C. R. LUND

W9-AQX-030

3. Energy Conversion Table

Three unusual but convenient "energy" quantities are temperature (K) representing the molecular energy kT at a given T, wavenumber (cm^{-1}) representing the molecular energy $hc\tilde{v}$ at a given frequency $\tilde{v}$ in cm^{-1} units (where c is the speed of light in $cm\,s^{-1}$), and electron volt (eV). All of these molecular energies are changed to molar energies on multiplication by Avogadro's number.

	K	cm^{-1}	kJ mol^{-1}	kcal mol^{-1}	eV
1 K =	1	0.69504	8.31451×10^{-3}	1.98722×10^{-3}	8.61739×10^{-5}
1 cm^{-1} =	1.43877	1	1.19627×10^{-2}	2.85914×10^{-3}	1.23984×10^{-4}
1 kJ mol^{-1} =	1.20272×10^2	8.35935×10^1	1	0.23901	1.03643×10^{-2}
1 kcal mol^{-1} =	5.03217×10^2	3.49755×10^2	4.18400	1	4.33641×10^{-2}
1 eV =	1.16044×10^4	8.06554×10^3	9.64853×10^1	2.30605×10^1	1

[a] E. R. Cohen and B. N. Taylor, CODATA Bulletin 63, Pergamon (1986).

[b] $F\,m^{-1} \equiv C^2\,m^{-1}\,J^{-1} \equiv C\,m^{-1}\,V^{-1}$.

[c] Note that $|c|$ is the *pure number* 29979245800 (i.e., it equals the magnitude of the speed of light in vacuum when it is expressed in $cm\,s^{-1}$ units).

Experiments in Physical Chemistry

Also Available from McGraw-Hill

Schaum's Outline Series in Science

Each outline includes basic theory, definitions and hundreds
of solved problems and supplementary problems with answers.

Current List Includes:

Analytical Chemistry
Applied Physics, 2/e
Biochemistry
College Chemistry, 6/e
College Physics, 7/e
Earth Sciences
Genetics, 2/e
Human Anatomy and Physiology
Lagrangian Dynamics
Modern Physics
Optics
Organic Chemistry
Physics for Engineering and Science
Zoology

Available at Your College Bookstore

DAVID P. SHOEMAKER
Emeritus Professor of Chemistry
Oregon State University

CARL W. GARLAND
Professor of Chemistry
Massachusetts Institute of Technology

JOSEPH W. NIBLER
Professor of Chemistry
Oregon State University

Experiments in Physical Chemistry

FIFTH EDITION

McGraw-Hill Book Company

New York   St. Louis   San Francisco   Auckland   Bogotá   Caracas
Colorado Springs   Hamburg   Lisbon   London   Madrid   Mexico   Milan
Montreal   New Delhi   Oklahoma City   Panama   Paris   San Juan
São Paulo   Singapore   Sydney   Tokyo   Toronto

This book was set in Times Roman.
The editors were Karen S. Misler and John M. Morriss;
the cover was designed by John Hite.
The production supervisor was Denise L. Puryear.
Project supervision was done by The Universities Press.
R. R. Donnelley & Sons Company was printer and binder.

EXPERIMENTS IN PHYSICAL CHEMISTRY

Copyright © 1989, 1981, 1974, 1967, 1962 by McGraw-Hill, Inc. All rights reserved.
Printed in the United States of America. Except as permitted under the United States
Copyright Act of 1976, no part of this publication may be reproduced or distributed in
any form or by any means, or stored in a data base or retrieval system, without the prior
written permission of the publisher.

1234567890 DOC DOC 89321098

ISBN 0-07-057007-8

Shoemaker, David P.
   Experiments in physical chemistry.
   Includes index.
   1. Chemistry, Physical and theoretical—Laboratory
manuals. I. Garland, Carl W. II. Nibler, Joseph W.
III. Title.
QD457.S56   1989   541.3'073   88-6863
ISBN 0-07-057007-8

CONTENTS

Preface   **xi**

I. Introduction   **1**

Organization of the Book   2
Preparation for an Experiment   4
Safety   5
Recording of Experimental Data   5
Literature Work   7
Reports   12
Sample Report   14
Special Projects   27

II. Treatment of Experimental Data   **29**

Errors   30
Statistical Treatment of Random Errors   39
Uncertainty in the Mean Value   43
Significance Testing   51
Propagation of Errors   55
Calculations   63
Graphs and Graphical Methods   72
Fundamental Limitations on Instrumental Precision   77
Summary   80
Exercises   82

III. Gases   **86**

1. Gas Thermometry   86
2. Joule–Thomson Effect   95
3. Heat-capacity Ratios for Gases   104

IV. Transport Properties of Gases   **119**

Kinetic Theory of Transport Phenomena   119
4. Viscosity of Gases   130
5. Diffusion of Gases   141

V. Thermochemistry   **153**

Principles of Calorimetry   153
6. Heats of Combustion   161
7. Strain Energy of the Cyclopropane Ring   170
8. Heats of Ionic Reaction   180

VI.   **Solutions**   **187**

    9.   Partial Molal Volume   187
   10.   Cryoscopic Determination of Molecular Weight   195
   11.   Freezing-point Depression of Strong and Weak Electrolytes   205
   12.   Chemical Equilibrium in Solution   211

VII.   **Phase Equilibria**   **219**

   13.   Vapor Pressure of a Pure Liquid   219
   14.   Binary Liquid-Vapor Phase Diagram   229
   15.   Binary Solid-Liquid Phase Diagram   238
   16.   Liquid-Vapor Coexistence Curve and the Critical Point   246

VIII.   **Electrochemistry**   **253**

   17.   Conductance of Solutions   253
   18.   Temperature Dependence of EMF   267
   19.   Activity Coefficients from Cell Measurements   270

IX.   **Chemical Kinetics**   **277**

   20.   Method of Initial Rates: Iodine Clock   277
   21.   Kinetics of a Hydrolysis Reaction   287
   22.   Enzyme Kinetics: Inversion of Sucrose   297
   23.   Kinetics of the Decomposition of Benzenediazonium Ion   311
   24.   Gas-phase Kinetics   316
   25.   Kinetics of a Fast Reaction   329

X.   **Surface Phenomena**   **339**

   26.   Surface Tension of Solutions   339
   27.   Physical Adsorption of Gases   350

XI.   **Macromolecules**   **361**

   28.   Osmotic Pressure   361
   29.   Intrinsic Viscosity: Chain Linkage in Polyvinyl Alcohol   370
   30.   Helix-Coil Transition in Polypeptides   380

XII.   **Electric and Magnetic Properties**   **390**

   31.   Dipole Moment of Polar Molecules in Solution   390
   32.   Dipole Moment of HCl Molecules in the Gas Phase   402
   33.   Magnetic Susceptibility   418
   34.   NMR Determination of Paramagnetic Susceptibility   430

XIII.   **Spectra and Molecular Structure**   **440**

   35.   Absorption Spectrum of a Conjugated Dye   440
   36.   Infrared Spectroscopy: Vibrational Spectrum of SO_2 446
   37.   Raman Spectroscopy: Vibrational Spectrum of CCl_4 451
   38.   Vibrational-Rotational Spectra of HCl and DCl   461
   39.   Vibrational-Rotational Spectrum of Acetylene   469

40. Spectrum of the Hydrogen Atom 482
41. Band Spectrum of Nitrogen 489
42. Absorption and Emission Spectra of Molecular Iodine 497
43. Electron Spin Resonance Spectroscopy 507
44. NMR Determination of Keto-Enol Equilibrium Constants 522

XIV. Solids **532**

45. Determination of Crystal Structure by X-ray Diffraction 532
46. Single Crystal X-ray Diffraction with the Buerger Precession Camera 550
47. Lattice Energy of Solid Argon 572
48. Statistical Thermodynamics of Iodine Sublimation 582

XV. Electrical Measurements **599**

Circuit Elements 599
Semiconductor Devices 604
Operational Amplifiers 615
Analog-to-Digital Conversion 621
Digital Multimeters 624
Potentiometer Circuits 626
Wheatstone Bridge Circuits 632
Capacitance Measurements 636

XVI. Temperature **640**

Temperature Scales 640
Triple-point and Ice-point Cell 643
Thermometers 645
Temperature Control 665

XVII. Vacuum Techniques **677**

Introduction 677
Theoretical Background 678
Pumping Speed for a System 678
Desorption and Virtual Leaks 685
Pumps 686
Vacuum Gauges 696
Miscellaneous Vacuum Components 703
Design of Vacuum Lines 707
Leak Detection 711
Safety Considerations 713

XVIII. Instruments **715**

Balances 715
Barometer 720
Cathetometer 722
Oscilloscope 722

pH Meter   725
Polarimeter   728
Radiation Counters   730
Recorders and Plotters   731
Refractometers   732
Signal-Averaging Devices   737
Spectroscopic Components   739
Spectroscopic Instruments   754
Timing Devices   763
Westphal Balance   764

XIX.   Miscellaneous Procedures   **766**

Volumetric Procedures   766
Purification Methods   773
Gas-handling Procedures   774
Electrodes for Electrochemical Cells   785
Materials for Construction   787
Solders and Adhesives   794
Tubing Connections   797
Shopwork   799

XX.   Least-squares Fitting Procedures   **801**

Introduction   801
Foundations of Least Squares   802
Weights   808
Rejection of Discordant Data   810
Goodness of Fit   811
Comparison of Models   815
Uncertainties in the Parameters   817
Summary of Procedures   820
Sample Least-squares Calculation   821

XXI.   Use of Computers   **828**

Types of Information Handled by Computers   830
Hardware   834
Software   841
Computer Programming   846
Interfacing the Computer with Experiments   853

Appendixes   **862**
A.   Glossary of Symbols   862
B.   International System of Units   866
C.   Concentration Units for Solutions   869
D.   Barometer Corrections   870
E.   Safety   871
F.   Research Journals   877

Index   **878**

Endpapers

Physical Constants and Conversion Factors   front
Energy Conversion Table   front
Relative Atomic Masses of the Elements   back

"I often say that when you can measure what you are speaking about, and express it in numbers, you know something about it; but when you cannot express it in numbers, your knowledge is of a meagre and unsatisfactory kind; it may be the beginning of knowledge, but you have scarcely, in your thoughts, advanced to the stage of Science, whatever the matter may be."

Sir William Thomson
(Lord Kelvin)

"It is much easier to make measurements than to know exactly what you are measuring."

J. W. N. Sullivan (1928)

PREFACE

This book is intended as a textbook for a junior-level laboratory course in physical chemistry. It is assumed that the student will be taking concurrently (or has previously taken) a lecture course in the principles of physical chemistry. The book contains 48 selected experiments which have been tested by extensive use. Four of the experiments are new, and one previous experiment has been greatly transformed. These five experiments involve the determination of dipole moments in the gas phase (Experiment 32), vibrational–rotational spectra of acetylenes (Experiment 39), visible absorption and emission spectra of I_2 (Experiment 42), ESR spectroscopy (Experiment 43), and a spectrophotometric/statistical thermodynamic study of I_2 sublimation (Experiment 48). All the other experiments have been carefully revised, and numerous changes have been made to bring both theory and experimental procedures up to date. The most extensive changes occur in Experiments 2, 17, 22, 31, and 36.

In addition to the experiments themselves, there are nine chapters containing material of a general nature. These chapters should be useful not only in undergraduate laboratory courses but also in special project work, graduate thesis research, and general research in chemistry. In Chapter I, the section on the use of scientific literature and reference sources has been expanded and modernized and other sections have been modified in smaller but still significant ways. In recognition of the vital importance of the evaluation and processing of experimental data in physical chemistry, Chapters II, XX, and XXI have been greatly strengthened. Chapters II and XXI have been rewritten completely. Chapter II, which contains a discussion of calculations and error analysis, has been expanded and numerous examples are now provided as an aid to the student. Chapter XXI, an introduction to the use of computers, has been brought up to date; it now focuses on the IBM PC/XT/AT family of microcomputers and includes some material on the interfacing of computers to experimental apparatus. Chapter XX on least-squares procedures has been reorganized for greater ease of use.

Chapters XV to XIX deal with basic experimental methods of broad value in chemistry. These chapters have been rewritten, modernized, and expanded to meet current needs. Of special significance are the revisions of Chapter XV (Electrical Measurements) and Chapter XVI (Temperature) and the inclusion of new material on spectroscopic components and instruments in Chapter XVIII.

The experiments in this book are not primarily concerned with "techniques" *per se* or with the analytical applications of physical chemistry. We believe that an experimental physical chemistry course should serve a dual purpose: (1) to illustrate and test theoretical principles and (2) to develop a research orientation by providing basic experience with physical measurements that yield quantitative results of important chemical interest.

Each experiment is accompanied by a theoretical development in sufficient detail to provide a clear understanding of the method to be used, the calculations required, and the significance of the final results. The depth of coverage is frequently greater than that which is available in introductory physical chemistry textbooks. Experimental procedures are described in considerable detail as an aid to the efficient use of laboratory time and teaching staff. Emphasis is given to the reasons behind the design and procedure for each experiment so that the student can learn the general principles of a variety of experimental techniques. Stimulation of individual resourcefulness through the use of special projects or variations on existing experiments is also desirable. We strongly urge that on some occasions the experiments presented here should be used as points of departure for work of a more independent nature.

To conserve space, six experiments from the fourth edition have been dropped. All of those experiments worked well but were less often used than those that have been retained. Anyone who is interested could still carry them out by following the procedures in the fourth edition. Indeed, there are valid experiments available in earlier editions as well, such as those on thermal conductivity of gases (second edition) and diffusion in liquid solutions (third edition).

It is important to note two major changes in notation and units. The definition of thermodynamic work has been changed to the work done *on* a system (e.g., $dw = -p\,dV$). As a result, the first law of thermodynamics has the form $dE = dq + dw$. Système International (SI) units are used, but not slavishly. Chemically convenient quantities like the gram (g), cubic centimeter (cm^3), and liter ($L \equiv dm^3 = 10^3\,cm^3$) are still used where useful—densities in $g\,cm^{-3}$, concentrations in $mol\,L^{-1}$, molecular masses in g. Conversions of such quantities into their SI equivalents is trivially easy. The situation with pressure is not so simple, since the SI pascal is a very awkward unit. Throughout the text, both bar and atmosphere are used. Generally bar ($\equiv 10^5\,Pa$) is used when a precisely measured pressure is involved, and atmosphere ($\equiv 760\,Torr \equiv 1.01325 \times 10^5\,Pa$) is used to describe casually the ambient air pressure, which is usually closer to 1 atm than to 1 bar. Standard states for all chemical substances are officially defined at a pressure of 1 bar; normal boiling points for liquids are still understood to refer to 1 atm values. The conversion factors given inside the front cover will help in coping with non-SI pressures.

McGraw-Hill and the authors would like to thank the following reviewers for their useful comments and suggestions: Alexander Amell, University of New Hampshire; Joseph Chaiken, Syracuse University; Edward Grant, Purdue

University; Doug Ridge, University of Delaware and Richard Wilde, Texas Technical University.

We gratefully acknowledge the assistance of students, teaching assistants, and faculty colleagues at the Massachusetts Institute of Technology and Oregon State University, as well as many helpful comments from faculties who have used this book at other universities. We are particularly grateful to Professors Walter H. Stockmayer of Dartmouth College and Clark C. Stephenson of M.I.T. for their interest and helpful suggestions spanning all five editions, including leads for new experiments; to Professor Michael W. Schuyler of Oregon State University for much helpful guidance in the rewriting of Chapter XXI (Use of Computers); to Professor John C. Decius of Oregon State University for contributions to the development of Experiment 48 (Statistical Thermodynamics of Iodine Sublimation); and to Professor Robert Scott of the University of California at Los Angeles for suggestions contributing to the revision of the introductory section on Principles of Calorimetry in Chapter V. We encourage and welcome feedback from those who are using the book, either as students or as instructors.

DAVID P. SHOEMAKER
CARL W. GARLAND
JOSEPH W. NIBLER

CHAPTER

I

INTRODUCTION

Physical chemistry deals with the physical principles underlying the properties of chemical substances. Like other branches of physical science, it contains a body of theory that has stood the test of experiment and is continually growing as a result of new experiments. In order to learn physical chemistry, one must become familiar with the experimental foundations on which the theoretical principles are based. Indeed, in many cases, the ability to apply the principles usefully requires an intimate knowledge of those methods and practical arts that are called "experimental technique."

For this reason, lecture courses in physical chemistry are usually accompanied by a program of laboratory work. Such experimental work should not just demonstrate established principles but should also develop research aptitudes by providing experience with the kind of measurements that can yield important new results. This book attempts to achieve that goal. Its aim is to provide a clear understanding of the principles of important experimental methods, the design of basic apparatus, the planning of experimental procedures, and the significance of the final results. In short, the aim is to train not laboratory technicians but future research scientists.

Severe limitations of time and equipment must be faced in presenting a set of experiments as the basis of a laboratory course that will provide a reasonably broad coverage of the wide and varied field of physical chemistry.

1

Although high-precision research measurements would require refinements in the methods described here and would often require more sophisticated and more elaborate equipment, each experiment in this book is designed so that meaningful results of reasonable accuracy can be obtained.

In the beginning, students will seldom have the skill and the time to plan detailed experimental procedures. Moreover, the trial-and-error method is not necessarily the best way to learn good experimental technique. Laboratory skills are developed slowly during a sort of apprenticeship. To enable an inexperienced student to make efficient use of available time, both the apparatus and the procedure for these experiments are described in considerable detail. *The student should keep in mind the importance of understanding why the experiment is done in the way described.* This understanding is a vital part of the experience necessary for planning special or advanced experiments of a research character. As the students become more experienced, it is desirable that they be required to plan more of their own procedure. This can easily be accomplished by introducing variations in the experiments described here. A change in the chemical system to be studied, the use of equipment that purposefully differs from that described, or the choice of a different method for studying the same system will force the student to work out modifications of the procedure. Finally, at the end of the course, it is recommended that some students do special projects that are completely independent of the experiments described in this book. A brief description of such special projects is given at the end of this chapter.

In addition to a general knowledge of laboratory techniques, creative research work requires the ability to apply two different kinds of theory. Many an experimental method is based on a special phenomenological theory of its own; this must be well understood in order to design the experiment properly and in order to calculate the desired physical property from the observed raw data. Once the desired result has been obtained, it is necessary to understand its significance and its interrelationship with other known facts. This requires a sound knowledge of the fundamental theories of physical chemistry (e.g., thermodynamics, statistical mechanics, and quantum theory). Considerable emphasis has been placed on both kinds of theory in this book.

In the final analysis, however, research ability cannot be learned merely by performing experiments described in a textbook; it has to be acquired through contact with inspired teachers and through the accumulation of considerable experience. The goal of this book is to provide a solid frame of reference for future growth.

ORGANIZATION OF THE BOOK

Before undertaking any experiments, the student should become familiar with the overall contents and structure of this book. Chapter I provides a general introduction to experimental work in physical chemistry. This includes advice on how to prepare for an experiment, how to record data in a laboratory notebook, and how to report the results. Chapter II describes in detail many mathematical procedures for analyzing data. Much of this material is concerned with the quantitative assessment of random errors and their effect on the uncertainty in the final result. These two chapters should be read at the very beginning, since they provide general information of value for all experiments in physical chemistry.

Experiments. Chapters III to XIV contain the descriptions of the experiments, which are numbered 1 to 48. In addition, Chapters IV and V each contain some separate introductory material that is pertinent to all the experiments in the given chapter. Each figure, equation, and table in a given experiment is identified by a single number: e.g., Fig. 1, Eq. (8), Table 2. For cross references, double numbering is used: e.g., Fig. 38-1 refers to Fig. 1 in Exp. 38 and Eq. (V-8) refers to Eq. (8) in the introductory part of Chapter V.

Every attempt has been made to write each experiment in sufficient detail that it can be intelligently performed without the necessity of extensive outside reading. However, it is assumed that the student will refer frequently to a standard textbook in physical chemistry for any necessary review of elementary theory. Literature sources are explicitly cited for those topics that are beyond the scope of a typical undergraduate textbook, and these numbered references are listed together at the end of each experiment. In addition, a selected list of reading pertinent to the general topic of each experiment is given under the heading General Reading. It is hoped that every student will do some reading in these books and journal articles since this is an excellent way to broaden one's scientific background.

A complete and very detailed list of equipment and chemicals[1] is given at the end of each experiment. The list is divided into two sections: Those items listed in the first paragraph are required for the exclusive use of a single team; those in the second paragraph are available for the common use of several teams performing the same experiment simultaneously. It is assumed that standardized stock solutions will be made up in advance and will be available for the student's use. The quantities indicated in parentheses† are for the use of the instructor and do *not* indicate the amounts of each chemical to be taken by a single team. In addition to the items included in these apparatus lists, it is assumed that the laboratory is equipped with analytical balances, a supply of distilled or deionized water, and a barometer as well as gas, water, and 110-V ac power lines. Also desirable, but not absolutely necessary, are gas-handling lines and a rough vacuum line.

---

† On the basis of the authors' experience, it is necessary to make available these amounts for each team that will do the experiment. They are scaled up from the amounts stated in the experiment to provide for possible wastage and to give a generous safety factor. It is hoped that they will be useful as a rough guide the first time an experiment is given.

Basic reference material. Chapters XV to XIX contain a variety of information on experimental procedures and devices. This includes the theory and practice of many types of electrical measurements, the principles and devices used in measurement and control of temperature, vacuum techniques, detailed descriptions of frequently used instruments, and a description of many miscellaneous procedures such as gas handling and the construction of equipment.

Chapter XX provides an introduction to least-squares fitting procedures and a discussion of how to assess the magnitude of random errors and how to judge the quality of any given fit to a set of data points. Chapter XXI presents a brief account of the use of computers in physical chemistry. This includes both data processing and data acquisition and instrument control.

In summary, the experiments are designed to provide significant "hands-on" experience in physical chemistry laboratory work and the general chapters (Chapters I–II and XV–XXI) are designed to present basic information of value not only in undergraduate courses but also in independent research work.

PREPARATION FOR AN EXPERIMENT

Although most of the experiments in this book can be performed by a single person, they have been written with the assumption that a pair of students will work together as a team. Such teamwork is advantageous, since it provides an opportunity for valuable discussion of the experiment between partners. The amount of experimental work to be assigned will be based on the amount of laboratory time available for each experiment. Many of the experiments can be completed in full during a single four-hour laboratory period. Many others are designed for six to eight hours of laboratory work but may be abridged so that meaningful results can be obtained in a single four-hour period. Some of the experiments require at least six hours and should not be attempted in a shorter time (in particular, Exps. 22, 24, 27, 28, 30).

Before the student arrives in the laboratory to perform a given experiment, it is essential that the experiment be studied carefully, with special emphasis on the method, the apparatus design, and the procedure. It will usually be necessary to make changes in the procedure whenever the apparatus to be used or the system to be studied differs from that described in this book. Planning such changes or even successfully carrying out the experiment as described requires a clear understanding of the experimental method.

Experimental work in physical chemistry requires many complex and expensive pieces of apparatus; many of these have been constructed especially for the student's use and cannot readily be replaced. Each team should accept complete responsibility for its equipment and should *check it over carefully before starting an experiment.*

SAFETY

Experimental work can be subject to hazards of many kinds, and every person working in a laboratory should be alert to possible safety problems. Once one is aware of the particular hazards involved in an experimental procedure, the instinct for self-preservation usually provides sufficient motivation for finding ways of avoiding them. The principal danger lies in ignorance of specific hazards and in forgetfullness.

A detailed analysis of all types of laboratory hazards and the procedures for dealing with them is beyond the scope of this book. Certain specific hazards are pointed out in connection with individual experiments. It is also assumed that the instructing staff will provide specific warnings and reminders where needed. Some general remarks on the kinds of safety hazards that should be kept in mind are given in Appendix E, and very complete treatments can be found elsewhere.[2,3] At this point, we wish to stress a few basic principles that apply to all laboratory work.

1. Determine the potential hazards and review the safety procedures appropriate for any experiment before beginning the work.
2. Know the location and proper use of safety and emergency equipment such as fire extinguishers and alarms, first-aid kits, safety showers, eyewash fountains, emergency telephone numbers, and emergency exits.
3. Call attention to any unsafe conditions you observe. Someone else's accident can be dangerous to you as well as to them.
4. Check all electrical equipment carefully before plugging it into the power line; unplug equipment before making any changes in electrical connections. Make sure that no part of the equipment has exposed high voltages.
5. Do not use mouth suction to draw up chemicals into a pipette; use a pipetting bulb instead.
6. Safety glasses or other appropriate eye protection must be worn in the laboratory at all times.
7. Avoid safety hazards and environmental contamination by following accepted waste-disposal procedures.
8. *Never work in the laboratory alone.* (For graduate research work and professional work, it is sometimes necessary to work alone. If so, make sure that someone—a security guard or an individual working in a nearby laboratory—will check on you every few hours.)

RECORDING OF EXPERIMENTAL DATA

The laboratory notebook is the essential link between the laboratory and the outside world and is the ultimate reference concerning what took place in the laboratory.[4] As such, it is not only the source book for the production of reports and publications but also a permanent record which may be consulted even after a great many years. It is standard practice in experimental research work to record *everything relevant* (data, calculations, notes and comments, literature surveys, and even some graphs) directly in a bound notebook with numbered pages. Spiral-bound notebooks are preferable since they lie flat and stay open at the desired page. Such notebooks are available with pages that are ruled vertically as well as horizontally to give a $\frac{1}{4}$-inch grid; this facilitates tabulation of columns of figures and permits rough plots of the data to be made directly on the notebook pages during the experiment. In an undergraduate laboratory course it is often convenient to make a carbon copy of the recorded data to include with the written report. (Notebooks with duplicate sets of numbered pages, in which alternate pages are perforated for removal, may be used; however, ordinary carbon paper and bond paper can be used with any style of research notebook.) In any case, the *original* is a part of a permanent notebook.

Whatever style of notebook is used, the principle is the same: *Record all data directly in your notebook.* Data may be copied into the notebook from a partner's notebook in those cases where it is clearly impossible for both partners to record data at the same time. Even then, a carbon copy or a photocopy of the original pages is better, since it avoids copying errors and saves time. In particular, *do not use odd scraps* of paper to record such incidental data as weights, barometer readings, and temperatures with the idea of copying them into the notebook at a later time. If anything must be copied from another source (calibration chart, reference book, etc.), identify it with an appropriate reference.

A ball-point pen is best for recording data, especially if carbon copies are required; otherwise, any pen with permanent ink is satisfactory. Pencils are unsuitable for recording primary data. If a correction is necessary, draw a single line through the incorrect number so as to leave it legible and then write the correct number directly above or beside the old one. If something happens to vitiate the data on an entire page, cross out these data and record the circumstances. *No original pages should ever be removed from a laboratory notebook.* Never leave blank notebook pages or any significant number of blank lines on a notebook page, except at the end of an experiment. Always record your data in serial (chronological) fashion except when you are recording data in tabular form. Neatness and good organization are desirable, but legibility, proper labeling, and completeness are absolute necessities.

For each experiment undertaken, make an entry for it in a table of contents at the beginning of the notebook, indicating the page number and the date on which work began. If a long experiment has logical subdivisions, these should also be entered in the table of contents.

What to record. Every data page for an experiment should have a clear heading that includes your name (and the name of any lab partner), the subject of the experiment, the date, and a page number.

The notebook should contain all the information that would be needed to permit someone else to perform the same experiment in the same way. In addition to the measurements and observations that constitute the results of the experiment, all other data that are relevant to the interpretation of the results should be recorded. It is not necessary to duplicate information that is conveniently available in some permanent record such as a journal article or laboratory text, provided that complete references are given.

The record of an experiment should begin with a brief statement of the experiment to be performed. The procedure being followed should be described in all essential detail. When the procedure is described elsewhere, the notebook entry may be abbreviated to a reference to the published description. However, be certain to record any variations from the published procedure and to specify the apparatus that was used (make and model for commercial equipment, otherwise a sketch). For each chemical substance used, record the name, formula, source, grade or stated purity, and concentration (in the case of a solution).

Data should be entered directly in the notebook, in tabular form whenever possible. Use complete, explicit headings; do not rely on your memory for the meaning of unconventional symbols and abbreviations. Be sure that all numerical values are accompanied by the appropriate units. In many cases, it is necessary or at least wise to record such laboratory conditions as the ambient temperature, the atmospheric pressure, or the relative humidity.

Instrumental records, such as spectra, should be dated and signed by the investigator. They may be taped or stapled into the notebook if they are of a convenient size or can be suitably reduced with a copying machine. This should not be done if there are many such records, as it makes the notebook clumsy and awkward to use. Records that are not fastened in the notebook should be numbered and placed in a laboratory file provided for the purpose (preferably a three-ring binder), and an appropriate reference should be made in the notebook.

LITERATURE WORK

As a matter of policy, very few of the experiments contain a direct reference to published values of the final result that is to be reported by the student. Students who wish to compare their results with accepted literature values are expected to do the necessary library work. As a general principle, it is best to refer directly to an original journal article rather than to some secondary source. It is commonly assumed that recent measurements are more precise than older ones; this assumption is based on the fact that methods and equipment are constantly being improved. But this does not mean that there is valid reason to reject or suspect a published result merely because it is old. The quality of research data depends strongly on the integrity, conscientious care, and patience of the research worker; much fine work done many years ago in certain areas of physical chemistry has not yet been improved upon. In evaluating results based on old but high-quality research data, one must, however, be alert to the possible need for corrections necessitated by more recent theoretical developments or by improved values of physical constants.

Literature search. It is convenient to distinguish between two types of literature search—one concerned with information about a single chemical species, and one concerned with a broad area of research, a method, a class of reactions, or a class of compounds. In the former case, it is usually best to begin directly with abstract journals. In the latter case, it is more efficient to begin by checking specialized monographs (consult the subject card catalog in the library), review journals, and annual series such as "Advances in ——."

The chemical literature of the review type consists of collections of articles or chapters on specific topics written by presumed experts. The introduction to these reviews usually specifies the period of time covered. Ideally, the reviews include critical evaluations of the work and progress in a specific area of research together with a rather complete bibliography of the important original research articles pertinent to that area. However, it should not be assumed that every reference to a particular subject is given or that all of the author's critical interpretations are necessarily correct. A selection of the review literature that is of special value in physical chemistry is listed below.

Review journals

Accounts of Chemical Research
Advances in Physics
Chemical Reviews
Chemical Society Reviews

Review series

Advances in Chemistry Series, American Chemical Society, Washington, D.C. (1950–) [more than 200 books issued at irregular intervals].
Advances in Chemical Physics, I. Prigogine and S. Rice (eds.), Wiley-Interscience, New York (1958–).
Advances in Magnetic Resonance, J. S. Waugh (ed.), Academic Press, New York (1965–).
Advances in Quantum Chemistry, P.-O. Löwdin (ed.), Academic Press, New York (1964–).
Annual Reviews of Physical Chemistry, B. S. Rabinovitch, J. M. Schurr and H. L. Strauss (eds.), Annual Reviews, Inc., Palo Alto, Calif. (1950–).
Molecular Structures and Dimensions, O. Kennard, D. G. Watson, F. H. Allen, and S. M. Weeds (eds.), Cambridge Crystallographic Data Centre and International Union of Crystallography (1970–).
Solid State Physics, H. Ehrenreich and D. Turnbull (eds.) [formerly F. Seitz and D. Turnbull], Academic Press, New York (1955–).
Structure and Bonding, C. K. Jorgensen *et al.* (eds.), Springer-Verlag, New York (1966–).
Structure Reports, W. B. Pearson and J. Trotter (eds.), Oosthoek, Utrecht, Vol. 8– (1940–) [vols. 1–7 were published in German under the title *Struckurbericht*].

The use of *Chemical Abstracts* (*CA*) and *Physics Abstracts* (*PA*) (Science Abstracts Series A) is the most effective way to make a detailed search of the physical chemical literature published from 1907 to the present time. Each of these abstract publications issues annual indexes, which are merged into cumulative indexes every four (*PA*) or five (*CA*) years. The index structures of *PA* and *CA* are somewhat different. *Physics Abstracts* provides an author index and a subject index. The latter contains both general topic entries, such as "dielectric relaxation," with specific compounds arranged as subentries, and chemical substance entries such as "argon" or "nickel compounds." *Chemical Abstracts* provides five different indexes: author, general subject, chemical substance, formula, and patent. If one is searching for information on a specific compound, the chemical substance index is naturally the most useful. It is larger and more complete than the subject index; furthermore, many convenient subheadings are used to display properties, reactions, and simple derivatives of the given compound.

Some practice is necessary to acquire a rapid and thorough search technique. The considerations outlined below should be helpful as an introduction to the use of abstracts. First, it is usually advisable to begin this type of literature search with the most recent abstract indexes available. In this way one may find recent review articles or monographs that are sufficiently complete in coverage that a search of the older literature can be avoided. Second, search under both the chemical name and key subject entries for the property of interest. Third, it is of obvious importance to use the correct name in searching for the desired compound. To ensure that the name being used is consistent with *Chemical Abstracts* usage, it is convenient to find the compound in the formula index and then use the name listed there or to refer to the most recent "Collective Index Guide to the Chemical Abstracts," which is issued every five years. Note also that each chemical compound has a unique Chemical Abstracts Service (CAS) registry number. In dealing with compounds with complicated structures, and thus complex names, it may be useful to cite the registry number when referring to the compound.

The various abstract indexes discussed above contain item numbers that refer directly to an abstract in one of the abstract volumes. These abstracts are brief summaries of the important contents of the original research publications, often identical to the abstracts given at the beginning of such publications. If the abstract indicates that the research results are of interest, the original journal article should be consulted whenever possible, even if the abstract gives the desired information. If a cited journal article is especially pertinent but not available locally, a reproduction can be obtained through interlibrary loan services. The availability of all journals abstracted in *Chemical Abstracts* is listed in "Chemical Abstracts Service Source Index," published by the American Chemical Society.

It is important to realize that there is a delay of several months between the time a journal article appears and the time its abstract is printed in *Chemical Abstracts* or *Physics Abstracts*. In addition, there is a long delay after an abstract volume is complete before the subject and substance indexes become available. Thus it is wise to check the most recent literature by scanning the title pages of likely journals. A convenient method for doing this is to scan through *Current Contents: Physical, Chemical, and Earth Sciences,* a weekly compilation of the latest tables of contents for many journals. A list of specific journals of greatest interest to physical chemists is given in Appendix F.

Another search method involves the use of the *Science Citation Index,* which is published bimonthly in three parts. The Source Index gives a listing of authors with the titles of their recent publications and the necessary bibliographic information. The Subject Index is organized under key words and gives the names of authors who used the given word in the title of a publication. The Citation Index is based on author names and their specific publications; it lists other publications that cite the given reference.

In recent years, a number of computer databases have been developed to aid in the search of current literature. The most important of these for physical chemists are *CAS Online* (Chemical Abstracts Service of the American Chemical Society) and *INSPEC* (derived from Physics Abstracts). Many chemistry departments and science libraries have modem links to these and other databases via some commercial vendor such as STN International or Dialog.

Reference books. The three series of books listed below describe important experimental techniques and apparatus for a variety of physical measurements. These books also deal with the phenomenological theory of of the methods described.

Techniques of Chemistry, Vol. I: Physical Methods of Chemistry, A. Weissberger and B. W. Rossiter (eds.), Wiley-Interscience, New York, part I (1971)–part VI (1977).
Physical Methods of Chemistry, 2d ed., vols. I and II, B. W. Rossiter and J. F. Hamilton (eds.), Wiley-Interscience, New York (1986).
Methods of Experimental Physics, L. Marton and C. Marton (original eds.), R. Celotta and J. Levine (current eds.), Academic Press, New York (1959–).

Techniques of Chemistry is a multivolume series dealing with a wide range of topics, with emphasis on organic and inorganic chemistry. Volume I on physical methods consists of eleven separate books: Part IA and IB—Components of Scientific Instruments; Part IIA and IIB—Electrochemical Methods; Part IIIA–IIID—Optical, Spectroscopic, and Radioactivity Methods; Part IV—Determination of Mass, Transport, and Electrical-Magnetic Properties; Part V—Determination of Thermodynamic and Surface Properties; Part VI—Supplement and Cumulative Index. The *Physical Methods of Chemistry* series is a direct extension and revision of *Techniques of Chemistry.* The two volumes that are presently available cover Components of Scientific Instruments (Vol. I) and Electrochemical Methods (Vol. II).

Handbooks and compilations of physical properties. The sources listed below are mainly devoted to numerical tabulations of various physical properties.

They are convenient, but some are distinctly secondary sources of information. It is often difficult to judge the quality of the data listed, since references to the original sources are sometimes inadequate and transcription errors can occur. If at all possible, it is wise to confirm important information by consulting the original literature. A particularly useful reference for obtaining a wide variety of routine physical data is the *Handbook of Chemistry and Physics*. Although a new edition of this handbook is issued every year, changes are introduced very slowly. Thus any recent edition is likely to be as useful as the newest one for almost all purposes.

American Institute of Physics Handbook, D. E. Gray *et al.* (eds.), 3d ed., McGraw-Hill, New York (1972).

Binary Alloy Phase Diagrams, T. B. Massalski *et al.* (eds.), ASM International, Metals Park, Ohio (1987).

Crystal Data, J. and G. Donnay (eds.), Amer. Cryst. Assoc. Monogr. no. 5 (1963) and no. 6 (1967); *Crystal Data*, J. Donnay and H. Ondik (eds.), 3d ed., US Natl. Bur. Stand. and Joint Comm. on Powder Diffraction Stand., Vol. 1 (1972), Vol. 2 (1973), Vols. 3 and 4 (1979).

Crystal Structures, R. W. G. Wyckoff (ed.), 2d ed., Wiley-Interscience, New York, vol. 1 (1963)–vol. 6 (1971).

Handbook of Chemistry and Physics, R. C. Weast (ed.), 69th ed., CRC Press, Boca Raton, Fla. (1988/89).

Journal of Physical and Chemical Reference Data, published for the U.S. Natl. Bur. Stand. by American Chemical Society and American Institute of Physics, Washington, D.C. (1972–) [regular issues plus many supplements].

JANAF Thermochemical Tables, 3d ed., supplement 1 to vol. 14 of *J. Phys. Chem. Ref. Data* [see above] (1985).

Landolt-Börnstein Numerical Data and Functional Relationships in Science and Technology. New Series, K.-H. Hellwege (ed.), Springer-Verlag, Berlin [especially Groups II–IV] (1965–).

Molecular Spectra and Molecular Structure IV. Constants of Diatomic Molecules, K. P. Huber and G. Herzberg, Van Nostrand-Reinhold, New York (1979).

National Standard Reference Data Series, U.S. Natl. Bur. Stand., U.S. Government Printing Office, Washington, D.C. (1964–71).†

National Standard Reference Data Series of the USSR, Hemisphere-Harper & Row, New York (1987) [a series of property tables for simple fluids].

NBS Tables of Chemical Thermodynamic Properties, supplement 2 to vol. 11 of *J. Phys. Chem. Ref. Data* [see above] (1982) [inorganic and C_1 and C_2 organic substances in SI units].

Physik Daten/Physics Data, Fachinformationszentrum, Karlsruhe, West Germany (1976–) [esp. vols. 3-1 to 3-4, 9-1, 21-1].

Selected Values of Chemical Thermodynamic Properties, U.S. Natl. Bur. Stand. Technical Notes 270-1 to 270-5, U.S. Government Printing Office, Washington, D.C. (1965–71) [updating of NBS Circular 500].

Selected Values of Electric Dipole Moments for Molecules in the Gas Phase, NSRD-NBS 10, U.S. Government Printing Office, Washington, D.C. (1967).

Selected Values of Properties of Chemical Compounds, Texas A & M University, Thermodynamics Research Center, College Station (1955–73).

---

† This NSRDS series consists of many different volumes, edited by different persons, on specialized topics; check a library catalog for further information. As of 1972, the *Journal of Physical and Chemical Reference Data* became the publication vehicle of NSRDS.

Solubilities of Inorganic and Organic Compounds, H. Stephen *et al.* (eds.), vols. 1–3, Pergamon, New York (1979).

Solubilities: Inorganic and Metal-organic Compounds, W. F. Linke (ed.), 4th ed., Van Nostrand, New York (1958–).

Tables of Chemical Kinetics: Homogeneous Reactions, U.S. Natl. Bur. Stand. Circ. 510, U.S. Government Printing Office, Washington, D.C. (1951).

Tables of Interatomic Distances and Configurations in Molecules and Ions, L. E. Sutton (ed.), The Chemical Society, London (1958, supplement 1965).

Tables of Thermal Properties of Gases, U.S. Natl. Bur. Stand. Circ. 564, U.S. Government Printing Office, Washington, D.C. (1955).

Thermophysical Properties of Matter: The TPRC Data Series, Y. S. Touloukian and C. Y. Ho (eds.), IFI/Plenum, New York, vols. 1–7 (1970).

Chemical nomenclature. In searching the chemical literature and in reading and writing research papers, a knowledge of systematic nomenclature for chemical compounds is indispensable. In 1957 the International Union of Pure and Applied Chemistry (IUPAC), Committee on Chemical Nomenclature, recommended the adoption of an international nomenclacture. The more important nomenclature rules, together with recommended chemical symbols and terminology, were published in *J. Amer. Chem. Soc.* **82,** 5517, 5523, 5545, 5575 (1960).

A more detailed set of IUPAC nomenclature rules for organic compounds is given in Refs. 5 and 6. The most convenient and useful source to consult if one has problems with nomenclature is the most recent issue of the *Collective Index Guide of Chemical Abstracts.*

REPORTS

The evaluation of any experimental work is based primarily on the contents of a written report. Indeed, the advancement of science depends heavily on the exchange of written information. Research work is not completed until the results have been properly reported. Such reports should be well organized and readable, so that anyone unfamiliar with the experiment can easily follow the presentation (with the aid of explicit references where necessary) and thereby obtain a clear idea as to what was actually done and what result was obtained.

An attempt should be made to use a scientific style comparable in quality to the literary style expected in an essay. Correct spelling and grammar should not be disregarded just because the report is to be read by a scientist instead of a literary editor. The report should be as concise and factual as possible without sacrificing clarity. In particular, mathematical equations should be accompanied by enough verbal material to make their meaning clear.

Two general sources of information dealing with proper literary usage are:

A Dictionary of Modern English Usage, H. W. Fowler, 2d ed., Oxford University Press, New York (1983).

The Elements of Style, W. Strunk, Jr. and E. B. White, 3d ed., Macmillan, New York (1979).

Four related sources dealing specifically with technical writing are:

The Craft of Scientific Writing, M. Alley, Prentice-Hall, Englewood Cliffs, N.J. (1987).
Technical Writing, J. M. Lannon, 3d ed., Little-Brown, Boston (1985).
Communicating Technical Information, R. R. Rathbone, 2d ed., Addison–Wesley, Reading, Mass. (1985).
Effective Writing for Engineers, Managers, Scientists, H. J. Tichy, Wiley, New York (1966).

Stylistic details for preparing journal articles (such as recommended symbols, nomenclature, abbreviations, and the proper presentation of formulas, equations, tables, and literature citations) are discussed in:

The ACS Style Guide, J. S. Dodd (ed.), American Chemical Society, Washington, D.C. (1985). [A general reference book on scientific writing with emphasis on manuscripts intended for one of the journals published by the American Chemical Society.]
Style Manual, 3d ed., American Institute of Physics, New York (1978). [Provides guidance in the preparation of papers for AIP journals.]

The preparation of journal articles will not be discussed here. The recommendations given below concern student reports written as part of a laboratory course, although the same advice might well serve for technical reports of any kind.

Most important of all, the report must be an original piece of writing. Copying or even paraphrasing of material from textbooks, printed notes, or other reports is clearly dishonest and must be carefully avoided. Brief quotations, enclosed in quotation marks and accompanied by a complete reference, are permissible where a real advantage is to be gained. Certainly there is no point in giving more than a brief summary of the theory or the details of experimental procedure if these are adequately described in some readily available reference. In part, a report is likely to be judged on how clearly it states the essential points without oscillating between minute detail on one topic and vague generalities on another.

Except for general physical and numerical constants or well-known theoretical equations, any data or material taken from an outside source must be accompanied by a complete reference to that source.

The content and length of any given report will depend on the subject matter of the experiment and on the standards established by the instructor. It is our belief that at least in some cases the report should be quite complete and should include a quantitative analysis of the experimental uncertainties and a detailed discussion of the significance of the results (see the sample report given below). For many experiments a brief report with only a qualitative treatment of errors and a short discussion may be considered adequate. In either case, a clear presentation of the data, calculations, and results is essential to every report on experimental work.

Format. Unless otherwise instructed, all reports should be prepared on $8\frac{1}{2}$- by

11-inch paper with reasonable margins on all sides. The pages should be stapled together or bound in a folder. Legibility is absolutely essential. Double-spaced typewritten or word-processed reports are best, but handwritten reports submitted in ink on wide-line ruled paper are perfectly satisfactory unless you are cursed with illegible handwriting. Crossing out and the insertion of corrections are permissible, but try to keep the report as a whole reasonably neat.

Presentation of graphs. A general discussion of the graphical treatment of experimental data is given in Chapter II. As part of that discussion, the proper technique for plotting data points and drawing lines or curves is fully described. We shall be concerned here only with the final steps necessary for the presentation of such figures as part of a report.

Vertical and horizontal axes should be drawn in, and the main divisions along each axis must be clearly marked and numbered. Each axis is then labeled with the appropriate symbol or words with the units indicated in parentheses [for example: T (°C), t (s), A (cm^2), density (g cm^{-3})]. The data points and the symbols surrounding them (usually small circles) are inked in so that the data will stand out prominently. Error bars should be added if either the vertical of horizontal uncertainties in the data values are greater than the size of the symbol used.

A smooth curve showing the variation of the data is desirable but this line should not actually be drawn through symbols surrounding the data points. If several lines or curves lie close together, distinguish them by using dashed lines as well as solid lines. When necessary, one can achieve further differentiation by varying the lengths of the dashes or alternating long and short dashes. If a curve is drawn to represent an equation, the calculated points should not be inked or encircled and the curve should be drawn so as to conceal these points. The equation itself or the number by which it is designated in the text should be written beside the curve. It is good practice to indicate clearly on the graph the numerical values of any slopes, intercepts, areas, maxima, or other features that are important in the calculations.

Each figure must have a figure number and a short legend prominently displayed, and it should be referred to by number in the body of the report.

SAMPLE REPORT

A sample undergraduate laboratory report on a very simple experiment is given here as a brief illustration of the typical structure to be used in technical reports. This example is intended to show the kind of material that each section should contain. It is not meant to provide a rigid model; the content of any given report will necessarily depend on the character of the experiment and the judgment of the individual student. In particular, it should be stressed that most reports will be longer and more complex than this example.

DETERMINATION OF THE DENSITY OF

CRYSTALLINE GERMANIUM

Maria Smith   Sept. 30, 1987

(Partner: John Klein)

Abstract

A pycnometer has been used to determine the density of two samples of germanium at 25°C. The values obtained, with their 95% confidence limits, are ρ(sample I) = 5.310 ± 0.003 g cm^{-3} and ρ(sample II) = 5.332 ± 0.003 g cm^{-3}. Both samples consisted of several pieces of crystalline material, with the pieces in sample I being larger and more irregular. It seems likely that sample I had hidden defects and voids that introduced a systematic error since the x-ray value of the density is 5.325 ± 0.002 g cm^{-3}. However, the density of sample II is larger than the x-ray value by an amount slightly greater than the sum of the cited uncertainties, which cannot be explained by voids in the sample.

The sample report is displayed as a facsimile of a typed and handwritten report on the upper part of the next twelve pages. The lower part of these pages contain a commentary on each section of the report. Although the given outline need not be followed exactly, the topics covered should be included somewhere in the report.

Title page and Abstract. The front page of the report should display a title, your name, the name of any experimental partners, the date on which the report is submitted, and an abstract. The abstract is typically 50 to 150 words long; the example above contains about 90 words. It should summarize the results of the experiment and state any significant conclusions. *Numerical results with confidence limits should be included.*

I. Introduction

The purpose of this experiment is to measure the density of crystals of germanium. Since the density ρ is defined by

$$\rho = W_S/V_S \tag{1}$$

it is desired to measure the volume V_S occupied by a known weight W_S of the metal sample.

The method involves the use of a pycnometer of known volume which is first weighed empty, then weighed containing the solid sample to be studied. The difference gives the weight of the solid, W_S. Finally the pycnometer (containing the solid sample) is filled with a liquid of known density and reweighed; the weight, and therefore the volume, of the liquid can be found by difference. Since the total volume of the pycnometer is known, one can then calculate the volume V_S which is occupied by the solid.

II. Experimental Method

The experimental method was similar to that described in the textbook (Aardvark and Zebra, 3rd ed., Exp. 13). The design of the pycnometer used, which differs from that described in the textbook, is shown in the following sketch:

Introduction. This section should state the purpose of the experiment and give a *very brief* outline of the necessary theory, which is often accomplished by citing pertinent equations with references wherever appropriate. Each equation should appear on a separate line and should be part of a complete sentence. Number all equations consecutively throughout the entire report, and refer to them by number. All symbols should be clearly identified the first time they appear. A very short description of the experimental method can also be included.

The introduction should cover the above topics as concisely as possible; the sample contains about 135 words. More complicated experiments will require longer introductions, but the normal length should be between 100 and 300 words. In the case of a journal article, the introduction is often much longer, since pertinent recent work in the field is cited and briefly summarized.

capillary top

ground-glass joint

The procedure was modified as follows: After distilled water had been added to the pycnometer containing the sample, the pycnometer (with capillary top removed) was completely immersed in a flask of distilled water and boiled under low pressure for 15 minutes to remove air trapped by the solid or dissolved by the water.[1] After this boiling, the pyconometer was equilibrated for 30 minutes in a 25°C thermostat bath before the top was inserted.

The weight of the solid sample is given by

$$W_S = W_2 - W_1 \tag{2}$$

where W_1 is the weight of the empty pycnometer and W_2 is that of the pycnometer plus the solid sample. The weight of water contained in

Experimental method. This section is usually *extremely brief* and merely cites the appropriate references which describe the details of the experimental procedure. If reference is made to the textbook and/or laboratory notes assigned for the course, an abbreviated title may be cited in lieu of a complete bibliographic entry. Any reference to other books or materials should be assigned footnote numbers and should be properly listed at the end of the experiment in the form illustrated by the references in this book. Description of experimental procedures should normally be given *only* for those features not described in or differing from the cited references. A simple sketch of apparatus is appropriate only when it differs from that described in the references. A summary statement or condensed derivation of the phenomeno-logical equations used to analyze the raw data should be given. NOTE: A statement of the number of runs made and the conditions under which they were carried out (concentration, temperature, etc.) should always be included at the end of this section.

the pycnometer, W_L, is

$$W_L = W_3 - W_2 \qquad (3)$$

where W_3 is the weight of pycnometer plus sample plus water. If the density of the liquid (water) is denoted by ρ_L, it follows from Eq. (3) that the volume of the solid sample is given by

$$V_S = V - V_L = \frac{\rho_L V + W_2 - W_3}{\rho_L} \qquad (4)$$

where V is the total volume of the pycnometer. From Eqs. (1), (2), and (4), we obtain

$$\rho = \frac{W_S}{V_S} = \frac{\rho_L(W_2 - W_1)}{\rho_L V + W_2 - W_3} \qquad (5)$$

Since the values of V and ρ_L are known, it is only necessary to determine W_1, W_2 and W_3 in order to calculate the density of the solid.

Two duplicate runs, carried out using the same procedure, were made on each of two different germanium samples. Sample I consisted of larger and somewhat more irregular pieces than did Sample II.

Results. This section should present the experimental results in full detail, making use of tables and graphs where appropriate. No result should be excluded merely because it is unexpected or inconsistent with other data or theoretical models. The cause for discrepancies, if known, can be pointed out in the Discussion section. Primary measurements ("raw data") should be given, as well as derived quantities. It is essential that the units be completely specified.

It is undesirable to present detailed computations in the main body of the report; however, a typical sample calculation should be given in an appendix to illustrate how the data analysis was carried out. If one or more computer programs have been used in processing the data, they should be cited in the Results section. If a program is not well known and documented elsewhere, provide a complete listing of the program in an appendix.

III. Results

The average values of the measured weights W_1, W_2, and W_3 are listed below together with the stated value of the pycnometer volume V and the literature value for the density of water ρ_L:

ρ_L = 0.99707 g cm^{-3} at 25°C (taken from the

Handbook of Chemistry and Physics[2])

V = 12.445 ± 0.003 cm^3 (given by instructor)

W_1 = 8.6309 g

Sample I: W_2 = 42.0301 g

W_3 = 48.1732 g

Sample II: W_2 = 45.8479 g

W_3 = 51.3036 g

The density of germanium can now be calculated by substitution of the above data into Eq. (5). The resulting densities are 5.315 g cm^{-3} for sample I and 5.337 g cm^{-3} for sample II. The weights used in these calculations have not been corrected for the effect of air buoyancy on the weighings. Rather than correct each weight to vacuum, we may use a simple formula given by Bauer and Lewin[1] for correcting the final calculated result. This formula gives for the corrected density ρ^*

$$\rho^* = \rho + 0.0012 \, (1 - \rho/\rho_L) \tag{6}$$

Tables should be numbered and given self-explanatory captions. The quantities displayed in a table should be clearly labelled with the units specified. More detailed suggestions for preparing tables of data are given in Ref. 7.

Many reports will require a graphical presentation of data or calculated results. Each figure should be numbered and given a legend. Figures serve several purposes: to supplement or replace tables as a means of presenting results; to display relationships between two or more quantities; to find values needed in the calculation of results (see Chapter II). Some advantages of graphical display are that the relationship between two variables is shown more clearly by graphs than by tables; deviations of individual results from expected trends are more readily apparent; "smoothing" of the data can be done better if it is necessary and appropriate; interpolation and extrapolation are easier. A disadvantage of figures is that graphical displays cannot always show the full

When Eq. (6) is applied to our results we obtain for ρ^* the following values:

$$\text{Sample I:} \quad 5.310 \text{ g cm}^{-3}$$

$$\text{Sample II:} \quad 5.332 \text{ g cm}^{-3} \tag{7}$$

$$\text{Average:} \quad 5.321 \text{ g cm}^{-3}$$

According to Eq. (5) the uncertainty in ρ will depend on the uncertainty in each of five variables; however, the value of ρ_L is known to five significant figures and its uncertainty may certainly be neglected in comparison to those of the other variables. We may take as reasonable 95% confidence limits for the weighings $\Delta W_1 = \Delta W_2 = 0.001$ g and $\Delta W_3 = 0.002$ g. The higher value for ΔW_3 includes the possible failure to attain an exact filling of the pycnometer with water. For ΔV we take 0.003 cm^3, the value given by the instructor, although such a value seems rather high. On the basis of these uncertainty values, it is clear that the major contributors to the limit of error $\Delta \rho$ are the uncertainty in V and, to a lesser extent, the uncertainty in the difference W_2-W_3. The contribution to $\Delta \rho$ due to the uncertainty in the difference $W_2 - W_1$ is much less (since $W_2 - W_1$ is about 5.5 times larger than $W_2 - W_3$ and the uncertainty

precision of the results. For a report, it is normally advisable to include both figures and tables. Some advice on the preparation of graphs is given on pp. 72–74. Indicate clearly the scales and units used. Error bars should be included when the uncertainties are greater than the size of the symbols enclosing the points.

An *error analysis* dealing with the uncertainty in the final result due to random errors in the measurements will normally be part of the Results section. The type of error analysis undertaken will depend a great deal on the nature of the experiment; see Chapters II and XX for more details. The analysis given in the sample report is typical of a straightforward propagation-of-errors treatment. If a long and complex propagation-of-errors treatment is required, this should be given in an appendix and the resulting uncertainties should be stated, or shown as error bars on a plot, in the main body of the report. It is important to combine and simplify all expressions as much as in $W_2 - W_1$ is less than that in $W_2 - W_3$) and can be neglected in obtaining an approximate value for $\Delta\rho$. With this simplification, the 95% confidence limit in ρ is approximately given by

$$(\Delta\rho)^2 \approx \frac{\rho^2}{(\rho_{LV} + W_2 - W_3)^2} [\rho_L^2(\Delta V)^2 + (\Delta W_2)^2 + (\Delta W_3)^2] \quad (8)$$

The resulting error limits are ±0.003 g cm^{-3} for both samples.

IV. Discussion

The values and limits of error obtained for the density of germanium at 25°C are

Sample I: 5.310 ± 0.003 g cm^{-3}

Sample II: 5.332 ± 0.003 g cm^{-3}

It is obvious that these two values deviate from each other by considerably more than the sum of the limits of error. Furthermore, this difference is even more disturbing when one considers the fact that any error in V will affect both densities in the same way and thus not contribute to the uncertainty in the difference. Our results suggest that the material examined may be somewhat inhomogeneous, yielding two samples of slightly different bulk densities. We suggest the possi-

possible in order to avoid obtaining unwieldy error equations. Since uncertainty values need not be calculated to better than about 15 percent accuracy, one should always try to find labor-saving approximations. Where the number of runs is so small that reliable limits of error cannot be deduced from statistical considerations, error limits must be assigned largely on the basis of experience and judgment. Try to make realistic assessments, avoiding excessive optimism (choosing error limits that are too small) or excessive pessimism (choosing limits that are unrealistically large). For a long and detailed report, a quantitative analysis of errors should always be derived and a numerical value of the uncertainty in the final result should be given. For a brief report, a qualitative discussion of the sources of error may suffice.

Discussion. This is the most flexible section of the entire report, and the student must depend heavily on his or her own judgment for the choice of topics for discussion. The final results of the experiment should be clearly presented. A comparison, often in a tabular or graphical form, between these results and theoretical expectations or experimental values from the literature bility that voids inaccessible to the liquid are present in Sample I, or perhaps in both samples to different degrees. On this assumption, the greater confidence would be placed in the higher value.

The "literature" value given in the Handbook of Chemistry and Physics[3] is 5.35 g cm^{-3} at 20°C, but no uncertainty figure or source is cited there. A more accurate value can be calculated from the atomic mass (72.59 g mol^{-1} for Ge) and the volume of the unit cell as determined by x-ray crystallography. Germanium has a cubic crystal structure with 8 atoms per unit cell, and the cubic unit cell edge is 5.65754 ± 0.00002 Å at 25°C.[4] Thus the x-ray density is

$$\rho_{x-ray} = \frac{8 \times 72.59}{6.022045 \times 10^{23} \ (5.65754 \times 10^{-8})^3} \tag{9}$$

$$= 5.3252 \text{ g cm}^{-3} \text{ at } 25°C$$

The uncertainty in this value is dominated by the uncertainty ±0.03 in the atomic mass of Ge.[3] The resulting uncertainty in the density is ±0.0022 g cm^{-3}.

Our density values are closer to this x-ray value than to the value of ρ cited in the handbook. The value of ρ for sample I is

is usually appropriate. A comment should be made on any discrepancies with the acccepted or expected values. In this sample discussion, comment is also made on "internal discrepancies," possible systematic errors and ways to reduce them, and the relative importance of various sources of random error; a brief suggestion is made for an improvement in the experimental method. Other possible topics include suitability of the method used compared with other methods, other applications of the method, mention of any special circumstances or difficulties that might have influenced the results, discussion of any approximations that were made or could have been made, suggestions for changes or improvements in the calculations, mention of the theoretical significance of the result. At the end of several of the experiments in this book there are questions that provide topics for discussion; however, the student should usually go beyond these topics and include whatever other discussion he or she feel feels to be pertinent.

lower than the x-ray value, which could be explained by the presence of voids as suggested above. However, the ρ value for sample II is larger than the x-ray value by an amount that is greater than the sum of the quoted uncertainties. We have no explanation for this difference, but it does suggest the presence of systematic errors. One possible source of error is the volume V of the pycnometer. The stated value of 12.445 cm^3 has a rather large error limit of ± 0.0003 cm^3. Indeed, Appendix B shows that the largest contribution to the overall error in ρ comes from the uncertainty in V. No information is available about how V was determined or why it has such a large uncertainty. Our experimental precision indicates that a more reliable value of V could be obtained by measuring the weight of the pycnometer filled with water alone. Such a recalibration of the pycnometer volume would reduce the uncertainty in ρ due to random errors and might also shift the ρ values by a significant amount, which could improve the agreement between our ρ(sample II) and the x-ray value. However, a change in the V value would not improve the agreement between the two samples.

Further work should be done to test the quality of the method (by studying a well-behaved reference solid of known density) and the reproducibility of measurements on germanium (by studying Ge samples prepared in different ways).

References. In citing a reference in the text of the report, use either a numerical superscript or a number in brackets to refer the reader to the appropriate entry in the reference list (e.g., . . . Smith[3] or . . . Smith [3]). An appropriate style for referring to a book is illustrated by entry 4. If the publisher's name is not well known, it should be given in full; if the city of publication is not well known, the state or country should also be given (see entry 2). The citation style for referring to a book containing chapters by several different authors is illustrated by entry 1. A recommended citation style for journal articles is shown in entry 4. Standard abbreviations for the titles of many important journals are given in Appendix F. For typewritten or handwritten reports, it is common practice to underline only the journal volume number.

References

 1. N. Bauer and S. Z. Lewin, Determination of Density, in A. Weissberger and B. W. Rossiter (eds.), "Techniques of Chemistry," Vol. I, Part IV, Chap. 2, esp. pp. 101-105, Wiley-Interscience, New York (1972).

 2. "Handbook of Chemistry and Physics," 67th ed., p. F-10, CRC Press, Boca Raton, Florida (1986-87).

 3. Ibid., p. B-92.

 4. A. Smakula and J. Kalnajs, Phys. Rev. 99, 1737 (1955); W. B. Pearson, "A Handbook of Lattice Spacings and Structures of Metals and Alloys," Vol. 2, p. 971, Pergamon Press, Oxford (1967).

Appendix A: Data sheet

            } see attached pages

Appendix B: Sample calculations

Appendices. Important material that is so detailed that inclusion in the main body of the report would break the continuity of the text should be assembled in appendices. Examples are a long mathematical derivation, extensive tables of primary data (e.g., temperature–time values in calorimetry or composition–time values in chemical kinetics), or a detailed listing of any nonstandard computer program.

 All experimental data must be recorded directly in a laboratory notebook along with any identifying numbers on special apparatus and all necessary calibration data; see pp. 5–7 for further details. For reports made in an undergraduate laboratory course, a carbon copy or a photocopy of all pertinent notebook pages should be included as an appendix to the report. The first page of this appendix should have a clear heading with the student's name, the name of any partner, and the dates on which the experimental work was performed.

 A sample calculation should also be presented as an appendix to an undergraduate laboratory report. This appendix should show how one obtains the final results starting from the raw data. In general, the numbers used in the computations should have more significant figures than are justified by the precision of the final result, in order to avoid mathematical errors due to roundoff. Units should be included with each step of the calculation. Also specify the source of raw data used (e.g., run 2 on page 14 of notebook).

Appendix A: Data for Exp. 13   work done Sept. 16, 1987

Xerox copy of p. 24 of lab notebook attached below.

24

Maria Smith   16 Sept. 1987
Partner: John Klein

Exp. 13: Density of Ge
Pycnometer #7 has $V = 12.445 \pm 0.003$ cm^3
Bath temperature = 24.96°C

	W_1 (empty)	W_2 (solid)	W_3 (solid + water)
Sample I			
Run 1	8.6313 g	42.0307 g	48.1749 g
Run 2	8.6308	42.0259 ~~0295~~	48.1715
Ave	—	42.0301 g	48.1732 g
Sample II			
Run 3	8.6316	45.8468	51.3055
Run 4	8.6299	45.8490	51.3017
Ave	—	45.8479 g	51.3036 g

Ave W_1 = 8.6309 g

estimate weighing uncertainty ± 0.001 g
temp variations of bath $\sim \pm 0.05$°C

Appendix B: Sample calculation for Exp. 13

The computation of ρ for germanium sample I is shown here in detail. Using Eq. (5) with weights in gram, V in cm^3, and ρ_L in $g\ cm^{-3}$, we obtain

$$\rho(I) = \frac{(0.99707)(42.0301 - 8.6309)}{(0.99707)(12.445) + 42.0301 - 48.1732}$$

$$= \frac{(0.99707)(33.3992)}{12.4085 - 6.1431} = \frac{33.3013}{6.2654}$$

$$= 5.3151\ g\ cm^{-3}$$

The correction for air buoyancy is given by Eq. (6), and we find

$$\rho^* = 5.3151 + 0.0012\left(1 - \frac{5.3151}{0.99707}\right)$$

$$= 5.3099\ gm\ cm^{-3}$$

for the corrected density ρ^* of sample I. The 95% confidence limit for the density of this sample is obtained from Eq. (8):

$$(\Delta\rho)^2 = \frac{(5.315)^2}{(6.265)^2}\left[(0.997)^2(.003)^2 + (.001)^2 + (.002)^2\right]$$

$$= 0.720\ [8.95 + 1 + 4] \times 10^{-6} = 10.0 \times 10^{-6}$$

$$\therefore \Delta\rho = 0.0032\ g\ cm^{-3}$$

SPECIAL PROJECTS

In order to become a creative and independent research scientist, one must acquire a complex set of abilities. It is often necessary to invent new experimental methods or at least to adapt old ones to new needs. New apparatus must be designed, constructed, and fully tested. Most important of all, an intelligent procedure must be established for the use of this apparatus in making precise measurements.

Performing assigned experiments that are described in a book such as this one is the first step in developing research ability. Once some basic experience in carrying out physical measurements has been acquired, individually supervised experimental work on an original research problem is an excellent way to develop independence and experience with advanced research techniques in a specialized field. In preparation for such research work, we have found it profitable to encourage interested students to perform a special project in lieu of two or three regular experiments.

These special projects are intended to provide experience in choosing an interesting topic, in designing an experiment with the aid of literature references, in building apparatus, and in planning an appropriate experimental procedure. At least 20 hours of laboratory time should be available for carrying out such a project. Although there are certain limitations which are imposed by the available time and equipment, challenging and feasible topics with a research flavor can be found in most branches of physical chemistry. Indeed, it is sometimes possible to make a significant start on an original research problem that will eventually lead to publishable results. The primary emphasis should, however, be placed on independent planning of the experimental work rather than on original proposals for new research.

The two (or possibly three) partners working on a given special project should plan the experiment together, starting three or four weeks in advance, and should discuss their ideas frequently with an instructor or a teaching assistant. All work done in the laboratory should be supervised by an experienced research worker in order to avoid safety hazards.

Some of the projects done in the laboratory courses at MIT and OSU are listed below as examples of the kind of work that might be attempted.

Rotational Raman spectra of N_2 and O_2
Resonance fluorescence spectrum of Br_2
Vibrational–rotational spectra of CD_3H and CH_3D
Spectrophotometric study of stability of metal ion–EDTA complexes
Kinetics of the $H_2 + I_2 = 2HI$ reaction in the gas phase
Weak-acid catalysis of BH_4^- decomposition
Photochemistry of the cis–trans azobenzene interconversion
Isotope effect on reaction-rate constants
Susceptibility of a paramagnetic solid as a function of temperature
Dielectric constant of polypropylene glycol

X-ray study of short-range order in liquid mercury
Fluorescence and phosphorescence of complex ions in solution
Infrared study of hydrogen bonding of CH_3OD with various solvents
Light scattering near the critical point in ethane
Polanyi dilute flame reaction; e.g., $K + Br_2$
EPR study of gas-phase hydrogen and deuterium atoms
EPR spectra of methyl semiquinones
Photodissociation of NO_2
Fluorescence quenching of excited K atoms
Shock-tube kinetics: recombination of I atoms
Dielectric dispersion in high-polymer solutions
Mössbauer spectroscopy of ferrous and ferric salts
Mass spectrometry of fragmentation patterns of $CHCl_3$ and CH_2Cl_2

Many of these projects were quite ambitious and required hard work and enthusiasm on the part of both students and staff. Not all were completely successful in terms of precise numerical results, but each one was instructive and enjoyable. Frequently they resulted in an excellent scientific rapport between the students and the instructing staff.

Safety. It has already been emphasized that safe laboratory procedures require thoughtful awareness on the part of both students and instructors. This is especially important in the planning and execution of special projects, where new procedures need to be developed and often modified as the work progresses. Appendix E on safety hazards and safety equipment should be read before beginning a course of experimental work in physical chemistry and reviewed carefully before beginning any special project.

REFERENCES

1. A convenient source of information about the commercial availability of some 100 000 chemicals is *Chem Sources-USA,* published annually by Directories Publishing Co., Clemson, S.C. 29633-1824. An international directory of 300 000 chemicals is published by Walter de Gruyter & Co., Berlin/New York under the general title *chemBUYdirect.* It consists of several volumes: *chemADDRESSbook, chemPRODUCTindex,* and *chemSUPPLIERSdirectory.* The most recent edition is 1976, with supplements issued annually.
2. N. V. Steere (ed.), "CRC Handbook of Laboratory Safety," 2d ed., CRC Press, Boca Raton, Fla. (1971).
3. "Prudent Practices for Handling Hazardous Chemicals in Laboratories," prepared by the National Research Council's Committee on Hazardous Substances, National Academy Press, Washington, D.C. (1981).
4. H. M. Kanare, " Writing the Laboratory Notebook," American Chemical Society, Washington, D.C. (1985).
5. "Nomenclature of Organic Chemistry, Definitive Rules for Section A. Hydrocarbons. Section B. Fundamental Heterocyclic Systems," International Union of Pure and Applied Chemistry, 2d ed., Butterworths, London (1966).
6. "Nomenclature of Organic Chemistry, Definitive Rules for Section C. Characteristic Groups Containing Carbon, Hydrogen, Oxygen, Nitrogen, Halogen, Sulfur, Selenium, and/or Tellurium," International Union of Pure and Applied Chemistry, Butterworths, London (1965).
7. E. Cortelyou, *J. Chem. Educ.* **34,** 590 (1954).

CHAPTER
II

TREATMENT OF EXPERIMENTAL DATA

The usual objective of performing an experiment in physical chemistry is to obtain one or more numerical results. Between the recording of measured values and the reporting of numerical results there are processes of arithmetical calculations, some of which may involve averaging or smoothing the measured values but most of which involve the application of formulas derived from theory. Part of this chapter is devoted to a discussion of general techniques for carrying out such calculations.

However, our concern with the treatment of experimental data does not end when we have obtained a numerical result for the quantity of interest. We must answer the question: "How good is the numerical result?" Without an answer to this question, the numerical result may be next to useless. The expression of how "good" the result may be is usually couched in terms of its *accuracy*, i.e., a statement of the degree of the uncertainty of the result. A related question, often to be asked before the experiment is begun, is "How good does the result need to be?" The answer to this question may influence important decisions as to the experimental design, equipment, and degree of effort required to achieve the desired accuracy.

Indeed, it may be useful to discuss the matter also in economic terms. The "economic value" of the numerical result of an experiment generally depends upon the degree of accuracy that is responsibly claimed for the result, since the user of the result—in economic terms, the "consumer"—may be obliged to make important economic decisions that depend on assurance that the true numerical value is within stated bounds, at least with a stated probability. Is a product being tested of a saleable standard? Will the quality of an experimental result justify a research proposal to a governmental agency for continued financial support? Are the results assuredly adequate to justify inclusion in an important publication, upon which scientific reputation (and perhaps future promotion) may depend? To claim too high an accuracy through ignorance, carelessness, or self deception is to cheat the "consumer," no less than deliberate deception would. To claim too low an accuracy through overconservatism or intellectual laziness is to sell one's work too cheap, lessening the value of the result to the "consumer" and wasting resources that have been employed to achieve the accuracy that could rightfully have been claimed.

ERRORS

Random errors. We will assume that the numerical result to which our discussion applies is obtained with an instrument that measures a physical quantity for which the *a priori* range of possible values constitutes a continuum. This may be a continuous-reading instrument with a scale divided with numbered marks; customarily the scale reading is estimated to some fraction of the smallest interval ("scale division") between marks, typically one fifth or one tenth. The reading of the scale by a person with normal visual acuity is subject to a possible error (typically one or two tenths of a scale division). Even if the observer is capable of reading the scale with great precision, the measurement at any particular moment is subject to unpredictable, random fluctuations of the environmental conditions (temperature variations, voltage fluctuations in the electrical supply, electrical noise in the equipment, mechanical vibrations, etc.) which affect the instrumental response. Thus, repetitions of the measurement generally yield different readings, distributed at random over a small range. The same may apply to a digital instrument when the smallest increment in the numerical output, corresponding to ± 1 in the final digit, is smaller than the differences expected between different readings of the same quantity. If the smallest increment of numerical output is *not* smaller than expected random fluctuations, the instrument itself is making a contribution to measurement error because of round-off, and is perhaps not suitable for its intended purpose. Whatever the measuring instrument, we may say that the measurements are subject to *random error*. A statement of the *precision* of a measurement concerns its reproducibility when the measurement is made repeatedly with the same instrumentation and is therefore an expression of the uncertainty due to random error.

Faced with the inevitability of random error, the observer customarily makes several measurements of a given physical quantity and then averages them to obtain a value that is expected to be more reliable than any single measurement. The average or mean of N measurements of an experimental variable x is

$$\bar{x} = \frac{1}{N} \sum_{i=1}^{N} x_i \tag{1}$$

where x_i is the result of the ith measurement.

The observer also customarily reports in some manner the scatter of the measurements that go into this expression in order to give an indication of the precision. A crude way is to state the *range* of measured values:

$$R = x_{\text{largest}} - x_{\text{smallest}} \tag{2}$$

For several reasons this does not give a very clear measure of the precision of the individual measurements but can be used as the basis of quick estimates of more sophisticated measures. Very commonly the average deviation (a.d.) is cited:

$$\text{a.d.} = \frac{1}{N} \sum_{i=1}^{N} |x_i - \bar{x}| \tag{3}$$

Traditionally this has been cherished for its arithmetic simplicity, a virtue of declining value in this age of computers. Its usefulness for quantitative comparisons is limited because under given experimental conditions its value depends sensitively on sample size (the value for a pair of measurements being, on the average, about 30 percent smaller than the value calculated with a large group of measurements).

A measure of precision that is unbiased by sample size is the *variance*, denoted by S^2 and defined by

$$S^2 \equiv \frac{1}{N-1} \sum_{i=1}^{N} (x_i - \bar{x})^2 \tag{4a}$$

This quantity has the property of additivity so that estimates from several sources of random error may be combined, and it is useful in probability calculations. An alternate form of S^2, algebraically equivalent, and often more convenient for use with a calculator or computer, is

$$S^2 = \frac{1}{N-1} \left(\sum x_i^2 - N\bar{x}^2 \right) = \frac{N}{N-1} (\overline{x^2} - \bar{x}^2) \tag{4b}$$

Particular attention should be paid to the divisor $(N-1)$, which is known as the number of "degrees of freedom" and is equal to the number of *independent* data on which the calculation of S^2 is based. The N values of $x_i - \bar{x}$ are at this stage of the game not all independent; one degree of freedom has been used up in calculating the mean $\bar{x}$. That is, one of the N deviations, say the Nth one $(x_N - \bar{x})$, is not an independent variable since it can be calculated from the other deviations:

$$(x_N - \bar{x}) = -\sum_{i=1}^{N-1} (x_i - \bar{x}) \tag{5}$$

The difference between a factor of $N-1$ or N in the denominators of Eqs. (4) and (5) is trivial if one has 100 measurements but becomes very important when N is small.

The square root of the variance,

$$S = \frac{1}{\sqrt{N-1}} \left[\sum_{i=1}^{N} (x_i - \bar{x})^2 \right]^{1/2} \tag{6}$$

is often called the *estimated standard deviation,* or *e.s.d.* The significance of this name will become clear later. This parameter is widely used to indicate the precision of individual measurements. More important is the precision of the mean of the measurements. It will be shown later that the *estimated standard deviation of the mean* of N values is

$$S_m = \frac{S}{\sqrt{N}} = \frac{1}{\sqrt{N(N-1)}} \left[\sum_{i=1}^{N} (x_i - \bar{x})^2 \right]^{1/2} \tag{7}$$

It is apparent that the precision of the mean can be increased by increasing the number of individual measurements. The gain in going from $N=2$ to $N=10$ is quite dramatic; that in going from $N=10$ to $N=20$ is less so; and going from $N=90$ to $N=100$ results in only a small increase in precision. Thus, while in principle the random error can be reduced to as small an amount as desired by repeated measurements, as a practical matter there comes rapidly a "point of diminishing returns" in continuing to repeat measurements, and it is only rarely useful to make more than about 10 or 20 measurements of the same physical quantity.

Systematic errors. There is an important reason why excessive efforts should not be made to reduce the magnitude of random errors by making a very large number of measurements of the same quantity. This reason is that measurements are also subject to other kinds of error, chiefly *systematic errors,* the magnitude of which can all too easily submerge random error. A systematic error is one that cannot be reduced or eliminated by any number of repetitions of a measurement because it is inherent in the method, the instrumentation, or occasionally the interpretation of data. While the precision of a result, as already stated, is an expression of uncertainty due to random error, the *accuracy* of a result is an expression of *overall uncertainty including that due to systematic error.* Accuracy is very much more difficult to assess than precision.

Examples of systematic error are a calibration error in the instrument, failure to establish properly a "zero" reading of the instrument scale, improper graduation or alignment of an instrumental scale, uncompensated instrumental drift, leakage of material (e.g. of gas in a pressure or vacuum system) or of electricity (in an electrical circuit), and incomplete fulfillment of necessary experimental conditions (e.g. incomplete reaction in a calorimeter, incomplete dehydration of a sample prior to weighing). An example of another kind is faulty theoretical treatment of the results of the measurements to obtain the desired result, perhaps through a faulty approximation in the phenomenological theory involved.

A particular kind of systematic error might be called "personal error" because it results from subjective judgements or personal idiosyncrasies on the part of the observer. Examples are consistent parallax errors in reading scales and setting cross-hairs, excessive reaction times in actuating a stopwatch or electronic timer, strong number prejudices that influence the reading of scales (tendency to favor 0s and 5s, or superstition in favor of 7 and against 13), and arbitrariness in rejecting or not rejecting readings that deviate somewhat from the mean of the others. Such errors can be particularly insidious if the observer has a personal stake in the outcome of the experiment (whether to confirm a theory and win a Nobel prize, or to get the result thought to be expected by the instructor and win an "A" on the experiment). Persons of the highest character may not be altogether immune from this tendency, since biases are often subconscious.

Systematic errors are difficult to eliminate because they do not make their presence known as random errors do. They must be painstakingly sought out and eliminated through careful analysis of the entire experiment and of all assumptions on which the experimental method is founded. Some will be relatively easy to find, such as calibration and alignment errors; others, such as errors in the theory or the intervention of totally unexpected phenomena, can defy all attempts by ordinary means to identify them. All too often, important systematic errors are never found at all, and without doubt many are imbedded in the permanent scientific literature. Fortunately, some become apparent when new results, obtained by other methods, depart from the old ones by much more than the stated limits of error. R. A. Millikan's final value of Avogadro's number N_0, which he calculated as the ratio of the accurately known Faraday constant to his value of the electronic charge from the oil-drop experiment, was $(6.064 \pm 0.006) \times 10^{23}$ mol^{-1}; the currently recognized value is $(6.022137 \pm 0.000007) \times 10^{23}$, based on X-ray diffraction measurements of unit cell volume in a crystal together with the crystal density and formula weight. (Millikan was a very careful experimenter; the systematic error in his value resided mainly in data reported by others that he used in his calculations.)

As a rule the best assurance of low systematic error is agreement among two or more entirely different methods of measurement. Further discussion of systematic errors can be found elsewhere.[1,2]

If all significant sources of systematic error cannot be eliminated, the experimentalist should at least seek to identify them and place limits on their possible magnitudes. However, to the maximum extent possible, the scientist should aim to reduce the limits of systematic error to magnitudes that are small in comparison with the random errors of the experiment, so that statistical estimates made with equations in this chapter will be meaningful.

Other errors. Another kind of error is important enough to mention. It is called by some "erratic error," which is a fancy name for "mistakes." Mistakes in arithmetic computation are an example, but these are inexcusable since all computations should be recorded in the laboratory notebook and carefully checked. One-time reading errors of instrument scales in which digits are improperly read (the numeral 5 being mistaken for a 6, for example), or in which pairs of digits are transposed or a decimal point is misplaced in recording, are examples of mistakes that can be very damaging because they usually cannot be traced without repeating the experiment and because they may be very much larger than any conceivable random error. One of the most frequent examples is weighing errors, since a sample of material being prepared for an experiment is weighed only once. Other mistakes are failure of the experimenter to execute some essential step in the procedure (such as turning on a heater or a circulating pump), or faulty execution of the procedure (such as use of a wrong optical filter or even a wrong chemical substance for investigation). These mistakes may be untraceable, since clues may not exist in the laboratory notebook if the experimenter has been less than meticulous. Serious mistakes may be revealed only when the consequent numerical result appears to be impossible or ridiculous or is in wide disagreement with a theoretical value or prior result.

Any measurement that will of its nature be done only once should provoke in the experimenter a strong sense of insecurity that will motivate him or her to be very deliberate in recording the data in the notebook and checking them thoroughly before going on to something else. Wherever possible, measurements should be done at least in duplicate. Duplicate measurements are NOT, as many suppose, for the purpose of enabling statistical treatment of random error! For that purpose, at least four measurements should be made, preferably more. Mistakes in procedure can only be avoided by exercise of due care in planning and executing the experiment. If the procedure is excessively complicated, prepare a checklist ahead of time. Make sure that illumination is adequate, labels are clear, scales are clean. Record in the notebook the identification numbers of optical filters, tared vessels, and all relevant data from labels of sample containers. Keep the experimental setup orderly and as uncomplicated as possible.

Rejection of discordant data. It occasionally happens in making multiple measurements that one value differs from the rest considerably more than they differ from one another. Should the discordant value be rejected before an average is taken? This question provides one of the most severe tests to which the scientific objectivity of the experimenter is exposed, for he may (indeed, he should) reject any measurement, whether concordant or discordant, if he has

TABLE 1
Critical Q values for rejection of a discordant value at 90 percent confidence level[3]

N	3	4	5	6	7	8	9	10
Q_c	0.94	0.76	0.64	0.56	0.51	0.47	0.44	0.41

valid reason to believe that it is defective. Thus, the first step would always be to check for evidence of a determinate error (used the wrong pipette, added the wrong reagent, inverted the digits in writing down the number, etc.). If none can be found, the suspect value must be included unless valid statistical arguments can be presented to show that such a large deviation on the part of a member of the hypothetical ideal population of similar measurements is highly improbable. There are a number of ways of going about this. Most of them—such as the traditional rule about discarding values which differ from the mean of the others by more than four times the average deviation of the others—are based on faulty application of large-number statistics to small-sample problems and are not to be trusted. There is a very simple approach, known as the Q test, which is both statistically sound and straightforward to use.

When one in a series of 3 to 10 measurements appears to deviate from the mean by more than seems reasonable, calculate the quantity Q, defined by

$$Q \equiv \frac{|\text{(suspect value)} - \text{(value closest to it)}|}{\text{(highest value)} - \text{(lowest value)}} \tag{8}$$

Compare this value of Q with the critical value Q_c in Table 1 corresponding to the number of observations in the series. If Q is equal to or larger than Q_c, the suspect measurement should be rejected. If Q is less than Q_c, this measurement must be retained.

Example. Five determinations of the baseline reading of a spectrophotometer at a standardizing wavelength were 0.32, 0.38, 0.21, 0.35, and 0.34 absorbance units. May the 0.21 be discarded?

$$Q = \frac{0.32 - 0.21}{0.38 - 0.21} = 0.65$$

From Table 1 the critical value Q_c for $N = 5$ is 0.64. The value of Q is higher, hence the 0.21 value may be discarded with 90 percent confidence.

It should be noted that only *one* value in a series may be discarded in this manner. If there is more than one divergent value in a small series, it means that the data really do show a lot of scatter. When one value in a small series is considerably off but not sufficiently to justify rejection by the Q test, there is good reason to consider reporting the median value (i.e., the middle value) instead of the mean. As an illustration of this, consider the situation if the five spectrophotometer readings had been 0.32, 0.38, 0.23, 0.35, and 0.34. The value of Q for this set (0.60) does not quite exceed the critical value Q_c, and the 0.23 value cannot be rejected. However, reporting the median value 0.34 avoids the large effect that 0.23 would have if included in calculating the mean, while still allowing it a "vote." The mean of all five values is 0.32, whereas the mean without the suspect value is 0.35.

When N exceeds 10 it is suggested that an observation be rejected if its deviation from the mean of the others exceeds $2.6S$, where S is the estimated standard deviation of the others from their mean; this corresponds to about a one percent probability that the suspect observation is valid.

Whether the above-recommended criteria for rejection are used, or alternate criteria are used on the basis of an objectively justified preference, *it is important that the chosen criteria be used consistently.* The criteria to be used must be decided upon ahead of time; the selection of criteria at the time the measurements are at hand, and the decision to reject or not to reject a discordant measurement must be made, runs the grave risk of introducing bias. If a series of measurements presents a problem that cannot be objectively treated with previously selected criteria, then if at all possible the entire series should be rejected and a new series of measurement made after careful review of the experimental procedure to minimize the possibility of large errors, mistakes, or bias.

A cautionary tale. A professor once received from a chemical supply house a specimen of a crystalline organic compound needed in his research. This compound was optically active, with a positive rotation (i.e., it was dextrorotatory). The professor felt the need to reassure himself that the sample was not contaminated by the other (levorotatory) optical isomer. Such contamination would decrease the magnitude of the positive rotation. Therefore he gave a small amount of the material to a graduate student with instructions to determine its *specific rotation* at 25°C with the highly precise polarimeter that he had in his laboratory (see Exp. 30 and Chapter XVIII).

The next day, the student came back to the professor bearing his laboratory notebook. This showed that he had first adjusted the thermostat bath that supplies water to the jacket of the polarimeter so that it would cycle nicely between 24.8°C and 25.3°C. He then weighted out 1.5220 g of the material (which he had stored overnight in a desiccator) and dissolved it in distilled water in a volumetric flask to make 25.00 mL of solution, being careful to equilibrate the flask in the thermostat bath before making the solution up to the mark. He then filled a 20.00 cm (2.000-dm) polarimeter tube with the solution and obtained the following values for the optical rotation with sodium-D light, taking care to approach the extinction point from alternate sides:

α (deg)	dev
+20.04	−0.004
+20.07	+0.026
+20.05	+0.006
(+20.09)	rejected
+20.04	−0.004
+20.02	−0.024
+20.04	−0.004
+20.03	−0.014
+20.06	+0.016
+20.05	+0.006

$$\bar{\alpha} = +20.044 \qquad \sum \text{dev}^2 = 0.001824$$

$$S = \left(\frac{0.001824}{9-1}\right)^{1/2} = 0.015° \qquad S_m = \frac{0.015°}{\sqrt{9}} = 0.005°$$

With Eq. (33) and Table 2 (see section on Confidence Limits: the Student t distribution, pp. 45–47) he calculated a 95 percent confidence limit of error, assuming eight degrees of freedom, of

$$2.31 \times 0.0050° = 0.012°$$

This means that if he were to repeat this procedure many times he would find that the mean value of α would differ from 20.044° by no more than 0.012° 95 percent of the time.

Using the formula for the specific rotation, equivalent to Eq. (30-12),

$$[\alpha] = \frac{V}{Lm}\alpha \qquad (9)$$

where V is the volume of the solution as made up, in cubic centimeters (at the precision of this experiment milliliters may be considered equivalent to cubic centimeters), L is the length of the polarimeter tube in decimeters, m is the mass of the solute in grams, and α is the rotation angle in degrees, the student calculated for the specific rotation at 25°C with sodium-D light

$$[\alpha]_D^{25} = \frac{25.00}{2.0000 \times 1.5220} \times (+20.044) = +164.62 \text{ deg dm}^{-1} (\text{g cm}^{-3})^{-1}$$

Estimating from his general experience 95 percent confidence error limits of 0.02 cm³ in V, 0.002 dm in L, and 0.0003 g in m, and using his 95 percent confidence limits of 0.012° in the mean value of α, the student calculated by propagation of errors (see later section, pp. 55–63) 95 percent confidence limits of ±0.22 in the $[\alpha]_D^{25}$ result.

When the professor saw the final result for the specific rotation, +164.62 ± 0.22, his jaw dropped in bewilderment. The reported value *exceeded* the literature value of the pure dextrarotatory compound, +152.70, by 11.92, about fifty times the claimed limit of error! This ridiculous result could only mean that a serious error or mistake had been made in this determination, unless the literature value itself were seriously in error, which the professor strongly doubted. He looked for, but could not find, a mistake in arithmetic. He examined the data and asked the student why the reading of 20.09° had been rejected. The student replied that the reading seemed too far out of line. After asking acidly why the 20.02° reading had not also been rejected for the same reason, the professor explained to the student how to perform a Q test. This test showed that the reading had been wrongfully rejected (calculated $Q = 0.29$; for $N = 10$, $Q_c = 0.41$; see Table 1 and accompanying discussion). With a revised average of +20.049° ($S = 0.020°$, $S_m = 0.006°$) the professor quickly recalculated the specific rotation, obtaining

$$[\alpha]_D^{25} = +164.66 \pm 0.23 \text{ deg dm}^{-1} (\text{g cm}^{-3})^{-1}$$

Clearly, the main problem had not yet been found.

He asked to be shown the experimental setup. On seeing the balance on which the student had weighed out the material he noticed that the numerical scale on one of the weight dials was smudged in a few places; one of the 6s looked a little like a 5. Could the student have recorded 1.5220 g when the true weight might have been 1.6220 g? That would be enough to account for most of the difference between the experimental and literature values. Under pressure, the student conceded that this was possible, even likely. The professor calculated, tentatively,

$$[\alpha]_D^{25} = +164.66 \times \frac{1.5220}{1.6220} = +154.51$$

This was now within shooting distance, although the occurrence of the weighing mistake could not be considered a certainty. (The professor did point out to the student that had the error gone the other way, he might have thrown out the sample and either tried to find new material or given up the research.)

The professor then inspected the polarimeter setup. He noticed that the circulating pump, which was to supply constant-temperature water from the thermostat bath to the water jacket, was not running. "Did you have the circulating pump on during your measurements?" he asked. "What circulating pump?" replied the student, who had not realized that water would not flow by itself through the tubing connecting the bath to the instrument, and had failed to notice the none-too-conspicuous pump and its on-off switch. Nor had he taken the trouble to read the thermometer that was attached to the polarimeter jacket, which now read 18.0°C rather than 25.0°C. The professor was momentarily stumped, for there were no available data on the temperature coefficient of rotation for this substance. At length he reasoned that the principal effect of temperature on the rotation would be through the temperature dependence of the density of the solution. Using density values for water at the two temperatures the professor calculated

$$[\alpha]_D^{25} = +154.51 \times \frac{0.998625(@18°C)}{0.997075(@25°C)} = +154.75$$

"Wait a minute," said the student. "If the polarimeter tube were at 25°C instead of 18°C, the solution should be less dense; thus α, and therefore the specific rotation, should be lower, not higher." With a red face the professor inverted the ratio and obtained +154.27. ("Scarcely significant," mumbled the professor.) The result was still more than a degree higher than the literature value.

One more thought occurred to him. Something his experience told him should be in the notebook was not there. He took the polarimeter tube, rinsed it thoroughly and filled it with distilled water, and placed it in the instrument (after turning the circulating pump on). Instead of a rotation reading of 0.00° he obtained a reading of +0.26°. The student had failed to check the zero adjustment on the polarimeter scale! The average of 10 readings gave +0.258° ($S_m = 0.006°$). Making the needed correction, the professor obtained $[\alpha]_D^{25} = +152.52 \ \text{deg dm}^{-1} \ (\text{g cm}^{-3})^{-1}$, which was now slightly *below* the literature value of 152.70.

Because of doubt about the weighing error and the temperature correction, it was necessary to repeat the determination. The resulting value of $[\alpha]_D^{25}$ was $+152.61 \pm 0.23 \ \text{deg dm}^{-1} \ (\text{g cm}^{-3})^{-1}$. Evidently the optical purity of the specimen was quite high; the difference from the literature value of +152.70 is evidently not significant at the 95 percent confidence level.

Besides illustrating various kinds of errors, this tale demonstrates that there is a good deal more to the management of experimental data than the identification and estimation of errors. Topics such as propagation of errors and significance testing have been touched upon; these will be discussed in detail later in this chapter.

STATISTICAL TREATMENT OF RANDOM ERRORS

Error frequency distribution[2,4,5]. Let it be supposed that a very large number of measurements x_i ($i = 1, 2, \ldots, N$) are made of a physical quantity x and that these are subject to random errors ε_i. For simplicity we shall assume that the true value x_0 of this quantity is known and therefore that the errors are known. We are here concerned with the frequency $n(\varepsilon)$ of occurrence of errors of size ε. This can be shown by means of a bar graph, like that of Fig. 1, in which the error scale is divided into ranges of equal width and the height of each bar represents the number of measurements yielding errors that fall

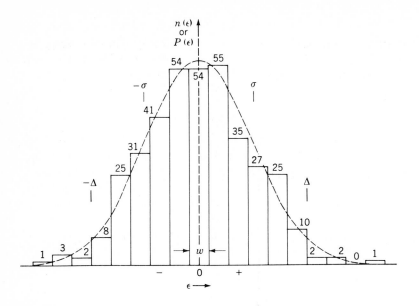

FIGURE 1

A typical distribution of errors. The bar graph represents the actual error frequency distribution $n(\varepsilon)$ for 376 measurements; the estimated normal error probability function $P(\varepsilon)$ is given by the dashed curve. Estimated values of the standard deviation σ and the 95-percent-confidence limit Δ are indicated in relation to the normal error curve.

within the respective range. The width w is chosen as a compromise between the desirability of having the numbers in each bar as large as possible and the desirability of having the number of bars as large as possible.

It is seen that even with as many as 376 measurements the graph shows irregularities, owing to the fact that the number of measurements represented by each bar is subject to statistical fluctuations that are not small in comparison with the number itself. From this graph we can make rough predictions concerning the probability that a measurement will yield an error of a given size. If we greatly increase the number of measurements represented (while perhaps decreasing the width w of the range more slowly than the rate of increase in the total number of measurements), the statistical fluctuations will become smaller in relation to the heights of the bars and our probability predictions are improved; we may even draw a smooth curve through the tops of the bars and assume it to represent an *error probability function $P(\varepsilon)$*. The vertical scale of the $P(\varepsilon)$ distribution should be adjusted by multiplication with an appropriate factor so that the function $P(\varepsilon)$ is normalized, i.e. so that

$$\int_{-\infty}^{\infty} P(\varepsilon)\,d\varepsilon = 1 \tag{10}$$

Its significance is that the probability that a single measurement will be in error by an amount lying in the range between ε and $\varepsilon + d\varepsilon$ is equal to $P(\varepsilon)\,d\varepsilon$.

A probability function derived in this way is approximate; the true probability function cannot be inferred from any finite number of measurements. However, it can often be assumed that the probability function is represented by a gaussian distribution called the *normal error probability function*,

$$P(\varepsilon) = \frac{1}{\sqrt{2\pi}\,\sigma}\, e^{-\varepsilon^2/2\sigma^2} \tag{11}$$

where σ is a parameter called the *standard deviation*. It is the root-mean-square error expected with this probability function:

$$\sigma \equiv (\overline{\varepsilon^2})^{1/2} = \left(\frac{1}{\sqrt{2\pi}\,\sigma} \int_{-\infty}^{\infty} \varepsilon^2 e^{-\varepsilon^2/2\sigma^2}\, d\varepsilon \right)^{1/2} \tag{12}$$

If the true value x_0 and thus the errors ε_i themselves are known, σ can be estimated from

$$\sigma = \left(\frac{1}{N} \sum_{i=1}^{N} \varepsilon_i^2 \right)^{1/2} \tag{13}$$

The dashed curve in Fig. 1 represents a normal error probability function, with a value of σ calculated with Eq. (13) from the 376 errors ε_i.

The usual assumptions leading to the normal error probability function are those required for the validity of the *central limit theorem*. This remarkable theorem states that if (1) an error in measurement is compounded of a large number of unpredictably variable contributions (which need *not* conform to a normal error probability distribution function or any other particular distribution function), if (2) these contributions are mutually independent, and if (3) all of them are small compared with the expected error itself, then the probability distribution function for the error itself will be approximately normal and will approach a normal distribution in the limit as the number of contributions increases indefinitely. The proof of this theorem, which we will not give here,[6,7] begins with an expression of the *joint probability* function for all of the individual contributions to the error (see Eq. (23)) and continues with manipulation of this expression by Fourier transformation and other techniques. The assumptions leading to this theorem are sufficient but not always altogether necessary; the normal error probability function may arise at least in part from circumstances different from those described. The factors which, in fact, determine the distribution are seldom known in detail. Thus, it is common practice to assume that the normal error probability function is applicable even in the absence of valid *a priori* reasons, and indeed the distributions obtained frequently conform to the normal error probability function to within ordinary statistical error, as in the case of the distribution shown in Fig. 1. However, even with 376 measurements the normal error probability function is not proved to be valid in all respects; with a much larger number of measurements it might become apparent that the true probability function is slightly skewed or flat-topped or double-peaked (bimodal), etc.

So far the discussion has dealt with the errors themselves, as if we knew their magnitudes. In actual circumstances we cannot know the errors ε_i by which the measurements x_i deviate from the true value x_0, but only the deviations $(x_i - \bar{x})$ from the mean $\bar{x}$ of a given set of measurements. If the random errors follow a gaussian distribution and the systematic errors are negligible, the best estimate of the true value x_0 of an experimentally measured quantity is the arithmetic mean $\bar{x}$. If an experimenter were able to make a very large (theoretically infinite) number of measurements, he could determine the *true mean* μ exactly, and the spread of the data points about this mean would indicate the precision of the observation. Indeed, the probability function for the deviations would be

$$P(x - \mu) = \frac{1}{\sqrt{2\pi}\,\sigma} \exp\left[-\frac{(x - \mu)^2}{2\sigma^2} \right] \tag{14}$$

where μ is the mean and σ the standard deviation of the distribution of the hypothetical *infinite population* of all possible observations. In the absence of systematic errors, μ should be equal to the true value, x_0.

If a large enough number of data points are available (at least 30 and preferably 100), classical probability calculations provide a very satisfactory description of the precision of the measurements in question. As a matter of fact one very seldom has anything like 30 measurements of the same quantity; most experiments are based on averages of 6 or less. Classical probability statistics are inadequate for the treatment of small numbers of observations, and techniques developed only within recent years are necessary to avoid large errors in the estimates of error. For a finite (and usually small) number of observations of a quantity, one obtains data that show a certain amount of spread. The true mean μ and the true spread of the hypothetical infinite population of measurements are what one wishes to have. Usually the best that we can actually have with a finite number N of measurements are *estimates* of μ and σ. These are respectively the mean $\bar{x}$, and the spread of the measurements for which we use the quantity S, the estimated standard deviation (e.s.d.), defined in Eq. (6). Clearly

$$\text{as } N \to \infty, \qquad \bar{x} \to \mu \qquad \text{and} \qquad S \to \sigma \tag{15}$$

and the *experimental* probability function may be written

$$P_{\exp}(x - \bar{x}) = \frac{1}{\sqrt{2\pi}\,S} \exp\left[-\frac{(x - \bar{x})^2}{2S^2} \right] \tag{16}$$

The normal probability functions as expressed by Eq. (14) (and approximated by Eq. (16)) is useful in theoretical treatments of random errors. For example, the normal probability distribution function is used to establish the probability P that an error is less than a certain magnitude δ, or, conversely, to establish the limiting width of the range, $-\delta$ to δ, within which the integrated probability P has a certain value:

$$P = \int_{-\delta}^{\delta} \frac{1}{\sqrt{2\pi}\,\sigma} \exp\left[-\frac{\varepsilon^2}{2\sigma^2}\right] d\varepsilon \tag{17}$$

If δ is specified as equal to σ, it is found that $P = 0.6826$; i.e., 68.26 percent of all errors do not exceed the standard deviation in magnitude. If P is specified as one-half, we find that $\delta_{1/2} = 0.6745\sigma$; this is called the *quartile deviation*, the name arising from the fact that one quarter of the area under the normal probability curve is contained between $\varepsilon = 0$ and $\varepsilon = \delta_{1/2}$. (Unfortunately, $\delta_{1/2}$ is also sometimes called the "probable error", a term that can be misleading.) If P is specified as 0.95 (95 percent), then

$$\delta_{0.95} = 1.96\sigma \simeq 2\sigma \tag{18}$$

This may be described as a "95 percent confidence limit" for an individual measurement. Confidence limits will be discussed more fully in the next section with application to mean values. It must be emphasized that Eq. (18) can be used only when σ is known. For σ to be known to a satisfactory approximation from the scatter of N measurements alone, N should be at least 20, preferably more.

UNCERTAINTY IN THE MEAN VALUE

Estimated standard deviation of the mean. The error ε_m in the mean of N observations (Eq. (1)) is the mean of the individual errors ε_i:

$$\varepsilon_m = \frac{1}{N} \sum_{i=1}^{N} \varepsilon_i \tag{19}$$

The mean square error is obtained by squaring both sides of this expression and averaging over an infinite population of sets of N measurements. The averaging is done by multiplying the square of each side of Eq. (19) by the probability distribution function $P(\varepsilon)$ and integrating from $-\infty$ to ∞.

$$S_m^2 = \overline{\varepsilon_m^2} = \frac{1}{N^2}\left[\sum_{i=1}^{N} \overline{\varepsilon_i^2} + \sum_{i=1}^{N}\sum_{j=1}^{N}{}' \overline{\varepsilon_i\varepsilon_j}\right] \tag{20}$$

where the prime indicates a sum over j values that are not equal to i. The mean of $\varepsilon_i\varepsilon_j$ vanishes because $\varepsilon_i\varepsilon_j$ is as often negative as positive. Therefore

$$S_m^2 = \frac{1}{N^2} N S^2$$

or

$$S_m = S/\sqrt{N} \tag{21}$$

in agreement with Eq. (7).† Thus, from Eq. (18), we have for a 95 percent confidence limit *in the mean,* when σ is at least approximately known,

$$\Delta \equiv \delta_{m,0.95} = 1.96\sigma/\sqrt{N} \simeq 2\sigma/\sqrt{N} \tag{22}$$

Where N is less than about 20 the Student t distribution described below should be used instead of the normal distribution.

The general results given above and some other important ones can also be proved through the use of the normal probability distribution, Eq. (11).[3-6] The *joint probability* that of N observations the first lies in the range x_1 to $x_1 + dx_1$, the second lies in the range x_2 to $x_2 + dx_2$, and so on is the product of the individual probabilities:

$$dP_{\text{joint}} = P_{\text{joint}}(x_1, x_2, \ldots, x_N) \, dx_1 \, dx_2 \ldots dx_N$$
$$= P_1(x_1) \, dx_1 \, P_2(x_2) \, dx_2 \ldots P_N(x_N) \, dx_N \tag{23}$$

On substituting the normal probability distributions into this equation, we obtain

$$dP_{\text{joint}} = (2\pi)^{-N/2} \sigma^{-N} \exp\left[-\sum_{i=1}^{N} \frac{(x_i - \mu)^2}{2\sigma^2} \right] dx_1 \, dx_2 \ldots dx_N \tag{24}$$

We now change from the set of variables $x_1, x_2, \ldots, x_N$ to a new set of N variables. Of these the only one we need to identify here is $\bar{x}$ (defined by Eq. (1)); we integrate over the rest of them and obtain (with procedures we omit here)[3]

$$dP_{\bar{m}} = P_m(\bar{x}) \, d\bar{x} \tag{25}$$

where

$$P_m(\bar{x}) = \frac{1}{\sqrt{2\pi} \, (\sigma/\sqrt{N})} \exp\left[-\frac{(\bar{x} - \mu)^2}{2(\sigma/\sqrt{N})^2} \right] \tag{26}$$

is the probability distribution function for the distribution of mean values $\bar{x}$ around the true mean of the distribution μ. Note that this equation resembles Eq. (11) except that $\sigma/\sqrt{N}$ replaces σ; therefore, in the case where σ is known, we may write

$$\sigma_m = \frac{\sigma}{\sqrt{N}} \tag{27}$$

In a similar way, using the *experimental* error distribution function of Eq. (16) instead of the true one of Eq. (11), we can derive Eq. (21).

---

† This is a quantitative statement of the intuitively obvious conviction that the mean of a group of N independent measurements of equal weight has a higher precision than any single one of the measurements.

Confidence limits: the Student t distribution. In most cases, as already stated, σ is not precisely known ahead of time. We need to find some kind of distribution function that will apply when σ is not known. The joint probability of Eqs. (23) and (24) can be integrated after a *different* change of variables from that mentioned above.[8] In this case the only variable that we do not integrate over is

$$\tau \equiv \frac{\bar{x} - \mu}{(S/\sqrt{N})} = \frac{\bar{x} - \mu}{S_m} \tag{28}$$

After integration over all new variables except τ we obtain

$$dP = P(\tau)\, d\tau \tag{29}$$

where

$$P(\tau) = \left[\frac{\Gamma(N/2)}{\sqrt{(N-1)\pi}\,\Gamma([N-1]/2)} \right]\left(1 + \frac{\tau^2}{N-1}\right)^{-N/2} \tag{30}$$

The factor in large square brackets, containing the gamma functions, is a normalization constant, with a value which assures that

$$\int_{-\infty}^{\infty} P(\tau)\, d\tau = 1 \tag{31}$$

The distribution represented in Eqs. (29) and (30) is known as the Student t distribution; "Student" is a pseudonym for W. S. Gosset who developed this distribution for application to biometrics. Note that σ does not appear in the expression. The shape of the Student t distribution, unlike that of the normal distribution, is variable, depending on N; when N is large it rapidly approaches the normal distribution. The distribution functions for several values of the number of degrees of freedom ($\nu = N - 1$) are shown in Fig. 2.

We may use this distribution in much the same way we use the normal distribution. Suppose we wish to find the (symmetrical) range of values of τ over which the integral of the Student probability function is a fraction P (say 0.95, or 95 percent). Let us define this range to be from $-t$ to t:

$$\int_{-t}^{t} P(\tau)\, d\tau = P \tag{32}$$

"Critical" values of t for a given number of degrees of freedom ν and a given P are given in Table 2 and in published tables.[9] Let us define a limit of error δ as the value of $\bar{x} - \mu$ that corresponds to the limit of integration t. Then from Eq. (28)

$$\delta = tS_m = t\frac{S}{\sqrt{N}} \qquad \Delta = t_{0.95}S_m = t_{0.95}\frac{S}{\sqrt{N}} \tag{33}$$

where again we use Δ to mean a 95 percent confidence limit in the mean. We emphasize here the use of ν rather than N in Fig. 2 and Table 2 because the

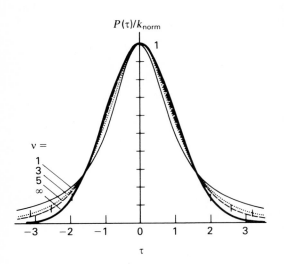

$P(\tau)/k_{norm}$

FIGURE 2

Student t distribution functions $P(\tau)$ for $v = 1, 3, 5, \infty$ (v = number of degrees of freedom). The quantities actually plotted are

$$P(\tau)/k_{norm} = \left[1 + \frac{\tau^2}{N-1}\right]^{-N/2}$$
$$= [1 + \tau^2/v]^{-(v+1)/2}$$

which are not normalized; they need to be multipled by the normalizing constant k_{norm}, which is the bracketed quantity in Eq. (30) with $N - 1 = v$. For $v = 1$ this is 0.3183, for $v = 3$ it is 0.3676, for $v = 5$ it is 0.3796, and for $v = \infty$ it is $1/\sqrt{2\pi} = 0.3989$. The curve for $v = \infty$ is the normal error probability curve. The short vertical bars mark the 95-percent confidence limits. These are not indicated for $v = 1$, as they are at $\tau = \pm 12.7$, far outside the plotting range.

TABLE 2
Critical values of t^a

v	P P'	0.50 0.75	0.80 0.90	0.90 0.95	0.95 0.975	0.98 0.99	0.99 0.995	0.999 0.9995
1		1.00	3.08	6.31	12.7	31.8	63.7	637.0
2		0.816	1.89	2.92	4.30	6.96	9.92	31.6
3		0.765	1.64	2.35	3.18	4.54	5.84	12.9
4		0.741	1.53	2.13	2.78	3.75	4.60	8.61
5		0.727	1.48	2.01	2.57	3.36	4.03	6.86
6		0.718	1.44	1.94	2.45	3.14	3.71	5.96
7		0.711	1.42	1.89	2.36	3.00	3.50	5.40
8		0.706	1.40	1.86	2.31	2.90	3.36	5.04
9		0.703	1.38	1.83	2.26	2.82	3.25	4.78
10		0.700	1.37	1.81	2.23	2.76	3.17	4.59
15		0.691	1.34	1.75	2.13	2.60	2.95	4.07
20		0.687	1.32	1.72	2.09	2.53	2.85	3.85
30		0.683	1.31	1.70	2.04	2.46	2.75	3.65
∞		0.674	1.28	1.64	1.96	2.33	2.58	3.29

[a] P is the probability that the mean μ of the population does not *differ from* the sample mean $\bar{x}$ by a factor of more than t, the tabular entry for a given number of degrees of freedom v. P' is the probability that μ does not *exceed* $\bar{x}$ (or, alternatively, the probability that $\bar{x}$ does not *exceed* μ) by a factor of more than t for a given number of degrees of freedom v.

figure and the table are then relevant to cases in which several parameters are determined from the N measurements, rather than only one (in the present case, the mean $\bar{x}$). Such a case is least-squares curve fitting (Chapter XX). If the number of parameters to be determined by least squares is n and the number of data points is N, then $v = N - n$.

As an example, suppose that in determining an optical rotation with a polarimeter (see "A Cautionary Tale", above) we have for $N = 10$, $v = 9$, a mean rotation angle α of 20.049°; the scatter of the data give $S = 0.020°$. For 95 percent confidence limits we look under the $P = 0.95$ column to the row corresponding to $v = 9$, and read 2.26 for the value of t. Therefore the 95 percent confidence limits are $\pm \Delta$, where

$$\Delta = 2.26 \times 0.020°/\sqrt{10} = 0.014°$$

and the average value of the optical rotation should be reported as

$$\alpha = +20.049° \pm 0.014° \qquad (95\%, \ v = 9)$$

Caution is required when using tables from some sources[9] since for those tables the integration of Eq. (32) is from $-\infty$ to t rather than from $-t$ to t. If we designate the corresponding integrated probability as P', then

$$P' = (P + 1)/2 \qquad \text{or} \qquad P = 2P' - 1 \tag{34}$$

Values of P' are also given in Table 2, and are useful in significance tests under certain circumstances to be described later.

Another caution. The Student t distribution is to be used when the numerical value to which it attaches is a mean of a definite number of direct observations, or is a numerical result calculated from such a mean by a procedure that introduces no uncertainties comparable in magnitude to the random errors of the direct observations. This is *not* the case for the specific rotation $[\alpha]_D^{25}$ in "The Cautionary Tale" related earlier; the uncertainty contribution to the final result that is due to random errors in the raw data on optical rotations α_i is *not* large compared to the contributions due to the *estimated* uncertainties in the other variables (particularly V and L) that are required for calculating the final result $[\alpha]_D^{25}$. The number of degrees of freedom applicable to the final result cannot be precisely known, or at least it would be an extremely difficult matter to determine. In addition, we must reiterate that the statistical treatment can mean little or nothing in the face of large systematic errors, except only to indicate the *precision* of the measurement method.

Short-cut methods based on the range. For small samples (N from 3 to about 10) the range, $R = x(\text{largest}) - x(\text{smallest})$, can be used to obtain rough uncertainty estimates. The range, when multiplied by the appropriate value of K_2 from Table 3, gives an approximate but useful estimate of the standard deviation. The factor J in Table 3 is equivalent to $K_2 t/\sqrt{N}$, where t is the critical value from the $P = 95$ percent column in Table 2 for the appropriate

TABLE 3
Factors for small-sample estimates based on the range R

N	K_2	$J(95\%)$
3	0.59	1.3
4	0.49	0.72
5	0.43	0.51
6	0.40	0.40
7	0.37	0.33
8	0.35	0.29
9	0.34	0.26
10	0.33	0.23

degree of freedom. Hence, to get a quick estimate of the 95 percent confidence limits of the mean of a few observations, simply calculate $\pm JR$. In summary,

$$S \simeq K_2 R \quad \text{and} \quad \Delta(95\% \text{ confidence}) \simeq JR \quad (35)$$

These techniques, which have developed out of the statistical methodology known as quality control, are becoming widely adopted because of their convenience. They are, however, decidedly inferior to the methods already described.

Estimation of limits of error. It must be kept in mind that the above discussion of random errors is based on the notion that one has no prior information about the precision of the measurements, in which case the uncertainty must be estimated from the scatter of the data in a given sample. In practice the number of independent measurements of an experimental quantity is often very small (sometimes two or only one) owing to time limitations, and it is unrealistic or impossible to estimate the confidence limits from the range. However, the experimenter frequently has a good idea of the precision of the instrument or method he is using. In this case, and also in cases where systematic errors must be estimated as well, the 95 percent confidence limits must be assigned on the basis of the individual judgement and experience of the experimenter. The ability to assign realistic limits of error, large enough to be safe but not so large as to detract unnecessarily from the value of the measurement, is one of the marks of a good experimentalist. It would be very difficult to formulate a set of rules that would be applicable in all cases; the confidence limits assigned in any given case will depend on the characteristics and capabilities of the instrument, its age and condition, the reproducibility of its reading, the quality of its calibration, and the user's experience and familiarity with it. A reasonable 95 percent confidence limit in weighing a small solid object on an analytical balance with weights calibrated to 0.05 mg might be 0.3 mg; in determining a short time interval with a manually actuated timer it might be 0.4 s; in reading an instrumental scale it might be 0.2 of the smallest scale division, or more depending on the instrument; etc.

Confidence limits assigned in this way will be represented in later sections by Δ on the rough assumption that they are 95 percent, based on long-term experience. However, one person's 95 percent estimate might be another's 90 percent, or 99 percent; this is unavoidable, and the experimenter simply has to do the best he or she can. Estimation of meaningful standard deviations in this manner is more difficult; if these are needed, rough values can be obtained as the estimated 95 percent confidence limits divided by 2.

Significant figures. Many numerical data in the literature and in everyday use carry no explicit statement of uncertainty, such as an estimated standard deviation (e.s.d.) S, or 95 percent confidence limits Δ. Some idea of the uncertainty in a numerical value is (or should be) conveyed by the number of digits employed to express that numerical value. The digits that are essential to conveying the numerical value to its full accuracy or precision are called significant figures. Thus the numbers 4521 and 6.784 each have four significant figures, provided all of the digits are meaningful. So also do 0.006784 and 4,521,000 since the leading and trailing zeros generally determine only the magnitude of the number and have nothing to do with the relative precision. In the latter case, however, there is some ambiguity because one or more of the trailing zeros may be significant. The ambiguity may be removed by expressing the number in exponential notation: 4.521×10^6 or 4.5210×10^6, etc., as the case may be.

This discussion applies to numerical values of variables whose possible range constitutes a continuum; in computer terminology these are called "real" numbers, as opposed to integers (or, of course, complex numbers). Integers should be identified as such and all of the digits necessary to express them should be given.

The use of too few digits in reporting numerical values robs the user of valuable information. The usual fault, however, is the reckless use of too many digits, which conveys a false sense of the accuracy or precision of the numerical value.

In reporting numerical values, certain rules regarding significant figures should be followed.

1. Numerical values are considered to be uncertain in the last digit by ± 3 or more, and perhaps slightly uncertain in the next-to-last digit. Ordinarily the next-to-last digit should not be uncertain by more than ± 2.
2. In rounding off numbers, (*a*) increase the last retained digit by 1 if the leftmost digit to be dropped is more than 5, or is 5 followed by any nonzero digits:

$$3.457 \rightarrow 3.46 \qquad 52.6502 \rightarrow 52.7$$

(b) leave the last retained digit unchanged if the leftmost digit to be dropped is less than 5:

$$0.3462 \rightarrow 0.346$$

(c) if the leftmost digit to be dropped is 5 followed by no digits except zero, then increase the last retained digit by 1 if it is odd, and leave it unchanged if it is even:

$$73.135 \rightarrow 73.14 \qquad 48.725 \rightarrow 48.72$$

For a discussion of the retention of digits in numbers during computation, see the later section on Calculations.

The presentation of numerical results and their uncertainties. At some point the final numerical result or results of an experiment and their uncertainties must be presented in a report or a publication, and it is important that this be done in a clear manner. Uncertainties are often presented (usually with "±") without making clear what kind of uncertainty estimates are being employed. Some conventions have arisen, which must be used with caution since they are by no means universally accepted; see Table 4.

Whatever the uncertainty figure, it is important that the number of decimal places given or implied by the uncertainty figure be the same as that of the numerical value to which it applies. Obviously, the same units apply to the uncertainty figure as to the numerical value. However, it is usually not necessary to state the units twice; it is sufficient to state them following the uncertainty figure. To be sure to avoid confusion, even with the confidence level at the usual 95 percent, it is becoming increasingly common to state the level of confidence and the number of measurements or the number of degrees of freedom, e.g.,

$$\alpha = -42.8 \pm 0.3° \qquad (95 \text{ percent}, N = 7)$$

or

$$\alpha = -42.8 \pm 0.3° \qquad (95 \text{ percent}, 6 \text{ degrees of freedom})$$

TABLE 4
Most common modes of uncertainty presentation

Result	Uncertainty	Presentation
In terms of the e.s.d., S, with parentheses around the uncertain digits		
3.42 bar	$S = 0.04$ bar	3.42(4) bar
0.451 nm	$S = 0.012$ nm	0.451(12) nm
In terms of the 95 percent confidence limit, Δ, with ±		
66.32 J K^{-1}	$\Delta = 0.21$ J K^{-1}	66.32 ± 0.21 J K^{-1}
5.02×10^{-4} V	$\Delta = 5 \times 10^{-6}$ V	$(5.02 \pm 0.05) \times 10^{-4}$ V

when the result presented is a mean of several independent measurements. Otherwise, when the limits are estimated on the basis of judgement and experience,

$$\alpha = -42.8 \pm 0.3° \qquad (95 \text{ percent})$$

is better than a presentation with no confidence limit explicitly stated.

As results are prepared for presentation the relationship between the 95-percent confidence limit and the number of significant figures retained in the final result should be recognized: generally the confidence limit should be expressed as a single digit equal to or greater than 3 or as a two-digit number not greater than 25. Correspondingly, if the uncertainty is expressed as an e.s.d., it should be expressed as a single digit equal to or greater than 2 or a two-digit number not greater than 15. These guides are not intended to be used inflexibly; departures may be made from them for good reasons.

SIGNIFICANCE TESTING

A frequent question that arises in quantitative experimentation is whether the observed difference between an experimentally measured quantity and the accepted, theoretical, or standard value is representative of a genuine difference or merely the natural result of random experimental error. Similarly, the question may arise whether two experimental studies differ in their results because of real differences either in the systems studied or in the methods used, or whether the apparent discrepancy is no greater than might be found by chance simply because of natural variability in the data.

The basic philosophy of significance testing is simple. Two series of measurements are made and their results are observed to differ by an amount we shall call D. The maximum difference that could be expected to occur if both series represent random samples of the same thing (i.e., no real difference in the quantity being measured in the two series) at a given probability level is calculated from probability theory. If the observed difference is smaller, it is concluded that there is no real evidence of difference. If it is larger, it is concluded that the difference is probably real (at whatever probability level was chosen). The conventionally accepted probability level for considering differences significant is 95 percent. It is based on the experience that if action is taken at lower levels of significance (e.g., 90 or 80 percent), there are too many embarrassing false alarms, whereas if the requirements for significance are made too stringent (e.g., at the 99-percent level), too many things get missed. Under circumstances in which it is very important not to claim a significant difference when none is really there, the 99 percent or even the 99.9 percent level may be set as the basis for taking action, but in ordinary scientific practice the 95 percent level has been found to be best.

We will here show how the the Student t statistic can be used to answer the question as to whether the above-mentioned difference D is statistically significant at a specified probability level. We begin by defining the effective uncertainty in the difference D between the two values being compared, using statistical data obtained in calculating mean values. This uncertainty, expressed as an e.s.d., S_D, is then converted to a confidence limit Δ_D with the help of Eq. (33) written in the form

$$\Delta_D = tS_D \tag{36}$$

where t is found in Table 2. If the difference D exceeds the confidence limit Δ_D, the difference between the two values is said to be significant at the specified confidence level; otherwise, it is said to be not significant.

An equivalent approach is to determine t from $t = D/S_D$ and find in Table 2 the values of t for the applicable value of v that bracket it. The heading of the column with the *lower* t entry represents the probability level at which the difference is *significant*; the heading of the column with the *higher* t entry represents the probability level at which the difference is *not significant*.

We will examine two different cases.

CASE 1. The mean $\bar{x}$ of N experimental observations is being compared with a theoretical, accepted, standard, or "true" value x_0 that may be regarded as exact. Is the difference D between $\bar{x}$ and x_0 statistically significant? The uncertainty in $\bar{x}$ is characterized by S_m as calculated with Eq. (7). Given the desired confidence level P (95 percent, or whatever) and the number of degrees of freedom v, t is found from Table 2 and Δ_D is calculated with Eq. (36). If $D \equiv |\bar{x} - x_0| > \Delta_D$, the difference D is significant. A related situation occurs when the question asked is whether $\bar{x}$ is significantly *greater than* x_0. (The same question could be asked with *less than*.) In this situation the value of t is found under P' rather than under P; if $D \equiv \bar{x} - x_0 > \Delta_D$, the difference D is significant.

Example. The accepted value of the weight percent of bromine in a standard KBr sample is 67.14, which may be regarded as exact. Four analyses of this sample by a new method gave a mean value $\bar{x}$ of 67.07 and an e.s.d. S of 0.04. Is the difference $D = 0.07$ significant at the 95 percent level? Here $S_D = S_m = 0.04/\sqrt{4} = 0.02$, and with $v = 3$ we find from Table 2 and $P = 0.95$ that $t = 3.18$. Thus:

$$\Delta_D = 3.18 \times 0.02 = 0.063$$

Since $D > \Delta_D$ we conclude that the difference D *is* significant at the 95 percent confidence level. (It is, however, *not* significant at the 97.5 percent level.) We therefore conclude with 95 percent confidence that the new method of analysis is subject to systematic error.

Example. The accepted value of the density of pure silver metal at 25°C is $x_0 = 10.501$ g cm^{-3}. This value, based on X-ray diffraction measurements, may be regarded as exact for the purposes of this example because the uncertainty, ± 0.0002 g cm^{-3}, is quite small compared to the random errors inherent in routine density measurements. It may also be regarded as an upper limit for pure silver metal. Low values can be given by a routine pycnometric determination if, for example, tiny air bubbles are entrained in the sample (see Sample Report, Chapter I). The density of a sample of finely divided pure silver metal was determined six times with a pycnometer. The mean density and the e.s.d. of the mean were $\bar{x} = 10.446 \text{ g cm}^{-3}$ and $S_m = 0.024 \text{ g cm}^{-3}$, thus $D' = -0.055 \text{ g cm}^{-3}$. Is the measured density significantly lower than the accepted value? Setting $S_D = S_m = 0.024 \text{ g cm}^{-3}$ we obtain, with $v = 5$, $P' = 0.95$

$$\Delta_D = 2.01 \times 0.024 = 0.048$$

Since the magnitude of D' is greater than this, the difference is significant at the 95 percent confidence level, and systematic errors in the method, perhaps due to the presence of tiny air bubbles, are to be suspected. Notice that if, instead of taking $P' = 0.95$, we were to take $P = 0.95$, we would obtain

$$\Delta_D = 2.57 \times 0.024 = 0.062$$

indicating erroneously that the difference is *not* significant at the 95 percent level.

CASE 2. The mean $\bar{x}_1$ of N_1 observations of a physical variable is being compared with the mean $\bar{x}_2$ of N_2 observations of the same physical variable made by a different method. Are the two means significantly different (suggesting that at least one may be subject to systematic error)? We need to have statistical estimates of uncertainty for both data sets; let the measurement e.s.d.'s be S_1 and S_2. The square root of the *pooled* variance of the two data sets is[1]

$$S_{12} = \left[\frac{(N_1 - 1)S_1^2 + (N_2 - 1)S_2^2}{N_1 + N_2 - 2} \right]^{1/2} \tag{37}$$

and the e.s.d. in the *difference between the two means, D,* is

$$S_D = S_{12}\left(\frac{1}{N_1} + \frac{1}{N_2} \right)^{1/2}$$
$$= S_{12}\left(\frac{N_1 + N_2}{N_1 N_2} \right)^{1/2} \tag{38}$$

Again, this is multiplied by the t value appropriate to the desired confidence level, but here the number of degrees of freedom v is $N_1 + N_2 - 2$. If $D \equiv |\bar{x}_1 - \bar{x}_2| > \Delta_D = t S_D$ then the difference D is significant.

If $N_1 = N_2 = N$, Eqs. (37) and (38) can be greatly simplified:

$$S_D = (S_{m1}^2 + S_{m2}^2)^{1/2} \tag{39}$$

and if in addition $S_{m1} = S_{m2} = S_m$, then

$$S_D = \sqrt{2}\, S_m \tag{40}$$

In either case the appropriate number of degrees of freedom v is $2N - 2$.

Again, the situation may instead be one in which the question to be asked is whether one mean is significantly *greater than* (or, alternatively, *less than*) the other mean at the stated probability level. Here again P' should be used instead of P.

Example. The heat of neutralization of an amine is measured four times in a large calorimeter and six times in a small one with the following results:

Large: $N = 4$, $\overline{\Delta H_1} = 30.5 \text{ kJ mol}^{-1}$, $S_1 = 1.3 \text{ kJ mol}^{-1}$

Small: $N = 6$, $\overline{\Delta H_2} = 28.9 \text{ kJ mol}^{-1}$, $S_2 = 1.7 \text{ kJ mol}^{-1}$

Is the difference $D = 1.6 \text{ kJ mol}^{-1}$ significant? The e.s.d. of the difference is, from Eqs. (37) and (38)

$$S_D = \left(\frac{3 \times 1.69 + 5 \times 2.89}{4 + 6 - 2} \times \frac{4 + 6}{4 \times 6} \right)^{1/2} = 1.01 \text{ kJ mol}^{-1}$$

For $v = 4 + 6 - 2 = 8$ and $P = 0.95$, t from Table 2 is 2.31. Therefore

$$\Delta_D = 2.31 \times 1.01 = 2.33 \text{ kJ mol}^{-1}$$

Since $\Delta_D > D$ the difference is not significant at the 95 percent level. If we set $t = D/S_D = 1.6/1.01 = 1.58$, we see that this is bracketed in Table 2 between 1.40 for $P = 0.80$ and 1.86 for $P = 0.90$. Therefore the difference is significant at the 80 percent confidence level but not at the 90 percent level. Thus there may be systematic errors in one or both calorimetric methods, but one would like to be more sure before one were to take serious action.

Example. A chemical company was trying to decide between Company A and Company B for a contract to provide a certain catalytic product at a set price. The chemical company was leaning strongly toward Company B for reasons not connected with the quality of the product (i.e., one of the directors had a brother-in-law working for Company B), but could not justify signing with that company if Company A could be shown to provide a product that was significantly better in its catalytic yield at the 95 percent confidence level. The crucial test agreed upon by all parties was to measure, in both products under identical conditions and under inspection by representatives of both suppliers, the value of the catalytic activity as measured by the conversion of CO and H_2 to methanol. In each case, 11 measurements of the yield variable were made. For Company A's product the mean was 53.23 (in the appropriate units, which we are not concerned with here) with $S_m = 0.75$; for Company B's product the mean was 50.87 with $S_m = 0.85$. Should Company A be selected to provide the product? Here $D = 2.36$, and $v = 20$. At $P = 0.95$, $t = 2.09$. Using Eqs. (39) and (36) we obtain

$$S_D = (0.75^2 + 0.85^2)^{1/2} = 1.25$$

$$\Delta_D = 2.09 \times 1.25 = 2.61 \qquad \text{(95 percent, } v = 20\text{)}$$

Since $D < \Delta_D$ the directors of the chemical company were happily preparing to award the contract to Company B, until a bright young PhD chemist recently added to their research division pointed out that P' should have been used instead of P. The question being asked was *not* "is A's product significantly *different* from B's product?" but rather "is A's product significantly *better* than B's product?," it having been presumed ahead of time that otherwise B's product would be the one chosen. With $t = 1.72$,

$$\Delta_D = 1.72 \times 1.25 = 2.15$$

Glumly, the directors signed the contract with Company A. The record does not reveal whether the bright young PhD was promoted or is now working for Company A.

The above discussion and examples are based on the application of statistical theory. Something should be said about the situations in which the limits of error for one or both of the values that are being compared have been estimated by the experimenter on the basis of judgement and experience rather than with equations derived from statistics. Here, the number of degrees of freedom is unknown or without meaning. In the situation analogous to Case 1, if the difference between the measured and "true" values exceeds the estimated Δ for the former, the difference is regarded as significant at the 95 percent confidence level. In the situation analogous to Case 2, the pooled uncertainty may be taken as the square root of the sum of the squares of the individual estimated confidence limits, weights being used if judged necessary:

$$\Delta_D = (\Delta_1^2 + \Delta_2^2)^{1/2} \tag{41}$$

or

$$\Delta_D = \left[\frac{w_1 \Delta_1^2 + w_2 \Delta_2^2}{w_1 + w_2} \right]^{1/2} \tag{42}$$

If the difference between the means exceeds Δ_D then the difference is regarded as significant at the 95 percent confidence level. In these situations there is little point in distinguishing between "different from" and "greater (less) than" in view of the roughness of the 95-percent-confidence estimate.

PROPAGATION OF ERRORS

An experiment has been performed: a variety of direct measurements have been made—e.g. weights, volumes, temperatures, measured e.m.f.'s, spectral absorbances, etc.—and uncertainties in all of them have been estimated, either from statistical data obtained by repeated measurements or from judgement and experience. From the values obtained by direct measurement, and with the aid of a phenomenological theory, a final numerical result is calculated. Let the desired numerical result by designated by F and the directly measured quantities be designated by $x, y, z, \ldots$. The latter quantities are assumed to be mutually independent. Let their uncertainties, usually in the form of 95-percent confidence limits, be designated $\Delta(x), \Delta(y), \Delta(z), \ldots$. The value of F is determined by substituting the experimentally determined values of

$x, y, z, \ldots$ into a mathematical formula, which we write symbolically as

$$F = f(x, y, z, \ldots) \tag{43}$$

The issue to be discussed here is how can one estimate the uncertainty of the final result F, usually in the form of a 95-percent confidence limit.

Infinitesimal changes dx, dy, etc., in the experimentally determined values will produce in F the infinitesimal change dF, where

$$dF = \frac{\partial F}{\partial x} dx + \frac{\partial F}{\partial y} dy + \frac{\partial F}{\partial z} dz + \cdots \tag{44}$$

If the changes are finite rather than infinitesimal, but are small enough that the values of the partial derivatives are not appreciably affected by the changes, we have approximately

$$\Delta F = \frac{\partial F}{\partial x} \Delta x + \frac{\partial F}{\partial y} \Delta y + \frac{\partial F}{\partial z} \Delta z + \cdots \tag{45}$$

This is equivalent to a Taylor expansion in which only the first-power terms have been retained. Now suppose that Δx represents the experimental error $\varepsilon(x)$ in the quantity x:

$$\Delta x = \varepsilon(x) \equiv x \text{ (measured)} - x \text{ (true)} \tag{46}$$

Such errors will produce an error in F,

$$\Delta F = \varepsilon(F) \equiv F \text{ (calculated from measured } x, y, z, \ldots) - F \text{ (true)} \tag{47}$$

the value of which is given by

$$\varepsilon(F) = \frac{\partial F}{\partial x} \varepsilon(x) + \frac{\partial F}{\partial y} \varepsilon(y) + \frac{\partial F}{\partial z} \varepsilon(z) + \cdots. \tag{48}$$

Propagation of systematic errors. Systematic or "determinate" errors are not governed by probability statistics and do not involve such concepts as the variance or standard deviation. If a systematic error in a quantity x is established, the sign and magnitude of $\varepsilon(x)$ is known, and Eq. (48) can be used for the propagation of the systematic error in F. Two important examples of this are:

$$F = x + y - z \qquad \varepsilon(F) = \varepsilon(x) + \varepsilon(y) - \varepsilon(z) \tag{49}$$

$$F = \frac{xy}{z} \qquad \frac{\varepsilon(F)}{F} = \frac{\varepsilon(x)}{x} + \frac{\varepsilon(y)}{y} - \frac{\varepsilon(z)}{z} \tag{50}$$

Of course, if systematic errors are known they should be corrected. Equation (48) provides one way of calculating the correction; normally it would be better simply to recalculate the result. Much more often, systematic errors are unknown; if suspected and of unknown sign, allowance for their possible magnitudes can be included in the estimated limits of "random" error.

Propagation of random errors. In the case of random errors, we do not know the actual values of $\varepsilon(x)$, $\varepsilon(y)$, etc., and we cannot determine the actual value of $\varepsilon(F)$. However, we have already assigned to each experimental variable an e.s.d. S or a confidence limit Δ. We now wish to deduce the corresponding uncertainty in the final result F, taking into consideration the high probability that the errors in the several variables will tend somewhat to cancel one another out. Let us square both sides of Eq. (48):

$$[\varepsilon(F)]^2 = \left(\frac{\partial F}{\partial x}\right)^2 [\varepsilon(x)]^2 + \left(\frac{\partial F}{\partial y}\right)^2 [\varepsilon(y)]^2 + \cdots + 2\left(\frac{\partial F}{\partial x}\right)\left(\frac{\partial F}{\partial y}\right)\varepsilon(x)\varepsilon(y) + \cdots$$

Now let us average this expression over all values expected for $\varepsilon(x)$, $\varepsilon(y)$, . . . , in accordance with the normal frequency distribution. Since the $\varepsilon(x)$ independently have average value zero, we expect the cross terms to vanish. However, the squared terms are always positive and will not vanish. If we replace the average of each squared error by the variance, we obtain

$$S^2(F) = \left(\frac{\partial F}{\partial x}\right)^2 S^2(x) + \left(\frac{\partial F}{\partial y}\right)^2 S^2(y) + \cdots \tag{51}$$

It can be seen immediately that the variance in the mean of N measurements $S^2(\bar{x}) = S^2(x)/N$, which was discussed previously, can be obtained as a trivial example of Eq. (51) by taking

$$F = \bar{x} = \frac{1}{N}\sum_{i=1}^{N} x_i$$

Equation (51), which is not limited to normal distributions, is an equation that can be used when all or substantially all of the experimental variables are means of sets of measurements from which statistical measures of variance are available. From $S(F)$ a confidence limit $\Delta(F)$ can be derived if the total number of degrees of freedom can be ascertained. In some cases this may be the sum of the number of degrees of freedom from all of the experimental variables; in general it will be larger than the number of degrees of freedom from any one variable. This complicated problem may be bypassed by assuming that the central limit theorem is at work: the error probability distribution in F is more closely normal than are the error probability distributions in the experimental variables. Thus to a satisfactory approximation, the value of t to be used can be taken to be that for $v = \infty$ (for $P = 95$ percent, $t = 1.96$).

However, we are usually working in situations where many or most of the uncertainties in the experimental variables are estimated on the basis of judgement and experience. On the assumption that the errors in such variables are distributed in a roughly normal manner, we may resort to the following approximate expression in terms of confidence limits:†

$$\Delta^2(F) = \left(\frac{\partial F}{\partial x}\right)^2 \Delta^2(x) + \left(\frac{\partial F}{\partial y}\right)^2 \Delta^2(y) + \left(\frac{\partial F}{\partial z}\right)^2 \Delta^2(z) + \cdots \tag{52}$$

In certain cases, the propagation of random errors can be carried out very simply.

1. For $F = ax \pm by \pm cz$,

$$\Delta^2(F) = a^2 \Delta^2(x) + b^2 \Delta^2(y) + c^2 \Delta^2(z) \tag{53}$$

2. For $F = axyz$ (or axy/z or ax/yz or a/xyz),

$$\frac{\Delta^2(F)}{F^2} = \frac{\Delta^2(x)}{x^2} + \frac{\Delta^2(y)}{y^2} + \frac{\Delta^2(z)}{z^2} \tag{54}$$

3. For $F = ax^n$,

$$\frac{\Delta^2(F)}{F^2} = n^2 \frac{\Delta^2(x)}{x^2} \tag{55}$$

In case 3, the fractional error limit in F, $\Delta(F)/F$, is simply n times the fractional error in x. Note also the differences between Eqs. (53) and (54) for random errors and the comparable expressions in Eqs. (49) and (50) for systematic errors.

Equation (53) includes the important special case $(x - y)$ which is applicable, for example, to differences between two readings of the same variable (e.g., the weight of a liquid plus container and the weight of the empty container, the initial and final readings of a thermometer). In many cases the limits of error in the initial and final readings may be taken as equal, in which case the limit of error in the difference between the two readings is equal to $\sqrt{2}$ times the limit of error in a single reading. Equation (54) can be simplified in the special case that the fractional error in each variable is approximately the same [i.e., $\Delta(x)/x \simeq \Delta(y)/y \simeq \Delta(z)/z$]. In this case, the fractional error in F is roughly equal to the sum of the fractional errors in the individual variables divided by the square root of the number of variables.

As an example, consider the use of Eq. (9) to calculate the specific rotation of a solute in solution (see "A Cautionary Tale"). Values of the independent variables and their 95 percent confidence limits are as follows:

$x = V$	$y = L$	$z = m$	$u = \bar{a}$
25.00 cm^3	2.000 dm	1.7160 g	+20.950°
$\Delta =$ 0.02 cm^3	0.002 dm	0.0003 g	0.016°

---

† The use of Δ as a limit of error, as in $\Delta(x)$, is not to be confused with its use as a difference, as in Δx.

The confidence limits for the first three variables were estimated on the basis of judgement and experience; the last was determined from a statistical analysis of data from ten individual measurements with Eq. (7). With

$$F = [\alpha]_D^{25} = \frac{V}{Lm}\, \tilde{\alpha} \tag{9}$$

we calculate

$$F = [\alpha]_D^{25} = \frac{25.00}{2.000 \times 1.7160} \times 20.950 = 152.61 \deg \mathrm{dm}^{-1}\,(\mathrm{g\,cm}^{-3})^{-1}$$

We observe that Eq. (9) is of the type for which Eq. (54) is applicable. Therefore,

$$\Delta^2([\alpha]) = [\alpha]^2 \left[\frac{\Delta^2(V)}{V^2} + \frac{\Delta^2(L)}{L^2} + \frac{\Delta^2(m)}{m^2} + \frac{\Delta^2(\alpha)}{\alpha^2} \right]$$

$$= 152.61^2 \left[\frac{0.02^2}{25.00^2} + \frac{0.002^2}{2.000^2} + \frac{0.0003^2}{1.7160^2} + \frac{0.016^2}{20.950^2} \right]$$

$$= 2.33 \times 10^4 (64 + 100 + 3 + 58) \times 10^{-8} = 0.0524$$

$$\Delta([\alpha]) = 0.23 \deg \mathrm{dm}^{-1}\,(\mathrm{g\,cm}^{-3})^{-1}$$

Many propagation-of-error treatments are not so simple as those illustrated by Eqs. (53) to (55). The expression for F is often complicated enough or the functional form sufficiently awkward to justify a breakdown of the procedure into steps:

$$F = f_0(A, B, \dots)$$

where

$$A = f_1(x_1, x_2, \dots) \qquad B = f_2(y_1, y_2, \dots), \text{ etc.} \tag{56}$$

We can then write

$$\Delta^2(F) = \left(\frac{\partial F}{\partial A}\right)^2 \Delta^2(A) + \left(\frac{\partial F}{\partial B}\right)^2 \Delta^2(B) + \cdots$$

$$\Delta^2(A) = \left(\frac{\partial A}{\partial x_1}\right)^2 \Delta^2(x_1) + \left(\frac{\partial A}{\partial x_2}\right)^2 \Delta^2(x_2) + \cdots \tag{57}$$

$$\Delta^2(B) = \left(\frac{\partial B}{\partial y_1}\right)^2 \Delta^2(y_1) + \left(\frac{\partial B}{\partial y_2}\right)^2 \Delta^2(y_2) + \cdots, \text{ etc.}$$

provided that $A, B, \dots$, are *independent*; if they are not independent (that is, in the event that some of the variables x_i are identical with some of the y_i), the calculated limit of error will be incorrect. For example, if we let

$$F = a(e^{kx} - 1) + b(y - cx) = A + B$$

it would be improper to write

$$\Delta^2(F) = \Delta^2(A) + \Delta^2(B)$$

$$\Delta^2(A) = (ake^{kx})^2 \, \Delta^2(x) \qquad \Delta^2(B) = b^2 \, \Delta^2(y) + b^2c^2 \, \Delta^2(x)$$

Treatment of the function as a whole gives the correct result

$$\Delta^2(F) = (ake^{kx} - bc)^2 \, \Delta^2(x) + b^2 \, \Delta^2(y)$$

which differs from the result first given by a term $-2abcke^{kx} \, \Delta^2(x)$.

When the expression for F is a complicated one, it is advisable to watch for opportunities for transforming the expression or parts of the expression so that single symbols representing known quantities can be substituted for groupings of several symbols. The most advantageous simplifications are usually obtained when the quantities substituted are the end results of the calculations, such as F itself, or at least quantities which represent the later stages of the calculation of F. This is advantageous not only because it results in the greatest economy of symbols but also because it presents the most frequent opportunities for cancellation and simplification. For example, the limit of error of

$$F = C\frac{x^2}{y}$$

can be obtained from

$$\Delta^2(F) = \left(\frac{2Cx}{y}\right)^2 \Delta^2(x) + \left(\frac{Cx^2}{y^2}\right)^2 \Delta^2(y)$$

but for computational purposes it is more convenient to divide this equation by F^2, since the value of F will already have been calculated; we obtain

$$\frac{\Delta^2(F)}{F^2} = 4\frac{\Delta^2(x)}{x^2} + \frac{\Delta^2(y)}{y^2}$$

Estimation of uncertainties is often more complicated than in the above instances because the functional form of F may be more complicated. Before one begins to differentiate F, which is always in principle possible but often yields masses of unwieldly expressions with ever-present possibility of mistakes, it is well to examine F carefully with two purposes in mind: (1) to see if the expression for F can be simplified to make differentiation easier, and (2) to see if there are any terms that can be neglected because their effect on the uncertainty is small. While F must be evaluated in its precise form to yield the desired numerical result to the maximum possible precision, it should be remembered that the value of the limit of error need not be known to great precision; thus a quantity can often be dropped in a limit-of-error treatment even if it cannot be neglected in the calculation of F itself. A term can ordinarily be dropped in the limit-of-error treatment if the effect produced in $\Delta(F)$ is no more than about 5 percent or even 10 percent. An overgenerous (overconservative) estimate of the limit of error is to be preferred to an insufficient (overoptimistic) estimate; if a term is to be dropped, it is better that dropping it make $\Delta(F)$ a little too large than a little too small. For example, in the cryoscopic determination of the molecular weight of an unknown substance, the molecular weight is given by Eq. (10-21) as

$$M = \frac{1000 g K_f}{G \, \Delta T_f} (1 - k_f \, \Delta T_f)$$

and since k_f is of the order of 0.01 K^{-1} the term $k_f \, \Delta T_f$ is itself very small compared with unity and will contribute negligibly to the uncertainty in comparison to ΔT_f itself; therefore, the treatment of uncertainties may be based on the simpler, approximate expression

$$M = \frac{1000 g K_f}{G \, \Delta T_f}$$

To take a more complicated example, consider the measurement of the temperature of a boiling organic liquid with the gas thermometer described in Exp. 1. The absolute temperature is given by

$$T = T_0 \frac{p(1 + pv/p_r V + 3\alpha t - B/\bar{V})}{p_0(1 + p_0 v/p_r V - B_0/\bar{V})}$$

This equation has been obtained by combining Eq. (1-8) with a similar equation in which T is replaced by T_0, which is the temperature at the ice point (273.15 K). For definitions of the quantities in this expression, see Exp. 1; quantities with subscript zero are those corresponding to T_0. The volume V of the thermometer bulb is not to be confused with $\bar{V} \approx RT_0/p_0$, which is the molar volume of the gas employed. The direct partial differentiation of this equation with respect to p or p_0 would yield a very messy expression, so we are motivated to look for ways to simplify the equation. Since the last two terms in parentheses in the denominator are small compared to unity, we can apply the series expansion expression for the reciprocal of the binomial $1 + x$,

$$\frac{1}{1+x} = 1 - x + x^2 - \cdots$$

keeping only the first two terms on the right. Thus we may write the expression for T, ignoring terms of second order in small quantities, as [cf. Eqs. (1–12), (1–13)]

$$T = T_0 \frac{p}{p_0} \left[1 + \frac{(p - p_0)}{p_r} \frac{v}{V} + 3\alpha t - \frac{(B - B_0)}{\bar{V}} \right]$$

$$= T_0 \left[\frac{p}{p_0} \left(1 + 3\alpha t - \frac{p_0}{RT_0}(B - B_0) \right) + \frac{(p^2 - pp_0)}{p_0 p_r} \frac{v}{V} \right]$$

This may be differentiated with respect to p, p_0, p_r, and v/V without much difficulty. Before going to the trouble of differentiating, we should examine the terms for magnitude. Using values of α, B, and B_0 given in Exp. 1, and noting that v/V is of the order of 0.01, we may surmise that for the purpose of deriving a limit-of-error estimate all terms except the leading one may be neglected;

$$T = T_0 \frac{p}{p_0}$$

Let us put in some numbers. Assume that the gas in the bulb is helium; then $B - B_0$ is certainly negligible (see Table 1-1). For $p = 510.1 \pm 1.0$ Torr, $p_0 = 418.2 \pm 1.0$ Torr, $T_0 = 273.15$ K (regarded as exact), T by the above simple equation is 333.2 K; $t = 60.0$°C. With, in addition, $p_r = 456.5 \pm 1.0$ Torr, $3\alpha t = 3(3.2 \times 10^{-6})60 = 0.00058$, and $v/V = 0.011 \pm 0.003$, the original precise expression gives $T = 333.94$ K. Using the simple expression for our limit-of-error treatment we have

$$\frac{\Delta^2(T)}{333.94^2} = \frac{1.0^2}{510.1^2} + \frac{1.0^2}{418.2^2}$$

or $\Delta(T) = 1.06$ K. Thus

$$T = 333.9 \pm 1.1 \text{ K} \quad t = 60.8 \pm 1.1\text{°C}$$

Suppose, however, that we are slightly unsure about the effect of the neglect of the v/V term on the limit of error. Let us disregard the extremely small terms involving α and $(B - B_0)$ and otherwise carry out a full-fledged limit-of-error treatment. The partial derivatives and their values are:

$$\frac{\partial T}{\partial p} = \frac{T_0}{p_0}\left[1 + \frac{(2p - p_0)}{p_r}\frac{v}{V}\right] = 0.662 \text{ K Torr}^{-1}$$

(instead of 0.653 K Torr^{-1} from the simple approximate expression),

$$\frac{\partial T}{\partial p_0} = -T_0\left[\frac{p}{p_0^2} + \frac{p^2}{p_0^2 p_r}\frac{v}{V}\right] = -0.807 \text{ K Torr}^{-1}$$

(instead of -0.798 K Torr^{-1} from the simple approximate expression),

$$\frac{\partial T}{\partial p_r} = -T_0\frac{p(p - p_0)}{p_0 p_r^2}\frac{v}{V} = -0.0016 \text{ K Torr}^{-1}$$

(instead of being neglected altogether), and

$$\frac{\partial T}{\partial(v/V)} = T_0\frac{p(p - p_0)}{p_0 p_r} = 67.1 \text{ K}$$

(instead of being neglected altogether). Then, we obtain from Eq. (52)

$$\Delta(T) = [1.0^2(0.662^2 + 0.807^2 + 0.0016^2) + 0.003^2(67.1)^2]^{1/2}$$
$$= (1.0895 + 0.0405)^{1/2}$$
$$= 1.063 \text{ K}$$

which is not significantly different from the result of the earlier calculation.

None of the experiments described in this book will involve mathematical treatments for limit-of-error estimation more complicated than this example. With careful judgement, most limit-of-error treatments can be made much simpler.

Another technique that may be used in cases where the formula for F is very complicated and difficult to differentiate is "computational differentiation," which is extremely tedious to do by hand or with a calculator but reasonably easy to do on a programmable calculator or a computer that has already been programmed to compute the result F. This consists of first computing the result F by substituting in the values of the x_i, then calculating the result $F + \Delta_1 F$ obtained by changing x_1 to $x_1 + \Delta x_1$ where Δx_1 is a very small increment, leaving the other x_i unchanged. Then the change $\Delta_1 F$ is obtained by subtraction. This is done successively with $x_2, x_3, \ldots$, yielding $\Delta_2 F, \Delta_3 F, \ldots$. Then

$$\frac{\partial F}{\partial x_i} \approx \frac{\Delta_i F}{\Delta x_i} \tag{58}$$

These values of partial derivatives may then be substituted into Eq. (52) to yield the desired limit-of-error value.

CALCULATIONS

Calculations may be performed by any method that does not introduce round-off errors that are significant in comparison with the experimental errors. A pocket calculator will suffice for most of the experiments in this book. For lengthy or repetitive calculations, the use of a computer is strongly recommended (see Chapter XXI). Finally, longhand arithmetic should not be regarded as a lost art on those occasions where calculators or computers fail or happen not to be available.

Precision and reliability of calculations. The rules given on pp. 49–50 for determining the number of significant figures to be retained in a numerical result should be kept in mind in doing calculations. However, these rules do not apply strictly to the retention of digits in intermediate numerical values obtained in the course of a computation. It is always wise to retain *at least* one additional digit during computations to avoid unnecessary accumulated round-off errors in the final result, but one must remember to adjust properly the number of significant figures when reporting the final result. The retention of too many digits during manual computation constitutes wasted computational effort and may contribute to the possibility of mistakes in arithmetic and in the transcription of numbers. These are not problems when a computer is being used. The computer programmer usually has the choice between two levels of computational precision: "normal precision" (approximately 7 significant digits in a typical personal computer), and "double precision" (typically 15 or 16 significant digits).

Computations, whether by hand, by pocket calculator, or by computer, should be done with awareness of certain principles concerning the effects of arithmetic operations. These follow directly from the principles underlying the propagation of errors, discussed in the preceding section.

When a number of numerical values are *added,* the precision of the result can be no greater than that of the least precise numerical value involved. Thus, generally, the number of decimal places in the result should be the same as the number of decimal places in the component with the fewest:

$$
\begin{array}{r}
32.7 \\
+\ \ 3.62 \\
+10.008 \\
\hline
\end{array}
$$

$$=46.328 \rightarrow 46.3$$

When the *difference* between two numbers yields a result that is relatively small, that difference not only has a precision limited to that of the less precise number but has also a *relative* precision much less than either number:

$$
\begin{array}{r}
673.425 \\
-672.91 \\
\hline
\end{array}
$$

$$0.515 \rightarrow 0.52$$

The less precise of the two numbers has a relative precision of about 1 in 67000 (0.0015 percent) while the difference has a relative precision of about 1 in 50 (2 percent). The effect can be particularly devastating in an intermediate calculation on a computer, where the user of the computer may be blissfully unaware of the loss of relative precision.

In *multiplication* and *division,* the *relative* precision of the result can be no greater than that of the least (relatively) precise number involved. Thus the number of *significant figures* in the result should be *approximately* that in the component having the fewest of them. In judging the appropriate number of significant figures in the result, bear in mind that the relative precision does not depend on the number of significant figures alone; 999, with 3 significant figures, is about as precise as 1001, with 4. When in doubt, employ the larger number of significant figures. For example:

$$346 \times 121 \times 900.0 = 37{,}679{,}400 \rightarrow 3.77 \times 10^7$$

The uncertainty in 121, the least relatively precise figure, is about 0.8 percent. One in 377 is about 0.27 percent, which implies a greater relative precision, but if the result is rounded to 3.8×10^7, with a relative precision of 2.6 percent, too much significance will be lost. The principles are no different when division is involved:

$$5.44 \times 0.132/8.092 = 0.0887395 \rightarrow 0.0887 \text{ or } 0.089$$

The uncertainty in 0.132 is about 0.7 percent, that in 0.089 is about 1.1 percent, implying a small but tolerable loss of precision. If this is an intermediate calculation, it would be better to keep the extra digit.

It should be kept in mind that in the case of a calculation with many numbers, round-off errors may accumulate to the extent that they reduce the precision significantly.

The above-mentioned considerations are not a substitute for a propagation-of-errors treatment to determine the uncertainty in the final result, especially when the calculations have been carried out on a computer and the result is given with an artifically large number of digits.

No matter what computational procedure has been used, there must be reliable assurance that mistakes in arithmetic and in the transcription of numerical data have not been made. Normally, an electronic calculator or computer can be trusted not to make mistakes in arithmetic, but mistakes in the manual transcription of input numbers and output numbers are all too easy to make. A printed record from a computer, which can be checked and double-checked, can provide satisfactory assurance against mistakes of this kind.

The best way to check a calculation is to repeat it, preferably in a different way, to the same precision. When an experiment has been performed by two persons as partners, the calculations can be done by each person independently, with cross-checking of intermediate and final numerical results.

Numerical methods. This term is commonly used in contradistinction to analytic methods for carrying out such mathematical procedures as differentiation, integration, solution of algebraic equations and of differential equations, data smoothing, etc. Analytic methods are exact, or at least capable of being carried to any arbitrary precision; numerical methods as applied to experimental data are necessarily approximate, being limited by the finite number of data employed and by their precision.

Given a set of experimental data y_i measured at a series of values x_i for the independent variable x, *which for convenience we shall take to be evenly spaced*, it is often desired to construct a function $y(x)$ to represent them. If there is reason to believe that this function should conform to some analytic form depending on a few parameters, these parameters can be adjusted to give the "best fit" to the data by the method of *least squares,* as described in Chapter XX. The analytic form may be an equation representing the applicable phenomenological theory, which in principle may be exact. Alternatively, one may use some general "curve-fitting" function (such as a polynomial) having no particular relation to the experiment at hand, which will provide an approximate representation. In either case the function can be differentiated, integrated, etc. directly by analytic methods, and there is no need for the application of numerical methods for these tasks. If no clear-cut analytic form is apparent and recourse is not taken to a general fitting function, these manipulations must be done numerically (or graphically, as described later).

Before the data can be interpreted, whether by analytic or numerical means, it is sometimes necessary to "smooth" the data so that differences from point to point do not show random fluctuations. This is the equivalent of "drawing a smooth curve" through the data points on a graph, but it can be done more objectively by numerical methods, several of which are described elsewhere.[10,11,12] We shall give here one method, based on fitting a third-degree polynomial to $2n + 1$ adjacent points. For $n = 2$, the "smoothed" value of y_i is

$$y_i' = \tfrac{1}{35}[17y_i + 12(y_{i+1} + y_{i-1}) - 3(y_{i+2} + y_{i-2})] \tag{59}$$

For $n = 3$,

$$y_i' = \tfrac{1}{21}[7y_i + 6(y_{i+1} + y_{i-1}) + 3(y_{i+2} + y_{i-2}) - 2(y_{i+3} + y_{i-3})] \tag{60}$$

For $n = 4$

$$y_i' = \tfrac{1}{231}[59y_i + 54(y_{i+1} + y_{i-1}) + 39(y_{i+2} + y_{i-2})$$
$$+ 14(y_{i+3} + y_{i-3}) - 21(y_{i+4} + y_{i-4})] \tag{61}$$

This procedure is equivalent to the Savitsky–Golay method,[12] algorithms for which have been included in computer software for scientific instruments such as the Fourier-Transform-Infra-Red (FTIR) spectrometer.

Having now the data x_i, y_i (which have been smoothed if necessary), we can now consider methods for such manipulations as differentiation and integration that do not require a fit to some analytic function or fitting function covering the entire range of x. For *differentiation*, a *local* construction of the function $y(x)$ can be based on a general interpolation formula,[13] which utilizes successive differences obtained as follows:

$$
\begin{array}{ccccccc}
x_0 = x_0 & y_0 & & & & & \\
& & \Delta y_0 & & & & \\
x_1 = x_0 + w & y_1 & & \Delta^2 y_0 & & & \\
& & \Delta y_1 & & \Delta^3 y_0 & & \\
x_2 = x_0 + 2w & y_2 & & \Delta^2 y_1 & & \Delta^4 y_0 & \\
& & \Delta y_2 & & \Delta^3 y_1 & & \Delta^5 y_0 \quad (62)\\
x_3 = x_0 + 3w & y_3 & & \Delta^2 y_2 & & \Delta^4 y_1 & \vdots \\
& & \Delta y_3 & & \Delta^3 y_2 & \vdots & \\
x_4 = x_0 + 4w & y_4 & & \Delta^2 y_3 & \vdots & & \\
& & \Delta y_4 & \vdots & & & \\
x_5 = x_0 + 5w & y_5 & & & & & \\
\vdots & \vdots & & & & &
\end{array}
$$

where $\Delta y_0 = y_1 - y_0$, $\Delta y_1 = y_2 - y_1$, $\Delta^2 y_0 = \Delta y_1 - \Delta y_0$, etc. The degree of smoothness of the data can be seen by inspection of the differences; there is no point in continuing the table beyond the point where the differences cease to vary in a reasonably smooth and regular manner. An expression for y as a function of x in the neighborhood of x_0 is given by the Gregory–Newton interpolation expression

$$y = y_0 + r\,\Delta y_0 + \frac{r(r-1)}{2!}\Delta^2 y_0 + \frac{r(r-1)(r-2)}{3!}\Delta^3 y_0 + \cdots \tag{63}$$

where

$$r \equiv \frac{x - x_0}{w} \tag{64}$$

By differentiation we obtain

$$\frac{dy}{dx} = \frac{1}{w}\left(\Delta y_0 + \frac{2r-1}{2!}\Delta^2 y_0 + \frac{3r^2 - 6r + 2}{3!}\Delta^3 y_0 \right.$$

$$+ \frac{4r^3 - 18r^2 + 22r - 6}{4!}\Delta^4 y_0$$

$$\left. + \frac{5r^4 - 40r^3 + 105r^2 - 100r + 24}{5!}\Delta^5 y_0 + \cdots \right) \tag{65}$$

$$\frac{d^2 y}{dx^2} = \frac{1}{w^2}\left[\Delta^2 y_0 + (r-1)\,\Delta^3 y_0 + \frac{6r^2 - 18r + 11}{12}\Delta^4 y_0 \right.$$

$$\left. + \frac{2r^3 - 12r^2 + 21r - 10}{12}\Delta^5 y_0 + \cdots \right] \tag{66}$$

On referring to Eq. (62), we see that the successive differences entering into Eqs. (63), (65), and (66) lie on a downward-slanting line. With as much justification an upward-slanting line may be used and, indeed, must be used if x_0 is at or very near the end of the table. No changes in the above expressions other than a replacement of Δy_0, $\Delta^2 y_0$, $\Delta^3 y_0$, ... by Δy_{-1}, $\Delta^2 y_{-2}$, $\Delta^3 y_{-3}$ and a change of all signs in Eqs. (63), (65), and (66) to plus are required if we continue to define r with Eq. (64). Indeed a good procedure is to employ both downward-slanting and upward-slanting Δ's, and to restrict r to the range $-\frac{1}{2} \le r \le \frac{1}{2}$.

For evaluation of the definite integral

$$Y(a, b) = \int_a^b y\,dx \tag{67}$$

it is often possible to utilize procedures that do not require smoothing of the data.[14] The simplest procedure, based on the "trapezoidal rule," makes use of the expression

$$Y(x_0, x_n) = \frac{w}{2}\left(y_0 + 2y_1 + 2y_2 + \cdots + 2y_{n-1} + y_n\right) \tag{68}$$

This is equivalent to approximating the function by linear segments between the tabular entries (data points). A better procedure makes use of "Simpson's one-third rule":

$$Y(x_0, x_n) = \frac{w}{3}(y_0 + 4y_1 + 2y_2 + 4y_3 + 2y_4 + \cdots + 4y_{n-1} + y_n) \tag{69}$$

This rule requires that n be even, i.e., that the number of values of y be odd. This procedure is exact if y is strictly a quadratic function of x. Its use in other cases is equivalent to fitting quadratic forms (parabolas) to three points at a time, joining them at the even-numbered points. Still another rule, "Simpson's three-eighths rule,"

$$Y(x_0, x_n) = \frac{3w}{8}[y_0 + 3(y_1 + y_2) + 2y_3 + 3(y_4 + y_5) + 2y_6$$

$$+ \cdots + 3(y_{n-2} + y_{n-1}) + y_n] \tag{70}$$

requires that n be divisible by 3 and is equivalent to fitting third-degree functions to four points at a time. More refined procedures are available, but their slight superiority seldom justifies the added complications.

In cases where n is divisible neither by 2 nor by 3, the range of integration may be split into two parts, one for Simpson's one-third rule, the other for Simpson's three-eighths rule. Alternatively, if the curve is approximately linear in one or two intervals, the trapezoidal rule may be used in these intervals.

For many purposes it is desirable to construct the function $y(x)$ over the entire range of x. There are several ways of doing this, of which two are very common. The first is to fit a polynomial

$$y(x) = \sum_{j=0}^{N} a_j x^j = a_0 + a_1 x + a_2 x^2 + \cdots + a_N x^N \tag{71}$$

to the experimental data by determining the "best" values of the constant coefficients a_j.[15] If the number m of data points is exactly equal to the number of terms $n = N + 1$ (N being the degree of the polynomial), then the coefficients are determined by the solution of a set of $m = n$ simultaneous linear equations (one for each data point x_i, y_i). The resulting polynomial will then fit each of the data points exactly, but between the data points the polynomial will provide only an approximate prediction of y. The predicted values of y', the first derivative of y with respect to x, are in general everywhere approximate, and particularly so near the ends of the range. If $m < n$, the system of equations is underdetermined, and a unique solution yielding the values of the coefficients is unobtainable. If $m > n$, the system of equations is overdetermined, and the coefficients may be determined by any of several methods, chief among which is the method of least squares, described in Chapter XX. Here the polynomial does not in general fit the data points exactly; indeed, if $m \gg n$ it should not be necessary to smooth the data beforehand.

The other widely used method of constructing the function $y(x)$ is by use of Fourier series:[16]

$$y(x) = a_0 + 2 \sum_{h=1}^{N} [a_h \cos(2\pi hx/L) + b_h \sin(2\pi hx/L)] \tag{72}$$

Here the coefficients to be determined are a_h and b_h. This function is periodic in x, with a period of arbitrary length L. The coefficients are determined by

$$a_0 = \frac{1}{L} \int_0^L y(x) \, dx \simeq \frac{\Delta x}{L} \sum_{i=1}^{m} y_i = \bar{y} \tag{73a}$$

$$a_h = \frac{1}{L} \int_0^L y(x) \cos(2\pi hx/L) \, dx \simeq \frac{\Delta x}{L} \sum_{i=1}^{m} y_i \cos(2\pi hx_i/L) \tag{73b}$$

$$b_h = \frac{1}{L} \int_0^L y(x) \sin(2\pi hx/L) \, dx \simeq \frac{\Delta x}{L} \sum_{i=1}^{m} y_i \sin(2\pi hx_i/L) \tag{73c}$$

If the period L is extended to infinity we obtain a *Fourier transform pair:*

$$y(x) = \int_{-\infty}^{\infty} F(h) e^{2\pi ihx} \, dx \tag{74a}$$

$$F(h) = \int_{-\infty}^{\infty} y(x) e^{-2\pi ihx} \, dx \tag{74b}$$

where i is here the square root of minus one, and where $y(x)$ and $F(h)$ are in general complex quantities but may be real in special circumstances. In practical usage, of course, the infinite limits are replaced by finite ones, taken as large as possible; generally the integrals are evaluated by numerical means, being approximated by sums with very narrow intervals between terms.

The Fourier series is particularly appropriate when the function $y(x)$ is known to be periodic, and is much used in problems involving waves (sound, light, ocean waves, diffraction phenomena, etc.). It can, however, be used for general fitting of data, with L set at some value greater than the range of x. Often it is desirable that the range covered by L should extend somewhat beyond the range of x at both ends, since troublesome peaks may occur at both ends of the range covered by L owing to the fact that the number of terms in the series is not infinite. Generally the number of terms should be as high as possible, but should not exceed a fraction (say, a tenth to a quarter) of the number of data points. The use of the Fourier series for general data fitting is sometimes criticized as artificial because the trigonometric functions, on a computer, are themselves provided in the form of approximating polynomials; thus, the Fourier series could itself be expressed as a polynomial constructed in a rather special way.

It is unrealistic to attempt the use of the Fourier series or of the Fourier integral transforms without the aid of a computer.

In recent years a "fast Fourier transform" (FFT) algorithm for computers has been developed.[17,18] This is particularly useful in certain kinds of chemical instrumentation, specifically nuclear magnetic resonance and infrared absorption spectrometers. In such instruments the experimental observations are obtained directly in the form of a Fourier transform of the desired spectrum; a computer that is built into the instrument performs the FFT and yields the spectrum (see Chapter XVIII).

Having obtained a fitting function $y(x)$ in the form of a polynomial, Fourier series or integral transform, or other form, we may differentiate it or integrate it as desired. We may also need to determine one or more zeros of the function.[19] The equation $y(x) = 0$ is in general nonlinear; analytic methods of solving such equations (other than polynomials of degree one or two) are generally extremely complicated or nonexistent. However, solutions to such equations can be obtained routinely by numerical methods to any desired precision with the aid of a computer and the use of convergent iterative methods. We will describe two methods here.

The equation $y(x) = 0$ may have several roots (if $y(x)$ is a polynomial of degree N it will have N roots, but some of them may be complex); we will seek only one root for this discussion. Assume that the function $y(x)$ varies slowly enough that it does not change sign more than once between two adjacent data points. Let us say that y_1 (at x_1) and y_2 (at x_2) are of opposite sign. Then, by linear interpolation between the two points, to the first approximation (first iteration) the root is given by

$$x_3 = \frac{|y_1| x_2 + |y_2| x_1}{|y_1| + |y_2|} = x_1 + \frac{|y_1|}{|y_1| + |y_2|} (x_2 - x_1) \qquad (75)$$

The value of $y_3 = y(x_3)$ is now calculated. It is extremely unlikely that it is zero, indicating that x_3 is the root we seek; the zero in x will presumably be on one side or the other of x_3. In the second iteration we repeat the above procedure, interpolating either between x_1 and x_3 or between x_3 and x_2 depending on which range includes the zero. The procedure is repeated in subsequent iterations to obtain x_4, x_5, and so on until the change between iterations becomes insignificant and we can say that the process has effectively converged on the root to the desired degree of precision. This method is known as the "false position" method.

Another method, closely related to the first, is the Newton–Raphson method. Here the derivative with respect to x, denoted y', is evaluated at x_1, a point judged to be close to the root. Then

$$x_2 = x_1 - \frac{y_1}{y'_1} \qquad (76)$$

for the first approximation (first iteration). This is equivalent, in graphical terms, to drawing a tangent to the curve at the point x_1, y_1 and extending it to intersect with the x axis. Then, for the second iteration, y_2 and y'_2 are evaluated to obtain x_3, and so on.

Example. By least squares a third degree polynomial is fitted to the data from an experiment:

$$y(x) = 0.331x^3 - 1.410x^2 + 2.050x - 4.015$$

We seek the value of a root of the equation

$$y(x) = 0$$

It is apparent from plotting the function that a zero lies between $x = 3.000$ and $x = 4.000$. From the equation for the function $y(x)$ it is found that

$$\text{for } x_1 = 3.000, \qquad y_1 = -1.618$$
$$\text{for } x_2 = 4.000, \qquad y_2 = +2.809$$

Application of Eq. (75) gives

$$x_3 = \frac{3.000 \times 2.809 + 4.000 \times 1.618}{1.618 + 2.809} = 3.366$$

$$\text{for } x_3 = 3.366, \qquad y_3 = -0.469 \qquad \text{(Iteration 1)}$$

Since y_3 is negative, the zero must lie between $x_3 = 3.366$ and $x_2 = 4.000$. In the same manner we obtain

$$x_4 = 3.456, \qquad y_4 = -0.107 \qquad \text{(Iteration 2)}$$

Continuing, we obtain

$$x_5 = 3.476, \qquad y_5 = -0.023 \qquad \text{(Iteration 3)}$$
$$x_6 = 3.480, \qquad y_6 = -0.005 \qquad \text{(Iteration 4)}$$
$$x_7 = 3.481, \qquad y_7 = -0.003 \qquad \text{(Iteration 5)}$$
$$x_8 = 3.481, \qquad y_8 = -0.003 \qquad \text{(Iteration 6)}$$

Apparently the process has converged to a satisfactory stopping point. The failure of y to decrease below 0.003 in magnitude results from round-off error in x; keeping additional significant figures during the calculation would reduce this magnitude in the same number of iterations. (Try it and see!) Let us now apply the Newton–Raphson method to the same problem:

$$y'(x) = 0.993x^2 - 2.820x + 2.050$$
$$\text{for } x_1 = 3.000, \qquad y_1 = -1.618, \qquad y_1' = 2.527$$

$$x_2 = 3.000 - \frac{(-1.618)}{2.527} = 3.640$$

$$x_2 = 3.640, \qquad y_2 = 0.730, \qquad y_2' = 4.943 \qquad \text{(Iteration 1)}$$

Continuing, we obtain

$$x_3 = 3.492, \qquad y_3 = 0.046, \qquad y_3' = 4.312 \qquad \text{(Iteration 2)}$$
$$x_4 = 3.481, \qquad y_4 = -0.003, \qquad y_4' = 4.226 \qquad \text{(Iteration 3)}$$
$$x_5 = 3.482, \qquad y_5 = 0.002, \qquad y_5' = 4.270 \qquad \text{(Iteration 4)}$$
$$x_6 = 3.482, \qquad y_6 = 0.002, \qquad y_6' = 4.270 \qquad \text{(Iteration 5)}$$

Satisfactory convergence has been obtained, somewhat faster than with the false position method. Usually the Newton–Raphson method is indeed the one that converges faster, but in some cases it can diverge in the early iterations. It may be helpful to use the false position method for the first one or two iterations, and then switch over to the Newton–Raphson method for further iterations leading to satisfactory convergence.

Many other numerical methods are available, which are beyond the scope of this brief introduction but are described in the works already referenced. Some are available in the form of published computer programs.[20,21,22] Some of the methods described in this section are not difficult to program for a computer and are recommended as programming exercises for interested students.

GRAPHS AND GRAPHICAL METHODS

By a graph we mean a representation of numerical values or functions by the positions of points and lines on a two-dimensional surface. A graph is inherently more limited in precision that a table of numerical values or an analytic equation, but it can contribute a "feel" for the behavior of data and functions that numerical tables and equations cannot. A graph reveals much more clearly such features as linearity or nonlinearity, maxima and minima, points of inflection, etc. Graphical methods of smoothing data and on differentiation and integration are usually easier than numerical methods. Graphs and graphical methods suffer, of course, from the limitations of a two-dimensional surface; thus, normally, a plotted point has only two degrees of freedom, which we assume here to be represented by one independent variable x (increasing from left to right) and one dependent variable y (increasing from bottom to top). The word "normally" is required in the above sentence because three-dimensional plots, representing for example points with coordinates x, y, and z, can be represented in oblique projection on a two-dimensional surface by special (mainly computer) techniques.

Wherever possible, the data from an experiment should be plotted at an early stage, even if numerical methods will be used subsequently for greater accuracy. This is particularly true when functions are to be fitted by least-squares or other methods, since a graph may make evident special problems or requirements that may otherwise be missed.

Manual preparation of graphs. For constructing a graph by hand, the following procedure is recommended.

1. *Design your graph first.* Plan it so as to make best use of the area available, and to present the data in a manner that will be clearest to the viewer. In most cases the entire useful area of a normal-sized sheet of graph paper should be dedicated to the graph. In some cases, the paper should be turned sideways (i.e., with binding margin on top) so that the independent variable, x, runs longways on the page, to provide an aspect more favorable for displaying the desired features. Determine the maximum and minimum values of x and y (remember, it is *not* necessary that they both be zero at the lower left corner!) so as to fit the points, lines, and curves into the available area. Be sure to make adequate allowance of space for borders all around, scale numbers and scale legends at least at the bottom and at the left, and a legend telling what the graph is about.

2. *Choose your graph paper.* The graph paper should be of good quality, accurately ruled with thin, lightweight lines. In most cases these will constitute a grid of lines equally spaced in two perpendicular directions; special papers are available for plotting logarithmic functions in one direction ("semilog paper") or in both directions ("log–log paper"), or for plotting in polar coordinates, etc. Most good graph papers are ruled in colors that are soft enough that they do not distract the viewer from the plotted lines and points, which are usually in black. (Graph paper ruled in very light blue is useful when it is desired to make photocopies of the graph in which the grid from the graph paper is not visible.) The most commonly used graph papers are ruled with a 1-mm spacing, the lines at 5 and 10 mm being heavier. The smallest divisions on the paper should preferably represent multiples of 1 or 2 or perhaps 5 in the plotted variable.

3. *Define the plotting area.* Draw, with heavy pencil lines, a rectangle within which all points and lines will be plotted, leaving sufficient margins all around, extra width for the binding margin, and also adequate allowance for the scale numbers and legends mentioned above. Indicate the major scale divisions with short, heavy pencil marks and label them with appropriate numerical values. Under the scale numbers for the independent variable enter the scale legend, stating the quantity that is varying, and its units, e.g.,

$$T, K$$

for absolute temperature in kelvins. To avoid scale numbers that are too large or too small for convenient use, multiply the quantity by a power of 10, e.g.,

$$\rho \times 10^4, \text{g cm}^{-3}$$

for the density of a gas. This is equivalent to multiplying all the scale numbers by 10^{-4} but is much more convenient. Similarly enter the scale legend for the dependent variable along the side scale on the left; if necessary turn the paper sideways to write the legend so that it will run vertically.

4. *Plot the points; draw the lines and curves.* Using a sharp, hard pencil, plot the experimental points as small dots as accurately as possible. One should endeavor to plot with an accuracy of one-fifth of the smallest grid division. Draw small circles (or squares, triangles, etc.) of uniform size (2 or 3 mm) around the points in ink to give them greater visibility.

If it is desired to draw a straight line to provide a good linear fit to the plotted points, draw a faint pencil line with a good straightedge such as a transparent ruler or draftsman's triangle, and do not hesitate to erase the line and try again. When you are satisfied that you have made the best possible visual fit, go over the line again with a softer pencil or a pen to make it sufficiently heavy; this time, do not allow the line to cross the circles or other symbols drawn around the experimental points. Smooth nonlinear curves may be drawn either freehand or with ships curves, splines, or other devices, with as much trial-and-error and erasing as required.

If a "theoretical curve" is to be drawn based on an analytic function representing predicted behavior, plot the function at as many points, closely spaced, as reasonably possible; do not draw symbols around them. Carefully draw a smooth curve through the points, lightly at first, and then "heavy it up" as desired.

5. *Add a legend.* Somewhere on the graph (if possible, at the bottom) enter a legend with an identifying figure number. The legend should state what the graph is about, identify symbols and line types used, and provide all needed information that will not be provided in the document to which the graph will be attached.

6. *Check the overall appearance.* After all points, lines, curves, scales, and legends are in place, check the overall clarity and legibility of the graph. If points or lines are too faint, make them heavier; if the graph is messy or confusing, attempt to improve it by erasing and redrawing the features concerned. If the end result falls seriously short in the expected qualities, *start over.* Your second attempt will probably take much less time to prepare, and it will look much better and will better suit its intended purpose. For examples of well prepared graphs representing experimental data, consult current scientific journals such as those listed in Appendix *F.*

Computer-generated plots. In many cases, particularly when the data to be plotted have been generated by a computer, one should take advantage of the ability of computers to generate graphs. These may be generated on the computer's monitor screen for inspection and editing and then produced on external graphic devices such as $X-Y$ plotters, or they may be printed out on a dot-matrix printer with a screen dump. The capabilities of spreadsheet programs such as Lotus 1-2-3 and Symphony to produce graphs should not be overlooked. Much of the above discussion of manual preparation of graphs applies also to computer graphs. Since computer graphs can be produced quickly, it may be worth while to make several tries to obtain the best results. The first plot may be a "default plot" to see the overall appearance of the plotted data; then user-chosen scales, symbols, aspect ratio, and legends may be introduced as needed.

Graphical differentiation and integration. The first derivative of a plotted function $y(x)$ with respect to x is given by the slope of a line drawn tangent to the curve at the point x, y concerned. Draw the tangent line long enough to maximize the accuracy of the slope determination. If two points x_1, y_1 and x_2, y_2 are accurately established at the two ends of the line, the slope and thus the derivative is given by

$$y'(x) = \frac{y_2 - y_1}{x_2 - x_1} \tag{77}$$

A good way to draw the tangent line is to hold a small rectangular mirror with its reflecting surface perpendicular to the paper, adjust it so that the reflection of the curve is tangent to and symmetrical with the curve itself, and then use the mirror as a straightedge to draw the tangent line. Another method is to use a compass to strike off arcs intersecting the curve on both sides and then draw a chord through the intersection points. The tangent can be drawn parallel to the chord if the curvature is uniform. If it is not, construct a second chord at a different distance in the same way and extend both lines to an intersection. Draw the (approximate) tangent line through this intersection point (see Fig. 3).

A definite integral, such as Eq. (67), can be determined by measuring the area under the curve between the desired limits (see Fig. 3). The area can be measured by use of an instrument known as a planimeter, by cutting out the area concerned with a pair of scissors and weighing it, by approximating it as

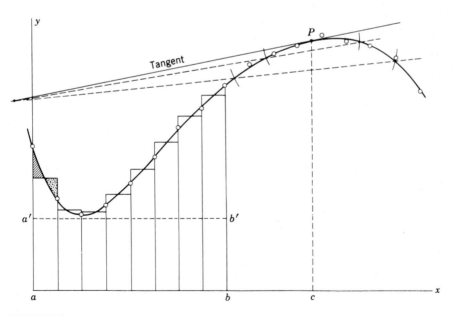

FIGURE 3

An illustration of two graphical procedures: (a) the construction of a tangent by the method of chords, and (b) the evaluation of a definite integral by approximating the curve by a bar graph. The top of each bar is drawn so that the two small areas thereby defined (e.g., the shaded and stippled areas on the first bar) appear equal.

well as possible by a bar graph, or by " counting squares." In the last method a count is made of the smallest squares of the graph-paper grid that lie wholly or *more than half* inside the area concerned. Whatever the method used, the area to be measured should be minimized as much as possible; there is no need to run the planimeter around the rectangular area below $a'b'$ in Fig. 3, for instance.

Uncertainties in graphically derived quantities. It frequently happens that an intermediate or final result of the calculations in a given experiment is obtained from the slope or an intercept of a straight line on a graph, say a plot of y against x. In such a case it is desirable to evaluate the uncertainty in the slope or in the position of the intercept. A rough procedure for doing this is based on drawing a rectangle with width $2\Delta(x_i)$ and height $2\Delta(y_i)$ around each experimental point (x_i, y_i), with the point at its center. The assignments of the limits of error $\Delta(x_i)$ and $\Delta(y_i)$ are made as described previously. The significance of the rectangle is that any point contained in it represents a possible position of the "true" point (x_i, y_i) and all points outside are ruled out as possible positions. Having already drawn the best straight line through the experimental points, and having derived from this line the slope or intercept, draw now two other (dashed) lines representing maximum and minimum values of the slope or intercept, consistent with the requirement that both lines pass through every rectangle, as shown in Fig. 4. (It should be borne in mind that the way in which the limiting lines are drawn will depend upon whether it is the slope or an interpolate intercept or an extrapolate intercept that is the quantity of interest.) Where there are a half dozen or more experimental points, it may be justifiable to neglect partially or completely one or more obviously "bad" points in drawing the original straight line and the limiting lines, provided that good judgment is exercised. The difference between the

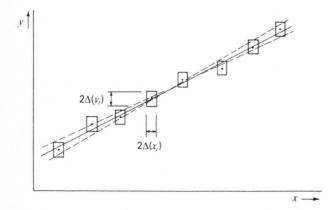

FIGURE 4
A graphical method of determining the limit of error in a slope.

two slopes or intercepts of the limiting lines can be taken as an estimate of twice the limit of error in the slope or intercept of the best straight line.

Where the number of points is sufficiently large, the limits of error of the position of plotted points can be inferred from their scatter. Thus, an upper bound and a lower bound can be drawn, and the lines of limiting slope drawn so as to lie within these bounds. Since the theory of least squares can be applied not only to yield the equation for the best straight line but also to estimate the uncertainties in the parameters entering into the equation (see Chapter XX), such graphical methods are justifiable only for rough estimates. In either case, the possibility of systematic error should be kept in mind.

FUNDAMENTAL LIMITATIONS ON INSTRUMENTAL PRECISION

In all experiments given in this book the precision of the measurements is governed by practical limitations of apparatus construction and operation; the precision is presumably capable of being increased by further refinement of the apparatus or the technique for using it, and it might well be imagined that there is no limit to the possible improvement. However, there are certain definite limitations imposed by physics. Although these limitations are not likely to be encountered by the beginning laboratory student, it is well that he or she should be aware of them.

Limitations due to thermal agitation. The classical law of equipartition of energy states that every degree of freedom of a system has an average kinetic energy equal to $kT/2$, where k is Boltzmann's constant. The student is used to thinking of this amount of energy (about 2×10^{-21} J at room temperature) as being of significance only for individual atoms and molecules, or at most for microscopic particles undergoing Brownian motion. However, Brownian motion may be observed in systems as large as laboratory instruments. An impressive example of this is a very sensitive galvanometer, commonly employed in the past for measuring extremely small currents. Even in the absence of an applied current, the spot of light on the galvanometer scale is observed to be in constant motion, wandering slowly and erratically due to Brownian motion, in the vicinity of its average position. An analogous phenomenon is observed in electrical circuits in the form of "electrical noise." This may, for example, impose an important limitation on the sensitivity of a very sensitive voltmeter or electrometer. It constitutes a very severe limitation on the reception of extremely faint radio waves, especially in radio astronomy, where increasing the amplification is futile because the electrical noise is amplified to the same degree as the desired signal.

An electrical measuring instrument contains electrical circuits incorporating capacitance, inductance, and resistance. In the absence of resistance, a circuit tends to oscillate with a definite frequency ν when disturbed. For optimum performance an amount of resistance is incorporated that is barely sufficient to damp the oscillations resulting from transient inputs; the circuit is then said to be "critically damped." For a critically damped circuit it can be shown[23] that the r.m.s. fluctuations in voltage V and in current I are given by

$$V_{rms} \equiv (\overline{V^2})^{1/2} = (\pi k T R v)^{1/2} \tag{78}$$

$$I_{rms} \equiv (\overline{I^2})^{1/2} = (\pi k T v / R)^{1/2} \tag{79}$$

where R is the total resistance of the circuit, including damping resistance. If the units used in the above equations are volts, amperes, ohms, and hertz (cycles per second), k must be given in joules per kelvin (1.38×10^{-23}).

The r.m.s. values given above effectively constitute estimated standard deviations S for *individual* measurements of V or I. These limitations on precision can be overcome to some degree by making many measurements and taking the mean, provided the conditions of measurement do not change during the time required. If a voltage V is automatically and continuously recorded on a chart recorder, a straight line or a smooth curve can be drawn through the wiggly curve traced by the recorder pen. However, as in cases discussed earlier, the improvement in precision varies as the square root of the number of measurements and is therefore severely limited in practice.

Since the fluctuations are thermal in origin, they can in principle, and also in practice in some cases, be reduced by reducing the temperature of the detection device. In radio astronomy the extremely faint radio signals may be first picked up by a "maser" (the microwave equivalent of the laser) maintained at liquid nitrogen or liquid helium temperature. This device amplifies the signal to a level where ordinary methods of amplification can take over.

Limitations due to particulate nature. An important limitation sometimes encountered is due to the particulate nature of electricity (electrons, ions) and of radiation (photons). The measurement of radiation intensities is in certain cases (e.g. X-rays) performed by counting particles or photons one at a time. The number N counted in a time interval of given magnitude is subject to statistical fluctuations; a count of N is subject to an estimated standard error given by

$$S = \sqrt{N} \tag{80}$$

The only way to overcome limitations imposed by the particulate nature of electricity and radiation is to scale up the experiment in physical size, time span, or intensity, all of which are subject to severe practical limitations. Doubling the source intensity in an X-ray experiment or doubling the counting time will (on the average) result in the number of counts being doubled, while the standard error is increased by a factor of $\sqrt{2}$ and the relative error is decreased in the same proportion. To improve the relative precision by a factor of 10 requires a scale-up factor of 100; thus, the square-root relationship imposes a severe limitation to improvement of precision.

Limitations due to the uncertainty principle of quantum mechanics. According to quantum mechanics, one cannot simultaneously measure both of two conjugate variables (e.g., position and momentum, or energy and time) to infinite precision; the product of the standard errors in the two variables must be greater than $h/2\pi$, where h is Planck's constant $(6.64 \times 10^{-34} \text{ J s})$. For instance, the energy of an excited state of an atom relative to the ground state, as determined by the wavelength of an emitted photon, is somewhat uncertain because the lifetime of the excited state is limited by collisions or other perturbations if not by the finite probability of spontaneous emission. The uncertainty principle applies to the results of any *one* experiment comprising simultaneous measurements of a pair of conjugate variables. Suppose that a single photon emitted by a system under study (say, a gas undergoing an electrical discharge) were to pass between the slits of a very-high-resolution spectrometer, pass through the dispersing element (prism or grating), pass between the very narrow slits of a detector, and then evoke from a photomultiplier tube a response which is recorded as a single event. The wavelength λ of the photon is determined by the angular position of the detector slits with respect to the dispersing element, and the energy of the photon is therefore determined $(E = h\nu = hc/\lambda$, where c is the velocity of light). The conjugate variable t is known, say, from the pressure of the light-emitting gas, which determines the r.m.s. time between deactivating molecular collisions. That is just the one photon; many others from the same source possess slightly different wavelengths and therefore miss the detector slit opening. If the experiment were to stop right there, the determination of the wavelength that characterizes the spectral transition under study would be truly limited in precision due to the uncertainty principle.

However, that is not the way we would conduct the wavelength determination in practice. Let the detector slit scan slowly over a narrow range bracketing the position corresponding to the expected wavelength; photons will begin to enter the slit opening at a very low rate at first, and then at a gradually increasing rate which goes through a maximum and declines; if the counting rate is automatically recorded on a chart, a more-or-less gaussian peak will be obtained. This constitutes in effect a large number (perhaps several thousand) of "individual experiments." The width of the peak is related to the estimated standard deviation S; the center of the peak can be measured with a precision corresponding to S_m. In any practical spectroscopic instrument, a peak of finite width is obtained even with spectral lines of essentially zero width, owing to the finite width of the detector slit and other geometrical factors; in fact

$$S = (S_{\text{qm}}^2 + S_{\text{inst}}^2)^{1/2} \tag{81}$$

where S_{qm} characterizes the quantum mechanical estimated standard deviation and S_{inst} characterizes the instrumental estimated standard deviation.) The same experiment can be done with a photographic plate as the detector. Let us suppose that each incoming photon exposes a single grain in the photographic emulsion; this is an "individual experiment." The resulting "line" on the plate is of finite width, and on microscopic examination is seen to be grainy. The density profile of the "line," which results from many thousands of "individual experiments," can be measured with a scanning microphotometer, and the width and center can be determined. Either with the scanning spectrophoto-meter or with the photographic spectrograph, the number of "individual experiments" is so large that the square-root relationship between precision and number of measurements almost never provides a severe limitation; in the experiments described above, the precision of the determination of the center of the spectral line is limited more by uncertainties in the geometrical dimensions of the apparatus than by quantum mechanics.

Although it is somewhat hard to find real examples in which the precision of a physical chemistry measurement is materially and primarily limited by the uncertainty principle, this principle should be kept in mind as a potential limiting factor in future experiments.

SUMMARY

The principal equations in this chapter and key text references are cited here as a practical aid in the treatment of experimental data. This summary is not intended to be a substitute for thoughtful judgement. Use the equations with care; if there is any doubt concerning the applicability of an equation, review the associated text material.

Rejection of a discordant value. For $3 \leq N \leq 10$ where N is the number of measurements of the same quantity, calculate

$$Q = \frac{|(\text{suspect value}) - (\text{value closest to it})|}{(\text{highest value}) - (\text{lowest value})} \tag{8}$$

and compare it with Q_c from Table 1 (p. 35). If $Q > Q_c$ then reject the suspect value; otherwise retain it. If $N > 10$, reject the suspect value if and only if it differs from the mean of the others by more than $2.6S_m$, where S_m is the e.s.d. of the mean of the others.

Mean, variance, estimated standard deviation (e.s.d.). For a data sample of N values of x, denoted x_i where $i = 1, 2, \ldots, N$, the unweighted mean is

$$\bar{x} = \frac{1}{N} \sum_{i=1}^{N} x_i \tag{1}$$

The variance S^2 is given by

$$S^2 = \frac{1}{N-1} \sum_{i=1}^{N} (x_i - \bar{x})^2 = \frac{N}{N-1} (\overline{x^2} - \bar{x}^2) \tag{4}$$

The estimated standard deviation (e.s.d.) is the square root of the variance:

$$S = \frac{1}{\sqrt{N-1}} \left[\sum_{i=1}^{N} (x_i - \bar{x})^2 \right]^{1/2} \tag{6}$$

E.s.d. of the mean; confidence limit of error of the mean. The e.s.d. of the mean is

$$S_m = \frac{S}{\sqrt{N}} = \frac{1}{\sqrt{N(N-1)}} \left[\sum_{i=1}^{N} (x_i - \bar{x})^2 \right]^{1/2} \tag{7}$$

The 95-percent confidence limit of error Δ is

$$\Delta = t_{0.95} S_m = t_{0.95} \frac{S}{\sqrt{N}} \tag{33}$$

The value of $t_{0.95}$ is given in Table 2 (p. 46) under $P = 0.95$ for the applicable number of degrees of freedom

$$\nu = N - 1$$

Significant figures. More digits than are required to express the precision of the numbers involved should be retained throughout the calculation. Before presentation of the final result, the excess digits should then be removed with appropriate rounding of the last digit retained. See pp. 49–50 for recommended roundoff procedures.

Presentation of results with uncertainties See Table 4, p. 50.

Significance testing. Determine the difference D between the two values being compared: $\bar{x}$ and x_0 (where x_0 is exact) in Case 1 (p. 52), or $\bar{x}_1$ and $\bar{x}_2$ in Case 2 (p. 53). For Case 1 determine $S_D = S_m$, the e.s.d. of the mean involved; for Case 2 determine the *pooled* e.s.d. S_D from S_{m1} and S_{m2} with Eqs. (37)–(38), or if applicable Eqs. (39) or (40). For the number of degrees of freedom ν given by

$$\nu = N - 1 \qquad \text{(Case 1)}$$
$$\nu = N_1 + N_2 - 2 \qquad \text{(Case 2)}$$

find t in Table 2 (p. 46) under $P = 0.95$ (or other selected level of confidence) if the question being asked is "Is the difference D significant?" Find t under $P' = 0.95$ if the question being asked is "Is $\bar{x}$ (or $\bar{x}_1$) significantly *greater than* (alternatively, *less than*) x_0 (or $\bar{x}_2$)?" Calculate

$$\Delta_D = t S_D \tag{36}$$

If $D > \Delta_D$ then the answer to the question is "Yes." If one (or both) of the values being compared is not a mean or an exact value, and its uncertainty has been estimated on the basis of judgement and experience, calculate Δ_D with Eq. (41) or Eq. (42).

Propagation of random errors. Let $x, y, z, \ldots$ be independent, directly measured quantities; let F be a result calculated from them with an equation symbolically represented by

$$F = f(x, y, z, \ldots) \tag{43}$$

$$\Delta^2(F) = \left(\frac{\partial F}{\partial x}\right)^2 \Delta^2(x) + \left(\frac{\partial F}{\partial y}\right)^2 \Delta^2(y) + \left(\frac{\partial F}{\partial z}\right)^2 \Delta^2(z) + \cdots \tag{52}$$

$F = ax \pm by \pm cz$:

$$\Delta^2(F) = a^2 \, \Delta^2(x) + b^2 \, \Delta^2(y) + c^2 \, \Delta^2(z) \tag{53}$$

$F = axyz \qquad$ or $\qquad axy/z \qquad$ or $\qquad ax/yz \qquad$ or $\qquad a/xyz$:

$$\frac{\Delta^2(F)}{F^2} = \frac{\Delta^2(x)}{x^2} + \frac{\Delta^2(y)}{y^2} + \frac{\Delta^2(z)}{z^2} \tag{54}$$

$F = ax^n$:

$$\frac{\Delta^2(F)}{F^2} = n^2 \frac{\Delta^2(x)}{x^2} \tag{55}$$

For complicated cases, see examples; pp. 58–62.

Numerical methods.
Smoothing: See Eqs. (59)–(61), pp. 55–66.
Differentiation by differences: See Eqs. (62)–(66), pp. 66–67.
Integration: See Eqs. (67)–(70), pp. 67–68.
Fitting functions (polynomials; Fourier series): See Eqs. (71)–(74), pp. 68–69.
Roots of nonlinear equations: See Eqs. (75) and (76) and examples, pp. 70–72.

EXERCISES

1. For each of the following series of measurement values, determine whether any measurements should be rejected at the 90-percent confidence level. After rejecting values where appropriate, determine S, S_m, and Δ (95 percent confidence limit) for each series.

(a) 2.8   2.7   2.7   2.5   2.9   2.6   3.0   2.6
(b) 97.13   97.10   97.20   97.35   97.10   97.19
(c)   0.134   0.120   0.109   0.124   0.131   0.119   0.135   0.132   0.132

2. A chemist dispenses a titrant, 0.1000 M HCl, from a burette at 25.0°C. The initial reading is taken six times; the values (in mL) are

6.79   6.78   6.79   6.77   6.76   6.78

The final reading is also taken six times; the values are

28.02   28.03   28.01   28.02   28.02   28.03

Calculate the number n of moles of HCl in the solution dispensed. Give S and S_m for the initial and final volumes, and give a limit of error (95 percent confidence) for n.

3. A crystallographic society, in a study of systematic errors in x-ray determinations of unit cell sizes of crystals by the powder diffraction method (see Exp. 46), sends a specimen of a standard cubic crystalline material to each of several laboratories. Each is to determine the unit cell parameter a_0 by its own technique in six independent experiments at 25.0°C, and report the mean value together with the estimated standard deviation of the mean, S_m. The first two laboratories to report gave the following results:

7.4132(18) Å   7.4065(20) Å

Are the two results significantly different at the 90-percent confidence level? at 95 percent? At 97.5 percent? At 99 percent?

4. Each of the following quantities is uncertain (95-percent confidence) by 0.20 percent. Express the quantities in the proper numbers of significant figures, with specification of the (absolute) uncertainty and units in proper form.
   (a) 2.05675 cm
   (b) 388.982 K
   (c) 1.21335×10^3 Torr

5. The heat of vaporization of a liquid may be obtained from the approximate integrated form of the Clausius–Clapeyron equation

$$\Delta \bar{H}_v = -R \frac{\ln(p_2/p_1)}{1/T_2 - 1/T_1} = R \frac{T_1 T_2}{T_2 - T_1} \ln \frac{p_2}{p_1}$$

(See Exp. 13 for the exact form of the Clapeyron equation. The integrated form given here involves certain reasonable approximations.) For water, the vapor pressure is measured to be 9.2 Torr at 10.0°C (283.15 K) and 55.3 Torr at 40.0°C (313.15 K). Taking R to be 8.3145 J K^{-1} mol^{-1}, calculate $\Delta \bar{H}_v$. Taking the limit of error (95-percent confidence) in a temperature measurement to be 0.1 K and that in a pressure measurement to be 0.1 Torr, calculate the uncertainty in $\Delta \bar{H}_v$. A handbook gives 43,893 J mol^{-1} for $\Delta \bar{H}_v$ at 25°C. Is the difference between this value and your value significant at the 95-percent confidence level?

6. A student determines the density of a solid with a pycnometer (see Sample Report, Chapter I) at 25.0°C. The weight of the empty, dry pycnometer is 6.2330 g. Some of the solid material is introduced; the weight of pycnometer plus solid is 39.4156 g. Water (density $\rho = 0.997048$ g cm^{-3} at 25.0°C) is added to fill all space not occupied by the solid; the weight of pycnometer plus solid plus water is 46.1357 g. The pycnometer is emptied and filled with water only; the weight is 19.7531 g. The student's estimated 95-percent confidence limit of error in any single weighing is 0.0020 g. Calculate the density and its 95-percent confidence limit of error, and present it and its limit of error in proper form with the correct number of significant figures and proper units.

7. The viscosity η of pure ethanol in centipoises was determined at eight different

temperatures at one atmosphere pressure with the following results:

t, °C	0	10	20	30	40	50	60	70
η, cP	1.78	1.45	1.17	1.10	0.83	0.71	0.60	0.49

Prepare a graph to display these data, carefully following the instructions given in this chapter. Draw a smooth curve through the points to represent the data. Determine graphically the temperature coefficient of viscosity $(\partial\eta/\partial t)_p$ at 25.0°C. (*Note to instructor*: Perhaps the laboratory budget can be stretched to provide a small prize, such as a pizza or a six-pack of the student's favorite soft drink, for the best graph.)

REFERENCES

1. D. A. Skoog and D. M. West, "Fundamentals of Analytical Chemistry," 4th ed., pp. 39–90, Holt, New York (1982).
2. E. B. Wilson, Jr., "An Introduction to Scientific Research," pp. 232–235, McGraw-Hill, New York (1952).
3. W. J. Dixon, *Ann. Math. Stat.*, **22**, 68 (1951).
4. P. R. Bevington, "Data Reduction and Error Analysis for the Physical Sciences," pp. 42–49, McGraw-Hill, New York (1969).
5. L. G. Parratt, "Probability and Experimental Errors in Science," Wiley, New York (1961).
6. F. Reif, "Fundamentals of Statistical and Thermal Physics," pp. 37–39, McGraw-Hill, New York (1965).
7. A. I. Khinchin, "Mathematical Foundations of Statistical Mechanics," p. 166, Dover, New York (1949).
8. E. B. Wilson, *op. cit.*, pp. 237–242.
9. CRC "Handbook of Tables for Probability and Statistics," 2d ed., pp. 24, 282–292, CRC Press, Cleveland, Ohio (1974).
10. P. R. Bevington, *op. cit.*, pp. 256–259.
11. W. R. Hamming, "Numerical Methods for Scientists and Engineers," 2d ed., pp. 567–573, 597–599, Dover, New York (1973).
12. R. Annino and R. D. Driver, "Scientific and Engineering Applications with Personal Computers," pp. 341–350, Wiley-Interscience, New York (1986).
13. R. W. Hamming, *op. cit.*, pp. 302–309.
14. *Ibid.*, pp. 577–581.
15. *Ibid.*, pp. 437–443.
16. *Ibid.*, pp. 503–538.
17. *Ibid.*, pp. 539–545.
18. R. Annino and R. D. Driver, *op. cit.*, pp. 352–364.
19. R. W. Hamming, *op. cit.*, pp. 59–77.
20. R. Annino and R. D. Driver, *op. cit.*, pp. 503–547.
21. "ASYST, a Scientific System" [in four volumes, with software], Macmillan Software Company, New York (1985).
22. V. H. Press, B. T. Flannery, J. A. Teukolsky, and W. T. Vettering, "Numerical Recipes—the Art of Scientific Computing," Cambridge University Press (1986). [A paperback supplement, accompanied by disks containing the programs given in the book in either FORTRAN or Pascal is available.]
23. H. J. J. Braddick, "The Physics of the Experimental Method," chap. IX, Chapman & Hall, London (1956).

GENERAL READING

D. C. Baird, "Experimentation," Prentice-Hall, Englewood Cliffs, N.J (1962).

P.R. Bevington, *op. cit.*

N.C. Barford, "Experimental Measurements: Precision, Error, and Truth," Addison-Wesley, Reading, Mass. (1967).

B. A. Barry, "Errors in Practical Measurements in Science, Engineering, and Technology," Wiley, New York (1978).

W. H. Beyer (ed.), "Standard Mathematical Tables," 27th ed., CRC Press, Baton-Roca, Fla. (1984).

W. C. Hamilton, "Statistics in Physical Science," Ronald Press, New York (1964).

R. W. Hamming, *op. cit.*

E. M. Pugh, and G. H. Winslow, "The Analysis of Physical Measurements," Addison-Wesley, Reading, Mass. (1966).

E. Whittaker and G. Robinson, "The Calculus of Observations," 4th ed., Blackie, Glasgow (1944).

E. B. Wilson, Jr. *op. cit.*

CHAPTER
III

GASES

EXPERIMENTS

1. Gas thermometry
2. Joule-Thomson effect
3. Heat-capacity ratios for gases

EXPERIMENT 1
GAS THERMOMETRY

A fundamental attribute of temperature is that for any body in a state of equilibrium the temperature may be expressed by a number on a *temperature scale,* defined without particular reference to that body. The applicability of a universal temperature scale to all physical bodies at equilibrium is a consequence of an empirical law (sometimes called the "zeroth law of thermodynamics"), which states that, if a body is in thermal equilibrium separately with each of two other bodies, these two will be also in thermal equilibrium with each other.

However, a temperature scale must be somehow defined, in order that each attainable temperature shall have a unique numerical value. A temperature scale may be defined over a certain range by the readings of a *thermometer,* which is a body possessing some easily measurable physical property that is for all practical purposes a sensitive function of the temperature alone. The specific volume of a fluid, the electrical resistivity of a metal, and the thermoelectric potential at the junction of a pair of different metals are examples of properties frequently used. Since these properties can be much more easily and precisely measured on a relative basis than on an absolute one, it is convenient to base a temperature scale, in large part, on certain reproducible "fixed points" defined by systems whose temperatures are fixed by nature. Examples of these are the triple point of water (the temperature at which ice, liquid water, and water vapor coexist in equilibrium) and the melting or boiling points of various pure substances under 1 atm pressure. Once the temperature has been fixed at two or more points, the temperatures at other points in the range of the thermometer can be defined in terms of the value of the physical property concerned. Thus, the thermometer may serve as a device for interpolating among two or more fixed points.

In the original centigrade scale of temperature the two fixed points were taken as the "ice point" (temperature at which ice and air-saturated water are in equilibrium under a total pressure of 1 atm), assigned a value of 0°C, and the 'steam point" (temperature at which pure water and water vapor are in equilibrium under a pressure of 1 atm), assigned a value of 100°C. A mercury thermometer with a capillary of uniform bore could be marked at 0°C and 100°C on the capillary stem by use of these fixed points, and the intervening range could then be marked off into 100 equal subdivisions. It is important to recognize that the scale thus defined is not identical with one similarly defined with alcohol (for example) as the thermometric fluid; in general, the two thermometers would give different scale readings at the same temperature because the thermal expansion coefficients of the fluids will have different temperature dependences.

Thus any one physical property of any arbitrarily specified substance would seem to define a temperature scale of a rather arbitrary kind. It would seem clearly preferable to define the temperature on the basis of some fundamental law. In the middle of the nineteenth century, Clausius and Kelvin stated the second law of thermodynamics and proposed the *thermodynamic temperature scale*, which is based on that law.[1] This scale is fixed at its lower end at the absolute zero of temperature. A scale factor, corresponding to the size of the degree, must be specified to complete the definition of the temperature scale; this can be accomplished by specifying the numerical value of the temperature of a reproducible fixed point or by specifying the numerical width of the interval between two fixed points. Kelvin adopted the latter procedure in order to make the degree coincide with the (mean) centigrade degree; accordingly, the interval between the ice point and the steam point was fixed at 100 K, and the absolute (or Kelvin) temperature of the ice point has been found by experiment to be approximately 273.15 K. However, by international agreement (1948, 1954) the former procedure is now used. The triple point of pure water is defined as exactly 273.15 K (Kelvin) and as exactly 0.01°C (degrees Celsius). Thus the relation of the Celsius (formerly centigrade) temperature t to the Kelvin (or absolute) temperature T is†

$$t(°C) \equiv T(K) - 273.15 \tag{1}$$

---

† The differences between the original and the present Kelvin scales and between the old centigrade and the thermodynamic Celsius scales are small (on the order of 10^{-3} K) and of importance only for refined measurements.

Practical difficulties arise in making very precise determinations of temperature on the thermodynamic scale; the precision of the more refined thermometric techniques considerably exceeds the accuracy with which the experimental thermometer scale may be related to the thermodynamic scale. For this reason, a scale known as the *International Practical Temperature Scale* has been devised, with several fixed points and with interpolation formulas based on practical thermometers (e.g., the platinum resistance thermometer between −259.34°C and 630.74°C). This scale is intended to correspond as closely as possible to the thermodynamic scale but to permit more precision in the measurement of temperatures. Further details about this scale are given in Chapter XVI.

METHOD

The establishment of the International Practical Temperature Scale has required that the thermodynamic temperatures of the fixed points be determined with as much accuracy as possible. For this purpose a device was needed that measures essentially the thermodynamic temperature and does not depend on any particular thermometric substance. On the other hand, since it was needed only for a few highly accurate measurements, it did not need to have the convenience of such instruments as resistance thermometers, mercury thermometers, and thermocouples. A device that has filled this need is the *gas thermometer*. It is based on the perfect-gaw law, expressed by

$$pV = nRT \tag{2}$$

where n is the number of moles, R is a universal constant called the gas constant, and T the "perfect-gas" temperature. At ordinary gas densities there are deviations from the perfect-gas law, but it is exact in the limit of zero gas density, where it is applicable to all gases. The "perfect-gas temperature scale" defined by Eq. (2) can be shown by thermodynamics to be identical with a thermodynamic temperature scale,[†] defined in terms of the second law of thermodynamics.

Gas thermometry measurements must, of course, be made with a real gas at ordinary pressures. However, it is possible to estimate the deviations from perfect-gas behavior and to convert measured pV values to "perfect-gas" pV values. For this purpose a virial equation of state for a real gas is often used:

$$\frac{p\bar{V}}{RT} = 1 + \frac{B}{\bar{V}} + \frac{C}{\bar{V}^2} + \cdots \tag{3}$$

where $B, C, \ldots$ are known as the second, third, ... *virial coefficients* and $\bar{V}$ is the molar volume of the gas at p and T. At ordinary pressures the series

---

[†] The perfect-gas properties required for this identity are (1) Boyle's law is obeyed: $p\bar{V} = f(T)$; and (2) the internal energy per mole is a function of temperature only: $\bar{E} = g(T)$.

converges rapidly, and for many purposes terms beyond the one containing B can be neglected. Values of second (and in some cases, third and fourth) virial coefficients are known over a wide range of temperatures from gas compressibility measurements; these values of B are useful in correcting the readings of a gas thermometer.

The establishment of the International Practical Temperature Scale has been accomplished largely with the aid of measurements made with the helium gas thermometer.[2] The most precise gas thermometry method is the constant-volume method, in which a definite quantity of the gas is confined in a bulb of constant volume V at the temperature T to be determined and the pressure p of the gas is measured. A problem is encountered, however, in measuring the pressure, which is usually done with a mercury manometer operating at room temperature; a way must be found to communicate between the bulb and the manometer. This is usually accomplished by connecting the bulb to the room-temperature part of the system by a slender tube and allowing a portion of the gas to occupy a relatively small, constant "dead-space" volume at room temperature. The pressure in the manometer system is then equalized to the gas pressure in this dead space by a null device, represented in the apparatus of the present experiment by the null manometer.

THEORY

The number of moles of gas n in the bulb of volume V at temperature T and in the dead space of volume v at room temperature is constant. Assuming the perfect-gas law, we have

$$\frac{pV}{RT} + \frac{pv}{RT_r} = n \tag{4}$$

where T_r is room temperature. The small region of large temperature gradient actually existing between the bulb and the dead space is replaced, without appreciable error, by a sharp division between two uniform temperatures T and T_r. Since the second term is small in comparison with the first, we shall introduce only a very small error (~ 0.05 percent with the present apparatus) by making the substitution

$$T_r = \frac{p_r}{p} T \tag{5}$$

where p_r is the pressure measured by the manometer when the bulb is at room temperature. We then obtain

$$\frac{pV}{RT}\left(1 + \frac{pv}{p_r V}\right) = n \tag{6}$$

Since only the pressure is being measured directly, the volume V must remain constant or else vary in a known way with the temperature (and pressure). If α

is the coefficient of linear thermal expansion of the material from which the bulb is constructed, the coefficient of volume expansion is 3α. If this is assumed constant with temperature, we can write

$$\frac{pV_0}{RT}\left(1 + \frac{pv}{p_rV} + 3\alpha t\right) = n \tag{7}$$

where V_0 is the bulb volume at 0°C and t is the Celsius temperature of the gas thermometer bulb. For very precise work, other variables, including the dependence of the bulb volume on the difference between internal and external pressure, must be taken into account. To the precision of the present experiment, this effect of pressure is inconsequential.

Finally, if we allow for gas imperfections by including the second virial term, we can write

$$\frac{pV_0}{RT}\left(1 + \frac{pv}{p_rV} + 3\alpha t - \frac{B}{\bar{V}}\right) = n \tag{8}$$

where

$$\frac{1}{\bar{V}} = \frac{n}{V} \cong \frac{p_0}{RT_0} \tag{9}$$

and p_0 and T_0 are the values of p and T at (say) the ice point.

The above equations contain approximations that are acceptable for the present experiment. For high-precision work a much more detailed treatment is required.[2]

EXPERIMENTAL

The object of the experiment described below is to set up a gas thermometer, calibrate it at the ice point (in lieu of the experimentally more difficult triple point), and use it to determine the temperatures of one or more other fixed points. These may include the steam point, the boiling point of liquid nitrogen (in lieu of liquid oxygen, which is more difficult to obtain and more hazardous to use), the sublimation temperature of solid carbon dioxide (Dry Ice), the transition temperature of sodium sulfate decahydrate to the monohydrate and saturated solution, etc. The experiment will be performed with an apparatus which resembles in principle a research gas thermometer but is very much simpler and somewhat less precise. The apparatus is shown in Fig. 1.

Gas thermometer bulb a. This is a Pyrex bulb with a volume V of about 100 cm^3, with an attached glass capillary passing through a large rubber stopper. The upper end of this capillary tube is connected, by a length of flexible stainless-steel capillary tubing, to the null manometer.

Null manometer b. When the mercury surfaces in the two arms are at the same level, the pressure in the bulb is equal to that in the rest of the system, including the manometer d; moreover, the "dead-space" volume v then has a

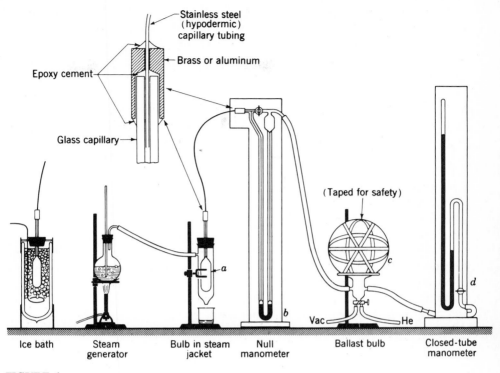

FIGURE 1
Gas thermometry apparatus.

fixed value. This dead space is the volume of gas at temperature T_r between the top of the bulb and the mercury in the left arm. The upper part of each arm consists of a long capillary tube. This design prevents loss of mercury in the event of any accidentally large pressure difference across the null manometer. The two arms are connected at the top by a capillary stopcock, which can be opened to admit or remove gas from the bulb. During measurements this stopcock is kept closed, and the quantity of gas in the bulb and dead space is thereby fixed.

Ballast bulb *c.* This is a bulb of large volume (at least 5 liters) which permits small adjustments in pressure to be made by the addition or removal of small quantities of gas or air through the three-way stopcock at the bottom. One of the two lower arms of this stopcock is connected to the source of gas to be used (or to room air); the other is connected to a laboratory vacuum line, an aspirator, or a vacuum pump.

Closed-tube mercury manometer *d.* This is the instrument used for the measurement of gas pressures in this experiment. The space in the closed arm is evacuated, so the difference between the two mercury levels is a direct measure of the pressure of the gas in the other arm. For precise work, the temperature of the manometer should be determined and the manometer readings corrected (to 0°C) for the differential expansion of the mercury and the scale; however, in this experiment, if the ambient temperature is reasonably constant, the corrections should ultimately cancel out and may be omitted.

The gas thermometer bulb is first filled with the gas to be used, preferably helium or nitrogen (free of water contamination). With the stopcock on the null manometer open, the entire system is evacuated and then filled to the desired pressure with the gas. If complete evacuation is not possible (as is the case when an aspirator or a laboratory vacuum line is used), the filling must be done several times, with a delay of 1 or 2 min between each repetition to allow for diffusive mixing. When the bulb has been filled with the gas to the desired pressure, the stopcock on the null manometer is closed. After this point any gas, including air, may be allowed in the ballast bulb and manometer.

When a temperature is being measured with the thermometer, gas or air is removed from or admitted to the ballast bulb until the null manometer is balanced. When it is evident that equilibrium has been attained and the null manometer is precisely balanced, the closed-tube manometer levels are read and recorded. It is suggested that four readings be taken, alternately bringing the pressure to the null value from above and below.

Procedure. Assemble the apparatus as indicated in Fig. 1. Make sure that the stopcock on the null manometer is properly greased. Then fill the apparatus with the gas to be used, as described above. At the final filling, adjust the pressure to about 600 Torr, and after 1 or 2 min close the null-manometer stopcock. After drift has ceased, make a reading with the bulb at room temperature. The difference between the two manometer levels is then the pressure of gas in the bulb at room temperature, p_r.

To check for possible leaks, it is advisable to restore the ballast bulb to atmospheric pressure for 10 or 15 min and then reduce the pressure so as to rebalance the null manometer. If p_r has increased significantly, a leak should be suspected.

Ice point. In the Dewar flask, prepare a "slushy" mixture of finely shaved clean ice and distilled water. This should be fluid enough to permit the gas thermometer bulb to be lowered into place and to allow a metal ring stirrer to be operated but should have sufficient ice to maintain two-phase equilibrium over the entire surface of the bulb. Mount the bulb in place, and commence stirring the mixture, moving the stirrer slowly up and down from the very top to the very bottom of the Dewar flask. Do not force the stirrer, since the bulb and its capillary stem are fragile. After equilibrium is achieved, take the ice-point pressure readings and record p_0.

Steam point. While the ice-point readings are being made, the water in the steam generator should be heated to boiling. When ready, the gas thermometer bulb should be positioned in the steam jacket so that it does not touch the wall at any point. Steam should be passed through the jacket at a rate slow enough to avoid overpressure in the jacket; steam should emerge from the bottom at low velocity. The rubber tubing connecting the steam generator to the steam jacket should run downhill all the way to permit steady drainage of any water condensed in the tubing. When drift ceases, the steam-point readings may be taken.

Repeat the ice-point measurements. The results should agree with those obtained previously to within experimental error; if the differences are larger, a leak must be suspected.

Measure the other fixed points assigned. If liquid nitrogen is used, insert the bulb into the Dewar slowly to prevent violent boiling and excessive loss of liquid.

If time permits, refill the gas thermometer bulb with another gas or reduce p_r to 300 Torr and repeat some or all of the above measurements. During the steam-point measurements, record the barometric pressure. (The barometer reading must be corrected for temperature. A discussion of the use of precision barometers is given in Chapter XVIII, and the necessary corrections are given in Appendix D.)

Also record the values of V and v for the apparatus used in the experiment.

CALCULATIONS

Equation (8) can be written in the form

$$T = Ap \tag{10}$$

where the proportionality factor

$$A = \frac{V_0}{nR}\left(1 + \frac{pv}{p_r V} + 3\alpha t - \frac{B}{\bar{V}}\right) \tag{11}$$

is nearly constant with temperature; but for the precision attainable with this experiment, the proportionality factor should be evaluated at each temperature. Making use of the known thermodynamic temperature of the ice point, we can write

$$T = \frac{273.15}{p_0}\frac{A}{A_0}p \tag{12}$$

$$\frac{A}{A_0} = 1 + \frac{p - p_0}{p_r}\frac{v}{V} + 3\alpha t - \frac{1}{\bar{V}}(B - B_0) \tag{13}$$

where p_0, A_0, and B_0 pertain to the ice point. Since the last two terms depend on the temperature, an approximate value of the temperature must be known before they can be evaluated; this can be obtained by setting A/A_0 equal to unity in Eq. (12). For Pyrex glass, $\alpha = 3.2 \times 10^{-6}\,\mathrm{K}^{-1}$. Second virial

TABLE 1
Second virial coefficients (in $cm^3 mol^{-1}$)

t, °C	He	Ar	N_2	CO_2
−250	~0			
−200	+10.4			
−150	11.4			
−100	11.7	−64.3	−51.9	
−50	11.9	−37.4	−26.4	
0	11.8	−21.5	−10.4	−154
50	11.6	−11.2	−0.4	−103
100	11.4	−4.2	+6.3	−73
150	11.0	+1.1	11.9	−51

coefficients for various gases are given in Table 1, and $\bar{V}$ can be calculated from Eq. (9).

The student should report the temperature determined for each fixed point on both the Kelvin and Celsius scales. In cases where the temperature is a boiling point or sublimation point, calculate and report also the temperature on both scales corrected to a pressure of 1 atm. In the neighborhood of 1 atm, the boiling point of water increases with pressure by 0.037 K $Torr^{-1}$. For liquid nitrogen, the increase is 0.013 K $Torr^{-1}$.

DISCUSSION

What property of helium makes it particularly suitable for gas thermometry over the temperature range covered by this experiment? Is the correction for gas imperfection in this experiment of significant magnitude in relation to the experimental uncertainty? If not, by how much must the precision of the pressure measurements be improved before gas imperfection corrections become significant? How does this depend on the choice of gas to be used?

APPARATUS

Closed-tube manometer; null manometer; properly taped and mounted ballast bulb; heavy-wall pressure tubing; Dewar flask; large ring stirrer; notched cover plate for Dewar with hole for mounting gas thermometer bulb; bunsen burner; steam generator with rubber connecting tubing; steam jacket; two ring stands; ring clamp and iron gauze; two clamp holders; one large and one medium clamp.

Cylinder of helium or dry nitrogen; pure ice (1 kg); ice grinder; liquid nitrogen (1 liter); boiling chips; stopcock grease; vacuum pump or water aspirator.

REFERENCES

1. G. W. Castellan, "Physical Chemistry," 3d ed., pp. 99, 160, Addison-Wesley, Reading, Mass. (1983).
2. J. A. Beattie and coworkers, *Proc. Am. Acad. Arts Sci.* **74,** 327 (1941); **77,** 255 (1949).

GENERAL READING

"Thermometers," New Encyclopaedia Britannica, 15th ed., Vol. 23, p. 734ff (1985).
T. J. Quinn, "Temperature," Academic Press, New York (1983).
R. P. Benedict, "Fundamentals of Temperature, Pressure, and Flow Measurements," 3d ed., Wiley-Interscience, New York (1984).

EXPERIMENT 2
JOULE–THOMSON EFFECT

The Joule–Thomson effect is a measure of the deviation of the behavior of a real gas from what is defined to be ideal-gas behavior. In this experiment a simple technique for measuring this effect will be applied to a few common gases.

THEORY

An ideal gas may be defined as one for which the following two conditions apply at all temperatures for a fixed quantity of the gas: (1) Boyle's law is obeyed; i.e.,

$$pV = f(T)$$

and (2) the internal energy E is independent of volume. Accordingly E is independent of pressure as well, and in the absence of other pertinent variables (such as applied fields) E of an ideal gas is therefore a function of the temperature alone:

$$E = g(T)$$

It is apparent that the enthalpy H of an ideal gas is also a function of temperature alone:

$$H \equiv E + pV = h(T)$$

Accordingly we can write for a definite quantity of an ideal gas at all temperatures

$$\left(\frac{\partial E}{\partial V}\right)_T = \left(\frac{\partial E}{\partial p}\right)_T = \left(\frac{\partial H}{\partial V}\right)_T = \left(\frac{\partial H}{\partial p}\right)_T = 0 \tag{1}$$

The absence of any dependence of the internal energy of a gas on volume was suggested by the early experiments of Gay-Lussac and Joule. They found that when a quantity of gas in a container initially at a given temperature was allowed to expand into another previously evacuated container without work or heat flow to or from the surroundings ($\Delta E = 0$), the final temperature (after the two containers came into equilibrium with each other) was the same as the initial temperature. However, that kind of experiment (known as the Joule experiment) is of limited sensitivity, because the heat capacity of the containers is large in comparison with that of the gases studied. Subsequently Joule and Thomson[1] showed, in a different kind of experiment, that real gases do undergo small temperature changes on free expansion. This experiment utilized continuous gas flow through a porous plug under adiabatic conditions. Because of the continuous flow, the solid parts of the apparatus come into thermal equilibrium with the flowing gas, and their heat capacities impose a much less serious limitation than in the case of the Joule experiment.

Let it be imagined that gas is flowing slowly from left to right through the porous plug in Fig. 1. To the left of the plug the temperature and pressure of the gas are T_1 and p_1, and to the right of the plug they are T_2 and p_2. The volume of a definite quantity of gas (say 1 mol) is V_1 on the left and V_2 on the right, and the internal energy is E_1 and E_2, respectively. When 1 mol of gas flows through the plug, the work done upon the system by the surroundings is

$$w = p_1V_1 - p_2V_2$$

Since the process is adiabatic, the change in internal energy is

$$\Delta E = E_2 - E_1 = q + w = w$$

Combining these two equations we obtain

$$E_1 + p_1V_1 = E_2 + p_2V_2$$

or

$$H_1 = H_2 \tag{2}$$

Thus this process takes place at constant enthalpy.

For a process involving arbitrary infinitesimal changes in pressure and temperature, the change in enthalpy is

$$dH = \left(\frac{\partial H}{\partial p}\right)_T dp + \left(\frac{\partial H}{\partial T}\right)_p dT \tag{3}$$

In the present experiment dH is zero and dT and dp cannot be arbitrary but

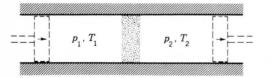

FIGURE 1

Schematic diagram of the Joule-Thomson experiment. The stippled area represents a porous plug.

are related by

$$\mu \equiv \left(\frac{\partial T}{\partial p}\right)_H = -\frac{(\partial H/\partial p)_T}{(\partial H/\partial T)_p} \tag{4}$$

The quantity μ defined by this equation is known as the *Joule–Thomson coefficient*. It represents the limiting value of the experimental ratio of temperature difference to pressure difference as the pressure difference approaches zero:

$$\mu = \lim_{\Delta p \to 0} \left(\frac{\Delta T}{\Delta p}\right)_H \tag{5}$$

Experimentally ΔT is found to be very nearly linear with Δp over a considerable range; this is in accord with expectations based on the theory given below.

The denominator on the right side of Eq. (4) is the heat capacity at constant pressure C_p. The numerator is zero for an ideal gas [see Eq. (1)]. Accordingly, for an ideal gas the Joule–Thomson coefficient is zero, and there should be no temperature difference across the porous plug. For a real gas, the Joule–Thomson coefficient is a measure of the quantity $(\partial H/\partial p)_T$ [which can be related thermodynamically to the quantity involved in the Joule experiment, $(\partial E/\partial V)_T$]. Using the general thermodynamic relation

$$\left(\frac{\partial H}{\partial p}\right)_T = -T\left(\frac{\partial V}{\partial T}\right)_p + V \tag{6}$$

it can be shown that, for an ideal gas satisfying the criteria already given,

$$pV = \text{const} \times T \tag{7}$$

where T is the absolute thermodynamic temperature. The coefficient $(\partial H/\partial p)_T$ is therefore a measure of the deviation from the behavior predicted by Eq. (7). On combining Eqs. (4) and (6), we obtain

$$\mu = \frac{T(\partial V/\partial T)_p - V}{C_p} \tag{8}$$

In order to predict the magnitude and behavior of the Joule–Thomson coefficient for a real gas, we can use the van der Waals equation of state,[2] which is

$$\left(p + \frac{a}{\bar{V}^2}\right)(\bar{V} - b) = RT \tag{9}$$

where $\bar{V}$ is the molar volume. We can rearrange this equation (with neglect of the very small second-order term $ab/\bar{V}^2$ and substitution of p/RT for $1/\bar{V}$ in a first-order term) to obtain

$$p\bar{V} = RT - \frac{ap}{RT} + bp$$

Thus

$$\left(\frac{\partial \bar{V}}{\partial T}\right)_p = \frac{R}{p} + \frac{a}{RT^2}$$

Combination of these two equations yields

$$\left(\frac{\partial \bar{V}}{\partial T}\right)_p = \frac{\bar{V} - b}{T} + \frac{2a}{RT^2} \tag{10}$$

which on substitution into Eq. (8) gives the expression

$$\mu = \frac{(2a/RT) - b}{\bar{C}_p} \tag{11}$$

This expression does not contain p or $\bar{V}$ explicitly, and the molar heat capacity $\bar{C}_p$ may be considered essentially independent of these variables. The temperature dependence of $\bar{C}_p$ is small, and accordingly that of μ is also small enough to be neglected over the ΔT obtainable with a Δp of about 1 bar (namely about 1 K or less for the gases considered here). Accordingly we may expect that μ will be approximately independent of Δp over a wide range, as stated previously.

For most gases under ordinary conditions, $2a/RT > b$ (the attractive forces predominate over the repulsive forces in determining the nonideal behavior) and the Joule–Thomson coefficient is therefore positive (gas cools on expansion). At a sufficiently high temperature the inequality is reversed, and the gas warms on expansion. The temperature at which the Joule–Thomson coefficient changes sign is called the *inversion temperature* T_I. For a van der Waals gas

$$T_I = \frac{2a}{Rb} \tag{12}$$

TABLE 1
Values of constants in equations of state[a]

	He	H$_2$	N$_2$	CO$_2$
van der Waals:[3]				
a	0.03457	0.2476	1.408	3.640
b	0.02370	0.02661	0.03913	0.04267
Beattie-Bridgeman:[4]				
A_0	0.0219	0.2001	1.3623	5.0728
a	0.05984	−0.00506	0.02617	0.07132
B_0	0.01400	0.02096	0.05046	0.10476
b	0.0	−0.04359	−0.00691	0.07235
$10^{-4}c$	0.0040	0.0504	4.20	66.00

[a] Units assumed are V in $\mathrm{dm^3\,mol^{-1}} \equiv \mathrm{L\,mol^{-1}}$, p in $\mathrm{bar} \equiv 10^5$ Pa, T in Kelvin. ($R = 0.083145 \, \mathrm{bar\,dm^3\,K^{-1}\,mol^{-1}}$.)

This temperature is usually several hundred degrees above room temperature. However, hydrogen and helium are exceptional in having inversion temperatures well below room temperatures. This results from the very small attractive forces in these gases (see Table 1 for values of the van der Waals constant a).

Other semiempirical equations of state can be used to predict Joule–Thomson coefficients. Perhaps the best of these is the Beattie–Bridgeman equation,[4,5] which can be written (for 1 mol) as

$$p = \frac{RT(1 - \varepsilon)}{\tilde{V}^2}(\tilde{V} + B) - \frac{A}{\tilde{V}^2} \tag{13}$$

where $A = A_0(1 - a/\tilde{V})$, $B = B_0(1 - b/\tilde{V})$, and $\varepsilon = c/\tilde{V}T^3$. In this equation of state there are five constants which are characteristic of the particular gas: A_0, B_0, a, b, and c. In terms of these constants and the pressure and temperature, the Joule–Thomson coefficient is given[4] by

$$\mu = \frac{1}{\tilde{C}_p}\left\{-B_0 + \frac{2A_0}{RT} + \frac{4c}{T^3} + \left[\frac{2B_0 b}{RT} - \frac{3A_0 a}{(RT)^2} + \frac{5B_0 c}{RT^4}\right]p\right\} \tag{14}$$

This equation predicts a small dependence on pressure not shown by Eq. (11), which is based on the van der Waals equation.

EXPERIMENTAL

The experimental apparatus is shown in Figs. 2 and 3. The "porous plug" is a fritted glass plate sealed in a 30-mm glass tube. A Corning plate of porosity $F =$ "fine" (4.0 to 5.5 μm pore size) is convenient. The lower end of this tube is drawn down to 6 to 8 mm and connected by a short length of pressure tubing (securely clamped or wired) to *at least* 100 ft of $\frac{1}{4}$-inch copper tubing wound in a coil of convenient shape. This coil and the lower end of the glass tube (up to the fritted disk) are submerged in a constant-temperature bath in order to bring the gas to the bath temperature. The other end of the coil is connected to a supply of gas at a pressure approximately 1 atm above atmospheric pressure. An open-tube mercury manometer (or a Bourdon pressure gauge) is used to measure Δp. The upper end of the glass tube (above the fritted disk) is always at atmospheric pressure but is thermally insulated with an outer jacket filled with an insulating medium such as vermiculite or Styrofoam.

The temperature difference ΔT between the gas above the fritted disk and the water in the bath is measured with a calibrated copper–constantan (Type T) thermocouple using the finest wires available. It is recommended that a cylindrical cardboard tube be placed in the cell to position the sample junction 5 to 10 mm above the fritted disk as shown in Fig. 3. The tube may be roughly centered with a few tufts of glass wool. The reference junction of the thermocouple should be embedded in a small amount of wax or oil at the bottom of a test-tube which is mounted in the constant-temperature bath.

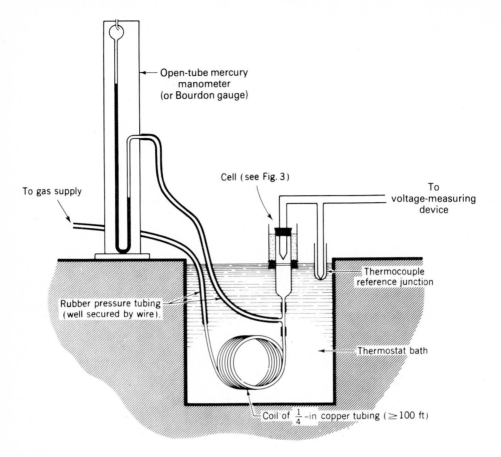

FIGURE 2
The Joule-Thomson apparatus.

Because ΔT is less than 1 K for the pressure range to be studied, a potentiometer or digital voltmeter with $\pm 0.1\,\mu V$ sensitivity is desirable for accurate measurements of thermocouple potential differences that will be less than $40\,\mu V$. Such small potentials can also be measured directly with a sensitive galvanometer.† In this case, the galvanometer-plus-thermocouple system must be calibrated in order to convert galvanometer deflections into temperature differences. As an alternative to a thermocouple, one can use two microbead thermistors and an appropriate resistance-bridge circuit (see Chapters XV and XVI). A calibration is also required in this instance.

---

† For temperature changes less than 1 K, the thermocouple can be attached directly to a galvanometer such as the Leeds & Northrup 2430d, having a sensitivity of $0.0005\,\mu A\ mm^{-1}$ and an internal resistance of $550\,\Omega$. For larger temperature differences, an appropriate resistance can be used in series with the thermocouple to reduce the sensitivity.

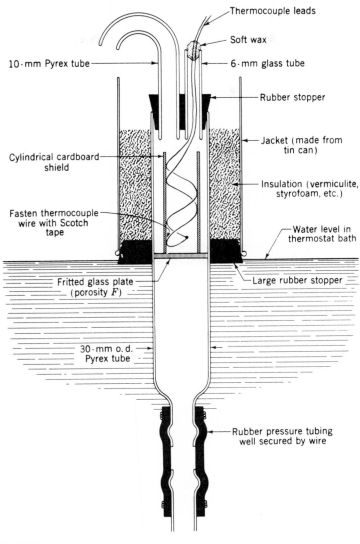

FIGURE 3
Detail of Joule-Thomson cell.

The cell depicted in Fig. 3 is a satisfactory design as long as care is taken to ensure that the high-pressure gas has reached temperature equilibrium with the water bath. Alternative designs are possible (see, for example, Ref. 6) in which the "reference" thermocouple junction is mounted in the high-pressure gas below the fritted disk. Such an arrangement would give a more direct measure of ΔT, but is less convenient if a calibration is required.

It should be noted that a careful *absolute* calibration of the thermocouple is not actually necessary since it is a change ΔT in temperature that is of primary interest in this experiment. In fact, to the accuracy feasible in this experiment, *no* calibration is needed if a potentiometer or digital voltmeter is used to measure the thermocouple potential. This is because the temperature coefficient (Seebeck coefficient) α of a copper–constantan thermocouple varies only slightly with temperature, from 39 to 43 μV K^{-1} over the temperature range from 0°C to 50°C. At 25°C, α is 40.6 μV K^{-1} and the variation in this coefficient is small from one thermocouple to another. For small temperature differences, a linear relation $\Delta V_{TC} = \alpha \, \Delta T = \alpha (dT/dp) \, \Delta p$ is a good approximation for the thermocouple potential difference between the two junctions.† Thus the slope of a plot of ΔV_{TC} versus Δp can be combined with an assumed value of $\alpha = 40.6 \,\mu$V K^{-1} to yield dT/dp and hence μ. This procedure is particularly desirable for a cell with the thermocouples mounted on both sides of the fritted disk, since a calibration is not easily carried out in this case.

Procedure. Set up the apparatus as shown in Figs. 2 and 3. The gas supply should be a cylinder or supply line equipped with a pressure regulator and a needle valve. Be sure that the supply pressure is constant; if there are other demands on the gas supply, careful planning is required. All gas connections should be securely wired or clamped. **Warning**: all changes in pressure must be carried out **very slowly**.

For the initial equilibration of the system, close the needle valve and adjust the main gas supply valve until the regulator gauge reads about 3 atm. Then **very slowly** open the needle valve until the pressure in the system (as indicated by the manometer or Bourdon gauge) is increasing at about a rate of 50 Torr min^{-1}. Continue to adjust the needle valve until Δp is about 750 Torr (this should take *at least* 15 min during which the system will be purged of residual air or any previous sample). If the gas pressure is increased too rapidly, the coil will not be able to warm the initial surge of gas to bath temperature and the porous frit will be cooled to below its steady-state value; this will cause a very slow (2 to 3 hours) attainment of a constant value for ΔT. With care in making pressure changes, a steady state should be achieved in about 40 minutes.

During this initial stabilization period, the thermocouple can be calibrated if a galvanometer is to be used as the measuring device. The sample junction should be removed from the Joule–Thomson cell and taped to the bulb of a Beckmann thermometer (or the body of a high-resolution platinum resistance thermometer), which is placed first in the bath and then in a Dewar

---

† In practice one often finds that $\Delta V_{TC} = \alpha \, \Delta T + \delta V_{TC}$, where δV_{TC} is a small offset voltage observed when both the reference and the measuring junction are at the same temperature. This "nonthermodynamic" result can occur if the thermocouple wire has regions of compositional variation or strain (e.g., from kinking) which are subject to a temperature gradient. δV_{TC} can be ignored in this experiment since it affects only the intercept of the plot of ΔV_{TC} versus Δp and not the slope.

flask containing water 0 to $-1\,K$ cooler than the bath temperature. The temperature of the water in the Dewar can be adjusted by adding small amounts of hot or cold water. Obtain several calibration points starting from small ΔT values with the thermocouple leads connected directly across the galvanometer terminals. If the galvanometer deflection goes off-scale for larger ΔT values, a second calibration curve should be obtained with a resistance added in series to reduce the sensitivity. (With the L & N 2430d, a 1500-Ω resistor is recommended.) After the calibration is complete, the wires should be dried very carefully, and the thermocouple junction re-installed in the cell.

Record values of ΔV_{TC} (or ΔT) and the pressure differential Δp until no significant change occurs over a 10- to 15-minute interval. Then *very slowly* close the needle valve so as to reduce the pressure drop Δp to approximately 600 Torr. This change should require *at least* 5 minutes! Record data as before until a steady state is achieved at this new setting (about 20 minutes). Repeat the procedure to obtain data at $\Delta p = 450$, 300, and 150 Torr. If time is available, make measurements on both nitrogen and carbon dioxide.

CALCULATIONS

For each gas studied, plot ΔV_{TC} (or ΔT) versus Δp. Draw the best straight line through these points; this line will of course only pass through zero if the thermocouple has been carefully calibrated. Use least squares to calculate the best slope along with its standard deviation. From the slope, evaluate the Joule–Thomson coefficient in units of $K\,bar^{-1}$. For comparison, calculate μ for these gases at 25°C from the van der Waals and/or the Beattie–Bridgeman constants given in Table 1. $\bar{C}_p$ values for N_2 and CO_2 at 25°C are 29.125 and 37.11 $J\,K^{-1}\,mol^{-1}$, respectively.

DISCUSSION

For the Joule experiment we can write

$$\eta \equiv -\left(\frac{\partial T}{\partial V}\right)_E = \frac{(\partial E/\partial V)_T}{(\partial E/\partial T)_V} = \frac{T(\partial p/\partial T)_V - p}{C_v} \tag{15}$$

This quantity is called the Joule coefficient. It is the limit of $-(\Delta T/\Delta V)_E$, corrected for the heat capacity of the containers, as ΔV approaches zero. With the van der Waals equation of state we obtain $\eta = a/\bar{V}^2\bar{C}_v$. The corrected temperature change when the two containers are of equal volume is found by integration to be $\Delta T = -a/2\bar{V}\bar{C}_v$, where $\bar{V}$ is the initial molar volume and $\bar{C}_v$ is the molar constant-volume heat capacity.

It is instructive to calculate this ΔT for a gas such as CO_2. In addition, the student may consider the relative heat capacities of 10 liters of the gas at a pressure of 1 bar and that of the quantity of copper required to construct two spheres of this volume with walls (say) 1 mm thick and then calculate the ΔT expected to be oberved with such an experimental arrangement.

APPARATUS

Joule–Thomson cell comprising glass tube with fritted disk, large rubber stopper with outer jacket, insulation medium such as vermiculite, rubber stopper with bent glass outlet tube and tube for thermocouple wires, and cardboard cylinder; clamp to hold cell in bath; thermocouple (insulated copper–constantan, No. 30) with test tube for reference junction; clamp for same; sensitive potentiometer (such as L & N K-3), voltmeter (such as Keithley 196), or galvanometer (such as L & N 2430d) with auxiliary resistance (e.g., $1500\,\Omega$); coil of $\frac{1}{4}$-inch copper tubing, *at least* 100 ft; open-tube manometer (80 cm) containing mercury or 0–1 atm pressure gauge; T tube; two 4-inch and two 36-inch (or longer) pieces of rubber tubing for making connections; Beckmann thermometer and thermometer clamp (or high-resolution platinum thermometer); 500-ml beaker.

   Constant-temperature bath regulated at 25°C; cylinders or supply lines for gases (N_2, CO_2) equipped with pressure regulators and needle valves; copper wire for securing connections of rubber tubing; Scotch tape; soft wax; paraffin wax or oil; glass wool.

REFERENCES

1. J. P. Joule and W. Thomson (Lord Kelvin), *Phil. Trans.* **143**, 357 (1853); **144**, 321 (1854). [Reprinted in "Harper's Scientific Memoirs I, The Free Expansion of Gases." Harper, New York (1898).]
2. G. W. Castellan, "Physical Chemistry," 3d ed., pp. 34ff., Addison-Wesley, Reading, Mass. (1983).
3. "Landolt-Börnstein Physikalisch-chimische Tabellen," 5th ed., p. 254. Springer, Berlin (1923) [reprinted by Edwards, Ann Arbor, Mich. (1943)]. [This is the source of van der Waals constants cited in the 1987 Edition of the CRC Handbook of Chemistry and Physics.]
4. J. A. Beattie and W. H. Stockmayer, The Thermodynamics and Statistical Mechanics of Real Gases, in H. S. Taylor and S. Glasstone (eds.), "A Treatise on Physical Chemistry," Vol. II, pp. 187ff., esp. pp. 206, 234, Van Nostrand, Princeton, N.J. (1951).
5. J. A. Beattie and O. C. Bridgeman, *J. Am. Chem. Soc.* **49**, 1665 (1927); *Proc. Am. Acad. Arts Sci.* **63**, 229 (1928).
6. C. E. Hecht and G. Zimmerman, *J. Chem. Educ.* **31**, 530 (1954); A. M. Halpern and S. Gozashti, *J. Chem. Educ.* **63**, 1001 (1986).

EXPERIMENT 3
HEAT-CAPACITY RATIOS FOR GASES

The ratio C_p/C_v of the heat capacity of a gas at constant pressure to that at constant volume will be determined by either the method of adiabatic expansion or the sound velocity method. Several gases will be studied, and the results will be interpreted in terms of the contribution made to the specific heat by various molecular degrees of freedom.

THEORY

In considering the theoretical calculation of the heat capacities of gases, we shall only be concerned with perfect gases. Since $\bar{C}_p = \bar{C}_v + R$ for an ideal gas (where $\bar{C}_p$ and $\bar{C}_v$ are the molar quantities C_p/n and C_v/n), our discussion can be restricted to C_v.

The number of "degrees of freedom" for a molecule is the number of independent coordinates needed to specify its position and configuration. Hence a molecule of N atoms has $3N$ degrees of freedom. These could be taken as the three cartesian coordinates of the N individual atoms, but it is more convenient to classify them as follows.

1. *Translational degrees of freedom*: Three independent coordinates are needed to specify the position of the center of mass of the molecule.

2. *Rotational degrees of freedom*: All molecules containing more than one atom require a specification of their orientation in space. As an example, consider a rigid diatomic molecule; such a model consists of two point masses (the atoms) connected by a rigid massless bar (the chemical bond). Through the center of mass, which lies on the rigid bar, independent rotation can take place about two axes mutually perpendicular to each other and to the rigid bar. (The rigid bar itself does not constitute a third axis of rotation under ordinary circumstances for reasons based on quantum theory, there being no appreciable moment of inertia about this axis.) Rotation of a *diatomic* molecule or any linear molecule can thus be described in terms of *two* rotational degrees of freedom. *Nonlinear* molecules for which the third axis has a moment of inertia of appreciable magnitude and constitutes another axis of rotation require *three* rotational degrees of freedom.

3. *Vibrational degrees of freedom*: One must also specify the displacements of the atoms from their equilibrium positions (vibrations). The number of vibrational degrees of freedom is $3N - 5$ for linear molecules and $3N - 6$ for nonlinear molecules. These values are determined by the fact that the total number of degrees of freedom must be $3N$. For each vibrational degree of freedom there is a "normal mode" of vibration of the molecule, with characteristic symmetry properties and a characteristic harmonic frequency. The vibrational normal modes for CO_2 and H_2O are illustrated schematically in Fig. 1.

Using classical statistical mechanics one can derive the theorem of the equipartition of energy. According to this theorem $kT/2$ of energy is associated with each quadratic term in the expression for the energy.[1] Thus there is associated with each translational or rotational degree of freedom for a molecule a contribution to the energy of $kT/2$ of kinetic energy and for each vibrational degree of freedom a contribution of $kT/2$ of kinetic energy and

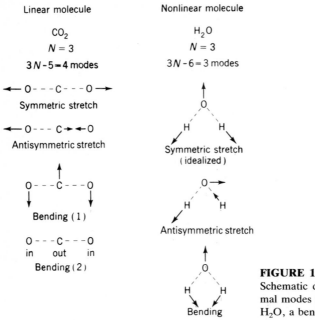

FIGURE 1
Schematic diagrams of the vibrational normal modes for CO_2, a linear molecule, and H_2O, a bent molecule.

$kT/2$ of potential energy. (The corresponding contributions to the energy per *mole* of gas are $RT/2$.)

Clearly, a monatomic gas has no rotational or vibrational energy but does have a translational energy of $\frac{3}{2}RT$ per mole. The constant-volume heat capacity of a *monatomic* perfect gas is thus

$$\bar{C}_v = \left(\frac{\partial \bar{E}}{\partial T}\right)_V = \tfrac{3}{2}R \tag{1}$$

For diatomic or polyatomic molecules, we can write

$$\bar{E} = \bar{E}(\text{trans}) + \bar{E}(\text{rot}) + \bar{E}(\text{vib}) \tag{2}$$

In Eq. (2), contributions to the energy from electronic states have been neglected, since they are not significant at room temperature for most molecules. Also any small intermolecular energies which occur for imperfect gases are not considered.

The equipartition theorem is based on classical mechanics. Its application to translational motion is in accord with quantum mechanics as well. At ordinary temperatures the rotational results are also in accord with quantum mechanics. (The greatest deviation from the classical result is in the case of hydrogen, H_2; at temperatures below 100 K the rotational energy of H_2 is significantly below the equipartition value, as predicted by quantum mechanics.)

The vibrational energy is, however, highly quantized and depends strongly on temperature; the various vibrational modes are at ordinary temperatures only partially "active," and the degree of activity depends strongly on the temperature. As a general rule, the heavier the atoms or the smaller the force constant of the bond (i.e., the lower the vibrational frequency), the more "active" is a given degree of freedom at a given temperature and the greater is the contribution to the heat capacity. Moreover, the frequencies of modes that are predominantly bending of bonds tend to be much lower than those that are predominantly stretching of bonds. In the case of most gaseous *diatomic* molecules (where the one vibrational mode is a pure stretch) the vibrational contribution to $\bar{C}_v$ is very small; for example, N_2 would have its classical equipartition value for $\bar{C}_v$ only above about 4000 K. Many polyatomic molecules, especially those containing heavy atoms, will at room temperature have significant *partial* vibrational contributions to $\bar{C}_v$.

We are now in a position to calculate for polyatomic molecules approximate or at least limiting values for $\bar{C}_v$ and for the ratio $\bar{C}_p/\bar{C}_v$, which, for a perfect gas, is given by

$$\gamma \equiv \frac{\bar{C}_p}{\bar{C}_v} = 1 + \frac{R}{\bar{C}_v} \tag{3}$$

For monatomic gases and all ordinary diatomic molecules (where vibration is not important at room temperature) definite values can be calculated. For a brief discussion of the calculation of $\bar{C}_v$ (vib) for polyatomic molecules, see Exp. 36.

A. ADIABATIC EXPANSION METHOD

For the reversible adiabatic expansion of a perfect gas, the change in energy content is related to the change in volume by

$$dE = -p\,dV = -\frac{nRT}{V}\,dV = -nRT\,d\ln V \tag{4}$$

Moreover, since E for a perfect gas is a function of temperature only, we can also write $dE = C_v\,dT$. Substituting this expression into Eq. (4) and integrating, we find that

$$\bar{C}_v \ln \frac{T_2}{T_1} = -R \ln \frac{\bar{V}_2}{\bar{V}_1} \tag{5}$$

where $\bar{C}_v$ and $\bar{V}$ are molar quantities (that is, C_v/n, V/n). It has been assumed that C_v is constant over the temperature range involved. This equation predicts the decrease in temperature resulting from a reversible adiabatic expansion of a perfect gas.

Consider the following two-step process involving a perfect gas denoted by A:

Step I: Allow the gas to expand adiabatically and reversibly until the pressure has dropped from p_1 to p_2.

$$A(p_1, \bar{V}_1, T_1) \rightarrow A(p_2, \bar{V}_2, T_2) \tag{6}$$

Step II: At constant volume, restore the temperature of the gas to T_1.

$$A(p_2, \bar{V}_2, T_2) \rightarrow A(p_3, \bar{V}_2, T_1) \tag{7}$$

For step I, we can use the perfect-gas law to obtain

$$\frac{T_2}{T_1} = \frac{p_2 \bar{V}_2}{p_1 \bar{V}_1} \tag{8}$$

Substituting Eq. (8) into Eq. (5) and combining terms in $\bar{V}_2/\bar{V}_1$, we write

$$\ln \frac{p_2}{p_1} = \frac{-(\bar{C}_v + R)}{\bar{C}_v} \ln \frac{\bar{V}_2}{\bar{V}_1} = -\frac{\bar{C}_p}{\bar{C}_v} \ln \frac{\bar{V}_2}{\bar{V}_1} \tag{9}$$

since for a perfect gas

$$\bar{C}_p = \bar{C}_v + R \tag{10}$$

For step II, which restores the temperature to T_1,

$$\frac{\bar{V}_2}{\bar{V}_1} = \frac{p_1}{p_3} \tag{11}$$

Thus

$$\ln \frac{p_1}{p_2} = \frac{\bar{C}_p}{\bar{C}_v} \ln \frac{p_1}{p_3} \tag{12}$$

This can be rewritten in the form

$$\gamma = \frac{\ln(p_1/p_2)}{\ln(p_1/p_3)} \tag{13}$$

Although in theoretical calculations of $\bar{C}_v$, we are only concerned with perfect gases, equations applying to reversible adiabatic expansion can be derived that are not limited to perfect gases. We will here derive a general expression for $(\partial p/\partial \bar{V})_S$ which we will apply to a perfect gas to obtain Eq. (9); we will also apply it to a van der Waals gas. The adiabatic expansion method may be incapable of sufficient precision to justify a van der Waals treatment, but the expressions obtained may be used with the more precise sound velocity method.

We may write

$$\left(\frac{\partial p}{\partial \bar{V}}\right)_S = \left(\frac{\partial T}{\partial \bar{V}}\right)_S \bigg/ \left(\frac{\partial T}{\partial p}\right)_S \tag{14}$$

From

$$dS̃ = \left(\frac{\partial S̃}{\partial Ṽ}\right)_T dṼ + \left(\frac{\partial S̃}{\partial T}\right)_V dT \tag{15a}$$

and

$$dS̃ = \left(\frac{\partial S̃}{\partial p}\right)_T dp + \left(\frac{\partial S̃}{\partial T}\right)_p dT \tag{15b}$$

we obtain, setting $dS̃ = 0$ in both cases,

$$\left(\frac{\partial T}{\partial Ṽ}\right)_S = -\left(\frac{\partial S̃}{\partial Ṽ}\right)_T \bigg/ \left(\frac{\partial S̃}{\partial T}\right)_V; \tag{16a}$$

$$\left(\frac{\partial T}{\partial p}\right)_S = -\left(\frac{\partial S̃}{\partial p}\right)_T \bigg/ \left(\frac{\partial S̃}{\partial T}\right)_p \tag{16b}$$

Now $(\partial S̃/\partial T)_V = C̃_V/T$ and $(\partial S̃/\partial T)_p = C̃_p/T$. On applying the appropriate Maxwell's relations (reciprocity relations arising from the perfect differentials $dÃ = -S̃\,dT - p\,dṼ$ and $dG̃ = -S̃\,dT + Ṽ\,dp$, respectively), namely

$$\left(\frac{\partial S̃}{\partial Ṽ}\right)_T = \left(\frac{\partial p}{\partial T}\right)_V; \qquad \left(\frac{\partial S̃}{\partial p}\right)_T = -\left(\frac{\partial Ṽ}{\partial T}\right)_p \tag{17a,b}$$

we obtain the relation

$$\left(\frac{\partial p}{\partial Ṽ}\right)_S = -\frac{C̃_p}{C̃_v}\left(\frac{\partial p}{\partial T}\right)_V \bigg/ \left(\frac{\partial Ṽ}{\partial T}\right)_p \tag{18}$$

The treatment is perfectly general up to this point; Eq. (18) can be applied to any gas law $f(p, Ṽ, T) = 0$. For the perfect gas law we easily obtain

$$\left(\frac{\partial p}{\partial Ṽ}\right)_S = -\gamma\frac{p}{Ṽ} \tag{19}$$

Integration of this equation at constant entropy yields Eq. (9). For a van der Waals gas we have

$$\left(p + \frac{a}{Ṽ^2}\right)(Ṽ - b) = RT \tag{20}$$

from which we obtain (using the perfect gas approximation in the small term involving a)

$$\left(\frac{\partial p}{\partial T}\right)_V = \frac{R}{Ṽ - b} \tag{21a}$$

$$\left(\frac{\partial Ṽ}{\partial T}\right)_p = \frac{Ṽ - b}{T} + \frac{2a}{RT^2} \tag{21b}$$

Combining these with Eq. (18) yields, on simplification (keeping only terms to

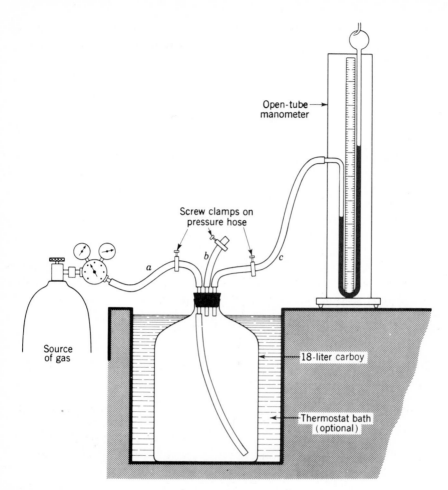

FIGURE 2
Apparatus for the adiabatic expansion of gases.

first order in the small quantities a and b),

$$\left(\frac{\partial p}{\partial \bar{V}}\right)_S = -\gamma \frac{p}{\bar{V}} \left(1 - \frac{a}{p\bar{V}^2} + \frac{b}{\bar{V}}\right) \tag{22}$$

The adiabatic expansion method, due to Clement and Desormes,[2] uses the very simple apparatus shown in Fig. 2. The change in state (6) is carried out by quickly removing and replacing the stopper of a large carboy containing the desired gas at a pressure initially somewhat higher than 1 atm pressure, so that the pressure of gas in the carboy momentarily drops to atmospheric pressure p_2. The change in state (7) consists of allowing the gas remaining in the carboy to return to its initial temperature. The initial pressure p_1 and the final pressure p_3 are read from an open-tube manometer.

The thermodynamic equations (4) to (13) apply only to that part of the gas that remains in the carboy after the stopper is replaced. We may imagine the gas initially in the carboy to be divided into two parts by an imaginary surface; the part above the surface leaves the carboy when the stopper is removed and presumably interacts irreversibly with the surroundings, but the part below the surface expands *reversibly* against this imaginary surface, doing work in pushing the upper gas out. The process is approximately adiabatic only because it is rapid; within a few seconds the gas near the walls will have received an appreciable quantity of heat by direct conduction from the walls, and the pressure can be seen to rise almost as soon as the stopper is replaced.

The approximate assumptions on which this experiment is based are not without controversy. An alternate assumption,[3] with which we do not agree, is that the whole of the gas in the container, initially at pressure p_1, upon removal of the stopper expands *irreversibly* against the surroundings at a *constant* external pressure p_2 (1 atm) until the internal pressure has been reduced to p_2. This assumption raises the question of where the boundary of the system is to be defined, since the external pressure p_2 is exerted only *outside* the neck of the carboy and since the treatment requires a closed system of uniform internal pressure and temperature. In any case, the alternate assumption results in the final equation

$$\gamma = \frac{p_1/p_2 - 1}{p_1/p_3 - 1} \tag{23}$$

The numerator and the denominator in this equation are the leading terms in the power series expansions of the logarithms in the numerator and the denominator in Eq. (13). Under the conditions of the experiment, the difference between the values of γ given by Eqs. (13) and (23) is only about one or two percent.

The experiment will presumably give a somewhat *low* result (low p_3 and therefore low $\bar{C}_p/\bar{C}_v$ ratio) if the expansion is to an appreciable degree irreversible, a somewhat *low* result if the stopper is left open so long that the conditions are not sufficiently adiabatic, and a somewhat *high* result if the stopper has not been removed for a long enough time to permit the pressure to drop momentarily to atmospheric. There should be no significant irreversibility if during the expansion there are no significant pressure gradients in the gas below the imaginary surface mentioned above, and such pressure gradients should not be expected if the throat area is small in comparison with the effective area of this imaginary surface. With an 18-liter carboy and a p_1 about 50 Torr above 1 atm, the volume of air forced out should be about 1 liter, which should provide a large enough surface area to fulfill this condition approximately if the throat diameter is not more than about 2 or 3 cm. The effect of the length of time the stopper is removed is much more difficult to estimate by calculation; some idea can be obtained from the duration of the sound produced when the stopper is removed and from the rate of rise of the manometer reading immediately after the stopper is replaced. For the purpose of this experiment the student may assume that, if the stopper is removed completely from the carboy to a distance of 2 or 3 in. away and replaced tightly as soon as physically possible, the desired experimental conditions are approximately fulfilled. Some additional assurance may be gained from the reproducibility obtained in duplicate runs.

The adiabatic expansion method is not the best method of determining the heat-capacity ratio. Much better methods are based on measurements of the velocity of sound in gases. One such method, described in Part B of this experiment, consists of measuring the wavelength of sound of an accurately known frequency by measuring the distance between nodes in a sonic resonance set up in a Kundt's tube. Methods also exist for determining the heat capacities directly,[4] although the measurements are not easy.

EXPERIMENTAL

The apparatus should be assembled as shown in Fig. 2. If desired, the carboy may be mounted in a thermostat bath; if so, it must be clamped securely to overcome buoyancy. The manometer is an *open-tube* manometer, one side of which is open to the atmosphere; the pressure that it measures, therefore, is the difference of pressure from atmospheric pressure. A suitable liquid for the manometer is dibutyl phthalate, which has a density of 1.046 g cm^{-3} at room temperature (20°C). To convert manometer readings (millimeters of dibutyl phthalate) to equivalent readings in millimeters of mercury, multiply by the ratio of this density to the density of mercury, which is 13.55 g cm^{-3} at 20°C. To find the total pressure in the carboy, the converted manometer readings should be added to atmospheric pressure as given by a barometer. It is unnecesary to correct all readings to 0°C, as all pressures enter the calculations as ratios.

Seat the rubber stopper firmly in the carboy and open the clamps on tubes *a* and *b*. Clamp off the tube *c*. The connections shown in Fig. 2 are based on the assumption that the gas to be studied is heavier than air (or the previous gas in the carboy) and therefore should be introduced at the bottom in order to force the lighter gas out at the top; in the event that the gas to be studied is lighter, the connections *a* and *b* should be reversed.

Allow the gas to be studied to sweep through the carboy for 15 min. The rate of gas flow should be about 6 liter min^{-1}, or 100 ml s^{-1} (measure roughly in an inverted beaker held under water in the thermostat bath). Thus, 5 volumes of gas (90 liters) will pass through the carboy.

Retard the gas flow to a fraction of the flushing rate by partly closing the clamp on tube *a*. **Carefully** open the clamp on tube *c,* and then cautiously (to avoid blowing liquid out of the manometer) clamp off the exit tube *b,* keeping a close watch on the manometer. When the manometer has attained a reading of about 600 mm of oil, clamp off tube *a.* Allow the gas to come to the temperature of the thermostat bath (about 15 min), as shown by a constant manometer reading. Record this reading; when it is converted to an equivalent mercury reading and added to the barometer reading, p_1 is obtained.

Remove the stopper *entirely* (a distance of 2 or 3 inches) from the carboy, and replace it *in the shortest possible time*, making sure that it is tight. As the gas warms back up to the bath temperature, the pressure will increase and finally (in about 15 min) reach a new constant value p_3, which can be determined from the manometer reading and the barometer reading. At some point in the procedure, a barometer reading (p_2) should be taken.

Repeat the steps above to obtain two more determinations with the same gas. For these repeat runs, long flushing is not necessary; one additional volume of gas should suffice to check the effectiveness of the original flushing.

Measurements are to be made on both helium and nitrogen. If there is sufficient time, study carbon dioxide also.

CALCULATIONS

For each of the three runs on He, N_2 (and CO_2), calculate $\bar{C}_p/\bar{C}_v$ using Eq. (13). Also calculate the theoretical value of $\bar{C}_p/\bar{C}_v$ predicted by the equipartition theorem. In the case of N_2 and CO_2, calculate the ratio both with and without vibrational contribution to $\bar{C}_v$.

DISCUSSION

Compare your experimental ratios with those calculated theoretically, and make any deductions you can about the presence or absence of rotational and vibrational contributions, taking due account of the uncertainties in the experimental values. For CO_2, how would the theoretical ratio be affected if the molecule were nonlinear (like SO_2) instead of linear? Could you decide between these two structures from the $\bar{C}_p/\bar{C}_v$ ratio alone?

B. SOUND VELOCITY METHOD

The heat-capacity ratio $\gamma \equiv C_p/C_v$ of a gas can be determined with good accuracy by measuring the speed of sound c. For an ideal gas,

$$\gamma = \frac{Mc^2}{RT} \qquad (24)$$

where M is the molecular weight. A brief derivation[5] of this equation will be sketched below.

For longitudinal plane waves propagating in the x direction through a homogeneous medium of mass density ρ, the wave equation is

$$\frac{\partial^2 \xi}{\partial t^2} = c^2 \frac{\partial^2 \xi}{\partial x^2} \tag{25}$$

where ξ is the particle displacement. Consider a layer AB of the fluid, of thickness δx and unit cross-sectional area, which is oriented normal to the direction of propagation. The mechanical strain on AB due to the sound wave is $\partial \xi / \partial x$:

$$\frac{\partial \xi}{\partial x} = \frac{\Delta V}{V} = -\frac{-\Delta \rho}{\rho} \tag{26}$$

where ΔV and $\Delta \rho$ are, respectively, the volume and density changes caused by the acoustic pressure (mechanical stress) change Δp in the ambient pressure p. From Hooke's law of elasticity,

$$-\Delta p = B_S \frac{\partial \xi}{\partial x} \tag{27}$$

where B_S is the adiabatic elastic modulus.† The net acoustic force acting on AB due to the sound wave is $-(\partial \Delta p / \partial x)\, \delta x$. According to Newton's second law of motion, this force is also equal to $(\rho\, \delta x)\, \partial^2 \xi / \partial t^2$. Thus

$$\frac{\partial \Delta p}{\partial x} = -\rho \frac{\partial^2 \xi}{\partial t^2} \tag{28}$$

or

$$\frac{\partial^2 \xi}{\partial t^2} = \frac{B_S}{\rho} \frac{\partial^2 \xi}{\partial x^2} \tag{29}$$

Comparison of Eqs. (25) and (29) leads to

$$\rho c^2 = B_S \tag{30}$$

and one can now rewrite Eq. (27) as $\Delta p = \rho c^2 (\Delta \rho / \rho) = c^2\, \Delta \rho$ which is equivalent to

$$c^2 = \left(\frac{\partial p}{\partial \rho} \right)_S \tag{31}$$

Now

$$\rho = M / \bar{V} \tag{32}$$

Therefore

$$c^2 = -\frac{\bar{V}}{\rho} \left(\frac{\partial p}{\partial \bar{V}} \right)_S = -\frac{\bar{V}^2}{M} \left(\frac{\partial p}{\partial \bar{V}} \right)_S \tag{33}$$

---

† Since the strain variations occur so rapidly that there is not time for heat flow from adjacent regions of compression and rarefaction, the adiabatic modulus is the appropriate one for sound propagation.[5]

If sound propagation is considered as a reversible adiabatic process we can use Eq. (19) to obtain for a perfect gas

$$c^2 = \gamma \frac{p}{\rho} = \gamma \frac{RT}{M} \tag{34}$$

which can be rewritten as Eq. (24). For a van der Waals gas we obtain (to first order in a and b)

$$c^2 = \gamma \frac{p}{\rho} \left(1 - \frac{a}{p\tilde{V}^2} + \frac{b}{\tilde{V}}\right)$$
$$= \gamma \frac{RT}{M} \left(1 - \frac{2a}{p\tilde{V}^2} + \frac{2b}{\tilde{V}}\right) \tag{35}$$

which may be readily solved for γ. For comparison, the value of γ for a van der Waals gas is (to first order in a)

$$\gamma = 1 + \frac{R}{\tilde{C}_v} \left(1 + \frac{2a}{p\tilde{V}^2}\right) \tag{36}$$

assuming that we are given a theoretical value for $\tilde{C}_v$. (If this is based on a perfect gas, the translational part may be slightly in error because of gas imperfections, but that error will be expected to be much less for $\tilde{C}_v$ than for $\tilde{C}_p$.)

METHOD

This experiment is based on a modified version of Kundt's tube, in which the wavelength λ of standing waves of frequency f are determined electronically. Figure 3 shows the experimental apparatus, which utilizes an audio oscillator driving a miniature speaker as the source and a microphone as the detector.

It is possible to connect the microphone directly to a sensitive voltmeter and watch the variation in the intensity of the received signal as the detector is moved along the tube. The intensity will decrease to a minimum when an antinode is reached and then increase to a maximum at a node. However, it is more precise and more convenient to use an oscilloscope to indicate the phase relationship between the speaker input and the microphone signal. The output of the audio oscillator should drive the horizontal sweep of the oscilloscope. The signal received by the microphone after propagation through the gas is applied (with amplification if necessary) to the vertical sweep of the oscilloscope. The phase difference is determined from the Lissajous figures which appear on the screen (see Chapter XVIII). When the pattern changes from a straight line tilted at 45° to the right (in phase, 0° phase angle) to another straight line tilted at 45° to the left (out of phase, 180° phase angle), the piston has moved one-half the wavelength of sound.

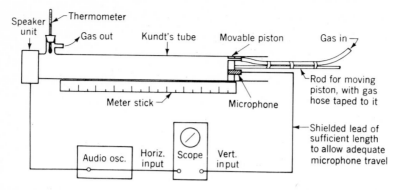

FIGURE 3

Apparatus for measuring the speed of sound in a gas. The Kundt's tube should be a glass tube ~150 cm long and ~5 cm in diameter. The close-fitting Teflon piston should move over a range of at least 100 cm. If a constant-temperature jacket is provided for the Kundt's tube, the position of the piston can be determined from the position of a fiducial mark on the rod.

EXPERIMENTAL

The apparatus should be assembled as shown in Fig. 3. Details of the procedure will depend on the particular electronic components used, especially the type of oscilloscope. Additional instructions may be provided by the instructor.

If the oscillator is not accurately calibrated, this can be done with a calibrated frequency meter or by (1) displaying the waveform on the screen if the oscilloscope has an accurate timebase; (2) obtaining a Lissajous figure with a standard frequency source (internal or external); or (3) holding a vibrating tuning fork near the microphone and obtaining Lissajous figures at the fundamental and several harmonics (disconnect the speaker unit during this calibration). The oscillator should be calibrated at several frequencies between 1 and 2 kHz.†

Move the piston to a position near the far end of its travel. Allow the gas to be studied to sweep through the Kundt's tube for about 10 min in order to displace all the air. Then reduce the flow rate until there is a very slow stream of gas through the tube; this will prevent air from diffusing in during the run. Adjust the gain controls so that the two signals are about equal in magnitude (i.e., a circle is seen when the signals differ in phase by 90°).

Set the oscillator at the desired frequency (1 kHz for N_2 and CO_2, 2 kHz for He), and slowly move the piston toward the speaker until a straight-line pattern is observed on the screen. Record the position of the piston and note whether the phase shift is 0° or 180°. Then move the piston inward again until

---

† The frequency unit cycles per second is called a Hertz (Hz). Thus 1 kHz = 1000 Hz = 1000 cps.

the next straight-line pattern is observed, and again record the position. Continue to record such readings as long as a satisfactory pattern can be observed. Record the temperature of the outlet gas several times during the run.

If necessary, repeat the entire procedure for a given gas until consistent results are obtained. If time permits, repeat the procedure using a different frequency. Measurements are to be made on helium (or argon), nitrogen, and carbon dioxide. At some point, obtain a reading of the barometric pressure.

CALCULATIONS

From the average value of the spacing $\lambda/2$ between adjacent nodes and the known frequency f, calculate the speed of sound c in each gas. Use Eq. (24) to calculate γ, and compare these experimental values with the theoretical values predicted by the equipartition theorem. In the case of N_2 and CO_2, calculate the theoretical ratio both with and without vibrational contribution to $\bar{C}_v$.

Using the van der Waals constants given for N_2 and CO_2 in Table 2-1, recalculate γ for those gases using Eq. (35) and compare them with the theoretical value given by Eq. (36).

DISCUSSION

Compare your experimental ratios with those calculated theoretically, and make any deductions you can about the presence or absence of rotational and vibrational contributions, taking due account of the uncertainties in the experimental values. For CO_2, how would the theoretical ratio be affected if the molecule were nonlinear (like SO_2) instead of linear? Could you decide between these two structures from the $\bar{C}_p/\bar{C}_v$ ratio alone?

Compare the speed of sound in each gas with the average speed of the gas molecules and with the average velocity component in a given direction. Explain why the speed of sound is independent of the pressure.

Does the precision of the measurements justify the van der Waals expression, Eq. (35), instead of the perfect-gas expression, Eq. (34)? Justify your answer in terms of your estimates of uncertainty.

APPARATUS

Adiabatic expansion method. Large-volume vessel (such as 18-liter glass carboy); three-hole stopper fitted with three glass tubes; open-tube manometer, with dibutyl phthalate as indicating fluid (may contain a small amount of dye for ease of reading); three long lengths and one short length of rubber pressure tubing; three screw clamps; cork ring and brackets, needed if vessel is to be mounted in a water bath; thermometer; 500-ml beaker.

Thermostat bath, set at 25°C (unless each vessel has an insulating jacket); cylinders of helium, nitrogen, and carbon dioxide.

Sound velocity method. Kundt's tube, complete with speaker unit (miniature cone or horn driver) and movable piston holding microphone; stable audio oscillator, perferably with calibrated dial; small crystal microphone; oscilloscope; audio-frequency amplifier, if scope gain is inadequate; electrical leads; rubber tubing; stopper; thermometer.

Cylinders of helium (or argon), nitrogen, and carbon dioxide.

REFERENCES

1. G. W. Castellan, "Physical Chemistry," 3d ed., pp. 74ff., Addison-Wesley, Reading, Mass. (1983).
2. Lord Rayleigh, "The Theory of Sound," 2d ed., Vol. II, pp. 15–23, Dover, New York (1945).
3. G. L. Bertrand and H. O. McDonald, *J. Chem. Educ.* **63,** 252 (1986).
4. K. Schell and W. Heuse, *Ann. Physik* **37,** 79 (1912); **40,** 473 (1913); **59,** 86 (1919).
5. J. Blitz, "Fundamentals of Ultrasonics," 2d ed., pp. 10–13, Butterworth, London (1967); S. M. Blinder, "Advanced Physical Chemistry," pp. 356–361, Macmillan, London (1969).

GENERAL READING

J. Blitz, *op. cit.*
J. R. Partington, "An Advanced Treatise on Physical Chemistry," Vol. I, pp. 792ff., Longmans, London (1949).

CHAPTER
IV

TRANSPORT
PROPERTIES
OF GAS

EXPERIMENTS

4. Viscosity of gases
5. Diffusion of gases

KINETIC THEORY OF TRANSPORT PHENOMENA

In this section we shall be concerned with a molecular theory of the transport properties of gases. The molecules of a gas collide with each other frequently, and the velocity of a given molecule is usually changed by each collision that the molecule undergoes. However, when a one-component gas is in thermal and statistical equilibrium, there is a definite distribution of molecular velocities—the well-known Maxwellian distribution.[1] Figure 1 shows how the molecular velocities are distributed in such a gas. This distribution is isotropic (the same in all directions) and can be characterized by a *root-mean-square speed u*, which is given by

$$u = \left(\frac{3kT}{m}\right)^{1/2} = \left(\frac{3RT}{M}\right)^{1/2} \tag{1}$$

where k is Boltzmann's constant, R is the gas constant, m is the mass of one molecule, and M is the molecular weight. Sometimes it is more convenient to use the *mean speed $\bar{c}$*:

$$\bar{c} = \left(\frac{8kT}{\pi m}\right)^{1/2} = \left(\frac{8}{3\pi}\right)^{1/2} u \tag{2}$$

119

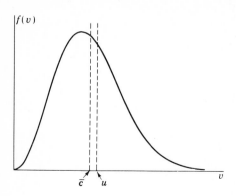

FIGURE 1

Schematic Maxwellian velocity distribution: $f(v)\,dv$ is the fraction of the molecules with velocities between v and $v + dv$. Values of the rms speed u and the mean speed $\bar{c}$ are shown.

Another very important concept in kinetic theory is the average distance a molecular travels between collisions—the so-called *mean free path*. On the basis of a very simple conception of molecular collisions, the following equation for the mean free path λ can be derived:

$$\lambda = \frac{1}{\pi d^2 \bar{N}}$$

where d is the "molecular diameter," or center-to-center collision distance, and $\bar{N}$ is the number of molecules per unit volume. When the fact is properly taken into account that molecules do not all have the same velocities, a different numerical coefficient is obtained:[1]

$$\lambda = \frac{1}{\sqrt{2}\,\pi d^2 \bar{N}} = \frac{1}{\sqrt{2}\,\pi d^2} \frac{kT}{p} = \left(\frac{1}{\sqrt{2}\,\pi N_0}\right) \frac{RT}{pd^2} = 3.738 \times 10^{-25} \frac{RT}{pd^2} \qquad (3)$$

Note that the ideal-gas law has been used for $\bar{N}$ and that the numerical coefficient in the final expression has units of mol. If SI units are used for all quantities, which means expressing the pressure in Pascals, the resulting value of λ will be in meters. However, for obtaining a feeling for the magnitude of λ compared to the scale of typical apparatus, centimeter units are more convenient. Some typical values of λ are given in Table 1.

We shall consider in detail the predictions of the hard-sphere model for the viscosity, thermal conductivity, and diffusion of gases; indeed, the kinetic theory treatment of these three transport properties is very similar. But first let us consider the simpler problem of molecular effusion.

Effusion of gases. The theory of effusion (Knudsen flow) is quite straightforward, since only molecular flow is involved; i.e., in the process of effusing, the molecules act independently of each other. For a Maxwellian distribution of velocities it can be shown[1,2] that the number Z of molecular impacts on a unit

TABLE 1
Mean free paths of various gases at 25°C

Gas	Mean free path, cm		
	At 760 Torr	At 0.1 Torr	At 10^{-4} Torr
He	19.0×10^{-6}	14.5×10^{-2}	145
H_2	12.1×10^{-6}	9.2×10^{-2}	92
Ar	6.8×10^{-6}	5.2×10^{-2}	52
O_2	6.9×10^{-6}	5.3×10^{-2}	53
N_2	6.4×10^{-6}	4.9×10^{-2}	49
CH_4	5.2×10^{-6}	4.0×10^{-2}	40
CO_2	4.2×10^{-6}	3.2×10^{-2}	32
SO_2	3.0×10^{-6}	2.3×10^{-2}	23

area of wall surface in unit time is

$$Z = \tfrac{1}{4}\bar{c}\bar{N} = \frac{1}{\sqrt{6\pi}} u\bar{N} \tag{4}$$

Combining Eq. (1) or Eq. (2) with Eq. (4), we obtain

$$Z = \left(\frac{RT}{2\pi M}\right)^{1/2} \bar{N}$$

The number z of *moles* colliding with 1 cm^2 of wall surface per second is

$$z = \left(\frac{RT}{2\pi M}\right)^{1/2} \frac{n}{V} \tag{5}$$

where n/V is the number of moles per unit volume. For a perfect gas Eq. (5) becomes

$$z = \frac{p}{\sqrt{2\pi MRT}} \tag{6}$$

Let two bodies of the same gas, at the same temperature but at two different pressures (p and p'), communicate through a small hole of area A, all dimensions of which are *small* in comparison with the mean free path of the gas molecules. It can then be assumed that the Maxwellian distribution of velocities is essentially undisturbed and that the same number of molecules will enter the hole as would otherwise have collided with the corresponding area of wall surface. It will also be assumed that all molecules entering the hole will pass through to the other side. The *net* rate of effusion through the hole will be the difference between the numbers of moles flowing through the hole in unit time in the two directions:

$$\frac{dn_{\text{eff}}}{dt} = \frac{A(p - p')}{\sqrt{2\pi MRT}} \qquad (7)$$

This equation assumes that the gas is perfect, but it should be emphasized that the rate of effusion calculated from it is independent of the size and shape of the molecules provided only that these are such as to give a sufficiently long mean free path at the given temperature and pressure.

Gas viscosity. Consider a body of gas under shear as shown in Fig. 4-1 and discussed in the introduction to Exp. 4. Assume that all the molecules in the gas are of one kind and that the gas undergoes laminar flow. By the mass velocity of the gas at each point, we understand the (vectorial) average velocity of all molecules passing through an infinitesimal region at that point. If the velocity gradient is substantially constant within one or a few mean free paths, we may also say that the mass velocity of the gas at a given point is the (vectorial) average velocity of all molecules that have suffered their most recent collision in an infinitesimal region at that point. We shall assume that, with respect to a system of coordinates moving at the mass velocity at a given point, the distribution of velocities of gas molecules which underwent their most recent collisions at that point is the same as it would be if the gas were not under shear (i.e., that the distribution is Maxwellian and therefore isotropic). We shall also assume, for simplicity, that all molecules travel the same distance λ between collisions.

Let us focus our attention on a laminar plane $x = x'$, with $v = v'$. Molecules coming from below and colliding with molecules on this layer may be assumed (in a rough approximation) to have suffered their most recent previous collision on a plane at a distance λ below the plane in question. The average momentum component in the shear direction parallel to the lamina (call it the y direction) for one of these molecules is therefore

$$mv = mv' - m\frac{dv}{dx}\lambda$$

where m is the mass of a molecule. Since by hypothesis the molecules have average momentum mv' after collision, an average amount of momentum

$$-m\frac{dv}{dx}\lambda$$

is transferred to the laminar plane x' from below, per collision. The number of molecules reaching a unit area of the plane from below per unit time is approximately[3]

$$\tfrac{1}{6}\bar{N}\bar{c}$$

where $\bar{N}$ is the number of molecules per unit volume and $\bar{c}$ is their mean

molecular speed. Therefore, the rate of upward transfer of momentum to a unit area of the plane from below is

$$- \tfrac{1}{6} m \bar{N} \bar{c} \lambda \frac{dv}{dx}$$

Similarly it can be shown that the rate of downward transfer of momentum from a unit area of the plane x' to the plane at $(x' - \lambda)$ is

$$+ \tfrac{1}{6} m \bar{N} \bar{c} \lambda \frac{dv}{dx}$$

A downward flow of momentum is equivalent to an upward flow of momentum of the opposite sign. Thus the net rate of *upward* flow of momentum through any given plane, per unit area, is

$$- \tfrac{1}{3} m \bar{N} \bar{c} \lambda \frac{dv}{dx}$$

This rate of change of momentum must be balanced by a force in accordance with Newton's second law:

$$\frac{f}{A} = \tfrac{1}{3} m \bar{N} \bar{c} \lambda \frac{dv}{dx} \tag{8}$$

We can write Eq. (4-1), the defining equation for the viscosity coefficient η, as

$$\frac{f}{A} = \eta \frac{dv}{dx} \tag{9}$$

Combining Eqs. (8) and (9) we obtain an expression for η:

$$\eta \cong \tfrac{1}{3} m \bar{N} \bar{c} \lambda \tag{10}$$

Making use of Eq. (2) for $\bar{c}$ and Eq. (3) for λ, we find

$$\eta = \frac{2}{3\pi^{3/2}} \frac{(mkT)^{1/2}}{d^2} = \frac{2}{3\pi^{3/2} N_0} \frac{(MRT)^{1/2}}{d^2} \tag{11}$$

This "mean-free-path treatment" involves several simplifying and somewhat crude approximations; a more sophisticated and rigorous hard-sphere theory[1] gives results identical with Eqs. (10) and (11) except for the numerical factors:

$$\eta = \frac{5\pi}{32} m \bar{N} \bar{c} \lambda \tag{12}$$

and

$$\eta = \frac{5}{16\pi^{1/2} N_0} \frac{(MRT)^{1/2}}{d^2} = 2.928 \times 10^{-25} \frac{(MRT)^{1/2}}{d^2} \tag{13}$$

It should be noted again that the numerical coefficient above has units of mol.

According to this equation, the gas velocity coefficient should be independent of pressure and should increase with the square root of the absolute temperature. The viscosities of gases are, in fact, found to be substantially independent of pressure over a wide range. The temperature dependence generally differs to some extent from $T^{1/2}$ because the effective molecular diameter is dependent on how hard the molecules collide and therefore depends somewhat on temperature. Deviation from hard-sphere behavior in the case of air (diatomic molecules, N_2 and O_2) is demonstrated by Eq. (4-19).

Equation (14) can also be used for the calculation of molecular diameters from experimental viscosity data. These can be compared with molecular diameters as determined by other experimental methods (molecular beams, diffusion, thermal conductivity, x-ray crystallographic determination of molecular packing in the solid state, etc.).

The SI unit for viscosity is $N\,m^{-2}\,s = Pa\,s = kg\,m^{-1}\,s^{-1}$, which has no special name. The cgs unit is the *poise* (P), which equals $1\,dyn\,cm^{-2}\,s = 0.1\,N\,m^{-2}\,s$; this unit is widely used and will probably continue to be used for the forseeable future. Remember to convert poise into SI units before carrying out calculations involving viscosities.

Thermal conductivity. The mechanism of thermal conduction is analogous to that of viscous resistance to fluid flow. In the case of fluid flow, where a velocity gradient exists, momentum is transported from point to point by gas molecules; in thermal conduction, where a temperature gradient exists, it is kinetic energy that is transported from point to point by gas molecules. If in the mean-free-path treatment of viscosity, the momentum is replaced by the average kinetic energy $\bar{\varepsilon}$ of a molecule, we obtain for one-dimensional heat flow in the x direction an equation analogous to Eq. (8):

$$\dot{q} = \tfrac{1}{3}\bar{N}\bar{c}\lambda\left(\frac{-d\bar{\varepsilon}}{dx}\right) \tag{14}$$

where $\dot{q}$ is rate of heat flow per unit area, $\bar{N}$ is the number of molecules per unit volume (in cubic centimeters), $\bar{c}$ is the mean speed, and λ is the mean free path. Now

$$\frac{d\bar{\varepsilon}}{dx} = \frac{d\bar{\varepsilon}}{dT}\frac{dT}{dx} = m\bar{C}_v\frac{dT}{dx} \tag{15}$$

where m is the mass of a molecule and $\bar{C}_v$ is the constant-volume specific-heat capacity (i.e., per gram) of the gas; thus we obtain

$$\dot{q} = \tfrac{1}{3}m\bar{N}\bar{c}\lambda\bar{C}_v\left(\frac{-dT}{dx}\right) \tag{16}$$

By comparison with $\dot{q} = -K(dT/dx)$, the defining equation for the coefficient of thermal conductivity K, we find that

$$K = \tfrac{1}{3}m\bar{N}\bar{c}\lambda\bar{C}_v \tag{17}$$

Using Eq. (10) for the coefficient of viscosity of a gas, we obtain the following relationship between coefficients of thermal conductivity and of viscosity:

$$\frac{K}{\eta} = \bar{C}_v \tag{18}$$

This simple expression, which is derived using the mean-free-path treatment, is correct in form, but the numerical coefficient is wrong. More advanced treatments[1] for a hard-sphere monatomic gas yield

$$K = \frac{25\pi}{64} m\bar{N}\bar{c}\lambda\bar{C}_v \qquad \text{monatomic} \tag{19}$$

From this result and Eq. (12), we find for monatomic gases

$$\frac{K}{\eta} = \frac{5}{2}\bar{C}_v = \frac{5}{2}\frac{\bar{C}_v}{M} = \frac{15R}{4M} \tag{20}$$

where the statistical mechanical value $3R/2$ has been used for the molar heat capacitiy $\bar{C}_v$. Interactions between diatomic and polyatomic molecules, involving transfer of rotational and vibrational energy, are much more difficult to treat theoretically[1,2] and will not be discussed here.

The SI unit for the thermal conductivity coefficient K is $J\,m^{-1}\,K^{-1}\,s^{-1}$, although the values are often cited in units of $J\,cm^{-1}\,K^{-1}\,s^{-1}$. Use of SI units for both K and η yields units of $J\,kg^{-1}\,K^{-1}$ for K/η as expected from R/M.

Diffusion. The theory of the diffusion of a gas is related to the theory of viscosity and of thermal conductivity in the sense that all three are concerned with a transport of some quantity from one point to another by the thermal motion of gas molecules. For diffusion, where a concentration gradient exists, chemical identity is transported.

Let there be two kinds of perfect gases, designated 1 and 2, present in a closed system of fixed total volume at concentrations of $\bar{N}_1$ and $\bar{N}_2$ molecules per cubic centimeter or $\bar{n}_1$ and $\bar{n}_2$ moles per cubic centimeter, respectively. These gases will have mean free paths of λ_1 and λ_2.

Let it be assumed that there is a concentration gradient in the x direction only. Consider a plane x_0, parallel to the y and z axes, and consider for a moment those molecules of kind 1 that pass through a small area $dA = dy\,dz$ on that plane. Assume for simplicity that these molecules suffered their most recent collision at or near the surface of an imaginary sphere of radius λ_1 with its center at dA. Now the mean distance from the plane x_0 up or down to the sphere is $2\lambda_1/3$. Therefore the mean concentration of these molecules, in the regions below the plane where they suffered their most recent collisions, is

$$\bar{N}_- = \bar{N}_1(x_0) - \frac{2\lambda_1}{3}\frac{d\bar{N}_1}{dx}$$

and the mean concentration similarly defined above the plane is

$$\bar{N}_+ = \bar{N}_1(x_0) + \frac{2\lambda_1}{3}\frac{d\bar{N}_1}{dx}$$

The number of molecules passing through a unit area of a given plane surface depends on the concentration of the molecules:

$$Z = \tfrac{1}{4}\bar{c}\bar{N} \tag{21}$$

where $\bar{c}$ is the mean speed. The *net* number of molecules of kind 1 crossing the x_0 plane in the upward direction is therefore

$$Z_1 = \tfrac{1}{4}\bar{c}_1(\bar{N}_- - \bar{N}_+) = -\tfrac{1}{3}\bar{c}\lambda_1\frac{d\bar{N}_1}{dx} \tag{22}$$

So far we have neglected the mass flow, if any, that is needed for the maintenance of constant and uniform pressure. If there is mass flow with a given velocity w_0 (which will be taken as positive when the mass flow is upward), the molecular velocities will be Maxwellian only with respect to a set of coordinate axes moving with that velocity, since the diffusing molecules suffered their recent collisions with molecules moving upward at the *mean* velocity. Equation (22) therefore represents the number of molecules per second crossing a unit area on a plane (parallel to the y and z axes) that is moving upward at velocity w_0. The number of molecules per second crossing a unit area on a *fixed* plane is then

$$\begin{aligned} Z_1 &= \bar{N}_1 w_0 - \tfrac{1}{3}\bar{c}_1\lambda_1\frac{d\bar{N}_1}{dx} \\ Z_2 &= \bar{N}_2 w_0 - \tfrac{1}{3}\bar{c}_2\lambda_2\frac{d\bar{N}_2}{dx} \end{aligned} \tag{23}$$

Now for maintenance of constant pressure we must have $Z_1 = -Z_2$; thus we obtain

$$\begin{aligned} w_0 &= \frac{1}{3(\bar{N}_1 + \bar{N}_2)}\left(\bar{c}_1\lambda_1\frac{d\bar{N}_1}{dx} + \bar{c}_2\lambda_2\frac{d\bar{N}_2}{dx}\right) \\ &= \frac{1}{3\bar{N}}\frac{d\bar{N}_1}{dx}(\bar{c}_1\lambda_1 - \bar{c}_2\lambda_2) \end{aligned} \tag{24}$$

and

$$Z_1 = -\frac{1}{3\bar{N}}(\bar{N}_1\lambda_2\bar{c}_2 + \bar{N}_2\lambda_1\bar{c}_1)\frac{d\bar{N}_1}{dx} = -Z_2 \tag{25}$$

where, since the pressure is constant, we have used the relation

$$-\frac{d\bar{N}_1}{dx} = \frac{d\bar{N}_2}{dx} \tag{26}$$

We can also write, in terms of moles,

$$z_1 = -\frac{1}{3\bar{N}}(\bar{n}_1\lambda_2\bar{c}_2 + n_2\lambda_1\bar{c}_1)\frac{d\bar{n}_1}{dx} = -z_2$$

$$= \tfrac{1}{3}(X_1\lambda_2\bar{c}_2 + X_2\lambda_1\bar{c}_1)\frac{d\bar{n}_1}{dx} \tag{27}$$

where X_1 and X_2 are mole fractions ($X_1 = \bar{n}_1/\bar{n}$, $X_2 = \bar{n}_2/n$; $n = \bar{n}_1 + \bar{n}_2$). By comparison with Eq. (5-2), the defining equation for the diffusion constant D_{12}, we find that

$$D_{12} = \tfrac{1}{3}(X_1\lambda_2\bar{c}_2 + X_2\lambda_1\bar{c}_1) \tag{28}$$

This equation appears to predict that D_{12} will be a function of the composition of the gas, although experimentally the diffusion constant is almost independent of composition. However, we must be careful in our definition of λ_1 and λ_2. We must take account of the fact that collisions of molecules of one species with one another can have no significant effect on the diffusion; such collisions do not affect the total momentum possessed by all the molecules of that species and thus do not affect the mean mass velocity of the species in its diffusion. Thus the total number of molecules of that species crossing the reference plane in a given period of time is not affected by such collisions and is the same as if such collisions did not take place at all. We should define λ_1 for our present purpose as the mean free path of molecules of species 1 between successive collisions with molecules of species 2 and *vice versa*. Accordingly we write [see Eq. (3)]

$$\lambda_1 = \frac{1}{\sqrt{2}\,\pi\bar{N}_2 d_{12}^2} \tag{29}$$

where by d_{12} we mean the center-to-center collision distance (assuming the molecules to be "hard spheres"). Equation (28) then becomes

$$D_{12} = \frac{1}{3\sqrt{2}\,\pi d_{12}^2\bar{N}}(\bar{c}_1 + \bar{c}_2) \tag{30}$$

which is independent of the composition. If we now introduce the kinetic theory expressions for $\bar{c}_1$ and $\bar{c}_2$ as given by Eq. (2) and make use of the perfect-gas expression for $\bar{N}$, namely,

$$\bar{N} = N_0\bar{n} = N_0\frac{p}{RT} \tag{31}$$

where N_0 is Avogadro's number, we obtain for the diffusion constant

$$D_{12} = \frac{2}{3\pi^{3/2}N_0}\frac{(RT)^{3/2}}{p}\frac{\sqrt{1/M_1} + \sqrt{1/M_2}}{d_{12}^2} \tag{32}$$

and for the *self-diffusion constant* (see Exp. 5)

$$D_1 = \frac{4}{3\pi^{3/2}N_0} \frac{(RT)^{3/2}}{p} \frac{\sqrt{1/M_1}}{d_1^2} \tag{33}$$

More advanced treatments[1] lead to a slightly different form of expression for D_{12} and different numerical coefficients for both D_{12} and D_1:

$$D_{12} = \frac{3}{8\sqrt{2\pi}\,N_0} \frac{(RT)^{3/2}}{p} \frac{\sqrt{1/M_1 + 1/M_2}}{d_{12}^2} \tag{34}$$

and

$$D_1 = \frac{3\pi}{16} \bar{c}_1 \lambda_1 = \frac{3}{8\sqrt{\pi}\,N_0} \frac{(RT)^{3/2}}{p} \frac{\sqrt{1/M_1}}{d_1^2} \tag{35}$$

where λ_1 is the mean free path as usually defined for a pure gas.

Equations (32) and (34), which are based on a "hard-sphere" model, are in agreement in predicting no dependence of the diffusion constant on gas composition. However, for real molecules a slight composition dependence should exist, which depends upon the form of the intermolecular potential.[1]

That Eqs. (32) and (34) differ in the dependence of D_{12} on the molecular weights should not be altogether surprising, since in our simple treatment we have assumed that the velocities of molecules after collisions are uncorrelated with their velocities before collisions, while, in fact, such correlations in general exist and are functions of the ratio of the masses of the colliding particles. To take this and other remaining factors properly into account would require a treatment that is beyond the scope of this book. It may suffice to point out here that the square-root quantity in Eq. (34) is equivalent to $\sqrt{1/\mu_{12}N_0}$, where

$$\mu_{12} = \frac{m_1 m_2}{m_1 + m_2}$$

is the "reduced mass" of a system comprising a molecule of kind 1 and a molecule of kind 2.

The diffusion constant is predicted by Eqs. (32) to (35) to be inversely proportional to the total pressure. Experimentally this is the case to roughly the degree to which the perfect-gas law applies. The equations appear to predict that the diffusion constant will be proportional to the three-halves power of the temperature; however, as in the case of viscosity, significant deviations from this behavior occur, as actual molecules are not truly "hard spheres" and have collision diameters that depend on the relative speeds with which molecules collide with one another.

By means of Eq. (34) it is possible to calculate center-to-center collision distances d_{12} from a measured diffusion constant D_{12}. If three diffusion constants D_{12}, D_{13}, and D_{23} can be measured, individual molecular diameters d_1, d_2, and d_3 can be obtained if it can be assumed that the collision distances are the arithmetic averages of the molecular diameters involved, as would be the case with hard spheres:

$$d_{ij} = \tfrac{1}{2}(d_i + d_j) \tag{36}$$

If a self-diffusion constant D_1 can be measured approximately through one of the approaches discussed in Exp. 5, a molecular diameter d_1 can be obtained directly from Eq. (35). Conversely, Eqs. (34) and (35) can be used to estimate a diffusion constant, if molecular diameters are known from some other source, or to calculate self-diffusion constants from binary diffusion constants (or vice versa) by calculating first the molecular diameters. In the latter case it can be argued that any remaining approximations in the treatment will largely cancel out.

   As expected, the self-diffusion constant can be related to other transport coefficients. From Eqs. (12) and (35), we find

$$\frac{D}{\eta} = \frac{6}{5\rho} = \frac{1.2}{\rho} \tag{37}$$

where ρ is the mass density of the gas. This is a hard-sphere model result, and more advanced treatments[1] show that the coefficient of $1/\rho$ is larger for soft-sphere models. For spherical molecules with a somewhat artificial inverse-fifth-power repulsion, the coefficient is about 1.5. Clearly D can also be related to K for monatomic gases. From Eqs. (20) and (37)

$$\frac{D}{K} = \frac{0.32M}{\rho R} \tag{38}$$

   The SI unit for D is $m^2\,s^{-1}$, but again it is common to cite values in $cm^2\,s^{-1}$. To avoid problems associated with calculation of ratios such as D/η and D/K, consistent use of SI units is recommended.

REFERENCES

1. W. Kauzmann, "Kinetic Theory of Gases," Benjamin, New York (1966); R. D. Present, "Kinetic Theory of Gases," McGraw-Hill, New York (1958).
2. I. N. Levine, "Physical Chemistry," 3d ed., Chaps. 15, 16, McGraw-Hill, New York (1988).
3. R. S. Berry, S. A. Rice and J. Ross, "Physical Chemistry," p. 1072, Wiley, New York (1980).

GENERAL READING

S. Chapman and T. G. Cowling, "The Mathematical Theory of Non-uniform Gases," 3d ed., Cambridge University Press, Cambridge (1970).

EXPERIMENT 4
VISCOSITY OF GASES

It is a general property of fluids (liquids and gases) that an applied shearing force that produces flow in the fluid is resisted by a force that is proportional to the gradient of flow velocity in the fluid. This is the phenomenon known as *viscosity*.

Consider two parallel plates of area A, a distance D apart, as in Fig. 1. It is convenient to imagine that D is small in comparison with any dimension of the plates in order to avoid edge effects. Let there be a uniform fluid substance between the two plates. If one of the plates is held at rest while the other moves with uniform velocity v_0 in a direction parallel to its own plane, under ideal conditions the fluid undergoes a pure shearing motion and a flow velocity gradient of magnitude v_0/D exists throughout the fluid. This is the simplest example of *laminar flow*, or pure viscous flow, which takes place under such conditions that the inertia of the fluid plays no significant role in determining the nature of the fluid motion. The most important of these conditions is that the flow velocities be small. In laminar flow in a system with stationary solid boundaries, the paths of infinitesimal mass elements of the fluid do not cross any of an infinite family of stationary laminar surfaces that may be defined in the system. In the simple example above, these laminar surfaces are planes parallel to the plates. When fluid velocities become high, the flow becomes *turbulent* and the momentum of the fluid carries it across such laminar surfaces so that eddies or vortices form.

In the above example, with laminar flow, the force f resisting the relative motion of the plates is proportional to the area A and to the velocity gradient v_0/D:

$$f = \eta A \frac{v_0}{D} \tag{1}$$

The constant of proportionality η is called the *coefficient of viscosity* of the fluid, or simply the *viscosity* of the fluid. A convenient unit for viscosity is the *poise,* which is the cgs unit equivalent to $10^{-1}\,\mathrm{N\,s\,m^{-2}} = 10^{-1}\,\mathrm{Pa\,s}$.

The viscosities of common liquids are of the order of a centipoise (cP); at 20.20°C, the viscosity of water is 1.000 cP, while that of ethyl ether is 0.23 cP,

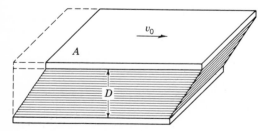

FIGURE 1
Ideal plane-parallel laminar flow.

and that of glycerin is 830 cP (8.3 P). The viscosity of pitch or asphalt at room temperature is of the order of 10^{10} P, and the viscosity of glass (a supercooled liquid) is many powers of 10 higher. The viscosities of gases are of the order of 100 or 200 μP. The viscosities of liquids and soft solids decrease with increasing temperature, while those of gases increase with increasing temperature. Viscosities of all substances are substantially independent of pressure at ordinary pressures but show change at very high pressures and, apparently in the case of gases alone, at very low pressures also.

METHOD

Among the various methods[1,2] that have been used for determining the viscosities of liquids and gases are several that involve the measurement of viscous drag on a rotating disk or cylinder immersed in the fluid or a sphere falling through it. The simplest and most commonly used methods, however, depend on measurement of the rate of viscous flow through a cylindrical capillary tube, in relation to the pressure gradient along the capillary.

For absolute viscosity measurements, the pressure should be constant at each end of the capillary. A simple but excellent method for gases has been described by A. O. Rankine[2] in which a constant-pressure difference is maintained across the ends of a capillary by a pellet of mercury falling in a parallel tube of considerably larger diameter (up to 3.5 mm). However, it is rarely necessary to make absolute determinations. Most viscosity measurements are made in "relative viscosimeters," in which the viscosity is proportional to the time required for the flow of a fixed quantity of fluid through the capillary, under a pressure differential that varies during the experiment *but always in the same manner.* If the pressure differential as a function of the volume of fluid passed through the capillary is accurately known, the viscosimeter can be used for absolute determinations. More commonly the unknown factors are lumped into a single apparatus constant, the value of which is determined by time-of-flow measurements on a reference substance of known viscosity. This is the principle of the Ostwald viscosimeter routinely used for measurements on liquids and solutions, in which capillary flow is induced by gravity (and accordingly the time of flow is proportional to the viscosity divided by the liquid density). It is also the principle of the method used for measurements on gases in this experiment, in which gas is caused to flow through the capillary by mercury displacement.

THEORY

Let us imagine a long cylindrical capillary tube, of length L and radius r. Let x denote the distance outward radially from the axis of the tube and z denote the distance along the axis. Under conditions of laminar flow, the laminar surfaces are a family of right circular cylinders coaxial with the tube.

Consider two such cylindrical surfaces, of radius x and $x + dx$ and length

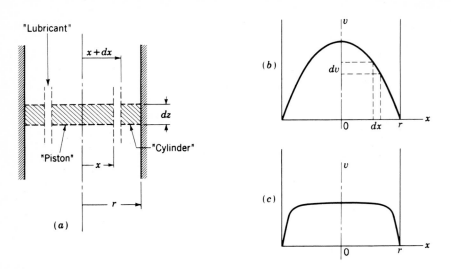

FIGURE 2
(a) Definition of variables. (b) Parabolic velocity distribution. (c) Abnormal (approximately uniform) velocity distribution existing near inlet.

dz as shown in Fig. 2(a). What is the difference in velocity dv between the two cylinders when the flow rate is such as to produce a pressure gradient dp/dz along the tube?

The force tending to cause the inner cylinder to slide past the outer one is the same as it would be for a *solid* inner piston of radius x sliding with the same relative velocity in a *solid* cylinder of radius $x + dx$, with the fluid playing the role of a "lubricant" between. Since the spacing dx is infinitesimal in comparison with the radius of curvature, we may imagine that the conditions existing between the two surfaces resemble those of Fig. 1. The force acting on an element of length dz of the tube is

$$df = \frac{dp}{dz} \pi x^2 \, dz$$

The corresponding cross-sectional area is

$$dA = 2\pi x \, dz$$

and the spacing D is dx. In order to determine the corresponding velocity difference dv, we substitute for f, A, D, and v_0 in Eq. (1) and rearrange to obtain

$$dv = \frac{\pi x^2}{\eta 2\pi x} \frac{dp}{dz} \, dx = -\frac{1}{2\eta} \left(-\frac{dp}{dz} \right) x \, dx \qquad (2)$$

To find the velocity at any value of x, we integrate from $x = r$, where the velocity vanishes, to $x = x$:

$$v(x) = \int_r^x dv = \frac{1}{4\eta}\left(-\frac{dp}{dz}\right)(r^2 - x^2) \tag{3}$$

This equation predicts that the distribution of velocities in the tube can be represented by a parabola, as in Fig. 2(b). The volume rate of flow past a given point (volume per unit time) is

$$\phi_V = \int_0^r v(x) 2\pi x\, dx = \frac{\pi r^4}{8\eta}\left(-\frac{dp}{dz}\right) \tag{4}$$

and the mean flow velocity $\bar{v}$ is†

$$\bar{v} = \frac{\phi_V}{\pi r^2} = \frac{1}{8\eta}\left(-\frac{dp}{dz}\right)r^2 = \tfrac{1}{2}v_{max} \tag{5}$$

If the fluid is incompressible, the volume rate of fluid flow must be constant along the tube and we find, where p_1 and p_2 are the pressures at the inlet and outlet of the tube, respectively, that

$$\phi_V = \frac{\pi r^4 (p_1 - p_2)}{8\eta L}$$

This equation, known as *Poiseuille's law,* can be applied under ordinary laminar-flow conditions to liquids, and also to gases in the limiting case where $(p_1 - p_2)$ is negligible in comparison with p_1 or p_2. For a more general treatment of gases, their compressibility must be taken into account. If we assume the perfect-gas law, we can write for the *molar* rate of flow (moles per unit time)

$$\phi_n = \frac{p}{RT}\phi_V = \frac{\pi r^4}{8\eta RT} p\left(-\frac{dp}{dz}\right) = \frac{\pi r^4}{16\eta RT}\left[-\frac{d(p^2)}{dz}\right] \tag{6}$$

Since conservation of matter requires that the mole rate of flow be constant along the tube, $d(p^2)/dz$ must be constant and accordingly we can write

$$-\frac{d(p^2)}{dz} = \frac{p_1^2 - p_2^2}{L} \tag{7}$$

We now obtain

$$\phi_n = \frac{\pi r^4 (p_1^2 - p_2^2)}{16\eta LRT} \tag{8}$$

In the apparatus to be used in this experiment (see Fig. 3) the gas contained in a bulb B is forced through a capillary tube by mercury flowing

---

† It should be emphasized that $\bar{v}$ is *not* the mean molecular speed $\bar{c}$ defined in Eq. (IV-2).

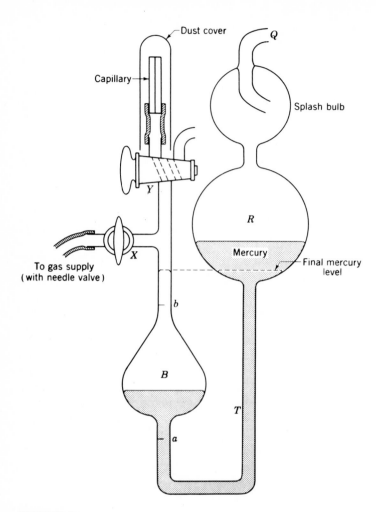

FIGURE 3
Gas viscosity apparatus.

into the bulb from an upper reservoir R. The time during which the mercury level rises from a lower fiducial mark a to an upper fiducial mark b is denoted as t_{ab}. For any such apparatus of given design and dimensions and containing a constant amount of mercury, the inlet pressure p_1 is a definite function of the volume V occupied by the gas in the bulb and tubing below the capillary; p_2 is a constant (atmospheric pressure). If n is the number of moles of gas below the capillary,

$$\frac{dn}{dt} = \frac{1}{RT}\frac{d(p_1 V)}{dt} = \frac{1}{RT}\left(V\frac{dp_1}{dV} + p_1\right)\frac{dV}{dt}$$

Setting $(-dn/dt)$ to ϕ_n we obtain[†]

$$-\frac{dV}{dt} = \frac{\pi r^4(p_1^2 - p_2^2)}{16\eta L[V(dp_1/dV) + p_1]} \tag{9}$$

On inverting and integrating, we find that the time required for the mercury meniscus to rise from mark a to mark b is

$$t_{ab} = \eta \frac{16L}{\pi r^4} \int_{V_b}^{V_a} \frac{V(dp_1/dV) + p_1}{p_1^2 - p_2^2} dV \equiv K\eta \tag{10}$$

where the "apparatus constant" K contains a part $(16L/\pi r^4)$ determined by the capillary tube and a part (the integral over V) determined by the shape and dimensions of the bulb and the mercury pressure system. While it is possible to determine K from capillary dimensions and measurements of p_1 as a function of V (as described later), more satisfactory results can be obtained by determining K from the time of flow of a gas of known viscosity.

Factors influencing apparatus design. Practical application of the theory presented above depends on the validity of a number of assumptions. One, which has already been mentioned, is the assumption that the flow is laminar. It has been shown by dimensional analysis that for any given type of hydrodynamic experiment the conditions for onset of turbulent flow depend upon the magnitude of a certain combination of pertinent experimental variables that is a pure number, called the Reynolds number R. For flow through a long, round, straight tube,

$$R = r\bar{v}\frac{\rho}{\eta} \tag{11}$$

where ρ is the density of the fluid. It is found empirically that laminar flow is always obtained in such a tube when R is less than 1000 regardless of the magnitude of any of the individual variables r, $\bar{v}$, ρ, η. Laminar flow can be obtained with a Reynolds number as high as 10 000 or even 25 000 if sufficiently careful attention is given to the smoothness of the walls and to the shape of the inlet, but ordinarily turbulent flow is obtained with Reynolds numbers in excess of a few thousand.[3]

Another important factor concerns the transition from a uniform velocity distribution at the inlet of the tube to the parabolic velocity distribution inside the tube (see Fig. 2(b) and (c)). The region in which this transition takes place may be of considerable length, and Eq. (8) will not apply very well over this region. Thus it is important that the length of the region be small in comparison with the length of the tube. Experiments have shown[3] that the

---

[†] The minus sign is necessary, since dn/dt is negative while ϕ_n is positive.

length L' of the transition region is given approximately by the equation

$$L' = \tfrac{1}{4}Rr \qquad (12)$$

where r is the radius of the tube and R is the Reynolds number.

It is important that the tube be very straight; the critical Reynolds number for onset of turbulence will be greatly reduced if the tube is appreciably curved. As a suitable criterion it might be specified that deviations from straightness should be negligible in comparison with r over distances of the order of L'.

Kinetic-energy correction. In deriving Eq. (10) it was assumed that "end effects" are negligible. Such end effects are most likely to be important at the inlet of the capillary, where a pressure drop occurs as a result of the necessity of imparting kinetic energy to the fluid.

By the law of Bernoulli, the pressure drop is equal to the "flow-average" *kinetic-energy density* of the fluid (that is, the rate of flow of kinetic energy past a given fixed point divided by the volume rate of flow of fluid past this point). For flow through a pipe or capillary,

$$\Delta p = \frac{1}{\pi r^2 \bar{v}} \int_0^r \rho \frac{v^2}{2} v 2\pi x \, dx \qquad (13)$$

Beyond the transitional region of length L' we may assume the parabolic velocity distribution given by Eqs. (3) and (5):

$$v = 2\bar{v}\left(1 - \frac{x^2}{r^2}\right) \qquad (14)$$

The overall kinetic-energy pressure drop from the inlet to the point where the distribution has become parabolic is obtained by substituting Eq. (14) into Eq. (13) and is

$$\Delta p = \rho(\bar{v})^2 = \frac{\rho}{\pi^2 r^4} \phi_v^2 \qquad (15)$$

This quantity is called the *kinetic-energy correction*. In practice, it is subtracted from the overall pressure difference $p_1 - p_2$, since no comparable pressure change need ordinarily be considered at the outlet end. This results from the fact that the kinetic energy of the fluid carries it in a jet stream far out into the body of fluid beyond the outlet, so that the kinetic energy is dissipated (as heat, rather than as potential energy) at a considerable distance from where it could otherwise be effective in producing a pressure change at the outlet.

To take into account the effect of kinetic energy on the time of flow t_{ab}, we can subtract Δp, as given by Eq. (15), from p_1 in Eqs. (8) and (9), then invert and integrate as before. We obtain the integral appearing in Eq. (10), plus additional terms which are small. To simplify the latter we assume that

$(p_1 - p_2)$ and $V\, dp_1/dV$ are negligible in comparison with p_1. We obtain

$$t_{ab} = K\eta + \frac{\rho V_{ab}}{8\pi\eta L} \tag{16}$$

where V_{ab} is the volume of the bulb between the two fiducial marks and K is the same apparatus constant as previously defined.†

Slip correction. It has been assumed throughout the above discussion that Eq. (1) is valid down to the smallest dimensions that are of any significant importance in the experiment. This assumptions appears well founded for laminar *liquid* flow, but for gases it breaks down when the mean free path is not negligibly small in comparison with the apparatus dimensions. Thus, at very low pressures or at ordinary pressures in capillaries of very small diameter, gas viscosities appear to be lower than when measured under ordinary conditions. When the mean free path of a gas is small but not negligible in comparison with the radius of the capillary, the gas behaves as if it were "slipping" at the capillary walls, rather than having zero velocity at the walls as shown in Fig. 2(b) and given in Eq. (3). Indeed, the mean velocity of the gas infinitesimally close to the wall is not zero, for although half the molecules in this region have suffered their most recent collision at the wall the other half have suffered their more recent collision at some distance from the wall, of the order of a mean free path λ. Therefore the mass velocity, instead of being zero at the wall, is a fraction approximating $4\lambda/r$ of the average mass velocity $\bar{v}$. Thus, if Eq. (16) yields the "apparent viscosity" η_{app}, the true viscosity can be obtained from the equations

$$\eta = \eta_{\text{app}}\left(1 + \frac{4\lambda}{r}\right) \tag{18}$$

Dependence on flow velocity. With regard to the tacit assumption that the viscosity coefficient is independent of flow velocities and gradients of flow

---

† One may ask whether the viscous drag and kinetic energy of the mercury in tube T of the apparatus shown in Fig. 3 will influence the experimental results. Simple calculations are likely to show that kinetic-energy effects (including possible turbulence) will be more important than viscous drag. A reasonable design criteria (for 0.1 percent maximum error) might be based on the application of Eq. (15) to the mercury in tube T:

$$3\frac{\rho_{\text{Hg}}}{\pi^2 r_T^4}\phi_{V,\text{max}}^2 \le 0.001(p_{1,\text{max}} - p_2) \tag{17}$$

where the factor 3 appears reasonable, since there are three straight portions of the tube. If the Reynolds number is high enough to indicate turbulent flow of the mercury, the factor 3 may be too small. This formula assumes a parabolic velocity distribution; a nearly uniform distribution would appear more likely in view of Eq. (12) but is a less conservative assumption, since it would imply an additional factor of $\frac{1}{2}$ on the left side of Eq. (17).

velocities, it must be remembered that the molecular velocities are essentially Maxwellian in a gas under shear only when variations in mass velocity over distances of the order of a mean free path are small in comparison with the molecular velocities themselves; if these differences are considerable, it is possible that the perturbed distribution of molecular velocities will manifest itself in an altered apparent viscosity coefficient. Without going into detail, it may be stated that if $\bar{v} \ll u$ (where u is the r.m.s. molecular velocity) and if $\lambda \ll r$, the error in η due to perturbation of the Maxwellian distribution will be negligible.

EXPERIMENTAL

Using the apparatus shown in Fig. 3, measure the flow times t_{ab} for dry air, helium, argon, carbon dioxide, and/or any other gases provided, at the same temperature in the same apparatus. Be careful not to disturb the leveling of the apparatus; the time of flow is rather sensitive to tipping of the apparatus, and it is important to have the same leveling for all gases studied. Moisture should be removed from the air by inserting a drying tube in the rubber hose line between the compressed-air supply and the viscosity apparatus. The drying tube (Fig. 4) should be full but not too tightly packed. It should be removed from the line when other gases are studied. If CO_2 is studied, be sure it is at room temperature before starting the run, as it cools considerably on expansion at the outlet of the cylinder.

Admit one of the gases to the bulb B by slowly opening stopcock X while

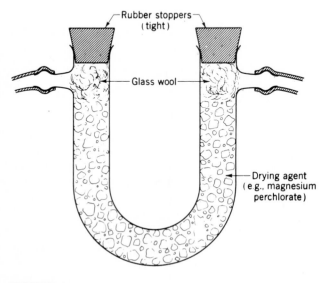

FIGURE 4
U tube for drying air.

stopcock Y remains closed; the mercury level in the bulb should drop quite slowly. Close stopcock X when the Hg level is about 2 cm below the lower fiducial mark. Carefully vent the bulb through the open outlet from the three-way stopcock Y to flush the system. Fill and flush a second time. Refill the bulb with gas, then open stopcock Y to the capillary and measure t_{ab} with a stopcock. Four or more runs should be made if time permits.

Repeat the above procedure for each of the gases to be studied. Also record the temperature and the corrected barometric pressure p_2. Record at least approximate values for all the pertinent apparatus factors—L, r, V_B, maximum value of $p_1 - p_2$, radius of tube T.

Optional: Make an absolute determination of the apparatus constant K. Measure the length L of the capillary tube with a millimeter scale. Fill the capillary with mercury, measure the length of the pellet of mercury contained in the capillary, expel the mercury into a weighing bottle, and weigh it. From the results calculate r for the capillary. To find p_1 as a function of V, attach to the outlet tube Q a gas burette with a parallel leveling bulb. Use water as the indicating fluid. By means of one of the stopcocks (say Y) permit the mercury to rise a little at a time in bulb B. After each change measure the quantity introduced by adjusting and reading the gas burette, and measure the pressure difference $(p_1 - p_2)$ by measuring the difference in mercury levels, perferably with a cathetometer. Alternatively, connect the gas inlet tube P of the apparatus directly to an open-tube manometer to measure $(p_1 - p_2)$ directly. Plot p_1 against V and draw a curve through the points. Evaluate dp_1/dV at a number of points and calculate the integrand of Eq. (10). Evaluate the integral graphically and compute K. Then compare the value of K with the value obtained by calibration with dry air.

CALCULATIONS

For calibration purposes dry air was chosen as the standard, since it is easily available and has been carefully studied by many investigators. Data from many sources can be summarized in the empirical Sutherland expression[4]

$$\eta = \frac{(145.8 \times 10^{-7})T^{3/2}}{T + 110.4} \tag{19}$$

which gives the viscosity of dry air in poise at 1 atm as a function of absolute temperature T. At 25°C, the value is 183.7 μP (0.01837 cP). For the temperature of your experiment, calculate η for dry air from Eq. (19).

For each gas studied, average the flow times for the individual runs to obtain an average t_{ab}.

If Eq. (10) were valid, the ratio of viscosities for two gases would equal the corresponding ratio of flow times. Assuming Eq. (10) and using t_{air} and η_{air}, find approximate values of the viscosities of the other gases from their flow times.

For each gas (including air) calculate the maximum value of the Reynolds number from Eq. (11) and verify that it is below 1000. Use Eq. (12) to verify that L', the length of the transition region, is small compared with the length L of the capillary. Estimate the largest value of $\bar{v}$ from Eq. (5) and compare it with the molecular r.m.s. speed u and mean speed $\bar{c}$.

If the required conditions have been met, kinetic-energy and slip corrections may now be made in order to get more precise values of η. First calculate λ for each gas by combining Eqs. (IV-2) and (IV-12) and using the ideal-gas law:

$$\lambda = \frac{32\eta}{5\pi} \frac{1}{m} \frac{1}{\bar{N}} \left(\frac{\pi m}{8kT}\right)^{1/2} = \frac{16\eta}{5p} \left(\frac{RT}{2\pi M}\right)^{1/2} \tag{20}$$

With Eq. (18) find η_{app} for dry air and use this value in Eq. (16) to find the value of K. Now calculate η_{app} for the other gases from Eq. (16) (one can use the approximate value of η in evaluating the kinetic-energy correction term). Finally apply the slip correction to obtain the true viscosity.

Report the final values obtained for η together with the temperature at which they were determined. Report also for each gas the calculated mean free path at 1 bar and the effective molecular diameter as calculated from the viscosity with Eq. (IV-13).

APPARATUS

Gas viscosity apparatus, complete with capillary tube and mercury; gas supply outlet(s) with needle valve(s); glass U tube with rubber stoppers and policemen; two lengths of gum-rubber tubing; 0 to 30°C thermometer; stopwatch; millimeter scale. If an absolute determination is to be made: additional mercury; weighing bottle; gas burette and leveling bulb; ring stands and clamps; cathetometer or an open-tube manometer.

Cylinders of argon, helium, and carbon dioxide; source of compressed air; glass wool; drying agent (such as magnesium perchlorate); stopcock grease.

REFERENCES

1. J. F. Golubev, "Viscosity of Gases and Gas Mixtures," Chap. 2 [translated from Russian by Israel Program for Scientific Translations, Jerusalem (1970)].
2. J. Reilly and W. N. Rae, "Physico-chemical Methods," 5th ed., Vol. I, pp. 667, 690 and Vol. III, p. 247, Van Nostrand, Princeton, N.J. (1953).
3. L. Prandtl, "Essentials of Fluid Dynamics," Blackie, Glasgow (1952).
4. "Tables of Thermal Properties of Gases," Table 1-B, U.S. Natl. Bur. Stand. Circ. 564, U.S. Government Printing Office, Washington, D.C. (1955).

GENERAL READING

S. Pai, "Viscous Flow Theory," Vol. I: Laminar Flow, Van Nostrand-Reinhold, Princeton, N.J. (1965).

J. F. Johnson, J. R. Martin, and R. S. Porter, "Determination of Viscosity," in A. Weissberger and B. W. Rossiter (eds.), "Techniques of Chemistry: Vol. I. Physical Methods of Chemistry," part VI, Chap. 2, Wiley-Interscience, New York (1977).

EXPERIMENT 5
DIFFUSION OF GASES

In this experiment, the mutual diffusion coefficients for the Ar–CO_2 and He–CO_2 systems are to be measured using a modified Loschmidt apparatus. These transport coefficients are then compared with theoretical values calculated with hard-sphere collision diameters.

When a component of a continuous fluid phase (a liquid solution or a gas mixture) is present in nonuniform concentration, at uniform and constant temperature and pressure and in the absence of external fields, that component diffuses in such a way as to tend to render its concentration uniform. For simplicity, let the concentration of a given substance $\bar{n}_1$ be a function of only one coordinate x; that is, the concentration of substance 1 varies only along the x direction, which we shall take as the upward direction. The net "flux" z_1 of the substance passing upward past a given fixed point x_0 (i.e., amount per unit cross-sectional area per unit time) is under most conditions found to be proportional to the negative of the concentration gradient:

$$z_1 = -D\left(\frac{d\bar{n}_1}{dx}\right)_{x_0} \tag{1}$$

where D is the *diffusion constant,* which is characteristic of the diffusing substance and usually also of the other substances present. This is called "Fick's first law of diffusion."

When only two substances are present in a closed gaseous system of fixed total volume at uniform and constant temperature, the distribution of one of the substances is completely determined when the distribution of the other is specified. The sum of the partial pressures of the two components is equal to the total pressure, and if the two components are perfect gases, the total pressure is constant and the sum of the two concentrations is also constant. Designating the two substances by the subscripts 1 and 2, we can write

$$z_1 = -D_{12}\frac{d\bar{n}_1}{dx} = D_{12}\frac{d\bar{n}_2}{dx} \tag{2}$$

$$z_2 = -z_1 = -D_{21}\frac{d\bar{n}_2}{dx} = D_{21}\frac{d\bar{n}_1}{dx} \tag{3}$$

from which it can be seen that D_{12}, the diffusion constant for the diffusion of component 1 into component 2, is equal to D_{21}, the diffusion constant for the diffusion of 2 into 1.

One can also define a diffusion constant for the diffusion of a substance into itself. Suppose that we have a single molecular species (substance 1, say) present in a vessel but that the molecules initially in a certain portion of the vessel can somehow be tagged or labelled (without changing their kinetic properties) so that later they can be distinguished from the others. After a period of time the distribution of the tagged molecules could then be determined experimentally and the *self-diffusion constant* D_1 can be calculated. In actuality, it is not possible to tag molecules without some effect on their kinetic properties. A close approach to such tagging can be obtained, however, with the use of different isotopes that may be distinguished with a mass spectrometer or a Geiger counter. Another close approach can be obtained in a few cases by the use of molecules that are very similar in size, shape, and mass but different in chemical constitution, for example, the pair N_2 and CO or the pair CO_2 and N_2O. For each of these pairs the diffusion constant D_{12} should constitute a reasonably good estimate of the self-diffusion constants D_1 and D_2.

In the case where the two gases are distinctly different in their diffusing tendencies, the mutual diffusion constant D_{12} lies somewhere between the two self-diffusion constants D_1 and D_2 in value. It is also found that, when the two self-diffusion constants are very different, the mutual diffusion constant is largely determined by the more rapidly diffusing substance rather than by the more slowly diffusing one; thus there is a greater difference between the diffusion constants for the systems Kr–He and Kr–H_2 than for Kr–He and CO_2–He. The gas having the high self-diffusion constant (He or H_2) diffuses rapidly into the gas having the low self-diffusion constant. But there must be a *mass flow* of the resulting mixture in the opposite direction to maintain uniform pressure, so the gas with the lower diffusion constant gives the appearance of diffusing, although in reality it is mainly undergoing displacement.

THEORY

Let the concentration of gas 1 be $\bar{n}(x)$, a function of x only, in a vertical tube of uniform cross section A (x increasing upward). Let us calculate the rate of accumulation of gas 1 in the region from x to $x + dx$. The net flow of the gas into the region from below, per square centimeter per second, is given by Fick's first law,

$$z_1(x) = -D_{12}\left(\frac{\partial \bar{n}_1}{\partial x}\right)_x$$

The net flow of the gas out of the region in the upward direction, per square centimeter per second, is

$$z_1(x + dx) = -D_{12}\left(\frac{\partial \bar{n}_1}{\partial x}\right)_{x+dx}$$

$$= -D_{12}\left[\left(\frac{\partial \bar{n}_1}{\partial x}\right)_x + \left(\frac{\partial^2 \bar{n}_1}{\partial x^2}\right)_x dx\right]$$

The rate of increase of the concentration $\bar{n}_1(x)$ is given by

$$\frac{\partial \bar{n}_1(x)}{\partial t} = \frac{z_1(x) - z_1(x + dx)}{dx}$$

and therefore

$$\frac{\partial \bar{n}_1}{\partial t} = D_{12} \frac{\partial^2 \bar{n}_1}{\partial x^2} \tag{4}$$

This is the differential equation for diffusion in one dimension (called "Fick's second law").

Let two perfect gases be placed in a *Loschmidt apparatus*[1]—a vertical tube of length $2L$, of uniform cross-section, closed at both ends, and containing a removable septum in the center. The heavier of the two gases (gas 1, say) is initially confined to the lower half and the lighter (gas 2) to the upper half in order to avoid convective mixing. Both gases are at initial concentration $\bar{n}_0$. At time $t = 0$ the septum is removed, so that the gases are free to interdiffuse. Let us define a quantity $\Delta \bar{n}$ as follows:

$$\Delta \bar{n} = \bar{n}_1 - \bar{n}_2 \equiv 2\bar{n}_1 - \bar{n}_0 \tag{5}$$

The differential equation to be solved,

$$\frac{\partial \Delta \bar{n}}{\partial t} = D \frac{\partial^2 \Delta \bar{n}}{\partial x^2} \tag{6}$$

is the differential equation (4) in $\bar{n}_1$ minus a similar one in $\bar{n}_2$, and

$$D = D_{12} = D_{21}$$

Taking the origin at the middle of the tube, we write the boundary conditions

$$t = 0: \quad \Delta \bar{n} = \bar{n}_0 \qquad -L \leq x < 0$$
$$\Delta \bar{n} = -\bar{n}_0 \qquad 0 < x \leq L \tag{7}$$
$$t = \infty: \quad \Delta \bar{n} = 0 \qquad -L \leq x \leq L \tag{8}$$

$$\text{All } t: \quad \frac{\partial \Delta \bar{n}}{\partial x} = 0 \qquad x = -L$$

$$\frac{\partial \Delta \bar{n}}{\partial x} = 0 \qquad x = L \tag{9}$$

The last boundary condition follows from the fact that there is no flow of gas through the two ends of the tube; see Eq. (1). Boundary condition (7) can be expressed in terms of a Fourier series expansion[2] of $\Delta \bar{n}$:

$$t = 0: \quad \Delta \bar{n} = -\frac{4}{\pi} \bar{n}_0 \left(\sin \frac{\pi}{2L} x + \tfrac{1}{3} \sin \frac{3\pi}{2L} x + \tfrac{1}{5} \sin \frac{5\pi}{2L} x + \ldots \right) \tag{7'}$$

This is the Fourier series for a "square wave" of amplitude $\bar{n}_0$ and wavelength $2L$.

We shall omit the procedure for finding the solution of Eq. (6) and simply give the result:

$$\frac{\Delta \bar{n}}{\bar{n}_0} = \frac{\bar{n}_1 - \bar{n}_2}{\bar{n}_1 + \bar{n}_2} = -\frac{4}{\pi} \left(e^{-\pi^2 Dt/4L^2} \sin \frac{\pi}{2L} x + \tfrac{1}{3} e^{-9\pi^2 Dt/4L^2} \sin \frac{3\pi}{2L} x \right.$$

$$\left. + \tfrac{1}{5} e^{-25\pi^2 Dt/4L^2} \sin \frac{5\pi}{2L} x + \dots \right) \quad (10)$$

The reader can easily verify that the solution satisfies the differential equation and the boundary conditions (7) or (7'), (8), and (9). The behavior of $\Delta \bar{n}$ with x and t is shown qualitatively in Fig. 1(a).

After time t, let us replace the removable septum and analyze the two portions of the tube for total quantity of one of the gases, say gas 1. The total amount of gas 1 in the upper part will be equal to the total amount of gas 2 in the lower part. Let W_1 and n_1 be the weight and number of moles of gas 1 in the lower part and W_1' and n_1' be the weight and the number of moles of gas 1 in the upper part, and let similar symbols with subscript 2 pertain to gas 2. Define a quantity f by

$$f \equiv \frac{n_1 - n_2}{n_1 + n_2} = \frac{(\bar{n}_1 - \bar{n}_2)_{\mathrm{av}}}{\bar{n}_0} = \frac{1}{L} \int_{-L}^{0} \frac{\Delta \bar{n}}{\bar{n}_0} \, dx \quad (11)$$

Since $n_1' = n_2$, we have also

$$f = \frac{n_1 - n_1'}{n_1 + n_1'} = \frac{W_1 - W_1'}{W_1 + W_1'} \quad (12)$$

This expression permits the calculation of f from the results of analyzing separately the upper and lower parts of the tube for gas 1. From Eqs. (10) and (11), we obtain

$$f = \frac{8}{\pi^2} \left(e^{-\pi^2 Dt/4L^2} + \tfrac{1}{9} e^{-9\pi^2 Dt/4L^2} + \tfrac{1}{25} e^{-25\pi^2 Dt/4L^2} + \dots \right) \quad (13)$$

At $t = 0$, the sum of the series in parentheses is equal to $\pi^2/8$, so the quantity f is initially unity. After a short time all terms except the first become negligible, and f therefore decays with time in a simple exponential manner.

At what time t should the run be stopped in order to obtain the most precise value of D, assuming that the only significant errors are those in determining f and that the uncertainty in f is constant (independent of the time)? Neglecting all but the first time of Eq. (13), we find, on solving for D and differentiating, that

$$\frac{\partial D}{\partial f} = -\frac{1}{t} \frac{4L^2}{\pi^2} \frac{1}{f}$$

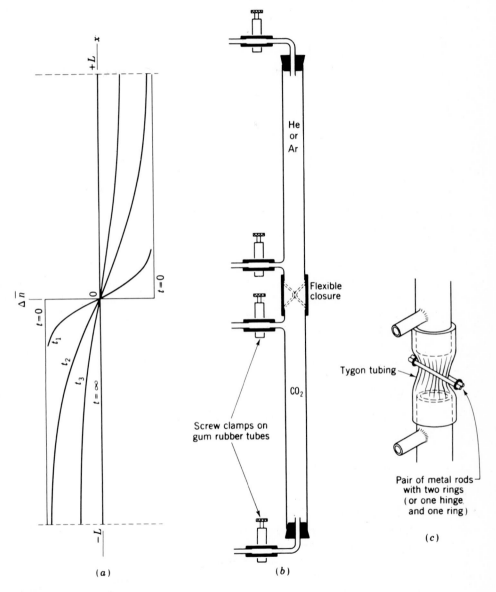

FIGURE 1
Loschmidt diffusion apparatus. (a) Curves showing schematically the quantity $\Delta\bar{n} = \bar{n}_1 - \bar{n}_2$. (b) Cell. (c) Detail of flexible closure.

We wish to find at what value of t the derivative $\partial D/\partial f$ is at a minimum:

$$0 = \frac{\partial^2 D}{\partial t \, \partial f} = \frac{4L^2}{\pi^2} \left(\frac{1}{t} \frac{1}{f^2} \frac{\partial f}{\partial t} + \frac{1}{t^2} \frac{1}{f} \right) = \frac{4L^2}{\pi^2 ft} \left(-\frac{\pi^2 D}{4L^2} + \frac{1}{t} \right)$$

thus the optimum time is

$$t_{\text{opt}} = \frac{4L^2}{\pi^2 D} \tag{14}$$

At this optimum time, Eq. (13) gives

$$f = 0.81057(0.367879 + 0.000014 + \cdots) = 0.2982$$

Note that neglect of the second and higher terms of Eq. (13) proves to be justified if the experiment is run for any time in the neighborhood of the optimum time or for any greater time.

METHOD

In this experiment we shall measure the diffusion constants for the systems CO_2–He and CO_2–Ar at room temperature. In each run, CO_2, the heavier gas, will be initially in the lower portion of a vertical tube. The gas in the two portions at the end of the run will be analyzed for CO_2 by sweeping it through absorption tubes (U tubes) filled with Ascarite (asbestos impregnated with NaOH, which reacts with CO_2) and with magnesium perchlorate (to absorb the water liberated in the first reaction); see Fig. 2. These tubes are weighed

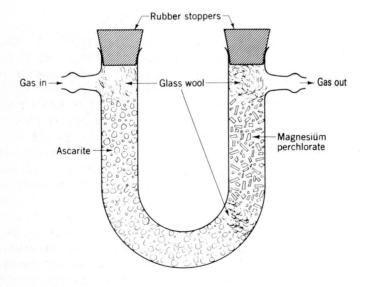

FIGURE 2
U-tube for absorbing CO_2 gas.

before and after sweeping the gas through them. The diffusion cell, a modified Loschmidt type, is shown in Fig. 1(b). In place of a removable septum at the center, the two halves of this tube are joined by a flexible plastic (Tygon) sleeve of approximately the same inner diameter as the glass, which can be pinched tight to achieve a closure as shown in Fig. 1(c).

A variant of the same basic method is shown in Fig. 3. In this case, a vacuum system is used for handling the gases, a large-bore stopcock is used in place of the flexible closure, and a different method is used for the quantitative analysis of CO_2. Instead of absorbing the CO_2 on Ascarite, the CO_2–Ar or CO_2–He mixture is slowly pumped through a copper coil immersed in liquid nitrogen. The CO_2 will be efficiently trapped and after all the Ar or He has been removed, the trap is allowed to warm to room temperature and the CO_2 pressure is determined.

In round numbers the length L in the apparatus to be employed is 50 cm, and the two diffusion constants are of the order of magnitude $0.6 \text{ cm}^2 \text{ s}^{-1}$ for CO_2–He and $0.15 \text{ cm}^2 \text{ s}^{-1}$ for CO_2–Ar. Using these figures the optimum diffusion times can be estimated from Eq. (14). The actual time should approximate this optimum time but in any case should not differ from it by more than thirty percent.

EXPERIMENTAL

Procedure A. In carrying out the experiment with the apparatus shown in Figs. 1 and 2, the student should follow a careful plan worked out with his partner, taking into account the following:

1. Even a tiny leak is serious; all stoppers and clamps *must be tight*. Rubber and Tygon tubing must be in good condition.
2. When one gas is being flushed out by another, the gas being introduced should enter at the lower end of the vessel if it is the heavier and at the upper end if it is the lighter.
3. Considerable heat is evolved in the reaction of CO_2 with Ascarite; a U tube that has absorbed a considerable amount of CO_2 should be allowed to cool for about 15 min before it is weighted.
4. Carbon dioxide gas emerges rather cool from the tank valve. Time should be allowed for it to warm to room temperature before the diffusion run is started. It is wise to vent the lower chamber, containing CO_2, momentarily at the side arm before the run to restore atmospheric pressure.
5. If the cylinder is not equipped with a regulator valve, turn off the gas supply at the cylinder before closing off the system. The inlet of the system should be clamped off before the outlet.
6. The CO_2-containing gas should be swept into the drying tubes no faster than 5 bubbles per second. Use a dibutyl-phthalate bubbler for metering

gas flow. One bubble is approximately 0.1 cm^3. This rate can be obtained by synchronization with the ticking of a watch. It is well to adjust the flow rate (with the outlet valve on the compressed air line) *before* attaching to the system. The sweeping air should go *first* through the bubbler and *then* through a purifying tube (to remove CO_2 and H_2O) before it enters the system. In case the flow rate is preadjusted, it is important to make sure that the system is open at both ends at the moment the air tube is attached.

7. Drying tubes should be closed off with rubber policemen when gas is not being passed through them. Take care that the drying tube always has the same two policemen.

8. Do not neglect the volumes of gas that are contained in any significant lengths of rubber tubing. Before using a rubber tube that may previously have contained CO_2, flush it out momentarily with air. Also, do not neglect the possibility of a heavy gas spilling out of the system if it is open at two places at different levels for more than a second or two.

Prepare as shown in Fig. 2 and weigh two U tubes, A and B. It is not necessary to correct weights to vacuum readings. A third U tube, which need not be weighed, should be prepared for purification of the air to be used for sweeping.

Flush out the lower half of the cell with CO_2, using a flushing rate of at least 10 to 15 bubbles per second through the bubbler. *Make sure the center partition is closed* before flushing. Pass at least six volumes of gas through to be sure of complete replacement.

Connect the lower half-cell to U tubes A and B in series, and sweep out the gas in the lower half-cell into the U tubes with three or four volumes of air (purified with the third U tube), at 5 bubbles per sec. Without stopping the sweeping, *quickly* pull off tube A and reconnect tube B. Immediately close the ends of tube A. After one or two more additional volumes have been swept through tube B, weigh tube B first, then tube A. Tube B should not have gained more than a few milligrams. Add the two gains in weight; this quantity is $W_1 + W_1'$, the total weight of CO_2 in the cell.

Now flush out the lower half-cell with CO_2 and the upper one with the other gas (He or Ar) to be studied, as above. This should be done during the above weighings to save time.

Begin the diffusion by *carefully* (not too suddenly) opening the center partition and simultaneously starting a stopwatch. After the estimated optimum time, close the partition.

Sweep the contents of the lower tube into U tubes A and B as before. After the weighings, repeat with the upper tube. The value of $W_1 + W_1'$ should agree with the value obtained above to within a very few milligrams; if greater disagreement is obtained, a leak must be suspected.

Repeat the procedure with the second gas combination. If time permits, repeat each of the above runs, with the diffusion cell inverted, and in each case take as your value of f the mean of the two values obtained.

Measure the length L as precisely as possible. Also record the barometric pressure and the ambient temperaure.

Procedure B. In carrying out the experiment with the apparatus shown in Fig. 3, the student should review the basic operation of a vacuum system (see Chapter XVII).

The diffusion tube was designed and constructed such that the length L of the upper section is equal to the length of the lower section *plus* the length of the bore of stopcock A. Hence the stopcock bore and the lower section of the tube are both filled with CO_2.

The CO_2 and He (or Ar) gas tanks are connected to the diffusion tube as shown in Fig. 3. Stopcocks A through F should be opened and the line should be pumped on for about 20 min. Close stopcock F and check for an increase in pressure. The rate of increase should be no greater than 5 Torr hr^{-1}. Then evacuate the entire apparatus again for a few minutes.

Close stopcock D and fill the system with CO_2 to a pressure between 750 and 770 Torr. Close stopcock B and read the CO_2 pressure exactly. After closing stopcock A, pump away the remaining CO_2. Fill the upper portion of the tube with He to a pressure *as close as possible* to the CO_2 pressure (the difference should be less than 1 Torr). Close stopcock C and evacuate the remaining portion of the apparatus. Begin the diffusion by *carefully* opening

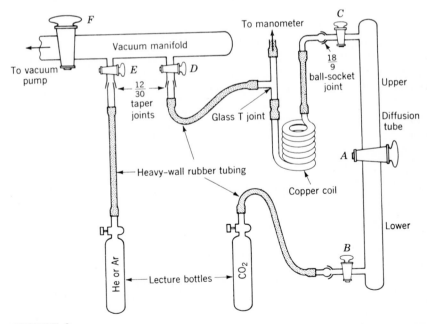

FIGURE 3
Modified Loschmidt apparatus, to be used according to procedure B. Stopcock A should have a very large bore which is as close as possible to the diameter of the diffusion tube. Stopcocks B to E should be high-vacuum type. As a safety precaution, each gas lecture bottle should be connected to the system via a reducing valve or a pressure-relief valve (not shown).

stopcock A and simultaneously starting a stopwatch or timer (be sure to align the stopcock bore with the vertical axis of the upper and lower tubes). After the estimated optimum time, close stopcock A and record the elapsed time.

Place a Dewar flask of liquid nitrogen around the copper coil. After the trap has been cooled (as indicated by an absence of violent boiling of the liquid nitrogen), *first* close stopcock D and then open C. After about 30 seconds, very slowly open stopcock D until the pressure (watch the manometer) just begins to decrease. If the stopcock is opened too much, the trapping of CO_2 will not be complete. Adjust stopcock D to give a pressure decrease of about 20 Torr every 15 seconds. It should take not more than a few minutes to remove the helium. Then close both stopcocks D and C. To remove any He entrapped in the solid CO_2, remove the Dewar flask from the coil and allow the CO_2 to vaporize. However, do not wait for the coil to reach room temperature. Place the Dewar flask around the coil again and when the solidification is complete, fully open stopcock D for about a minute (*leave stopcock C closed*). If the above operations were performed properly, than at least 99 percent of the CO_2 will have been trapped.

With stopcock D closed, remove the Dewar flask and allow the bulk of the CO_2 to vaporize. Then place a liter beaker containing water at room temperature around the coil (this will hasten the warming up of the coil). After about a minute, remove the beaker and dry the coil with a towel. Wait a few minutes for the coil to come to thermal equilibrium with the room and then measure the CO_2 pressure. This pressure in the upper half of the tube is denoted by p_1'.

Now evacuate all of the system above stopcock A. Then close stopcock C and open A. Repeat the procedure described above in order to determine the CO_2 pressure associated with the gas in the lower half of the tube.

The quantity f in Eq. (12) is defined in terms of pressures by

$$f = (p_1 - p_1')/(p_1 + p_1').$$

Finally repeat the complete procedure with Ar as the second gas. In addition to determining the diffusion coefficients at atmospheric pressure, measure one of them at half an atmosphere.

Measure the length L as precisely as possible, and also record the ambient temperature.

CALCULATIONS

Your experimental data should be summarized in a table that includes for each run the initial pressure, the temperature, the length of the run, and the final CO_2 weights (or pressures) from the upper and lower parts of the cell.

By use of Eqs. (12) and (13) neglecting all but the first term in the series, calculate D for each gas combination. Then, with this value of D, calculate the second term in the series; if it is more than 1 percent of the first term, recalculate D. The resulting values of D are to be designated D_{12} for CO_2–He, D_{13} for CO_2–Ar.

Gas-viscosity measurements yield molecular diameters of 2.58 Å (angstroms) for helium, 3.42 Å for argon, and 4.00 Å for carbon dioxide.[3] With these values, calculate d_{12} and d_{13} from Eq. (IV-36), and calculate D_{12} and D_{13} using Eq. (IV-34). Compare with your experimental values.

The molecular diameters cited above were obtained from an analysis of viscosity data in terms of the Lennard–Jones "6,12 potential":[3]

$$u(r) = 4\varepsilon\left[\left(\frac{\sigma}{r}\right)^{12} - \left(\frac{\sigma}{r}\right)^{6}\right] \tag{15}$$

where $u(r)$ is the potential energy of two molecules at a distance r. The diameters correspond to σ values (the distance at which the potential energy equals zero). Although these values are often called effective hard-sphere diameters (why?), one should remember that they do not result from the solution of the hard-sphere expression, Eq. (IV-13). Alternatively, if one wishes to calculate an *accurate* diffusion coefficient, it is meaningless to use such σ values in conjunction with Eq. (IV-34).

Optional. At 273 K, D_{23} (for He–Ar) is 0.653 cm² s⁻¹. From D_{12}, D_{13}, and D_{23} at *room temperature* calculate d_{12}, d_{13}, and d_{23}. (To correct diffusion constants from one temperature to another, assume a $T^{3/2}$ dependence if the temperature change is small. This is only approximate, since the d's may vary in some degree with the temperature.) Then obtain d_1, d_2, and d_3 and use these to calculate D_1, D_2, and D_3 from Eq. (IV-35). Determine the ratios of the self-diffusion constants to their respective viscosities. How do these compare with the theoretical ratios discussed in the introductory section of Chapter IV?

DISCUSSION

The flexible closure or the large-bore stopcock, though simple and convenient, is admittedly less satisfactory than a removable diaphragm closure would be. Discuss its disadvantages from the points of view of (1) mixing produced in the opening and closing operations; (2) inexactness of the matching of its diameter with that of the tubes; and (3) any other defects that occur to you. Try wherever possible to employ arguments that are not wholly qualitative; estimate orders of magnitude where you can. If possible, suggest an improved closure design. What additional techniques might be used with this experimental method to enable D_{23} (for He–Ar) to be measured directly? Using your value for D_{13} estimate the time necessary for a diffusion run on CO_2–Ar in the present apparatus to achieve a state where $W_1 = 1.01W_1'$ (i.e., where the concentration of CO_2 is only 1 percent greater in the lower half than in the upper half).

APPARATUS

Procedure A. Diffusion cell (Loschmidt type) with sturdy support; clamping device for flexible closure; two one-hole rubber stoppers, with short bent tubes, for ends of cell; four 1-in. lengths of rubber tubing; four screw clamps; four short glass tubes; three glass U tubes with rubber stoppers and policemen; bubbler, containing dibutyl phthalate; three long and two short lengths of rubber tubing; needle valve; two ring stands; four clamps with clamp holders; 0 to 30° thermometer; stopwatch; meter stick.

Cylinders of helium, argon, and carbon dioxide; source of compressed air; glass wool; Ascarite; drying agent (such as magnesium perchlorate).

Procedure B. Vacuum system with mechanical pump and manifold having provision for two attachments via taper joints; Loschmidt diffusion cell with stopcocks attached at top and bottom; two standard taper joints to fit manifold; two ball-socket joints to fit diffusion cell; glass (or metal) T joint; five long lengths and one short length of heavy-wall rubber pressure tubing; tightly wound copper coil; large Dewar flask; closed-tube manometer; 1-liter beaker; towel; stopwatch or electrical timer; meter stick; 0 to 30°C thermometer.

Liquid nitrogen; cylinders or lecture bottles of helium, argon, and carbon dioxide.

REFERENCES

1. P. J. Dunlap, B. J. Steel and J. E. Lane, "Experimental Methods for Studying Diffusion in Liquids, Gases, and Solids," in A. Weissberger and B. W. Rossiter (eds.), "Techniques of Chemistry: Vol. I. Physical Methods of Chemistry," part IV, Chap. 4, Wiley-Interscience, New York (1972).
2. Any text on advanced calculus, such as R. Courant, "Differential and Integral Calculus," Wiley-Interscience, New York (1937); or F. B. Hildebrand, "Advanced Calculus for Applications," 2d ed., Prentice-Hall, Englewood Cliffs, N. J. (1976).
3. J. O. Hirschfelder, C. F. Curtiss and R. B. Bird, "Molecular Theory of Gases and Liquids," 2d ed., pp. 1035–1044, 1110, Wiley, New York (1967).

GENERAL READING

W. Kauzmann, "Kinetic Theory of Gases," Benjamin, New York (1966).
R. D. Present, "Kinetic Theory of Gases," McGraw-Hill, New York (1958).
R. E. Cunningham, "Diffusion in Gases and Porous Media," Plenum, New York (1980).

CHAPTER
V

THERMOCHEMISTRY

EXPERIMENTS

6. Heats of combustion

7. Strain energy of the cyclopropane ring

8. Heats of ionic reaction

PRINCIPLES OF CALORIMETRY

We are concerned here with the problem of determining experimentally the enthalpy change ΔH or the energy change ΔE accompanying a given isothermal change in state of a system, normally one in which a chemical reaction occurs. We can write the reaction schematically in the form

$$A(T_0) + B(T_0) = C(T_0) + D(T_0) \tag{1}$$
$$\underset{\text{initial state}}{} \underset{\text{final state}}{}$$

In practice we do not actually carry out the change in state isothermally; this is not necessary because ΔH and ΔE are independent of the path. In calorimetry we usually find it convenient to use a path composed of two steps:

Step I. A change in state is carried out *adiabatically* in the calorimeter vessel to yield the desired products but in general at another temperature:

$$A(T_0) + B(T_0) + S(T_0) = C(T_1) + D(T_1) + S(T_1) \tag{2}$$

where S represents those parts of the system (e.g., inside wall of the calorimeter vessel, stirrer, thermometer, solvent) that are always at the same temperature as the reactants or products because of the experimental arrangement; these parts, plus the reactants or products, constitute the system under discussion.

Step II. The products of step I are brought to the initial temperature T_0 by adding heat to (or taking it from) the system:

$$C(T_1) + D(T_1) + S(T_1) = C(T_0) + D(T_0) + S(T_0) \tag{3}$$

As we shall see, it is often unnecessary to carry out this step in actuality since the associated change in energy or enthalpy can be calculated from the known temperature difference.

By adding Eqs. (2) and (3) we obtain Eq. (1) and verify that these two steps describe a complete path connecting the desired initial and final states. Accordingly, ΔH or ΔE for the change in state (1) is the sum of the values of this quantity pertaining to the two steps:

$$\Delta H = \Delta H_{\mathrm{I}} + \Delta H_{\mathrm{II}} \tag{4a}$$

$$\Delta E = \Delta E_{\mathrm{I}} + \Delta E_{\mathrm{II}} \tag{4b}$$

The particular convenience of the path described is that the heat q for step I is zero, while the heat q for step II can be either measured or calculated. It can be measured directly by carrying out step II (or its inverse) through the addition to the system of a measurable quantity of heat or electrical energy, or it can be calculated from the temperature change $(T_1 - T_0)$ resulting from adiabatic step I if the heat capacity of the product system is known. For step I,

$$\Delta H_{\mathrm{I}} = q_p = 0 \qquad \text{constant pressure} \tag{5a}$$

$$\Delta E_{\mathrm{I}} = q_v = 0 \qquad \text{constant volume} \tag{5b}$$

Thus, if *both* steps are carrried out at constant pressure,

$$\Delta H = \Delta H_{\mathrm{II}} \tag{6a}$$

and if *both* are carried out at constant volume,

$$\Delta E = \Delta E_{\mathrm{II}} \tag{6b}$$

Whether the process is carried out at constant pressure or at constant volume is a matter of convenience. In nearly all cases it is most convenient to carrry it out at constant pressure; the experiment on heats of ionic reaction is an example. An exception to the general rule is the determination of a heat of combustion, which is conveniently carried out at constant volume in a "bomb calorimeter." However, we can easily calculate ΔH from ΔE as determined from a constant-volume process (or ΔE from ΔH as determined from a constant-pressure process) by use of the equation

$$\Delta H = \Delta E + \Delta(pV) \tag{7}$$

When all reactants and products are condensed phases, the $\Delta(pV)$ term is negligible in comparison with ΔH or ΔE, and the distinction between these two quantities is unimportant. When gases are involved, as in the case of combustion, the $\Delta(pV)$ term is likely to be significant in magnitude, though still small in comparison with ΔH or ΔE. Since it is small, we can employ the perfect-gas law and rewrite Eq. (7) in the form

$$\Delta H = \Delta E + RT \, \Delta n_{\text{gas}} \tag{8}$$

where Δn_{gas} is the *increase* in the number of moles of gas in the system.

We must now concern ourselves with procedures for determining ΔH or ΔE for step II. We might envisage step II being carried out by adding heat to the system or taking heat away from the system and measuring q for this process.[†] Usually, however, it is much easier to measure work than heat. In particular, electrical work done on the system by a heating coil (often referred to as "Joule heating") can be used conveniently to carry out either step II or its inverse, whichever is endothermic. Since the work is dissipated *inside* the system, the work is positive:

$$w_{\text{el}} = \int V_h \, dQ = \int V_h I \, dt \tag{9a}$$

where V_h is the voltage drop across the heater, Q is the electric charge, and I is the electric current. For precise work, one should measure both V_h and I during the heating period. In many instances, however, it is possible to assume that the resistance of the heating coil R_h is constant and make only measurements of I by determining the potential drop V_s across a standard resistance R_s in series with the heating coil. In such a case, one can write

$$w_{\text{el}} = \int I^2 R_h \, dt = \frac{R_h}{R_s^2} \int V_s^2 \, dt \tag{9b}$$

Note that the electrical work given by Eq. (9) is in joules when resistance is in ohms, potential (voltage) is in volts, and time is in seconds. If the electrical heating is done adiabatically,

$$\Delta H_{\text{II}} = w_{\text{el}} \qquad \text{constant pressure} \tag{10a}$$

$$\Delta E_{\text{II}} = w_{\text{el}} \qquad \text{constant volume} \tag{10b}$$

Our discussion so far has been limited to determining ΔH_{II} or ΔE_{II} by directly carrying out step II (or its inverse). However, it is often not necessary to carry out this step in actuality. If we know or can determine the heat capacity of the system, the temperature change $(T_1 - T_0)$ resulting from step I provides all the additional information we need:

$$\Delta H_{\text{II}} = \int_{T_1}^{T_0} C_p(\text{C} + \text{D} + \text{S}) \, dT \tag{11a}$$

$$\Delta E_{\text{II}} = \int_{T_1}^{T_0} C_v(\text{C} + \text{D} + \text{S}) \, dT \tag{11b}$$

---

† This could be done by placing the system in thermal contact with a heat reservoir (such as a large water bath) of known heat capacity until the desired change has been effected and calculating q from the measured change in the temperature of the reservoir.

The heat capacities ordinarily vary only slightly over the small temperature ranges involved; accordingly we can simplify Eqs. (11) and combine them with Eqs. (6) to obtain the familiar expressions

$$\Delta H = -C_p(C + D + S)(T_1 - T_0) \tag{12a}$$

$$\Delta E = -C_v(C + D + S)(T_1 - T_0) \tag{12b}$$

where C_p and C_v are average values over the temperature range.

The heat capacity must be determined if it is not known. A direct method, which depends on the assumption of the constancy of heat capacities over a small range of temperature, is to measure the adiabatic temperature rise $(T_2' - T_1')$ produced by the dissipation of a measured quantity of electrical energy. We then obtain

$$\left.\begin{array}{c} C_p \\ C_v \end{array}\right\} = \frac{w_{el}}{T_2' - T_1'} \qquad \begin{array}{c} (13a) \\ (13b) \end{array}$$

at constant pressure or at constant volume, respectively. This method is exemplified in the experiment on heats of ionic reaction.

An indirect method of determining the heat capacity is to carry out another reaction altogether, for which the heat of reaction is known, in the same calorimeter under the same conditions. This method depends on the fact that in most calorimetric measurements on chemical reactions the heat-capacity contributions of the actual product species (C and D) are very small, and often negligible, in comparison with the contribution due to the parts of the system denoted by the symbol S. In a bomb calorimeter experiment the reactants or products amount to a gram or two, while the rest of the system is the thermal equivalent of about 2.5 kg of water. In calorimetry involving dilute aqueous solutions, the heat capacities of such solutions can in a first approximation be taken equal to that of equivalent weights or volumes of water. Thus we may write, in place of Eqs. (12a) and (12b),

$$\left.\begin{array}{c} \Delta H \\ \Delta E \end{array}\right\} = -C(S)(T_1 - T_0) \qquad \begin{array}{c} (14a) \\ (14b) \end{array}$$

for constant-pressure and constant-volume processes, respectively. In Eqs. (14a) and (14b) we have omitted any subscript from the heat capacity as being largely meaningless, since only solids and liquids, with volumes essentially independent of pressure, are involved. The value of $C(S)$ can be calculated from the heat of the known reaction and the adiabatic temperature change $(T_2' - T_1')$ produced by it, as follows:

$$C(S) = \begin{cases} \dfrac{-\Delta H_{known}}{T_2' - T_1'} & \text{constant pressure} \qquad (15a) \\[3mm] \dfrac{-\Delta E_{known}}{T_2' - T_1'} & \text{constant volume} \qquad (15b) \end{cases}$$

This method is exemplified in the experiment on heats of combustion by bomb calorimetry.

Let us now consider step I, the adiabatic step, and the measurement of the temperature difference $(T_1 - T_0)$ which is the fundamental measurement of calorimetry. If this step could be carried out in an *ideal* adiabatic calorimeter, the temperature variation would be like that shown in Fig. 1(a). In this case there would be no difficulty in determining the temperature change $\Delta T = T_1 - T_0$ since $(dT/dt) = 0$ before the time of mixing the reactants and after the products achieve thermal equilibrium. The only cause of temperature change here is the chemical reaction. However, it is an unrealistic idealization to assume that step I is truly adiabatic; as no thermal insulation is perfect, some heat will in general leak into or out of the system during the time required for the change in state to occur and for the thermometer to come into equilibrium with the product system.

In addition, we usually have a stirrer present in the calorimeter to aid in the mixing of reactants or for hastening thermal equilibration. The mechanical work done on the system by the stirrer results in the continuous addition of energy to the system at a small, approximately constant rate. During the time required for the change in state and thermal equilibration to occur, the amount of energy introduced can be significant. A typical temperature–time variation is shown in Fig. 1(b), where a greatly expanded temperature scale has been used with an interrupted temperature axis in order to show clearly the pre- and post-reaction temperature changes. In this case, a more detailed analysis is needed in order to extract the adiabatic ΔT value from the observed data points.

Consider Fig. 2, which shows a schematic plot of $T(t)$ associated with step I as carried out in a typical calorimeter. This plot is based on the assumptions that the heat leakage through the calorimeter walls is small and

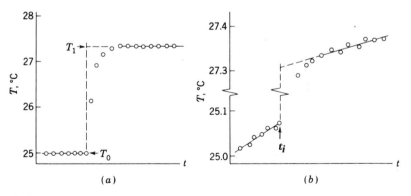

(a)   (b)

FIGURE 1
(a) Plot of temperature T vs. time t for an ideal adiabatic calorimeter. (b) Plot with interrupted temperature axis and greatly expanded temperature scale for a typical calorimeter run. Such a plot is designed to show clearly the temperature variation before and after the time t_i when the reaction is initiated.

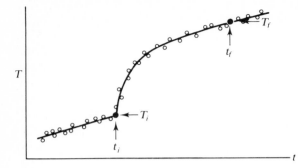

FIGURE 2
Schematic plot of temperature T vs. time t observed in a typical calorimeter experiment. The temperature-time points (T_i, t_i) and (T_f, t_f) can be chosen somewhat arbitrarily as long as t_i is prior to or at t_{react} and t_f is a sufficiently long time after t_{react} (see text).

the energy input due to the stirrer is moderate. These assumptions are appropriate for the thermochemistry experiments described in this book. The initial temperature T_i is conveniently taken to be the value at time t_i when the reaction is initiated following a period of essentially linear T, t readings that serve to establish the initial (pre-reaction) drift rate $(dT/dt)_i$. The temperature is followed after reaction for a period long enough to achieve a roughly linear variation and to establish the drift rate $(dT/dt)_f$ at an arbitrary point T_f, t_f.

The analysis of step I will be given for a constant-volume calorimeter run, but analogous results hold at constant pressure. The effect of heat leakage is evaluated using Newton's law of cooling,

$$\frac{dq}{dt} = -k(T - T_s) \tag{16}$$

where T_s is the temperature of the surroundings (air temperature in the laboratory) and k is a thermal rate constant that depends on the thermal conductivity of the calorimeter insulation. The mechanical power input to the system (dw/dt) will be denoted by P, which is assumed to be a constant independent of t. Thus

$$C_v \frac{dT}{dt} = \frac{dE}{dt} = P - k(T - T_s) \tag{17}$$

or

$$\left(\frac{dT}{dt}\right)_{leak+stir} = \frac{1}{C}[P - k(T - T_s)] \tag{18}$$

where C is C_v for constant-volume reactions and C_p for constant-pressure reactions. The temperature difference $T_f - T_i$ shown in Fig. 2 is given by

$$T_f - T_i = \Delta T + \int_{t_i}^{t_f} \left(\frac{dT}{dt}\right)_{leak+stir} dt \tag{19}$$

where $\Delta T = T_1 - T_0$ is the change due to an adiabatic chemical reaction and the integral is the net change due to heat leak and stirrer power input. Thus it follows that

$$T_1 - T_0 \equiv \Delta T = (T_f - T_i) - \frac{1}{C} \int_{t_i}^{t_f} \{P - k[T(t) - T_s]\} \, dt \qquad (20)$$

The following are the most important experimental approaches aimed at minimizing the effect of nonadiabatic conditions and the effect of the stirrer; they can be used separately or in combination.

1. The calorimeter may be built in such a way as to minimize heat conduction into or out of the system. A vessel with an evacuated jacket (Dewar flask), often with silvered surfaces to minimize the effect of heat radiation, may be used. This corresponds to making the thermal rate constant k very small.

2. One may interpose between system and surroundings an "adiabatic jacket," reasonably well insulated from both. This jacket is so constructed that its temperature can be adjusted at will from outside, as, for example, by supplying electrical energy to a heating circuit. In use, the temperature of the jacket is continuously adjusted so as to be as close as possible to that of the system, so that no significant quantity of heat will tend to flow between the system and the jacket. This corresponds to making $T - T_s$ very small.

3. The stirring system may be carefully designed to produce the least rate of work consistent with the requirement of thorough and reasonably rapid mixing. This corresponds to making P small.

4. We may assume that the rate of gain or loss of energy by the system resulting from heat leakage and stirrer work is reasonably constant with time at any given temperature. We may therefore assume that the temperature variation as a function of time should be initially and finally almost linear.

One simple possibility would be $k(T - T_s) = 0$ and $P = $ constant, in which case $(dT/dt) = P/C$ has the same value at all t and Eq. (19) gives

$$\Delta T = T_f - T_i - \frac{dT}{dt}(t_f - t_i) \qquad \text{if } k(T - T_s) = 0 \qquad (21)$$

As shown in Fig. 3(a), ΔT in this case is just the vertical distance between two parallel T–t lines. Naturally, this distance can be evaluated at any t. A more realistic possibility is that $k(T - T_s)$ is small but not negligible and $P = $ constant. In this case, the T–t plot looks like Fig. 3(b), where the initial leak + stir drift rate $(dT/dt)_i$ at t_i differs from the final rate $(dT/dt)_f$ at t_f. It can be shown[1] in this case that ΔT is given by

$$\Delta T = (T_f - T_i) - \left(\frac{dT}{dt}\right)_i (t_d - t_i) - \left(\frac{dT}{dt}\right)_f (t_f - t_d) \qquad (22)$$

where t_d is chosen so that the two shaded regions in Fig. 3(b) are of equal area. The derivation of Eq. (22) from Eq. (20) requires a bit of sleight-of-hand

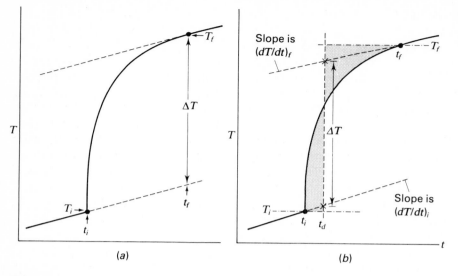

FIGURE 3

Determination of the "adiabatic" temperature change ΔT from experimental $T(t)$ measurements made in non-ideal calorimeters: (a) case of no heat leak but a constant stirring power input, (b) case of a small heat leak as well as stirrer energy input. The value of t_d is chosen so that the two shaded regions have equal areas; see text for details.

manipulation. Let us define a new variable $T_c \equiv T_s + (P/k)$, in terms of which Eq. (18) gives

$$\left(\frac{dT}{dt}\right)_i = -\frac{k}{C}(T_i - T_c) \quad \text{and} \quad \left(\frac{dT}{dt}\right)_f = -\frac{k}{C}(T_f - T_c) \tag{23}$$

and Eq. (20) becomes

$$\Delta T = (T_f - T_i) + \frac{k}{C}\int_{t_i}^{t_f}(T - T_c)\,dt \tag{24}$$

Note that

$$\int_{t_i}^{t_f}(T - T_c)\,dt = \int_{t_i}^{t_d}(T - T_i)\,dt + (T_i - T_c)(t_d - t_i)$$

$$+ \int_{t_d}^{t_f}(T - T_f)\,dt + (T_f - T_c)(t_f - t_d) \tag{25}$$

By the definition given above for t_d, the two integrals in Eq. (25) cancel out. Substituting Eq. (25) into Eq. (24) and making use of Eq. (23) leads directly to Eq. (22). It is clear from this derivation that Eq. (22) will still be valid even if the T variation prior to t_i and after t_f is not linear. In such a case, one merely uses the tangents to the T–t curves at t_i and t_f.

In summary, one should follow the temperature variation before and after reaction for a period long enough to allow a good evaluation of $(dT/dt)_i$ and $(dT/dt)_f$. Then make a plot like Fig. 3(b), choose the point (T_f, t_f), determine the best value for t_d, and calculate ΔT from Eq. (22). In practice, t_f should be chosen as close to t_i as is consistent with a well-characterized final drift rate. Note that t_d will lie much closer to t_i than to t_f for reactions which go to completion quickly. Whenever the time required for a change in state is long, as in the case of determining the heat capacity by electrical heating, the position of t_d will lie near the middle of the t_i to t_f range. As an approximate procedure for handling the analysis of electrical heating plots, one can choose t_i as the time the heater is turned on and t_f as the time it is turned off and choose t_d as the midpoint of this heating period.

In the experiments on heats of combustion we make use of approaches 2 (optionally), 3, and 4. In the experiment on heats of ionic reaction we make use of 1, 3, and 4.

In the above discussion, we have been concerned with determining ΔH or ΔE for a chemical change, with arbitrary amounts of reactants. The quantities in which we are interested are the molar quantities $\Delta \bar{H}$ or $\Delta \bar{E}$ corresponding to one mole of reaction. The number of moles of reaction n can be calculated from the quantity of the limiting reactant. Thus we have

$$\Delta \bar{H} = \frac{\Delta H}{n}, \qquad \Delta \bar{E} = \frac{\Delta E}{n} \tag{26}$$

REFERENCE

1. H. C. Dickinson, *Bull. Natl. Bur. Stand.* **11,** 189 (1914); S. R. Gunn, *J. Chem. Thermodyn.* **3,** 19 (1971).

GENERAL READING

J. M. Sturtevant, "Calorimetry," in A. Weissberger and B. W. Rossiter (eds.), "Techniques of Chemistry: Vol. I. Physical Methods of Chemistry," part V, chap. VII, Wiley-Interscience, New York (1971).

M. L. McGlashan, "Chemical Thermodynamics," Academic, New York (1979).

EXPERIMENT 6
HEATS OF COMBUSTION

In this experiment a bomb calorimeter is employed in the determination of the heat of combustion of an organic substance such as naphthalene, $C_{10}H_8$. The heat of combustion of naphthalene is $-\Delta \bar{H}$ for the reaction at constant temperature and pressure:

$$C_{10}H_8(s) + 12O_2(g) = 10CO_2(g) + 4H_2O(l) \tag{1}$$

THEORY

A general discussion of calorimetric measurements is presented in the section Principles of Calorimetry. That material should be reviewed as background for this experiment. It should be noted that no specification of pressure is made for the reaction in Eq. (1) other than that it is constant. (Indeed, in this experiment the reaction is not actually carried out at constant pressure, though the results are corrected to constant pressure in the calculations.) In fact, energy and enthalpy changes attending physical changes are generally small in comparison with those attending chemical changes, and those involved in pressure changes on condensed phases (owing to their small molal volumes and low compressibilities) and even on gases (owing to their resemblance to perfect gases) at constant temperature are very small. Thus, energy and enthalpy changes accompanying chemical changes can be considered as being independent of pressure for nearly all practical purposes.

EXPERIMENTAL

The calorimeter (Emerson design), shown in Fig. 1, consists of a high-pressure stainless-steel bomb which sits in a calorimeter pail containing 2 L of water. Projecting downward into this pail through a hole in the cover is a motor-driven stirrer and a precision thermometer. All these parts constitute what has been denoted by the letter S in the Principles of Calorimetry. The bomb contains a pair of electrodes to which are attached a short length of fine iron wire in contact with the specimen. Electrical ignition of this wire provides the means of initiating combustion of the sample, which burns in a small metal pan. The bomb is constructed in two parts, which are put together by means of a large ring-nut and a lead gasket. A needle valve at the top is provided for filling the bomb with oxygen to about 300 psi.

The system as shown in Fig. 1 is surrounded by the Daniels adiabatic jacket,[1] which consists of two concentric metal cans with a space between that can be filled with water. Adjustment of the temperature of the jacket is done by passage of electric current between the two cans, by conduction through the water. The water must have a certain degree of electrical conductivity. Often ordinary tap water or a mixture of tap water and distilled water can be used; more predictable performance can be obtained by adding a small quantity of electrolyte to distilled water (0.05 g NaCl/liter of water). The electric current is controlled by a tapping key to maintain the jacket temperature as nearly as possible at the temperature registered by the calorimeter thermometer. The use of this adiabatic jacket should in principle eliminate the need to plot and

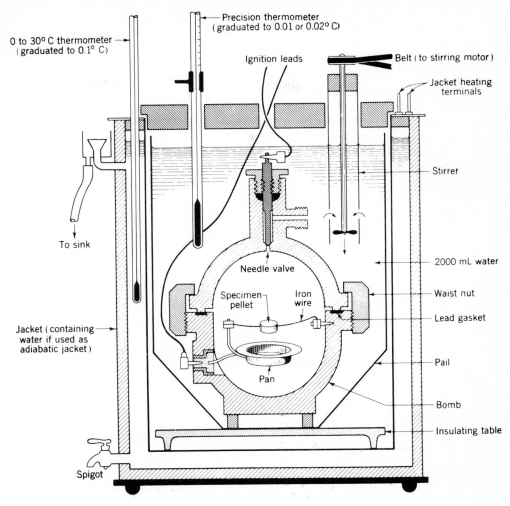

FIGURE 1
Bomb calorimeter (Emerson design), shown with Daniels adiabatic jacket which may also be used empty as an air jacket. The mercury thermometer can be replaced by a high-resolution resistance thermometer.

extrapolate temperature values. However, there are inevitable errors (and some experimental awkwardness) in adjusting the jacket temperature during a run. It is also possible to operate the calorimeter without controlling the temperature of the Daniels adiabatic jacket. In this case the space between the two cans is left empty to provide thermal insulation. The heat leakage that does occur can be compensated for very well by the extrapolation technique illustrated in Fig. V-3. This insulating air-jacket technique is strongly recommended, and a detailed procedure for the use of the adiabatic jacket will not be given.

Calorimeters of the Parr design are also widely used. They are available with insulating air jackets and with adiabatic water jackets in which water is circulated through the cover as well as the body of the jacket. The temperature of the adiabatic jacket can be adjusted manually by varying the flow through hot and cold water inlets. Detailed instructions for the operation of the Parr adiabatic jacket are available elsewhere,[2] but it is recommended that an air-jacket technique be used. The principal advantage of the Parr calorimeter lies in the design of the self-sealing nickel-alloy bomb. This bomb, shown in Fig. 2, has a head that is precision-machined to fit into the top of the cylindrical body of the bomb. Closure is achieved by compressing a rubber ring that seals against the cylinder wall. No bench socket and long-handled wrench are necessary to seal the bomb. The retaining cap is merely screwed down hand-tight; on pressurizing the bomb, a seal develops automatically. Releasing the pressure will break the seal and allow the cap to be unscrewed manually. The head shown in Fig. 2 has a single spring-loaded check valve that closes

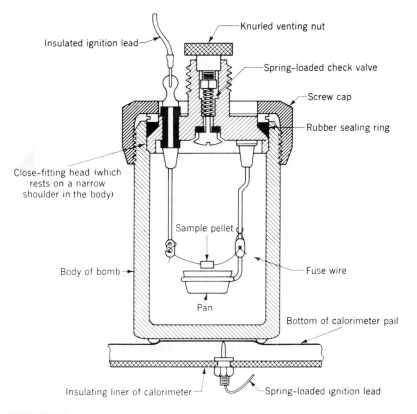

FIGURE 2

Parr single-valve bomb, shown in contact with bottom of calorimeter pail. One electrical contact is made automatically through the pail to the body of the bomb.

automatically when the bomb is pressurized. To vent this type of bomb, the knurled nut on the top of the valve is unscrewed one turn (to reduce the spring tension) and pressed down. A double-valve head, which has a check valve for filling and a needle valve for venting the bomb, is also available. Another difference between the Emerson and Parr designs involves the electrical lead connections to the bomb. In the Parr bomb, one lead goes directly to an insulated terminal and the other lead makes contact automatically through the oval calorimeter pail that holds the bomb.

Following the procedure given below, two runs should be made with benzoic acid to determine the heat capacity of the calorimeter and two runs should be made with naphthalene or another substance (perhaps an unknown).

Procedure. The successful operation of this experiment requires close attention to detail, as there are possible sources of trouble that may prevent the experiment from working properly. It is for this reason, and because laboratory scheduling usually does not allow the student to work out his own technique by trial and error, that the procedure is presented in considerable detail. The basic procedure is very similar for both the Emerson and the Parr calorimeters. The text below is written for the Emerson calorimeter. *Changes required when using the Parr calorimeter are given in square brackets.* Further details on the operation of the Parr calorimeter are available in a manual published by the manufacturer.[2]

The student's attention is particularly drawn to the following **warning**: There is a hazard of electric shock and of short circuit from exposed terminals and metal parts of the calorimeter, especially when outer-jacket heating is employed with the Emerson calorimeter. Keep the working space dry. The bomb is expensive and should be handled carefully. In particular, be very careful not to scratch or dent the closure surfaces. When the bomb is dismantled its various parts should be placed gently on a clean, folded towel.

Condition of apparatus. The bomb must be clean and dry with no bits of iron wire in the terminals. Make sure that the jacket is completely empty and that all switches on the control box are off. Check to see that the inside of the calorimeter is dry. By testing in a beaker of water, verify that the strirrer is operating so as to impel the water *downward* [not important for the Parr].

Filling of the bomb. Cut the iron wire, free from sharp bends or kinks, to the correct length and weigh it accurately. Press pellets of the substance concerned, of weight 0.5 ± 0.1 g for naphthalene and 0.8 ± 0.1 g for benzoic acid. Shave them to the desired weight with a knife or spatula if necessary. Fuse the wire into the pellet by heating the wire with current provided by a 1.5-V dry cell. Blow gently on the wire to prevent it from overheating. The pellet should be at the center of the wire. Weight it accurately. The weight of the pellet alone is obtained by difference. Handle it very carefully after weighing.

Install the pellet and wire in the bomb. The pellet should be over the pan, and the wire should touch *only* the terminals. Carefully assemble the bomb and tighten the waist nut with a wrench. [Screw down the retaining cap hand-tight for the Parr; *never* use a wrench.]

Attach the bomb to the oxygen-filling apparatus. [The knurled nut on the check valve of the Parr must be removed first.] Line up the fitting carefully before tightening, as misalignment may result in damaged threads. Open the needle valve *one turn* only. *Carefully* open the supply valve, fill the bomb to 300 psi, then close the needle valve hand-tight only and remove the bomb. [The Parr has an automatic check valve for filling; admit oxygen *slowly* until the pressure reaches 400 psi.] Release the pressure to flush out most of the atmospheric nitrogen originally present in the bomb. Retighten and refill to the same pressure. If trouble is encountered in filling the bomb, check the condition of the gasket on the filling apparatus. Check the bomb for leaks by immersion in water. If a leak is found around the waist, retighten the waist nut with the wrench and try again. [Leaks are less likely to occur with the Parr. If a leak does occur, vent the bomb, loosen and rotate the head slightly, and then retighten the screw cap.] An occasional bubble—one every 5 or 10 seconds—is inconsequential.

Assembly of calorimeter. Dry the bomb. Make electrical connections *tightly* to the top and side; the side connector *must not touch the waist ring nut.* Place the bomb in the dry pail. Check to see that the side connector does not touch the pail. Set the pail in the calorimeter, making sure that it is centered and does not touch the inner wall of the calorimeter. [For the Parr, the bomb and the pail must be in position in the calorimeter before the single connection to the top of the bomb can be made.]

Fill a 2-L volumetric flask with water at 25°C. A convenient way of doing this is to use both hot and cold water; add the two as required, swirling and checking with a thermometer until the flask is almost full, then make up to the mark. Pour the water from the flask carefully into the pail; avoid splashing. Allow the flask to drain a minute.

Put the calorimeter lid in place and introduce the stirrer as far as it will go. Clamp the precision thermometer in place as low as it will go without obscuring the scale. The clamp should be as low as possible and *not too tight.* The jaws of the clamp should be rubber-covered to protect the thermometer. The thermometer should register within half a degree of 25°C.

If a test meter or an improvised tester consisting of a flashlight bulb and a dry cell is available, it is advisable to check the electrical continuity between the two prongs of the plug on the ignition cord. With all switches off, plug the stirrer and ignition cords into the control box (see Fig. 3). Recheck the wiring and then plug the control box into a 110-V ac outlet. Turn on the stirrer and make sure that it runs smoothly. [The Parr stirrer motor operates directly on 110-V, 60-cycle ac; no series resistor such as that on the control box is required.]

Instead of the ignition circuit shown in the control box of Fig. 3, a low-voltage transformer can be used. Transformer ignition circuits are avail-

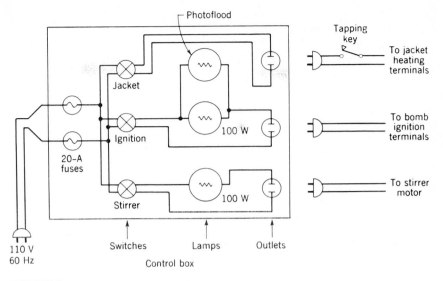

FIGURE 3
Electrical circuit for Emerson bomb calorimeter.

able commercially as accessories from calorimeter manufacturers or can be easily constructed. A suitable design is shown in Fig. 4.

Making the run. Begin time–temperature readings, reading the precision thermometer once every 30 s and recording both the time and the temperature. Estimate the temperature to thousandths of a degree if feasible. Tap the thermometer *gently* before each reading. *Do not interrupt the time–temperature readings until the run is over.* The calorimeter pail temperature should change at a very slow linear rate (of the order of 0.001 K min^{-1}). After this steady rate has persisted for at least 5 min, the bomb may be ignited.

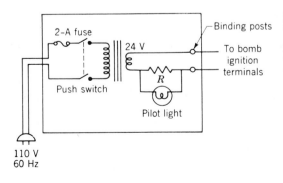

FIGURE 4
Transformer ignition circuit for bomb calorimeter. The switch is a momentary-contact push switch which is only on when it is held down. Release the switch if ignition does not occur after about 5 s (the pilot light will indicate when the fuse wire burns through). The choice of resistor R will depend on the type of calorimeter and the length of fuse wire to be used (a current of 3 to 4 A is desired).

To ignite, turn the ignition switch on and then immediately off. Record the exact time. A dull brief flash may be observable in the light bulbs, indicating the passage of current for the instant required to burn the wire through. In nearly all such cases the burning wire will ignite the pellet, and after a period of 10 or 15 s the temperature will begin to rise. After a few minutes the pail temperature should again show a slow, steady rate of change. Do not discontinue readings; continue them until the time since ignition has been at least four times the period required for attainment of this steady rate. The purpose of this is to provide a valid basis for extrapolation. When readings are completed, turn all switches off.

Disassemble the apparatus, release the bomb pressure, and open the bomb. Remove and weigh any unburned iron wire; ignore "globules" unless attempts to crush them reveal that they are fused metal rather than oxide. Subtract the weight of unburned iron wire from the initial iron-wire weight to obtain the net weight of iron burned. If the inside of the bomb is found to be coated with soot, the amount of oxygen present at the time of ignition was presumably insufficient to give complete combustion and the run should be discarded. Wipe dry all bomb parts.

CALCULATIONS

For each run, plot temperature versus time using an expanded, interrupted temperature scale as shown in Fig. V-1(b) and determine the initial and final drift rates $(dT/dt)_i$ and $(dT/dt)_f$. Then make an overall T versus t plot like Fig. V-3(b), choose t_f, and determine t_i and t_d. Finally, the "adiabatic" temperature change $\Delta T \equiv T_1 - T_0$ can be calculated from Eq. (V-22). This determination of ΔT could be made directly from the overall T versus t plot as shown in Fig. V-3(b), but the procedure described above provides greater precision.

The heat capacity $C(S)$ is found by determining the adiabatic temperature rise $(T_2' - T_1')$ obtained in the combustion of a known weight of benzoic acid (and, of course, a known weight of iron wire) and by making use of Eq. (V-15b). For calculating the energy change produced, the specific (i.e., per gram) energies of combustion of benzoic acid (BA) and iron wire (Fe) given below can be used:[3]

$$\Delta \bar{E}_{BA} = -26.42 \text{ kJ g}^{-1}$$
$$\Delta \bar{E}_{Fe} = -6.68 \text{ kJ g}^{-1} \tag{2}$$

In this experiment ΔE for the combustion of a weighted specimen of naphthalene or other substance is determined from the rise in temperature $(T_1 - T_0)$ and the heat capacity $C(S)$ by use of Eq. (V-14b). The value of ΔE obtained includes a contribution for the combustion of the iron wire; this must be subtracted to yield the contribution from the naphthalene alone. The molar energy change $\Delta \bar{E}$ is then obtained from the number of moles of the reactant that is present in the least equivalent amount (in this case, naphthalene) in accordance with Eq. (V-26). The molar enthalpy change $\Delta\bar{H}$ can then be obtained by use of Eq. (V-8).

Report the individual and average values of the heat capacity $C(S)$ and the individual and average values of the molar enthalpy change $\Delta\bar{H}$ for the combustion of naphthalene, calculated using the average $C(S)$.

DISCUSSION

How does the order of magnitude of the error introduced into the experimental result by the assumption of the perfect-gas law in Eq. (V-8) compare with the uncertainties inherent in the measurements in this experiment? What is the magnitude of the uncertainty introduced by lack of knowledge of the specific heat of the sample? Does your ΔH value pertain to the initial or the final temperature?

APPARATUS

Bomb calorimeter complete with the bomb, pail, jacket, insulating table, cover, and stirring motor with power cord; one set of electrical leads which connect to bomb; 110-V power source unit, equipped with fuse and bulbs; precision calorimeter thermometer covering range from 19°C to 35°C or a resistance thermometer with ±0.005°C resolution; magnifying thermometer-reader; 0 to 30°C thermometer; stopwatch; 2-L volumetric flask; 1.5-V dry cell; 500-mL beaker. If adiabatic jacket is to be used: tapping key and set of leads for jacket heating; water of proper conductivity for filling jacket.

Bomb-filling apparatus on table securely bolted to the floor; long-handled wrench for closing and opening bomb; cylinder of oxygen with appropriate fittings; large pail for leak-testing bomb; benzoic acid (5 g); naphthalene (5 g) or other solid to be studied; pellet press; spatula; 0.004-inch-diameter iron wire (50 cm); device for checking electrical continuity (optional).

REFERENCES

1. F. Daniels, *J. Amer. Chem. Soc.* **38,** 1473 (1916); T. W. Richards, *J. Amer. Chem. Soc.* **31,** 1275 (1909).
2. "Oxygen Bomb Calorimetry and Combustion Methods," Tech. Manual 130, Parr Instrument Co., Moline, Ill. (1960).
3. "Handbook of Chemistry and Physics," 68th ed., CRC Press, Boca Raton, Fla. (1987/8).

GENERAL READING

J. M. Sturtevant, "Calorimetry", in A. Weissberger and B. W. Rossiter (eds.), "Techniques of Chemistry: Vol. I. Physical Methods of Chemistry," part V, chap. VII, Wiley-Interscience, New York (1971).

EXPERIMENT 7

STRAIN ENERGY OF THE CYCLOPROPANE RING

The strain energy of the cyclopropane ring is determined from thermochemical measurements on two closely related compounds: the n-butyl ester of cyclopropanecarboxylic acid and the methyl ester of cyclohexanecarboxylic acid. Both compounds can be synthesized from the appropriate acid chlorides, and their heats of combustion are to be measured with a bomb calorimeter.

THEORY

The molar enthalpy change $\Delta\bar{H}_d$ for dissociation of an organic compound into separated atoms in their gaseous state

$$C_mH_nO_pN_q \cdots = mC(g) + nH(g) + pO(g) + qN(g) + \cdots \tag{1}$$

is determined principally by the number and kinds of chemical bonds in the molecule. In many cases, the heat of dissociation $\Delta\bar{H}_d$ can be closely approximated by the sum of so-called "bond energies" (more properly, bond enthalpies) B_i for all the bonds in the molecule:†

$$\Delta\bar{H}_d \cong \sum_i B_i \tag{2}$$

For all conventional chemical bonds, the B_i values are positive; $\Delta\bar{H}_d$ must have a positive sign for a stable molecule.

Certain molecules are made less stable by the presence of a *strain energy* **S** corresponding to the bending or stretching of bonds from their normal state as a result of geometrical requirements, or are made more stable by the presence of a *resonance energy* **R** corresponding to aromatic character, conjugation, hyperconjugation, etc.:

$$\Delta\bar{H}_d = \sum_i B_i - \mathbf{S} + \mathbf{R} \tag{3}$$

The quantities **S** and **R** are defined so as to be positive in sign. We shall be concerned here with the strain energy of the cyclopropane ring, in which the C—C—C bond angles are constrained by geometry to be 60° rather than their normal values which are in the neighborhood of the tetrahedral angle 109.5°.

Consider the compounds n-butylcyclopropanecarboxylate (I) and methylcyclohexanecarboxylate (II), which have the same molecular composi-

---

† Strictly speaking, a heat of dissociation calculated from bond energies applies to a compound in the vapor state, but the heat of vaporization from the liquid or solid state (typically 20 to 50 kJ mol⁻¹) is ordinarily smaller in magnitude than the errors inherent in the bond-energy treatment.

tion $C_8H_{14}O_2$:

I   II

The quantity ΣB_i should have the same value for both compounds, since each
has 7 C—C, 1 C=O, 2 C—O, and 14 C—H bonds. However, molecule I
possesses appreciable strain energy while molecule II does not. (The cyclo-
hexyl ring is staggered, so that the angles are close to $109.5°$.) Neither
molecule would be expected to have appreciable resonance energy; in any
case, resonance effects would be about the same for both molecules and would
cancel out. Hence

$$S(\text{I}) = \Delta \bar{H}_d(\text{II}) - \Delta \bar{H}_d(\text{I}) \tag{4}$$

The quantity $-\Delta \bar{H}_d$ is closely related to, but should be carefully
distinguished from, the heat of formation of a compound, which is the molar
enthalpy change resulting from the formation of the compound from the
elements *in their standard states* at the given temperature and pressure (usually
298 K, 1 bar). The standard state of carbon at 298 K, 1 bar is graphite, $C(s)$;
those of hydrogen, oxygen, and nitrogen are the gases $H_2(g)$, $O_2(g)$, and
$N_2(g)$. $\Delta \bar{H}_f$ for the change in state

$$mC(s) + \tfrac{1}{2}nH_2(g) + \tfrac{1}{2}pO_2(g) + \tfrac{1}{2}qN_2(g) \rightarrow C_mH_nO_pN_q \qquad (298 \text{ K}, 1 \text{ bar}) \tag{5}$$

is related to $\Delta \bar{H}_d$ by

$$-\Delta \bar{H}_f = \Delta \bar{H}_d - m\, \Delta \bar{H}_d^0[C(s)] - \tfrac{1}{2}n\, \Delta \bar{H}_d^0[H_2(g)] - \tfrac{1}{2}p\, \Delta \bar{H}_d^0[O_2(g)]$$
$$- \tfrac{1}{2}q\, \Delta \bar{H}_d^0[N_2(g)]$$
$$= \Delta \bar{H}_d - 716.68m - 217.97n - 249.17p - 472.70q \tag{6}$$

where the numerical values are given in kJ mol^{-1}. The quantities $\Delta \bar{H}_d^0$ are the
heats of dissociation of the elements *in their standard states* into separated
gaseous atoms; their values have been obtained from a variety of calorimetric
and spectroscopic measurements.[1] The sign of $\Delta \bar{H}_f$ is usually, but not always,
negative.

Since the change in state (5) will seldom take place as a single reaction in
a laboratory calorimeter, the heat of formation is not measured directly.
However, by Hess's law, the desired enthalpy change can be obtained by a
suitable algebraic combination of the enthalpy changes for other changes in
state corresponding to actual reactions. As an example of the kind of reaction
that is useful for this purpose, we consider combustion with high-pressure
oxygen gas:

$$C_mH_nO_p(s \text{ or } l) + (m + \tfrac{1}{4}n - \tfrac{1}{2}p)O_2(g) \rightarrow mCO_2(g) + \tfrac{1}{2}nH_2O(l)$$
$$(298 \text{ K}, 1 \text{ bar}) \tag{7}$$

We will here restrict ourselves to the combustion of solids and liquids; combustion of gases requires a very different experimental technique. Moreover, we will not consider compounds containing elements other that C, H, and O, since the combustion products of compounds containing other elements may be somewhat variable; nitrogen, for example, may appear in the product as $N_2(g)$, oxides of nitrogen, or nitric acid. Note that in Eq. (7) the products are gaseous CO_2 and liquid water. The heat of formation for the organic compound $C_m H_n O_p$ is related to the heat of combustion $\Delta \bar{H}_c$, which is the enthalpy change for the change in state (7), by

$$-\Delta \bar{H}_f = \Delta \bar{H}_c + 395.51m + 142.92n \qquad (8)$$

where, again, all quantities are in kJ mol^{-1} at 298 K, 1 bar. The sign of $\Delta \bar{H}_c$ is, for all practical purposes, invariably negative. The second and third terms on the right arise from the heats of combustion of graphite and gaseous hydrogen. The heat of dissociation of the compound is then given by

$$\Delta \bar{H}_d = \Delta \bar{H}_c + 1112.19m + 360.89n + 249.17p \qquad (9)$$

From Eqs. (4) and (9) we see that

$$\mathbf{S}(I) = \Delta \bar{H}_c(II) - \Delta \bar{H}_c(I) \qquad (10)$$

all other terms having canceled out.

Equation (9) may be used to calculate $\Delta \bar{H}_d$ from the experimentally determined heat of combustion of compound II for comparison with the sum of tabulated bond energies [see Eq. (2)]. A short list of bond energies[2] is given in Table 1. These bond energies are mostly averages of values derived from heats of formation of several compounds. The values for double and triple bonds are subject to greater variability than those for single bonds and should be used with some caution. Values of $\Delta \bar{H}_d$ calculated with bond energies are typically in error by 1 or 2 percent in favorable cases and by as much as 5 percent (or even more) in bad cases. It should be no surprise if the value obtained from Eq. (2) for compound II (which is negligibly affected by strain) differs from that calculated with Eq. (9) by an amount that exceeds the small difference between the enthalpies of I and II due to strain. However, it may be expected that since molecules I and II are very similar in all respects other than strain, the errors inherent in the application of Eq. (3) and Table 1 will be essentially the same for both molecules. Accordingly we may expect that Eq. (10) will give a fairly good measure of the strain energy $\mathbf{S}$ in molecule I.

It must be kept in mind that the strain energy $\mathbf{S}$ represents a small difference between two large quantities, $\Delta \bar{H}_c(I)$ and $\Delta \bar{H}_c(II)$, obtained from measurements on two different substances; the difference is only of the order of a few percent. Thus it is clear that all physical measurements involved in the calorimetry must be carried out with the highest possible precision (one or two parts per thousand) and also that the two different substances must be highly pure. The principal impurity to be feared in compound I is *n*-butanol, which boils only about 45°C lower than compound I at a pressure of 90 Torr. The

TABLE 1
Selected bond enthalpies at
298 K (in kJ mol^{-1})

Bond	B	Bond	B
C—C	348	C=C	612
C—H	413	C=O	733
C—O	356	C=N	613
C—N	299	O=O	497
C—Cl	332	N=N	413
C—Br	276		
C—I	238	C≡C	820
O—H	463	C≡N	890
N—H	388	N≡N	945
H—H	436		

replacement of a small percentage of compound I by an impurity such as butanol will produce a much smaller percentage error in $\Delta\bar{H}_c$; the error will depend on the difference between the heats of combustion per gram of the two substances. Water, which has no heat of combustion, will cause a percentage error in $\Delta\bar{H}_c$ equal to its percentage as an impurity.

EXPERIMENTAL

Since the two esters I and II are not available commercially, they will be synthesized by a procedure that is basically that of Jeffery and Vogel.[3] The starting materials for the syntheses are the free acids, cyclopropanecarboxylic acid and cyclohexanecarboxylic acid. Each is first converted to the acid chloride with thionyl chloride, and the acid chloride is then allowed to react with the appropriate alcohol to form the ester. Alternatively, the acid chlorides can be obtained commercially from the Aldrich Chemical Co., Inc., Milwaukee, Wisconsin, and the first part of the syntheses can be dropped. The purity of the two esters is checked by refractometry, infrared spectroscopy, and vapor-phase chromatography, and the heats of combustion are determined with a bomb calorimeter. It is suggested that each student in the team synthesize and check the purity of one of the esters and that both work together on the calorimetric measurements.

Synthesis. In following the directions given below, one should adhere closely to the amounts of material used in each step. The yields given for the two esters are those obtained routinely and correspond to a competent execution of each preparation. Substantially lower yields may require repetition of the preparation in order to obtain sufficient material for the calorimetry work.

Both HCl and SO$_2$ are evolved in parts of this experiment and all operations where either of these gases are evolved must be carried out in a hood.†

Preparation of acid chlorides

Cyclopropanecarboxylic acid chloride is prepared according to the reaction

$$\triangleright\!\!-\!\!\overset{\overset{\displaystyle O}{\|}}{C}\!-\!OH + SOCl_2 \longrightarrow \triangleright\!\!-\!\!\overset{\overset{\displaystyle O}{\|}}{C}\!-\!Cl + HCl(g) + SO_2(g) \qquad (11)$$

Assemble in a hood an apparatus consisting of a single-neck 50-mL round-bottom flask, a Claissen adapter, a vertical water-cooled condenser for reflux (joined to the main stem of the adapter), and a dropping funnel (joined at the side arm of the adapter). The flask should contain a small Teflon-covered stirring bar and should sit in a heating mantle on top of a magnetic stirring unit.

Allow 13.1 g (0.11 mol) of freshly distilled or reagent-grade thionyl chloride to run into the flask from the dropping funnel; close the stopcock as soon as the thionyl chloride has run into the flask. Start the magnetic stirrer and the flow of water through the condenser. Heat the flask until the thionyl chloride has begun to reflux. Add 6.25 g (0.073 mol) of cyclopropanecarboxylic acid dropwise from the dropping funnel to the flask. As soon as an evolution of gas more vigorous than that due to the refluxing thionyl chloride becomes apparent, turn off the external source of heat. The rate of addition of acid should be adjusted so that constant evolution of gas is obtained—the addition usually requires about 15 min. After the addition of acid has been completed, close the stopcock and allow the system to stir at room temperature until the evolution of gas has slowed considerably. Reflux the mixture for 20 min. Then allow the system to cool and rearrange the apparatus to a conventional apparatus for simple distillation. In carrying out this rearrangement, close each entry to the reaction flask as each piece of equipment (dropping funnel, condenser) is removed from the flask since the acid chloride decomposes slowly in contact with moist air. The small amount of residual thionyl chloride (bp 79°C/760 Torr) is then distilled from the acid chloride. At this point, one may continue and isolate the acid chloride by distillation (bp 119°C/760 Torr), or one may use the undistilled residue in the next step of the synthesis. The final yields of ester are comparable from both routes.

The procedure for preparing cyclohexanecarboxylic acid chloride is exactly the same as that described above for the preparation of cyclopropane-carboxylic acid chloride. In this case 11.9 g (0.10 mol) of freshly distilled or reagent-grade thionyl chloride and 7.66 g (0.06 mol) of cyclohexanecarboxylic acid are used. Cyclohexanecarboxylic acid is a solid melting at 31°C but is

---

† Reactions and distillations carried out in a hood are subject to strong drafts of air and should be protected. In particular, distillation apparatus should be carefully wrapped in glass wool and aluminum foil or both, to prevent excessive and uneven cooling.

easier to handle as a liquid. As an effective and safe (under normal precautions) method to melt the acid for weighing and addition to the thionyl chloride, apply steam to the exterior of the bottle or the dropping funnel. As with the previous synthesis, one may distill this acid chloride (bp 125°C/760 Torr), or one may use it without distillation in the next step.

Preparation of n-butyl cyclopropanecarboxylate

$$\underset{\text{S}}{\triangleright}\!\!-\!\!\overset{\overset{\displaystyle O}{\|}}{C}\!\!-\!\!Cl + n\text{-}C_4H_9OH \longrightarrow \underset{\text{S}}{\triangleright}\!\!-\!\!\overset{\overset{\displaystyle O}{\|}}{C}\!\!-\!\!OC_4H_9 + HCl(g) \qquad (12)$$

Assemble in a hood the same apparatus as before, except that the condenser is replaced with a $CaCl_2$ drying tube, and the heating mantle is omitted. Start the magnetic stirrer after 5.40 g (0.073 mol) of anhydrous n-butanol has been placed in the 50-mL flask. Pour the cyclopropanecarboxylic acid chloride into the dropping funnel, and stopper the dropping funnel with another $CaCl_2$ drying tube. [If commercially prepared acid chloride is used, add 7.63 g (0.073 mol).] Since the reaction is exothermic, before allowing the acid chloride to react with the alcohol *prepare an ice bath* for use in case heat is evolved too rapidly. Add the acid chloride dropwise at a rate that maintains a vigorous evolution of HCl gas; the addition will usually take about 20 min. After the addition period, the ester reaction mixture is stirred for 40 min at room temperature.

After the solution is cool, add about 15 mL of ethyl ether, pour the mixture *carefully* into about 15 mL of water (**caution**: the reaction is exothermic), and mix well. Separate the aqueous layer, extract it twice with 15-mL portions of ether. Wash the combined layer twice with 15 mL portions of saturated aqueous $NaHCO_3$ solution. Extract the washings twice with 15-mL portions of ether; in each case, use the ether layer from the extraction of the first portion of $NaHCO_3$ solution to extract the second. Dry the combined organic layers for 20 to 30 min with $MgSO_4$, swirling to hasten equilibrium. Filter into a 100-mL round-bottom flask, attach a Vigreux column, and drive off the ether with a steam bath (in a hood).

The ester should preferably be distilled at reduced pressure (bp 110 to 112°C/90 Torr) to minimize thermal decomposition. Simple distillation is likely to yield a product with a higher-than-acceptable contamination by butanol, so fractionation is recommended. A Vigreux column should be wrapped with glass wool and aluminum foil for thermal insulation and fitted with an Ace vacuum distilling head having Teflon greaseless stopcocks. This head has a "cold-finger" condenser, rotatable so that the reflux ratio can be adjusted, and has stopcocks to permit "fraction cutting" (change of receiving vials without releasing the vacuum). The stopcock through which the distillate flows is greaseless to avoid contamination of the product.

Before distilling the reduced liquid, transfer it (together with a 5-mL ether washing from the previous flask) to a 25-mL round-bottom flask. The distillation should be carried out with a reflux ratio of at least 4:1. Three fractions should be collected: a low-boiling one containing most of the ether and residual butanol, one boiling above the boiling point of butanol but below the stable boiling point of the ester, and the last containing the ester boiling at constant temperature. The usual yield in the last fraction is 6 to 6.5 g.

Take refractive indexes of the last two fractions (Jeffrey and Vogel[3] report $n_D = 1.42446$; at MIT, we have found $n_D^{25} = 1.4259$), and analyze both fractions by vapor-phase chromatography (VPC).† No impurities should show on the VPC trace when the peak height is about 90 percent full deflection. Decreased attenuation of the VPC may result in trace amounts of impurities being shown. To calibrate for the effect of butanol on the VPC spectrum, run also a sample consisting of $200 \, \mu L$ of the final ester fraction and $2 \, \mu L$ of n-butanol.

The infrared absorption spectrum of a 5 percent solution of the final fraction in carbon tetrachloride should be taken. Verification that the cyclopropyl group did not isomerize to propenyl-2 to any significant extent under the synthesis conditions should be found in the absence of any significant vinyl C—H bending band at $990 \, cm^{-1}$. (Other such bands at 909 and $1320 \, cm^{-1}$ would be more or less obscured by other structural features of the spectrum. The vinyl C–H stretching bands at 3090 and $3030 \, cm^{-1}$ would be obscured by the cyclopropyl C–H stretching band at about $3060 \, cm^{-1}$.) If there is any suspicion of such isomerization, a bromine test may be performed. Dilute about 0.5 g of the ester with 1 or 2 mL of CCl_4 and add one drop of a 5 percent solution of bromine in CCl_4; immediate fading of the bromine color would confirm that unsaturation is present. The presence of any significant quantity of butanol or free carboxylic acid would be indicated by absorption bands at $3625 \, cm^{-1}$ or 2700 to $2500 \, cm^{-1}$, respectively.

Preparation of methyl cyclohexanecarboxylate

$$\langle S \rangle - \overset{\overset{\textstyle O}{\|}}{C} - Cl + CH_3OH \longrightarrow \langle S \rangle - \overset{\overset{\textstyle O}{\|}}{C} - O - CH_3 + HCl(g) \quad (13)$$

The procedure for this synthesis is essentially the same as that for the n-butyl cyclopropanecarboxylate described above. In the present case 3.8 g (0.12 mol) of methanol is used. [If commercially prepared acid chloride is used, add 8.80 g (0.06 mol).] Since methanol boils substantially lower than butanol, simple distillation may be used instead of fractionation, but it is advisable to do it at reduced pressure. A conventional setup with a "pig"-type adapter for fraction

---

† A silicone column (e.g., 10 percent 550 Dow-Corning Silicone fluid on Chrom P) at 100°C and a flow rate of $60 \, mL \, s^{-1}$ is sufficient to show any significant impurities, although tailing of the ester peaks generally occurs. A Carbowax 20M column gives less tailing under the same conditions.

cutting may be used. As with the other ester, the liquid to be distilled should be transferred, with an ether washing, to a 25-mL round-bottom flask. Distill the ester at about 35 Torr (bp 90—92°C/35 Torr) and collect three fractions: one low boiling, one from 50°C to 90°C at 35 Torr, and the last fraction after the boiling point of the ester is stable. The usual yield in the final fraction is 6.0 to 8.5 g of the ester.

Record the refractive indexes of the last two fractions ($n_D^{25} = 1.4410$), and analyze them by vapor-phase chromatography. Only minor impurity peaks should show on the VPC trace at low attenuation with the ester peak off scale. Obtain an infrared spectrum of a 5 percent solution of the final fraction in carbon tetrachloride. Isomerization of the ring to a terminally unsaturated straight chain is much less likely than with the cyclopropane ring. Check for absorption in the medium-strength $3085 \, \text{cm}^{-1}$ and $3030 \, \text{cm}^{-1}$ vinyl C–H stretching bands; vinyl C–H bending bands would tend to be obscured by other features.

Calorimetry. The procedures for the operation of the bomb calorimeter are, with minor exceptions noted below, those described in detail in Exp. 6. That experiment and the general discussion presented in the section Principles of Calorimetry should be studied carefully.

The heat capacity of the calorimeter is to be determined with duplicate runs on solid benzoic acid, C_6H_5COOH. Two runs are then made on each liquid ester. Any run with unsatisfactory features in the time–temperature plot or indications of incomplete combustion (soot, etc.) must be rejected. If any pair of runs fails to give reasonably concordant results (agreement within 0.5 percent or better), additional runs on the same substance should be made.

For the present experiment the adiabatic jacket is to be used empty as an air jacket. The procedure is experimentally simpler and capable of high accuracy provided that the ambient temperature in the laboratory is reasonably steady.

Benzoic acid. A pellet of benzoic acid, $0.8 \pm 0.1 \, \text{g}$, is accurately weighed and placed in the sample pan in the bomb. Weigh also an 8-cm piece of iron ignition wire. Instead of fusing the wire into the pellet as described in Exp. 6, you may connect the wire to the terminals in such a way that it is held by light "spring action" against the surface of the pellet. In shaping the wire for this purpose avoid kinks and sharp bends and make certain that the wire does not touch the sample pan itself.

Seal the bomb, flush and fill with oxygen, test for leaks, and assemble the calorimeter as described in Exp. 6. Be sure to adjust the temperature of the water in the 2-L volumetric flask to 25°C ± 1°C before introducing it into the calorimeter pail. When all is ready, turn on the stirring motor and begin time–temperature readings. Read the precision thermometer every 30 s, and record both time and temperature. Tap the thermometer gently before each reading, and estimate the temperature to thousandths of a degree. A temperature-time curve on an expanded scale (see Fig. V-1(b)) should be plotted concurrently. The readings should fall on a straight line, with a small positive slope due to energy input from the stirrer. When the readings have shown a straight-line behavior continuously for at least 5 min, ignite the bomb. Record the time of ignition and continue temperature readings until the total elapsed time following ignition is at least four times the period required to achieve a slow, steady rate of change in the upper temperature. Then turn off the control box switches, release the bomb pressure, open the bomb, remove and weight any unburned iron wire. The presence of any soot indicates incomplete combustion, probably owing to insufficient oxygen; a leak is to be suspected. In such a case, the result of the run must be rejected. A run should also be rejected if the time–temperature plot contains any unusual features which make extrapolation uncertain.

Esters. About 0.7 to 0.9 g of one of the esters is introduced into the sample pan from a 2-mL disposable hypodermic syringe. The syringe is filled nearly half full, weighed on an analytical balance, discharged into the sample pan, and reweighed. Before each weighing, withdraw the plunger so as to pull a small quantity of air into the needle. This will prevent loss of liquid through dribbling from the needle tip. Naturally a drop hanging from the tip should not be wiped off at any time between the two weighings. Although the 8-cm iron wire described above may be used, a double wire obtained by attaching the middle of a 16-cm length to one terminal and the two ends to the other terminal may provide better insurance against incomplete combustion. The two wires should be carefully shaped to dip into the liquid without touching each other or the metal of the pan. Care should be taken not to wet the walls of the pan above the meniscus any more than necessary, as liquid clinging to the walls may possibly fail to burn. Handle the bomb very carefully; avoid bumping or tipping. The run is performed as described above for benzoic acid.

CALCULATIONS

Use the plotting and calculation procedures described in Exp. 6 in order to determine the adiabatic temperature change associated with each combustion run. The same extrapolation procedure should be used for both esters, and used in as consistent a manner as possible so that any systematic errors inherent in the procedure will cancel out in the calculation of the strain energy.

Determine $C(S)$, the heat capacity of the calorimeter system, from

$$C(S) = \frac{26.43 w_{BA} + 6.69 w_{Fe}}{\Delta T'} \qquad \text{kJ K}^{-1} \qquad (14)$$

where w_{BA} and w_{Fe} are the weights in grams of benzoic acid and iron wire *burned* (be sure to correct for any unburned wire), and $\Delta T'$ is the extrapolated temperature difference for the benzoic acid run. The value of $C(S)$ should be approximately 10.5 kJ K^{-1}. The molar energy of combustion of each ester is calculated with the equation

$$\Delta \bar{E}_c = -\frac{M}{w}[C(S)\,\Delta T - 6.69 w_{Fe}] \qquad \text{kJ mol}^{-1} \qquad (15)$$

where $\Delta T = T_1 - T_0$ is the extrapolated temperature difference for the adiabatic ester combustion, w is the weight in grams of the ester, and M is the ester molecular mass. It is the change in internal energy that is obtained directly in this experiment, since combustion in a bomb is a constant-volume process. The molar enthalpy change $\Delta \bar{H}_c$ for combustion at constant pressure can be calculated from $\Delta \bar{E}_c$ on the assumption that the gases O_2 and CO_2 obey the ideal-gas law reasonably well and that the enthalpies of all substances concerned are independent of pressure. Both of these assumptions are valid to an accuracy acceptable for our purposes. Thus from Eq. (V-8) we have

$$\Delta \bar{H}_c = \Delta \bar{E}_c + RT\,\Delta \bar{n}_{gas} = \Delta \bar{E}_c - (\tfrac{1}{4}n - \tfrac{1}{2}p)RT \qquad (16)$$

where $\Delta \bar{n}_{gas}$ is the change in the number of moles of gas per mole of solid or liquid $C_mH_nO_p$ compound burned. Another necessary assumption, also valid to an accuracy sufficient for our purposes, is that the effect on $\Delta \bar{H}_c$ of the presence of a small fraction of the product water in the vapor phase rather than the liquid phase is negligible. This may be checked by a simple calculation based on the volume of the bomb interior (about 200 mL), the vapor pressure of water at room temperature (about 25 Torr), and the molar heat of vaporization of water (about 42 kJ mol^{-1}).

The strain energy **S** may be calculated from the heats of combustion for the two esters by use of Eq. (10).

Report the yields, boiling ranges, and refractive indexes of the two esters. If available, the VPC and IR curves should be attached to the report. The identification number of the calorimeter and bomb, the individual and average values of the calorimeter heat capacity $C(S)$ and the heats of combustion $\Delta \bar{H}_c$ for the two esters, and the apparent strain energy for the cyclopropane ring should be reported. An estimate of the uncertainties in $C(S)$, the two $\Delta \bar{H}_c$ values, and **S** should be given.

DISCUSSION

Calculate $\Delta \bar{H}_d$ for compound II with Eq. (9), and compare this value with the sum of bond energies obtained with Eq. (2) and Table 1. Comment on the agreement between these values. Also compare your **S** value with a literature estimate of the strain energy.[4]

APPARATUS

Synthetic work. One 25-, one 50-, and one 100-mL single-neck round-bottom flask; Claissen adapter; water-cooled condenser and rubber hoses; dropping funnel; Vigreux column; Ace vacuum distillation head; 0 to 150°C thermo-meter; magnetic stirring unit; heating mantle; two $CaCl_2$ drying tubes; ice bath; glass wool; aluminum foil.

Refractor; vapor-phase chromatograph; infrared spectrometer; fume hood; steam bath; pump for reduced pressure distillation; thionyl chloride (30 g); cyclopropanecarboxylic acid (7 g) and cyclohexanecarboxylic acid (8 g) *or* cyclopropanecarboxylic acid chloride (8 g) and cyclohexanecarboxylic acid chloride (9 g); n-butanol (6 g); methanol (4 g); ethyl ether (175 mL); saturated aqueous $NaHCO_3$ solution (40 ml); $MgSO_4$; CCl_4.

Calorimetry. See detailed list at the end of Exp. 6. An additional item required here is a 2-mL disposable hypodermic syringe.

REFERENCES

1. L. Pauling, "Nature of the Chemical Bond," 3d ed., pp. 64–107, Cornell University Press, Ithaca, N.Y. (1960).
2. L. Pauling, "General Chemistry," 3d ed., p. 913, Freeman, New York (1970); P. W. Atkins, "Physical Chemistry," 3d ed., table 4.5, Freeman, New York (1986).
3. G. H. Jeffrey and A. I. Vogel, *J. Chem. Soc.* **1948,** 1804 (1948).
4. R. A. Nelson and R. S. Jessup, *J. Res. Natl. Bur. Stand.* **48,** 206 (1952).

GENERAL READING

L. Pauling, "Nature of the Chemical Bond," *loc. cit.*
J. M. Sturtevant, "Calorimetry," in A. Weissberger and B. W. Rossiter (eds.), "Techniques of Chemistry: Vol. I. Physical Methods of Chemistry," part V, chap. VII, Wiley-Interscience, New York (1971).

EXPERIMENT 8
HEATS OF IONIC REACTION

It is desired in this experiment to determine the heat of ionization of water:

$$H_2O(l) = H^+(aq) + OH^-(aq) \qquad \Delta \bar{H}_1 \qquad (1)$$

and the heat of the second ionization of malonic acid ($HOOC—CH_2—COOH$, hereinafter designated H_2R):

$$HR^-(aq) = H^+(aq) + R^{2-}(aq) \qquad \Delta \bar{H}_2 \qquad (2)$$

These will be obtained by studying experimentally with a solution calorimeter the heat of reaction of an aqueous HCl solution with an aqueous NaOH solution, for which the ionic reaction is

$$H^+(aq) + OH^-(aq) = H_2O(l) \qquad \Delta \bar{H}_3 \qquad (3)$$

and that of the reaction of an aqueous NaHR solution with an aqueous NaOH solution,

$$HR^-(aq) + OH^-(aq) = R^{2-}(aq) + H_2O(l) \qquad \Delta\bar{H}_4 \qquad (4)$$

The last two equations can be written in the general form

$$A + B = \text{products} \qquad (5)$$

where A represents the acid ion and B the basic ion.

THEORY

A general discussion of calorimetric measurements is presented in the section Principles of Calorimetry, which should be reviewed in connection with this experiment. We shall not here consider the concentration dependence of these enthalpy changes. Such concentration dependence is generally a small effect, since the heats of dilution involved are usually much smaller than the heats of chemical reaction (indeed they are zero for perfect solutions). Since we are here dealing with solutions of moderate concentration, particularly in the case of the NaOH solution, it may be useful to make parallel determinations of heats of dilution of the solutions concerned by a procedure similar to that described here if time permits.

EXPERIMENTAL

In this experiment 500 mL of solution A, with a precisely known concentration in the neighbourhood of 0.25 M, is reacted with 50 mL of solution B at a concentration sufficient to provide a slight excess over the amount required to react with solution A. The reaction is carried out in the solution calorimeter shown in Fig. 1. The calorimeter is a vacuum Dewar flask containing a motor-driven stirrer, a heating coil of precisely known electrical resistance, and a precision thermometer. Various methods of mixing the two solutions might be employed; here we use an inner vessel having an outlet hole plugged by stopcock grease which can be blown out by applying air pressure at the top.

For determining the heat capacity, a dc electric current of about 1.5 A is passed through a heating coil of ~6 Ω resistance during a known time interval. The magnitude of the current is measured as a function of time by determining the potential difference developed across a standard resistance in series with the coil. The electrical circuit is shown in Fig. 2.

It is recommended that two complete runs be made with HCl and NaOH and two with NaHR and NaOH.

Procedure. It is advisable to carry out a test of the electrical heating procedure in advance of the actual runs. This will also provide a chance to check on the heating coil resistance R_h under actual operating conditions (i.e., when the coil is hot). Place roughly 550 mL of water in the calorimeter and introduce the

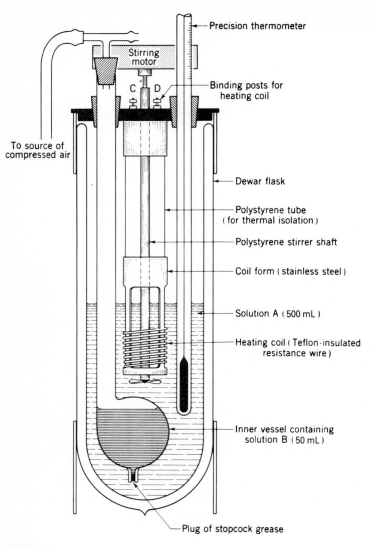

— Precision thermometer

Stirring motor

C D — Binding posts for heating coil

To source of compressed air

— Dewar flask

— Polystyrene tube (for thermal isolation)

— Polystyrene stirrer shaft

— Coil form (stainless steel)

— Solution A (500 mL)

— Heating coil (Teflon-insulated resistance wire)

— Inner vessel containing solution B (50 mL)

— Plug of stopcock grease

FIGURE 1

Solution calorimeter. The mercury thermometer can be replaced by a high-resolution resistance thermometer.

stirrer-heater unit and thermometer. Make the electrical connections as shown in Fig. 2. Turn on the stirring motor and then turn on the current to the heater; **never** pass current through the heating coil when it is not immersed in a liquid, as it will overheat and may burn out. Measure the potential difference V_h across the heating coil by throwing the DPDT switch to the right; i.e., measure the voltage drop between binding posts C and D on the stirrer-heater unit (see Fig. 1). Then measure V_s across the standard resistor in series with

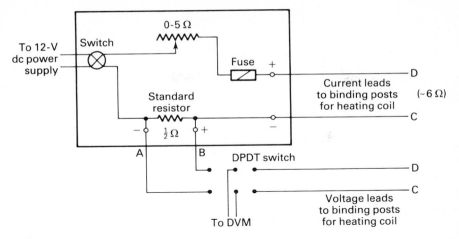

FIGURE 2
Electrical heating circuit for solution calorimeter. The standard resistor should be a wire-wound resistor with a low temperature coefficient and rated for 2 W. If a heating coil of higher resistance (say ~60Ω) is to be used, the current can be reduced to ~0.5 A and a 1-Ω standard resistor can be used.

the heater by throwing the DPDT switch to the left; i.e., measure the voltage drop between A and B on the heater supply unit. The heater resistance can then be calculated from the known value of R_s and the measured ratio V_h/V_s. Also note the rate of temperature rise during heating. All of these observations will be useful during the actual runs.

Voltage measurements are best made with a good-quality digital voltmeter (DVM) having a high impedance. Such a meter, which can be read easily and rapidly, will facilitate measurements during the actual heating runs. Voltage measurements can also be made with a potentiometer; see Chapter XV for a description of both digital multimeters and potentiometers. Either of two alternate procedures can be used to determine the electrical work dissipated inside the calorimeter during a heating run: (a) measure both V_h and V_s as a function of time and use Eq. (V-9a); or (b) measure only V_h as a function of time, assume that R_h is constant with the value determined above, and use Eq. (V-9b). If a DVM is used, procedure (a) is recommended. If a potentiometer is used, procedure (b) is preferable since potentiometer readings are somewhat time consuming and, furthermore, a high-resistance potential divider would be needed in order to measure V_h. In this case, the value of R_h will be provided by the instructor.

Fill a 500-mL volumetric flask with solution A, the temperature of which should be within a few tenths of a degree of 25.0°C. (Adjust the temperature by swirling under running hot or cold water before the flask has been entirely filled, then make up to the mark.) Pour the solution into the clean and reasonably dry calorimeter and allow the flask to drain for a minute. Work a plug of stopcock grease into the capillary hole in the bottom of the inner vessel. The glass must be absolutely dry, or the grease will not stick. Pipette in 50 mL of solution B. Place the inner vessel in the calorimeter carefully.

Introduce the stirrer-heater unit and thermometer into the calorimeter, making sure that the thermometer bulb is completely immersed (level with or just below the heating coil). The inner vessel should be held in a hole in the calorimeter cover by a split stopper so that it does not rest on the bottom of the Dewar flask. Check the electrical connections to the heating coil. Connect the T tube to a compressed-air supply; turn on the air, adjust to barely audible flow, and attach the T tube to the top of the inner vessel. Turn on the stirring motor.

Start temperature–time measurements. Read the thermometer every 30 s, estimating to thousandths of a degree if feasible. Tap the thermometer stem *gently* before each reading. After a slight but steady rate of temperature change due to stirring and heat leak has been observed for 5 min, initiate the reaction by blowing the contents of the inner vessel into the surrounding solution. This is done by placing a finger over the open end of the T tube. Release the pressure as soon as bubbling is heard. Record the time. After 15 s, blow out the inner vessel again to ensure thermal equilibrium throughout all the solution.

After a plateau with a slight, steady rate of temperature change has prevailed for 5 min, turn on the current to the heater and *record the exact time*. Immediately measure the potential difference across the standard resistor in series with the heater. Continue to make voltage measurements approximately every 30 s, alternating between readings of V_s and V_h (or readings of V_s at least once every minute if a potentiometer is being used, in which case the potentiometer should be standardized and set at the expected potential before the current is turned on). Note the time at which each voltage reading is taken.

When a temperature rise of about 1.5°C has been obtained by electrical heating, turn off the current, again noting the exact time. Simultaneously blow out the inner vessel again to mix the contents with the surrounding solution. After a final plateau with a slight, steady rate of temperature change has been achieved for about 7 min, the run may be terminated.

CALCULATIONS

For each run, plot temperature versus time using a greatly expanded temperature scale with an interrupted temperature axis (see Fig. V-1b). Determine the drift rates $(dT/dt)_i$ and $(dT/dt)_f$ before and after the chemical reaction took place and the rate $(dT/dt)_h$ after electrical heating was completed. Carry out appropriate extrapolations to determine the temperature differences $(T_1 - T_0)$ and $(T_2' - T_1')$ shown in Fig. 3. In the case of $(T_1 - T_0)$ associated with the adiabatic reaction, use the method described in the section Principles of Calorimetry and illustrated in Fig. V-3(b). If $(dT/dt)_f$ and $(dT/dt)_h$ have reasonably similar values, one can simplify the determination of

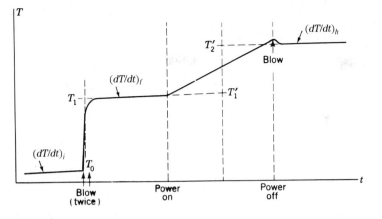

FIGURE 3
Schematic plot of temperature versus time: heat of ionic reaction.

$(T_2' - T_1')$ by choosing t_d as the midpoint of the electrical heating period as illustrated in Fig. 3.

The heat capacity is determined from Eqs. (V-9) and (V-13a). The integral in either Eq. (V-9a) or (V-9b) should be evaluated graphically unless the integrand is essentially constant during the heating period. The enthalpy change of the ionic reaction is calculated from Eq. (V-12a). Calculate the number of moles of reaction from the number of moles of A (the limiting reactant) present and calculate the molar enthalpy change $\Delta \bar{H}_3$ or $\Delta \bar{H}_4$ from Eq. (V-26).

From the average values of ΔH_3 and ΔH_4 obtained above, calculate $\Delta \bar{H}_1$ and $\Delta \bar{H}_2$.

DISCUSSION

Using literature data for the appropriate heats of formation,[1] estimate the difference between the $\Delta \bar{H}_3$ value for the concentrations employed in this experiment and the value which would apply at infinite dilution. Compare this effect with a qualitative estimate of the experimental uncertainty in the measured $\Delta \bar{H}_3$. Indicate why there should be a large percentage error in $\Delta \bar{H}_4$ as determined here.

APPARATUS

Solution calorimeter (1-liter Dewar); stirrer-heater unit complete with motor and power cord; inner vessel; precision calorimeter thermometer covering range from 19 to 35°C or a resistance thermometer with a resolution of ±0.005°C; magnifying thermometer reader; 0 to 30°C thermometer; two split rubber stoppers; glass T tube with stopper and rubber tubing; stopwatch; dc power-supply unit for heater (see Fig. 2); double-pole, double-throw switch; digital multimeter or potentiometer setup (see Chapter XV); 15 electrical leads with lugs attached; 250-mL beaker; 500-mL volumetric flask; a 50- and a 10-mL pipette; 100-mL beaker; rubber pipetting bulb.

Solutions. NaOH, concentration slightly greater than $2.5\,M$ (0.4 liter); $0.25\,M$ HCl (1.5 liters); $0.25\,M$ NaHC$_3$H$_2$O$_4$ (1.5 liters; the sodium acid malonate solution can be prepared by neutralizing malonic acid with sodium hydroxide to a point just past the NaHC$_3$H$_2$O$_4$ end point); stopcock grease; source of compressed air.

REFERENCES

1. D. D. Wagman, *et al.*, "The NBS Tables of Chemical Thermodynamic Properties: Selected Values for Inorganic and C$_1$ and C$_2$ Organic Substances in SI Units," *J. Phys. Chem. Ref. Data* **11**, Suppl. 2 (1982).

CHAPTER
VI

SOLUTIONS

EXPERIMENTS

9. Partial molar volume
10. Cryoscopic determination of molecular weight
11. Freezing-point depression of strong and weak electrolytes
12. Chemical equilibrium in solution

EXPERIMENT 9
PARTIAL MOLAR VOLUME

In this experiment the partial molar volumes of sodium chloride solutions will be calculated as a function of concentration from densities measured with a pycnometer.

THEORY

Most thermodynamic variables fall into two types. Those representing *extensive* properties of a phase are proportional to the amount of the phase under consideration; they are exemplified by the thermodynamic functions V, E, H, S, A, G. Those representing *intensive* properties are independent of the amount of the phase; they include p and T. Variables of both types may be regarded as examples of homogeneous functions of degree l, that is, functions having the property

$$f(kn_1, \ldots, kn_i, \ldots) = k^l f(n_1, \ldots, n_i, \ldots) \tag{1}$$

where n_i represents for our purposes the number of moles of component i in a phase. Extensive variables are functions of degree one and intensive variables are functions of degree zero.

Among intensive variables important in thermodynamics are *partial molar quantities*, defined by the equation

$$\bar{Q}_i = \left(\frac{\partial Q}{\partial n_i}\right)_{p,T,n_{j\neq i}} \tag{2}$$

where Q may be any of the extensive quantities already mentioned. For a phase of one component, partial molar quantities are identical with so-called molar quantities, $\tilde{Q} = Q/n$. For an ideal gaseous or liquid solution, certain partial molar quantities ($\bar{V}_i$, $\bar{E}_i$, $\bar{H}_i$) are equal to the respective molar quantities for the pure components while others ($\bar{S}_i$, $\bar{A}_i$, $\bar{G}_i$) are not. For nonideal solutions, all partial molar quantities differ in general from the corresponding molar quantities, and the differences are frequently of interest.

A property of great usefulness possessed by partial molar quantities derives from Euler's theorem for homogeneous functions, which states that, for a homogeneous function $f(n_1, \ldots, n_i, \ldots)$ of degree l,

$$n_1 \frac{\partial f}{\partial n_1} + n_2 \frac{\partial f}{\partial n_2} + \cdots + n_i \frac{\partial f}{\partial n_i} + \cdots = lf \tag{3}$$

Applied to an extensive thermodynamic variable Q, for which $l = 1$, we see that

$$n_1 \bar{Q}_1 + n_2 \bar{Q}_2 + \cdots + n_i \bar{Q}_i + \cdots = Q \tag{4}$$

Equation (4) leads to an important result. If we form the differential of Q in the usual way,

$$dQ = \frac{\partial Q}{\partial n_1} dn_1 + \cdots + \frac{\partial Q}{\partial n_i} dn_i + \cdots + \frac{\partial Q}{\partial p} dp + \frac{\partial Q}{\partial T} dT$$

and compare it with the differential derived from Eq. (4),

$$dQ = \bar{Q}_1 \, dn_1 + \cdots + \bar{Q}_i \, dn_i + \cdots + n_1 \, d\bar{Q}_1 + \cdots + n_i \, d\bar{Q}_i + \cdots$$

we obtain

$$n_1 \, d\bar{Q}_1 + \cdots + n_i \, d\bar{Q}_i + \cdots - \left(\frac{\partial Q}{\partial p}\right)_{n_i, T} dp - \left(\frac{\partial Q}{\partial T}\right)_{n_i, p} dT = 0 \tag{5}$$

For the important special case of constant pressure and temperaure,

$$n_1 \, d\bar{Q}_1 + \cdots + n_i \, d\bar{Q}_i + \cdots = 0 \quad (\text{const } p \text{ and } T) \tag{6}$$

This equation tells us that changes in partial molar quantities (resulting of necessity from changes in the n_i) are not all independent. For a binary solution one can write

$$\frac{d\bar{Q}_2}{d\bar{Q}_1} = -\frac{X_1}{X_2} \tag{7}$$

where the X_i are *mole fractions,* $X_i = n_i / \sum n_i$. In application to free energy this equation is commonly known as the Gibbs–Duhem equation.

We are concerned in this experiment with the partial molar volume $\bar{V}_i$, which may be thought of as the increase in the volume of an infinite amount of solution (or an amount so large that insignificant concentration change will result) when 1 mole of component i is added. This is by no means necessarily equal to the volume of 1 mole of pure i.

Partial molar volumes are of interest in part through their thermo-dynamic connection with other partial molar quantities such as partial molar Gibbs free energy, known also as chemical potential. An important property of chemical potential is that for any given component it is equal for all phases that are in equilibrium with each other. Consider a system containing a pure solid substance (e.g., NaCl) in equilibrium with the saturated aqueous solution. The chemical potential of the solute is the same in the two phases. Imaging now that the pressure is changed isothermally. Will solute tend to go from one phase to the other, reflecting a change in solubility? For an equilibrium change at constant temperature involving only expansion work, the change in the Gibbs free energy G is given by

$$dG = V\, dp \tag{8}$$

Differentiating with respect to n_2, the number of moles of solute, we obtain

$$d\bar{G}_2 = \bar{V}_2\, dp \tag{9}$$

where the partial molar free energy (chemical potential) and partial molar volume appear. For the change in state

$$NaCl(s) = NaCl(aq)$$

we can write

$$d(\Delta \bar{G}_2) = \Delta \bar{V}_2\, dp$$

or

$$\left[\frac{\partial (\Delta \bar{G}_2)}{\partial p} \right]_T = \Delta \bar{V}_2 \tag{10}$$

Thus if the partial molar volume of solute in aqueous solution is greater than the molar volume of solid solute, an increase in pressure will increase the chemical potential of solute in solution relative to that in the solid phase; solute will then leave the solution phase until a lower, equilibrium solubility is attained. Conversely, if the partial molar volume in the solution is less than that in the solid, the solubility will increase with pressure.

Partial molar volumes, and in particular their deviations from the values expected for ideal solutions, are of considerable interest in connection with the theory of solutions, especially as applied to binary mixtures of liquid components where they are related to heats of mixing and deviations from Raoult's law.

METHOD[1]

We see from Eq. (4) that the total volume of an amount of solution containing 1 kg (55.51 mol) of water and m mole of solute is given by

$$V = n_1 \bar{V}_1 + n_2 \bar{V}_2 = 55.51 \bar{V}_1 + m \bar{V}_2 \tag{11}$$

where the subscripts 1 and 2 refer to solvent and solute, respectively. Let $\bar{V}_1^0$ be the molar volume of pure water ($= 18.016/0.997044 = 18.069$ cm^3 mol^{-1} at 25.00°C). Then we define the *apparent molar volume* ϕ of the solute by the equation

$$V = n_1 \bar{V}_1^0 + n_2 \phi = 55.51 \bar{V}_1^0 + m \phi \tag{12}$$

which can be rearranged to give

$$\phi = \frac{1}{n_2} (V - n_1 \bar{V}_1^0) = \frac{1}{m} (V - 55.51 \bar{V}_1^0) \tag{13}$$

Now

$$V = \frac{1000 + m M_2}{d} \text{ cm}^3 \tag{14}$$

and

$$n_1 \bar{V}_1^0 = \frac{1000}{d_0} \text{ cm}^3 \tag{15}$$

where d is the density of the solution and d_0 is the density of pure solvent, both in units of g cm^{-3}, and M_2 is the solute molecular weight in grams. Substituting Eqs. (14) and (15) into Eq. (13), we obtain

$$\phi = \frac{1}{d} \left(M_2 - \frac{1000}{m} \frac{d - d_0}{d_0} \right) \tag{16}$$

$$= \frac{1}{d} \left(M_2 - \frac{1000}{m} \frac{W - W_0}{W_0 - W_e} \right) \tag{17}$$

In Eq. (17), the directly measured weights of the pycnometer—W_e when empty, W_0 when filled to the mark with pure water, and W when filled to the mark with solution—are used. This equation is preferable to Eq. (16) for calculation of ϕ, as it avoids the necessity of computing the densities to the high precision that would otherwise be necessary in obtaining the small difference $d - d_0$.

Now by the definition of partial molar volumes and by use of Eqs. (11) and (12),

$$\bar{V}_2 = \left(\frac{\partial V}{\partial n_2} \right)_{n_1, T, p} = \phi + n_2 \frac{\partial \phi}{\partial n_2} = \phi + m \frac{d\phi}{dm} \tag{18}$$

Also

$$\text{\Large *} \quad \bar{V}_1 = \frac{1}{n_1} \left(n_1 \bar{V}_1^0 - n_2^2 \frac{\partial \phi}{\partial n_2} \right) = \bar{V}_1^0 - \frac{m^2}{55.51} \frac{d\phi}{dm} \tag{19}$$

We might proceed by plotting ϕ versus m, drawing a smooth curve through the points, and constructing tangents to the curve at the desired concentrations in order to measure the slopes. However, for solutions of simple electrolytes, it has been found that many apparent molar quantities such as ϕ vary linearly with $\sqrt{m}$, even up to moderate concentrations.[2] This behavior is in agreement with the prediction of the Debye–Hückel theory for dilute solutions.[3] Since

$$\frac{d\phi}{dm} = \frac{d\phi}{d\sqrt{m}}\frac{d\sqrt{m}}{dm} = \frac{1}{2\sqrt{m}}\frac{d\phi}{d\sqrt{m}} \tag{20}$$

we obtain from Eqs. (18) and (19)

$$\text{❉}\quad \bar{V}_2 = \phi + \frac{m}{2\sqrt{m}}\frac{d\phi}{d\sqrt{m}} = \phi + \frac{\sqrt{m}}{2}\frac{d\phi}{d\sqrt{m}} = \phi^0 + \frac{3\sqrt{m}}{2}\frac{d\phi}{d\sqrt{m}} \tag{21}$$

$$\text{❉}\quad \bar{V}_1 = \bar{V}_1^0 - \frac{m}{55.51}\left(\frac{\sqrt{m}}{2}\frac{d\phi}{d\sqrt{m}}\right) \tag{22}$$

where ϕ^0 is the apparent molar volume extrapolated to zero concentration. Now one can plot ϕ versus $\sqrt{m}$ and determine the best *straight* line through the points. From the slope $d\phi/d\sqrt{m}$ and the value of ϕ^0, both $\bar{V}_1$ and $\bar{V}_2$ can be obtained.

EXPERIMENTAL

Make up 200 mL of approximately $3.2\,m$ $(3.0\,M)$ NaCl in water. Weigh the salt accurately and use a volumetric flask; then pour the solution into a dry flask. If possible, prepare this solution in advance (since the salt dissolves slowly). Solutions of $\frac{1}{2}$, $\frac{1}{4}$, $\frac{1}{8}$, and $\frac{1}{16}$ of the initial molarity are to be prepared by successive volumetric dilutions; for each dilution pipette 100 mL of solution into a 200-mL volumetric flask and make up to the mark with distilled water.

The pycnometer is rinsed with distilled water and thoroughly dried before each use. Use an aspirator, and rinse and dry by suction; use a few rinses of acetone to expedite drying. The procedure given here is for the Ostwald–Sprengel type pycnometer; the less accurate but more convenient stopper type can be used with a few obvious changes in procedure.† To fill, dip one arm of

---

† The filling of a Weld-type pycnometer must be carried out with great care. The temperature of the laboratory must be below the temperature at which the determination is to be made. Fill the pycnometer body with the liquid, and seat the capillary stopper firmly. Wipe off excess liquid around the tapered joint, cap the pycnometer, and immerse it in the thermostat bath to a level just below the cap. When temperature equilibrium has been reached, remove the cap, wipe off the excess liquid from the capillary tip, being careful not to draw liquid out of the capillary, and remove the pycnometer from the bath. As the pycnometer cools to room temperature, the liquid column in the capillary will descend; be careful not to heat the pycnometer with your hands since this will force liquid out of the capillary. Carefully clean and wipe off the whole pycnometer, including the cap, but not including the tip of the stopper. Cap the pycnometer, place it in the balance, and allow it to stand about 10 minutes before weighing.

the pycnometer into the vessel containing the solution (preferably at a temperature *below* 25°C) and apply suction by mouth with a piece of rubber tubing attached to the other arm. Hang the pycnometer in the thermostat bath (25.0°C) with the main body below the surface but with the arms emerging well above. Allow at least 15 min for equilibration. While the pycnometer is still in the bath, adjust menisci to fiducial marks with the aid of a piece of filter paper. Remove the pycnometer from the bath and quickly but thoroughly dry the outside surface with a towel and filter paper. Weigh the pycnometer, hanging it from the hook above the balance pan.

The pycnometer should be weighed empty and dry (W_e), and also with distilled water in it (W_0), as well as with each of the solutions in it (W). It is advisable to repeat W_e and W_0 as a check, inasmuch as the results of all runs depend upon them.

All weighings are to be done on an analytical balance to the highest possible precision. Record the values of all weights, then apply any weight calibration corrections. It is unnecessary to correct the weights to vacuum readings in this experiment, although for very precise work this must be done.

As an alternative and very convenient procedure, one can use a Cassia volumetric flask instead of a pycnometer. Although the precision of density measurements made with this flask is not as good as that obtainable with the Ostwald–Sprengel pycnometer, it is adequate for the present purposes. The Cassia flask, shown in Fig. 1(c), is a special glass-stoppered volumetric flask with 0.1-mL graduations between 100 mL and 110 mL, calibrated to contain the indicated volume to within ±0.08 mL. Since this flask is 26 cm high and will weigh over 120 g when full, it must be weighed on a top-loading balance. A

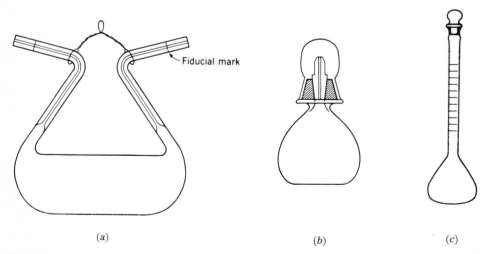

(a) (b) (c)

FIGURE 1

Pycnometers: (a) Ostwald-Sprengel type; (b) Weld (stopper) type; (c) Cassia volumetric flask (not drawn to scale).

high-quality balance of this type, such as the Ohauf balance, capable of weighing to within ± 0.01 g, is required.

The Cassia flash should be weighed empty and dry (W_e) and twice with distilled water in it, once with the level near 100 mL (W_0) and once with the level near 105 mL (W_0'). It should then be weighed with each of the solutions in it (W) to a level just above 100 mL. In every case, record the liquid volume reading to within ± 0.05 mL (V_0, V_0', V for each solution). As with the pycnometers, the Cassia flask must be rinsed well and dried prior to each filling. The filled flask is then thermally equilibrated in a constant-temperature bath (25°C) for at least 15 min. On removing the flask from the bath, dry the outside thoroughly before weighing. Since more solution is required to fill a Cassia flask than a pycnometer, you will need *at least* 225 mL of the 3 M stock solution. Using two 200-mL volumetric flasks and two pipettes (one 50-mL and one 100-mL), you can prepare all the dilutions required.

CALCULATIONS

The success of this experiment depends greatly upon the care with which the computations are carried out. For this reason a simple computer program for analyzing the data is strongly recommended. This computer program may be either written by the student or made available by the instructor. If computer facilities are unavailable, all calculations should be carefully checked. If the work is done by two or more students, the partners are encouraged to work together, performing the same calculations independently and checking results after each step.

Calculate the density d of every solution to within an accuracy of at least one part per thousand:

$$d = \frac{W_{\text{soln}}}{V} = \frac{W - W_e}{V} \tag{23}$$

If a pycnometer was used, determine its volume from $W_0 - W_e$ and the density d_0 of pure water at 25°C (0.997044 g cm^{-3}). If a Cassia flask was used, carry out this volume calculation for both fillings with water and compare your results with the direct volume readings V_0 and V_0'. If necessary, devise a calibration procedure that can be applied to correct the Cassia volume readings obtained on the solutions.

The molalities m (concentration in mole per kg of solvent) that are needed for the calculations can be obtained from the molarities M (concentration in mole per liter of solution) obtained from the volumetric procedures by using the equation

$$m = \frac{1}{1 - (M/d)(M_2/1000)} \frac{M}{d} = \frac{1}{(d/M) - (M_2/1000)} \tag{24}$$

where M_2 is the solute molecular weight (58.45 g mol^{-1}) and d is the experimental density in g cm^{-3} units.

Calculate ϕ for each solution using Eq. (17) for pycnometer data or Eq. (16) for Cassia flask data. Plot ϕ versus $\sqrt{m}$. Determine the slope $d\phi/d\sqrt{m}$ and the intercept ϕ^0 at m equal zero from the best straight line through these data points. This can be done graphically or with a linear least-squares fitting procedure.

Calculate $\bar{V}_2$ and $\bar{V}_1$ for $m = 0$, 0.5, 1.0, 1.5, 2.0, and 2.5. Plot them against m and draw a smooth curve for each of the two quantities.

In your report, present the curves (ϕ versus $\sqrt{m}$, $\bar{V}_2$ and $\bar{V}_1$ versus m) mentioned above. Present also in tabular form the quantities d, M, m, $(1000/m)(W - W_0)/(W_0 - W_e)$, and ϕ for each solution studied. Give the values obtained for the pycnometer volume V_p and ϕ^0 and $d\phi/d\sqrt{m}$.

DISCUSSION

The density of $NaCl(s)$ is 2.165 g cm^{-3} at 25°C. How will the solubility of NaCl in water be affected by an increase in pressure?

Discuss qualitatively whether the curves of $\bar{V}_1$ and $\bar{V}_2$ versus m behave in accord with Eq. (7).

APPARATUS

Pycnometer (approximately 70 mL) with wire loop for hanging in bath or Cassia flask; one or two 200-mL volumetric flasks; 100-mL pipette, and 50-mL pipette if a Cassia flask is used; pipetting bulb; 250-mL Erlenmeyer flask; one 250- and one 100-mL beaker; large weighing bottle; short-stem funnel; spatula; filter paper and gum-rubber tube (1 to 2 ft long) if an Ostwald–Sprengel pycnometer is used.

Constant-temperature bath set at 25°C; bath hardware for holding flasks and pycnometer; reagent-grade sodium chloride (35 g of solid or 200 mL of solution of an accurately known concentration, 50 g or 285 mL if a Cassia flask is used); acetone to be used for rinsing; cleaning solution.

REFERENCES

1. F. T. Gucker, Jr., *J. Phys. Chem.* **38,** 307 (1934).
2. D. O. Masson, *Phil. Mag.* **8,** 218 (1929).
3. O. Redlich and P. Rosenfeld, *Z. Phys. Chem.* **A155,** 65 (1931).

GENERAL READING

G. N. Lewis and M. Randall (revised by K. S. Pitzer and L. Brewer), "Thermodynamics," 2d ed., chap. 17, McGraw-Hill, New York (1961).

N. Bauer and S. Z. Lewin, "Determination of Density", in A. Weissberger and B. W. Rossiter (eds.), "Techniques of Chemistry: Vol. I. Physical Methods of Chemistry," part IV, chap. 2, Wiley-Interscience, New York (1972).

EXPERIMENT 10
CRYOSCOPIC DETERMINATION OF MOLECULAR WEIGHT

When a substance is dissolved in a given liquid solvent, the freezing point is nearly always lowered. This phenomenon constitutes a so-called "colligative" property of the substance—a property that depends in its magnitude primarily on the number of moles of the substance that are present in relation to a given amount of solvent. In the present case, the amounts by which the freezing point is lowered, called the *freezing-point depression* ΔT_f, is approximately proportional to the number of moles of solute dissolved in a given amount of the solvent. Other colligative properties are vapor-pressure lowering, boiling-point elevation, and osmotic pressure.

THEORY

In Fig. 1, T_0 and p_0 are, respectively, the temperature and vapor pressure at which the pure solvent freezes. Strictly speaking, T_0 is the temperature at the "triple point" with a total pressure equal to p_0 rather than the freezing point at 1 atm. However, the pressure effect is small—of the order of a few thousandths of a degree—and moreover, as it is essentially the same for a dilute solution as for the pure solvent, it will cancel out in the calculation of ΔT_f. The curve labeled p_l is the vapor pressure of the pure liquid solvent as a function of temperature, and that labeled p_s is the vapor pressure ("sublimation pressure") of the pure solid solvent. The dashed curve labeled p_x is the vapor pressure of a solution containing solute in mole fraction X. We now introduce Assumption 1: *The solid in equilibrium with a solution at its freezing point is essentially pure solid solvent.* The validity of this assumption results primarily from the crystalline structure of the solid solvent, in which the molecules are

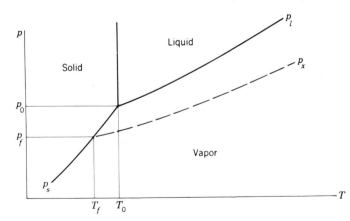

FIGURE 1
Pressure-temperature diagram for solvent and solution.

packed in a regular manner closely dependent upon their shape and size. The substitution of one or more solute molecules for solvent molecules, to yield a solid solution, is usually accompanied by a very large increase in free energy, for the solute molecule usually has a size and shape poorly adapted to filling suitably the cavity produced by removing a solvent molecule. Exceptions occur when solute and solvent molecules are very similar, as in the toluene–chlorobenzene, benzene–pyridine, and other similar systems, and in some inorganic salt mixtures and many metal alloys.

By the second law of thermodynamics, when a solution is in equilibrium with solid solvent, the solvent vapor pressures of the two must be equal. Therefore for a solution containing solute at mole fraction X, the freezing point T_f and the corresponding equilibrium vapor pressure p_f are determined by the intersection of the sublimation-pressure curve p_s with the solution vapor-pressure curve p_x. To determine the point of intersection we write the equations for the two curves and solve them simultaneously for p and T.

The equation of the sublimation curve is conveniently obtained from the integrated Clausius–Clapeyron equation:

$$\ln \frac{p_s}{p_0} = \frac{\Delta \bar{H}_s}{R} \left(\frac{1}{T_0} - \frac{1}{T} \right) \tag{1}$$

where $\Delta \bar{H}_s$ is the molar heat of sublimation of the solid solvent. In order to use this equation we must introduce Assumption 2: *All assumptions required in the derivation of the integrated Clausius–Clapeyron equation may be here assumed.* These are that (*a*) the vapor obeys the perfect-gas law, (*b*) the volume of a condensed phase is negligible in comparison with that of an equivalent amount of vapor, and (*c*) the enthalpy change accompanying vaporization or sublimation is independent of temperature. Part (*c*) of this assumption may be dispensed with, as we shall see later. The vapor-pressure curve p_x for the solution may be related to that for the liquid p_l, the equation for which we may take in accordance with the above assumption as

$$\ln \frac{p_l}{p_0} = \frac{\Delta \bar{H}_v}{R} \left(\frac{1}{T_0} - \frac{1}{T} \right) \tag{2}$$

where $\Delta \bar{H}_v$ is the molar heat of vaporization of the liquid solvent. To obtain the equation for the p_x curve we introduce Assumption 3: *Raoult's law applies.* Raoult's law is applicable to ideal solutions and, in the limit of infinite dilution, to real solutions. It is usually a fairly good approximation for moderately dilute real solutions. According to this law,

$$\frac{p_x}{p_l} = X_0 = 1 - X \tag{3}$$

where X_0 is the mole fraction of solvent and X is the mole fraction of solute in the solution. Combining Eqs. (2) and (3) we obtain, as the equation for p_x as a function of T,

$$\ln\frac{p_x}{p_0} = \ln\frac{p_x}{p_l} + \ln\frac{p_l}{p_0} = \ln(1-X) + \frac{\Delta\tilde{H}_v}{R}\left(\frac{1}{T_0} - \frac{1}{T}\right) \tag{4}$$

Now we are ready to solve for the coordinates of the intersection point p_f, T_f. Setting p_x equal to p_s, we obtain from Eqs. (1) and (4)

$$\ln(1-X) = \frac{\Delta\tilde{H}_s - \Delta\tilde{H}_v}{R}\left(\frac{1}{T_0} - \frac{1}{T_f}\right)$$

or

$$\ln(1-X) = -\frac{\Delta\tilde{H}_f}{RT_0 T_f}\Delta T_f \tag{5}$$

where

$$\Delta T_f \equiv T_0 - T_f \tag{6}$$

and $\Delta\tilde{H}_f$ is the molar heat of fusion of the liquid solvent.

It is worth noting here that Eq. (5) may be derived more elegantly on a somewhat different set of assumptions, in which those pertaining to properties of the solvent vapor do not appear. Those assumptions, together with Raoult's law, may be replaced by the single assumption that the *activity of the solvent is directly proportional to the mole fraction of the solvent in the solution.*

We wish to derive expressions for calculating the molality m of the solution (or the molecular weight M of the solute) from a measured freezing-point depression ΔT_f. The molality of the solution is given by

$$m = \frac{1000}{M_0}\frac{X}{1-X} \tag{7}$$

where M_0 is the molecular weight of the solvent in grams per mole. If g is the weight of solute and G is the weight of solvent,

$$\frac{X}{X_0} = \frac{g/M}{G/M_0}$$

thus

$$M = M_0\frac{g}{G}\frac{X_0}{X} = M_0\frac{g}{G}\frac{1-X}{X}$$

Combining this with Eq. (7) we obtain

$$M = \frac{g}{G}\frac{1000}{m} \tag{8}$$

The expressions commonly used for determination of molalities or molecular weights from freezing-point depressions are derived with the following approximations:

$$\ln(1 - X) \cong -X$$
$$T_0 T_f \cong T_0^2 \tag{9}$$
$$1 - X = X_0 \cong 1$$

With the aid of these approximations, Eq. (5) becomes

$$X = \frac{\Delta \tilde{H}_f}{RT_0^2} \Delta T_f \tag{10}$$

Combining Eq. (10) with Eqs. (7) and (8) we obtain

$$m = \frac{1000 \, \Delta \tilde{H}_f}{M_0 RT_0^2} \Delta T_f = \frac{\Delta T_f}{K_f} \tag{11}$$

and

$$M = \frac{M_0 g RT_0^2}{G \, \Delta \tilde{H}_f \, \Delta T_f} = 1000 \frac{g}{G} \frac{K_f}{\Delta T_f} \tag{12}$$

where

$$K_f \equiv \frac{M_0 RT_0^2}{1000 \, \Delta \tilde{H}_f} \tag{13}$$

K_f is called the *molal freezing-point depression constant,* since it is equal to the freezing-point depression predicted by Eq. (11) for a 1-molal solution.

It will often be the case that no further refinement of these expressions is justified. In some cases, however, the use of a higher order of approximation is worthwhile. In the treatment that follows, we shall retain terms representing no more than two orders in ΔT.

Let the right-hand side of Eq. (5) be called y,

$$y \equiv -\frac{\Delta \tilde{H}_f}{RT_0 T_f} \Delta T_f$$

and express e^y in a Maclaurin series, so that Eq. (5) becomes

$$1 - X = e^y = 1 + y + \frac{y^2}{2!} + \cdots$$

Since y is small in comparison with unity,

$$X \cong -y \left(1 + \frac{y}{2}\right) \cong -\frac{y}{1 - y/2} \tag{14}$$

The other approximations of Eq. (9) are replaced by

$$T_0 T_f = T_0^2 \frac{T_f}{T_0} = T_0^2 \left(1 - \frac{\Delta T_f}{T_0}\right) \cong \frac{T_0^2}{1 + \Delta T_f / T_0} \tag{15}$$

and

$$1 - X = X_0 = 1 + y \cong \frac{1}{1-y} \tag{16}$$

In addition, if it is not desired to retain Assumption 2(c) that $\Delta \bar{H}_f$ is strictly constant, we can replace it in the expression for y by its mean value over the temperature range ΔT_f:

$$\overline{\Delta \bar{H}_f} = \Delta \bar{H}_f^0 - \Delta \bar{C}_p \frac{\Delta T_f}{2} = \Delta \bar{H}_f^0 \left(1 - \frac{\Delta \bar{C}_p}{2\,\Delta \bar{H}_f^0} \Delta T_f \right) \tag{17}$$

where $\Delta \bar{H}_f^0$ is the molar heat of fusion of the pure solvent at its freezing point.

In combining the above equations we follow the usual rules for multiplying and dividing binominals and reject terms of second and higher order in ΔT_f in comparison with unity. We obtain

$$\frac{X}{1-X} = \frac{\Delta \bar{H}_f^0}{RT_0^2} \Delta T_f (1 + k_f\,\Delta T_f) \tag{18}$$

where

$$k_f = \frac{1}{T_0} + \frac{\Delta \bar{H}_f^0}{2RT_0^2} - \frac{\Delta \bar{C}_p}{2\,\Delta \bar{H}_f^0} \tag{19}$$

We further obtain, by use of Eqs. (7) and (8),

$$m = \frac{\Delta T_f}{K_f}(1 + k_f\,\Delta T_f) \tag{20}$$

$$M = 1000\,\frac{g}{G}\frac{K_f}{\Delta T_f}(1 - k_f\,\Delta T_f) \tag{21}$$

In Table 1 are given values of the pertinent constants for two common solvents: cyclohexane and water. It will be seen that for freezing-point depressions of about 2 K, omission of the correction term $k_f\,\Delta T_f$ leads to errors of the order of 1 percent in m or M.

Where Raoult's law fails, we may expect in the case of a nondissociating

TABLE 1

		Cyclohexane	Water
Molecular weight (g)	M_0	84.16	18.02
Celsius freezing point (°C)	t_0	6.68	0.00
Absolute freezing point (K)	T_0	279.83	273.15
Molar heat of fusion at T_0 (J mol^{-1})	$\Delta \bar{H}_f^0$	2678	6008
Molar heat-capacity change on fusion (J K^{-1} mol^{-1})	$\Delta \bar{C}_p$	15.1	38.1
Molal freezing-point depression constant (K molal^{-1})	K_f	20.4	1.855
Correction constant (K^{-1})	k_f	0.003	0.005

and nonassociating solute that the experimental conditions will at least lie in a concentration range over which the deviation in vapor pressure can be expressed fairly well by a quadratic term in the solute mole fraction X. Within this range the main effect on the above equations will be to change k_f by a small constant amount. Thus we may expect that a plot of M [calculated either with Eq. (12) or with Eq. (21)] against ΔT_f, with a number of experimental points obtained at different concentrations, should yield an approximately straight line which, on extrapolation to $\Delta T_f = 0$, should give a good value for M. In the event of failure of the assumption of no solid solution, however, the limiting value of M itself should be expected to be in error.†

METHOD

In this experiment the freezing point of a solution containing a known weight of an "unknown" solute in a known weight of cyclohexane is determined from cooling curves. From the result at each of two concentrations, the molecular weight of the unknown is determined.

The apparatus is shown in Fig. 2. The inner test tube, containing the solution, stirrer, and thermometer, is partially insulated from a surrounding ice–salt cooling bath through being suspended in a larger test tube with an air space between them. To provide additional insulation, the space between may be filled by a hollow plug of expanded polystyrene foam. The thermometer is either a special cryoscopic mercury thermometer of appropriate range, with graduations every 0.01 or 0.02°C, or a digital resistance thermometer with a resolution of ±0.01°C.

Under the conditions of the experiment, heat flows from the inner system, at temperature T, to the ice–salt bath, at temperature T_b, at a rate which is approximately proportional to the temperature difference:

$$-\frac{dH}{dt} = A(T - T_b) \tag{22}$$

where H is the enthalpy of the inner system and A is a constant incorporating shape factors and thermal-conductivity coefficients. If $(T - T_b)$ is sufficiently large in comparison with the temperature range covered in the experiment, we can write

$$-\frac{dH}{dt} \cong \text{const} \tag{23}$$

In the absence of a phase change, the rate of change of the temperature is

---

† This can easily be shown by a treatment parallel to the derivation here given, taking account of the facts that the equilibrium concentration of solute in the solid phase increases (and that of solvent decreases) with increasing solute concentration in the liquid phase and that the solvent vapor pressure of the solid decreases as the solvent concentration in the solid decreases. Thus for the solid a new vapor-pressure curve should be drawn below p_s in Fig. 1, and its intersection point with p_x is to the right of that shown.

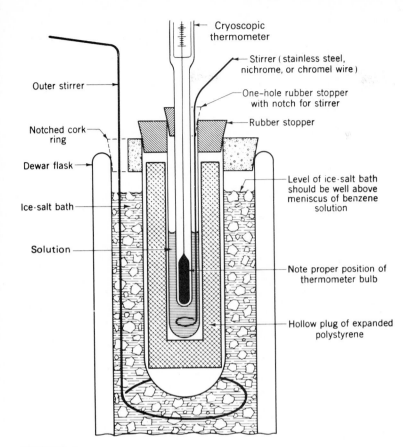

FIGURE 2
Apparatus for cryoscopic determination of molecular weight. The cryoscopic thermometer can be replaced by a resistance thermometer.

given by

$$-\frac{dT}{dt} = \frac{1}{C}\left(-\frac{dH}{dt}\right) \tag{24}$$

where C is the heat capacity of the inner system. When a pure liquid freezes, dT/dt vanishes as long as two phases are present and we have a "thermal arrest." When pure solid solvent separates from a liquid solution on freezing, the temperature does not remain constant because the solution becomes continually more and more concentrated and the freezing point T_f correspondingly decreases. It can be shown that in this case

$$-\frac{dT}{dt} = \frac{1}{(N_0\,\Delta\bar{H}_f/\Delta T_f) + C}\left(-\frac{dH}{dt}\right) \tag{25}$$

where N_0 is the number of moles of solvent present in the liquid phase. Thus, when solid solvent begins to freeze out of solution on cooling, the slope changes discontinuously from that given by Eq. (24) to the much smaller slope

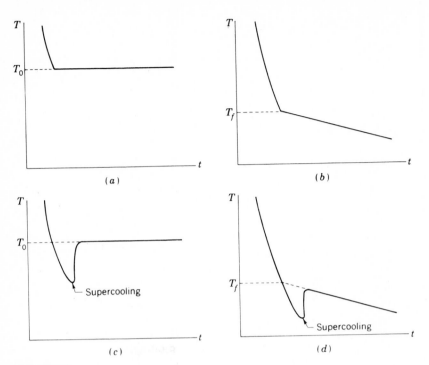

FIGURE 3
Schematic cooling curves: (*a*) and (*c*) show the cooling curves for pure solvent; (*b*) and (*d*) show the cooling curves for a solution.

given by Eq. (25), and we have what is called a "break" in the cooling curve.

Figure 3 shows schematically the types of cooling curves that are expected as a result of these considerations. It will be noted that the solutions may "supercool" before solidification of solvent takes place. In the present experiment, the extent of supercooling rarely exceeds about 2 K and is best kept below 1 K by seeding—introducing a small crystal of frozen solvent.

When supercooling occurs, the recommended procedure for estimating the true freezing point of the solution (the temperature at which freezing would have started in the absence of supercooling) is to extrapolate back that part of the curve that corresponds to freezing out of the solvent until the extrapolate intersects the cooling curve of the liquid solution. To a good approximation the extrapolation may be taken as linear.†

---

† To justify this extrapolation procedure, consider two experiments starting at the same temperature and the same time under conditions identical in all respects except that in one case supercooling is allowed to take place and in the other it is somehow prevented. By Eq. (23) the enthalpies of the two systems remain identical throughout the experiment. Except during the interval of supercooling, both systems are in equilibrium, and therefore when their enthalpies are equal, they are identical in all respects, including temperature. Therefore, except for the interval of supercooling, the two curves when superimposed are congruent throughout their entire length. The dashed line representing the extrapolation in Fig. 3(*c*) or (*d*) is therefore part of the curve for the hypothetical experiment in which supercooling was prevented.

This procedure is valid only when the extent of supercooling is small in comparison with the difference in temperature between the system and the ice–salt cooling bath. When this is not the case, Eq. (22) must be used in place of Eq. (23), and it is then seen that the rate of decrease of enthalpy is dependent in significant degree on whether or not supercooling takes place. Therefore the amount of supercooling must be kept small, and the bath must be kept as cold as possible.

EXPERIMENTAL

The success of this experiment depends primarily on a careful experimental technique. The solvent to be used is cyclohexane, which must be of reagent-grade purity and scrupulously dry. Pour about 25 mL of cyclohexane into a clean, dry, glass-stoppered flask for your own use. During the experiment, take care to keep the inner part of the apparatus dry and to expose the inner test tube to the air as little as possible in order to avoid condensation of moisture inside.

The ice–salt mixture should be made up freshly for each run. Mix about one part by volume of coarse rock salt with about four parts by volume of finely crushed ice in a beaker or battery jar. Allow it to stand a few minutes to become slushy, and then pour the desired amount into the Dewar flask. (This ice–salt mixture can be replaced by a mixture of 10 parts ice with 1 part denatured alcohol.)

Fill a small beaker with crushed ice and a little water, and place in it a test tube containing 1 or 2 mL of cyclohexane and a very thin glass rod. Stopper the tube with cotton or glass wool to keep out moist air from the room. This frozen cyclohexane is to be used for seeding. When it is desired to seed the system during a run, remove the glass rod, making sure it carries a small amount of frozen cyclohexane, and carefully insert it into the solution with the least possible disruption of the experiment.

Cyclohexane should be introduced into the inner test tube from a weighing bottle (weighed before and after delivery) or from a pipette. *Do not pipette cyclohexane by mouth*; use a rubber bulb. The quantity normally required is 15 mL. If a pipette is used, record the ambient temperature.

The thermometer and stirrer should be inserted with the thermometer carefully mounted in such a way that the sensing element (bulb of a mercury thermometer or resistance coil of a platinum resistance thermometer) is about halfway between the bottom of the test tube and the upper surface of the liquid and concentric with the tube so that the stirrer can easily pass around it. If the thermometer is too close to the bottom, a bridge of frozen solvent can easily form which will conduct heat away from the thermometer and result in low readings.

Much depends on the technique of stirring the solution. The motion of the stirrer should carry it from the bottom of the tube up to near the surface of the liquid; before assembling the apparatus, it is well to observe carefully how high the stirrer can be raised without too frequent splashing. The stirring should be continuous throughout the run and should be at the rate of about 1 stroke per second. The outer bath should be stirred a few times per minute.

The test tube containing cyclohexane, stopper, thermometer, and stirrer is held in the beaker of ice water, and the cyclohexane is stirred until it visibly starts to freeze. The outside of the tube is wiped dry, and the tube is allowed to warm up at least 1°C above the freezing point. Then the tube is placed in the assembled apparatus and temperature readings are taken every 30 s. If a mercury cryoscopic thermometer is being used, tap the thermometer gently before each reading, and estimate the temperature readings to tenths of the smallest scale division. If the temperature falls 0.5°C below the normal freezing point without evidence of freezing, seed the liquid. Once freezing has occurred, continue temperature readings for about 5 min. If the cyclohexane and apparatus are suitably dry, the temperature should remain constant or fall by no more than about 0.02°C during that time.

The inner test tube is removed and warmed (with stirring) in a beaker of water until the cyclohexane has completely melted. The stopper is lifted, and an *accurately weighed* pellet of the unknown compound, of about 0.6 g, is introduced. The pellet is dissolved by stirring, the inner tube is cooled with stirring in an ice bath to about the freezing temperature of pure cyclohexane and then dried on the outside, and the apparatus is reassembled. Temperature readings are taken every 30 s throughout the run. After a depression of about 2°C has been reached, it is advisable to seed the system at intervals of about 0.5°C until freezing starts. After freezing has begun, temperature readings should be continued for a period at least four times as long as the estimated period of supercooling to provide data for an adequate extrapolation.

The inner tube is removed, the cyclohexane is melted as before, and a second weighed pellet of the unknown is added and stirred into solution. Before this pellet is added, it is well to make a rough calculation, based on the results of the first run, to make sure that the temperatures in the second run will remain on scale and, if necessary, to modify the weight of the pellet accordingly. The second run is carried out as before, after cooling in an ice bath to about the temperature at which freezing was obtained in the first run.

If the results from these two runs are not in agreement and time allows, a repeat of both runs should be made using somewhat different pellet weights.

CALCULATIONS

If the cyclohexane was introduced with a pipette, its weight G can be calculated using the density:

$$\rho(\text{cyclohexane}) = 0.779 - 9.4 \times 10^{-4}(t - 20) \text{ g cm}^{-3}$$

where t is the Celsius temperature. Plot the cooling-curve data for pure cyclohexane and for each solution studied. If supercooling took place, perform the extrapolations as shown in Fig. 3. Report the weight g of solute and the freezing-point depression ΔT_f for each run.

Calculate the molecular weight from both Eqs. (12) and (21) using the appropriate constants in Table 1. An extrapolated value of M can be obtained from the calculated values from either equation by plotting the calculated molecular weight against the depression ΔT_f. If the elementary analysis or empirical formula of the unknown is given, deduce the molecular formula and the exact molecular weight.

APPARATUS

Dewar flask; notched cork ring to fit top of Dewar; large test tube with polystyrene-foam insert; large ring stirrer; inner test tube which fits into polystyrene insert; large rubber stopper with hole for inner test tube; medium stopper with hole for thermometer and a notch for small ring stirrer; 1-quart battery jar or 1000-mL beaker; 250-mL beaker; 125-mL glass-stoppered flask for storing benzene; small test tube; glass rod (3 mm diameter and 20 cm long); 15-mL pipette; small rubber pipetting bulb; precision cryoscopic thermometer and magnifying thermometer reader, or a digital resistance thermometer with ±0.01°C resolution; stopwatch.

Dry reagent-grade cyclohexane (50 mL); glass wool; naphthalene or an unknown solid; pellet press; acetone for rinsing; ice (3 lb); ice grinder or shaver; coarse rock salt (2 lb).

GENERAL READING

E. L. Skau and J. C. Arthur, Jr., "Determination of Melting and Freezing Temperatures", in A. Weissberger and B. W. Rossiter (eds.), "Techniques of Chemistry: Vol. I. Physical Methods of Chemistry," part V, chap. 3, Wiley-Interscience, New York, (1971).

EXPERIMENT 11
FREEZING-POINT DEPRESSION
OF STRONG AND WEAK
ELECTROLYTES

In this experiment, the freezing-point depression of aqueous solutions is used to determine the degree of dissociation of a weak electrolyte and to study the deviation from ideal behavior that occurs with a strong electrolyte.

THEORY

Use will be made of the theory developed in Exp. 10 for the freezing-point depression ΔT_f of a given solvent containing a known amount of an ideal solute; this material should be reviewed.

In the case of a dissociating (or associating) solute, the molality given by Eq. (10-11) or (10-20) is ideally the *total* effective molality—the number of moles of all solute species present, whether ionic or molecular, per 1 kg of solvent. As we shall see, ionic solute species at moderate concentrations do not form ideal solutions and, therefore, do not obey these equations. However, for a weak electrolyte the ionic concentration is often sufficiently low to permit treatment of the solution as ideal.

Weak electrolytes. As an example, let us discuss a weak acid HA with nominal molality m. Owing to the dissociation

$$HA = H^+ + A^-$$

the equilibrium concentrations of HA, H^+, and A^- will be $m(1 - \alpha)$, $m\alpha$, and $m\alpha$, respectively, where α is the fraction dissociated. The total molality m' of all solute species is

$$m' = m(1 + \alpha) \tag{1}$$

This molality m' can be calculated from the observed ΔT_f using Eq. (10-20). Thus, freezing-point measurements on weak electrolyte solutions of known molality m enable the determination of α.

The equilibrium constant in terms of concentrations can be calculated from

$$K_c = \frac{(H^+)(A^-)}{(HA)} = m\frac{\alpha^2}{(1 - \alpha)} \tag{2}$$

In this experiment α and K_c are to be determined at two different molalities. Since these will be obtained at two different temperatures, the values of K_c should be expected to differ slightly.

Strong electrolytes. These are now generally believed to be completely dissociated in ordinary dilute solutions. However, their colligative properties when interpreted in terms of ideal solutions appear to indicate that the dissociation is a little less than complete. This fact led Arrhenius to postulate that the dissociation of strong electrolytes is indeed incomplete. Subsequently, this deviation in colligative behavior has been demonstrated to be an expected consequence of interionic attractions.

For a nonideal solution, Eq. (10-5) is replaced by

$$\ln a_0 = -\frac{\Delta \bar{H}_f}{RT_0 T_f}\Delta T_f \cong -\frac{\Delta \bar{H}_f}{RT_0^2}\Delta T_f \tag{3}$$

where a_0 is the *activity* of the solvent and is related to the mole fraction of solvent X_0 by

$$a_0 = \gamma_0 X_0 \tag{4}$$

The quantity γ_0 is the activity coefficient for the solvent and in electrolytic solutions differs from unity even at moderately low concentrations. Let us write $\ln a_0$ as $(\ln \gamma_0 + \ln X_0)$ in Eq. (3) and divide both sides by $\ln X_0$ to obtain

$$-\frac{\Delta \tilde{H}_f}{RT_0^2} \frac{\Delta T_f}{\ln X_0} = 1 + \frac{\ln \gamma_0}{\ln X_0} \equiv g_1 \tag{5}$$

where g_1 is called the *osmotic coefficient* of the solvent.[1,2] Now

$$X_0 = \frac{n_0}{n_0 + vn_1} \tag{6}$$

where n_0 is the number of moles of solvent and vn_1 is the total number of moles of ions formed from n_1 moles of solute (for example, $v = 2$ for HCl). Thus

$$-\ln X_0 \equiv \ln\left(1 + \frac{vn_1}{n_0}\right) \cong \frac{vn_1}{n_0} \tag{7}$$

for dilute solutions. Substituting this expression for $\ln X_0$ into Eq. (5), we have

$$\frac{\Delta \tilde{H}_f}{RT_0^2} \frac{n_0}{vn_1} \Delta T_f = g_1 \tag{8}$$

For a solution of molality m in a solvent with molecule weight M_0 in g mol^{-1}, we can replace (n_0/vn_1) by $(1000/M_0 vm)$. Equation (8) can then be written as

$$\left(\frac{\Delta \tilde{H}_f}{RT_0^2} \frac{1000}{M_0}\right) \frac{\Delta T_f}{vm} = \frac{\Delta T_f}{vmK_f} = g_1 \tag{9}$$

where K_f is the molal freezing-point depression constant defined by Eq. (10-13).

For an ideal solution, $\gamma_0 = 1$ and g_1 is unity. Then Eq. (9) is consistent with Eq. (10-11), since the total molality of all solute species is vm for a completely dissociated solute of molality m. For ionic solutions, the Debye–Hückel theory predicts a value of γ_0 different from unity and therefore a deviation of g_1 from unity. A treatment of this aspect of the Debye–Hückel theory is beyond the scope of this book, and we shall merely state the result. The osmotic coefficient g_1 at 0°C for dilute solutions of a single strong electrolyte in water is given[1] by

$$g_1 = 1 - 0.376\sigma |z_+ z_-| I^{1/2} \tag{10}$$

where z_+ is the valence of the positive ion, z_- is the valence of the negative ion, and I is the ionic strength:

$$I \equiv \tfrac{1}{2} \sum_i m_i z_i^2 \tag{11}$$

the sum being taken over all ionic species present. The quantity σ is a function of κa where, for aqueous solutions at 0°C, κa is given by MacDougall[1] as

$$\kappa a = 0.324 \times 10^8 a I^{1/2}$$

TABLE 1

κa	0.20	0.25	0.30	0.35	0.40	0.45	0.50	0.55
σ	0.7588	0.7129	0.6712	0.6325	0.5988	0.5673	0.5376	0.5108

when the effective ionic diameter a is expressed in centimeters. Values of σ are given for several values of κa in Table 1.

For a small ion, a is approximately 3×10^{-8} cm and we can take $\kappa a \cong I^{1/2}$. For a uni-univalent electrolyte such as HCl, this becomes simply

$$\kappa a \cong m^{1/2} \tag{12}$$

Therefore Eq. (10) can be written as

$$g_1 \cong 1 - 0.38 \sigma m^{1/2} \tag{13}$$

For a known value of m, κa for a uni-univalent electrolyte is given by Eq. (12) and one can find the appropriate value of σ by interpolating in Table 1. Use of this value in Eq. (13) allows one to calculate g_1. It should be emphasized that Eq. (13) is an approximation based on the Debye–Hückel theory and thus valid only in dilute solution.[2]

EXPERIMENTAL

The apparatus used in this experiment is shown in Fig. 1. The thermometer is either a special cryoscopic mercury thermometer of appropriate range, with graduations every 0.01 or 0.02°C, or a resistance thermometer with a resolution of ±0.01°C. In this experiment, an aqueous solution of a weak or strong acid is mixed with crushed ice until equilibrium is attained. The temperature is recorded, and two or more aliquots of the liquid phase are withdrawn for titration to determine the equilibrium nominal concentration m_0. The ice to be used should preferably be distilled-water ice.

The most difficult part of the experimental technique is the achievement of thorough mixing. This difficulty is aggravated by the fact that water has a maximum density near 4°C and solution at that temperature tends to settle to the bottom of the Dewar flask while colder solution tends to float near the top with the ice. The stirring must therefore be *vigorous and prolonged*. A good technique is to work the stirrer frequently above the mass of ice and with a vigorous downward thrust propel the ice all the way to the bottom of the flask. *It must not be assumed that equilibrium has been obtained until the temperature shown by the thermometer has become quite stationary and does not change when the stirring is stopped or when the manner or vigor of stirring is changed.*

Procedure. The solutions to be studied are the strong electrolyte HCl and weak electrolyte monochloroacetic acid, each at two concentrations—roughly 0.25 and 0.125 m. About 150 mL of each solution will be needed, and the

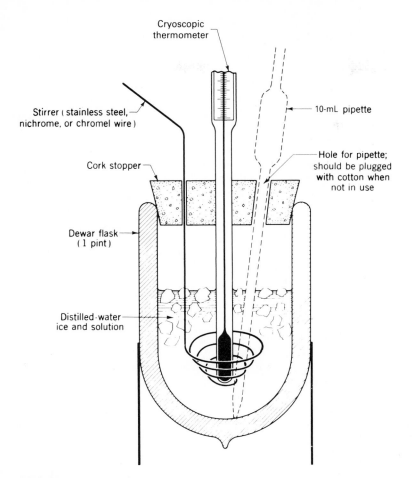

Cryoscopic
thermometer

Stirrer (stainless steel,
nichrome, or chromel wire)

10-mL pipette

Cork stopper

Hole for pipette;
should be plugged
with cotton when
not in use

Dewar flask
(1 pint)

Distilled-water
ice and solution

FIGURE 1
Apparatus for measuring the freezing point of aqueous solutions. The cryoscopic thermometer can
be replaced by a resistance thermometer.

$0.125\,m$ solutions required should be prepared by diluting the $0.25\,m$ stock
solutions with distilled water. Place each solution in a clean, glass-stoppered
flask packed in crushed ice. A flask of distilled water should also be packed in
ice.

Wash about $300\,\text{mL}$ of crushed, distilled-water ice with several small
amounts of the chilled distilled water, then fill the Dewar flask about one-third
full with this washed ice. About $100\,\text{mL}$ of the chilled distilled water is added
to the Dewar, the apparatus is assembled, and the mixture is well stirred to
achieve equilibrium. The temperature T_0 is recorded. If a cryoscopic mercury
thermometer is being used, tap it gently before reading and estimate to tenths
of the smallest division. It should be realized that a cryoscopic thermometer

may not read exactly 0°C in this ice-water bath and may deviate by as much as a ±0.1°C without coming under suspicion of being defective. The important function of a cryoscopic thermometer is to measure temperature *differences*. For this experiment, absolute accuracy in temperature values is not important but high resolution (i.e., precision in ΔT values) is important.

The water is poured off and replaced by 100 mL of chilled 0.25 m HCl solution. After stirring to achieve equilibrium, as described previously, the temperature is read and a 10-mL aliquot withdrawn with a pipette. Introduce the pipette quickly, while blowing a gentle stream of air through it to prevent solution from entering it until the tip touches the bottom of the flask. This will prevent small particles of ice from being drawn into the pipette. Alternatively, attach to the tip of the pipette a filter consisting of a short length of rubber tubing containing a wad of cotton or glass wool. This aliquot is discharged into a clean weighing bottle, warmed to room temperaure, and accurately weighed. It is then quantitatively transferred to a flask and titrated with 0.1 M NaOH to a methyl red or phenolphthalein end point.† Stirring should be resumed vigorously for about 5 min, then another temperature reading taken and a second aliquot withdrawn. If these results are not consistent with each other, a third aliquot should be taken after further stirring.‡

The solution is poured off and the experiment is repeated with 0.125 m HCl. Be sure that an adequate amount of ice is present. If time permits, a run should be made with 0.0625 m HCl also.

Similar runs are carried out with the solutions of monochloroacetic acid, preferably repeating the measurement of T_0.

CALCULATIONS

Calculate the equilibrium molality m (in mol/kg of water) for each aliquot. If the results for the two aliquots from a given run are consistent, the average values of m and ΔT_f may be used in further calculations. For monochloroacetic acid, a weak electrolyte, calculate the effective total molality m' from Eq. (10-20) using the appropriate constants in Table 10-1. Then calculate α and K_c for each of the two concentrations studied.

For hydrochloric acid, a strong electrolyte, calculate an experimental value of g_1 with Eq. (9) for each of the concentrations studied. In addition, use Eq. (13) to obtain a value of the osmotic coefficient g_1 based on the

---

† This NaOH solution should be prepared from carbonate-free NaOH as described in Chapter XIX. The solution is standardized by titrating, with phenolphthalein as an indicator, a weighed quantity of dry potassium acid phthalate, $KH(C_8H_4O_4)$.

‡ The temperatures and the concentrations of the two aliquots may differ slightly, owing to some melting of ice, but the differences should be consistent. If they are not consistent, at least one of the aliquots was presumably not withdrawn at equilibrium.

Debye–Hückel theory for each concentration. Compare these experimental and theoretical values.

DISCUSSION

The colligative behavior of strong electrolytes is sometimes expressed in terms of the Van't Hoff factor $i \equiv m_{app}/m$, where m_{app} is the "apparent" total molality as deduced from any colligative property when the solution is treated as ideal. Using the expressions for freezing-point depression, show the relation between i and g_1.

What additional data would be required in order to compare the values of K_c obtained for monochloroacetic acid at two different temperatures? Can you predict the direction of the change in K_c with T; i.e., the sign of dK_c/dT?

APPARATUS

Dewar flask (short, wide mouth, 1 pt); cork stopper with three holes; coil-type stirrer; stopwatch; two 10-mL weighing bottles; 50-mL burette; burette clamp and stand; two 100-mL volumetric flasks; one 10-, one 25-, and one 50-mL pipette; pipetting bulb; battery jar; wash bottle; precision cryoscopic thermometer and magnifying thermometer reader, or a digital resistance thermometer with ±0.01°C resolution.

Distilled- or deionized-water ice (500 g); 0.1 M sodium hydroxide solution (500 mL); 0.25 m HCl solution (400 mL) and 0.25 m monochloroacetic acid solution (400 mL); phenolphthalein indicator; stopcock grease.

REFERENCES

1. F. H. MacDougall, "Thermodynamics and Chemistry," 3d ed., Wiley, New York (1939).
2. R. S. Berry, S. A. Rice and J. Ross, "Physical Chemistry," pp. 929, 1004, Wiley, New York (1980).

GENERAL READING

E. L. Skau and J. C. Arthur, Jr., "Determination of Melting and Freezing Temperatures," in A. Weissberger and B. W. Rossiter (eds.), "Techniques of Chemistry: Vol. I. Physical Methods of Chemistry," part V, chap. 3, Wiley-Interscience, New York (1971).

EXPERIMENT 12
CHEMICAL EQUILIBRIUM
IN SOLUTION

In this experiment a typical homogeneous equilibrium in aqueous solution is investigated and the validity of the law of mass action is demonstrated. The equilibrium to be studied is that which arises when iodine is dissolved in aqueous KI solutions:

$$I_2 + I^- = I_3^-$$ (1)

In determining the concentration of the I_2 species present at equilibrium in the aqueous solution, use is made of a heterogeneous equilibrium: the distribution of molecular iodine I_2 between two immiscible solvents, water and carbon tetrachloride.

THEORY

Homogeneous equilibrium. For the change in state in aqueous solution given in (1), the thermodynamic equilibrium constant K_a, which must be expressed in terms of the activities of the chemical species involved, is given to a good approximation by the analogous expression involving concentrations:

$$K_a = \frac{a_{I_3^-}}{a_{I_2} a_{I^-}} = \frac{(I_3^-)}{(I_2)(I^-)} \frac{\gamma_{I_3^-}}{\gamma_{I_2} \gamma_{I^-}} \cong \frac{(I_3^-)}{(I_2)(I^-)} = K_c$$ (2)

where the a's are activities, the γ's are activity coefficients, and K_c is the equilibrium constant in terms of concentration. The parentheses denote dimensionless quantities equal in magnitude to the concentrations of the species in mol L^{-1}; i.e., $(X) = c_X/c_0$ where c_X is the molar concentration of species X and $c_0 = 1\ M$. The approximation $K_a = K_c$ follows from the fact that $\gamma_{I^-} = \gamma_{I_3^-}$ and $\gamma_{I_2} = 1$, as we show below.

A commonly used approximate form of the Debye–Hückel theory for the activity coefficients of ionic species at 25°C is[1]

$$\log \gamma_i = \frac{-0.509 z_i^2 \sqrt{I}}{1 + \sqrt{I}}$$ (3)

Since both I^- and I_3^- have the same charge z and are influenced by the same ionic strength I, the activity coefficients γ_{I^-} and $\gamma_{I_3^-}$ are equal within the accuracy of Eq. (3). It should, however, be pointed out that Eq. (3) does not take account of variations in the size and shape of ions. For the I_2 species, it will be observed that the I_2 concentration in aqueous solution is very small. Thus γ_{I_2} is approximately unity, since the activity coefficient of a dilute neutral solute species does not deviate greatly from its limiting value at infinite dilution. Therefore, the approximation given in Eq. (2) should be good to within 1 percent or better in moderately dilute solutions, and K_c at any specified temperature should be a constant independent of the concentrations of the individual species.

If, then, one starts with a solution of KI whose concentration C is known accurately and dissolves iodine in it, the iodine will be present at equilibrium partly as neutral I_2 molecules and partly as I_3^- ions (formed when I_2 reacts with some of the I^- ions initially present). The total iodine concentration,

$T = (I_2) + (I_3^-)$, can be found by titration with a thiosulfate solution. If there is an independent way to measure (I_2) at equilibrium, then (I^-) and (I_3^-) can be obtained from

$$(I_3^-) = T - (I_2) \tag{4}$$

and

$$(I^-) = C - (I_3^-) \tag{5}$$

We can measure (I_2) at equilibrium by taking advantage of the fact that CCl_4, which is immiscible with aqueous solutions, dissolves molecular I_2 but not any of the ionic species involved.

Heterogeneous equilibrium. We are concerned here with the distribution of molecular iodine I_2 as the solute between two immiscible liquid phases, aqueous solution and CCl_4. At equilibrium, the concentrations of I_2 in the two phases, $(I_2)_w$ and $(I_2)_{CCl_4}$, are related by a distribution constant k:

$$k = (I_2)_w/(I_2)_{CCl_4} \tag{6}$$

This distribution law applies only to the distribution of a definite chemical species, as does Henry's law. The distribution constant k is not a true thermodynamic equilibrium constant, since it involves concentrations rather than activities. Thus it may vary slightly with the concentration of the solute (particularly because of the relatively high concentration of I_2 in the CCl_4 phase); it is therefore advantageous to determine k at a number of concentrations. It can be determined directly by titration of both phases with standard thiosulfate solution when I_2 is distributed between CCl_4 and pure water. Once k is known, (I_2) in an aqueous phase containing I_3^- can be obtained by means of a titration of the I_2 in a CCl_4 layer that has been equilibrated with this phase. The use of a distribution constant in this manner depends upon the assumption that its value is unaffected by the presence of ions in the aqueous phase.

EXPERIMENTAL

Distribution ratio. Measure the distribution constant $k = (I_2)_w/(I_2)_{CCl_4}$, using CCl_4 and pure H_2O as solvents and I_2 as solute. Distilled water (200 mL) and solutions of I_2 in CCl_4 (50 mL) should be put into 500-mL glass-stoppered Erlenmeyer flasks and equilibrated at 25°C. The quantities to use are given in Table 1 (runs 1 to 3); three sets of conditions are used in order to check the variation of the distribution constant with concentration. The flasks containing the solutions should be vigorously shaken for 5 min and clamped in a thermostat bath. After 10 min of thermal equilibrium, remove the flasks one at a time, wrap them in a dry towel, shake them vigorously for 3 to 5 min, and then return them to the thermostat bath. Repeat this procedure for at least an hour. Before removing samples for analysis let the flasks remain in the bath for

TABLE 1[a]

Run no.	CCl$_4$ (50 mL), molarity I$_2$	Aqueous layer (200 mL), molarity KI	CCl$_4$ layer (use 20-mL pipette)		Aqueous layer (use 50-mL pipette)	
			Burette size, ml	Molarity S$_2$O$_3^{2-}$	Burette size, mL	Molarity S$_2$O$_3^{2-}$
1	0.080	0.0	50	0.1	10	0.01
2	0.040	0.0	50	0.1	10	0.01
3	0.020	0.0	10	0.1	10	0.01
4	0.080	0.15	10	0.1	50	0.1
5	0.040	0.15	10	0.1	10	0.1
6	0.080	0.03	50	0.1	10	0.1

[a] The table shows the nominal *initial* concentrations and volumes of the aqueous KI solutions and of the solutions of I$_2$ in CCl$_4$ to be used. Also given are the nominal concentrations of the thiosulfate solutions to be used for titrating and the sizes of burettes and pipettes to be used. The *actual*, precise concentrations of the solutions used should be read from the labels on the respective bottles.

10 min after the last shaking to allow the liquid layers to separate completely. After equilibration is complete, remove one flask at a time from the thermostat bath and place in a battery jar containing water at 25°C as shown in Fig. 1. A sample of the aqueous layer and a sample of the CCl$_4$ layer are removed with pipettes, **using a pipetting bulb** (see Appendix E for safety precautions for potentially hazardous chemicals). The flask is then stoppered and returned to the bath for an additional 30 min of equilibration with shaking as described above, after which a second sample of each phase is removed for titration. Additional information is given in Table 1.

The purpose of taking a second sample after further equilibration is to verify that equilibrium has been achieved. If the two titrations are in substantial agreement, the average value can be used. If the results of the titrations indicate that equilibrium had not been reached when the first samples were taken, the results of the second titrations should be used although proof of equilibrium is in this case lacking.

In removing the samples for titration, it is important to avoid contamination of the sample by drops of the other phase, especially when the other phase is very much more concentrated, as is the CCl$_4$ layer in this case. When pipetting samples of the CCl$_4$ solutions, blow a slow stream of air through the pipette while it is being introduced into the solution in order to minimize the contamination by the aqueous layer. It is necessary to use a rubber bulb with the pipette for these solutions; **do not pipette by mouth**. Between samples rinse the pipette well with acetone and dry it before taking the next sample.

Each sample is transferred from the pipette to a 250-mL Erlenmeyer flask containing 10 mL of 0.1 M KI and titrated with the appropriate thiosulfate solution (see Table 1). The iodide is added to reduce loss of I$_2$ from the aqueous solution by evaporation during the titration, by forming the non-

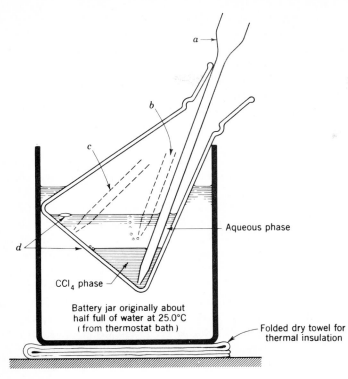

FIGURE 1

Arrangement for withdrawing samples. Use a rubber bulb for drawing up liquid in the pipette. (*a*) Position of pipette for withdrawing sample of CCl_4 phase. Tip should touch bottom at deepest point and should not be moved around. (*b*) On inserting pipette, blow *small* stream of air to keep aqueous phase from entering tip. (*c*) Position of pipette for withdrawing aqueous sample. (*d*) On inserting pipette, avoid contact with drops of CCl_4 phase (spheres on bottom or lenses floating on top).

volatile I_3^-. The reaction taking place during the titration is

$$2S_2O_3^{2-} + I_3^- \rightarrow S_4O_6^{2-} + 3I^-$$

Near the end point, as indicated by a very light yellow color of aqueous iodine solution, Thyodene† or 1 mL of soluble starch solution is added; this acts as an indicator by adsorbing iodine and giving a deep blue color. At the end point this blue color disappears sharply. If the starch has been added too soon, the

---

† Thyodene is a proprietary starch derivative which gives very sharp and reliable end points. It may be added directly as a powder and is therefore much more convenient for occasional use than the traditional "soluble starch" suspension. Should Thyodene be unavailable, a starch indicator solution may be prepared as follows. Mull 1 g of soluble starch powder in a mortar with several mL of boiling water. Pour the paste into 200 mL of boiling water, boil for 2 to 3 min, and allow to cool. When cool, the indicator is ready to use (about 2 mL/titration).

color may redevelop after an apparent end point owing to diffusion of iodine from the interior of the colloidal starch particles; the titration should be continued until no blue color reappears. However, the blue color may reappear on prolonged standing because of air oxidation of the iodide ion, and this effect should be disregarded. In the two-phase titration, shake the flask vigorously after adding each portion of thiosulfate solution.

If desired, an excellent end point for the two-phase titration can be taken as the disappearance of the reddish-violet iodine color from the CCl_4 layer. This may even be done in the titration of the aqueous sample by adding 1 or 2 mL of pure CCl_4. In this case, do not add the starch indicator.

It is recommended that a practice titration can be performed with one of the stock solutions of I_2 in CCl_4.

Since a thiosulfate solution is susceptible to attack by sulfur-metabolizing bacteria, it may be wise to check the standardization of the stock solution with a standard solution of KIO_3.[2] Place 1.3 to 1.4 g of KIO_3 in a weighing bottle, dry it for several hours at 110°C, cool in a desiccator, and weigh it exactly on an analytical balance. Transfer this salt to a clean 100-mL volumetric flask and make up to the mark with distilled water. Rinse a clean 5-mL pipette with several small portions of the iodate solution and then carefully deliver a 5-mL sample into a 250-mL Erlenmeyer flask. Add about 20 mL of distilled water, 5 mL of a 0.5 g/mL KI solution, and 10 mL of 1 M HCl. Titrate at once with the thiosulfate solution until the reddish color turns orange and then yellow and becomes pale. At this point add about 0.5 g of Thyodene indicator, mix well, and titrate until the blue color disappears.

Equilibrium constant. To determine the concentrations of I_2 and I_3^- in equilibrium in aqueous solutions we equilibrate aqueous KI solutions (200 mL) with solutions of I_2 in CCl_4 (50 mL) as shown in Table 1 (runs 4 to 6). The equilibration and titration procedures are identical with those described above; additional information is given in Table 1. With the distribution constant determined above and the initial I^- concentration, the equilibrium concentrations of I_2, I_3^-, and I^- can be calculated.

Adequate time for equilibration is of great importance; the flasks should be placed in the thermostat bath as soon as possible. The temperature of the thermostat bath should be checked several times throughout the equilibration.

WARNING. Like many other chlorinated hydrocarbons, CCl_4 is a toxic substance that can cause liver and/or kidney damage if ingested or inhaled.[3] In many of the cases of CCl_4 poisoning, the victims were chronic alcoholics or heavy drinkers, so there appears to be a synergistic effect of alcohol and CCl_4. The prescribed short-term exposure limit is 20 ppm CCl_4 vapor in the air ($\sim$125 mg m^{-3}).[3] In order to reach this level in a typical laboratory room of 750 m^3 volume, 100 g of liquid CCl_4 would have to evaporate into the atmosphere. Since CCl_4 liquid has a moderately low vapor pressure and as used in the present procedure lies below an aqueous layer in which it is not soluble, there should be no serious hazard in carrying out this experiment. The most important safety precautions are: (1) *do not pipette by mouth,* as already stressed in the procedure above; (2) *store CCl$_4$ liquid and solutions of I$_2$ in CCl$_4$ in a fume hood* and carry out all transfers from the stock bottle to a stoppered flask in this hood; and (3) *dispose of waste materials properly,* i.e., place used CCl$_4$ in a storage bottle kept in the hood.

CALCULATIONS

Calculate the distribution constant k from the results of runs 1 to 3. Plot k versus the iodine concentration in the CCl$_4$ solution, $(I_2)_{CCl_4}$. If k is not constant, discuss its variation with concentration.

Using Eqs. (4), (5), and (6) and the results of runs 4 to 6, calculate the equilibrium concentrations of I$_2$, I$_3^-$, and I$^-$ in each of the aqueous solutions. In each case use the appropriate value of k as read from the smooth curve of k versus $(I_2)_{CCl_4}$. Calculate the equilibrium constant K_c for each run. If any variation of K_c with concentration is found, do you regard it as experimentally significant?

APPARATUS

Three to six 500-mL glass-stoppered Erlenmeyer flasks; 1-qt battery jar or 1500-mL beaker; three pipettes—20, 50, 100 (or 200) mL; one 10- and one 50-mL burette; six 250-mL Erlenmeyer flasks; 10-mL graduated cylinder; three 250-mL beakers; burette clamp and stand; large rubber bulb for pipetting; wash bottle.

Constant-temperature bath set at 25°C; bath clamps for holding 500-mL Erlenmeyer flasks; pure CCl$_4$; acetone for rinsing; large carboy for waste liquids. Solutions: 0.08 M solution of I$_2$ in CCl$_4$ (300 mL); 0.04 M I$_2$ in CCl$_4$ (250 mL); 0.02 M I$_2$ in CCl$_4$ (100 mL); 0.15 M KI solution (800 mL); 0.03 M KI solution (400 mL); 0.1 M Na$_2$S$_2$O$_3$ solution (500 mL); 0.01 M Na$_2$S$_2$O$_3$ solution (100 mL); approximately 0.1 M KI solution (500 mL); Thyodene or 0.2 percent soluble starch solution, containing a trace of HgI$_2$ as preservative (50 mL). See the *warning* given above about the storage and handling of CCl$_4$.

REFERENCES

1. P. W. Atkins, "Physical Chemistry," 3d ed., pp. 242–245, Freeman, New York (1986).
2. D. A. Skoog and D. M. West, "Fundamentals of Analytical Chemistry," 4th ed., Holt, New York (1982).
3. "Documentations of the Threshold Limit Values," Amer. Conf. of Governmental Industrial Hygienists, Cincinnati, Ohio (issued annually).

GENERAL READING

G. N. Lewis and M. Randall (revised by K. S. Pitzer and L. Brewer), "Thermodynamics," 2d ed., McGraw-Hill, New York (1961).

E. G. Schiebel, "Liquid–Liquid Extraction" in E. S. Perry and A. Weissberger (eds.), "Techniques of Chemistry: Vol. XII. Separation and Purification," 3d ed., chap. 3, Wiley-Interscience, New York (1978).

CHAPTER VII

PHASE EQUILIBRIA

EXPERIMENTS

13. Vapor pressure of a pure liquid
14. Binary liquid–vapor phase diagram
15. Binary solid–liquid phase diagram
16. Liquid–vapor coexistence curve and the critical point

EXPERIMENT 13
VAPOR PRESSURE OF A PURE LIQUID

When a pure liquid is placed in an evacuated bulb, molecules will leave the liquid phase and enter the gas phase until the pressure of the vapor in the bulb reaches a definite value which is determined by the nature of the liquid and its temperature. This pressure is called the vapor pressure of the liquid at a given temperature. The equilibrium vapor pressure is independent of the quantity of liquid and vapor present as long as both phases exist in equilibrium with each other at the specified temperature. As the temperature is increased, the vapor pressure also increases up to the critial point, at which the two-phase system becomes a homogeneous, one-phase fluid.

If the pressure above the liquid is maintained at a fixed value (say by having the bulb containing the liquid open to the atmosphere), then the liquid may be heated up to a temperature at which the vapor pressure is equal to the external pressure. At this point, vaporization will occur by the formation of bubbles in the interior of the liquid as well as at the surface; this is the boiling point of the liquid at the specified external pressure. Clearly the temperature of the boiling point is a function of the external pressure; in fact, the variation

219

of the boiling point with external pressure is seen to be identical with the variation of the vapor pressure with temperature.

In this experiment the variation of vapor pressure with temperature will be measured and used to determine the molar heat of vaporization.

THEORY

We are concerned here with the equilibrium between a pure liquid and its vapor:

$$X(l) = X(g) \qquad (p, T) \tag{1}$$

It can be shown thermodynamically[1] that a definite relationship exists between the values of p and T at equilibrium as given by

$$\frac{dp}{dT} = \frac{\Delta S}{\Delta V} \tag{2}$$

In Eq. (2) dp and dT refer to infinitesimal changes in p and T for an equilibrium system composed of a pure substance with both phases always present; ΔS and ΔV refer to the change in S and V when one phase transforms to the other at constant p and T. Since the change in state (1) is isothermal and ΔG is zero, ΔS may be replaced by $\Delta H/T$. The result is

$$\frac{dp}{dT} = \frac{\Delta H}{T\,\Delta V} \tag{3}$$

Equation (2) or (3) is known as the Clapeyron equation. It is an exact expression which may be applied to phase equilibria of all kinds, although it has been presented here in terms of the one-component liquid–vapor case. Since the heat of vaporization ΔH_v is positive and ΔV is positive for vaporization, it is seen immediately that the vapor pressure must increase with increasing temperature.

For the case of vapor–liquid equilibria in the range of vapor pressures less than 1 atm, one may assume that the molar volume of the liquid $\tilde{V}_l$ is negligible in comparison with that of the gas $\tilde{V}_g$, so that $\Delta \tilde{V} = \tilde{V}_g$. This assumption is very good in the low-pressure region, since $\tilde{V}_l$ is usually only a few tenths of a percent of $\tilde{V}_g$. Thus we obtain

$$\frac{dp}{dT} = \frac{\Delta \tilde{H}_v}{T\tilde{V}_g} \tag{4}$$

Since $d \ln p = dp/p$ and $d(1/T) = -dT/T^2$, we can rewrite Eq. (4) in the form

$$\frac{d \ln p}{d(1/T)} = -\frac{\Delta \tilde{H}_v}{R}\frac{RT}{p\tilde{V}_g} = -\frac{\Delta \tilde{H}_v}{RZ} \tag{5}$$

where we have introduced the *compressibility factor* Z for the vapor:

$$Z = \frac{p\tilde{V}_g}{RT} \tag{6}$$

Equation (5) is a convenient form of the Clapeyron equation. We can see that *if* the vapor were a perfect gas ($Z \equiv 1$) and $\Delta \bar{H}_v$ were independent of temperature, then a plot of $\ln p$ versus $1/T$ would be a straight line the slope of which would determine $\Delta \bar{H}_v$. Indeed, for many liquids $\ln p$ is almost a linear function of $1/T$, which implies at least that $\Delta \bar{H}_v/Z$ is almost constant.

Let us now consider the question of gas imperfections, i.e., the behavior of Z as a function of temperature for the saturated vapor. It is difficult to carry out p–V–T measurements on gases close to condensation and such data are scarce, but data are available for water,[2] and theoretical extrapolations[3] have been made for the vapor of "normal" liquids based on data obtained at higher temperatures. Figure 1 shows the variation of the compressibility factor Z for a saturated vapor as a function of temperature in the case of water and two normal liquids, benzene and n-heptane. For the temperature axis, a "reduced" temperature T_r is used; $T_r = T/T_c$, where T_c is the critical temperature. This has the effect of almost superimposing the curves of many different substances; indeed, by the law of corresponding states such curves would be exactly superimposed. In general, it is clear that Z decreases as the temperature increases. Water, owing to its high critical temperature, is a reasonably ideal gas even at 100°C where Z equals 0.986. But n-heptane at its 1-atm boiling point of 98°C has a value of Z equal to 0.95 and is relatively nonideal. For many substances, sizable gas imperfections are present even at pressures below 1 atm.

Next we must consider the variation of $\Delta \bar{H}_v$ with temperature. For a

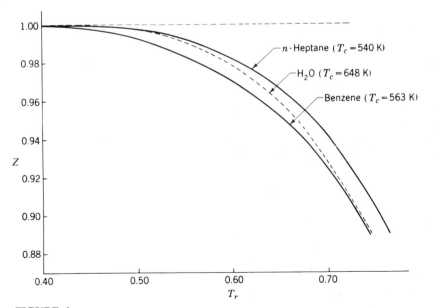

FIGURE 1
The compressibility factor Z of saturated vapor as a function of reduced temperature T_r for water, benzene, and n-heptane.

change in state such as Eq. (1),

$$\Delta H_{T_2} = \Delta H_{T_1} + \int_{T_1}^{T_2} \Delta C_p \, dT + \int_{p_1}^{p_2} \left(\frac{\partial \Delta H}{\partial p} \right)_T dp \tag{7}$$

Since the final term is zero for a perfect gas and small for most real gases, it is possible to approximate Eq. (7) by

$$\Delta H_{T_2} \simeq \Delta H_{T_1} + \overline{\Delta C_p}(T_2 - T_1) \tag{8}$$

where $\overline{\Delta C_p}$ is the average value over the temperature interval. For $\Delta \tilde{H}_v$ to be independent of temperature, the average value of $\Delta \tilde{C}_p$ must be very close to zero, which is generally not true. Heat capacities for water, benzene, and n-heptane are given below as typical examples.[4,5] For n-heptane, the specific heat of both gas and liquid changes rapidly with temperature; use of average values will give only an order-of-magnitude result. In general, the value of $\Delta \tilde{H}_v$ will decrease as the temperature increases.

Compound	Temp. range, °C	Average values, J K^{-1} mol^{-1}		
		$\tilde{C}_p(g)$	$\tilde{C}_p(l)$	$\Delta \tilde{C}_p$
Water	25–100	33.5	75	−41.5
Benzene	25–80	92	146	−54
n-Heptane	25–100	~197	~243	−46

Since both $\Delta \tilde{H}_v$ and Z decrease with increasing temperature, it is possible to see why $\Delta \tilde{H}_v / Z$ might be almost constant, yielding a nearly linear plot of $\ln p$ versus $1/T$.

METHODS

There are several experimental methods of measuring the vapor pressure as a function of temperature.[6] In the *gas-saturation method* a known volume of an inert gas is bubbled slowly through the liquid, which is kept at a constant temperature in a thermostat. The vapor pressure is calculated from a determination of the amount of vapor contained in the outcoming gas or from the loss in weight of the liquid. A common *static method* makes use of an *isotensicope,* a bulb with a short U tube attached. The liquid is placed in the bulb, and some liquid is placed in the U tube. When the liquid is boiled under reduced pressure, all air is swept out of the bulb. The isoteniscope is then placed in a thermostat. At a given temperature the external pressure is adjusted so that both arms of the U tube are at the same height. At this setting, the external pressure, which is equal to the pressure of the vapor in the isoteniscope, is measured with a mercury manometer. A common *dynamic method* is one in which the variation of the boiling point with external applied pressure is measured. The total pressure above the liquid can be varied and maintained at a given value by use of a large-volume ballast bulb; this pressure is then measured with a mercury manometer. The liquid to be studied is heated until boiling occurs, and the temperature of the refluxing vapor is measured in order to avoid any effects of superheating. The experimental procedure for both the isoteniscope method and a boiling-point method is given below.

EXPERIMENTAL

1   Boiling-Point Method

The apparatus should be assembled as shown in Fig. 2. A Claissen distilling flask can be used in this experment by closing off the side-arm with a rubber policeman. Use pressure tubing for connecting the condenser and the

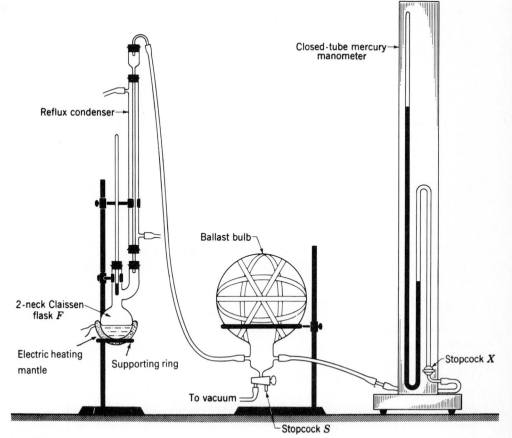

Closed-tube mercury manometer

Reflux condenser→

Ballast bulb

2-neck Claissen flask F

Electric heating mantle

Supporting ring

To vacuum

Stopcock X

Stopcock S

FIGURE 2
Boiling-point apparatus.

manometer to the ballast bulb. Fill the flask about one-third full (just above the level of the baffle) with the liquid to be studied. A few carborundum boiling chips should be added to reduce "bumping." Be sure that the thermometer scale is visible over a range of at least 50°C below the boiling point at 1 atm. Heating should be accomplished with an electrical heating mantle.

Use the manometer with the greatest care; be careful not to allow liquid to condense in it. Stopcock X should be open only when the manometer is to be read and closed immediately after each reading. Stopcock X should not be open at the same time as stopcock S is open, either to vacuum or to the air.

To make a reading, adjust the heating so as to attain steady boiling of the liquid, but avoid heating too strongly. When conditions appear to be steady, open stopcock X carefully and watch the manometer. When manometer and thermometer appear to be as steady as they can be maintained, read them as nearly simultaneously as possible. The thermometer scale should be read to the nearest 0.1°C, and vapor should be condensing on and dripping from the thermometer bulb to ensure that the equilibrium temperature is obtained. In reading the manometer, record the position of each meniscus (h_1 and h_2), keeping your line of sight level with the meniscus to avoid parallax error. Estimate your readings to the nearest 0.1 mm. Record the ambient air temperature at the manometer several times during the run.

To change the pressure in the system between measurements, first remove the heating mantle, then check to make sure that stopcock X is closed. After a short time admit some air or remove some air by opening stopcock S for a few seconds. Be especially careful in removing air from the ballast bulb to avoid strong bumping. Then open stopcock X carefully to measure the pressure on the manometer. Repeat as often as necessary to attain the desired pressure. Close stopcock X and restore the heating mantle to its position under the flask.

Take readings at approximately the following pressures:

Pressure descending: 760, 600, 450, 350, 260, 200, 160, 130, 100, 80 Torr
Pressure ascending: 90, 110, 140, 180, 230, 300, 400, 520, 670 Torr

2   Isoteniscope Method

In this technique much of the equipment shown in Fig. 2 is used, but the distilling flask and reflux condenser are replaced by an isoteniscope mounted in a glass thermostat as shown in Fig. 3. The heater is operated from a Variac voltage control, and the water bath should be vigorously stirred to ensure thermal equilibrium. The liquid to be studied is placed in the isoteniscope so that the bulb is about two-thirds full and there is no liquid in the U tube. Then the isoteniscope is placed in the thermostat (which should be at room temperature) and is connected to the ballast bulb and manometer (which are assembled and connected as in Fig. 2).

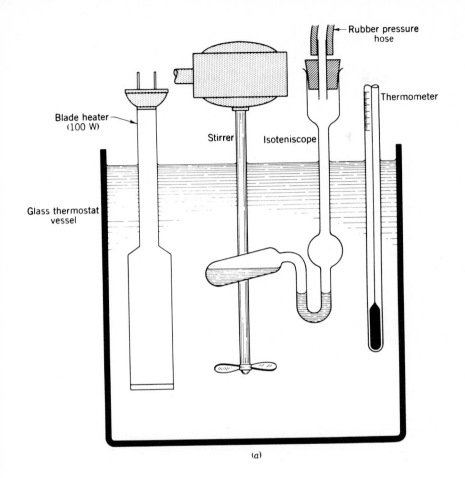

(a)

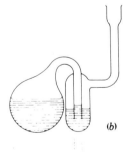

(b)

FIGURE 3
Isoteniscope: (a) Schematic diagram of the apparatus which must be connected to the ballast bulb and manometer shown in Fig. 2; (b) an alternate design for the isoteniscope. (An even more elaborate type of isoteniscope is described by Arm et al.[7])

Air is swept out of the bulb by **cautiously** and slowly reducing the pressure in the ballast bulb until the liquid boils very gently. Continue pumping for about 4 min, but be careful to avoid evaporating too much of the liquid. Then tilt the isoteniscope so that some liquid from the bulb is transferred into the U tube. Carefully admit air to the ballast bulb through stopcock S until the levels of the liquid in both arms of the U tube are the same. Read and record the temperature and the pressure on the mercury manometer. (The procedure for operating and reading this manometer is described in Section 1.) To establish that all the air has been removed from the isoteniscope, **cautiously** reduce the pressure in the ballast bulb until a few bubbles are observed passing through the U-tube liquid. Then determine the equilibrium pressure reading. Repeat the above procedure, if necessary, until successive vapor-pressure readings are in good agreement.

Once the air is removed and a good pressure reading at room temperature is obtained, heat the thermostat bath to a new temperature about 5°C above room temperature. Keep the liquid levels in the U tube approximately equal at all times. When the bath temperature is steady at its new value, adjust the pressure in the ballast bulb until the levels in the U tube are equal and record both temperature and pressure.

Take readings at approximately 5°C intervals until the bath is at about 75°C, and then take readings at decreasing temperatures which are between the values obtained on heating.

CALCULATIONS

Correct all manometer pressure readings $(h_2 - h_1)$ for the fact that the mercury is not at 0°C by multiplying by $(1 - 1.8 \times 10^{-4}t)$, where t is the Celsius temperature of the manometer. If t has been reasonably constant during the experiment, use an average value and apply the same correction factor to all pressures.

Convert all Celsius temperature readings to absolute temperatures T and plot $\ln p$ versus $1/T$. If there is no systematic curvature, draw the best straight line through the points. If there is noticeable curvature, draw a smooth curve through the points and also draw a straight line tangent to the curve at about the midpoint. Determine the slope of the straight line or tangent. From Eq. (5) it follows that this slope is $-\Delta \bar{H}_v / RZ$.

In addition to this graphical analysis, carry out a least-squares fit to your p versus T data. If visual inspection of your $\ln p$ versus $1/T$ plot does not indicate any systematic curvature, make a direct linear fit of $\ln p$ as a function of $1/T$. If systematic curvature is observed, make a least-squares fit of $\ln p$ with the empirical power-series form $a + b/T + c/T^2$. Then differentiate the resulting expression to obtain $\Delta \bar{H}_v / RZ$.

Report both the graphical and the least-squares values of $\Delta \bar{H}_v / Z$. Estimate the value of Z for the saturated vapor at the appropriate temperature from Fig. 1 and calculate $\Delta \bar{H}_v$ in $J\,mol^{-1}$. Report the value of the heat of vaporization and the vapor pressure for the liquid at the applicable temperature (corresponding to the midpoint of the range studied).

DISCUSSION

Using Fig. 1 and Eq. (8), estimate the variation in $\Delta \tilde{H}_v / Z$ expected over the range of temperatures studied. Indicate clearly whether $\Delta \tilde{H}_v / Z$ should increase or decrease with increasing temperature. Does your $\ln p$ versus $1/T$ plot show a curvature of correct sign?

Evaluate a quantitative uncertainty in your value of $\Delta \tilde{H}_v / Z$ by the method of "limiting slopes" and compare this with the standard deviation obtained from the least-squares fit to your data. Comment on this uncertainty in relation to the variation with temperature calculated above. Discuss possible sources of systematic errors.

If water or some other compound with a simple molecular structure has been studied, it is possible to combine the entropy of vaporization, $\Delta \tilde{S}_v = \Delta \tilde{H}_v / T$, with the third-law calorimetric entropy of the liquid to obtain a thermodynamic value for the entropy of the vapor. The statistical mechanical value of $\tilde{S}_g$ can be calculated using the known molecular weight and the spectroscopic parameters for the rotation and vibration of the gas-phase molecule. A comparison of $\tilde{S}_g$ (thermodynamic) with $\tilde{S}_g$ (spectroscopic) provides a test of the validity of the third law of thermodynamics. The case of H_2O is particularly interesting since ice has a nonzero residual entropy at 0 K due to frozen-in disorder in the proton positions.[8]

The pertinent thermodynamic data needed to carry out the calculation of $\tilde{S}_g$ for water vapor are as follows: $\tilde{S}_l$ (1 bar, 298.15 K) = 66.69 J K^{-1} mol^{-1}, which is the calorimetric value obtained by assuming (erroneously) that the third law is valid for ice;[9] $\bar{C}_p(l)$, which is essentially independent of temperature over the range 298–355 K, has the average value 75.36 J K^{-1} mol^{-1}.[5] You may neglect the effect of pressure on the value of $\tilde{S}_l$ and may also treat $H_2O(g)$ as an ideal gas at the vapor pressure p_m associated with the temperature T_m at which $\Delta \tilde{H}_v$ has been determined. These two simplifying assumptions do not have much effect of the final result, and they facilitate the calculation of $\tilde{S}_g(p_m, T_m)$.

The experimental data needed to calculate the statistical entropy $\tilde{S}_g$ are the rotational temperatures $\Theta_r = hcB/k$ and the vibrational temperatures $\Theta_v = hv/k = hc\bar{v}/k$, where B ($=h/8\pi^2 Ic$) and $\bar{v}$ are frequencies in wavenumber units (cm^{-1}), and c is the speed of light in cm s^{-1}. Since the nonlinear H_2O molecule is an asymmetric top, the three principal moments of inertia are all different. The three rotational temperatures are $\Theta_A = 40.1$ K, $\Theta_B = 20.9$ K, and $\Theta_C = 13.4$ K.[10] There are three nondegenerate normal modes of vibration (one bend, one symmetric stretch, and one asymmetric stretch), and the vibrational temperatures are $\Theta_{v1} = 2290$ K, $\Theta_{v2} = 5160$ K, and $\Theta_{v3} = 5360$ K.[10] The statistical-mechanical entropy of an ideal gas of *nonlinear* molecules is a sum of a translational contribution (the so-called Sackur–Tetrode term), a rotational contribution for three degrees of rotational freedom, and a vibrational contribution for $3N - 6$ vibrational degrees of freedom, where N is the number of atoms in the molecule. These three contributions are given by [10]

$$\bar{S}(\text{trans}) = R[1.5 \ln M + 2.5 \ln T - \ln p - 1.15171] \tag{9}$$

where M is the molecular mass in grams and p is the pressure in bar,

$$\bar{S}(\text{rot}) = R\left[1.5 + \ln \frac{\pi^{1/2} T^{3/2}}{\sigma(\Theta_A \Theta_B \Theta_C)^{1/2}}\right] \tag{10}$$

where σ is the symmetry number, and

$$\bar{S}(\text{vib}) = R\left\{\sum_i \left[\frac{\Theta_{vi}}{T} \frac{1}{(e^{\Theta_{vi}/T} - 1)} - \ln(1 - e^{-\Theta_{vi}/T})\right]\right\} \tag{11}$$

with $i = 1$ to 3 in the case of H_2O.

Calculate a statistical/spectroscopic value of $\bar{S}_g(p_m, T_m)$, compare this with your thermodynamic value, and report the discrepancy. For most substances, these two $\bar{S}_g$ values agree and the third law is valid for the solid as T approaches 0 K. However, you should find that $\bar{S}_g(\text{spectroscopic}) > \bar{S}_g(\text{thermodynamic})$ for H_2O, which means that ice does not have the perfect order at 0 K required by the third law. The explanation for this in terms of the disorder in the proton configurations in ice was first given by Pauling[11] and is well described by Davidson.[8]

APPARATUS

Ballast bulb (5 liter); closed-tube manometer; double-neck distillation flask (if Claissen flask is used, a rubber policeman is needed to close off side-arm); a Celsius thermometer to cover the desired range, with one-hole rubber stopper to fit flask; reflux condenser with large-hole rubber stopper to fit flask; bent glass tube in a rubber stopper to fit top of condenser; electrical heating mantle; two long pieces of rubber tubing for circulating water through condenser; three long pieces of heavy-wall, rubber pressure tubing; two condenser clamps and clamp holders; ring stand; iron ring (and clamp holder, if necessary); 0 to 30°C thermometer.

If isoteniscope is used: a glass thermostat (e.g., large battery jar); mechanical stirrer; electrical blade heater; Variac; isoteniscope. (An isoteniscope similar to that shown in Fig. 3(a) can be purchased from Kontes Glass Co., Vineland, N.J. 08360.)

Liquid, such as water, n-heptane, cyclohexane, or 2-butanone.

REFERENCES

1. Any standard text, such as I. N. Levine, "Physical Chemistry," 3d ed., p. 201, McGraw-Hill, New York (1988).
2. J. H. Keenan, "Steam Tables," Wiley-Interscience, New York (1978).

3. K. S. Pitzer, D. Z. Lippmann, R. F. Curl, Jr., C. M. Huggins and D. E. Petersen, *J. Amer. Chem. Soc.* **77,** 3433 (1955).
4. "Selected Values of Properties of Hydrocarbons," *Natl. Bur. Stand. Circ.* C461, U.S. Government Printing Office, Washington, D.C. (1947).
5. "Handbook of Chemistry and Physics," 68th ed., CRC Press, Boca Raton, Fla. (1987/8).
6. G. W. Thomson and D. R. Douslin, "Determination of Pressure and Volume," in A. Weissberger and B. W. Rossiter (eds.), "Techniques of Chemistry: Vol. I. Physical Methods of Chemistry," part V, chap. 2, Wiley-Interscience, New York (1971).
7. H. Arm, H. Daeniker and R. Schaller, *Helv. Chim. Acta* **48,** 1772 (1966).
8. N. Davidson, "Statistical Mechanics," p. 371*ff.* and chap. 11, McGraw-Hill, New York (1962).
9. W. F. Giauque and J. W. Stout, *J. Amer. Chem. Soc.* **58,** 1144 (1936).
10. D. A. McQuarrie, "Statistical Thermodynamics," chap. 8, Harper & Row, New York (1973).
11. L. Pauling, *J. Amer. Chem. Soc.* **57,** 2680 (1935).

GENERAL READING

G. W. Thomson and D. R. Douslin, *loc. cit.*

EXPERIMENT 14
BINARY LIQUID–VAPOR PHASE DIAGRAM

This experiment is concerned with the heterogeneous equilibrium between two phases in a system of two components. The particular system to be studied is cyclohexanone–tetrachloroethane at 1 atm pressure. This system exhibits a strong negative deviation from Raoult's law, resulting in the existence of a maximum boiling point.

THEORY

For a system of two components (A and B) we have from the phase rule[1]

$$F = C - P + 2 = 4 - P \tag{1}$$

where C is the number of *components* (minimum number of chemical constituents necessary to define the composition of every phase in the system at equilibrium), P is the number of *phases* (number of physically differentiable parts of the system at equilibrium), and F is the *variance* or number of *degrees of freedom* (number of intensive variables pertaining to the system that can be independently varied at equilibrium without altering the number or kinds of phases present).

When a single phase is present, the pressure p, the temperature T, and the composition X_B (mole fraction of component B) of that phase can be independently varied; thus a single-phase two-component system at equilibrium is defined, except for its size,† by a point in a three-dimensional plot in which the coordinates are the intensive variables p, T, and X_B (see Fig. 1). When two phases, e.g., liquid L and vapor V, are present at equilibrium, there are four variables but only two of them can be independently varied. Thus, if p and T are specified, X_{BL} and X_{BV} (the mole fractions of B in L and V) are fixed at their *limiting* values (X_{BL}^0 and X_{BV}^0) for the respective phases at this p and T. The loci of points $X_{BL}^0(p, T)$ and $X_{BV}^0(p, T)$ constitute two surfaces, shown in Fig. 1. The shaded region between them may be interpreted as representing the coexistence of two phases L and V if in this region X_B is interpreted as a mole fraction of B *for the system as a whole*. Within the two-phase region X_B is *not* to be regarded as one of the intensive variables constituting the variance (although in a single-phase region it is indeed one of these variables). In a two-phase region the value of X_B determines the relative proportions of the two phases in the system; as X_B varies from X_{BL}^0 to X_{BV}^0, the molar proportion x_V of vapor phase varies from zero to unity:

$$x_V = 1 - x_L = \frac{X_B - X_{BL}^0}{X_{BV}^0 - X_{BL}^0} \tag{2}$$

Figure 1 is drawn for the special case of two components that form a complete range of *ideal solutions*, i.e., solutions that obey Raoult's law with respect to both components at all compositions. According to this law the vapor pressure (or partial pressure in the vapor) of a component at a given temperature T_1 is proportional to its mole fraction in the liquid. Thus, in Fig. 1 the light dashed lines representing the partial pressures p_A and p_B and the total vapor-pressure line (L, joining p_A^0 and p_B^0) are straight lines when plotted against the liquid composition. However, the total vapor pressure as plotted against the *vapor* composition is not linear. The curved line V, joining p_A^0 and p_B^0, is convex downward, lying *below* the straight line on the constant-temperature section; its slope has everywhere the same sign as the slope of L.

The vapor pressures p_A^0 and p_B^0 of the pure liquids increase with temperature (in accord with the Clapeyron equation) as indicated by the curves joining p_A^0 with $p_A'^0$ and p_B^0 with $p_B'^0$. At a constant pressure, say 1 atm, the boiling points of the pure liquids are indicated as T_A^0 and T_B^0. The boiling point of the solution, as a function of X_{BL} or X_{BV}, is represented by the curve L^* or V^* joining these two points. Neither curve is in general a straight line. If Raoult's law is obeyed, both are convex upward in temperature, the vapor curve lying above the liquid curve in temperature and being the more convex.

In most binary liquid–vapor systems Raoult's law is a good approximation for a component only when its mole fraction is close to unity. Large deviations from this law are commonplace for the dilute component, or for

---

† The complete definition of the system would, of course, include also its shape, description of surfaces, specification of fields, etc.; ordinarily these have negligible effects as far as our present discussion is concerned.

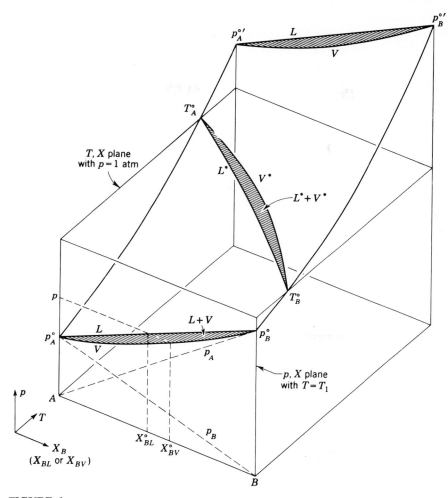

FIGURE 1
Schematic three-dimensional vapor-liquid equilibrium diagram for a two-component system obeying Raoult's law.

both components when the mole fraction of neither is close to unity. If at a given temperature the vapor pressure of a solution is higher than that predicted by Raoult's law, the system is said to show a *positive deviation* from that law. For such a system the boiling-point curve L^* at constant pressure is usually convex downward in temperature. If at a given temperature the vapor pressure of the solution is lower than that predicted by Raoult's law, the system is said to show a *negative deviation*; in this case the curve L^* is more convex upward. These deviations from Raoult's law are often ascribed to differences between "heterogeneous" molecular attractions ($A - - - B$) and "homogeneous" attractions ($A - - - - A$ and $B - - - - B$). Thus, the existence of a positive deviation implies that homogeneous attractions are stronger than heterogeneous attractions, and a negative deviation implies the reverse. This interpretation is consistent with the fact that positive deviations are usually associated with positive heats of mixing and volume expansions on mixing while negative deviations are usually associated with negative heats and volume contractions.

In many cases the deviations are large enough to result in maxima or minima in the vapor-pressure and boiling-point curves, as shown in Fig. 2. Systems for which the boiling-point curves have a maximum include acetone–chloroform and hydrogen chloride–water; systems with a minimum include methanol–chloroform, water–ethanol, and benzene–ethanol. At a maximum or a minimum, the compositions of the liquid and of the vapor are the same; accordingly there is a *point of tangency* of the curves L and V and of the curves L^* and V^* at the maximum or minimum. At every value of X_B the slope of V (or V^*) has the same sign as the slope of L (or L^*); one is zero where and only where the other is zero, at the point of tangency. (A common error in curves of this kind, found even in some textbooks, is to draw a cusp—point of discontinuity of slope—in one or both curves at the point of tangency; both curves are, in fact, smooth and have continuous derivatives.)

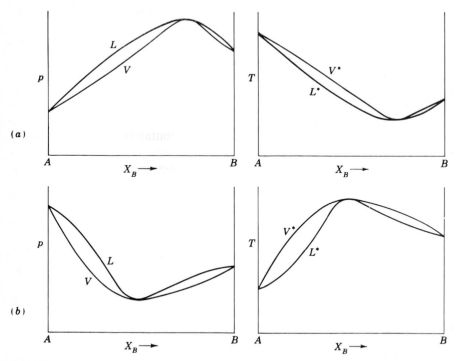

FIGURE 2
Schematic vapor-pressure and boiling-point diagrams for systems showing (a) a strong positive deviation and (b) a strong negative deviation from Raoult's law.

If the homogeneous attractions are very much stronger than the heterogeneous ones, phase separation may occur in the liquid; i.e., there is a limited mutual solubility of the two liquid components over certain pressure and temperature ranges. If the pressures and temperatures at which two liquid phases coexist include those at which equilibrium also exists with the vapor phase, a boiling-point diagram of a type similar to that shown in Fig. 15-1 is found. (That figure is a *melting-point* diagram, showing two solid phases in equilibrium with a liquid phase, but the principles are the same. It will be noted that boundaries of fields show discontinuities in slope at points representing the coexistence of three phases.)

Liquid–vapor phase diagrams, and boiling-point diagrams in particular, are of importance in connection with *distillation,* which usually has as its object the partial or complete separation of a liquid solution into its components.[2] Distillation consists basically of boiling the solution and condensing the vapor into a separate receiver. A simple "one-plate" distillation of a binary system having no maximum or minimum in its boiling-point curve can be understood by reference to Fig. 3. Let the mole fraction of B in the initial solution be represented by X_{BL1}. When this is boiled and a small portion of the vapor is condensed, a drop of distillate is obtained, with mole fraction X_{BV1}. Since this is richer in A than is the residue in the flask, the residue becomes slightly richer in $B,$ as represented by X_{BL2}. The next drop of distillate X_{BV2} is richer in B than was the first drop. If the distillation is continued until all the residue has boiled away, the last drop to condense will be virtually pure B. To obtain a substantially complete separation of the solution into pure A and B by distillations of this kind it is necessary to separate the distillate into portions by changing the receiver during the distillation, then subsequently to distill the separate portions in the same way, and so on, a very large number of successive distillations being required. The same result can be achieved in a single distillation by use of a fractionating column containing a large number of "plates"; discussion of the operation of such a column is beyond the scope of this book. If there is a maximum in the boiling-point curve (Fig. 2b) the

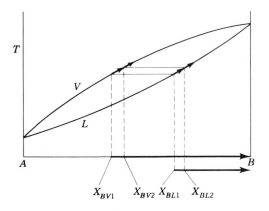

FIGURE 3
Variation of liquid and vapor compositions during distillation.

compositions of vapor and residue do not approach pure A or pure B, but rather the composition corresponding to the maximum. A mixture with this composition will distill without change in composition and is known as a "constant-boiling mixture" or "azeotrope." These terms are also applied to a mixture with a minimum boiling point. Azeotropes are important in chemical technology. Occasionally they are useful (as in constant-boiling aqueous hydrochloric acid, used as an analytical standard); often they are nuisances (as in the case of the azeotrope of 95 percent ethanol with 5 percent water, the existence of which prevents preparation of absolute ethanol by direct distillation of dilute solutions of ethanol in water). Extensive lists of azeotropes have been compiled.[3]

METHOD

A boiling-point curve can be constructed from data obtained in actual distillations in an ordinary "one-plate" distilling apparatus. Small samples of the distillate are taken directly from the condenser, after which small samples of the residue are withdrawn with a pipette. The samples of distillate and residue are analyzed, and their compositions are plotted on a boiling-point diagram against the temperatures at which they were taken. In the case of the distillate the temperature to be plotted for each sample should be an average of the initial and final values during the taking of the sample. In the case of the residue the temperature to be plotted should be that recorded at the point where the distillation is stopped to take the sample of residue.

For analysis of the samples, a physical method is often preferable to chemical methods. Chemical analysis usually is appropriate only when a simple tritration of each sample is involved, as in the case of the system $HCl-H_2O$. If a physical property is chosen as the basis for an analytical method, it should be one which changes significantly and smoothly over the entire composition range to be studied. The refractive index is one property that can be used for the cyclohexanone–tetrachloroethane system. The values of n_D^{20} are 1.4507 for cyclohexanone and 1.4942 for 1,1',2,2'-tetrachloroethane, and $\log n_D^{20}$ is almost a linear function of the weight percent of cyclohexanone. Thus one can interpolate linearly between the values listed in Table 1, and then convert weight percentages into mole fractions.

Another property that could also be used is the density, which varies in a nonlinear way between 0.9478 g/mL for cyclohexanone and 1.600 g/mL for tetrachloroethane at 20°C. In this case, a calibration curve should be constructed from known solutions or provided by the instructor. The density of each distillation sample can be measured by pipetting 1 mL into a small, previously weighed vial and then weighing again (to the nearest 0.1 mg).

Warning. These solutions will decompose slowly at room temperature ($\sim$1 day) and more rapidly during distillation at high temperatures. Impure solutions become yellowish, which can interfere with the refractive index measurements. Use well-purified starting materials and do not prolong the

TABLE 1
Logarithm of refractive index for cyclohexanone–tetrachloroethane mixtures

$\log n_D^{20}$	W% $C_6H_{10}O$	$\log n_D^{20}$	W% $C_6H_{10}O$	$\log n_D^{20}$	W% $C_6H_{10}O$
0.17441	0	0.16864	40	0.16360	80
0.17298	10	0.16719	50	0.16256	90
0.17155	20	0.16582	60	0.16158	100
0.17010	30	0.16473	70		

distillations unnecessarily. It should also be noted that many chlorinated hydrocarbons, including tetrachloroethane, are toxic chemicals. Chronic exposure to tetrachloroethane can cause liver damage. The toxic oral dose is quite high (~0.5 g per kilogram of body weight), but smaller doses can cause short-term medical problems. **Do not pipette by mouth**.

EXPERIMENTAL

A simple distilling apparatus that can be used for this experiment is shown in Fig. 4. The thermometer bulb should be about level with the side-arm to the condenser. Except when samples of distillate are being taken for analysis, an adequate receiving flask should be placed at the lower end of the condenser.

Before beginning the distillations, prepare twenty 5-mL shell vials for taking samples. Write on the corks the designations $1L, 1V, 2L, \ldots, 10V$ ($L = liquid$ residue; $V = $ condensed *vapor* or distillate). The samples to be taken are about 2 mL in size.

When the distillation is proceeding at a normal (not excessive) rate at about the desired temperature, quickly replace the receiver with a vial and read the thermometer. After about 2 mL has been collected, read the thermometer again, replace the receiver, and cork the vial tightly. Turn off and lower the heating mantle to halt the distillation. At the point where the temperature just begins to fall, record another thermometer reading. After the flask has cooled about 15°C, remove the stopper at the top of the flask and insert a 2-mL pipette equipped with a rubber bulb. Fill the pipette, discharge it into the appropriate vial, and stopper the vial.

The following procedure is recommended for economical use of materials in carrying out this experiment. The paragraph numbers correspond to sample numbers. A graduated cylinder is adequate for measuring liquids. The temperatures recommended are those appropriate for 760 Torr; at ambient pressure differing markedly from this value, the temperatures should be adjusted accordingly. For example, at Denver (altitude 1609 m) the average atmospheric pressure is 836 mbar = 627 Torr. Since the enthalpies of vaporization of cyclohexane and tetrachloroethane are both near 40 kJ mol^{-1}, this

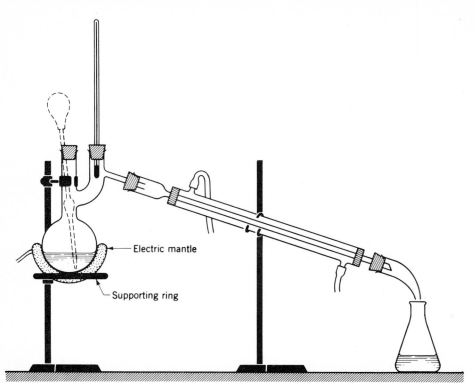

FIGURE 4
Distillation apparatus.

pressure change will reduce the boiling points of the pure materials by ~7°C compared with the values at 760 Torr. Comparable changes are expected for the azeotrope and for intermediate compositions.

1. Pure tetrachloroethane: Introduce 125 mL (~200 g) of 1,1',2,2'-tetra-chloroethane into the flask. Distill enough to give a constant temperature (should be near 146°C at 760 Torr). Collect samples (1V and 1L) for analysis.

2. 149°C (tetrachloroethane-rich side of azeotrope): Cool the distilling flask, and return the excess distillate of paragraph 1 to the flask. Add 38 mL (~36 g) of cyclohexanone. Begin distillation. When the temperature reaches 149°C, collect about 2 mL of distillate (2V) and 2 mL of residue (2L).

3. 151°C: Resume the distillation. Distill until the temperature reaches 151°C (this may take some time) and collect samples (3V, 3L).

4. 154°C: Resume the distillation. When the temperature reaches 154°C, collect samples (4V, 4L).

5. 157°C: Cool the flask somewhat, and add 35 mL of tetrachloroethane and 25 mL of cyclohexanone. Resume the distillation. When the temperatures reaches approximately 157°C, collect samples (5V, 5L).

6. Azeotrope: Cool the flask somewhat and add 36 mL of tetrachloroethane and 54 mL of cyclohexanone. Resume the distillation until the boiling point ceases to change significantly, and take samples (6V, 6L). (If the boiling point does not become sufficiently constant, analyze the remaining residue, and make up 100 mL of solution to the composition found. Distill to constant temperature and take samples.)

7. Pure cyclohexanone: Introduce 105 mL of cyclohexanone into the clean flask and determine the boiling point as in paragraph 1. (The temperature should be near 155°C at 760 Torr.) Collect samples (7V and 7L).

8. 156.5°C (cyclohexanone-rich side of azeotrope): Cool the distilling flask, return the excess distillate of paragraph 7, and add 20 mL of tetrachloroethane. Resume the distillation, and take samples (8V, 8L) at about 156.5°C.

9. 157°C: Cool the flask somewhat, and add 50 mL cyclohexanone and 17 mL tetrachloroethane. Resume the distillation and collect samples (9V, 9L) at about 157°C.

10. Azeotrope: Resume the distillation, continue to constant boiling temperature, and take samples (10V, 10L).

The indexes of refraction should be measured and recorded as soon as possible (the samples decompose on standing). The refractometer and the procedure for its use are described in Chapter XVIII. (If the experiment is being done by several teams using the same refractometer, it is wise to take samples to the refractometer as soon as six or eight samples are ready, or fewer if the instrument happens to be free.) If careful attention is given to the proper technique of using the refractometer, it should be possible to take readings at the rate of 1 sample per minute.

At the end of the experiment all cyclohexanone–tetrachloroethane mixtures should be poured into a designated waste vessel.

At some time during the laboratory period, the barometer should be read. The ambient temperature should be recorded for the purpose of making thermometer stem corrections.

CALCULATIONS

Interpolate in Table 1 and then convert the weight percentages to mole fractions. Plot the temperatures (after making any necessary stem corrections; see Chapter XVI) against the mole fractions. Draw one smooth curve through the distillate points V and another through the residue points L. Label all fields of the diagram to indicate what phases are present. Report the azeotropic composition and temperature, together with the atmospheric pressure (i.e., the properly corrected barometer reading).

APPARATUS

Claissen distilling flask; mercury thermometer, graduated to 0.1°C or a digital resistance thermometer with a resolution of ±0.1°C; one-hole cork stopper for thermometer to fit flask; solid cork stopper; straight-tube condenser with one-hole stopper to fit distilling side arm; two lengths of rubber hose for condenser cooling water; distilling adapter with one-hole stopper to fit end of condenser; two clamps and clamp holders; two ring stands; one iron ring; electrical heating mantle (or steam bath); 20 small vials (screw-top or with corks to fit); 100-mL graduated cylinder; two wide-mouth 250-mL flasks; 2-mL pipette; pipetting bulb; two 500-mL glass-stoppered Erlenmeyer flasks.

Refractometer, thermostated at 25°C; sodium-vapor lamp (optional); eye droppers; *clean* cotton wool; acetone wash bottles; pure 1,1′,2,2′-tetra-chloroethane (300 mL) and pure cyclohexanone (350 mL); acetone for rinsing; large bottle for disposal of waste solutions.

REFERENCES

1. P. W. Atkins, "Physical Chemistry," 3d ed., chap. 9, Freeman, New York (1986).
2. A. J. Teller, *Chem. Eng.* **61,** 168 (1954).
3. L. H. Horsley, "Azeotropic Data," ("Advances in Chemistry," no. 6), American Chemical Society, Washington, D.C. (1952).

GENERAL READING

A. Findlay, A. N. Campbell, and N. O. Smith, "The Phase Rule and its Applications," 9th ed., Dover, New York (1951).
A. Reisman, "Phase Equilibria: Basic Principles, Applications, Experimental Techniques," Academic Press, New York (1970).

EXPERIMENT 15
BINARY SOLID–LIQUID PHASE DIAGRAM

In this experiment we are concerned with the heterogeneous equilibrium between solid and liquid phases in a two-component system. From the many systems[1,2,3] that are suitable for study, the system naphthalene–diphenylamine has been selected for this experiment because of the simplicity of its phase diagram and the convenience of its temperature range.

THEORY

The principles underlying this experiment are identical with those discussed in Exp. 14. For our discussion here we can merely replace L (liquid) in that experiment by S (solid) and V (vapor) by L (liquid). Solid–liquid equilibria differ from liquid–vapor equilibria in being essentially independent of pressure changes of the order of a few atmospheres. This is a consequence of the Clapeyron equation, Eq. (13-3), owing to the small molar volume change associated with fusion. Accordingly we shall be concerned only with temperature–composition diagrams at 1 atm pressure.

All types of phase diagrams that have been found for liquid–vapor equilibria are also possible for solid–liquid equilibria. In some binary systems (particularly metal systems) the two components may form *solid solutions,* sometimes covering the entire composition range from pure A to pure B. Examples of systems with solid solutions covering the entire range include copper–nickel,[4] which has a phase diagram of the type shown in Fig. 14-3, with a nearly linear dependence of melting point on composition; d- and l-carvoxime $(C_{10}H_{14}NOH)$,[5] which has a maximum melting point (analogous to the maximum boiling point shown in Fig. 14-2b); and bromobenzene–iodobenzene,[6] which has a minimum melting point (analogous to Fig. 14-2a). An important condition for the existence of a complete range of solid solutions is that A and B have the same type of crystal structure. The solid solutions must also have the same crystal structure as A and B, the atomic sites being occupied largely at random by the two kinds of atoms or molecules.

More often, when solid solutions exist, they are limited in the range of their compositions. This may result from a difference between the crystal structures of pure A and pure B, or from differences in atomic or molecular size and shape resulting in "lattice incompatibility," or from factors analogous to those resulting in strong positive deviations from Raoult's law in liquids, or from any combination of these. Thus we may have two solid solutions at equilibrium, one (α) consisting predominantly of component A and the other (β) consisting predominantly of component B. These phases are often described as a solid solution of B in A and one of A in B, respectively.

Solid–liquid equilibria in a system of limited solid solubilities may yield a phase diagram of the type shown in Fig. 1. For this diagram, the compositions of the two phases (α and β, α and L, or β and L) coexisting at equilibrium at a given temperature and 1 atm pressure are found in the same way as in Exp. 14. A new feature is the possibility of the coexistence, at equilibrium, of three phases: α, β, and L. According to the phase rule, Eq. (14-1), with two components and three phases there is but one degree of freedom, and that is taken up in the arbitrary specification of the pressure as 1 atm. Thus the temperature is fixed (T_E) and the compositions of all phases are fixed $(X_B = X_\alpha^0, X_\beta^0, X_E)$. A system of three phases coexisting at equilibrium is represented by any point on the heavy solid horizontal line of Fig. 1. The point T_E, X_E is called the "eutectic point"; its significance will be discussed later.

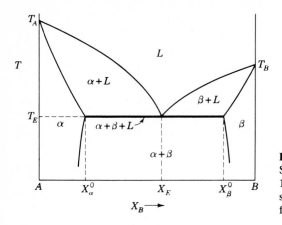

FIGURE 1
Schematic solid-liquid phase diagram at 1 atm for a binary system with limited solid solubilities and no compound formation.

Diagrams of the kind shown in Fig. 1, exhibiting limited solid solubility, exist for many systems, including azobenzene–azoxybenzene,[7] bismuth–tin, and lead–tin.

Very commonly, particularly in organic systems, where a high degree of lattice incompatibility is almost always present, solid solubility is so small that it may be regarded as negligible. Here the solid-solution regions α and β shrink to the vertical lines A and B and the type of phase diagram shown in Fig. 2(a) is obtained. The principal features of this type of diagram can be understood at least semiquantitatively from the theory of freezing-point depression. (It will be recalled that an essential requirement for the theory of

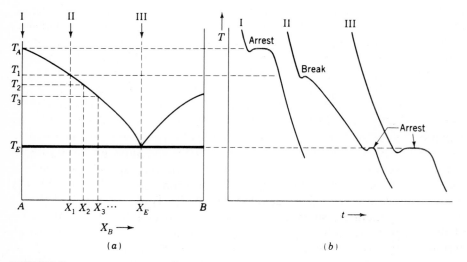

(a)   (b)

FIGURE 2
(a) Schematic solid-liquid phase diagram at 1 atm for a binary system with negligible solid solubilities and no compound formation. (b) Schematic cooling curves for this system at various overall compositions.

freezing-point depression developed in Exp. 10 is the absence of appreciable solid solubility.) From Eq. (10.5) we obtain the following equation for the solid–liquid curve ("liquidus curve") starting from the left at T_A, the melting point of A:

$$T \cong T_A + \frac{RT_A^2}{\Delta \bar{H}_A} \ln(1 - X_B) = T_A - \frac{RT_A^2}{\Delta \bar{H}_A}\left(X_B + \frac{X_B^2}{2} + \cdots\right) \qquad (1)$$

where $\Delta \bar{H}_A$ is the heat of fusion of A. Clearly this theoretical curve starts with a finite negative slope determined by the melting point and heat of fusion of pure component A; the slope increases in steepness with increasing X_B, so that the curve is concave downward. Similarly, for the liquidus curve starting from the right at T_B we obtain

$$T \cong T_B + \frac{RT_B^2}{\Delta \bar{H}_B} \ln X_B = T_B - \frac{RT_B^2}{\Delta \bar{H}_B}\left[(1 - X_B) + \frac{(1 - X_B)^2}{2} + \cdots\right] \qquad (2)$$

The eutectic composition and euctectic temperature are given by the intersection of the two liquidus curves and can be estimated by solving simultaneously Eqs. (1) and (2), on the assumption that the liquid represents an ideal solution with respect to both components over its entire composition range. This assumption is often not even roughly valid, especially in metal systems.

With the aid of Fig. 2(a) we can predict the general nature of the cooling curves in a system of this kind. These curves, examples of which are shown in Fig. 2(b), are plots of temperature against time obtained when liquid solutions of various compositions are allowed to cool by slow leakage of heat to the surroundings; some features of such curves are discussed in Exp. 10. When a liquid consisting of pure A is cooled, the temperature falls until solid A begins to form; the temperature then remains constant until solidification is complete, whereupon it falls again. Thus, the curve shows a "thermal arrest." While two phases are present in this one-component system, there is only one degree of freedom, which is taken up in the arbitrary specification of the pressure, and the temperature is fixed. When a liquid having the eutectic composition is cooled, the behavior is similar in that a thermal arrest is obtained. Although the number of components is increased from one to two, the number of phases at the eutectic point is increased from two to three, and again we have a single degree of freedom which is taken up in the arbitrary specification of the pressure. When a liquid of some other composition—say X_1 in Fig. 2(a)—is cooled, solid A begins to form at temperature T_1. This tends to deplete the liquid of component A, so that its composition passes through $X_2, X_3, \ldots$, and the temperature falls as long as solid A alone continues to come out of solution. With two components and two phases there are two degrees of freedom; thus, at constant pressure the temperature is not fixed but varies with the composition of the liquid. However, the slope of the cooling curve is much less than that for cooling of a single phase, owing to the heat liberated by the formation of solid A; see Eqs. (10-24) and (10-25). The abrupt change in slope, which occurs when solid A begins to form, is called a "break." When the composition of the solution finally reaches X_E, solid B begins to form together with solid A, and the two solids continue to separate from solution at the temperature T_E until no liquid remains; thus we have an arrest.

It is beyond the scope of this book to consider all possible types of solid–liquid binary phase diagrams. It will suffice here to mention two other important features, which arise when A and B can combine to form a solid compound of some stoichiometric composition A_mB_n or a solid solution varying in some degree from such a composition. Generally this compound or solid solution will differ in crystal structure both from A and from B. In Fig. 3 the phase diagram for the system Mg–Ni is shown. There are two solid compounds, Mg_2Ni and $MgNi_2$, with no appreciable solid solubility. The compound $MgNi_2$ has a sharp melting point and melts to give a liquid of the same composition. The liquidus curve has a maximum (horizontal tangent, no cusp) at this composition. The compound Mg_2Ni does not melt in this way; instead it undergoes a decomposition, or "peritectic transformation," to another solid phase, namely, $MgNi_2$, and to a liquid poorer in Ni than was the original compound. During this transformation, three phases are present— liquid, Mg_2Ni, and $MgNi_2$—and the temperature (at 1 atm) is therefore fixed. Some binary phase diagrams, particularly in metal systems, are exceedingly complicated and show many distinct compounds or phases.[1]

METHOD

The relation of cooling curves to phase diagrams (as illustrated by Fig. 2) forms the basis of "thermal analysis," an important technique for determining phase diagrams. This technique consists of cooling liquid mixtures of various compositions, making plots of temperature against time, and drawing inferences from features of these curves such as thermal arrests and breaks. A thermal arrest indicates the solidification temperature of a pure component, a compound, or a eutectic mixture (under favorable circumstances it may also indicate a peritectic transformation). A break indicates a point on the liquidus

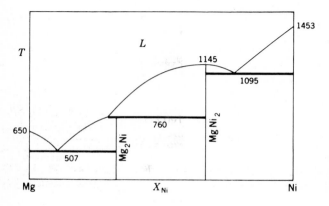

FIGURE 3
Phase diagram for the system
Mg–Ni at 1 atm.

curve corresponding to the temperature at which solid first appears at the given composition.

The interpretation of such curves is often subject to experimental difficulties. A frequent difficulty is supercooling (see Fig. 10-3), which tends to obscure breaks or make them difficult to distinguish from short arrests. When the initial composition is far from the eutectic, the thermal arrest at the eutectic temperature is often hard to observe, since the amount of liquid is small relative to the amount of solid already formed and stirring becomes very difficult. When solid solubility exists, it is often very difficult to obtain the true "solidus" curve from cooling curves. This is due to the difficulty of maintaining solid–liquid equilibrium, which requires a continuous variation in the composition of the solid. As such a variation requires either diffusion in the solid or continuous dissolution and reprecipitation of the solid, it is ordinarily very slow or nonexistent except at elevated temperatures. Owing to these difficulties, thermal analysis must often be supplemented by other techniques in elucidating a phase diagram; these include microscopic examination of solidified specimens (to determine the number of solid phases present or to distinguish regions of small crystals that form at the eutectic temperature from regions of larger crystals that form while only one solid phase is precipitating) and x-ray diffraction analysis (which enables identification of the crystalline phases that are present).

For the system naphthalene–diphenylamine, a reasonable number of cooling curves can easily be obtained with the use of either a mercury thermometer or a thermocouple. Other systems that may be studied conveniently with little or no change in apparatus or technique include naphthalene–p-dichlorobenzene,[8] naphthalene–p-nitrotoluene,[9] biphenyl–benzophenone, o-nitrophenol–p-toluidine, and acetanilide–phenacetin. At lower temperatures, with a refrigerant bath (see Chapter XVI for details) systems such as chlorobenzene–cyanobenzene and dimethylsulfoxide–water[10] can be studied;† an alcohol thermometer or a thermocouple may be used. A thermocouple is recommended also for study at higher temperatures of such systems as bismuth–tin and lead–tin.[11] Systems with phase diagrams of greater complexity may require a larger number of cooling curves; it may be appropriate to assign different composition ranges to several groups who will pool their data.

EXPERIMENTAL

The materials to be used should be of reagent grade. Diphenylamine is subject to discoloration resulting from air oxidation; it should preferably be obtained

---

† **Safety note:** All of the organic compounds mentioned here are **toxic** and must be handled with care; minimize skin contact and inhalation of vapors.

TABLE 1
Suggested compositions for thermal analysis ($A =$ diphenylamine; $B =$ naphthalene)

Run no.	Wt. % B	Prepare by adding . . .		. . . to sample used in run no.
1	100	5 g	B	
2	83.3	1 g	A	1
3	66.7	1.5 g	A	2
4	50.0	2.5 g	A	3
5	33.3	5 g	A	4
6	0	5 g	A	
7	16.7	1 g	B	6
8	25.0	0.67 g	B	7
9	Eutectic	(See text)		

in containers packed in an argon atmosphere and should be taken from a freshly opened container.

The apparatus consists of an inner test tube with a 20-mL capacity, a thermometer or thermocouple mounted in a notched one-hole rubber stopper fitting the test tube, a wire stirrer, a larger test tube to serve as an outer jacket, a large one-hole rubber stopper to support the inner test tube in the outer one, and a bath of cold water or ice and water. If a 0 to 100°C thermometer is used, it should have a stem long enough to permit all readings above 25°C to be visible above the stopper. If a thermocouple (such as copper–constantan or Chromel–Alumel) is used, the junction should be immersed in wax or oil at the bottom of a closed glass tube 6 to 8 mm in diameter. This tube can be mounted in the stopper in the same way as a thermometer. The reference junction should be in a similar tube immersed in a slushy mixture of ice and distilled water contained in a Dewar flask, and the thermoelectric potential can be measured with a digital voltmeter or a potentiometer (see Chapter XV).

Make up mixtures in accordance with Table 1. Weigh materials to 0.01 g on a triple-beam balance. Note that in most instances the mixture to be studied is made by adding A or B to the previous mixture in order to minimize the quantities of materials required.

To obtain a cooling curve, heat the inner test tube containing the mixture in a beaker of hot water until the solid is completely melted, wipe the tube dry, place it in the outer jacket, and place the assembly in cold water or ice and water. Stir continuously, and read the temperature at regular intervals (say 30 s). Continue reading to below 30°C if possible.†

Plot each temperature reading against time as soon as it is obtained.

---

† If time is limited, discontinue the cooling on all runs except 4, 8, and 9 shortly after the first definite break has occurred.

After each run determine break and/or arrest temperatures and plot them against weight percent B. From the results of runs 1 through 8 draw the liquidus curves and extrapolate them to an intersection at a point on the eutectic line. Determine the eutectic composition, and make up a mixture having this composition. Run a cooling curve on this mixture.

At the end of the experiment dispose of the mixtures and clean the glassware thoroughly. First liquefy the mixture by warming the test tube in a beaker of warm water, and then pour the contents into a designated disposal container. Remove the large rubber stopper, and wash the test tube, thermometer, and stirrer with petroleum ether or toluene. Use several small portions in succession in order to minimize the amount of solvent required, and pour the used solvent into the designated container. Dry the apparatus.

CALCULATIONS

Convert weight-percentage compositions to mole fractions. Plot the break and arrest temperatures against the overall composition X_B. Draw the eutectic line and the liquidus curves. Label all fields to show the phases present.

From the limiting slopes of the liquidus curves, estimate the heats of fusion of A and B, assuming that there is no appreciable solid solubility. With these values for the heats of fusion plot ideal curves using Eqs. (1) and (2). Compare the calculated intersection temperature and composition with the eutectic temperature and composition found by experiment.

DISCUSSION

Why does the liquidus curve have a horizontal tangent at the melting point of a pure compound while it has a finite slope at the melting point of either of the two pure components?

A method sometimes used for obtaining the liquidus curve in a system of sufficient optical transparency is the determination of the "clear point" (the temperature at which the last solid disappears on warming) and of the "cloud point" (the temperature at which the first solid appears on cooling) for each of a number of compositions. What would you expect to be the limitations of this procedure?

APPARATUS

Test tube with notched one-hole stopper; 0 to 100°C thermometer (all the stem above 25°C should be above the stopper); wire ring stirrer; outer test tube with large stopper (see Exp. 11 for above items); two 500-mL beakers; bunsen burner; tripod and wire gauze. A thermocouple with potentiometer and a small Dewar flask for the reference junction may be provided in place of the thermometer.

Diphenylamine (30 g); naphthalene (15 g); petroleum ether or toluene (for washing glassware); ice (if needed for cold junction); waste vessel.

REFERENCES

1. M. Hansen, "Constitution of Binary Alloys," McGraw-Hill, New York (1958).
2. J. Timmermans, "Les Solutions concentrées," Masson, Paris (1936).
3. International Critical Tables (I.C.T.), vols. II (pp. 400–455) and IV (pp. 22–215), McGraw-Hill, New York (1928).
4. M. Hansen, *op. cit.*, p. 602.
5. J. Timmermans, *op. cit.*, p. 30.
6. *Ibid.*, p. 88.
7. *Ibid.*, p. 348.
8. P. P. Blanchette, *J. Chem. Educ.*, **64**, 267 (1987).
9. J. M. Wilson, R. J. Newcombe, A. R. Denaro, and R. M. W. Rickett, "Experiments in Physical Chemistry," 2d ed., p. 46, Pergamon, Oxford (1968).
10. F. A. Betelheim, "Experimental Physical Chemistry," pp. 241–249, W. B. Saunders, Philadelphia (1971).
11. F. Daniels, J. W. Williams, P. Bender, R. A. Alberty, C. D. Cornwell, and J. E. Harriman: "Experimental Physical Chemistry," 7th ed., pp. 123–124, McGraw-Hill, New York (1970).

GENERAL READING

P. Gordon, "Principles of Phase Diagrams in Materials Systems," McGraw-Hill, New York (1968).

EXPERIMENT 16
LIQUID–VAPOR COEXISTENCE CURVE AND THE CRITICAL POINT

The object of this experiment is to measure the densities of coexisting carbon dioxide liquid and vapor near the critical temperature. The critical density of carbon dioxide and a value of the critical temperature will also be obtained from a Cailletet–Mathias curve constructed from the data.

THEORY

At its critical point, the liquid and vapor densities of a fluid become equal. Since this point is very difficult to determine directly, an extrapolation procedure is often used. This procedure is based on the "law of rectilinear diameters,"[1,2] which states that

$$\rho_{av} \equiv \tfrac{1}{2}(\rho_l + \rho_v) = \rho_0 - cT \tag{1}$$

where ρ_{av} is the average of the densities of the liquid and vapor that are in coexistence with each other at temperature T, and ρ_0 and c are constants characteristic of the fluid. The critical point is taken to be that point at which the straight line drawn through ρ_{av} as a function of T intersects the coexistence curve (locus of the liquid and vapor density values as a function of temperature). Such a plot is known as a Cailletet–Mathias curve.

METHOD

Six to eight heavy-wall glass capillary tubes of uniform bore and known cross-sectional area are filled with different, known amounts of CO_2. The diameters of the capillary tubes can be obtained prior to filling by weighing a thread of mercury of measured length contained in them. The capillaries are then filled on a vacuum system by freezing into them CO_2 gas contained in a known volume at a known pressure and sealing them off.[3] Residue corrections for the amount of CO_2 left behind in the vacuum system after sealing off the tubes should be made if necessary. Heavy-walled capillary tubing with a nominal bore of 1 mm is convenient for use in this experiment. Lengths of 10 to 30 cm are suitable; perhaps 20 cm is the best length to use. (It may be desirable to prepare a few tubes of different lengths which are filled to the same pressure as a way of testing the reproducibility of the method.) The amounts of CO_2 to be loaded into the tubes should be chosen to yield a range of pressures between 18 and 30 bar at 25°C. *The data on the calibration and loading of the capillaries should be made available at the beginning of the laboratory period.*

The filled capillary tubes are mounted in a well-stirred, temperature-regulated water bath which can be warmed or cooled and whose temperature can be measured accurately. If a tube contains less than a critical amount of CO_2, the meniscus between the liquid and vapor will fall toward the bottom of the tube on warming. At a certain temperature, say T_0, the meniscus in a given tube will reach the bottom; thus the filling density in that tube will equal the density of the saturated vapor at the temperature T_0. This density is the total mass of CO_2 in the tube (which is known) divided by the internal volume of the tube (= cross-sectional area × length of capillary).

At temperature T_0, all of the other tubes *that still have both liquid and vapor present* will contain vapor whose density is the same as that just calculated. The volume of vapor in each of these tubes can be determined (= cross-sectional area × length of tube occupied by vapor). Therefore, the mass of CO_2 present as vapor in each of these tubes can be found (= density of vapor × volume of vapor). The mass of liquid CO_2 in each tube can then be determined (= total CO_2 – vapor). Since the volume of the liquid in each tube can also be determined (= cross-sectional area × length of tube occupied by liquid), the density of the liquid phase can be found at temperature T_0.

EXPERIMENTAL

A glass-walled constant-temperature bath with an adjustable thermoregulator and an efficient stirrer is required. This bath should also be provided with a precision thermometer and a light to illuminate the capillary tubes. A cathetometer is used to measure the height of the menisci in the capillaries, which are completely immersed in the thermostat bath. If the cathetometer is not level, it should be adjusted using the spirit levels on the base and on the telescope. Focus the telescope on one of the capillaries in the bath. The vertical cross-hair should be lined up parallel to the side of the capillary. This may be achieved, if necessary, by rotating the eyepiece. When reading the height of an object, such as a meniscus, always use the same position on the horizontal cross-hair—usually near the center of the field. Note that the image is inverted by the telescope.

Now check the temperature of the bath. If it is above 24°C add a little ice to cool it to that temperature; while the bath is coming to equilibrium, one can begin to measure the dimensions of the tubes (see below). If the temperature of the bath is constant at 24°C or below, record the temperature and measure the positions of the menisci in all the tubes. The readings of the cathetometer should be recorded to the nearest 0.005 cm, and one should measure one of the tubes several times to establish the reproducibility of these readings. During the period in which these cathetometer measurements of the menisci levels are being made, the bath temperature should be determined several times and recorded to the nearest 0.01°C. The temperature should not vary over a range of more than 0.04°C during the measurements, and the average of these temperature readings should be used in the subsequent calculations.

It will be convenient to number the tubes in sequence from left to right, in order to simplify the tabulation of data. Once the first set of data are recorded, begin heating the bath to the next desired temperature by making a suitable adjustment to the temperature controller (see Chapter XVI). Recommended bath temperatures are ~24, 26.0, 28.0, 29.2, 29.7, 30.1, 30.5°C, and higher by increments of 0.3°C (or less) until the critical temperature is reached. The bath should regulate at each temperature for at least 10 min (preferable more near T_c) before the meniscus levels are measured.

Once you have begun making readings, be very careful not to jar the cathetometer or to make any more leveling adjustments. During the course of the entire experiment the only permissible movements of the cathetometer are the raising, lowering, and traversing of the telescope.

While the bath is heating up from one temperature setting to the next, carry out measurements of the dimensions of the capillary tubes. Since the top and bottom ends of these tubes are tapered or rounded, it is necessary to make two cathetometer readings at each end. As shown in Fig. 1, readings are made at the extreme tips of the inside space (a and d) and at levels where the tapering begins (b and c). The center section (bc) is a cylinder of uniform cross-sectional area; the small ends (ab and cd) can be described by cones or hemispheres.

If possible, try to observe the *critical opalescence* in carbon dioxide near the critical temperature. This milky opalescence is caused by large spatial fluctuations in the density and can only be seen if the filling density is close to

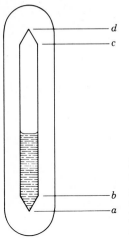

FIGURE 1
Sketch of a capillary tube with conical ends. The internal volume is
given by $V = \pi r^2(h_{bc} + \frac{1}{3}h_{ab} + \frac{1}{3}h_{cd})$, where $2r$ is the inner diameter of
the tube and h_{ij} represents the vertical distance between levels i and j.
If the ends look more hemispherical than conical, one should find
$h_{ab} \simeq h_{cd} \simeq r$; in this case, $V = \pi r^2(h_{bc} + \frac{4}{3}r)$.

the critical density of CO_2. In order to look for the opalescence, darken the
room (or at least shield the bath from bright lights) and shine a flashlight on
the capillary tube whose meniscus is closest to the center of the tube. Look for
the scattered light at roughly 90° from the direction of the flashlight beam.

Do not be surprised if, during the experiment, you find that some menisci
move towards the top of the tube with increasing temperature. Some of the
menisci will become difficult to see as the critical temperature is approached,
but do not give up looking too soon because only a few will disappear until just
before the critical temperature.

PRECAUTION. Do not raise the temperature of the thermostat bath appreciably above the critical temperature as the pressure increase in the
capillaries may cause them to explode (the critical pressure of CO_2 is
73.8 bar = 72.9 atm). Although the possibility of explosion is remote, the
wearing of *safety glasses*, which should be mandatory practice in all laboratory
work, should be particularly emphasized in this experiment.

CALCULATIONS

First, calculate the total internal volume V^i of each of the $i = 1, 2, \ldots, N$
capillaries. The drawn-out ends can be assumed to be conical (or hemispherical) in shape. Calculate the volume of liquid V_l^i in each capillary at each
temperature, and obtain the volume of the coexisting vapor V_v^i from

$$V_v^i = V^i - V_l^i \tag{2}$$

The liquid and vapor densities can be determined either by a graphical
method, or, preferably, by a numerical least-squares analysis as described in
Chapter XX. Both of these methods are described below.

Graphical method. Plot V_l versus temperature for each capillary in which the meniscus has reached the bottom, and extrapolate to zero volume. At this temperature, call it T_0, you know the density of the vapor in each of the capillaries still containing liquid and vapor. Obtain by interpolation the volume of vapor and liquid, V_v and V_l, for each of these capillaries at T_0. From the known total mass of CO_2 in each tube, several values of the liquid density at T_0 will be obtained. Not all of these values are equally reliable. You should take this into account in arriving at the "best" value for the liquid density at T_0. If the meniscus reaches the top of any of the capillaries, a similar extrapolation of vapor volume V_v versus temperature to zero volume gives a temperature at which the liquid density in all capillaries containing both liquid and vapor is known. Several values for the vapor density can be obtained from capillaries containing liquid and vapor at this temperature, and a "best" value of the vapor density can be chosen.

Least-squares analysis. In this experiment there are two phases, and only one component is present in the tubes. Thus the phase rule tells us that there is only one degree of freedom,

$$F = C - P + 2 = 1$$

and by fixing the temperature, the vapor pressure and the densities of the liquid and vapor are also fixed. At a given temperature T, there will be N equations of the form

$$m_i = \rho_l V_l^i + \rho_v V_v^i \tag{3}$$

where m_i is the total mass of CO_2 in the ith tube. V_l^i and V_v^i are the volumes occupied by liquid and vapor in the ith tube at T. In this experiment, the number of tubes N is six or more, while only two are needed to find the two unknowns $\rho_l(T)$ and $\rho_v(T)$ in Eq. (3). Owing to experimental errors, no pair of density values will satisfy all N equations exactly. Thus we will rewrite Eq. (3) in terms of the residual r_i for the ith tube:

$$r_i \equiv \rho_l V_l^i + \rho_v V_v^i - m_i \tag{4}$$

The method of least squares tells us that the best possible fit to all the data is obtained when the sum of the squares of the residuals is a minimum. Thus

$$S = \sum r_i^2 = \sum_{i=1}^{N} (\rho_l V_l^i + \rho_v V_v^i - m_i)^2 \tag{5}$$

should be a minimum. This quantity can be minimized with respect to the values of the two parameters ρ_l and ρ_v by using the equations

$$\frac{\partial S}{\partial \rho_l} = 0 = 2 \sum (\rho_l V_l^i + \rho_v V_v^i - m_i) V_l^i \tag{6a}$$

$$\frac{\partial S}{\partial \rho_v} = 0 = 2 \sum (\rho_l V_l^i + \rho_v V_v^i - m_i) V_v^i \tag{6b}$$

These equations reduce to two equations in two unknowns:

$$\rho_l[\sum (V_l^i)^2] + \rho_v[\sum (V_v^i V_l^i)] = \sum m_i V_l^i \tag{7a}$$

$$\rho_l[\sum (V_l^i V_v^i)] + \rho_v[\sum (V_v^i)^2] = \sum m_i V_v^i \tag{7b}$$

A least-squares analysis can then be used to solve these equations for the best values of ρ_l and ρ_v at each temperature.

Using the resulting pairs of density values, construct a Cailletet–Mathias plot (liquid and vapor densities versus temperature). From this plot, obtain the critical temperature T_c, test the "law" of rectilinear diameters (see Eq. (1)), and obtain the critical density ρ_c.

DISCUSSION

The critical temperature and molar volume can be related to the constants of the van der Waals' equation of state:

$$\left(p + \frac{n^2 a}{V^2}\right)(V - nb) = nRT \tag{8}$$

The resulting expressions are[2]

$$\tilde{V}_c = 3b \qquad T_c = \frac{8a}{27bR} \tag{9}$$

Using these equations and your values of T_c and $\tilde{V}_c$, calculate a and b. How well do your values agree with values tabulated in the literature? For a van der Waals' fluid, the critical pressure is given by

$$p_c = \frac{a}{27b^2} = \frac{3RT_c}{8\tilde{V}_c} \tag{10}$$

Calculate p_c from Eq. (10) and your T_c and $\tilde{V}_c$ values, and compare it with the accepted literature value. From the empirical corresponding-states behavior of many fluids, one would expect that $p_c \tilde{V}_c / RT_c \approx 0.28$. Will this yield a better p_c value than Eq. (10)?

Explain why the meniscus in some tubes moved down and disappeared at the bottom while the meniscus in other tubes may have moved up and disappeared at the top. Is it possible for the meniscus to remain stationary as the temperature is increased?

In recent years, it has been established experimentally and theoretically that the variation of many thermodynamic properties near a critical point can be described by the use of *critical exponents*.[4] In the case of fluids, the density difference $\rho_l(T) - \rho_v(T)$ is well represented by

$$\frac{\rho_l(T) - \rho_v(T)}{\rho_c} = B\left(\frac{T_c - T}{T_c}\right)^\beta \tag{11}$$

where the exponent β has a value of 0.325. Thus a log–log plot of $(\rho_l - \rho_v)$ versus $(T_c - T)$ should yield a straight line of slope β. Make such a plot and determine the value of β either graphically or from a least-squares fit of your $\Delta\rho(\Delta T)$ data with Eq. (11).

Binary-liquid option. As an alternative to this study of critical behavior in a pure fluid, one can use quite a similar technique to investigate the coexistence curve and critical point in a binary–liquid mixture.[5] Many mixtures of organic liquids (call them A and B) exhibit an upper critical point, which is also called a consolute point. In this case, the system exists as a homogeneous one-phase solution for all compositions if T is greater than T_c. For temperatures below T_c, the system separates into two coexisting liquid phases—one with $X_A < X_c$ and one with $X_A > X_c$, where X_c denotes the mole fraction of component A at the critical point. Since this binary-liquid phase separation can be studied at a constant pressure of 1 atm, the cells are much simpler and easier to fill than those required for a pure fluid.

APPARATUS

Thermostat bath provided with a heater, adjustable thermoregulator, and stirrer; thermometer with a resolution of at least ± 0.01 K (a platinum resistance thermometer, thermocouple or quartz thermometer could be used; see Chapter XVI); illuminating light and cathetometer for reading the vertical position of the menisci; number of sealed capillaries of uniform bore containing varying amounts of carbon dioxide, prepared in advance.

REFERENCES

1. J. S. Rowlinson, "Liquids and Liquid Mixtures," 2d ed., pp. 90–94, Plenum, New York (1969).
2. R. S. Berry, S. A. Rice, and J. Ross, "Physical Chemistry," p. 741, Wiley, New York (1980); M. W. Zemansky and R. Dittman, "Heat and Thermodynamics," 6th ed., McGraw-Hill, New York (1981).
3. M. S. Banna and R. D. Mathews: *J. Chem. Educ.* **56,** 838 (1979).
4. H. E. Stanley, "Introduction to Phase Transitions and Critical Phenomena," pp. 9–12, 39–49, Oxford University Press, New York (1971); M. R. Moldover, *J. Chem. Phys.* **61,** 1766 (1974).
5. S. B. Ngubane and D. T. Jacobs, *Am. J. Phys.* **54,** 542 (1986).

GENERAL READING

J. S. Rowlinson, *op. cit.,* chap. 3.

H. E. Stanley, *op. cit.,* chaps. 1–3.

H. N. V. Temperley, J. S. Rowlinson, and G. S. Rushbrooke (eds.), "Physics of Simple Liquids," chap. 7, North-Holland, Amsterdam (1968).

CHAPTER
VIII

ELECTROCHEMISTRY

EXPERIMENTS

17. Conductance of solutions
18. Temperature dependence of emf
19. Activity coefficients from cell measurements

CONDUCTANCE OF SOLUTIONS

In this experiment we shall be concerned with electrical conduction through aqueous solutions. Although water is itself a very poor conductor of electricity, the presence of ionic species in solution increases the conductance considerably. The conductance of such electrolytic solutions depends on the concentration of the ions and also on the nature of the ions present (through their charges and mobilities), and conductance behavior as a function of concentration is different for strong and weak electrolytes. Both strong and weak electrolytes will be studied at a number of dilute concentrations, and the ionization constant for a weak electrolyte can be calculated from the data obtained.

THEORY

Electrolyte solutions obey Ohm's law just as metallic conductors do. Thus the current I passing through a given body of solution is proportional to the applied potential difference V. The resistance R of the body of solution in ohms (Ω) is given by $R = V/I$, where the potential difference is expressed in volts and the current in amperes. The *conductance* L is defined as the reciprocal of the resistance,

$$L = \frac{1}{R} \tag{1}$$

and is expressed in units of Ω^{-1}. The conductance of a homogeneous body of uniform cross section is proportional to the cross-sectional area A and inversely proportional to the length l:

$$L = \frac{\kappa A}{l} \qquad \text{or} \qquad \kappa = \frac{1}{R}\frac{l}{A} = k/R \tag{2}$$

where κ is the *conductivity* with units $\Omega^{-1}\,m^{-1}$. The conductivity of a given solution in a cell of arbitrary design and dimensions can be obtained by first determining the cell constant k (the "effective" value of l/A) by measuring the resistance of a solution of known conductivity. The standard solution used for this purpose will be 0.02000 M potassium chloride. Once the cell constant k has been determined, conductivities can be calculated from experimental resistances by using Eq. (2).

The conductivity κ depends on the equivalent concentrations and the mobilities of the ions present. Consider a single electrolyte $A_{v_+}B_{v_-}$ giving ions A^{z+} and B^{z-} and having fractional ionization α at a solute concentration c in *moles per liter*. The concentration in equivalents per liter for the positive ions (and also for the negative ions) is $v\alpha c$, where $v = v_+ z_+ = v_-|z_-|$ denotes the number of equivalents of positive or negative ions per mole of the electrolyte. The conductivity is given by[1]

$$\kappa = 1000\alpha c \mathscr{F}(v_+ z_+ U_+ + v_-|z_-|\,U_-)$$
$$= 1000\alpha c v \mathscr{F}(U_+ + U_-) \tag{3}$$

The quantity $1000\alpha c$ is the concentration of A^+ ions (and also of B^- ions) in mol m^{-3}, $\mathscr{F} = N_0 e$ is the Faraday constant (96485.31 C equiv^{-1}), and U_i is the ionic mobility of charged species i. Note that the mobility is defined as the migration speed of an ion under the influence of unit potential gradient and hence has the units $m^2\,s^{-1}\,V^{-1}$. It is now convenient to define a new quantity, the *equivalent conductance* Λ, by

$$\Lambda = \kappa/1000vc = \alpha \mathscr{F}(U_+ + U_-) \tag{4}$$

Note that Λ has units of $\Omega^{-1}\,m^2\,equiv^{-1}$. In this experiement, we are concerned with simple one-one electrolytes A^+B^-. In this case, where $v = 1$ equiv mol^{-1}, there is no distinction between equivalents and moles and the equivalent conductance is the same as the *molar conductance* Λ_m.[1]

For a strong electrolyte, the fraction ionized is unity at all concentrations; thus Λ is roughly constant, varying to some extent owing to changes in mobilities with concentration but approaching a finite value Λ_0 at infinite dilution. The effect of ionic attraction reduces the mobilities, and it can be shown theoretically[2] for strong electrolytes in dilute solution that

$$\Lambda = \Lambda_0(1 - \beta\sqrt{c}) \tag{5}$$

Using this relation, Λ_0 for strong electrolytes can be obtained experimentally from measurements of conductance as a function of concentration. At infinite dilution the ions act completely independently, and it is then possible to express Λ_0 as the sum of the limiting conductances of the separate ions:

$$\Lambda_0 = \lambda_0^+ + \lambda_0^- \tag{6}$$

for a one-one electrolyte A^+B^-, where $\lambda_0^+ = \mathscr{F}U_+^0$ and $\lambda_0^- = \mathscr{F}U_-^0$.

For a weakly ionized substance, Λ varies much more markedly with concentration because the degree of ionization α varies strongly with concentration. The equivalent conductance, however, must approach a constant finite value at infinite dilution, Λ_0, which again corresponds to the sum of the limiting ionic conductances. It is usually impractical to determine this limiting value from extrapolation of Λ values obtained with the weak electrolyte itself, since to obtain an approach to complete ionization the concentration must be made too small for effective measurement of conductance. However, Λ_0 for a weak electrolyte can be deduced from Λ_0 values obtained for *strong electrolytes* by the use of Eq. (6). As an example, let us consider acetic acid (CH_3COOH, denoted as HAc) as a typical weak electrolyte. It follows from Eq. (6) that

$$\Lambda_0(HAc) = \Lambda_0(HX) + \Lambda_0(MAc) - \Lambda_0(MX) \tag{7}$$

where M^+ is any convenient univalent positive ion such as K^+ or Na^+ and X^- is a univalent negative ion such as Cl^- or Br^-. The only restriction on M and X is the requirement that HX, MAc, and MX must all be strong electrolytes so that their Λ_0 values can be obtained by extrapolation using Eq. (5).

For sufficiently weak electrolytes, the ionic concentration is small and the effect of ion attraction on the mobilities is slight; thus we may assume the mobilities to be independent of concentration and obtain the approximate expression

$$\alpha \cong \frac{\Lambda}{\Lambda_0} \tag{8}$$

If one measures Λ for a weak electrolyte at a concentration c and calculates Λ_0 from the conductivity data for strong electrolytes as described above, it is possible to obtain the degree of ionization of the weak electrolyte at concentration c.

Equilibrium constant for weak electrolyte. Knowing the concentration c of the weak electrolyte, say HAc, and its degree of ionization α at that concentration, the concentrations of H^+ and Ac^- ions and of un-ionized HAc can be calculated. Then the equilibrium constant in terms of concentrations K_c can be calculated from

$$K_c = \frac{(\text{H}^+)(\text{Ac}^-)}{(\text{HAc})} = c\frac{\alpha^2}{1-\alpha} \tag{9}$$

The equilibrium constant K_c given by Eq. (9) using α values obtained from Eq. (8) differs from K_a, the true equilibrium constant in terms of activities, owing to the omission of activity coefficients $(\gamma_\pm^2)$ from the numerator of Eq. (9) and the approximations inherent in Eq. (8). At the very low ionic concentrations encountered in the dissociation of a weak electrolyte, a simple extrapolation procedure can be developed to obtain K_a from the values of K_c. Since $K_a = K_c\gamma_\pm^2$ is an excellent approximation, it follows that

$$\log K_c = \log K_a - 2\log \gamma_\pm \tag{10}$$

According to Debye–Hückel theory,[3] the mean activity coefficient at low and moderate ionic concentrations is given by

$$\log \gamma_\pm = -|z_+ z_-| \frac{A\sqrt{I}}{1 + B\sqrt{I}} \tag{11}$$

The ionic strength I is defined by the expression

$$I = \tfrac{1}{2} \sum_i m_i z_i^2 \tag{12}$$

where m_i is the concentration of the ith ionic species in moles per kg of solvent (i.e., the molality), z_i is the charge on ion i in units of the electron charge, and the sum is taken over *all* ionic species present in the solution. For dilute aqueous solutions at 25°C, the quantity A in Eq. (11) has the value $0.509 \text{ kg}^{1/2}\text{ mol}^{-1/2}$ and the quantity B is found to be close to unity for many salts.[4] Furthermore, the difference between molarity and molality is sufficiently small in dilute aqueous solutions that one can approximate m_i by the molar concentration c_i in mol L^{-1}.

For dilute solutions of HAc, $I = \alpha c$ has a very small value and Eq. (10) is well approximated by

$$\log K_c = \log K_a + 2(0.509)\sqrt{\alpha c} \tag{13}$$

Thus if K_c has been determined from conductance measurements at a number of low HAc concentrations, one can plot $\log K_c$ against $\sqrt{\alpha c}$ and make a linear extrapolation to $c = 0$ to obtain K_a.

METHOD

The use of dc circuitry is impractical for determining ionic conductance from measurements of the resistance of the solution in a conductivity cell, since the electrodes quickly become "polarized"; that is, electrode reactions take place which set up an emf opposing the applied emf, leading to a spuriously high apparent cell resistance. Polarization can be prevented by (1) using a high (audio) frequency *alternating current,* so that the quantity of electricity carried during one half-cycle is insufficient to produce any measurable polarization, and at the same time by (2) employing platinum electrodes covered with a colloidal deposit of "platinum black," having an extremely large surface area, to facilitate the adsorption of the tiny quantities of electrode reaction products produced in one half-cycle so that no measurable chemical emf is produced.[5]

The resistance of a conductivity cell filled with an ionic solution can be measured accurately by use of a *Wheatstone-bridge* circuit employing high-frequency alternating current, where an audio oscillator is used as the source and where the detector is an oscilloscope. A schematic diagram of an ac Wheatstone bridge is given in Fig. 1. The condition of balance for an ac bridge requires that the alternating potential at points B and D be of equal amplitude and exactly in phase. This corresponds to a balance condition

$$\frac{Z_1}{Z_2} = \frac{Z_3}{Z_4} \tag{14}$$

where Z is the impedance. Since the cell has both resistance and capacitance, it represents a complex impedance Z_1; therefore, at least one of the other arms must be complex in order to satisfy Eq. (14). In other words, it is necessary to include some variable reactance in order to balance both the real (resistive) and imaginary (reactive) components and achieve a null signal through the detector. In the bridge circuit shown in Fig. 1, arm 2 is complex owing to the presence of an adjustable parallel capacitor C_2. Arms 3 and 4 are noninductively-wound pure resistance ratio arms (i.e., $Z_3 = R_3$ and $Z_4 = R_4$).

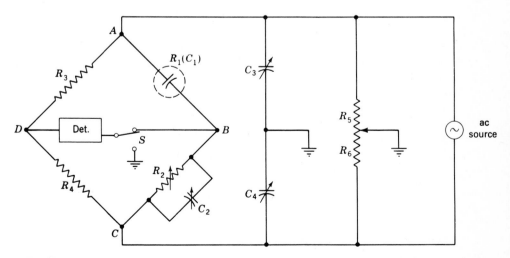

FIGURE 1
Schematic diagram of an ac Wheatstone bridge (with Wagner earthing device) suitable for conductance measurements.

Once a balance condition has been achieved, there is still the problem of relating the cell resistance R_1 to the known resistance R_2. The principal difficulty lies in the need to adopt an equivalent electrical circuit for the cell. This question is discussed in detail elsewhere,[5,6] but the simplest realistic model is

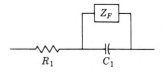

where the *series* capacitance C_1 is due to the electrical double layer at the interface between the electrode and the electrolyte and the frequency-dependent faradaic impedance Z_F is associated with the transfer of charge through the interface. Note that capacitances in parallel with R_1 have been neglected in this model. For "ideally polarized" electrodes, one can neglect the faradaic leakage (set $Z_F = \infty$). The resulting approximation for the impedance of the cell is a series combination of a resistance R_1 and a capacitance C_1, yielding

$$Z_1 = R_1 + iX_1 = R_1 + \frac{1}{i\omega C_1} = R_1 - i\frac{1}{\omega C_1} \tag{15}$$

where $i = \sqrt{-1}$, X is the capacitive reactance, and $\omega = 2\pi f$. The impedance of the parallel combination of resistance R_2 and capacitance C_2 used in arm 2 is given by

$$\frac{1}{Z_2} = \frac{1}{R_2} + \frac{1}{iX_2} = \frac{1}{R_2} + i\omega C_2 \tag{16}$$

Thus Eq. (14) becomes

$$\left(\frac{R_1}{R_2} + \frac{C_2}{C_1}\right) + i\left(\omega R_1 C_2 - \frac{1}{\omega R_2 C_1}\right) = \frac{R_3}{R_4} \tag{17}$$

where $\omega R_1 C_2 - (1/\omega R_2 C_1)$ must equal zero since R_3/R_4 is real. Using this condition, one finds from the real part

$$R_2 \frac{R_3}{R_4} = R_1\left(1 + \frac{1}{\omega^2 R_1^2 C_1^2}\right) \tag{18}$$

Typical $R_1 C_1$ values are so large that the second term above can be neglected,†

---

† Since the double layer is very thin (~ 1 Å), large capacitances are possible. A platinized electrode may have an effective surface area 10 to 100 times the area of a bright metal electrode of the same size. Thus the capacitance of the double layer for a 1 cm^2 platinized electrode can be 100 μF or greater. Taking as typical conditions $R_1 > 10^3\ \Omega$, $f = 10^3$ Hz and $C_1 > 100\ \mu$F, the quantity $1/\omega^2 R_1^2 C_1^2 < 2.5 \times 10^{-6}$, which is negligible compared to 1 in Eq. (18).

and one should find at all frequencies that

$$R_1 = R_2 \frac{R_3}{R_4} \tag{19}$$

This is fortunately the behavior observed for cells having platinum electrodes with a heavy deposit of platinum black such as those to be used in this experiment. The expression that is valid for bright platinum electrodes is considerably more complicated.[6]

The use of a parallel combination of R_2 and C_2, as in Fig. 1, is attractive because the value of C_2 needed to compensate for the capacitance of the cell is quite low, and small capacitors are cheaper, more accurate, and less frequency-dependent than large ones.[6] By adjusting C_2 and thus compensating for the phase shift in the cell, one can improve the balance of the bridge. However, the best achievable balance may still be a minimum rather than a null point if B and D are not at ground potential, since some current can still flow through the detector to ground via distributed capacity. Detection of the balance point can be improved for precise work by using a Wagner earthing circuit to bring point D (and thus point B) close to ground potential.[7] After an initial bridge balance is achieved, the detector is switched from B to ground and the Wagner circuit (C_3, C_4, R_5, R_6 in Fig. 1) is adjusted to minimize the signal through the detector. The bridge is then rebalanced, and this procedure is repeated until the best possible null point is achieved. The ideal detector is an oscilloscope, which can be used in a way that permits separate observation of both the capacitive and the resistive balance[8] and can also reveal any waveform distortion that might occur if there is serious polarization of the electrodes.

As described in Chapter XVIII, a 1:1 Lissajous figure can be generated on an oscilloscope screen when the imbalance signal between B and D is applied to the vertical input and a sine-wave signal from the audio oscillator is applied to the horizontal input. This configuration is shown in Fig. 2, where the details of the Wagner earthing device have been omitted for clarity. If all the bridge arms were pure resistance elements, bridge imbalance would be indicated by an oscilloscope trace that is a tilted straight line whose vertical projection is proportional to the voltage difference between B and D. Thus a horizontal trace would indicate bridge balance. If the unknown arm is a conductivity cell, the oscilloscope trace will in general be an ellipse since the cell is not a pure resistance and the potentials at B and D are generally not in phase. A capacitive balance of the bridge will close the ellipse to produce a tilted line; resistive balance is then indicated by a horizontal line. In practice, the balance procedure is more complex than this, and the Wagner earthing circuit is usually needed to obtain a good null pattern. The bridge is considered balanced when the minor axis of the ellipse is as short as possible and the major axis is as nearly horizontal as possible. Under the conditions encountered in this experiment, very good null patterns should be achieved for all but the most dilute solutions studied.

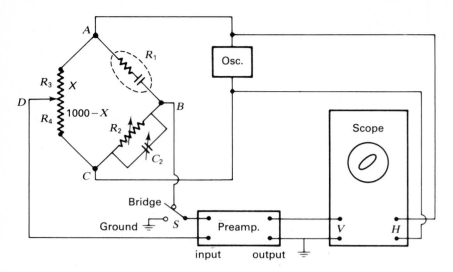

FIGURE 2

Conductance bridge. The Wagner earthing circuit has not been shown here. The slide-wire reading X is on a scale from 0 to 1000. The oscillator should be isolated from the bridge circuit by a good-quality transformer, and another transformer should be used to isolate the bridge from the preamplifier. The oscillator and cables should be well shielded.

EXPERIMENTAL

Set up the circuit shown in Fig. 2, including a Wagner earthing device like that illustrated in Fig. 1. Make sure that the leads or lugs on all hook-up wires are clean before making connections; if necessary, use fine emery paper to remove any dirt or oxide coating. All connections must be *tight* to avoid any unwanted contact resistances. A ten-turn Helipot variable resistance slide-wire can be used for the ratio arms 3 and 4. Alternatively, the slide-wire in a Leeds & Northrup Student Potentiometer can also be used; see Chapter XV for more details. A high-quality four- or five-decade resistance box with a range from $0\,\Omega$ to $100\,\mathrm{k}\Omega$ is used as R_2, while a variable or decade capacitor serves as C_2. Note that it is not necessary to know the value of C_2, but it must be variable over a sufficient range ($0–0.5\,\mu\mathrm{F}$) to allow compensation of the phase shift due to the cell.

The bridge should be operated with as low a power level and for as short a time as possible to prevent polarization effects.† A preamplifier is used to increase the imbalance signal before it is applied to the vertical input terminals of the oscilloscope. Be careful that the output leads ot the preamplifier are properly connected to the input terminals of the scope, one of which is usually

---

† In the unlikely event that significant polarization occurs, very poor balance patterns will be observed because the severe distortion of the waveform cannot be compensated by varying capacitor C_2.

grounded. In order to optimize the accuracy of the measurements, choose R_2 values whenever possible so that the variable terminal D is near the center of the slide-wire, i.e., the reading X is near 500 and $R_3/R_4 \simeq 1$).

Two methods of using the oscilloscope as a null detector are described below.

Sine wave. Set the selector switch on the scope to "internal sweep". As the null point is approached, the amplitude of the sinusoidal imbalance signal will diminish. At the maximum gain setting, the scope pattern near the balance point will be a 60-Hz leakage signal with a small 1-kHz bridge signal superimposed. Do not attempt to balance out the 60-Hz signal; merely balance for minimum 1-kHz ripple.

Lissajous figure. Set the selector switch on the scope to "horizontal input". Now both the horizontal and the vertical inputs come from the oscillator. Generally, an ellipsoidal pattern will be seen on the scope. The width (minor axis) of the ellipse can be reduced by balancing the reactance. The null point occurs when the major axis of the ellipse is as nearly horizontal as possible.

This balance point can now be improved by using the Wagner earthing network. Turn the selector switch S to the Ground position and minimize the vertical input to the scope by adjusting the elements in the Wagner circuit. Then return the selector switch to the Bridge position, and rebalance the bridge with C_2 and the slide-wire to obtain a better null. For low-resistance cells, the reactance has little effect on the bridge balance (i.e., sharp nulls are easily achieved). High-resistance cells, however, require repeated sequences of balancing the bridge, then the Wagner circuit, and then the bridge again in order to achieve good nulls.

Calibration. There may be small systematic errors in the slide-wire due to nonlinearity of the wire or failure of the indicator to correspond precisely to the position of the movable contact. In this case, the reading X on a scale from 0 to 1000 may yield a value of $X/(1000 - X)$ that does not equal the true value of R_3/R_4. This can be checked, and corrected if necessary, by carrying out calibration measurements using a high-precision resistance box in place of the cell in arm 1.

Set both the resistances R_1 and R_2 at 3000 Ω and balance the bridge. If there are no errors, the X reading should be exactly 500. To obtain a correction figure, make four readings of the balance approaching the null twice from each side. Then reverse the positions of the resistance boxes and take four more readings. The average of all eight readings will be called X^*. A correction quantity $(500 - X^*)$ should be added to all future X readings that are close to the center of the dial.

A calibration check is also needed for bridge balances obtained when X is substantially greater than 500. To achieve this, increase the resistance in arm 1 to 6000 Ω so that $R_3/R_4 = R_1/R_2 = 2.00$. Balance the bridge and record the value of X several times. Calculate the ratio $X/(1000 - X)$, which should equal

R_3/R_4 if there are no errors. Now calculate the correction term Δ defined by

$$\frac{X}{1000 - X}(1 + \Delta) \equiv \frac{R_3}{R_4} = \frac{R_1}{R_2} \tag{20}$$

Repeat this procedure for $R_1 = 9000\ \Omega$ and $12\,000\ \Omega$; then plot the values of Δ versus $X/(1000 - X)$. Use an interpolated value of Δ to correct any measured ratios $X/(1000 - X)$ obtained from X readings substantially above 500.

Conductivity cell. A satisfactory conductivity cell design is shown in Fig. 3. This cell is fragile and should be handled with care. The leads should be arranged to avoid placing any strain on the cell while it is mounted in the thermostat bath. The electrodes are sensitive to poisoning if very concentrated electrolytes are placed in the cell and will deteriorate if allowed to become dry. (Details concerning the preparation of "platinized" platinum electrodes are given in Chapter XIX.) After each emptying, the cell must *immediately* be rinsed or filled with water or solution. While the cell is in the bath, the filling arms must be stoppered by placing rubber policemen over the ends. If the cell is allowed to stand for any long time between runs, leave it filled with conductivity water rather than solution.

Draw liquid into the cell through the tapered tip as you would with a pipette. Empty the cell by allowing it to drain out the other end. The procedure for filling the cell with a new solution is to empty the cell, rinse it at least *four times* with aliquots of the desired solution using about enough to fill the cell one-fourth full, shake well on each rinsing, then fill the cell, being careful to avoid bubbles on the electrodes. Be careful to avoid loss of mercury from the contact arms.

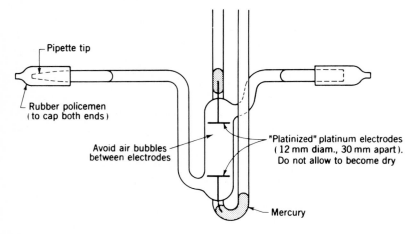

FIGURE 3
Conductivity cell.

Conductivity water. Ordinary distilled water is not suitable for use in this experiment, since it has too high a conductance. Much of this conductivity results from dissolved CO_2 gas from the air, which can be removed by boiling distilled water and capping a full bottle while it is still hot. Better conductivity water can be prepared by special distillation[5] or by passing distilled water through an ion-exchange resin to "deionize" it. Store conductivity water in screw-cap polyethylene bottles to avoid contamination by ionic substances leached from glass.

Rinse thoroughly and fill the cell with conductivity water. Place the cell in arm 1 of the bridge and determine its resistance. High precision is not required, but the resistance should be high enough to yield a water conductivity κ_w less than $2 \times 10^{-4} \, \Omega^{-1} \, m^{-1}$. For a cell like that shown in Fig. 3, which will have a cell constant k of roughly $250 \, m^{-1}$, this means a resistance $R_w > 1.3 \times 10^6 \, \Omega$. If the cell to be used has significantly different dimensions, estimate k from roughly measured values of the spacing l and the area A of the electrodes and determine the minimum resistance corresponding to the upper bound $2 \times 10^{-4} \, \Omega^{-1} \, m^{-1}$ on κ_w. If the cell resistance is too low, rinse the cell a few more times and repeat the measurement. Record the value of R_w obtained.

Standardization. Rinse and fill the cell with the $0.02000 \, M$ KCl solution.† Then mount the cell in a 25°C thermostat bath and connect it to the bridge circuit. Balance the bridge, adjusting the value of R_2 so that the slide-wire reading is roughly in the middle of its range (i.e., X near 500). If necessary, improve the null point by adjusting the value of the capacitor C_2 and adjusting the Wagner earthing device. If excessive capacitance C_2 appears to be required, polarization is probably taking place and an instructor should be consulted.

Once the best possible null point has been achieved, do not disturb any of the settings except the slide-wire contact. Final measurements on the solution consist of a set of four readings, the balance being approached twice from each side. The value of X on the slide-wire scale is recorded for each balance. There should be no appreciable drift in these values; if drift is observed, the cell is probably not in thermal equilibrium with the bath. Wait a few minutes and repeat the measurements. Be sure to record R_2 for the resistance box. The measurement on this KCl solution should be made with great care, since it determines the cell constant k and will affect all subsequent calculations.

Procedure. Determine the resistance R of the cell containing acetic acid solutions at concentrations $c_a = 0.05 \, M$, $c_a/4$, $c_a/16$, and $c_a/64$. Also determine

---

† It is recommended that a *precisely* $0.02000 \, M$ KCl solution be used for standardization. If this is not available, record the actual concentration used and make the appropriate correction as described in the Calculations section.

R for solutions of the strong electrolyte KAc at $c_b = 0.02\ M$, $c_b/4$, $c_b/16$, and $c_b/64$.

If time permits, determine R for KCl and/or HCl solutions at c_b, $c_b/4$, $c_b/16$, and $c_b/64$ as well.

The procedure for each solution is the same as that described above for the standardization measurement on $0.02\ M$ KCl. For several of the HAc solutions and for the most dilute strong electrolytes, the resistance may be so high that it is not possible to balance the bridge with X near 500. One can either use two decade resistance boxes in series for arm 2 or merely take slide-wire readings at larger X values. In either case, be sure that your bridge calibration procedure will provide any necessary correction factors.

For dilute solutions with high resistance, the bridge balance is more difficult to achieve. Iterative adjustments of R_2, C_2, and the Wagner device will be required to achieve a good null point.

Dilutions. The quality of your results depends on accurate dilutions and scrupulous care in avoiding contamination of the dilute solutions by stray electrolytes. Follow the dilution procedure described below.

Clean thoroughly a 100-mL volumetric flask, a 125-mL wide-mouth flask, and a 25-mL pipette. Rinse these with conductivity water. Rinse the 125-mL flask with two or three *small* aliquots of the stock solution, and then take not more than 100 mL of the solution, noting the concentrations given on the stock bottle.

Rinse the pipette with two or three small aliquots from the flask, and then pipette exactly 25 mL of the solution from the flask into the volumetric flask. Make up to the mark with conductivity water and mix thoroughly. Set this aside for the next measurement and dilution. Use the solution remaining in the 125 mL flask for rinsing and filling the conductivity cell. When the measurement of resistance of the cell containing that solution has been completed, discard the remaining solution from the 125-mL flask and rinse with two or three aliquots of the new solution from the volumetric flask. Pour the contents of the volumetric flask into the 125-mL flask, and rinse the volumetric flask with conductivity water.

Repeat the process described above as many times as necessary to obtain the dilutions needed. Care is important in this work, since dilution errors are cumulative.

Make sure that the thermostat bath is regulating properly at all times; record the bath temperature several times during the experiment. It is a good idea to carry out some rough calculations while you are taking the data, in order to verify that everything is working properly. For example, the product of the cell resistance times the dilution factor $(1, \frac{1}{4}, \frac{1}{16}, \frac{1}{64})$ should be almost constant for strong electrolytes but should decrease for HAc as the solution becomes more dilute.

At the end of each laboratory session, rinse the cell well, fill it with conductivity water, and store it with filling arms stoppered.

CALCULATIONS

The cell resistance R is obtained from Eq. (19). If the calibration procedure showed that significant slide-wire corrections are needed, apply these corrections during the process of converting X into the desired ratio (R_3/R_4).

From the result of the $0.02000\ M$ KCl measurement, calculate the cell constant k from Eq. (2) using $\kappa = 0.27653\ \Omega^{-1}\,m^{-1}$ at 25°C.[9] If the KCl solution was not precisely $0.02000\ M$, an appropriate κ value can be calculated from the following empirical equation, valid at 25°C over the range $c = 10^{-4}\ M$ to $0.04\ M$,[9]

$$\kappa(\Omega^{-1}\,m^{-1}) = 14.984c - 9.484c^{3/2} + 5.861c^2 \log c$$
$$+22.89c^2 - 26.42c^{5/2} \tag{21}$$

Calculate the conductivity κ for each solution studied. If the conductivity κ_w of the water used in the dilutions is significant compared to the measured conductivity κ_{obs} of any solution, a correction should be made:

$$\kappa = \kappa_{obs} - \kappa_w \tag{22}$$

Using Eq. (4), calculate the equivalent conductance Λ for each solution. The values of R, κ and Λ should be tabulated for all solutions studied. Cite the units for each quantity, and also give the value of the cell constant k (with units) in the table caption.

For the KAc solutions and any other strong electrolytes studied as a function of concentration, plot Λ versus $\sqrt{c}$ and extrapolate linearly to $c = 0$ in order to obtain Λ_0. In making these extrapolations, beware of increasingly large experimental uncertainties at the lowest concentrations and also of systematic errors due to conducting impurities or dilution errors at low concentrations. If you are making a least-squares fit with Eq. (5) rather than a graphical extrapolation, be sure to estimate the relative weights associated with each data point. If runs have not been made on one or more of the strong electrolytes, the following data[5,10] may be used, where Λ_0 is given in units of $\Omega^{-1}\,equiv^{-1}\,m^2$.

Solution	Λ_0 at 25°C	$d\Lambda_0/dT$
HCl	0.04262	6.4×10^{-4}
KCl	0.014986	2.8×10^{-4}
KAc	0.01144	2.1×10^{-4}

Combine your extrapolated Λ_0 values and/or the data cited above to obtain Λ_0 for acetic acid. If the temperature of any of your runs was different from 25°C, the $d\Lambda_0/dT$ data above will allow a correction to be made.

For each dilution of acetic acid, calculate α from Eq. (8) and then K_c from Eq. (9). Present in tabular form Λ, α, c, and K_c for each dilution of acetic acid. Plot $\log K_c$ against $\sqrt{\alpha c}$. If an extrapolation to zero concentration is feasible with a slope of correct sign and magnitude (see Eq. (13)), make it and obtain a value for K_a. If the data do not appear to be sufficiently good to warrant an extrapolation, report an average or best value for K_c. Report the temperature at which the HAc measurements were made.

DISCUSSION

How does the cell constant k compare with the geometric value l/A obtained from an approximate measurement of the dimensions of your cell? Why is the equivalent conductance Λ_0 so large for an HCl solution? How do the slopes of your Λ versus $\sqrt{c}$ plots for strong electrolytes compare with literature values and the values expected from Onsager's theory?[2] Find a literature (or textbook) value for the equilibrium constant K_a for HAc ionization. Using this value and Eq. (13), draw a dashed literature/theory line on your plot of log K_c versus $\sqrt{\alpha c}$. Are the deviations of your data points from this line reasonable in view of the experimental errors expected in this work? What is the limiting factor in the accuracy of your K_c measurements?

APPARATUS

AC Wheatstone-bridge circuit; oscillator (1 kHz, with power supply); pre-amplifier; oscilloscope; precision resistance decade box or plug box; decade or adjustable capacitor; Wagner earthing network; conductivity cell, filled with conductivity water and capped off with clean rubber policemen; holder for mounting cell in bath; two leads to connect cell to bridge; electrical leads with lugs or soldered tips; electrical switch; 100-mL volumetric flask; 25-mL pipette; pipetting bulb; two 100- or 250-mL beakers; two 125-mL Erlenmeyer flasks; 500-mL glass-stoppered flask for storing conductivity water.

Constant-temperature bath set at 25°C; emery paper; conductivity water (1500 mL); precisely $0.02000\,M$ KCl solution (300 mL); $0.02\,M$ HCl solution (200 mL); $0.02\,M$ potassium acetate solution (200 mL); $0.05\,M$ acetic acid (250 mL).

REFERENCES

1. S. I. Smedley, "The Interpretation of Ionic Conductivity in Liquids," Plenum, New York (1980).
2. L. Onsager, *Phys. Z.* **28**, 277 (1927); P. W. Atkins, "Physical Chemistry," 3d ed., chap. 27, Freeman, New York (1986).
3. P. W. Atkins, *op. cit.*, p. 237*ff*.
4. G. Scatchard, *Chem. Rev.* **19**, 309 (1936).
5. M. Spiro, "Conductance and Transference Determinations," in B. W. Rossiter and J. F. Hamilton (eds.), "Physical Methods of Chemistry," 2d ed., vol. II. Electrochemical Determinations, chap. 8, Wiley–Interscience, New York (1986).
6. R. A. Robinson and R. H. Stokes, "Electrolyte Solutions," 2d ed., pp. 88–95, Butterworth, London (1959); J. Braunstein and G. D. Robbins: *J. Chem. Educ.* **48**, 52 (1971).
7. F. E. Terman, "Electronic Measurements," 2d ed., pp. 40–44, McGraw-Hill, New York (1952); G. Jones and R. C. Josephs, *J. Amer. Chem. Soc.* **50**, 1049 (1928).

8. D. Edelson and R. M. Fuoso: *J. Chem. Educ.* **27**, 610 (1950).
9. M. Spiro, *op. cit.,* pp. 718–719; E. Juhasz and K. N. Marsh: *Pure Appl. Chem.* **53**, 1841 (1981).
10. H. S. Harned and B. B. Owen, "The Physical Chemistry of Electrolytic Solutions," 2d ed., Appendix A, Reinhold, New York (1950).

GENERAL READING

S. I. Smedley, *op. cit.*
M. Spiro, *op. cit.*

EXPERIMENT 18
TEMPERATURE DEPENDENCE OF EMF

In this experiment the following electrochemical cell is studied:

$$\text{Cd}(s) \mid \text{Cd}^{2+}\text{SO}_4^{2-}(aq, c) \mid \text{Cd}(\text{Hg}, X_2) \qquad \text{1 bar, } T \qquad (1)$$

The change in state accompanying the passage of 2 faradays of *positive* electricity from left to right through the cell is given by

$$\text{Anode: Cd}(s) = \text{Cd}^{2+}(aq, c) + 2e^-$$

$$\text{Cathode: } 2e^- + \text{Cd}^{2+}(aq, c) = \text{Cd}(\text{Hg}, X_2)$$

$$\text{Net: Cd}(s) = \text{Cd}(\text{Hg}, X_2) \qquad \text{1 bar, } T \qquad (2)$$

In the above, (s) refers to the pure crystalline solid, (aq, c) refers to an aqueous solution of concentration c and (Hg, X_2) represents a liquid (single-phase) cadmium amalgam in which the mole fraction of Cd is X_2.

From the emf of this cell and its temperature coefficient, the changes in free energy, entropy, and enthalpy for the above change in state are to be determined.

THEORY

When the cell operates reversibly at constant pressure and temperature, with no work being done except electrical work and expansion work,

$$\Delta G = -n\mathscr{E}\mathscr{F} \qquad (3)$$

where ΔG is the increase in free energy of the system attending the change in state produced by the passage of n faradays of electricity through the cell, $\mathscr{F}$ is the Faraday constant, and $\mathscr{E}$ is the emf (electromotive force) of the cell.

From the Gibbs–Helmholtz equation, we have for a change in state at constant pressure and temperature

$$\left(\frac{\partial \Delta G}{\partial T}\right)_p = -\Delta S \tag{4}$$

Combining Eqs. (3) and (4), we find that

$$\Delta S = n\mathscr{F}\left(\frac{\partial \mathscr{E}}{\partial T}\right)_p \tag{5}$$

Knowing both ΔG and ΔS, one can find ΔH by using

$$\Delta G = \Delta H - T\,\Delta S \tag{6}$$

METHOD

The emf of a cell is best determined by measurements with a potentiometer, since this method gives a very close approach to reversible operation of the cell. The cell potential is opposed by a potential drop across the slide-wire of the potentiometer, and at balance only very small currents are drawn from the cell (depending on the sensitivity of the galvanometer or detector used and the fineness of control possible in adjusting the slide-wire). Alternatively, a high-resolution digital voltmeter with a large internal impedance can also be used. Potentiometers and digital voltmeters are described and compared in Chapter XV.

EXPERIMENTAL

The cell consists of a small beaker with a top that will accommodate two electrodes as shown in Fig. 1. The *cadmium electrode* is made by plating

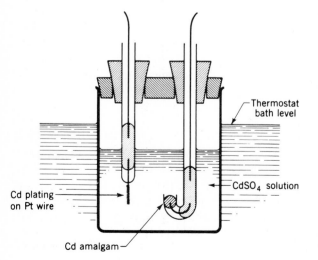

Thermostat
bath level

CdSO$_4$ solution

Cd plating
on Pt wire

Cd amalgam

FIGURE 1
Cadmium amalgam cell.

cadmium onto a platinum wire that is sealed through the bottom of a small glass tube. Electrical contact is made to mercury contained in this tube. The *amalgam electrode* is made by placing a small quantity of the cadmium amalgam in the cup of a special J-shaped glass tube with a platinum wire sealed in to make electrical contact with mercury contained in the long arm of the tube.

Procedure. A cadmium–mercury amalgam containing 2 percent Cd by weight will be used. This amalgam can be prepared as follows. Remove the oxide coating from a thin rod of pure Cd with dilute acid, rinse the rod thoroughly in distilled water, dry, and weigh. Then dissolve the rod in the proper (weighed) amount of triple-distilled mercury. Using a medicine dropper, fill the cup of the J electrode with this amalgam. Carefully insert this electrode into the cell, which previously has been rinsed and filled with a 0.1 M $CdSO_4$ solution. The cell should be filled only half full to prevent contact of the solution with the cover.

The platinum wire of the other electrode is plated with a cadmium deposit by immersing the electrode in a beaker containing a 0.1 M solution of $CdSO_4$ and a pure cadmium rod. The electrode is connected to the negative terminal of a 1.5-V dry cell, and the cadmium rod is connected to the positive terminal. Current is passed through this plating bath until a heavy deposit of cadmium is visible on the electrode. This cadmium electrode is then transferred to the cell. Handle the electrode with care to avoid flaking off any of the cadmium deposit. It is advisable to keep the electrode wet at all times.

Mount the assembled cell in a thermostat bath. Assemble the potentiometer circuit and connect the cell. Consider the sign of ΔG for the change in state to decide on the proper connection. Have an instructor check your circuit.

After the cell has been in the thermostat bath for at least 10 min, measure the emf and repeat this measurement at least three times at 5-min intervals to verify that there is no systematic drift in the emf. The potentiometer circuit should be checked against the standard cell immediately before each reading. The emf should be determined as a function of temperature at four more points in the range from 0 to 40°C. The same potentiometer must be used in making all these measurements.

CALCULATIONS

Plot the emf of the cell $\mathscr{E}$ versus the absolute temperature T. Report the value of $\mathscr{E}$ and the slope $(\partial \mathscr{E}/\partial T)_p$ at 25.0°C (298.15 K) from the best smooth curve through your experimental points. Using Eqs. (3), (5), and (6), calculate ΔG, ΔS, and ΔH in J mol^{-1}, and give the change in state to which they apply.

From the known weight percentage of Cd in the amalgam, compute the mole fraction X_2 of cadmium. Assuming for this concentration an activity coefficient of unity (so that the activity a_2 is equal to X_2), compute the standard free-energy change $\Delta G°$ for the change in state

$$Cd(s) = Cd(Hg)$$

by means of the equation

$$\Delta G = \Delta G° + RT \ln a \tag{7}$$

DISCUSSION

Explain why $\Delta G°$ is not equal to zero, as would be expected if the standard state were an "amalgam" with $X_2 = 1$.

APPARATUS

Complete potentiometer setup or high-resolution digital voltmeter (see Chapter XV); cell (50-mL beaker or weighing bottle); platinum electrode; J electrode for holding amalgam; two leads for connections to cell; two beakers; battery jar; large ring stirrer; ring stand; large clamp and clamp holder.

   Provision for electroplating cadmium from $0.1\,M$ $CdSO_4$ onto platinum electrode; dilute (2 percent) cadmium amalgam (1 mL); eye dropper; mercury for making contact to electrodes; constant-temperature baths set at 15, 25, and 35°C; approximately $0.1\,M$ $CdSO_4$ solution (150 mL); ice (1 kg).

GENERAL READING

F. E. Smith, *Phil. Mag.* **19**, 250 (1910).
C. E. Teeter, Jr., *J. Amer. Chem. Soc.* **53**, 3927 (1931).
V. K. LaMer and W. G. Parks: *J. Amer. Chem. Soc.* **56**, 90 (1934).

EXPERIMENT 19
ACTIVITY COEFFICIENTS FROM
CELL MEASUREMENTS

We shall be concerned with the electrochemical cell

$$Pt + H_2(g,\, p\ bar) \mid H^+Cl^-(aq,\, c) \mid AgCl(s) + Ag(s) \tag{1}$$

Measurements of emf are to be made with this cell under reversible conditions at a number of concentrations c of HCl. From these measurements, relative values of activity coefficients at different concentrations can be derived. To obtain the activity coefficients on such a scale that the activity coefficient is unity for the reference state of zero concentration, an extrapolation procedure based on the Debye–Hückel limiting law is used. By this means, the standard electrode emf of the silver–silver chloride electrode is determined and activity coefficients are determined for all concentrations studied.

THEORY

When 1 faraday of positive electricity passes reversibly through cell (1) from left to right, the overall change in state will be

$$AgCl(s) + \tfrac{1}{2}H_2(g) = H^+Cl^-(aq) + Ag(s) \tag{2}$$

Cell (1) may conveniently be considered as a combination of two half-cells or "electrodes", which are conventionally written as undergoing reduction reactions.

Hydrogen electrode. This can be written as

$$H^+(aq, c) \mid H_2(g, p \text{ bar}) + Pt \tag{3}$$

With the reversible passage of 1 faraday of positive electricity from left to right through this electrode, the change in state is

$$H^+(aq, c) + e^- = \tfrac{1}{2}H_2(g) \tag{4}$$

and the electrode emf is accordingly

$$\mathscr{E}_3 = \mathscr{E}_3^0 - \frac{RT}{\mathscr{F}} \ln \frac{f_{H_2}^{1/2}}{a_{H^+}} \tag{5}$$

where f_{H_2} is the fugacity of the hydrogen gas, a_{H^+} is the activity of the aqueous hydrogen ion, and $\mathscr{F}$ is the Faraday constant. By convention, the standard potential for the hydrogen electrode is taken to be zero. Thus

$$\mathscr{E}_3^0 = \mathscr{E}_{H^+/H_2}^0 = 0 \tag{6}$$

Silver–silver chloride electrode. This can be written as

$$Cl^-(aq, c) \mid AgCl(s) + Ag(s) \tag{7}$$

When 1 faraday of positive electricity passes reversibly from left to right through this electrode, the change in state is

$$AgCl(s) + e^- = Ag(s) + Cl^-(aq, c) \tag{8}$$

and the electrode emf is

$$\mathscr{E}_7 = \mathscr{E}_7^0 - \frac{RT}{\mathscr{F}} \ln a_{Cl^-} \tag{9}$$

where $\mathscr{E}_7^0 = \mathscr{E}_{AgCl/Ag^+Cl^-}^0$ is the standard electrode potential for the silver–silver chloride electrode (a reduction potential) and a_{Cl^-} is the activity of the chloride ion in the aqueous solution.

The cell. Noting that cell (1) corresponds to half-cell (7) minus half-cell (3) and, of course, that change in state (2) is Eq. (8) minus Eq. (4), the emf of the cell is given by

$$\mathscr{E} = \mathscr{E}_7 - \mathscr{E}_3 = \mathscr{E}^0 - \frac{RT}{\mathscr{F}} \ln \frac{a_{H^+}a_{Cl^-}}{f_{H_2}^{1/2}} \tag{10}$$

where

$$\mathscr{E}^0 = \mathscr{E}_7^0 - \mathscr{E}_3^0 = \mathscr{E}_{AgCl/Ag^+Cl^-}^0 \tag{11}$$

Activity coefficients. Let us write

$$f_{H_2} = \gamma' p \qquad a_{H^+}a_{Cl^-} = \gamma_\pm^2 c^2 \tag{12}$$

where γ' is the activity coefficient for $H_2(g)$ and $\gamma_\pm$ is the mean activity coefficient for $H^+Cl^-(aq)$. Equation (10) can now be written

$$\mathscr{E} = \mathscr{E}^0 - \frac{2.303RT}{\mathscr{F}} \log \frac{c^2}{p^{1/2}} - \frac{2.303RT}{\mathscr{F}} \log \frac{\gamma_\pm^2}{\gamma'^{1/2}} \tag{13}$$

where p is in bars and c is in moles per liter.† From this equation it is clear that, if emf measurements are made on two or more cells which differ only in the concentrations c of HCl, the ratios of the corresponding activity coefficients $\gamma_\pm$ can be determined from the differences in emf. For the determination of the individual values of these activity coefficients, it is necessary to know the values of $\mathscr{E}^0$ and γ'. At 25°C and 1 bar, the activity coefficient γ' is 1.0006, which for the purpose of this experiment may be taken as unity. To determine $\mathscr{E}^0$, however, requires a procedure equivalent to determining the emf with the solute in its "reference state," at which the activity coefficient $\gamma_\pm$ is unity. However, the reference state for a solute in solution is the limiting state of zero concentration, which is inaccessible to direct experiment. However, the Debye–Hückel theory predicts the limiting behavior of $\gamma_\pm$ as the concentration approaches zero, and we can make use of this predicted behavior in devising an extrapolation procedure for the determination of $\mathscr{E}^0$. According to the Debye–Hückel limiting law,[1]

$$\log \gamma_\pm \cong -A\sqrt{I} \tag{14}$$

In the present case, the ionic strength I can be set equal to the HCl molar concentration c. The value of the constant A given by the Debye–Hückel

---

† In the treatment given here, the standard-state pressure has been taken to be 1 bar $\equiv 10^5$ Pa, as recommended by the IUPAC. Prior to 1983, the standard-state pressure was 1 atm. Fortunately, the effect on standard free energies and therefore on standard electrode potentials is very small since 1 atm = 101325 Pa is close to 1 bar. For an electrode involving only condensed phases, like the AgCl/Ag$^+$Cl$^-$ electrode, one finds that $\mathscr{E}^0(\text{atm}) - \mathscr{E}^0(\text{bar}) = +0.169$ mV, which corresponds to a free energy difference of 16.4 J mol^{-1}. This difference arises from the new choice of $\mathscr{E}_{H^+/H_2}^0 = 0$ when hydrogen gas is in its standard state of 1 bar rather than the old choice of $\mathscr{E}_{H^+/H_2}^0 = 0$ when hydrogen was at a pressure of 1 atm. For the standard potential of some arbitrary half-cell $\mathscr{E}^0(\text{atm}) - \mathscr{E}^0(\text{bar}) = -0.338(\Delta n - 0.5)$ mV, where Δn is the number of moles of gas produced in the half-cell change in state.

theory for a uni-univalent electrolyte in aqueous solution at 25°C is 0.509. While knowledge of this value may be helpful to the extrapolation, it is not necessary.

Let us rearrange Eq. (13) and set γ' equal to unity:

$$\mathscr{E}^0 = \mathscr{E} - \frac{2.303RT}{\mathscr{F}} \log \frac{p^{1/2}}{c^2} + \frac{2.303RT}{\mathscr{F}} \log \gamma_{\pm}^2 \tag{15}$$

Now define $\mathscr{E}^{0\prime}$ by

$$\mathscr{E}^{0\prime} \equiv \mathscr{E} - \frac{2.303RT}{\mathscr{F}} \log \frac{p^{1/2}}{c^2} - 2\frac{2.303RT}{\mathscr{F}}(0.509)\sqrt{c} \tag{16}$$

The value of $\mathscr{E}^{0\prime}$ should be close to that of $\mathscr{E}^0$ [depending on the validity of Eq. (14) as an approximation for $\log \gamma_{\pm}$]; in any case $\mathscr{E}^{0\prime}$ will approach $\mathscr{E}^0$ as the concentration approaches zero. A plot of $\mathscr{E}^{0\prime}$ versus c should be approximately linear and have only a small slope, thus permitting a good extrapolation to zero concentration.[2]

Alternatively, one could define a quantity $\mathscr{E}^{0\prime\prime}$ by

$$\mathscr{E}^{0\prime\prime} \equiv \mathscr{E} - \frac{2.303RT}{\mathscr{F}} \log \frac{p^{1/2}}{c^2} \tag{17}$$

which would also equal $\mathscr{E}^0$ at infinite dilution. If this quantity $\mathscr{E}^{0\prime\prime}$ is plotted against $\sqrt{c}$, we expect to obtain a curve that will give a limiting straight-line extrapolation to zero concentration.

In either case, the extrapolated intercept is the desired value of $\mathscr{E}^0$. Often it is best to make both extrapolations; with only a few points, use of $\mathscr{E}^{0\prime}$ is recommended.

With the value of $\mathscr{E}^0$ the activity coefficients for the various concentrations studied can be determined individually by use of Eq. (13). They can then be compared with those predicted by a more complete form of the Debye–Hückel equation, such as

$$\log \gamma_{\pm} = -0.509 |z_+ z_-| \frac{\sqrt{I}}{1 + B\sqrt{I}} \tag{18}$$

where B is approximately equal to unity.[1]

EXPERIMENTAL

The cell vessel consists of a small beaker with a stopper or cover with holes through which the hydrogen electrode assembly and the silver–silver chloride electrode assembly may be introduced. The assembled cell is shown in Fig. 1.

The hydrogen electrode consists of a mounted platinum gauze square contained within a glass sleeve having large side-holes at about the level of the gauze and a side-arm for admission of hydrogen near the top. The platinum gauze is "platinized," that is, coated with a deposit of platinum black by

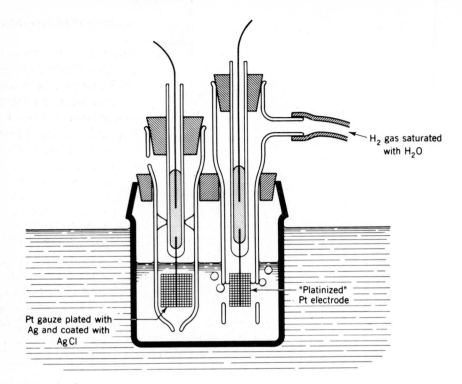

H₂ gas saturated
with H₂O

"Platinized"
Pt electrode

Pt gauze plated with
Ag and coated with
AgCl

FIGURE 1
Electrochemical cell with H_2 and Ag AgCl electrodes.

electrolytic deposition from a solution containing platinic chloride and a trace
of lead acetate. This deposit should be removed with warm aqua regia and
renewed if the electrode has been allowed to dry out or if there is evidence
that the deposit has been "poisoned."

The silver–silver chloride electrode consists of a mounted platinum
screen that has been heavily plated with silver from a cyanide bath, rinsed,
aged in an acidified silver nitrate solution, rinsed, coated with a thin layer of
silver chloride by anodizing in a dilute HCl solution (preferably no more than a
few days before use), and kept in dilute HCl pending use. This is mounted in a
glass sleeve with a small hole in the bottom to admit the cell solution and a
small side-hole near the top for passage of air; this sleeve protects the
electrode from mechanical damage and also prevents the attainment of any
significant concentration of dissolved H_2 in the solution in contact with the
electrode. (Further details concerning electrode preparation can be found in
Chapter XIX.)

Both electrode assemblies should be furnished ready for use. Care should
be exercised to prevent the electrodes from becoming dry. The use of
hydrogen in any significant quantity (as might occur when many students were

carrying out this experiment simultaneously) is attended by explosion hazard. Hydrogen usage must be kept at the minimum necessary for the experiment, and wastage and leakage must be guarded against. **Smoking and the use of flames for any purpose cannot be permitted**.

The hydrogen to be used should be oxygen-free. Oxygen is most conveniently removed by passing tank hydrogen through a commercial catalytic purifier. The hydrogen should also be saturated with water vapor at room temperature (about 25°C) by passing it through a fritted-glass bubbler filled with water.

The partial pressure p of the hydrogen at the electrode should be determined from the atmospheric pressure in the laboratory by subtracting the partial pressure of water vapor at 25°C and adding the mercury equivalent of the "water head" (the average difference in liquid levels inside and outside the hydrogen electrode shell).

Procedure. Fill the cell about half full with the appropriate HCl solution. Rinse the two electrodes in separate small portions of the same solution, and insert them. When the liquid level in the silver–silver chloride electrode shell has reached its equilibrium level, pass hydrogen through the cell at a moderate rate to sweep out the air and saturate the solution. After about 10 min, slow down the rate to a few bubbles per second and begin to make emf readings. Make at least four readings at 5-min intervals. These should show no significant drift.

Readings of emf are obtained with a potentiometer circuit. If a null galvanometer reading cannot be obtained with any setting of the potentiometer dial, reverse the leads and try again. The sign of the emf will depend on whether the null reading is obtained with the right-hand terminal of the cell (Ag + AgCl) connected to the positive or to the negative terminal of the potentiometer. Alternatively, a high-resolution digital voltmeter with a large internal impedance can also be used very successfully. See Chapter XV for a description and comparison of potentiometers and digital multimeters.

Care should be exercised to prevent the cell from becoming polarized by accidental shorting or by passage of excessive currents during the emf measurements.

Runs should be made with the following concentrations of HCl: 0.1, 0.05, 0.025, 0.0125, and 0.00625 M, obtained by successive volumetric dilution of 0.1 M HCl stock solution. Make up 200 mL of each solution.

CALCULATIONS

For each run calculate $\mathscr{E}^{0\prime}$ from Eq. (16) using the measured $\mathscr{E}$, the partial pressure p of H_2 in bars, and the concentration c of HCl. The value of $2.303RT/\mathscr{F}$ at 25°C is 0.05916 V.

Plot $\mathscr{E}^{0\prime}$ versus c and extrapolate to zero concentration to obtain $\mathscr{E}^0$ (which is equal to the standard electrode potential for the silver–silver chloride electrode). If you wish, also calculate $\mathscr{E}^{0'''}$ from Eq. (17), plot it versus $\sqrt{c}$, and extrapolate to obtain another value of $\mathscr{E}^0$ as a check.

With this value of $\mathscr{E}^0$ calculate mean activity coefficients for all concentrations studied by use of Eq. (15). For each concentration, calculate also a theoretical mean activity coefficient by use of the Debye–Hückel equation in the form given in Eq. (18). Present your experimental and theoretical activity coefficients in tabular form.

APPARATUS

Complete potentiometer setup or high-resolution digital voltmeter (see Chapter XV); 50-mL weighing bottle as cell; special three-hole stopper to fit cell and hold electrodes; two leads for connections to cell; hydrogen electrode, and Ag–AgCl electrode; 200-mL volumetric flask; 100-mL pipette; pipetting bulb; 250-mL flasks; two 250-mL beakers; small gas bubbler; two pieces of gum-rubber tubing; large clamp and clamp holder.

Cylinder of hydrogen gas, regulator fitted with a Deoxo purifier to remove any oxygen and a flow reducer to limit flow to $5\,\text{ft}^3\,\text{hr}^{-1}$; large fritted-glass bubbler to saturate hydrogen with water vapor; constant-temperature bath set at 25°C; 0.1 M HCl solution (350 mL).

REFERENCES

1. I. N. Levine, "Physical Chemistry," 3d ed., pp. 280–282, McGraw-Hill, New York (1988).
2. H. S. Harned and R. W. Ehlers: *J. Amer. Chem. Soc.* **55,** 2179 (1933).

GENERAL READING

R. A. Alberty, "Physical Chemistry," 6th ed., Wiley, New York (1983).
H. Eyring, "Electrochemistry," Academic Press, New York (1970).

CHAPTER
IX

CHEMICAL KINETICS

EXPERIMENTS

20. Method of initial rates: iodine clock

21. Kinetics of a hydrolysis reaction

22. Enzyme kinetics: inversion of sucrose

23. Kinetics of the decomposition of benzenediazonium ion

24. Gas-phase kinetics

25. Kinetics of a fast reaction

EXPERIMENT 20
METHOD OF INITIAL RATES: IODINE CLOCK

The homogeneous reaction in aqueous solution

$$IO_3^- + 8I^- + 6H^+ \rightarrow 3I_3^- + 3H_2O \tag{1}$$

like virtually all reactions involving more than two or three reactant molecules, takes place not in a single molecular step but in several steps. The detailed system of steps is called the *reaction mechanism*. It is one of the principal aims of chemical kinetics to obtain information to aid in the elucidation of reaction mechanisms, which are fundamental to our understanding of chemistry.

THEORY

The several steps in a reaction are usually consecutive and tend to proceed at different speeds. Usually, when the overall rate is slow enough to measure at all, it is because one of the steps tends to proceed so much more slowly than all the others that it effectively controls the overall reaction rate and can be designated the *rate-controlling* step. A steady state is quickly reached in which the concentrations of the reaction intermediates are controlled by the intrinsic speeds of the reaction steps by which they are formed and consumed. A study of the rate of the overall reaction yields information of a certain kind regarding the nature of the rate-controlling step and closely associated steps. Usually, however, rate studies supply only part of the information needed to formulate uniquely and completely the correct reaction mechanism.

When the mechanism is such that steady state is quickly attained, the rate law for Eq. (1) can be written in the form

$$-\frac{d(IO_3^-)}{dt} = f[(IO_3^-), (I^-), (H^+), (I_3^-), (H_2O), \ldots] \tag{2}$$

where parenthesized quantities are concentrations. In the general case the brackets might also contain concentrations of additional substances, referred to as catalysts, whose presence influences the reaction rate but which are not produced or consumed in the overall reaction. The determination of the rate law requires that the rate be determined at a sufficiently large number of different combinations of the concentrations of the various species present to enable an expression to be formulated which accounts for the observations and gives good promise of predicting the rate reliably over the concentration ranges of interest. The rate law can be written to correspond in form to that predicted by a theory based on a particular type of mechanism, but basically it is an empirical expression.

The most frequently encountered type of rate law is of the form [again using the reaction of Eq. (1) as an example]

$$-\frac{d(IO_3^-)}{dt} = k(IO_3^-)^m (I^-)^n (H^+)^p \cdots \tag{3}$$

where the exponents $m, n, p, \ldots$ are determined by experiment. Each exponent in Eq. (3) is the *order* of the reaction with respect to the corresponding species; thus, the reaction is said to be mth order with respect to IO_3^-, etc. The algebraic sum of the exponents, $m + n + p$ in this example, is the *overall order* (or, commonly, simply the *order*) of the reaction. Reaction orders are usually, but not always, positive integers within experimental error.

The order of a reaction is determined by the reaction mechanism. It is related to, and is often (but not always) equal to, the number of reactant molecules in the rate-controlling step—the "molecularity" of the reaction. Consider the following proposed mechanism for the hypothetical reaction $3A + 2B = $ products:

a.	$A + 2B = 2C$	(fast, to equilibrium, K_a)
b.	$A + C = $ products	(slow, rate controlling, k_b)

The rate law predicted by this mechanism is

$$-\frac{d(A)}{dt} = \tfrac{3}{2}k_b(A)(C) = \tfrac{3}{2}k_b(A)K_a^{1/2}(A)^{1/2}(B) = k(A)^{3/2}(B)$$

The overall reaction involves five reactant molecules, but it is by no means necessarily of fifth order. Indeed, the rate-controlling step in this proposed mechanism is bimolecular, and the overall reaction order predicted by the mechanism is $\tfrac{5}{2}$. It is also important to note that this mechanism is not the only one that would predict the above $\tfrac{5}{2}$-order rate law for the given overall reaction; thus experimental verification of the predicted rate law would by no means constitute proof of the validity of the above proposed mechanism.

It occasionally happens that the observed exponents deviate from integers or simple rational fractions by more than experimental error. A possible explanation is that two or more simultaneous mechanisms are in competition, in which case the observed order should lie between the extremes predicted by the individual mechanisms. A possible alternative explanation is that no single reaction step is effectively rate controlling.

We now turn our attention to the experimental problem of determining the exponents in the rate law. Except in first- and second-order reactions it is usually inconvenient to determine the exponents merely by determining the time behavior of a reacting system in which many or all reactant concentrations are allowed to change simultaneously and comparing the observed behavior with integrated rate expressions. A procedure is desirable which permits the dependences of the rate on the concentrations of the different reactants to be isolated from one another and determined one at a time. In one such procedure, all the species but the one to be studied are present at such high initial concentrations relative to that of the reactant to be studied that their concentrations may be assumed to remain approximately constant during the reaction; the apparent reaction order with respect to the species of interest is then obtained by comparing the progress of the reaction with that predicted by rate laws for first order, second order, and so on. This procedure would often have the disadvantage of placing the system outside the concentration range of interest and thus possibly complicating the reaction mechanism.

In another procedure, which we shall call the *initial rate method*, the reaction is run for a time small in comparison with the "half-life" of the reaction but large in comparison with the time required to attain a steady state, so that the actual value of the initial rate [the initial value of the derivative on the left side of Eq. (3)] can be estimated approximately. Enough different combinations of initial concentrations of the several reactants are employed to enable the exponents to be determined separately. For example, the exponent m is determined from two experiments which differ only in the IO_3^- concentration.

In the present experiment the rate law for the reaction shown in Eq. (1)

will be studied by the initial rate method, at 25°C and a pH of about 5. The initial concentrations of iodate ion, iodide ion, and hydrogen ion will be varied independently in separate experiments, and the time required for the consumption of a definite small amount of the iodate will be measured.

METHOD

The time required for a definite small amount of iodate to be consumed will here be measured by determining the time required for the iodine produced by the reaction (as I_3^-) to oxidize a definite amount of a reducing agent, arsenious acid, added at the beginning of the experiment. Under the conditions of the experiment, arsenious acid does not react directly with iodate at a significant rate but reacts with iodine as quickly as it is formed. When the arsenious acid has been completely consumed, free iodine is liberated which produces a blue color with a small amount of soluble starch that is present. Since the blue color appears rather suddenly after a reproducible period of time, this series of reactions is commonly known as the "iodine clock reaction."

The reaction involving arsenious acid may be written, at a pH of about 5,

$$H_3AsO_3 + I_3^- + H_2O \rightarrow HAsO_4^{2-} + 3I^- + 4H^+ \tag{4}$$

The overall reaction, up to the time of the starch end point, can be written, from reactions (1) and (4),

$$IO_3^- + 3H_3AsO_3 \rightarrow I^- + 3HAsO_4^{2-} + 6H^+ \tag{5}$$

Since with ordinary concentrations of the other reactants hydrogen ions are evidently produced in quantities large in comparison to those corresponding to pH 5, it is evident that buffers must be used to maintain constant hydrogen-ion concentration. As is apparent from the method used, the rate law will be determined under conditions of essentially zero concentration of I_3^-; the dependence of the rate on triiodide, which in fact has been shown to be very small,[1] will not be measured. Under these conditions, Eq. (3) is an appropriate expression for the rate.

A constant initial concentration of H_3AsO_3 is used in a series of reacting mixtures having varying initial concentrations of IO_3^-, I^-, and H^+. Since the amount of arsenious acid is the same in each run, the amount of iodate consumed up to the color change is constant, and related to the amount of arsenious acid by the stoichiometry of Eq. (5). The initial reaction rate in mole liter^{-1} sec^{-1} is thus approximately the amount consumed (per liter) divided by the time required for the blue end point to appear. From the initial rates of two reactions in which the initial concentration of only one reactant is varied and all other concentrations kept the same, it is possible to infer the exponent in the rate expression associated with the reactant which is varied. This is most conveniently done by taking logarithms of both sides of Eq. (3) and subtracting the expressions for the two runs.

EXPERIMENTAL

Solutions. Two acetate buffers, with hydrogen-ion concentrations differing by a factor of 2, will be made up by the student from stock solutions. Use will be made of the fact that at a given ionic strength the hydrogen-ion concentration is proportional to the ratio of acetic acid concentration to acetate ion concentration:

$$(H^+) = K \frac{(HAc)}{(Ac^-)} \frac{1}{\gamma_\pm^2} \tag{6}$$

where at 25°C, $K = 1.753 \times 10^{-5}$ when concentrations are expressed in units of mol liter^{-1}. The experiments will all be carried out at about the same ionic strength (0.16 ± 0.01), and accordingly the activity coefficient is approximately the same in all experiments, by the Debye–Hückel theory. It will also be seen that within wide limits the amount of buffer solution employed in a given total volume is inconsequential, provided the ionic strength of the resultant solution is always kept about the same. The solutions required are as follows:

Buffer A. Pipette 100 mL of 0.75 M NaAc solution, 100 mL of 0.22 M HAc solution, and about 20 mL of 0.2 percent soluble starch solution into a 500-mL volumetric flask, and make up to the mark with distilled water [yields $(H^+) \cong 10^{-5}\ M$].

Buffer B. Pipette 50 mL of 0.75 M NaAc solution, 100 mL of 0.22 M HAc solution, and about 10 mL of 0.2 percent soluble starch solution into a 250-mL volumeric flask, and make up to the mark with distilled water [yields $(H^+) \cong 2 \times 10^{-5}\ M$].

H_3AsO_3, 0.03 M. Should be made up from $NaAsO_2$ and brought to a pH of about 5 by addition of HAc.

KIO_3, 0.1 M.

KI, 0.2 M.

Suggested sets of initial volumes of reactant solutions, based on a final volume of 100 mL, are given in the table below.

Solution	Pipette sizes (mL)	Initial volumes of solutions, mL			
		1	2	3	4
H_3AsO_3	5	5	5	5	5
IO_3^-	5	5	10	5	5
Buffer A	20, 25	65	60	40	
Buffer B	20, 25				65
I^-	25	25	25	50	25

Two or three runs should be made on each of the four sets. Two or more runs should also be made on a set with proportions chosen by the student in which the initial compositions of *two* reacting species differ from those in set 1. In each case, the amount of buffer required is that needed to obtain a final volume of 100 mL.

It is convenient to use each pipette only for a single solution, if possible, to minimize time spent in rinsing. The pipettes should be marked to avoid mistakes.

The buffer solutions and the iodide solution should be equilibrated to 25°C by clamping flasks containing them in a thermostat bath set at that temperature. Two vessels of convenient size (ca. 250 mL) and shape (beakers or Erlenmeyer flasks), rinsed and drained essentially dry, should also be clamped in the bath. One of them should have a white-painted bottom surface (or have a piece of white cloth taped under the bottom) to aid in observing the blue end point unless other means are available to obtain a light background.

To make a run, pipette all the solutions *except* KI into one of the vessels and the KI solution into the other. Remove both vessels from the bath, and begin the reaction by pouring the iodide rapidly but quantitatively into the other solution, simultaneously starting the stopwatch. Pour the solution back and forth once or twice to complete the mixing, and place the vessel containing the final solution back into the bath. Stop the watch at the appearance of the first faint but definite blue color.

CALCULATIONS

The student should construct a table giving the actual initial concentrations of the reactants IO_3^-, I^-, and H^+. The H^+ concentrations should be calculated from the actual concentrations of NaAc and HAc in the stock solutions employed, with an activity coefficient calculated by use of the Debye–Hückel theory for the ionic strength ($I = 0.16$) of the reacting mixtures.

Using the known initial concentration of H_3AsO_3, calculate the initial rate for each run. From appropriate combinations of sets 1, 2, 3, and 4, calculate the exponents in the rate expression (3). Also calculate a value of the rate constant k from each run, and obtain an average value of k from all runs.

Write the rate expression, with the numerical values of the rate constant k and the experimentally obtained values of the exponents. Beside it write the temperature and ionic strength at which this expression was obtained.

Write another rate expression, in which those exponents which appear to be reasonably close (within experimental error) to integers are replaced by the integral values. Use this expression to calculate values for the initial rates of all sets studied, and compare them with the observed initial rates.

DISCUSSION

The kinetics of this reaction have been the subject of much study, and the mechanism is not yet completely elucidated with certainty. The following is an incomplete list of the mechanisms that have been proposed.

1. $IO_3^- + 2I^- + 2H^+ \xrightarrow{k} 2HIO + IO^-$ (slow)

Followed by fast reactions (Dushman[1])

2. $IO_3^- + H^+ \underset{}{\overset{K}{\rightleftharpoons}} HIO_3$ (fast, to equil.)

$I^- + H^+ \underset{}{\overset{K'}{\rightleftharpoons}} HI$ (fast, to equil.)

$HIO_3 + HI \xrightarrow{k} HIO + HIO_2$ (slow)

Followed by fast reactions (at low iodide concentrations; Abel and Hilferding[2])

3. $IO_3^- + I^- + 2H^+ \underset{}{\overset{K}{\rightleftharpoons}} H_2I_2O_3$ (fast, to equil.)

$H_2I_2O_3 \underset{}{\overset{K'}{\rightleftharpoons}} I_2O_2 + H_2O$ (fast, to equil.)

$I_2O_2 + I^- \xrightarrow{k} I_3O_2^-$ (slow)

Followed by fast reactions (Bray and Liebhafsky[3])

4. $IO_3^- + I^- + 2H^+ \underset{}{\overset{K}{\rightleftharpoons}} H_2I_2O_3$ (fast, to equil.)

$H_2I_2O_3 \xrightarrow{k} HIO + HIO_2$ (slow)

Followed by fast reactions (at low iodide concentrations; Bray and Liebhafsky[3])

5. $IO_3^- + H^+ \underset{}{\overset{K}{\rightleftharpoons}} IO_2^+ + OH^-$ (fast, to equil.)

$H^+ + OH^- \underset{}{\overset{1/K_w}{\rightleftharpoons}} H_2O$ (fast, to equil.)

$IO_2^+ + I^- \underset{}{\overset{K'}{\rightleftharpoons}} IOIO$ (fast, to equil.)

$IOIO + I^- \xrightarrow{k} I^+ + 2IO^-$ (slow)

Followed by fast reactions (Morgan, Peard, and Cullis[4])

6. $IO_3^- + H^+ \underset{}{\overset{K}{\rightleftharpoons}} IO_2^+ + OH^-$ (fast, to equil.)

$H^+ + OH^- \underset{}{\overset{1/K_w}{\rightleftharpoons}} H_2O$ (fast, to equil.)

$IO_2^+ + I^- \underset{}{\overset{K'}{\rightleftharpoons}} IOIO$ (fast, to equil.)

$IOIO \xrightarrow{k} IO^+ + IO^-$ (slow)

Followed by fast reactions (at low iodide concentrations; Morgan, Peard, and Cullis[4])

The student should discuss the above mechanisms in connection with his experimentally determined rate law.

The oxidation of iodide ion by chlorate ion ClO_3^- has also been studied.[5,6]

Although the reaction appears to be attended by complications that make it difficult to study, under certain conditions it can be carried out as an "iodine clock" experiment. (For the interested student, suggested concentration ranges for 20 to 25°C are ClO_3^-, 0.05 to 0.10 M; I^-, 0.025 to 0.10 M; H^+, 0.02 to 0.04 M. A sulfate–bisulfate buffer may be used.) The resulting rate law is not identical with that for the reaction with iodate, but appears to be compatible with mechanisms analogous to several of those given above.

A different type of "clock" reaction, suitable for student investigation, is the reaction of formaldehyde with bilsulfite ion.[7,8] The reaction involves a single slow step

$$HCHO + HSO_3^- \longrightarrow HOCH_2SO_3^- \qquad (7)$$

followed by the rapid buffer equilibrium

$$HSO_3^- \underset{}{\overset{K_a}{\rightleftharpoons}} H^+ + SO_3^{2-} \qquad (8)$$

As bisulfite ion is used up in reaction (7), the hydrogen-ion concentration adjusts itself according to the buffer equation (8). If an indicator such as phenolphthalein is added to the mixture, it will undergo a sudden color change when the pH of the solution reaches the pK_i of the indicator. The time τ required for the color change is related to the rate constant for reaction (7) by the equation[7]

$$\frac{1}{F_0 - B_0}\left[\log_{10}\frac{B_0(F_0 - B_0)K_a}{F_0 S_0 K_i'}\right] \cong 0.43 k_7 \tau \qquad (9)$$

where B_0 is the initial bisulfite concentration, S_0 is the initial sulfite concentration, F_0 is the initial formaldehyde concentration, K_a is the bisulfite equilibrium constant $(6.7 \times 10^{-8}$ at 25°C when using $mol\,L^{-1}$ concentration units), $K_i' = RK_i$, where K_i is the indicator equilibrium constant, and R is the indicator color ratio at the end point for the indicator concentration used.

"Flowing clock" modification of the experiment. In conventional "clock" reactions, it is difficult to obtain reliable results for reaction times of less than 20 or 30 s. Since much of modern chemical kinetics is concerned with reactions occurring on much faster time scales, it is worthwhile to introduce a modification of such experiments which makes use of a simple flow system to explore reaction times of the order of a few seconds or less.[9] Both the iodide oxidation and formaldehyde–bisulfite reactions can be investigated in this way.

The simple apparatus is shown in Fig. 1. The reactants are contained in two thermostated 250-mL graduated cylinders, each with a siphon tube connected to the T-shaped mixing chamber A, which is a 1-mm bore 3-way stopcock. The flow tube is a 2-mm bore Pyrex capillary which lies between the mixing chamber and an inline stopcock B used to start and stop the flow. A water aspirator or a house vacuum line protected with a 1-liter trap is connected to stopcock B. During tests of the apparatus carried out using

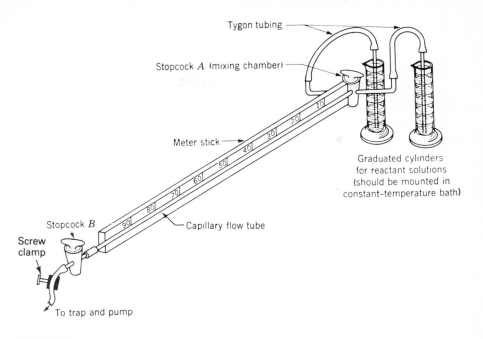

Tygon tubing

Stopcock A (mixing chamber)

Meter stick

Graduated cylinders
for reactant solutions
(should be mounted in
constant-temperature bath)

Stopcock B

Screw
clamp

Capillary flow tube

To trap and pump

FIGURE 1
Schematic diagram of the "flowing clock" apparatus.

water, stopcock A should be carefully adjusted so that equal volumes of fluid are removed from each of the two cylinders in a given time interval. Stopcock A is then left at that setting throughout all the runs.

A volume flow rate f in the range 1 to $10\,\mathrm{mL\,s^{-1}}$ should be suitable for carrying out the experiments described below, and this rate can be fine-tuned with an adjustable screw clamp on the tube between stopcock B and the trap. The best rate of flow for a given experiment will depend on the reaction being studied and the concentrations being used for the reactants. It is assumed that this value, determined from preliminary trial runs, has been provided by the instructor. The student should measure and adjust the flow rate using water in the graduated cylinders. Apply suction and open stopcock B *briefly* in order to fill the tubing and capillary with water. Then read the initial level of the water in each cylinder. Fully open stopcock B for a measured time interval τ_0, close B and read the final water levels. The flow rate f is given by $f = V/\tau_0$, where V is the total volume of water removed from the cylinders.

The cylinders should now be emptied of water, rinsed with the solutions to be studied, and then filled with the appropriate solutions. Apply suction and fully open stopcock B to carry out a run.

The reaction time τ for a given experiment is determined by measuring the distance x along the flow tube from stopcock A at which the appropriate color change occurs, and using the relationship

$$\tau = \frac{xA}{f} \tag{10}$$

where A is the cross-sectional area of the flow tube and f is the previously determined volume flow rate.

The iodine clock experiment can be carried out using peroxide in place of iodate and sodium thiosulfate in place of arsenious acid. The rate law by analogy to Eq. (3) is

$$-\frac{d(H_2O_2)}{dt} = k(H_2O_2)^m (I^-)^n (H^+)^p \tag{11}$$

A set of experimental conditions suitable for determining the exponents in Eq. (11) is given in the table below.

	Reservoir 1, vol in mL			Reservoir 2, vol in mL			
Run no.	4 M H$_2$O$_2$	2 M HCl	Distilled water	0.05 M KI	0.01 M Na$_2$S$_2$O$_3$	Distilled water	Starch solution
1	125	125	0	100	100	0	50
2	62.5	125	62.5	100	100	0	50
3	125	62.5	62.5	100	100	0	50
4	125	125	0	50	100	50	50

For the formaldehyde–bisulfite reaction, a set of runs is given in the following table, where the buffer is 0.3 M in sulfite ion and 0.05 M in bisulfite ion.

	Reservoir 1, vol in mL			Reservoir 2, vol in mL	
Run no.	Buffer solution	Phenol-phthalein	Distilled water	1 M Formaldehyde solution	Distilled water
1	120	2	128	120	130
2	60	2	188	60	190
3	30	2	218	30	220
4	15	2	233	15	235

A plot of $(F_0 - B_0)^{-1}$ versus τ will test Eq. (9) since the quantity in brackets is constant. Assuming that $K_a/K_i' = 79,$[7] one can estimate k_7.

APPARATUS

Three 200-mL beakers (bottom painted white); two 250-mL and one 100-mL beakers; two 5-, one 20-, two 25-, and one 50-mL pipettes; pipetting bulb; one 250- and one 500-mL volumetric flask; four 250-mL Erlenmeyer flasks with four corks to fit; one 10-mL graduated cylinder; glass-marking pencil; stopwatch.

Constant-temperature bath (set at 25°C) with provision for mounting beakers and flasks; 0.75 M NaAc solution (300 mL); 0.22 M HAc solution (300 mL); 0.03 M H_3AsO_3 solution (150 mL); 0.1 M KIO_3 solution (150 mL); 0.2 M KI solution (500 mL); 0.2 percent soluble starch solution, with trace of HgI_2 as preservative (75 mL).

For the flowing clock modification: two 250-mL graduated cylinders; 3-way stopcock; inline stopcock; 1 m of 2-mm bore Pyrex capillary tubing; meter stick; connecting tubing; water aspirator or other rough-vacuum source; 1-liter trap; assorted clamps and ring stands; stopwatch; thermostat bath; appropriate solutions.

REFERENCES

1. S. Dushman, *J. Phys. Chem.* **8,** 453 (1904).
2. E. Abel and K. Hilferding: *Z. Phys. Chem.* **136A,** 186 (1928).
3. W. C. Bray, *J. Amer. Chem. Soc.* **52,** 3580 (1930).
4. K. J. Morgan, M. G. Peard, and C. F. Cullis: *J. Chem. Soc.* **1951,** 1865.
5. W. C. Bray, *J. Phys. Chem.* **7,** 92 (1903).
6. A. Skrabal and H. Schreiner: *Monatsh. Chem.* **65,** 213 (1934).
7. P. Jones and K. B. Oldham: *J. Chem. Educ.* **40,** 366 (1963).
8. R. L. Barrett, *J. Chem. Educ.* **32,** 78 (1955).
9. M. L. Hoggett, P. Jones, and K. B. Oldham: *J. Chem. Educ.* **40,** 367 (1963).

GENERAL READING

I. N. Levine, "Physical Chemistry," 3d ed., chap. 17, McGraw-Hill, New York (1988).
K. L. Laidler and K. J. Laidler, "Chemical Kinetics," Harper-Row, New York (1986).

EXPERIMENT 21
KINETICS OF A HYDROLYSIS REACTION

In the presence of hydrogen ions, diethyl acetal in aqueous solution hydrolyzes as follows, the reaction going substantially to completion:

$$CH_3CH(OC_2H_5)_2 + H_2O \xrightarrow{H^+} CH_3CHO + 2C_2H_5OH \qquad (1)$$

In this experiment the reaction will be shown to be first order with respect to acetal and with respect to hydrogen ion, the rate constant will be determined, and the temperature dependence of the rate constant will be used to calculate the activation energy for the reaction.

Other acetals, as well as numerous other organic compounds such as esters and epoxides, also undergo acid-catalyzed hydrolysis with similar kinetics.[1] Use has been made of kinetics measurements on reactions of this kind as a means of measuring hydrogen-ion concentration independently of hydrogen-ion activity.[2]

THEORY

A first-order reaction is ordinarily understood to be one in which the rate of disappearance of a single species is proportional to the concentration of that species in the reaction mixture. This terminology is often used even when, strictly speaking, the total reaction order is different from unity owing to the participation of additional species in the reaction which are themselves not consumed (i.e., catalysts—H^+ in the present instance) or the participation of substances which are present in such large amounts that their concentrations do not undergo significant percentage change during the reaction (H_2O in the present instance). It would be more correct in such cases to state that the reaction is first order with respect to a given disappearing species.

The rate law for a first-order reaction may be expressed by the differential equation

$$-\frac{dc_A}{dt} = kc_A \tag{2}$$

where c_A is the instantaneous concentration of the disappearing reactant (here acetal) and k is the *specific reaction rate*, or simply *rate constant*, applicable to the specified conditions (temperature and, in the present case, hydrogen-ion concentration) which are constant throughout the course of the reaction. If the dependence of the rate of hydrolysis of acetal on hydrogen-ion concentration is known, it may be included in the expression for the rate law. In the present experiment it will be demonstrated that this dependence is first order, and accordingly we can write

$$-\frac{dc_A}{dt} = k'c_{H^+}c_A \tag{3}$$

where k' is another rate constant at the specified temperature, independent of the hydrogen-ion concentration.

If Eq. (2) is integrated with the initial condition $c = c_0$ at $t = 0$, we obtain

$$\ln \frac{c}{c_0} = -kt \qquad c = c_0 e^{-kt} \tag{4}$$

A reaction may be shown to be first order by a variety of methods, some of which include the following.

1. The logarithm of the concentration c is plotted against time and the experimental points are shown to fit a straight line.

2. The time interval required for any definite fraction of the reactant to disappear is shown to be a constant, independent of when the interval is taken during the reaction. In the special case where the fraction is one-half, the time interval is called the "half-life" of the reactant.

3. The *rate* of the reaction is studied as a function of time. A well-known example is that of radioactive decay, where the rate of decay is given by a radiation intensity as measured, say, by a Geiger counter.

4. The *initial rate* of the reaction, $-(dc/dt)_{t=0}$, is determined at a number of initial concentrations c_0, and the initial rate is shown to be directly proportional to the initial concentration (see Exp. 20).

All the above methods require that the actual concentration or its instantaneous time derivative be known at least to within a multiplicative constant. However, there are methods useful for cases where only a linear function of the concentration

$$f = Ac + B \tag{5}$$

is available (A and B being unknown constants):

5. The reaction is allowed to go substantially to completion by waiting a very long time or by temperature change or catalysis, and a value of f_∞ is obtained. Then we can use the equation, derived from Eqs. (4) and (5),

$$\ln \frac{f - f_\infty}{f_0 - f_\infty} = -kt \tag{6}$$

in the manner of method 1 above.

6. *Guggenheim method.* In the present experiment we shall use a method[3] which eliminates the need for an f_∞ value. From Eqs. (4) and (5) we have

$$f(t) = Ac_0\, e^{-kt} + B$$

and at time $t + \Delta t$

$$f(t + \Delta t) = Ac_0\, e^{-k\,\Delta t}\, e^{-kt} + B$$

Taking the difference we obtain

$$\Delta f \equiv f(t + \Delta t) - f(t) = Ac_0(e^{-k\,\Delta t} - 1)e^{-kt} \tag{7}$$

Thus, *for constant Δt, Δf* varies exponentially in a way similar to c. If we plot its logarithm against t as in method 1, we should obtain a straight line, from the slope of which the rate constant can be obtained:

$$\ln \Delta f = 2.303 \log \Delta f = -kt + \text{const} \tag{8}$$

The Guggenheim treatment and its results are in principle independent of the particular constant value chosen for Δt. The choice of a value for Δt depends on practical considerations: If it is too small, the scatter of points on the logarithmic plot will be too great, and if it is too large, the number of points that can be plotted becomes too small. For a clear demonstration of first-order kinetics and a good determination of the rate constant, the points plotted should cover at least one natural logarithmic cycle (one power of e). A good procedure would be to run the experiment over a total time sufficient to obtain about $1\frac{1}{2}$ cycles (i.e., until increments between successive readings have decreased to about one-fifth of the initial value) and to take about one-third of this total time for Δt.

This method may be seen to be very closely related to method 3 above. It uses two experimental readings separated by a constant time difference Δt to determine, not the actual rate itself, but rather a quantity which on the assumption of first-order kinetics must be directly proportional to the actual instantaneous rate. When the appropriate logarithmic plot gives a straight line, the assumption of first-order kinetics is confirmed. The method may be said to degenerate into method 3 in the limit of $\Delta t = 0$, and into method 5 in the limit $\Delta t = \infty$.

Dependence on hydrogen-ion concentration. In this experiment the rate constant k will be demonstrated to be proportional to the first power of the hydrogen-ion concentration. This suggests a mechanism which can be written

$$A + H_2O + H^+ = X^+ \qquad \text{(fast equilibrium, } K^*)$$
$$X^+ \rightarrow H^+ + \text{products} \qquad \text{(rate determining, } k_X) \tag{9}$$

where A is acetal and X^+ is an intermediate complex ion. This predicts for the reaction rate

$$-\frac{dc_A}{dt} = k_X c_{X^+} = k_X K^* c_{H^+} c_A \tag{10}$$

Comparison with Eqs. (2) and (3) gives

$$k = k' c_{H^+} = k_X K^* c_{H^+} \tag{11}$$

It is worthy of note that the rate constant is proportional to the *concentration*, rather than to the thermodynamic *activity*, of hydrogen ion. This was explained by Brønsted in terms of the transition-state theory.[4] It arises from the fact that the equilibrium constant K^* contains an activity coefficient for the ion X^+ in the numerator and one for the ion H^+ in the denominator while all other activity coefficients present are for uncharged species. For dilute solutions of given ionic strength the Debye–Hückel theory predicts that the activity coefficients of all univalent ions are very nearly equal, and those for H^+ and X^+ accordingly should cancel out. In the same order of approximation, the activity coefficients for uncharged species are unity. (In certain other cases activity coefficients do not cancel out, and the rate constants are then found to depend on the ionic strengths of the solutions.)

As a matter of convenience, acetate buffer solutions are to be used as a means of obtaining low but stable hydrogen-ion concentrations. The hydrogen-ion concentration in the reaction mixture can be calculated directly from the buffer composition and the acetic acid ionization constant K:

$$c_{H^+} = \frac{K}{\gamma_\pm^2} \frac{c_{HAc}}{c_{Ac^-}} \tag{12}$$

At 25°C, $K = 1.753 \times 10^{-5}$ with concentrations given in mol L^{-1}. Alternatively, the hydrogen-ion activity of the buffer can be determined with a pH meter, and the concentration calculated from

$$c_{H^+} = \frac{a_{H^+}}{\gamma_{H^+}} \tag{13}$$

Activity coefficients can be obtained with sufficient accuracy from the Debye–Hückel expression

$$\log \gamma_{H^+} = \log \gamma_\pm = -0.509 \frac{\sqrt{I}}{1 + \sqrt{I}} \tag{14}$$

where I is the ionic strength ($= $ NaAc concentration).

Dependence on temperature. The Arrhenius activation energy E_a is related to the temperature variation of the rate constant by the equation

$$\frac{d \ln k'}{dT} = \frac{E_a}{RT^2} \tag{15}$$

If this is integrated with the assumption of constant E_a, we obtain

$$\ln k' = 2.303 \log k' = -\frac{E_a}{RT} + \text{const} \tag{16}$$

which permits us to obtain E_a from the slope of a plot of logarithm of the rate constant against the reciprocal of the absolute temperature. If the rate constant is available at only two temperatures, we can use the equation

$$\ln \frac{k_2'}{k_1'} = 2.303 \log \frac{k_2'}{k_1'} = \frac{E_a}{R} \frac{(T_2 - T_1)}{T_1 T_2} \tag{17}$$

Since in this experiment a buffer solution is being used at more than one temperature, we should examine the effect of possible temperature variation of the hydrogen-ion concentration in the buffer, since, in general, the ionization constants of weak acids are functions of temperature. Failure to take this into account would have the effect of adding the heat of ionization of the weak acid to the activation energy. In the case of acetic acid, the heat of ionization happens to be very small (-0.8 kJ mol^{-1}) in comparison with the expected energy of activation, and this small quantity is partly offset by a very small correction $(+0.4 \, \text{kJ mol}^{-1})$ for the temperature variation of the activity coefficient. Therefore with an acetate buffer the effect of temperature on hydrogen-ion concentration can be neglected for the present experiment.

METHOD

As a measurable physical quantity which may be taken as a linear function of the concentration of the disappearing reactant, the *total volume V* of a given quantity of the reaction mixture might be used. To a good approximation, the volume is a linear function of the numbers of moles n_i of the component species i present:

$$V = \sum_i n_i \bar{V}_i$$

since the partial molar volumes $\bar{V}_i$ are approximately constant for both solvent and solutes in a dilute solution. In turn the numbers of moles n_i are linear functions of the concentrations of the disappearing reactants. Similarly we might use the electrical conductance of the reacting mixture in the event that the species concerned are ionic, or the absorbance for a given wavelength of light if the species have suitable absorption bands.

In the present experiment we shall make use of the total volume of the solution as measured with a dilatometer, which is a vessel to which a graduated capillary of small diameter is attached so that small changes in the volume of the contained liquid can be measured by movement of the meniscus of the liquid in the capillary. The capillary need not be graduated to correspond to the total volume of solution; it is sufficient that the graduations give the volume above any fixed reference point in the capillary in any arbitrary units. The capillary reading h is then a linear function of the total volume and thus, by the above argument, is also a linear function of the concentration of the disappearing reactant.

EXPERIMENTAL

The dilatometer is shown in Fig. 1. It must be handled with great care, as it is fragile. Special care should be exercised in cleaning it and in putting it into and taking it out of its holder.

Thermostat baths at 20.0, 25.0, and 30.0°C are required, with temperature regulation of ±0.002°C or better. It should be stressed that excellent temperature control is essential since the volume changes produced by even small temperature changes are of the same order of magnitude as those produced by the reaction being studied. A high-sensitivity thermometer should be used for careful and frequent monitoring of the bath regulation.

Materials. The acetal should be of high purity, dry, and free of acidic

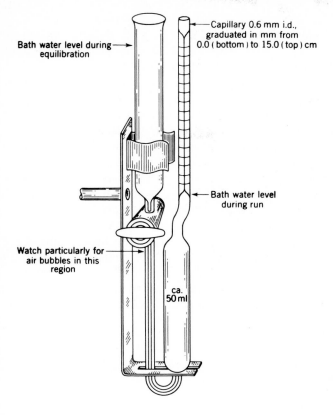

FIGURE 1
Dilatometer.

substances. If necessary the commercially available product may be refluxed over calcium oxide and fractionated.

Water used in the preparation of the stock solutions and water to be used for making up the final buffer solutions should be as free of dissolved gases as possible. Boiled distilled water may be used.

The following buffer solutions will be needed and should be made up volumetrically from the stock solutions provided (0.1 M NaAc, 1.5 M HAc):

1. 0.02 M in NaAc and 0.3 M in HAc ($c_{H^+} \cong 3.4 \times 10^{-4}\ M$; pH $\cong 3.5$)
2. 0.02 M in NaAc and 0.6 M in HAc ($c_{H^+} \cong 6.7 \times 10^{-4}\ M$; pH $\cong 3.2$)

Make up 250 mL of each of these solutions at the beginning of the experiment. If a pH meter is available, measure the pH of each buffer solution. At least 15 min before it will be needed, 100 mL of buffer solution should be placed in a volumetric flask and this flask should be put in thermostat bath to equilibrate. At the same time, a clean, dry 250-mL flask should be clamped in the bath for later use in preparing the reaction mixture. Flasks and solutions should be brought close to bath temperature before being placed in the bath if someone else is performing a run in the same bath.

Cleaning the dilatometer. The dilatometer must be cleaned very carefully. If it has been constructed with a stopcock having a Teflon plug, all use of stopcock grease is eliminated. In this case, the cleaning and reassembly procedures are simplified. However, a dilatometer with a standard glass stopcock will also perform well if attention is paid to the details concerning stopcock grease given below. In either case, the crucial step is the thorough cleaning of the capillary tube. First, the stopcock plug should be removed and set aside for thorough cleaning. After the stopcock barrel has been scrubbed out with a test-tube brush and detergent solution and any excess stopcock grease from the stopcock leads cleaned out with a pipe cleaner, the two ends of the barrel may be plugged with corks. To facilitate further cleaning and rinsing, a one-hole rubber stopper equipped with a short length of tubing is affixed to the mouth of the dilatometer and connected to a water aspirator with pressure tubing. With the aspirator on, the end of the capillary is dipped into a battery jar or lipless beaker containing a detergent solution (see Fig. 2) until the dilatometer is partly filled. It should then be carefully shaken for a few minutes. By the same technique it is repeatedly and very thoroughly rinsed with distilled water and dried by sucking air through it. It need not be rigorously dry except in the capillary tube and in the stopcock barrel and leads. The latter can be dried

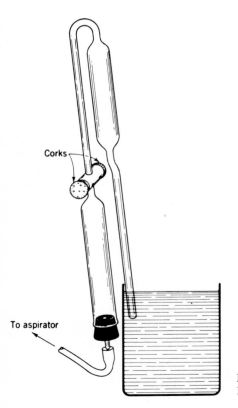

Corks

To aspirator

FIGURE 2
Method of washing and rinsing the dilatometer.

with the help of a towel and pipe cleaner. The stopcock plug should be scrubbed with detergent, and its bore should be cleaned with a pipe cleaner. It should be rinsed and dried thoroughly, carefully greased as described in Chapter XVII, and reinserted into the barrel. The dilatometer can then be installed in its holder and placed in the bath to equilibrate. Between and after runs it is not necessary to clean the dilatometer so thoroughly or to regrease the stopcock. The used solution should be removed by suction, and the dilatometer then thoroughly rinsed and dried.

Procedure. Make up the reacting mixture by pouring 100 mL of buffer solution into the clean dry 250-mL flask, and then pipetting in 2 mL of acetal. The mixture should be swirled in the bath until the acetal has completely dissolved.

With the stopcock of the dilatometer closed, introduce a sufficient quantity of the reacting mixture. After all air bubbles have risen to the surface, the dilatometer can be mounted higher in the bath (so that the entire capillary is above the water surface) and the stopcock can be opened to admit the solution into the bulb. When the bulb is partly filled, the one-hole stopper with affixed tube should be attached and suction applied by mouth to reverse the flow; the purpose of this is to dislodge any air bubbles which may be trapped in the tube below the stopcock. After a few repetitions of this suction, close the stopcock and raise the dilatometer from the bath for a few seconds to permit inspection for the presence of air bubbles; if there are none, the solution may be allowed to fill the bulb completely. *The rate of filling should be reduced drastically toward the end in order that it can be cut off suddenly as soon as the meniscus is on scale in the capillary.* It is inadvisable to draw the meniscus down in the capillary by suction unless absolutely necessary, for drainage from the capillary wall is likely to form liquid plugs which drain slowly and unreliably during the run.

Readings of the position of the meniscus (estimated to one-tenth of the smallest scale division) should be made at 30-s intervals, starting as soon as possible after the stopcock has been closed. They should be continued until differences between successive readings are averaging about one-fifth of the differences observed near the beginning of the run (discounting the first half dozen readings). The bath temperature should be frequently read to within 0.001°C and recorded.

Runs should be made under the following sets of conditions:

1. Buffer solution 1 ($c_{H^+} \cong 3.4 \times 10^{-4} M$) at 25.0°C
2. Buffer solution 2 ($c_{H^+} \cong 6.7 \times 10^{-4} M$) at 25.0°C
3. Buffer solution 2 ($c_{H^+} \cong 6.7 \times 10^{-4} M$) at 20.0°C
4. Buffer solution 1 ($c_{H^+} \cong 3.4 \times 10^{-4} M$) at 30.0°C

If time is limited, it is advisable to perform runs 2 and 3 or 1 and 4 in order to determine the activation energy E_a from data obtained at constant hydrogen-ion concentration.

CALCULATIONS

For each run a table should be constructed with the following entries: t, $h(t)$, and Δh, where $\Delta h = h(t + \Delta t) - h(t)$; Δt should be chosen in accord with the discussion of the Guggenheim method. A plot of $\log \Delta h$ against t is then constructed, and the best straight line drawn through the points. Semilog paper (one decimal cycle) is convenient. The plot is likely to show some initial curvature, which should be disregarded in drawing the straight line; it is presumably due to drift toward temperature equilibrium following the filling operation. If the plotted points tend to show a slight systematic oscillation about the line, a periodic variation in bath temperature due to thermoregulator cycling is probably responsible; if possible, compare the period of the oscillation with that of the thermoregulator cycling. Draw your line so it evens out these fluctuations. For each run, the rate constant k should be calculated from the slope of the best straight line.

In addition to this graphical determination of k values, a least-squares fit with Eq. (8) should also be carried out for each run. Delete from your data set any entries $h(t)$ at small t that give rise to Δh values that seem to be systematically in error due to a drift toward temperature equilibrium in the early stages of the run. It is also proper to delete individual data points that give rise to obviously erratic Δh values (presumably due to an erroneous reading). See Chapters II and XX for discussions of rejecting discordant data. Determine k and its standard deviation from each of a series of least-squares fits using different Δt values spanning the value chosen for your initial plot. If the best fit corresponds to a Δt value significantly different from the one you choose initially, prepare another plot of $\log \Delta h$ versus t using this better Δt value.

The hydrogen-ion concentrations should be calculated by use of either Eq. (12) or Eq. (13). If Eq. (12) is used, the changes in Ac^- and HAc concentrations due to the small ionization of the acetic acid should not be neglected; they may be taken account of in a second approximation, the magnitude of the ionization having been estimated in a first approximation using the concentration values calculated directly from those for the stock solutions.

The results of runs 1 and 2 should be used to demonstrate that the reaction is first order with respect to hydrogen ion, and a value of k' should be calculated for all four runs.

The activation energy E_a should be calculated from the variation of k' with temperature.

APPARATUS

Dilatometer; special clamp for mounting dilatometer; one-hole rubber stopper to fit top of dilatometer, with short length of glass tubing affixed; magnifier (optional); clock or stop-watch; two clamps; three clamp holders; one 100- and two 250-mL volumetric flasks; 250-mL Erlenmeyer flask; 2-, 50-, and 100-mL pipettes; pipetting bulb; battery jar; glass-stoppered bottle for storing acetal; medium length (~0.5 m) of gum-rubber tubing; corks to fit stopcock barrel.

Constant-temperature baths set at 20, 25, and 30°C with high-sensitivity thermometers mounted in each; water aspirator, equipped with rubber hose; stopcock grease; pipe cleaners; acetone for rinsing; reagent-grade acetal (15 mL); 0.1 M sodium acetate (200 mL); 1.5 M acetic acid (300 mL).

REFERENCES

1. J. N. Brønsted and W. F. K. Wynne-Jones: *Trans. Faraday Soc.* **25,** 59 (1929).
2. M. Kilpatrick, Jr., and E. F. Chase: *J. Amer. Chem. Soc.* **53,** 1732 (1931).
3. E. A. Guggenheim, *Phil. Mag.* **2,** 538 (1926).
4. I. N. Levine, "Physical Chemistry," 3d ed., sec. 23.8, McGraw-Hill, New York (1988).

GENERAL READING

J. W. Moore and R. G. Pearson, "Kinetics and Mechanism," 3d ed., Wiley-Interscience, New York (1973).

EXPERIMENT 22
ENZYME KINETICS: INVERSION OF SUCROSE

A very important aspect of chemical kinetics is that dealing with the rates of enzyme-catalyzed reactions. Enzymes are a class of proteins that catalyze virtually all biochemical reactions. In this experiment, we shall study the inversion of sucrose, as catalyzed by the enzyme invertase (β-fructofuranidase) derived from yeast. The rate of the enzyme-catalyzed reaction will then be compared to that of the same reaction catalyzed by hydrogen ions.

THEORY

The basic mechanism for enzyme-catalyzed reactions was first proposed by Michaelis and Menten and confirmed by a study of the kinetics of the sucrose inversion. A simple reaction mechanism by which an enzyme converts a reactant S, usually called a substrate, into products P is

$$E + S \underset{k_{-1}}{\overset{k_1}{\rightleftharpoons}} ES \tag{1}$$

$$ES \underset{k_{-2}}{\overset{k_2}{\rightleftharpoons}} E + P \tag{2}$$

The initial steps involving the enzyme–substrate complex ES are rapid (k_1 and

k_{-1} large) and the decomposition of the complex to form products is relatively slow. The back-reaction in which E and P recombine to form complex will be ignored here for two reasons. The rate constant k_{-2} is generally small, and the concentration of P is also small since we will be concerned with the initial stages of product formation. In some enzyme-catalyzed reactions, more complicated mechanisms involving several different complexes may be required.[1] In the present experiment, however, the simple mechanism given in Eqs. (1) and (2) is adequate to describe the kinetics.

The rate of formation of products when one ignores the back-reaction in Eq. (2) is

$$r = +\frac{d(\mathrm{P})}{dt} = k_2(\mathrm{ES}) \tag{3}$$

where parentheses indicate the concentration of the specified species in moles per liter. In order to proceed further, we need an expression for (ES) in terms of enzyme and substrate concentrations. The key innovation made by Michaelis and Menten was to use the steady-state approximation for the complex ES. That is, it is assumed that $d(\mathrm{ES})/dt$, the net rate of change of ES, is very small compared to the rates of formation and destruction of ES. Thus, we have

$$\frac{d(\mathrm{ES})}{dt} = k_1(\mathrm{E})(\mathrm{S}) - k_{-1}(\mathrm{ES}) - k_2(\mathrm{ES}) \cong 0 \tag{4}$$

which yields

$$(\mathrm{ES}) = \frac{k_1(\mathrm{E})(\mathrm{S})}{k_{-1} + k_2} = \frac{(\mathrm{E})(\mathrm{S})}{K_m} \tag{5}$$

where the Michaelis–Menten constant K_m is defined by $K_m = (k_{-1} + k_2)/k_1$. Note that if $k_2 \ll k_{-1}$, the quantity K_m has a simple interpretation as the equilibrium constant for the dissociation of the ES complex. The instantaneous concentration of enzyme, (E), is not known as a function of time but will not be needed. Since the enzyme is conserved, its total concentration in the form of free enzyme and enzyme–substrate complex must be constant and equal to the initial enzyme concentration $(\mathrm{E})_0$:

$$(\mathrm{E})_0 = (\mathrm{E}) + (\mathrm{ES}) \tag{6}$$

Combining Eqs. (3), (5) and (6), we obtain

$$r = \frac{k_2(\mathrm{E})_0(\mathrm{S})}{K_m + (\mathrm{S})} \tag{7}$$

A schematic plot of r versus (S) is shown in Fig. 1. When the substrate concentration is sufficiently low, $(\mathrm{S}) \ll K_m$, the kinetics are first order with respect to S: $r = [k_2(\mathrm{E})_0/K_m](\mathrm{S})$. When (S) is very high, $(\mathrm{S}) \gg K_m$, the reaction becomes zero order with respect to S and the rate approaches the limiting

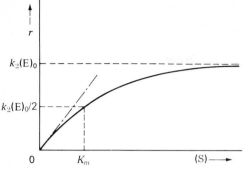

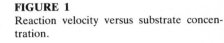

FIGURE 1
Reaction velocity versus substrate concentration.

value $r = k_2(E)_0$. The rate will be equal to half of this maximum value when $(S) = K_m$. Thus a determination of the dependence of r on the substrate concentration (S) will give both $k_2(E)_0$ and K_m values. If the molar concentration $(E)_0$ is known, the rate constant k_2 can also be determined. This quantity, often referred to as the *turnover number*, is of special interest because of its simple interpretation. The value of k_2 corresponds to $r_{max}/(E)_0$; thus the turnover number is the number of sucrose molecules hydrolyzed per second per molecule of enzyme when the enzyme is saturated with substrate (i.e., virtually all the enzyme is in the form of the complex ES).

In this experiment, you will measure the initial rate r_0 as a function of $(E)_0$ and $(S)_0$. The most practical way of analyzing such data is to plot the results in a manner that should yield a straight-line relationship. Equation (7) can be rewritten in two useful forms:

$$\text{L-B:} \qquad \frac{1}{r_0} = \frac{1}{k_2(E)_0} + \frac{K_m}{k_2(E)_0}\frac{1}{(S)_0} \qquad (8)$$

$$\text{E-H:} \qquad \frac{r_0}{(S)_0} = \frac{k_2(E)_0}{K_m} - \frac{r_0}{K_m} \qquad (9)$$

For a plot of $1/r_0$ versus $1/(S)_0$, called a Lineweaver–Burke plot, the intercept determines $k_2(E)_0$ and K_m can then be determined from the slope. For a plot of $r_0/(S)_0$ versus r_0, called an Eadie–Hofstee plot, the slope determines K_m and $k_2(E)_0$ can then be determined from the intercept. The classic plot is the Lineweaver–Burke plot, which has the advantage of displaying the variables separately on different axes. However, the Eadie–Hofstee plot has some practical advantages as demonstrated by a statistical comparison of several different methods of data analysis.[2]

Acid-catalyzed reaction. A number of enzyme-catalyzed reactions, including the sucrose inversion to be studied in this experiment, can also be carried out under nonphysiological conditions by using H^+ ions as a less efficient catalyst.

In the present case, the acid-catalyzed reaction rate has the form

$$r = +\frac{d(P)}{dt} = -\frac{d(S)}{dt} = k_H(H^+)(S) \tag{10}$$

This rate law will be studied in the absence of the enzyme and at H^+ concentrations in the range 0.1–$0.5\ M$, which are found to give suitable values for the rate. On the other hand, the enzyme-catalyzed rate will be studied in a buffer solution that maintains (H^+) at $\sim 10^{-5}\ M$, which effectively eliminates the acid-catalyzed path during those runs.

Activation energy. Rate constants are frequently represented by an Arrhenius equation,

$$k = A\,e^{-E_a/RT} \tag{11}$$

where the activation energy E_a represents an "energy threshold" which must be overcome before the reaction can take place. A conceptual model for the action of an enzyme, or any other catalyst, is that it lowers the activation barrier, thus permitting the reaction to take place more rapidly. One of the objectives of this experiment is to determine the activation energies of both the enzyme-catalyzed and the H^+-catalyzed reactions from measurements of the temperature dependences of the rates. This will allow you to decide whether the higher efficiency of the enzyme catalyst can be attributed simply to a lowering of the energy barrier for the reaction.

METHOD

The enzyme yeast invertase catalyzes the conversion of sucrose to fructose and glucose according to the hydrolysis reaction

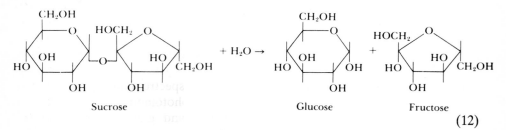

$$\text{Sucrose} \qquad\qquad \text{Glucose} \qquad\qquad \text{Fructose} \tag{12}$$

For this example of the simple Michaelis–Menten mechanism, we hypothesize that an intermediate complex ES is formed between one molecule of sucrose and one molecule of invertase. This complex then reacts with a molecule of water to yield the products P_1 and P_2 and to regenerate a molecule of the free enzyme. Thus the rate of product formation is given by

$$r = \frac{d(P_1)}{dt} = \frac{d(P_2)}{dt} = k_2'(H_2O)(ES) \tag{13}$$

where k_2' is the rate constant of the rate-determining step. For reactions occurring in dilute aqueous solutions, the water concentration does not change significantly during the reaction. Thus $k_2'(H_2O)$ can be replaced by a new constant k_2, and Eq. (13) becomes identical to Eq. (3).†

One of the most striking consequences of the sucrose hydrolysis is a large change in the optical-rotatory power of the solution as the sucrose is progressively converted to glucose and fructose.[3] This conversion actually produces a change in the direction in which the plane of polarization of light is rotated by the solution, and it is from this "inversion" that the enzyme derives its name. The inversion of sucrose in the presence of yeast was noted as early as 1832 by Persoz; and Michaelis and Menten used enzyme extracted from yeast in their classic study of the enzyme–substrate complex, in which the course of the reaction was monitored by observation of the change in optical-rotatory power accompanying it.

In the present experiment we shall use a rather simpler, although equally accurate, way of following the progress of the reaction. This method hinges on the fact that glucose and fructose, like all other monosaccharides, are active reducing agents, while the sucrose from which they are formed is not a reducing sugar. Thus any measure of the reducing capacity of a reaction mixture becomes, in effect, a measurement of the extent to which the conversion of sucrose to glucose and fructose has proceeded in that mixture. In this experiment we shall use a 3,5-dinitrosalicylic acid anion reagent in assaying the reducing capacity of various reaction mixtures.[4] Since the oxidized and reduced form of this reagent possess significantly different visible absorption spectra, measurement of the absorbance at a suitable wavelength (540 nm) can be used to measure the extent of hydrolysis that has occurred.

The procedure we will use is first to prepare a particular reaction system containing appropriate amounts of sucrose, enzyme, and buffer solution. This mixture is hydrolyzed for some specified time, and then "quenched" by addition of sodium hydroxide (contained in the 3,5-dinitrosalicylate reagent added at this point). The solution is heated to "develop" the color produced by the reagent in the presence of reducing activity in the solution, and the intensity of that color is determined with a spectrophotometer. After a calibration has been carried out, the spectrophotometric reading directly establishes the number of moles of fructose and glucose present in the solution. Then knowing the time during which the reaction proceeded, we can readily establish the average rate at which sucrose was converted to glucose and fructose.

---

† It should be kept in mind that there are other enzymatic reactions with the overall stoichiometry $S + B \rightarrow P$ where the concentration (B) of the second reactant does vary. In such cases, the simple mechanism given above must be modified.

EXPERIMENTAL

There are a large number of runs described in this section, and a given team may not be required to carry out all of them. In any event, it is recommended that students work in teams of two or three with the work divided equitably among the partners. To avoid any needless expenditure of time, each partner should carry out several runs concurrently. Do not, however, undertake so many that you are unable to stop each run at precisely the correct time. In making several runs of equal duration, a staggered schedule is recommended. For a given set of runs, set up a series of labelled assay tubes each containing all but the last solution—the one that will initiate the reaction. Add the prescribed amount of this last solution to the first tube and note the time; exactly one minute later, add the prescribed amount of the last solution to the second tube; and so on. You will then be able to quench these reactions at one-minute intervals with all the times precisely controlled.

Most of the runs are carried out at room temperature without any constant-temperature control. Note the ambient air temperature near your laboratory workbench at the beginning of each set of runs.

Procedure. The standard assay procedure to be used (with variations as indicated elsewhere) is as follows.

1. Pipette into a clean test tube—*in the order given*—the appropriate amounts of enzyme solution, distilled water, buffer solution, and sucrose solution. Swirl the assay tube vigorously before and *immediately* after the addition of the sucrose.

2. Begin timing when the sucrose solution is added and the tube swirled to mix the reagents. Let the reaction proceed for *exactly* 5.0 min, or whatever other period of time is specified in a particular run.

3. At the end of the allotted period, quench the reaction by adding 2.0 mL of dinitrosalicylate reagent (containing NaOH as the active quenching agent), and once again swirl the tube vigorously to mix the reagents.

4. Immerse test tube in boiling water for 5 min; cover *loosely* with Parafilm to prevent the escape of water vapor.

5. Cool the reaction tube by holding it in a slanted position and running cold tap water down the outside of the tube.

6. Dilute the solution with 15.0 mL of distilled water from a graduated pipette. Cover *tightly* and shake the tube to ensure thorough mixing.

7. Using a suitable spectrophotometer, determine the absorbance of the solution at 540 nm with distilled water as the reference. Read carefully the operating instructions provided for the instrument to be used. A general discussion of spectrophotometers and absorbance measurements is given in Chapter XVIII.

Enzyme activity. Carry through the standard assay procedure described above with 1.0 mL of enzyme solution, 0.5 mL of distilled water, 0.5 mL of buffer, and 1.0 mL of 0.3 M sucrose solution. The observed absorbance should lie in the range 0.6–0.9. If it is greater than 0.9, dilute the enzyme solution with sterile, previously boiled distilled water to reduce the activity to an appropriate level. If the absorbance is less than 0.6, the enzyme solution is too weak and a new solution must be made up.

A Blank runs. Our first concern is to establish that the dinitrosalicylate reagent gives no appreciable test in the absence of reducing sugars. To establish this point, one should systematically examine the results of the action of the reagent on each of the other substances present in the reaction mixture. The solutions to be tested are given below. In run A0, add the enzyme, water

Run	Enzyme, mL	Water, mL	Buffer, mL	0.3 M Sucrose, mL
A0	1.0	0.5	0.5	*1.0
A1	0.0	2.5	0.5	0.0
A2	0.0	1.5	0.5	1.0
A3	1.0	1.5	0.5	0.0

and buffer, then add 2.0 mL of the dinitrosalicylate reagent *before adding* the 1.0 mL of sucrose solution. Pick up the standard assay procedure at step 4. In the above table and in subsequent tables, an asterisk (*) is placed before the volume of sucrose to remind you to add dinitrosalicylate reagent before adding the sucrose. Run A0 constitutes a so-called "zero-time blank"—since the sucrose is added only after the reagent that should completely halt the enzymatic reaction—and constitutes a particularly sensitive blank run. In runs A1 to A3, mix the indicated reagents, wait 5.0 min, and then pick up the standard assay at step 3. None of these blank runs should develop an absorbance greater than about 0.05 at 540 nm.

B Standardization runs. We must establish the absorbance at 540 nm produced by the action of our reagent in the presence of known amounts of glucose and fructose. The reagent offers an exceedingly delicate test, and to standardize such a reagent we use a very dilute aqueous glucose–fructose solution (hereafter denoted by GFS) containing only 0.9 g of glucose and 0.9 g of fructose per liter. Five standardization runs are recommended.

In each case, make up the indicated solution and then pick up the standard assay procedure at step 3. The results of blank A2 should have shown that, in the absence of glucose and fructose, the mixture of reagents used in the standardization runs will yield no significant color. The absorbances obtained in runs B1 to B5 can thus be attributed to the action of the reagent on

Run	Enzyme, mL	Water, mL	Buffer, mL	0.3 M Sucrose, mL	GFS, mL
B1	0.0	1.1	0.5	1.0	0.4
B2	0.0	0.8	0.5	1.0	0.7
B3	0.0	0.5	0.5	1.0	1.0
B4	0.0	0.3	0.5	1.0	1.2
B5	0.0	0.0	0.5	1.0	1.5

the GFS. These data will be used to construct a calibration curve for the spectrophometric assay.

C–D   Progress of the reaction with time. For the second of the following two sets of runs, prepare some $0.03 M$ sucrose solution by diluting 1 mL of the $0.3 M$ stock solution with 9 mL of distilled water. Runs in series C use the concentrated sucrose solution, and those in series D use the dilute solution.

Run	Enzyme, mL	Water, mL	Buffer, mL	Sucrose, mL	Time, min
C0	0.2	1.3	0.5	*1.0 (0.3 M)	0
C1	0.2	1.3	0.5	1.0 (0.3 M)	1
C3	0.2	1.3	0.5	1.0 (0.3 M)	3
C5	0.2	1.3	0.5	1.0 (0.3 M)	5
C10	0.2	1.3	0.5	1.0 (0.3 M)	10
C20	0.2	1.3	0.5	1.0 (0.3 M)	20
D0	1.0	0.5	0.5	*1.0 (0.03 M)	0
D1	1.0	0.5	0.5	1.0 (0.03 M)	1
D3	1.0	0.5	0.5	1.0 (0.03 M)	3
D5	1.0	0.5	0.5	1.0 (0.03 M)	5
D10	1.0	0.5	0.5	1.0 (0.03 M)	10
D20	1.0	0.5	0.5	1.0 (0.03 M)	20

Runs C0 and D0 are zero-time blanks in which the sucrose addition is made only *after* inactivation of the enzyme by the addition of 2.0 mL of dinitrosalicylate reagent (denoted by * in the tables). For these two runs, pick up the standard assay procedure at step 4. In all other runs, begin timing when the sucrose solution is added, and let the reaction proceed for *precisely* the time indicated in each case. Then pick up the standard assay procedure at step 3.

E   Dependence of initial rate on substrate concentration. Prepare from the $0.3 M$ stock solution of sucrose, another 10 mL of $0.03 M$ sucrose solution. Taking care to use the sucrose solution called for in each case, proceed to the following runs—all of which are allowed to proceed for the standard time of

5.0 min except run E0, which is a zero-time blank. This set of runs will be analyzed to obtain k_2 and K.

Run	Enzyme, mL	Water, mL	Buffer, mL	Sucrose, mL
E0	1.0	0.5	0.5	*1.0 (0.3 M)
E1	1.0	0.5	0.5	1.0 (0.3 M)
E2	1.0	1.3	0.5	0.2 (0.3 M)
E3	1.0	0.5	0.5	1.0 (0.03 M)
E4	1.0	1.0	0.5	0.5 (0.03 M)
E5	1.0	1.3	0.5	0.2 (0.03 M)
E6	1.0	1.4	0.5	0.1 (0.03 M)

F   Non-enzymatic hydrolysis of sucrose.[3,4] The basic procedure for carrying out hydrolysis runs with H^+ ions as the catalyst is as follows. Pipette the indicated volume of 0.3 M sucrose solution into an assay tube. Add sufficient distilled water so that the final volume of the reaction mixture will be 4 mL (*not 3 mL, as in all preceding runs*). Add the indicated volume of 1 M HCl solution to initiate the catalytic hydrolysis, starting your timing as the acid is added. After exactly 5.0 min, stop the reaction by adding 5 mL of 1 M NaOH, which immediately neutralizes all of the catalytically active H^+ present. Add 2.0 mL of dinitrosalicylate reagent, and heat the reaction tube in a boiling-water bath for 5 min. Cool the reaction tube, and dilute by adding 9 mL of distilled water (*not 15 mL as in earlier runs*). Thus the final volume is, as before, 20 mL. Measure the absorbance of the solution at 540 nm.

A set of suggested concentrations at which this procedure can be carried out are given below. Run F0 is a blank, which will allow you to see how much (if any) hydrolysis takes place in the absence of any catalyst.

Run	0.3 M Sucrose, mL	Water, mL	1 M HCl, mL
F0	2.0	2.0	0.0
F1	1.0	2.0	1.0
F2	1.0	1.0	2.0
F3	2.0	1.0	1.0
F4	2.0	0.0	2.0

G–H   Dependence of the rate on temperature. In order to determine the temperature dependence of the enzyme-catalyzed rate, the reaction is carried out at several different temperatures which are held constant to within ±0.3°C or better. Choose from runs E a set of initial concentrations that gives an absorbance of ~0.5 at room temperature after the standard assay. Convenient temperatures for these runs, denoted as runs G, are 0°C (ice bath), 12, 25, 35, and 45°C. In each case, prepare the enzyme–water–buffer mixture in the assay tube, put about 2 mL of the sucrose stock solution in another test tube, and immerse both tubes in the thermostat bath for a few minutes to achieve temperature equilibrium. Next, pipette 1.0 mL of the equilibrated sucrose solution into the assay tube and read the bath temperature. Leave the assay tube in the constant-temperature bath until, 5.0 min after the addition of the sucrose solution, the reaction is terminated by the addition of 2.0 mL of the dinitrosalicylate reagent. Then complete the assay as usual.

Finally, choose from runs F a set of initial concentrations that gives a moderate rate of reaction. Then carry out a study of the temperature dependence of the acid-catalyzed rate over the range 0°C to 50°C. It is convenient to do these measurements, denoted as runs H, at the same temperatures and at the same times as runs G are carried out.

Precautions. In order to obtain successful results from this experiment, careful volumetric technique and judicious handling of reagents are required. The following points should be noted.

1. The pipettes, especially those to be used for dispensing the enzyme, should be cleaned carefully and rinsed thoroughly with boiled distilled water. When first using any pipette, rinse it with a small portion of the solution to be used and make sure that the pipette drains well. If possible, always use the same pipette for a given solution; *it is imperative that the pipettes used to dispense enzyme never be used for any other solution.* The pipettes needed are:

> Two 1-mL transfer pipettes, used to dispense the "standard" amounts of sucrose and enzyme solution
>
> Two 2-mL graduated pipettes, used for dispensing "nonstandard" amounts of sucrose and enzyme solution
>
> One 2-mL transfer pipette, used exclusively for dispensing the dinitrosalicylate reagent
>
> One 5-mL transfer pipette, used exclusively for dispensing the NaOH solution required in parts F and H of the experiment
>
> One 1-mL graduated pipette, used for dispensing buffer solution and distilled water in step 1 of the assay procedure
>
> One 25-mL graduated pipette, used for dilution in step 6

2. The enzyme, a protein, is highly susceptible to attack by airborne microorganisms. All glassware that comes in contact with the enzyme must be cleaned and rinsed very carefully. Place about 50 mL of the enzyme solution in a glass-stoppered Erlenmeyer flask, and keep this flask chilled with ice throughout the experiment. Any dilution of the enzyme stock solution must be done with chilled water that has been previously sterilized by boiling.

3. Sucrose solutions are also susceptible to attack by microorganisms. In this instance, it is sufficient to use a freshly prepared solution each day.

Solutions. A variety of comments will be given here about the preparation of the required solutions. As noted above, a new sucrose solution must be prepared each day by the student team doing the experiment. The enzyme solution must be prepared and tested in advance, presumably by the teaching staff. Other solutions may be made available to the students or be prepared by them, depending upon the policy of the instructor. A complete list of all solutions is given in the Apparatus section.

    Enzyme solution. Powdered invertase obtained from yeast is available from many biochemical supply houses, such as Sigma Chemical Co., P.O. Box 14508, St. Louis, MO 63178. A sterile solution of invertase must be prepared with great care. All apparatus used to make up this solution must be sterilized prior to use. More than 1 liter of distilled water is boiled for 10 min, covered with Al foil, and chilled in an ice bath. Then add a precisely weighed amount of invertase (approx. 5–10 mg) to *cold* water in a 1-liter volumetric flask and make up to the mark. *The enzyme solution should be kept tightly stoppered and chilled at all times.* The correct invertase concentration is very sensitive to the specific activity of the enzyme as purchased; it is necessary to carry out the standard assay and adjust the solution to an appropriate final concentration.

    Reducing-sugar reagent. This reagent is made up by mixing two solutions: a solution of about 25.0 g of 3,5-dinitrosalicylic acid in 1000 mL of warm $1\,M$ NaOH and one of 750 g of sodium potassium tartrate, $NaKC_4H_4O_6\cdot4H_2O$, in 750 mL of warm distilled water. These solutions are mixed slowly and diluted to a total volume of 2500 mL with warm distilled water.

    Glucose–fructose standard solution (GFS). Each individual team will need 100 mL of this solution. Weigh out exactly 0.090 g of glucose and 0.090 g of fructose, dissolve both solids in distilled water contained in a 100-mL volumetric flask, and make up to the mark. Note that fructose is somewhat hygroscopic; it may be necessary to handle it in a dry-box if the humidity is unusually high.

    Acetate buffer. A suitable buffer with pH 4.8–5.0 can be prepared by dissolving 4.10 g of sodium acetate in 1000 mL of distilled water and then adding 2.65 mL of glacial acetic acid.

Optional investigations. There are a number of additional investigations that can be carried out on this system. Three suggestions are given below.

1. Confirm the first-order dependence of the rate on the initial enzyme concentration $(E)_0$. A set of recommended runs based on the use of the $0.3\,M$ stock sucrose solution is given below. Each run is allowed to proceed for the standard time of 5.0 min except run I0, which is a zero-time blank.

Run	Enzyme, mL	Water, mL	Buffer, mL	0.3 M Sucrose, mL
I0	1.0	0.5	0.5	*1.0
I1	1.5	0.0	0.5	1.0
I2	1.0	0.5	0.5	1.0
I3	0.5	1.0	0.5	1.0
I4	0.2	1.3	0.5	1.0
I5	0.1	1.4	0.5	1.0

2. By a series of runs at different pH values, determine the pH at which invertase functions most effectively as a catalyst. Is the reaction rate highly dependent on pH? Why should pH have anything to do with the catalytic activity of invertase?

3. Investigate the inhibition of the enzymatic reaction by p-mercuribenzoate at concentrations of about 10^{-5} M. Show how an investigation of inhibition or "poisoning" of the catalyst might be used to establish the number of catalytically active sites on an invertase molecule.

DATA ANALYSIS

In order to convert the observed absorbance A into the desired product concentration (P) in the reaction mixture, a calibration must be carried out using the results of runs A2 and B1–B5. According to the Beer–Lambert law,[5] commonly called Beer's law,

$$A = \log(I_0/I) = \varepsilon cd \tag{14}$$

where I/I_0 is the fraction of the light transmitted through an absorbing solution of path length d and c is the concentration of the absorbing species (in this case, the reduced form of the dinitrosalicylate reagent in the final 20 mL solution). The absorption coefficient ε is a constant at any specified wavelength, 540 nm in this case; and d will also be a constant if matched cuvettes with the same path length are used.

The value of c is *proportional* to $n_1 + n_2$, where n_1 and n_2 are the numbers of moles of glucose and of fructose formed in the reaction up to the time it was quenched. It is obvious from the stoichiometry of Eq. (12), that $n_1 = n_2 = x$, where x is the number of moles of sucrose hydrolyzed. Thus, c is proportional to x, and we can rewrite Eq. (14) as

$$A = \alpha x \tag{15}$$

where α is a calibration constant. The product concentration $(P) = (P_1) = (P_2)$ is given by

$$(P) = \frac{x}{V} = \frac{A}{\alpha V} \tag{16}$$

where V is the volume of the reaction mixture (3 mL for enzyme runs and 4 mL for acid runs).

For a determination of α, plot the absorbance measured for runs A2 and B1–B5 versus the value of $(n_1' + n_2')/2$, where n_1' and n_2' are the numbers of moles of glucose and fructose added to the assay tube in the standardization runs. Since the GFS solution is made up by weighing, n_1' may not be precisely equal to n_2' and it is appropriate to use the average value. The molecular weight of both glucose and fructose is $180.16 \text{ g mol}^{-1}$. Make a graphical or least-squares evaluation of the slope and give its units.

Using the value of α determined above, the results of the standard assay made initially to check the enzyme activity, the assay in part C, and the given concentration of the enzyme stock solution in g L^{-1}, calculate the *specific activity* of the enzyme—that is, the number of micromoles of sucrose hydrolyzed per minute per gram of enzyme present. (The specific activity of an enzyme preparation is, of course, a function of the purity of the enzyme. As inactive protein is removed from the preparation, the specific activity will rise. When the specific activity can no longer be increased by any purification method, a homogeneous enzyme preparation may have been achieved; but proof of this depends on other criteria.) The exact chemical composition of invertase is still unknown, but its molecular weight has been estimated at $100\,000 \text{ g mol}^{-1}$. Combining this datum with your calculated specific activity, estimate the turnover number for the enzyme.

From the data of runs C1–C20 and D1–D20, calculate x, the number of moles of sucrose hydrolyzed in each time interval. If the reaction were zero order in sucrose, then we would expect that $(x/0.003) = k_0 t$, where $x/0.003$ is the concentration of either of the product species in mol L^{-1} units. Prepare a graph of the results obtained in these two series of runs, plotting x versus t, and indicate whether the data are consistent with the hypothesis that the reaction is zero order in sucrose. Note that even if a reaction starts out being zero order in sucrose, this cannot continue indefinitely. Indeed, we expect the inversion reaction to become first order in sucrose when (S) becomes sufficiently small.

If the reaction were first order in sucrose then, letting a stand for the number of moles of sucrose originally present, we should find that $\ln[a/(a - x)] = k_1 t$. Prepare a graph of the results obtained in the above two series of runs, plotting $\log[a/(a - x)]$ versus t, and indicate whether your data are consistent with the hypothesis that the reaction is first order in sucrose.

Show that the average rate measured over the first 5 min of reaction provides an acceptable approximation to the true initial rate r_0 by estimating the percent error associated with this approximation.

If the optional runs I1–I5 were carried out, determine the dependence of the initial rate on the initial concentration of the enzyme $(E)_0$.

Using the data from runs E, prepare a Lineweaver–Burke plot of $1/r_0$ versus $1/(S)_0$ and an Eadie–Hofstee plot of $r_0/(S_0)$ versus r_0. Determine the values at room temperature of $k_2(E)_0$ and K_m from both of these plots. Using the nominal molecular weight of $100\,000$ g mol^{-1} for invertase, calculate $(E)_0$ in mol L^{-1} units and obtain the value of k_2 in clearly stated units.

From the data of part F, estimate the rate constant k_H for the acid-catalyzed reaction. Also calculate a turnover number for the acid-catalyzed reaction, i.e., the number of molecules of sucrose hydrolyzed per second per hydrogen ion present. Note that Eq. (10) predicts that this turnover number depends on (S). Do your data confirm this?

From the data of part G, determine the initial rate r_0 for each temperature and plot log r_0 versus $1/T$. From the slope of the line, determine the activation energy E_a from

$$\frac{d \log r_0}{d(1/T)} = -\frac{E_a}{2.303R} \tag{17}$$

Finally, use the data from runs H for the temperature dependence of r_0 to determine E_a for the acid-catalyzed reaction.

DISCUSSION

There are a number of questions that might be addressed in the discussion of the results. How reproducible are the initial rate measurements? (Note that runs D5 and E3 are duplicates; also runs E1 and the standard assay for enzyme activity have identical initial concentrations.) Are the enzyme-catalyzed data compatible with the Michaelis–Menten mechanism? Do the data from both runs C and D follow apparent zero-order kinetics, and how does this agree with expectations based on comparing (S) with K_m? Which of the two types of analysis, Lineweaver–Burke or Eadie–Hofstee, seems to give the better results and why? How does k_2 agree with the estimate of the turnover number based on the specific activity? Are the acid-catalyzed data consistent with the rate law given in Eq. (10)?

Compare the enzyme turnover number with a typical hydrogen-ion turnover number. What does this tell you about the efficiency of enzymes? Compare the activation energy for the enzyme-catalyzed reaction with that for the acid-catalyzed reaction. Does the difference in E_a values completely account for the ratio of turnover numbers for the enzyme and H$^+$?

APPARATUS

Sterile preparation of yeast invertase (50 mL per team); pH 4.8–5 acetate buffer; 1 M sodium hydroxide; 3,5-dinitrosalicylate reagent solution; 1 M HCl; glucose–fructose standard solution; sucrose.

Two 1-mL transfer pipettes; one 2-mL transfer pipette; one 5-mL transfer pipette; one 1-mL graduated pipette; two 2-mL graduated pipettes; one 25-mL graduated cylinder or pipette; pipetting bulb; glass-stoppered flask; volumetric flasks, beakers, test tubes.

Boiling-water bath; stopwatch or other timer; spectrophotometer (Bausch & Lomb Spectronic 20, Sequoia-Turner model 340, or other suitable type); 0°C to 50°C constant-temperature baths.

REFERENCES

1. F. J. Kedzy and M. L. Bender, *Biochem.* **1,** 1097 (1962); M. L. Bender, F. J. Kezdy, and F. C. Wedler: *J. Chem. Educ.* **44,** 84 (1967).
2. H. B. Dunford, *J. Chem. Educ.* **61,** 129 (1984).
3. F. A. Bettelheim, "Experimental Physical Chemistry," pp. 269–277, Saunders, Philadelphia (1971).
4. J. G. Dawber, D. R. Brown, and R. A. Reed, *J. Chem. Educ.* **43,** 34 (1966).
5. I. R. Levine, "Physical Chemistry," 3d ed., pp. 705–6, McGraw-Hill, New York (1988).

GENERAL READING

M. L. Bender and L. J. Brubacker, "Catalysis and Enzyme Action," McGraw-Hill, New York (1973).
K. J. Laidler and P. S. Bunting, "The Chemical Kinetics of Enzyme Action," 2d ed., Oxford University Press, New York (1973).
I. Tinoco, Jr., K. Sauer, and J. C. Wang, "Physical Chemistry: Principles and Applications in Biological Sciences," 2d ed., chap. 8, Prentice-Hall, Englewood Cliffs, N.J. (1985).
G. G. Hammes, "Enzyme Catalysis and Regulation," Academic Press, New York (1982).

EXPERIMENT 23
KINETICS OF THE DECOMPOSITION
OF BENZENEDIAZONIUM ION

In an acidic aqueous solution, benzenediazonium ion ($C_6H_5N_2^+$) will decompose[1] to form nitrogen and phenol:

$$C_6H_5N_2^+ + H_2O \longrightarrow C_6H_5OH + N_2(g) + H^+ \qquad (1)$$

In this experiment you are to follow the course of the above reaction by a spectrophotometric measurement of the unreacted diazonium ion.[2] The reaction order with respect to $C_6H_5N_2^+$, the rate constant k, and the activation energy for the reaction are to be determined.

THEORY

The rate of reaction (1), $-d(C_6H_5N_2^+)/dt$, can be written in terms of the concentrations of the reacting species as

$$-\frac{d(C_6H_5N_2^+)}{dt} = k'(C_6H_5N_2^+)^n(H_2O)^m \qquad (2)$$

As written above, the reaction would be nth order with respect to $C_6H_5N_2^+$, mth order with respect to water, and would have an overall order of $n + m$. The reaction is not acid-catalyzed, although very high acid concentrations (say 12 M HCl) seem to increase the rate slightly.[1] Thus the effect of changes in the H^+ concentration due to reaction (1) can be completely neglected. Since the present experiment will be performed in a dilute aqueous solution, the concentration of water, (H_2O), will be very nearly constant throughout the reaction. Thus the factor $(H_2O)^m$ can be absorbed into the rate constant and Eq. (2) can be rewritten as

$$-\frac{dc}{dt} = kc^n \tag{3}$$

where c represents the instantaneous concentration $(C_6H_5N_2^+)$.

Equation (3) can readily be integrated to give

$$c = c_0\, e^{-kt} \qquad \text{for } n = 1 \tag{4a}$$

$$c^{1-n} - c_0^{1-n} = (n-1)kt \qquad \text{for } n \neq 1 \tag{4b}$$

where c_0 is the concentration at $t = 0$. In other words, if the reaction is first order, a plot of $\log c$ versus t should give a straight line of slope $-k/2.303$ and intercept $\log c_0$. If the reaction is nth order where n (which need not be an integer) is not equal to unity, a plot of c^{1-n} versus t should give a straight line of slope $(n-1)k$ and intercept c_0^{1-n}. By making such plots for various trial values of n and determining which gives the best straight-line dependence over a wide range of concentration, one can obtain the order n of the reaction and then determine the appropriate rate constant k_n.

If the value of a rate constant is measured at several different temperatures, it is almost always found that the temperature dependence can be represented by

$$k = \mathscr{A}\, e^{-E_a/RT} \tag{5}$$

where the factor $\mathscr{A}$ is independent of temperature. The Arrhenius activation energy E_a can be easily determined by plotting $\log k$ versus $1/T$. This should give a straight line of slope $-E_a/2.303R$.

METHOD

Any physical variable giving an accurate measure of the extent to which a reaction has gone toward completion can be used to obtain rate data. In this experiment, we shall monitor the concentration of unreacted diazonium ion by measuring the absorption of ultraviolet light by the solution.

According to the Beer–Lambert law,[3] the intensity of light I transmitted by an absorbing medium is given by

$$I = I_0\, e^{-c\varepsilon' d} \tag{6}$$

where c is the concentration of absorbing molecules, d is the path length, I_0 is the intensity of incident light, and ε' is an "extinction coefficient." The *absorbance A,* defined as $\log(I_0/I)$, thus gives a direct measure of the concentration:

$$A \equiv \log \frac{I_0}{I} = \frac{c\varepsilon'd}{2.303} = c\varepsilon d \qquad (7)$$

where ε is called the *molar absorption coefficient* when the concentration is expressed in moles per liter. When the sample is a solute in solution, I is the intensity of light transmitted by a cell filled with the solution and I_0 is the intensity transmitted by the cell filled with pure solvent.

For first-order reactions, the quantity εd will cancel out of Eq. (4a) and k can be obtained directly from a plot of $\log A$ versus t. However, for orders different from first order, εd does not cancel out of Eq. (4b). In such cases, εd must be evaluated by measuring the absorbance of a solution of known concentration in order to permit k to be calculated in concentration units.

A description of spectrophotometers and their use in determining the absorbance is given in Chapter XVIII.

EXPERIMENTAL

Kinetic runs are to be made at 25, 30, and 40°C. Well-regulated thermostat baths operating at about these temperatures will be required. The actual temperatures of each of these baths should be measured with a good thermometer and recorded.

The concentration of benzenediazonium ion can be determined by the absorbance at wavelengths between 295 and 325 nm. Below 295 nm products of the reaction produce interfering absorption, and above 325 nm the molar absorption coefficient is too small to permit effective measurement of changes in the benzenediazonium ion concentration.[4] The absorbance will be measured using a suitable spectrophotometer such as a Beckman Model DU-20 UV; detailed instructions for operating the instrument will be provided in the laboratory. A wavelength of 305 nm should be used, with a slit width of 0.3 mm. Make two absorbance readings on each sample.

The diazonium salt that should be used in this experiment is benzenediazonium fluoborate ($C_6H_5N_2BF_4$, M.W. 191.9). The great majority of diazonium salts are notoriously unstable solids and can decompose with explosive violence. The fluoborates are by far the safest to use and are not known to explode; however, reasonable caution should be used in preparing the compound. Since even benzenediazonium fluoborate will decompose slowly, it should not be prepared too far in advance, and it must be stored in a refrigerator. A simple high-yield procedure for its preparation has been given by Dunker, Starkey, and Jenkins.[5] Recrystallization of the product from 5 percent fluoboric acid yields white needle-like crystals which can be dried by vacuum pumping at 1 Torr for several hours.†

The diazonium salt should be available at the beginning of the experiment. Remove a *small* quantity from the refrigerator and warm it rapidly to room temperature. Prepare approximately $10^{-3} M$ solutions by placing an accurately weighed 15- to 20-mg sample of the salt into each of three 100-mL volumetric flasks, and then make up to the mark with a $0.2 M$ HCl solution. Label the flasks and be sure to record the exact weight of salt placed in each flask. Return the unused diazonium salt to the refrigerator at once. Be careful not to waste it.

Using a hypodermic syringe, withdraw a sample of about 5 mL of each solution *as soon as possible* after it is made up. Chill these samples by placing them in test tubes set in crushed ice, then put them aside for adsorbance measurments (to be made as soon as conveniently possible). These measurements will provide a value for εd in case it is needed later on.

Suspend one of the volumetric flasks in each of the three constant-temperature baths, and record the time. Allow about 20 min for the solutions to achieve thermal equilibrium. After thermal equilibrium has been attained, you may begin taking samples for spectrophotometric analysis.

When you take a sample, use the following procedure: Withdraw about 1 mL of the solution with the hypodermic syringe and use this to rinse out the syringe. Then quickly withdraw a 5-mL sample and place it in a labeled test tube set in crushed ice for chilling. Record the time of discharge into the test tube. After the sample is chilled (3 to 5 min) use about 1 mL of it to rinse out the spectrophotometer cell. Fill the cell with the remaining sample and measure its absorbance, using the $0.2 M$ HCl solution as a blank. These cells are fragile and expensive; handle them with care.

The reason for chilling the sample is to slow down the reaction rate so that a negligible amount of reactant will decompose between the time you chill the sample and the time you make the absorbance measurement. Since it is impossible to eliminate completely errors due to reaction after withdrawal and cooling, it is important to try to perform the sample removal and chilling

---

† After storage at 0°C in a vacuum desiccator for several weeks, these crystals tend to stick together. After six months there is clear evidence of decomposition; fortunately, the main impurity is phenol (2 to 3 percent) which does not interfere much with rate studies in aqueous solution. By a very careful purification, it is possible to obtain a benzenediazonium fluoborate sample which can be stored at 0°C for several months without any signs of decomposition. The sample is dissolved in acetone, and then chloroform is added until a few crystals are formed. When the solution is then chilled to −20°C for 30 min, the compound will crystallize out in the form of tiny white needles. After three such recrystallizations, the sample can again be dried by pumping, and can be stored in a vacuum desiccator.

procedure in as reproducible a fashion as possible. This will lead to a partial cancellation of such errors.†

Take as many measurements as you reasonably can without rushing. Since the runs at the higher temperatures will proceed more rapidly, it will be necessary to take samples from them at more frequent intervals. A suggested interval might be once every 25 min for the 25°C bath, every 15 min for the 30°C bath, and every 10 min for the 40°C bath.

At the end of the experiment, wash the cells thoroughly with distilled water and dry them very carefully. Store in a safe place.

CALCULATIONS

Using all the data points from any single run, prepare plots of the following: (1) $A^{1/2}$ versus t (order $\frac{1}{2}$), (2) $\log A$ versus t (order 1), (3) A^{-1} versus t (order 2), and (4) any additional plots you feel to be necessary in order to establish the order of the reaction.

Having determined the order of the reaction, make the appropriate plot for each of the runs and determine the value of k (in concentration units) for each temperature. It is convenient to make these plots using A values rather than c values. The value of the slope can then be corrected, if necessary, to obtain k in concentration units by utilizing the value of εd. You should also carry out a least-squares analysis of your data using the appropriate form of Eq. (4). Report both sets of k values, together with the standard deviations for the least-squares values. Finally, plot $\log k$ versus $1/T$ and determine E_a.

Report your result for the order of the reaction and list the best values obtained for $k(T_1)$, $k(T_2)$, and $k(T_3)$. Report also the activation energy E_a. Give the correct units for each quantity, using seconds as the unit of time, mol L^{-1} for concentration, and kJ for energy.

DISCUSSION

From your results, calculate the extent of reaction that would occur in a sample of the initial solution after 30 min at 0°C. Does this indicate that

---

† If a spectrophotometer with a cell compartment that can be temperature controlled is available, the procedure can be greatly improved and simplified. Adjust the bath temperature to 30°C, turn on the circulating pump, and wait until the cell compartment (with cell holder in place) has become stable. Then prepare a solution as described above and fill two spectrophotometer cells as soon as possible. (A third cell should already have been filled with 0.2 M HCl solution, for use as a blank.) Place the cells in the cell holder, and record the time at which the cell holder is returned to the cell compartment. Obtain absorbance readings on both samples as soon as possible; these initial readings will provide values of εd. Allow about 20 min for the solutions to achieve thermal equilibrium, and then begin measuring the absorbance every 15 min. Since the runs are made sequentially rather than simultaneously, it is advisable to replace the slow run at 25°C with a run at 45°C (use 5-min intervals). Fresh solutions are needed for each temperature studied.

varying lengths of time between the chilling of a sample and its absorbance measurement would introduce serious or negligible errors in your data?

APPARATUS

Spectrophotometer, such as one of the Beckman Model DU series; two or more quartz sample cells; constant-temperature baths set at 25, 30, and 40°C; lens tissue; three 100-mL volumetric flasks; clock, stopwatch, or other timer; hypodermic syringe; about 20 test tubes; plastic pail or large battery jar; beaker.

Benzenediazonium fluoborate (75 mg), stored in a refrigerator; 0.2 M hydrochloric acid (400 mL); crushed ice.

REFERENCES

1. E. A. Moelwyn-Hughes and P. Johnson, *Trans. Faraday Soc.* **36,** 948 (1940); M. L. Crossley, R. H. Kienle, and C. H. Benbrook: *J. Amer. Chem. Soc.* **62,** 1400 (1940).
2. J. E. Sheats, Ph.D. Thesis, Dept. of Chemistry, M.I.T. (1965).
3. P. W. Atkins, "Physical Chemistry," 3d ed., p. 464, Freeman, New York (1986).
4. A. Wohl, *Bull. Soc. Chim. Fr.* **6,** 1319 (1939).
5. M. F. W. Dunker, E. B. Starkey, and G. L. Jenkins, *J. Amer. Chem. Soc.* **58,** 2308 (1936).

GENERAL READING

F. Grum, "Visible and Ultraviolet Spectrophotometry," in A. Weissberger and B. W. Rossiter (eds.), "Techniques of Chemistry: Vol. I. Physical Methods of Chemistry," part IIIB, chap. 3, Wiley-Interscience, New York (1972).
J. F. Bunnett, "Kinetics in Solution," in E. S. Lewis (ed.), "Techniques of Chemistry: Vol. VI. Investigation of Rates and Mechanisms of Reactions," 3d ed., part I, chap. 4, Wiley-Interscience, New York (1974).

EXPERIMENT 24
GAS-PHASE KINETICS

Although a vast majority of important chemical reactions occur primarily in liquid solution, the study of simple gas-phase reactions is very important in developing a theoretical understanding of chemical kinetics. A detailed molecular explanation of rate processes in liquid solution is extremely difficult. At the present time reaction mechanisms are much better understood for gas-phase reactions; even so, this problem is by no means simple. This experiment will deal with the unimolecular decomposition of an organic compound in the vapor state. The compound suggested for study is cyclopentene or di-*t*-butyl peroxide, but several other compounds are also suitable; see, for example, Table XI.4 of Ref. 1.

THEORY

A discussion of bimolecular reactions (which usually give rise to second-order kinetics) can be based quite naturally on collision theory or on transition-state theory.[1,2] We shall assume a general knowledge of these treatments as background for the discussion of unimolecular reactions (which usually give rise to first-order kinetics). The crucial question is: How does a molecule acquire the necessary energy to undergo a spontaneous unimolecular decomposition? If the "activation" occurs via a collision, one could normally expect second-order kinetics. However, Lindemann pointed out that this would not be true if the time between collisions is short compared with the average lifetime of an activated molecule.

Let us consider a simple unimolecular decomposition† in the gas phase

$$A \rightarrow B + C$$

The proposed reaction mechanism will consist of three steps: (1) collisional activation, (2) collisional deactivation, and (3) spontaneous decomposition of the activated molecule. Thus,

$$A + A \rightarrow A + A^* \qquad k_1 \tag{1}$$

$$A^* + A \rightarrow A + A \qquad k_2 \tag{2}$$

$$A^* \rightarrow B + C \qquad k_A \tag{3}$$

where A^* is an activated A molecule with a definite internal energy ε (above its ground state) which is greater than a certain critical value ε^* necessary for reaction to occur, k_1 and k_2 are second-order rate constants, and k_A is a first-order rate constant. The overall rate of reaction is given by

$$+\frac{d(B)}{dt} = +\frac{d(C)}{dt} = -\frac{d(A)}{dt} = k_A(A^*) \tag{4}$$

where parentheses indicate concentrations of the species. As discussed below, the value of k_A will depend on the particular value of ε involved. If the concentration of A^* is small, one can make the steady-state approximation, $d(A^*)/dt = 0$, and obtain

$$+\frac{d(A^*)}{dt} = k_1(A)^2 - k_2(A^*)(A) - k_A(A^*) = 0 \tag{5}$$

Solving Eq. (5) for (A^*) and substituting the resulting expression into Eq. (4) gives

$$-\frac{d(A)}{dt} = \frac{k_1 k_A(A)^2}{k_A + k_2(A)} \tag{6}$$

---

† In the following treatment several details and difficulties will be overlooked; more advanced developments are given by Benson[1] and Trotman–Dickenson.[3]

High-pressure limit. At high concentrations deactivation is much more probable than decomposition, since the collision frequency is high, and deactivation should occur at the first collision suffered by A^*. Thus at high pressure, $k_2(A) \gg k_A$ and the reaction becomes first order:

$$-\frac{1}{(A)}\frac{d(A)}{dt} = \frac{k_1}{k_2} k_A \tag{7}$$

Also

$$\frac{k_1}{k_2} = \frac{(A^*)}{(A)} \cong f_A \tag{8}$$

where f_A is the fraction of A molecules in an activated state with some internal energy ε. Obviously f_A depends on the value of ε, as does k_A. Thus,

$$-\frac{1}{(A)}\frac{d(A)}{dt} = \int_{e^*}^{\infty} k_A(\varepsilon) f_A(\varepsilon)\, d\varepsilon = k_\infty \tag{9}$$

where k_∞, the value of the integral, corresponds to the experimental first-order rate constant k_{exp} determined at high pressures.

Rice and Ramsperger[4] and Kassel[5] have given an expression for k_A as a function of ε:

$$k_A(\varepsilon) = \begin{cases} 0 & \text{for } \varepsilon < \varepsilon^* \\ v\left(\dfrac{\varepsilon - \varepsilon^*}{\varepsilon}\right)^{n-1} & \text{for } \varepsilon \geqslant \varepsilon^* \end{cases} \tag{10}$$

Equation (10) is based on a model (called the RRK model) in which the molecule contains a total internal energy ε distributed among n weakly coupled harmonic oscillators. One of these oscillators is localized in a weak bond which will break when energy ε^* is concentrated in it; the other $n-1$ oscillators are assumed to act as a reservoir of freely available energy. The probability that an energy at least as large as ε^* is localized in one oscillator is given by $(1 - \varepsilon^*/\varepsilon)^{n-1}$, and v is the frequency of energy transfer among the various oscillators in the molecule. For such a model, one would expect v to be of about the same order of magnitude as molecular vibrational frequencies— namely, $10^{13}\,s^{-1}$. (A more realistic and sophisticated molecular model proposed by N. B. Slater[1,3] gives results very similar to those based on Eq. (10) and also predicts a value for v of about $10^{13}\,s^{-1}$.)

To get a rough approximation for $f_A(\varepsilon)$ we can use a classical model in which the normal modes of vibration of the molecule A are represented by n classical harmonic oscillators. With this drastic assumption, one can use Boltzmann statistics to obtain[6]

$$f_A(\varepsilon) = \frac{1}{(n-1)!\, kT}\left(\frac{\varepsilon}{kT}\right)^{n-1} e^{-\varepsilon/kT} \tag{11}$$

When Eqs. (10) and (11) are substituted into Eq. (9) we find that

$$k_\infty = \frac{\nu}{(n-1)!} e^{-\varepsilon^*/kT} \int_0^\infty x^{n-1} e^{-x} \, dx \qquad (12)$$

where $x = (\varepsilon - \varepsilon^*)/kT$. The definite integral is the gamma function $\Gamma(n) = (n-1)!$. Thus

$$k_\infty = \nu e^{-\varepsilon^*/kT} = \nu e^{-E^*/RT} \qquad (13)$$

where $E^* = N_0\varepsilon^*$. Equation (13) can be compared directly with the empirical Arrhenius equation which is used to obtain the experimental activation energy:

$$k = \mathscr{A} e^{-E_a/RT} \qquad (14)$$

The temperature-independent factor $\mathscr{A}$ has now been given a physical significance and an estimated magnitude.

The theory presented above can be improved by using a quantum-mechanical model for the oscillations of the activated complex; this will permit a consideration of the effect of possible structural changes in the activated complex. The final result from this more sophisticated model[7] is

$$k_\infty = \nu^* e^{\Delta S^*/R} e^{-\Delta H^*/RT} \qquad (15)$$

where ν^* is an average frequency of the vibrational modes in the activated complex and ΔS^* and ΔH^* are the entropy and enthalpy changes for forming the activated complex (transition state) from a normal species. Comparing Eqs. (14) and (15) one can identify $\mathscr{A}$ with $\nu^* e^{\Delta S^*/R}$, but ν^* is not well known, since it depends on the properties of an activated molecule (which are themselves not well known). Marcus[8] has proposed a modification of the RRK model, termed the RRKM treatment, in which ν^* is calculated by quantum statistical-mechanical methods. Vibrational frequencies of reactive species and activated complexes are explicitly considered, along with the effects of rotation and zero-point energies. On the whole, this theory has been quite successful in describing unimolecular reactions but the details (given in the general reading references) are beyond the scope of the present discussion. For simplicity, we will assign ν^* a value of 10^{13} s^{-1} (the value of $\mathscr{A}$ in many reactions). Then, when the experimental value of $\mathscr{A}$ is less than 10^{12} or greater than 10^{14}, we can look for a possible explanation in terms of a value of ΔS^* which differs appreciably from zero.†

Low-pressure limit. We now wish to comment on the behavior of Eq. (6) at low concentrations where the collision frequency is so low that deactivation

---

† Transition-state theory gives the same result as Eq. (15) except that ν^* is replaced by a universal factor kT/h ($= 6 \times 10^{12}$ s^{-1} at 300 K). Although this would appear to resolve the uncertainty as to the value of ν^* (and thus allow better calculation of ΔS^* from $\mathscr{A}$ values), the theory involves several assumptions which are open to question on physical grounds.[7]

will be very slow compared with decomposition. At low pressures, $k_2(A) \ll k_A$ and we find second-order kinetics. It is convenient to write the rate law in a pseudo first-order form:

$$-\frac{1}{(A)}\frac{d(A)}{dt} = k_{exp} = k_1(A) \tag{16}$$

A detailed treatment of the change with pressure of the apparent first-order rate constant k_{exp} is quite complicated,[2,3] but we can easily see that

$$k_{exp}(\text{low } p) = k_1(A) \ll \frac{k_1 k_A}{k_2} \cong k_\infty = k_{exp}(\text{high } p) \tag{17}$$

Thus at low initial pressures, if one analyzes the data according to a first-order rate law, the "apparent first-order rate constant" k_{exp} should decrease as the pressure is decreased (i.e., the half-time for the reaction at constant temperature is independent of pressure at high pressure but rises as the pressure approaches zero). For reactions where $\mathscr{A} \gg 10^{13}\,\text{s}^{-1}$, this change has been observed to occur, but normally (i.e., where $\mathscr{A} \sim 10^{13}\,\text{s}^{-1}$) the change will occur at very low pressure and is not observed experimentally.

Chain mechanisms.[9] The fact that first-order kinetics are observed for a gas-phase reaction does not prove that the unimolecular mechanism described above must be involved. Indeed, very many organic decompositions that experimentally are first order have complicated free-radical chain mechanisms.

As an example, let us consider a simple *hypothetical* chain mechanism for the dehydrogenation of a hydrocarbon:

$$\begin{aligned}
M_a &\rightarrow H + R & k_1 \\
H + M_a &\rightarrow H_2 + R & k_2 \\
R &\rightarrow H + M_b & k_3 \\
H + R &\rightarrow M_c & k_4
\end{aligned} \tag{18}$$

where the M's are stable molecules and R is a free radical. The chain is initiated by the first step, propagated by many repetitions of the second and third steps (to form the stable products H_2 and M_b), and terminated by the last step. By writing the equations for $d(H)/dt$ and $d(R)/dt$ and making the steady-state assumption that (H) and (R) are constant, one can obtain

$$(H) = \left(\frac{k_1 k_3}{k_2 k_4}\right)^{1/2} \qquad (R) = \left(\frac{k_1 k_2}{k_3 k_4}\right)^{1/2}(M_a) \tag{19}$$

Now the rate of disappearance of M_a is given by

$$-\frac{d(M_a)}{dt} = k_1(M_a) + k_2(H)(M_a)$$

$$= \left[k_1 + \left(\frac{k_1 k_2 k_3}{k_4}\right)^{1/2}\right](M_a) \tag{20}$$

and we see that the overall rate is first order. (When the chain is long, $k_2 \gg k_4$ and $k_3 \gg k_1$, so that k_1 is negligible compared with $k_1 k_2 k_3 / k_4$ in Eq. (20).)

The presence of a chain reaction can usually be detected by adding small quantities of an inhibitor such as nitric oxide or propylene, which will markedly reduce the rate by reacting with the radicals and greatly shortening the chain length.[10] Another technique is to add small amounts of a compound that is known to provide relatively large concentrations of free radicals and see if the overall rate is increased. A detailed treatment of chain mechanisms will not be given here, but it should be pointed out that the overall rate law is often very complex and the apparent order of the reaction often changes considerably with pressure. To obtain a complete picture of the kinetics may require measurements over the pressure range from a few Torr to 10 bar or more.

Wall effects. In the above discussion we have assumed that the reaction is homogeneous (i.e., no catalytic reaction at the walls of the reaction bulb). The fact that the data give first-order kinetics is not a proof that wall effects are absent. This point can be checked by packing a reaction bulb with glass spheres or thin-wall tubes and repeating the measurements under conditions where the surface-to-volume ratio is increased by a factor of 10 to 100. This will not be done in this experiment, but the system chosen for study must be free from serious wall effects or it may not be possible to discuss the experimental results in terms of the theory of unimolecular reactions.

METHOD[11]

The three basic experimental features of gas-phase kinetic studies are temperature control, time measurement, and the determination of concentrations. Of these, the principal problem is that of following the composition changes in the system. Perhaps the most generally applicable technique is the chemical analysis of aliquots; however, continuous methods are much more convenient. By far the easiest method is to follow the change in total pressure. This technique will be used in the present experiment. Obviously, the pressure method is possible only for a reaction that is accompanied by a change in the number of moles of gas. Also, the stoichiometry of the reaction should be straightforward and well understood so that pressure changes can be directly related to extent of reaction.

For the simple apparatus designs shown in Figs. 1 and 2 both the reactant and the products must be sufficiently volatile to avoid condensation in the manometer, which is much cooler than the reaction flask. (Heating wire may be wound around the connecting tubing between the furnace and the manometer to permit warming this section to about 50°C if necessary.) Also there should be no chemical reaction between the vapors and glass or mercury. If conventional stopcocks are used, the vapors must not react with or dissolve in the stopcock grease; this concern can be eliminated by using greaseless vacuum stopcocks. The furnace in Fig. 1 should be capable of operating at

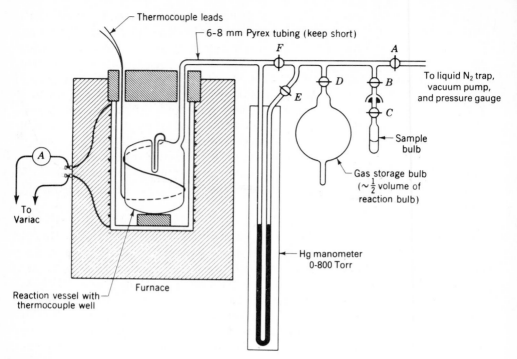

FIGURE 1
Apparatus for high-temperature gas-phase kinetics experiment.

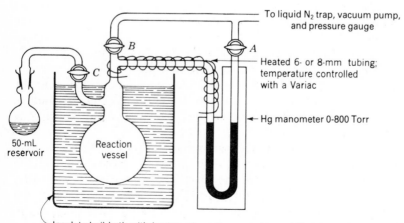

FIGURE 2
Apparatus for moderate-temperature gas-phase kinetics experiment.

temperatures up to 500°C, and the temperature should remain constant to within less than 1°C during a run. A large, well-insulated furnace has the disadvantage of heating up very slowly, but it will maintain a reasonably constant temperature without regulation. The Pyrex reaction bulb should be mounted in the center of the furnace. A large bulb (about 1000 mL) with a well to hold the thermocouple junction is preferable. However, a small bulb (as small as 100 mL) can be used, and the thermocouple can be wound around the outside of the bulb and then covered with aluminum foil. A two-junction Chromel–Alumel thermocouple, with the reference junction immersed in an ice bath, is used to measure the high temperatures achieved with this apparatus (see Chapter XVI for more information on thermocouples). The oil bath shown in Fig. 2 should be capable of operating in the range 140 to 160°C with a stability of ±0.5°C during a run. In the case of an oil bath, there is no problem in achieving thermal contact between the reaction vessel and the thermometer. The connecting tubing between the reaction bulb and the mercury in the manometer should be short and of as small bore as feasible. The rate of the reaction should be slow enough so that the time for filling the reaction bulb and for the gas to warm up to the furnace temperature is small compared with the length of the run. Since this initial period is about 30 to 60 s, the temperature should be chosen to give a half-life of at least 30 min.

We must finally derive an expression for the rate law in terms of the total pressure of the system. As an example, consider the gas-phase decomposition $A \rightarrow B + C$, and assume that the ideal-gas law holds for all species. Let the initial total pressure be p^0; this will also be the partial pressure of species A at zero time (p_A^0). At some arbitrary time t, $p_B = p_C = p_A^0 - p_A$, and since $p = p_A + p_B + p_C$,

$$p_A = 2p^0 - p \tag{21}$$

For a first-order reaction $\ln[(A)_0/(A)] = kt$; therefore

$$\ln p^0 - \ln(2p^0 - p) = \ln \frac{p^0}{2p^0 - p} = kt \tag{22}$$

Thus a plot of $\ln(2p^0 - p)$ versus t should be a straight line. For a reaction $A \rightarrow B + C + D$ or $A \rightarrow B + 2C$ in which three product species are created, it can easily be shown that $\ln p^0 - \ln \frac{1}{2}(3p^0 - p) = kt$. Thus a plot of $\ln(3p^0 - p)$ versus t should be a straight line in this case.

For a moderately fast reaction it is often difficult to make a direct measurement of p^0. It is then best to back-extrapolate the early pressure data to zero time to obtain a value of p^0. Another possibility is to allow the reaction to go to completion (assuming there is no back reaction) and measure p^∞, which is equal to $2p^0$ or $3p^0$ for the particular reactions under consideration. This method is usually not so reliable, since side-reactions that have little effect on the early stages of the reaction may influence the final pressure. (The experimental value of p^∞ is often found to be slightly less than the expected value.) Another approach is to use the Guggenheim method as in Exp. 21; this eliminates the need for a p^0 value.

In deriving Eq. (22) the dead-space volume V_D in the manometer and connecting tube has not been considered. Since V_D changes during the reaction owing to the change of the mercury level in the manometer, $p^0/(2p^0 - p)$ in Eq. (22) must be multiplied by a correction factor for very precise work.[12] With $V_D \sim 20$ mL and $V_B \sim 1000$ mL, this correction factor varies linearly with pressure from 1.00 to ~ 0.98 during the entire run and can be neglected.

Once the order of the reaction is established by a plot of $\ln(2p^0 - p)$ versus t, there is a rapid method of evaluating specific rate constants which utilizes the time for the reaction to proceed to a given fraction reacted. The half-time and third-time for a first-order reaction are given by

$$kt_{1/2} = \ln 2 = 0.691$$
$$kt_{1/3} = \ln 1.5 = 0.406$$

(23)

For the reaction $A \rightarrow B + C$, $t_{1/2}$ is the time required for the total pressure p to reach $3p^0/2$ and $t_{1/3}$ is the time required to reach $4p^0/3$. For the reaction $A \rightarrow B + 2C$, the total pressure will equal $2p^0$ at $t_{1/2}$ and $5p^0/3$ at $t_{1/3}$. Either (or both) times can be used to obtain k; it is also possible to check that the pressure data are consistent with first-order kinetics by calculating $t_{1/2}/t_{1/3}$. This ratio should be 1.70 for a first-order reaction; it is 1.50 and 2.00 for zero-order and second-order reactions, respectively.

EXPERIMENTAL

1   High-Temperature Reaction

The reaction suggested for study is the decomposition of cyclopentene to cyclopentadiene and hydrogen:

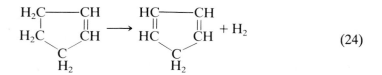

(24)

This reaction is quite clean-cut, and at least 95 percent of the products are accounted for by Eq. (24). Vanas and Walters[13] have studied the reaction in the gas phase and find that p^∞/p^0 is about 1.9, which indicates the occurrence of some side-reactions. However, they carried out a careful chemical analysis of the products and showed that the partial pressures of cyclopentadiene and hydrogen are equal to each other and to the increase in total pressure $\Delta p = p - p^0$ over at least the first half of the reaction. With the exception of the last stages of the decomposition, this reaction is very well suited to the pressure method. In addition, it has been shown that the reaction is homogeneous and does not involve a chain mechanism.

The rate of reaction should be studied at a single temperature for several different initial pressures in the range 50 to 200 Torr. A furnace temperature of about 510°C will give a convenient half-life (~30 min). Ordinarily it is only necessary to follow the reaction until Δp exceeds $0.5p^0$. However, data should be taken on one of the runs until Δp is at least $0.75p^0$. (If possible, let this reaction mixture stand overnight and obtain a value of p^∞.) If NO gas is available, add about 2 to 3 percent of NO to the cyclopentene for one of the runs and check for any indication of inhibition; consult the instructor regarding any necessary changes in the procedure. If more than one day is available for experimental work, raise the temperature of the furnace by about 10°C and make several more runs.

Procedure. The furnace should be turned on the day before measurements are to be made so that it will achieve a steady temperature. Set the power supply to a predetermined voltage appropriate for the desired reaction temperature. The heating current should be measured with a series ammeter and recorded periodically. At the start of the experiment, place the reference thermocouple junction in a Dewar flask filled with ice and distilled water (see Chapter XVI) and connect the thermocouple to a potentiometer or digital voltmeter (see Chapter XV). Measure the thermocouple emf and check to see if the furnace is at the proper temperature. If the temperature shows appreciable long-term drift (greater than $\pm 0.5 \, \mathrm{K \, h^{-1}}$) adjustment of the heater current will be necessary. On making any change in voltage setting, be careful to note the time response of the furnace as an aid in making later adjustments.

While the temperature stability of the furnace is being checked, carefully open all of the stopcocks in Fig. 1 *except C*, which isolates the sample bulb containing liquid cyclopentene. Evacuate the system until the pressure is 10^{-2} Torr or lower. Then close stopcocks E and F and check for leaks in the reaction bulb by monitoring the manometer reading over a period of ~10 min. Next close stopcock D, slowly freeze the cyclopentene with liquid nitrogen, and then open stopcock C in order to pump off any air. Close C, remove the liquid nitrogen, and allow the solid to melt so that it will liberate any dissolved gases. Repeat this degassing process at least once more, and then allow the cyclopentene to warm to room temperature. Close stopcock A and open D and then C to permit the storage bulb to be filled with vapor. The vapor pressure at 25°C is about 350 Torr.

To start a run, close stopcock C and then **slowly** open stopcock F and watch the manometer. When the pressure has reached about 95 percent of the desired value for p^0, close F and start a timer. Record pressure and time readings every minute during the early stages of the reaction and every few minutes thereafter until the end of the run.

Analysis of products. It is possible to design an apparatus which will permit the efficient removal of the entire reaction mixture for analysis. With the apparatus shown in Fig. 1, a qualitative analysis of the products can be made if desired. Analytical details are given by Vanas and Walters.[13]

2   Moderate-Temperature Reaction

The reaction suggested for study is the decomposition of *tert*-butyl peroxide (*t*BP).† Batt and Benson[14] have shown that the stoichiometry of this reaction is three (i.e., 3 mol of product for each mol of reactant) with 90 percent of the reaction accounted for by

$$(CH_3)_3COOC(CH_3)_3 \rightarrow C_2H_6 + 2(CH_3)_2CO \tag{25}$$

The mechanism of this reaction is given by the following scheme:

$$t\text{BP} \rightarrow 2t\text{BO}\cdot \tag{26}$$

$$t\text{BO}\cdot \rightarrow \text{Me}\cdot + (\text{Me})_2\text{CO} \tag{27}$$

$$\text{Me}\cdot + \text{Me}\cdot \rightarrow C_2H_6 \tag{28}$$

where the cleavage of the O—O bond in the first step is rate determining. The other 10 percent of the *t*BP decomposes according to

$$t\text{BP} \rightarrow CH_4 + (\text{Me})_2\text{CO} + \text{MeCOEt}$$
$$t\text{BP} \rightarrow 2CH_4 + \text{biacetonyl} \tag{29}$$

both of which involve intermediate $\text{MeCOCH}_2\cdot$ radicals. The reaction is homogeneous and chain contributions are less than 2 percent.[14] Thus, this reaction is quite well suited to the pressure method if two minor precautions are observed. (1) Fresh stopcock grease may absorb *t*BP, leading to a p^∞/p^0 ratio of ~2.9. Greaseless stopcocks are therefore recommended; if these are not available, use the minimum amount of grease necessary and, if possible, do not regrease stopcocks just before a run. (2) The side-arm to the manometer must be heated to avoid condensation of acetone.

The apparatus shown in Fig. 2 has a 500-mL reaction vessel which is connected to a 50-mL reservoir for liquid *t*BP via a stopcock and a standard-taper joint.‡ The connecting tubing between the reaction vessel and the mercury in the manometer must be wrapped with commercial heating tape or wound carefully with heating wire. The temperature of this section, which should be as short and of as small bore as feasible, is controlled with a variable-voltage power supply. It is not necessary to control this temperature accurately, but it should be close to the bath temperature (say within 20°C) and *must* be above the boiling point of acetone (56°C).

---

† No special handling procedures are required for *t*BP since it is one of the most stable organic peroxides known. It can be distilled at 1 atm without decomposition.

‡ This apparatus is very similar in design to an apparatus described by Ellison;[15] see also the design of Trotman–Dickenson.[16]

The rate of the reaction should be studied at a single temperature for several initial pressures in the range 30 to 100 Torr. A reaction temperature of about 160°C will give a convenient half-life. At least one run should be made at a lower temperature (say 150°C). In general, it is only necessary to follow the reaction until p exceeds $2p^0$, but data should be taken on one run until p is at least $2.5p^0$.

Procedure. If necessary, vacuum-distill the tBP. Place about 10 mL of tBP in the side reservoir and close stopcock C. Open stopcocks A and B and evacuate the system to a pressure of 10^{-2} Torr or lower while the oil bath is being heated to the desired temperature. Note that the pumping system should include a liquid nitrogen trap to prevent condensation of tBP or acetone in the pump. Check for leaks by closing stopcocks A and B; then open them again and continue pumping.

Heating of the oil bath should be started well in advance of the measurements so that the temperature will have stabilized at about 160°C. The temperature can be monitored with a thermometer (or with a thermocouple as described in the procedure for the high temperature reaction) and should not drift by more than ±0.5°C/h.

When the reaction vessel is at constant temperature, close stopcock A and **slowly** open stopcock C. Pump down the reservoir as low as possible, but do not waste tBP by prolonged pumping. Next close stopcock C and place a beaker of hot water (60 to 65°C) around the reservoir. Stopcock B is then closed and C is opened until the desired value of p^0 is achieved. Finally, close stopcock C and start a stopwatch or timer. Record the pressure and time readings every minute during the early stages of the reaction and every few minutes thereafter.

CALCULATIONS

Plot the pressure reading for each run versus time and extrapolate to zero time to obtain a value for p^0. For each run determine $t_{1/3}$ and $t_{1/2}$, and check their ratio with that expected for first-order kinetics. Use Eqs. (23) to calculate k, in units of s^{-1}, from each of these times. Analyze the data from your longest run by carrying out a linear least-squares fit of $\ln(np^0 - p)$ as a function of t, where $n = 2$ or 3 depending on the stoichiometry and p^0 is taken to be the value determined previously by extrapolation. The early data points should be weighted more heavily than the late ones. Plot $\ln(np^0 - p)$ versus t and add the best straight-line fit to the graph. Report the value of the resulting specific rate constant k. If possible, carry out a *nonlinear* least-squares fit with Eq. (22), or its analog for the case of three product species, taking both p^0 and k as freely adjustable fitting parameters. Do the resulting values of the parameters differ much from the previous values? If so, use the new p^0 value to make another plot of $\ln(np^0 - p)$ versus t.

Tabulate T, p^0, $t_{1/3}$, and k for each run, and list your best overall value of

k for each temperature studied. If data were obtained at two different temperatures, use the integrated form of Eq. (14) to calculate a value of the activation energy. Compare your result with the appropriate literature value. For cyclopentene, E_a is reported to be 246 kJ mol^{-1} on the basis of a linear plot of ln k versus $1/T$ for data over the range 485 to 545°C.[13] For *tert*-butyl peroxide, Batt and Benson[14] report 156.5 kJ mol^{-1} for the temperature range 130 to 160°C.

DISCUSSION

Calculate the frequency factor v and discuss its value in terms of the theory of unimolecular decompositions.

In the case of tBP, there is a complication owing to the fact that reaction (25) is exothermic enough to cause temperature gradients in a spherical reaction vessel. Batt and Benson have shown that such gradients increase with increasing temperature, pressure, and reaction volume. At 160°C there is a gradient of roughly 0.5°C during the first half of the reaction.[14] Estimate the effect that such a gradient will have on the rate constant determined in this experiment.

APPARATUS

High-temperature reaction. Gas kinetics apparatus as shown in Fig. 1, including vacuum line with greaseless stopcocks, sample bulb, gas storage bulb, manometer, Pyrex reaction bulb; high-temperature furnace; variable-voltage power supply; ammeter for measuring heater current; two-junction Chromel–Alumel thermocouple; two 1-qt Dewar flasks; millivolt-range poten-tiometer setup or digital voltmeter (see Chapter XV); thermocouple calibra-tion table; stopwatch or other timer.

Cyclopentene; ice (3 lb); liquid nitrogen (2 liters); cylinder of NO gas, with needle valve and pressure tubing (optional).

Moderate-temperature reaction. Gas kinetics apparatus as shown in Fig. 2, including vacuum line with greaseless stopcocks, sample reservoir bulb, reaction vessel, heated mercury manometer; regulated oil bath; thermometer; 500-mL beaker; 1-qt Dewar; stopwatch or other timer.

tert-Butyl peroxide (15 mL); liquid nitrogen (2 liters).

REFERENCES

1. S. W. Benson, "Foundation of Chemical Kinetics," chaps. X and XI, McGraw-Hill, New York (1960). [Reprinted by Krieger, Malabar, Florida (1982).]
2. Any standard physical chemistry text, such as P. W. Atkins, "Physical Chemistry," 3rd ed., chap. 30, Freeman, New York (1986); or I. N. Levine, "Physical Chemistry," 3rd ed., chap. 23, McGraw-Hill, New York (1988).

3. A. F. Trotman-Dickenson, "Gas Kinetics," secs. 2.3, 2.4, 3.2, Butterworth, London (1955).
4. H. C. Ramsperger, *Chem. Rev.* **10,** 27 (1932).
5. L. S. Kassel, "Kinetics of Homogeneous Gas Reactions," chap. 5, Reinhold (ACS Monograph), New York (1932).
6. S. W. Benson, *op. cit.,* pp. 222–223.
7. *Ibid.,* pp. 250–252.
8. R. A. Marcus, *J. Chem. Phys.* **20,** 359 (1952).
9. E. W. R. Steacie, "Atomic and Free Radical Reactions," pp. 82–86, Reinhold (ACS Monograph), New York (1946).
10. A. F. Trotman-Dickenson, *op. cit.,* pp. 153–160.
11. H. W. Melville and B. G. Gowenlock, "Experimental Methods in Gas Reactions," Macmillan, London (1963).
12. A. O. Allen, *J. Amer. Chem. Soc.* **56,** 2053 (1934).
13. D. W. Vanas and W. D. Walters, *J. Amer. Chem. Soc.* **70,** 4035 (1948).
14. L. Batt and S. W. Benson, *J. Chem. Phys.* **36,** 895 (1962).
15. H. R. Ellison, *J. Chem. Educ.* **48,** 205 (1971).
16. A. F. Trotman-Dickenson, *J. Chem. Educ.* **46,** 396 (1969).

GENERAL READING

S. W. Benson, *op. cit.*
J. W. Moore and R. G. Pearson, "Kinetics and Mechanism," 3rd ed., especially pp. 121–129, Wiley-Interscience, New York (1981).
K. J. Laidler, "Chemical Kinetics," 3rd ed., especially sec. 5.3, Harper and Row, New York (1987).
P. J. Robinson and K. A. Holbrook, "Unimolecular Reactions," Wiley-Interscience, New York (1972).

EXPERIMENT 25
KINETICS OF A FAST REACTION

Although conventional kinetic methods have proved very useful in studying a wide range of chemical reactions, special techniques are needed for investigating fast reactions (i.e., reactions with half-lives less than a few seconds). Such techniques may also provide a way to elucidate the rapid steps of reaction mechanisms for which only the slow rate-controlling step can be investigated by classical methods. In recent years, a wide variety of new methods have been developed for fast reactions in solution.[1,2] The earliest of these are the various flow methods (both continuous and stop-flow), which are suitable for reactions with $\sim 1\,\text{s} > t_{1/2} > 10^{-3}\,\text{s}$.[3] For half-lives below $10^{-3}\,\text{s}$, it is necessary to use methods that avoid the mixing of reactants, and several sophisticated techniques (such as relaxation methods) are now available for studying very fast reactions with $t_{1/2}$ values as short as $10^{-9}\,\text{s}$.[1,2]

In this experiment, the simplest fast-reaction technique—the continuous-flow method—will be used to study the kinetics of the formation of the ferric thiocyanate complex $FeSCN^{2+}$.

THEORY

For the fast reaction between ferric and thiocyanate ions in an acid solution of constant pH, the observed behavior is consistent with the simple mechanism[4]

$$\text{Fe}^{3+} + \text{SCN}^- \underset{k_r}{\overset{k_f}{\rightleftharpoons}} \text{FeSCN}^{2+} \tag{1}$$

where k_f is the bimolecular forward rate constant and k_r is the unimolecular reverse rate constant.† The rate law resulting from mechanism (1) is

$$\frac{d(\text{FeSCN}^{2+})}{dt} = k_f(\text{Fe}^{3+})(\text{SCN}^-) - k_r(\text{FeSCN}^{2+}) \tag{2}$$

Let us recall that the equilibrium constant K is related to the rate constants by

$$K = \frac{k_f}{k_r} = \frac{(\text{FeSCN}^{2+})_\infty}{(\text{Fe}^{3+})_\infty(\text{SCN}^-)_\infty} \tag{3}$$

where the subscript ∞ denotes the equilibrium $(t = \infty)$ value. Let us also note that

$$(\text{FeSCN}^{2+}) + (\text{SCN}^-) = (\text{FeSCN}^{2+})_\infty + (\text{SCN}^-)_\infty \tag{4}$$

at any time t. Using these relations, we can rewrite Eq. (2) in the form

$$\frac{d(\text{FeSCN}^{2+})}{dt} = k_f(\text{Fe}^{3+})[(\text{FeSCN}^{2+})_\infty + (\text{SCN}^-)_\infty]$$
$$- k_f[(\text{Fe}^{3+}) + K^{-1}](\text{FeSCN}^{2+}) \tag{5}$$

In order to simplify the integration of Eq. (5), let us choose the experimental conditions such that $(\text{Fe}^{3+}) \gg (\text{SCN}^-)$. This will allow us to assume that (Fe^{3+}) is essentially constant during the course of the reaction. If, in addition, the initial conditions are chosen so that $(\text{FeSCN}^{2+}) = 0$ at $t = 0$, we find

$$\ln \frac{(\text{FeSCN}^{2+})_\infty - (\text{FeSCN}^{2+})}{(\text{FeSCN}^{2+})_\infty} = -[(\text{Fe}^{3+}) + K^{-1}]k_f t \tag{6}$$

This is an approximate solution which becomes exact only when (Fe^{3+}) is constant (you can check Eq. (6) by direct differentiation). In actual practice, $(\text{Fe}^{3+})_0$ will be chosen to be ten times larger than $(\text{SCN}^-)_0$, so that (Fe^{3+}) will vary by about 10 percent during the reaction.‡

---

† In fact, the detailed mechanism is more complex, since both a direct and a base-catalyzed path exist.[4] The value of k_f as used in mechanism (1) is found to be dependent on the H^+ concentration: $k_f = k_1 + k_2/(\text{H}^+)$. Since this experiment is restricted to a single value of (H^+), the more general mechanism is omitted.

‡ With this choice of an initial ferric concentration, which is much larger than the thiocyanate concentration, any formation of $\text{Fe}(\text{SCN})_2^+$ species can be neglected. Also we shall work at a low pH value so that the hydrolysis of ferric ions can be neglected.

If a plot of $\ln[(FeSCN^{2+})_\infty - (FeSCN^{2+})]$ versus t is linear, then the first-order dependence on (SCN^-) and $(FeSCN^{2+})$ is confirmed. The rate dependence on (Fe^{3+}) has been established as first order[4] and will not be tested in this experiment.

METHOD

The continuous-flow method[3,5] is a simple and straightforward technique. Two reactant solutions are forced under pressure into a T-shaped mixing chamber, and then the reacting mixture flows down a long capillary tube. It is important that the solutions mix well and rapidly ($\sim 10^{-3}$ s). When the mixing time is much shorter than the half-life of the reaction, a steady state is set up such that various positions along the capillary tube correspond to different reaction times. A quantitative correlation between the reaction time t and the distance x from the mixing chamber to a given point along the tube can be achieved by measuring the time interval $\Delta\tau$ needed for a known volume ΔV of solution to flow through the tube. Assuming a constant cross-sectional area $\mathscr{A}$ for the capillary, the average linear flow velocity $\bar{v}$ is given by $\bar{v} = \Delta V/(\mathscr{A} \Delta\tau)$. Thus, on the average, the reaction time is related to x by

$$t = x \frac{\mathscr{A} \Delta\tau}{\Delta V} \tag{7}$$

It is important that the flow of the reacting solution should be as close to "mass flow" as possible, so that all the fluid arriving at a small volume element located at some fixed point along the tube will have travelled for the same length of time after leaving the mixing chamber. Laminar flow is undesirable since it is fast near the center of the capillary and quite slow near the walls (see Exp. 4). Since turbulent flow is much closer to mass flow, we wish to use flow rates that are greater than $1000\, \eta/\rho r$ [the minimum value for turbulent flow; see Eq. (4-11)]. Deviations from mass flow will blur the time value appropriate to a given distance x. Fortunately, the resulting error in the rate constant is no more than 2 to 3 percent for turbulent flow; but it may be as high as 10 percent for streamline flow.[3,5]

In this experiment, the reaction is to be followed using a spectrophotometer with a rather slow response, several seconds being required to make an absorbance reading. Since the reaction to be studied is rapid (half-life of less than 1 s), it is not possible to follow the progress of the reaction directly as in Exp. 23. However, the continuous-flow method is an appropriate way† to study fast reactions with a slowly responding instrument. The design described

---

† The principal disadvantage of this method is the fact that large quantities of solution are required. Although the present chemicals are cheap and easily obtained, this would be a serious limitation in studying reactions involving rare or costly reactants. In those cases, one would need to use a stop-flow method with a rapidly responding spectrophotometer.[3]

in the experimental section is based directly on that first used by Dalziel; excellent detailed descriptions of his equipment are available in the literature.[3,5]

The complex $FeSCN^{2+}$ has an absorption maximum near 450 nm, whereas neither Fe^{3+} nor SCN^- absorbs in that region.[6] Therefore,

$$(A_\infty - A) = \varepsilon d[(FeSCN^{2+})_\infty - (FeSCN^{2+})] \tag{8}$$

where A is the absorbance, ε is the molar absorption coefficient, and d is the optical path length. From Eqs. (6) and (8) we see that a plot of $\log(A_\infty - A)$ versus t should give a straight line of slope $-k_f[(Fe^{3+}) + K^{-1}]/2.303$. Note that A_∞ values obtained at different points along the tube may not all agree with each other, since there may be variations in the capillary tube and its placement with respect to the light beam. Also note that errors in $(A_\infty - A)$ caused by a drift in the zero setting of the spectrophotometer could cause appreciable errors in the rate constant. Readings corresponding to a large percent reaction are most sensitive since the $(A_\infty - A)$ values are then small.

Finally, we must remember that there will be a salt effect[7] on the rate of this reaction: both the rate constant k_f and the equilibrium constant K are functions of the ionic strength I. An (approximately constant) ionic strength of $0.40\,M$ will be chosen here to facilitate comparison with previous work.[4]

EXPERIMENTAL

A sketch of the complete assembled apparatus is shown in Fig. 1, and more detailed drawings of the capillary support frame are given in Fig. 2. There are two important differences between this design and that of Dalziel:[5] (1) the carboys of reactant solutions are mounted in a stationary thermostat bath and are connected to the movable capillary support frame by rubber pressure tubing, and (2) the mixing chamber is machined from a block of Lucite instead of being fabricated from glass. A detailed drawing of this mixing chamber is given in Fig. 3.

The spectrophotometer setup shown in Fig. 1 is based on the use of a Beckman Model DU instrument in which the cell holder unit (sample chamber) has been replaced by a special housing that allows the capillary support frame to slide back and forth. The Beckman Model DU is an old and simple design. Many other spectrophotometers are available that will provide excellent absorbance measurements, and any of these can be used as long as suitable adaptation of the sample chamber can be made. A modular spectrophotometer PTR model SP-100, available from PTR Optics Corp., is one example of a modern instrument with a flexible geometric design; similar components for modular instruments are available from Oriel Corp. and other manufacturers. No matter what instrument is being used, make sure that the alignment is correct (so that the capillary frame moves smoothly) before bolting the special adapter housing to the spectrophotometer. It is convenient to have the spectrophotometer sitting on a base with adjustable legs to

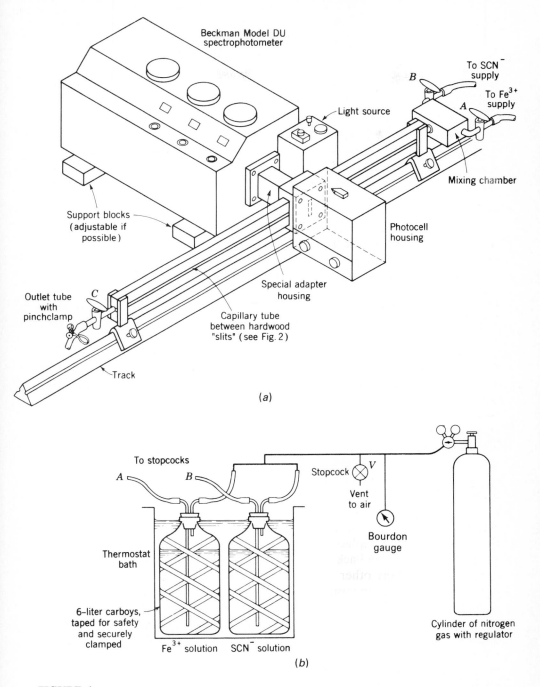

FIGURE 1
Flow-kinetics apparatus: (*a*) spectrophotometer setup; (*b*) schematic diagram of system for driving reactant solutions.

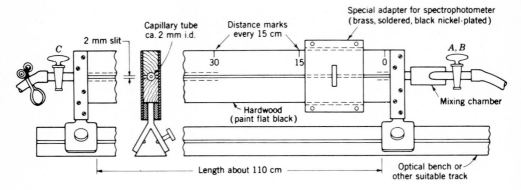

FIGURE 2
Detail of reaction capillary tube and support frame.

facilitate this alignment. Finally, attach the photodetector unit onto the rear of the adapter housing.

Next, fill the two carboys† with the appropriate solutions. Solution A should be $0.02\,M$ in $Fe(NO_3)_3$, $0.2\,M$ in $HClO_4$, and $0.14\,M$ in $NaClO_4$; solution B should be $0.002\,M$ in $NaSCN$, $0.2\,M$ in $HClO_4$, and $0.14\,M$ in $NaClO_4$. Be careful in handling these solutions; perchloric acid will give you dishpan hands. Now mount the carboys in the thermostat bath and connect them by a long piece of flexible pressure tubing to the appropriate stopcocks,

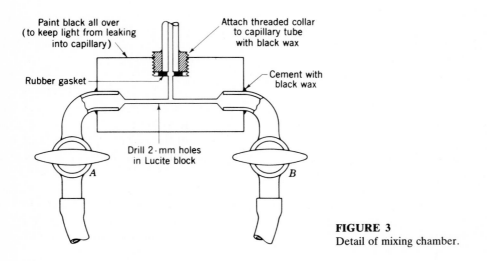

FIGURE 3
Detail of mixing chamber.

---

† Before being used in this experiment, each carboy should be pressure tested to verify that there are no flaws that would make it unsafe. It is recommended that the empty *but wet* carboy be placed behind a protective barrier and subjected to an internal gas pressure of 2 atm.

which are sealed into the mixing chamber. Again, test that the capillary frame will move smoothly back and forth through its housing. Do not force it. If there is any difficulty, ask an instructor to check the alignment. Proper alignment is crucial. Along the top of the capillary support frame, there should be a series of fiducial marks (say six or seven) which can be lined up with a reference mark on the special housing. This enables one to set the spectrophotometer at reproducible points along the capillary that are at known distances x away from the mixing chamber.

To complete the setup, connect the reagent carboys to each other via a T tube that leads to an open-tube mercury manometer and a regulated supply of a suitable driver gas (such as air or N_2).

It is important that stray light should not reach the capillary tube, since it may be reflected down the walls of the tube and be detected by the spectrophotometer. The support frame around the capillary is designed to reduce such stray light by providing a deep, narrow slit-like aperture; but one must still avoid horizontal light leaks. It may be worthwhile to cover both ends of the entire support frame with blackout cloths. Do not cover the source housing on the spectrophotometer, as it gets hot.

Success in this experiment depends on good technique; the procedure should be followed carefully. Before beginning the measurements, you should look over the general discussion of spectrophotometers given in Chapter XVIII and the instruction manual for the spectrophotometer to be used. Read carefully the detailed information given in the latter for the proper operation of the instrument.

Procedure. Turn on the spectrophotometer and allow at least 20 min for it to warm up prior to use. The wavelength setting should be 455 nm throughout the entire experiment. With both reagent stopcocks A and B and the vent stopcock V closed, slowly increase the gas pressure on the reagent solutions until Bourdon pressure gauge indicates about 500 Torr pressure above 1 atm. With the outlet stopcock C *open,* open and close the reagent stopcocks A and B (one at a time) several times to make sure that both solutions are flowing smoothly and to remove any air bubbles from the system. Use a beaker to catch the outflow from the capillary tube.† Then set the capillary frame at the first fiducial mark (one nearest to the mixing chamber) and carry out the three following steps.

1. Open stopcock A and allow the Fe^{3+} solution to flow for a sufficient time to remove from the capillary tube any solution containing $FeSCN^{2+}$ species (until the outflow is clear). Then close stopcock A and the outlet stopcock C. In accordance with the spectrophotometer operating instructions, zero

---

† As an optional step, you can record the time required for a given volume (say 250 mL) of each separate solution to flow through the capillary tube. The times for solutions A and B should agree within a few percent.

the instrument with this Fe^{3+} solution (i.e., adjust the slit width and sensitivity controls so that the instrument reads 100 percent transmission or $A = 0$).

2. To begin a run, open the outlet stopcock C and then turn both stopcocks A and B to their fully open positions. Catch the outflow of solution from the capillary in a beaker until the flow becomes stable (as indicated by a constant manometer reading, usually before 100 mL of solution is collected). Then *quickly* switch the outlet tube from the beaker to a 250-mL volumetric flask and simultaneously start a stopwatch. When this flask is full, stop the watch and record the elapsed time (~50 s). Return the outlet tube to the beaker. While one student is carrying out the above flow-rate measurement, his partner should determine the absorbance A of the reaction mixture and record that value together with the distance x from the mixing chamber. Work quickly to avoid any unnecessary waste of the reagent solutions.

3. When both the flow and absorbance measurements are complete, close the outlet stopcock C and then as soon as possible close both stopcocks A and B. This is a crucial step in the procedure; if A and B are left open, solution may siphon from one carboy to the other. After about 2 min, determine the absorbance again to obtain A_∞ (the infinite-time value). Verify that this value does not change after one more minute.

For the next run, move the capillary support frame so as to line up the second fiducial mark and repeat steps 1 to 3 at this new distance setting. Be careful in moving the capillary support frame.

Make two runs at each of the six or seven positions along the capillary tube. Use special care in making the absorbance readings at large values of x (corresponding to large percent reaction and a small $A_\infty - A$ value). If time permits, you should also take data at a different driving pressure. Either increase or decrease the gas pressure depending on whether you need more data at low percent reaction or at high, but it may not be safe to exceed about 700 Torr overpressure.

During the course of the experiment, more of solution A will be used up than solution B if the Fe^{3+} solution is always used in step 1 to make the zero adjustment of the spectrophotometer at each distance setting. The resulting change in the liquid level for solution A relative to that for solution B may change the relative flow rates of these solutions. This can be avoided by alternating the use of solutions A and B for making the zero adjustments (or, if necessary, corrected by merely running enough of solution B out through the capillary periodically to equalize the levels).

Before leaving the laboratory, record the concentrations of the reactant solutions, the temperature of the thermostat bath, the radius of the capillary tube, and the distances x corresponding to the various fiducial marks along the capillary support frame.

CALCULATIONS

Using Eq. (7), calculate the reaction time t corresponding to each run; tabulate these times together with the appropriate $(A_\infty - A)$ values. Plot $\log(A_\infty - A)$ versus t, and determine the slope of the best straight line through the data points. From the value of this slope, calculate the rate constant k_f. The literature value[4] of K is $146 \pm 5 \text{ L mol}^{-1}$ at 25°C and an ionic strength of 0.40, and you should take the (Fe^{3+}) value to be an average of the initial value ($t = 0$, but *after* mixing of the two solutions) and the equilibrium value ($t = \infty$, as calculated with K). Also report the back-reaction rate constant k_r. Give the proper units for each rate constant.

DISCUSSION

What fraction of the Fe^{3+} in solution A is hydrolyzed to $Fe(OH)^{2+}$ if the hydrolysis constant K_h is known[4] to be 2×10^{-3}? What are the principal sources of error in this experiment? How fast a reaction do you think could be measured with this apparatus? What is the limiting design factor in this method?

APPARATUS

Suitable spectrophotometer, such as a Beckman model DU or PTR model SP-100; 6-V battery and battery charger (or regulated dc power supply); adjustable support for spectrophotometer (optional); movable capillary support frame with mixing chamber and special housing attached; set of four retaining bolts for mounting housing on the spectrophotometer; two carboys (~6 liters) for reactant solutions; Bourdon pressure gauge (0–700 Torr overpressure); cylinder of nitrogen gas with pressure regulator (or source of regulated compressed air); pressure tubing and hose clamps; stopcock for venting the carboys; gum-rubber tubing for outlet of capillary; a 250-mL volumetric flask; a beaker (~500 mL); stopwatch; blackout cloths (optional).

Constant-temperature bath (set at 25°C) with provision for mounting carboys; solution A which is $0.0200\,M$ in $Fe(NO_3)_3$, $0.20\,M$ in $HClO_4$, and $0.14\,M$ in $NaClO_4$ (5 liters); solution B which is $0.00200\,M$ in $NaSCN$, $0.20\,M$ in $HClO_4$, and $0.14\,M$ in $NaClO_4$ (5 liters).

REFERENCES

1. E. F. Caldin, "Fast Reactions in Solution," Wiley, New York (1964).
2. C. F. Bernasconi (ed.), "Techniques of Chemistry: Vol. VI. Investigation of Rates and Mechanisms of Reactions," 4th ed., part II, Wiley-Interscience, New York (1986).
3. B. Chance, "Rapid Flow Methods," in G. G. Hammes (ed.), "Techniques of Chemistry: Vol. VI. Investigation of Rates and Mechanisms of Reactions," 3d ed., part II, chap. 2, Wiley-Interscience, New York (1974).
4. J. F. Below, Jr., R. E. Connick, and C. P. Coppel, *J. Amer. Chem. Soc.* **80,** 2961 (1958).

5. K. Dalziel, *Biochem. J.* **55,** 79 (1953).
6. H. S. Frank and R. L. Oswalt, *J. Amer. Chem. Soc.* **69,** 1321 (1947).
7. G. W. Castellan, "Physical Chemistry," 3d ed., pp. 862–864, Addison-Wesley, Reading, Mass. (1983).

GENERAL READING

E. F. Caldin, *op. cit.*
C. F. Bernasconi, *op. cit.*

CHAPTER X

SURFACE
PHENOMENA

EXPERIMENTS

26. Surface tension of solutions
27. Physical adsorption of gases

EXPERIMENT 26
SURFACE TENSION OF SOLUTIONS

The capillary-rise method is used to study the change in surface tension as a function of concentration for aqueous solutions of *n*-butanol and sodium chloride. The data are interpreted in terms of the surface concentration using the Gibbs isotherm.

THEORY

If a body of material is homogeneous, the value of any *extensive* property is directly proportional to the quantity of matter contained in the body:

$$Q = \bar{Q}_V V = \bar{Q}_m m = \bar{Q}_n n \tag{1}$$

where Q is the extensive quantity; V, m, and n are, respectively, the volume, mass, and number of moles of the substance involved. $\bar{Q}_V$, $\bar{Q}_m$, and $\bar{Q}_n$ are, respectively, the specific values of Q per unit volume, per unit mass, and per mole and have *intensive* magnitudes.

It is known, however, that the values of extensive properties of bodies often thought of as homogeneous are not always independent of the surface area. In fact, liquid bodies with surfaces are in general not entirely homogeneous, for the value of a given intensive quantity (say Q_V) in the region of the surface may at equilibrium deviate from the value that this quantity has in the bulk of the solution ($\bar{Q}_V$). We can write

$$Q = \int_V Q_V \, dV = \bar{Q}_V V + \int_\tau (Q_V - \bar{Q}_V) \, dV \tag{2}$$

where the second integral needs to be taken only over a region τ in the neighborhood of the surface, within which Q_V is different from $\bar{Q}_V$, the *bulk* value that prevails through the body except near the surface. We can replace the volume element dV in the second integral by $dA \, dx$, where dA is an element of surface area and dx an element of distance inward from the surface of the body. The integration over dA extends over the surface of the body, and that over dx extends to a distance τ inside the body beyond which Q_V is substantially equal to $\bar{Q}_V$. If $(Q_V - \bar{Q}_V)$ is independent of position on the surface of the body, the integration over dA can be carried out at once, and we have

$$Q = \bar{Q}_V V + \bar{Q}_A A \tag{3}$$

where we have introduced a new quantity, called a specific surface quantity,

$$\bar{Q}_A \equiv \int_\tau (Q_V - \bar{Q}_V) \, dx \tag{4}$$

This quantity represents the *excess,* per unit area of surface, of the quantity Q over what Q would be for a perfectly homogeneous body of the same magnitude with $Q_V = \bar{Q}_V$ throughout; see Fig. 1. In some cases $\bar{Q}_A$ may be negative, corresponding to a deficiency rather than an excess.

If the body has surfaces of several different kinds (like a crystal with different kinds of crystal faces or a liquid with part of its surface in contact with

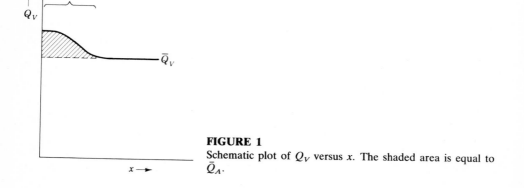

FIGURE 1

Schematic plot of Q_V versus x. The shaded area is equal to $\bar{Q}_A$.

the air and the remainder of its surface in contact with solid or other liquid phases), the parenthesized quantity in Eq. (2) is independent of surface positional coordinates only within the boundaries of each kind of surface. Each kind of surface will have in general a different $\bar{Q}_A$, and we write in such a case

$$Q = \bar{Q}_V V + \sum_i \bar{Q}_{A_i} A_i \tag{5}$$

Surface concentration. When the extensive quantity concerned is the number n of moles of solute, the corresponding specific bulk property is $n_V \equiv c$, the bulk concentration (concentration in the interior of the solution). The corresponding specific surface quantity $n_A \equiv \Gamma$ is called *surface concentration* and represents excess of solute per unit area of the surface over what would be present if the internal (bulk) concentration prevailed all the way to the surface. It may be expressed in moles per square centimeter.

Surface free energy or surface tension. Under conditions of constant temperature and pressure the equilibrium state sought by any system is that of lowest free energy G, and the maximum work (other than expansion work) done on the system in any change of state under these conditions is equal to the free-energy increase ΔG. A body of liquid with a free surface will tend to assume the shape which gives it the lowest possible free energy at the given temperature and pressure prevailing.

The free energy of a system containing variable surface areas can be written, from Eq. (5), as follows:

$$G = G_0 + \sum_i \gamma_i A_i \tag{6}$$

where $\bar{G}_{A_i} \equiv \gamma_i$ is the specific surface free energy or *surface tension* of surface i, which may be a free surface (exposed to air or vapor or vacuum) or an interface with another liquid or solid. In the event that the surface is an interface, this quantity is called *interfacial tension*. For a stable free liquid surface it clearly must be positive. The surface tension γ_i is equal to the expenditure of work required to increase the net area of surface i by one unit of area. If the increase in surface area is accomplished by moving a line segment of unit length in a direction perpendicular to itself, γ_i is equal to the force (or "tension") opposing this motion. Accordingly, the SI units for surface tension are $N\,m^{-1}$ or $J\,m^{-2}$; cgs units, which are still widely used in many tabulations, are $dyn\,cm^{-1}$ or $erg\,cm^{-2}$ ($1\,N\,m^{-1} = 1000\,dyn\,cm^{-1}$).

The variation of surface tension with temperature will not be discussed here, except to remark that surface tension decreases with temperature and that the rate of decrease is large enough to require that the temperature of measurement of surface tension be kept constant, to the order of 0.1°C, by means of a thermostat.

The Gibbs isotherm. It is found that the surface tensions of solutions are in general different from those of the corresponding pure solvents. It has also been found that solutes whose addition results in a decrease in surface tension tend to concentrate slightly in the neighborhood of the surface (positive surface concentration); those whose addition results in an increase in surface tension tend to become less concentrated in the neighborhood of the surface (negative surface concentration). The migration of solute either toward or away from the surface is always such as to make the surface tension of the solution (and thus the free energy of the system) lower than it would be if the concentration of solute were uniform throughout (surface concentration equal to zero). Equilibrium is reached when the tendency for free-energy decrease due to lowering surface tension is balanced by an opposing tendency for free-energy increase due to increasing nonuniformity of solute concentration near the surface.

Let us imagine a body of solution of volume V, surface area A, bulk concentration c, and bulk osmotic pressure Π at constant temperature T and external pressure p. For arbitrary changes dA and dV in the area and volume† of the solution, the free-energy change can be written

$$dG = \gamma\, dA - \Pi\, dV \tag{7}$$

This is an exact differential; from the well-known reciprocity relation[1] we find that

$$-\left(\frac{\partial \gamma}{\partial V}\right)_A = \left(\frac{\partial \Pi}{\partial A}\right)_V \tag{8}$$

This can be rewritten

$$-\frac{d\gamma}{dc}\left(\frac{\partial c}{\partial V}\right)_A = \frac{d\Pi}{dc}\left(\frac{\partial c}{\partial A}\right)_V \tag{9}$$

where the presence of total derivatives is justified, since at constant pressure and temperature both the osmotic pressure and the surface tension are determined completely by the concentration.

Now we can write for the total number of moles of solute $n = cV + \Gamma A$ and rearrange this to give

$$c = \frac{n - \Gamma A}{V} \tag{10}$$

Differentiating, we obtain

$$\left(\frac{\partial c}{\partial V}\right)_A = -\frac{c}{V} \qquad \left(\frac{\partial c}{\partial A}\right)_V = -\frac{\Gamma}{V} \tag{11}$$

---

† The volume change may be thought of as resulting from the motion of a piston, containing a semipermeable membrane, against the osmotic pressure Π.

For a perfect solute

$$\Pi = \frac{n}{V}RT = cRT$$

and

$$\frac{d\Pi}{dc} = RT \tag{12}$$

We finally obtain, on combining Eqs. (9), (11), and (12),

$$x_G \simeq \frac{\Gamma}{c} = -\frac{1}{RT}\frac{d\gamma}{dc} \tag{13}$$

This equation was first derived by Willard Gibbs and is called the *Gibbs isotherm*. We can also write it in the form†

$$\Gamma = -\frac{1}{RT}\frac{d\gamma}{d(\ln c)} = -\frac{1}{2.303RT}\frac{d\gamma}{d(\log c)} \tag{14}$$

Surface-active substances: surface adsorption. Many organic solutes in aqueous solution, particularly polar molecules and molecules containing both polar and nonpolar groupings, considerably reduce the surface tension of water. Such solutes tend to accumulate strongly at the surface where, in many cases, they form a unimolecular film of adsorbed molecules.

Usually the film is substantially complete and can accept no more solute molecules when the bulk concentration attains some small value, and hence the surface concentration undergoes little change when the bulk concentration is increased further over a wide range (see Fig. 2). From Eq. (14) we can therefore expect that over this range a plot of the surface tension of the solution against the logarithm of the bulk concentration should be linear. It will deviate from linearity both at low and at high concentrations, in both cases yielding a slope of smaller absolute magnitude.

Solutions of electrolytes. In solutions of certain electrolytes, among them NaCl and KCl, we find that the surface tension *increases* with concentration, indicating a *negative* surface concentration. This is a result of interionic electrostatic attraction, which tends to make the ions draw together and away from the surface. Measurements have shown that in dilute solution the surface tension increases linearly with concentration; thus we see from Eq. (13) that the surface concentration must be directly proportional to the bulk concentra-

---

† Equation (14) has been derived for a perfect solute and applies to real solutions only at low concentrations. In general, $d\gamma = -\Gamma RT\, d(\ln a)$, where a is the activity of the solute. Activity coefficients for *n*-butanol in aqueous solution at 0°C (which to fairly good approximation can be used at 25°C) are:[2] 0.1 *M*, 0.943; 0.2 *M*, 0.916; 0.4 *M*, 0.882; 0.6 *M*, 0.856; 0.8 *M*, 0.838; and 1.0 *M*, 0.823.

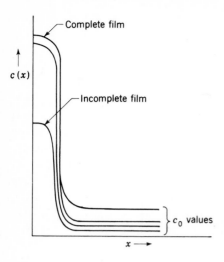

FIGURE 2

Variation of the solute concentration near a free surface for the case of positive surface adsorption. The bulk concentration is represented by c_0, and x is the distance measured inward from the surface.

tion. In actual practice the surface tension often drops initially at low concentrations and then rises linearly; the initial drop is presumably due to impurities and can be eliminated by careful work.

The ratio Γ/c has dimensions of length and is a measure of the "effective thickness" of the region at the surface in which the actual concentration is significantly less than the bulk concentration. Indeed, we may consider a very crude picture in which the actual concentration $c(x)$ may be supposed to be zero from the surface inward to a distance x_0 and to be equal to the bulk concentration c_0 from that distance inward. It will easily be seen from Eq. (4)

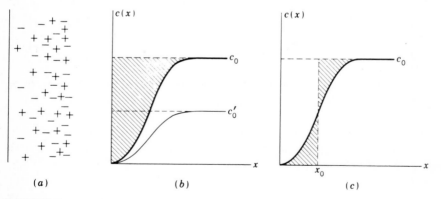

FIGURE 3

Variation of ionic solute concentration near a free surface: (a) schematic diagram of the distribution of ions, showing negative surface adsorption; (b) concentration versus distance for two different bulk concentrations (the shaded area is equal to $-\Gamma$ for a solution of bulk concentration c_0); (c) the "equivalent empty-layer thickness" x_0 chosen such that the two shaded areas will be equal.

that

$$x_0 = \frac{\Gamma}{c_0} \tag{15}$$

This distance, by the above argument, is independent of bulk concentration.

A model of a liquid surface in which there is no negative or positive surface concentration would have the solute uniformly distributed right out to the surface. In electrolyte solutions an ion tends to be surrounded by ions of the opposite charge, but at the surface there are ions that are not uniformly surrounded. There is, therefore, an unbalanced electrostatic force tending to draw these surface ions into the body of the solution. In Fig. 3(a) a more nearly correct model is shown, in which the ions have largely drawn away from the surface. The shaded area above the curve labeled c_0 in Fig. 3(b) represents the negative of the surface concentration; the lower curve corresponds to a different bulk concentration c_0' and shows the proportionality of surface concentration to bulk concentration. Figure 3(c) shows the significance of the "equivalent empty-layer thickness" x_0. The two shaded areas are equal.

METHOD

There are many experimental techniques for measuring the surface tension of liquids. In the *ring method* one determines the force necessary to pull a metal ring free from the surface of a liquid. The Du Nouy tensiometer, in which the ring is hung from the beam of a torsion balance, is often used, but it cannot easily be thermostated. In calculating the surface tension, it is necessary to apply a correction factor which takes into account the shape of the liquid held up by the ring. In the *drop-weight method* one determines the weight of a drop which falls from a tube of known radius. Again, a correction factor is needed because the drop that actually falls does not represent all the liquid that was supported by surface tension. In the *bubble-pressure method* one determines the maximum gas pressure obtained in forming a gas bubble at the end of a tube of known radius immersed in the liquid. Detailed descriptions of these and other methods may be found elsewhere.[3]

Capillary rise. In the absence of external forces a body of liquid tends to assume a shape of minimum area. It is normally prevented from assuming spherical shape by the force of gravity, as well as by contact with other objects. When a liquid is in contact with a solid surface, there exists a specific surface free energy for the interface, or interfacial tension γ_{12}. A solid surface itself has a surface tension γ_2, which is often large in comparison with the surface tensions of liquids. Let a liquid with surface tension γ_1 be in contact with a solid with surface tension γ_2, with which it has an interfacial tension γ_{12}. Under what circumstances will a liquid film freely spread over the solid surface and "wet" it? This will happen if, in creating liquid–solid interface and an equal area of liquid surface at the expense of an equal area of solid surface, the free energy of the entire system decreases:

$$\gamma_1 + \gamma_{12} - \gamma_2 < 0 \tag{16}$$

If we have a vertical capillary tube that dips into a liquid, a film of the liquid will tend to run up the capillary wall if condition (16) is obeyed. Then, in order to reduce the surface of the liquid, the meniscus will tend to rise in the tube. It will rise until the force of gravity on the liquid in the capillary above the outside surface, $\pi r^2(h + r/3)\rho g$, exactly counterbalances the tension at the circumference, which $2\pi r\gamma_1$. In these expressions ρ is the density of the liquid, g is the acceleration of gravity, h is the height of the liquid above the outside surface, r is the radius of the cylindrical capillary, and $r/3$ is a correction for the amount of liquid above the bottom of the meniscus, assuming it to be hemispherical (see Fig. 4). Thus we obtain

$$\gamma_1 = \frac{1}{2}\left(h + \frac{r}{3}\right)r\rho g \tag{17}$$

If Eq. (16) is not obeyed, but if instead

$$\gamma_1 \cos\theta + \gamma_{12} - \gamma_2 = 0 \tag{18}$$

for some value of θ, the liquid will not tend to spread indefinitely on the solid surface but will tend instead to give a *contact angle* θ (see Fig. 5). This may be the case with aqueous solutions or water on glass surfaces that are not entirely

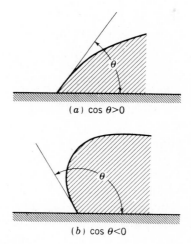

(a) cos θ>0

(b) cos θ<0

FIGURE 5

Contact angle between a liquid and solid surface: (a) cos θ > 0 (e.g., water on a glass surface that is not completely clean); (b) cos θ < 0 (e.g., Hg on glass or water on paraffin).

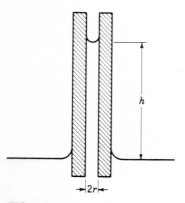

2r

FIGURE 4

Capillary rise h in a tube of radius r.

clean. We shall then have, instead of Eq. (17),

$$\gamma_1 \cos \theta = \frac{1}{2}\left(h + \frac{r}{3}\right)r\rho g \qquad (19)$$

However, in practice, there is usually some "hysteresis"; that is, the contact angle finally attained is somewhat variable, depending on whether the liquid has been advancing over the solid surface or receding from it. Thus two different capillary rise heights are to be expected. If the *same* height is obtained regardless of whether the liquid was allowed to rise from below or fall from above in the capillary, it may be assumed that Eqs. (16) and (17) are almost certainly valid. This is nearly always true of aqueous solutions in carefully cleaned glass capillary tubes.

EXPERIMENTAL

If the capillary tube has not been recently cleaned, it should be soaked in hot nitric acid for several minutes and rinsed copiously with distilled water. A clean capillary is essential to obtaining good results. When not in use, the capillary should be stored by immersing it in a tall flask of distilled water. Assemble the apparatus as shown in Fig. 6. Adjust the capillary tube upward or downward until the outside liquid level is at or slightly above the zero position on the scale.

Determine the height h of capillary rise for pure water at 25°C. Take at least four readings, alternately allowing the meniscus to approach its final position from above and below. Also be sure to read the position of the outside level. If there is not good agreement among these readings, reclean the capillary and repeat the measurements.

Repeat the above procedure with $0.8\,M$ n-butanol solution, dilute to precisely three-quarters the concentration and repeat, and so on until eight concentrations have been used (the last being $0.11\,M$). Rinse the apparatus and capillary with one or two small aliquots of fresh solution at each concentration change.

If time permits, repeat the above procedure with NaCl solutions. Use solutions of concentrations approximately 4, 3, 2, and 1 M.

CALCULATIONS

From the data obtained for pure water, calculate the capillary radius r and use it in calculating the surface tension of each solution studied. Use Eq. (17) for these calculations. For the butanol solutions one may assume that the density is equal to that of pure water; however, for NaCl solutions it is necessary to use values of ρ obtained by interpolation in Table 1. At 25.0°C for pure water, the surface tension is $72.0\,\mathrm{dyn\,cm^{-1}}$ ($0.072\,\mathrm{N\,m^{-1}}$) and the density is $0.9970\,\mathrm{g\,cm^{-3}}$. Tabulate all your γ values and give the units in which they are expressed.

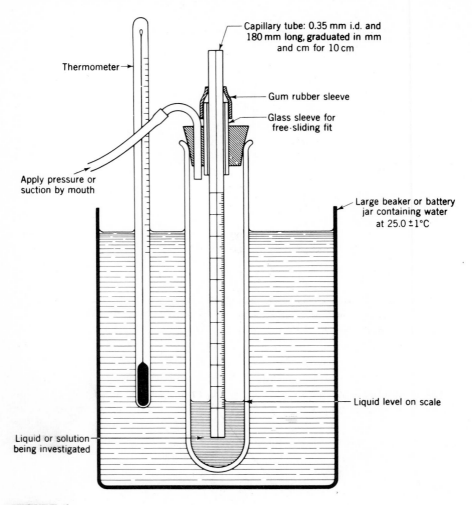

FIGURE 6
Apparatus for measuring surface tension by the method of capillary rise.

For the runs on n-butanol solutions, plot the surface tension of the solution γ versus the logarithm of the bulk concentration c and determine the slope. The surface concentration (in units of $mol\,cm^{-2}$ or $mol\,m^{-2}$) can be calculated from Eq. (14). Express this surface concentration in molecules per square angstrom, and obtain the "effective cross-sectional area" per molecule of adsorbed butanol in $Å^2$.

For the runs on NaCl solutions, plot γ versus the bulk concentration c expressed in moles per cubic centimeter. As seen by Eq. (13) the ratio Γ/c can be determined from the slope of the straight line obtained. Calculate the "effective empty-layer thickness" x_0 in $Å$ for NaCl. Since the laws of ideal solutions do not hold at the NaCl concentrations studied, the results obtained

TABLE 1
Density of NaCl solutions at 25°C[a]

Percentage by wt.	NaCl, g liter^{-1}	Mol liter^{-1}	Density, g cm^{-3}
0	0	0	0.9971
1	10.05	0.1719	1.0045
2	20.22	0.3460	1.0108
4	41.00	0.7015	1.0250
6	62.36	1.067	1.0394
8	84.31	1.443	1.0539
10	106.9	1.829	1.0686
12	130.0	2.224	1.0836
14	153.8	2.632	1.0986
16	178.2	3.049	1.1139
18	203.3	3.479	1.1295
20	229.1	3.920	1.1453
22	255.5	4.372	1.1614
24	282.6	4.836	1.1777
26	310.5	5.313	1.1944

(handwritten annotations in right margin):

1.0   1.0365
2.0   1.0740
3.0   1.1115
4.0   1.1490

[a] Calculated from data given in Landolt-Bornstein, New Series, Group IV, vol. Ib, p. 80, Springer-Verlag, Berlin/New York (1977).

with NaCl have only qualitative significance and the x_0 obtained represents only a rough order of magnitude value.

APPARATUS

One (or two) graduated capillary tubes (cleaned with concentrated HNO_3, rinsed thoroughly with distilled water, and stored in distilled water); ring stand; battery jar; test tube, with two-hole stopper assembly, to hold capillary; 0 to 30°C thermometer; one large clamp and one thermometer clamp; two clamp holders; 20-in. length of gum-rubber tubing; 200-mL volumetric flask; 50-mL pipette; pipetting bulb; 250-mL beakers.

Water aspirator with attached clean gum-rubber tube; $0.8\,M$ aqueous solution of n-butanol (250 mL); $4\,M$ solution of NaCl (400 mL).

REFERENCES

1. P. W. Atkins, "Physical Chemistry," 3d ed., pp. 62, 123, Freeman, New York (1986).
2. W. D. Harkins, and R. W. Wampler, *J. Amer. Chem. Soc.* **53**, 850 (1931).
3. A. E. Alexander and J. B. Hayter, "Determination of Surface and Interfacial Tension," in A. Weissberger and B. W. Rossiter (eds.), "Techniques of Chemistry: Vol. I. Physical Methods of Chemistry," part V, chap. 9, Wiley-Interscience, New York (1971).

GENERAL READING

N. K. Adams, "The Physics and Chemistry of Surfaces," 3d ed., Oxford University Press, London (1941).
R. Defay and I. Prigogine, "Surface Tension and Adsorption," Longmans, London (1966).

EXPERIMENT 27
PHYSICAL ADSORPTION OF GASES

A brief, general description of adsorption on the surface of a solid is given in Exp. 31, which deals primarily with monolayer adsorption from solution. The present experiment is concerned with the multilayer physical adsorption of a gas (the *adsorbate*) on a high-area solid (the *adsorbent*). Since such adsorption is caused by forces very similar to those that cause the condensation of a gas to a bulk liquid, appreciable adsorption occurs only at temperatures near the boiling point of the absorbate. The adsorption of N_2 gas on a high-area solid will be studied at 77.4 K (the boiling point of liquid nitrogen), and the surface area of the solid will be obtained.

THEORY

We shall be concerned here with the adsorption isotherm, i.e., the amount adsorbed as a function of the equilibrium gas pressure at a constant temperature. For physical adsorption (sometimes referred to as van der Waals' adsorption) five distinct types of isotherms have been observed;[1] we shall discuss the theory for the most common type, a typical example of which is shown in Fig. 1. It has unfortunately become standard practice to express the

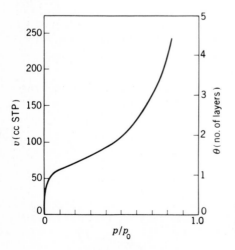

FIGURE 1
A typical isotherm for the physical adsorption of nitrogen on a high-area solid at 77 K.

amount adsorbed by v, the volume of gas in cubic centimeters at STP (standard pressure and temperature: $p = 1$ atm, $T = 273.15$ K) rather than in moles. If the adsorbed volume required to cover the entire surface with a complete *monolayer* is denoted by v_m, one can describe the isotherm in terms of the coverage θ:

$$\theta = \frac{v}{v_m} \tag{1}$$

As shown in Fig. 1, adsorption increases rapidly at high pressures and several layers of adsorbate are present even at a relative pressure $p/p_0 = 0.8$. The pressure p_0 is the saturation pressure of the gas (i.e., the vapor pressure of the liquid at that temperature). When $p/p_0 = 1.0$, bulk condensation will occur to form a liquid film on the surface $\theta \rightarrow \infty$). One can also see from Fig. 1 that there is no clear, sharp indication of the formation of a first layer; indeed, the second and higher layers usually begin to form before the first layer is complete. Clearly we cannot obtain a good value of v_m from the adsorption isotherm without the aid of a theory that will explain the shape of the isotherm.

Brunauer, Emmett, and Teller[2] were the first to propose a theory for multilayer adsorption (BET theory). Since the behavior of adsorbed molecules is even more difficult to describe in detail than that of molecules in the liquid state, the BET theory contains some rather drastic assumptions. In spite of this, it is still a generally useful theory of physical adsorption. The BET theory gives a correct semiquantitative description of the shape of the isotherm and provides a good means of evaluating v_m (which is then used to estimate the surface area of the solid).

The usual form of the BET isotherm is derived for the case of a free (exposed) surface, where there is no limit on the number of adsorbed layers that may form. Such an assumption is clearly not good for a very porous adsorbent, such as one with deep cracks having a width of a few monolayer thicknesses, since the surface of these cracks can hold only a few layers even when the cracks are filled. Fortunately, the BET theory is most reliable at low relative pressures (0.05 to 0.3) at which only a few complete layers have formed, and it can be applied successfully to the calculation of v_m even for porous solids.

The original derivation of the BET equation[1,2] was an extension and generalization of Langmuir's treatment of monolayer adsorption. This derivation is based on kinetic considerations—in particular on the fact that at equilibrium the rate of condensation of gas molecules to form each adsorbed layer is equal to the rate of evaporation of molecules from that layer. In order to obtain an expression for θ as a function of p (the isotherm) it is necessary to make several simplifying assumptions. The physical nature of these assumptions in the BET theory can be seen most clearly from a different derivation based on statistical mechanics.[3] Neither derivation will be presented here, but we shall discuss the physical model on which the BET equation is based. A detailed review of this model has been given by Hill;[4] the important assumptions are as follows.

1. The surface of the solid adsorbent is uniform; i.e., all "sites" for adsorption of a gas molecule in the first layer are equivalent.
2. Adsorbed molecules in the first layer are localized; i.e., they are confined to sites and cannot move freely over the surface.
3. Each adsorbed molecule in the first layer provides a site for adsorption of a gas molecule in a second layer, each one in the second layer provides a site for adsorption in a third, with no limitation on the number of layers.
4. There is no interaction between molecules in a given layer. Thus, the adsorbed gas is viewed as many independent stacks of molecules built up on the surface sites.
5. All molecules in the second and higher layers are assumed to be like those in the bulk liquid. In particular, the energy of these molecules is taken to be the same as the energy of a molecule in the liquid. Molecules in the first layer have a different energy owing to the direct interaction with the surface.

In brief, the statistical derivation is based on equilibrium considerations—in particular on finding that distribution of the heights of the stacks which will make the free energy a minimum.

The BET isotherm obtained from either derivation is

$$\theta = \frac{N}{S} = \frac{cx}{(1-x)[1+(c-1)x]} \tag{2}$$

where S is the number of sites, N is the number of adsorbed molecules, x is the relative pressure (p/p_0), and c is a dimensionless constant greater than unity and dependent only on the temperature. Note that θ equals zero when $x = 0$ and approaches infinity as x approaches unity, in agreement with Fig. 1. Making use of Eq. (1), we can rearrange Eq. (2) to the usual form of the BET equation:

$$\frac{x}{v(1-x)} = \frac{1}{v_m c} + \frac{(c-1)x}{v_m c} \tag{3}$$

Thus a plot of $[x/v(1-x)]$ versus x should be a straight line; in practice, deviations from a linear plot are often observed below $x = 0.05$ or above $x = 0.3$. From the slope s and the intercept I, both v_m and c can be evaluated:

$$v_m = \frac{1}{s+I} \qquad c = 1 + \frac{s}{I} \tag{4}$$

The volume adsorbed (in cm^3 at STP) is related to n, the number of moles adsorbed, by

$$v = nRT_0 \tag{5}$$

where $T_0 = 273.15$ K and $R = 82.06$ cm^3 atm. The total area of the solid is

$$A = N_0 n_m \sigma \tag{6}$$

where N_0 is Avogadro's number and σ is the cross-sectional area of an adsorbed molecule.

Although we shall not be concerned experimentally with measuring heats of adsorption, it is appropriate to comment that ΔH for the physical adsorption of a gas is always negative, since the process of adsorption results in a decrease in entropy. The *isoteric heat of adsorption* (the heat of adsorption at constant coverage θ) can be obtained by application of the Clausius–Clapeyron equation if isotherms are determined at several different temperatures; the thermodynamics of adsorption have been fully discussed by Hill.[5]

METHOD

Adsorption from the gas phase can be measured by either gravimetric or volumetric techniques. In the gravimetric method, first developed by McBain and Bakr, the weight of adsorbed gas is measured by observing the stretching of a helical spring from which the adsorbent is hung (see Fig. 2).† Alternatively, an electrostatic balance may be used. In the volumetric method, the amount of adsorption is inferred from pV measurements that are made on the gas before and after adsorption takes place. An excellent review of the many apparatus designs in common use has been given by Joy;[6] we shall consider only a conventional volumetric apparatus similar to one that has been very completely described by Barr and Anhorn.[7] This apparatus is shown in Fig. 3. The central feature in this design is a gas burette connected to a manometer that can be adjusted to maintain a constant volume in the arm containing the gas.

The gas burette shown in Fig. 3 is constructed from two 50-mL bulbs, large-bore capillary tubing, and a stopcock (*F*). Fiducial marks are engraved above the top bulb, between the bulbs, and below the bottom bulb. Before the burette is attached to the vacuum system, the volume between fiducial marks

---

† A suitable spring can be made by winding 0.010-in beryllium–copper wire on a 8- to 10-mm carbon rod for a length of four to six inches, heating it on the rod in a furnace to about 450°C, and then slowly cooling to room temperature. The total load on the spring (bucket plus contents) should not exceed about 5 g, for which a total deflection of about 10 to 15 cm should be obtained; the length of the spring should be adjusted accordingly, and the apparatus shown in Fig. 2 must be dimensioned to accommodate the maximum deflection. The spring should be tested for the presence of any hysteresis; if any is found, the maximum load must be decreased enough to eliminate it. The spring must be protected from mercury vapor if a mercury manometer, McLeod gauge, or mercury diffusion pump is used. Adequate protection is obtained by placing a wad of fine copper wool in the vacuum line ahead of the spring balance; this should be replaced each time the experiment is performed. (Alternatively, springs made of fused silica may be used; their performance is excellent and they are chemically inert, but they are very fragile.) The sample bucket can be made from a thin blown glass or silica bubble, or from aluminum foil.

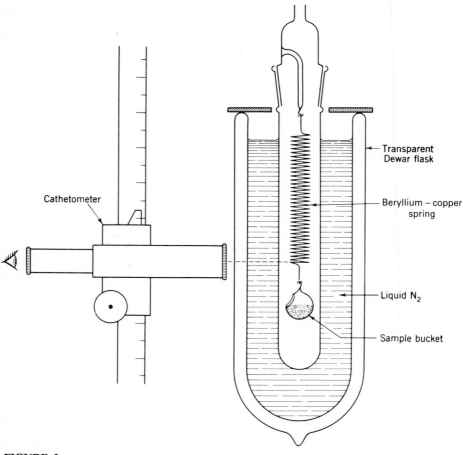

FIGURE 2
Gravimetric adsorption apparatus.

(V_1 and V_2) can be accurately measured by weighing the mercury required to fill each bulb. More elegant designs[6] achieve greater flexibility in operation by using five or six bulbs, but two are sufficient for the purpose of evaluating V_3, the volume of the tubing between the top of the gas burette and the zero level in the manometer. This is accomplished by measuring the pressure p of gas with both bulbs in use, then filling the lower bulb with mercury and determining the new pressure value p'. From the perfect-gas law

$$V_3 = \frac{pV_1 - (p' - p)V_2}{p' - p} \qquad (7)$$

When an isotherm is being determined, the temperature of the gas in the burette system must be known and should remain constant. Although a water jacket is often used, an air jacket is sufficient for the present experiment.

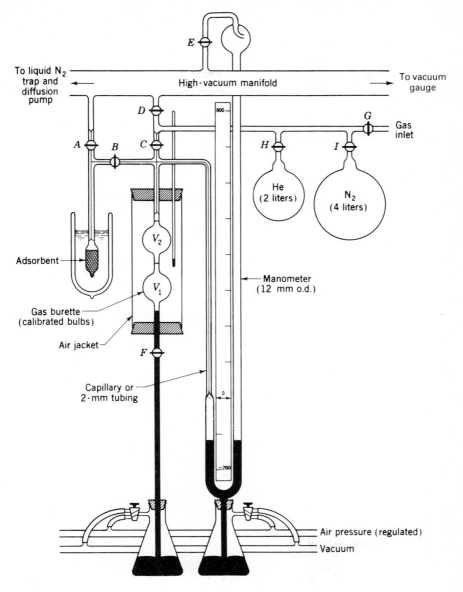

FIGURE 3
Apparatus for measuring adsorption isotherms by the volumetric method.

Adsorbent. For the apparatus described here, a sample of a high-area solid adsorbent which has a total surface area of 150 to 250 $m^2 g^{-1}$ should be used. Of the adsorbents most commonly used (charcoal, silica, alumina), silica is recommended as an excellent choice. The sample bulb should be small enough so that it is almost completely filled with the powder. Figure 4 shows a type of

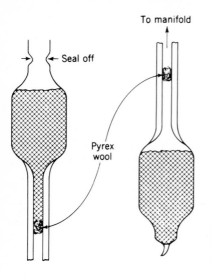

To manifold

Seal off

Pyrex
wool

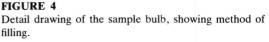

FIGURE 4
Detail drawing of the sample bulb, showing method of filling.

bulb that can easily be filled; a loose plug of Pyrex wool in the capillary will prevent loss of powder during filling and later during degassing. The exact weight of the sample should be determined before the bulb is attached to the system. After it is attached to the vacuum line, the sample must be degassed (i.e., pumped on to remove any physically adsorbed substances, principally water). This can be done by mounting a small electrical tube furnace around the sample bulb and heating to 200 to 250°C for several hours. Before heating, stopcock A should be opened to pump out all the air (stopcock B remains closed); this must be done cautiously to avoid forming a tight plug of powder in the capillary tube (gentle tapping will help also). After this initial degassing, a shorter degassing period will suffice between adsorption runs.

Adsorbate gas. For physical adsorption, any pure gas that does not react with the solid adsorbent can be used. However, for quantitative area measurements, a molecule of known adsorption area is required. Nitrogen gas is a very common choice and will be used in this experiment.

   Liquid nitrogen is the most convenient constant-temperature bath for isotherm measurements with N_2 gas. For high-precision work, the temperature of this bath should be measured with a thermocouple so that an accurate value of p_0 can be obtained, or p_0 should be measured directly with a separate nitrogen vapor-pressure manometer.[6,7] For this experiment, it is adequate to assume that p_0 is equal to the atmospheric pressure in the room.

   We can now develop the necessary equations for calculating the amount adsorbed from the pV data obtained during a run. Let us denote the known volume of the gas burette by V_B ($= V_1 + V_2 + V_3$ or $V_2 + V_3$ as the case may be) and its temperature by T_B. Since the sample bulb is not completely filled by the solid adsorbent, we must also consider the so-called dead-space volume occupied by gas. Let us denote this volume by V_s and consider it to be at a single temperature T_s. If the burette is filled with gas at an initial pressure p_1^0 and then stopcock B is opened, the pressure will drop to a new equilibrium value p_1. The number of moles adsorbed is

$$n_1 = \frac{p_1^0 V_B}{RT_B} - \left(\frac{p_1 V_B}{RT_B} + \frac{p_1 V_s}{RT_s}\right) \tag{8}$$

where typically the V's are expressed in cm^3, the p's in atm, and $R = 82.06\ cm^3$ atm. Equation (8) can be rewritten in terms of v_1 (the amount adsorbed in cm^3 at STP) by using Eq. (5):

$$v_1 = p_1^0 V_B^* - p_1(V_B^* + V_s^*) \tag{9}$$

where the p's are now in Torr and

$$V_B^* = \frac{273.15 V_B}{760 T_B} \qquad V_s^* = \frac{273.15 V_s}{760 T_s} \tag{10}$$

The value of V_B^* can be calculated from V_B and T_B, which are known. V_s^* is constant but unknown; however, it can be determined if a run is made using helium gas. Since He is not significantly adsorbed at liquid-nitrogen temperature, $v_1 = 0$ and V_s^* can be calculated from Eq. (9).

If N_2 gas is admitted to the bulb, adsorption does occur and Eq. (9) yields a value of v_1, the "volume" adsorbed at the equilibrium pressure p_1. Now stopcock B is closed and more N_2 gas is added to the burette to achieve a new initial pressure p_2^0. When b is opened, more adsorption will occur and the pressure will change to an equilibrium value p_2. The total volume adsorbed at p_2 is

$$v_2 = [p_1^0 V_B^* + (p_2^0 - p_1)V_B^*] - p_2(V_B^* + V_s^*) \tag{11}$$

The generalization of Eq. (11) for v_j, the volume adsorbed at p_j after j additions of gas to the burette, is obvious; for example,

$$v_3 = [p_1^0 V_B^* + (p_2^0 - p_1)V_B^* + (p_3^0 - p_2)V_B^*] - p_3(V_B^* + V_s^*)$$

Note that the bracket contains an expression for the *total* amount of gas added to the system and the second term represents the amount remaining in the gas phase at equilibrium. Since the effect of errors is cumulative in this method, care should be exercised in measuring all pressures.

Normally the entire gas burette is used ($V_B = V_1 + V_2 + V_3$), but on occasion it may be convenient to obtain two points on the isotherm from a single filling of the burette; this can be accomplished after the first point is obtained, by raising the mercury level so as to fill bulb V_1. Equation (11) is still applicable, but the value of the last term is changed, since V_B now equals $V_2 + V_3$ (in that term only) and a new equilibrium pressure must be used.

EXPERIMENTAL

Details of the operation of the vacuum system (Fig. 3) should be reviewed carefully in consultation with an instructor before beginning the experiment. In the following it is assumed that both the helium and nitrogen storage bulbs have already been filled to a pressure slightly over 1 atm, the diffusion pump is operating, and the sample has been degassed.

Stopcocks B, C, D, E, and, of course, G should be closed and the pressure determined with the vacuum gauge; it should be 10^{-5} Torr or less. Now set the mercury level exactly at the lower mark in the gas burette and close stopcock F. Also raise the mercury in the manometer to approximately the zero level. After closing stopcock A and opening C, D, and E, read the vacuum gauge again and also verify that the mercury level is the same in both arms of the manometer.

Slowly raise a narrow-mouth Dewar flask filled with liquid nitrogen into place, and clamp it so that the sample bulb is completely immersed. Then plug the mouth of the Dewar loosely with glass wool or a piece of clean towel to retard condensation of water and oxygen from the air. (Liquid O_2 dissolved in the liquid N_2 would raise the temperature of the bath.)

Close stopcocks D and E, and **very slowly** turn the stopcock on the helium storage bulb while watching the manometer. As soon as the mercury levels begin to move, cease turning the stopcock and let the gas slowly fill the gas burette to a pressure of about 300 Torr. When this pressure is reached, immediately close the stopcock on the helium storage bulb and then close C. Adjust the mercury level in the manometer so that the left arm is exactly at the zero level. Wait a few minutes for equilibrium to be achieved, readjust to the zero level if necessary, and record the pressure. Now carefully raise the mercury level in the gas burette to the center mark (i.e., fill bulb V_1 with Hg) and again adjust the manometer to the zero level. After a short wait, record the new pressure. Using the known values of V_1 and V_2 (given by the instructor), one can now calculate V_3 from Eq. (7).

Next open stopcock B slowly and allow He to enter the sample bulb. Wait for at least 3 min, then reset the manometer to the zero level and read the pressure. After several minutes more, readjust this level (if necessary) and read the pressure again. When the pressure is constant, record its value and also record T_B, the ambient temperature at the gas burette. The quantity V_s^* can be calculated from these data using Eq. (9) with $v_1 = 0$. Note that, in calculating V_B^* from Eq. (10), the appropriate value of V_B is $V_2 + V_3$.

Slowly open stopcock A and pump off the He gas. Lower the Dewar flask and allow the sample to warm up to room temperature. Open stopcocks C and D and continue pumping for at least 15 min. During this time, reset the Hg level in the gas burette to the lower mark and lower the mercury in the manometer to approximately the zero level. (Do the two arms still read the same? Open stopcock E and see if there is any change.) Now close stopcocks A and B and replace the liquid-nitrogen Dewar around the sample bulb. If time is limited and values of V_3 and V_s^* have been provided, this helium run may be omitted and the following procedure is then carried out directly after first cooling the sample.

Procedure for measuring the N_2 isotherm. Close stopcocks D and E, and fill the gas burette with N_2 gas to a pressure of about 300 Torr. Follow the procedure given above in filling the burette and measuring the pressure. Record this pressure p_1^0. *Slowly* open stopcock B and allow N_2 to enter the sample bulb. Again follow the procedure given previously for obtaining the new equilibrium pressure p_1. In the case of N_2, a longer period may be necessary to achieve equilibrium, since adsorption is now taking place. The volume adsorbed v_1 can be calculated from Eq. (9). [In this case V_B in Eq. (10) is $V_1 + V_2 + V_3$.]

Now close stopcock B and add enough N_2 gas to the burette to bring the pressure up to about 100 Torr. Record this pressure p_2^0; then open B and obtain the next equilibrium pressure p_2. The volume v_2 is calculated from Eq. (11). Continue to repeat this process so as to obtain about seven points in the pressure range from 0 to about 250 Torr. Each time start with an initial pressure p_j^0 somewhat greater (by say 15 to 25 percent) than the desired equilibrium pressure p_j. If time permits, obtain additional points somewhat more widely spaced over the range 250 to 650 Torr. At all times be sure that the liquid-nitrogen level completely covers the sample bulb.

Several times during the period, record the barometric pressure and T_B, the temperature at the gas burette.

CALCULATIONS

From your helium data and the known values of V_1 and V_2, calculate V_3 from Eq. (7). If the temperature at the gas burette has been fairly constant throughout the experiment, use the average value as T_B in Eq. (10) and calculate values of V_B^* when $V_B = V_1 + V_2 + V_3$ and when $V_B = V_2 + V_3$. These values can then be used in all further calculations. If T_B has varied by more than $\pm 0.5°C$, appropriate changes in V_B^* should be made where necessary. Now use Eq. (9) to calculate V_s^*.

For each equilibrium point on the isotherm, calculate a value of v, the volume adsorbed at pressure p. If the barometric pressure has been almost constant, its average value may be taken as p_0, the vapor pressure of nitrogen at the bath temperature. For each isotherm point, calculate $x = p/p_0$.

Plot the isotherm (v versus x) at 77 K and compare it qualitatively with the one shown in Fig. 1. Finally, calculate $x/v(1-x)$ for each point and plot that quantity versus x; see Eq. (3). Draw the best straight line through the points between $x = 0.05$ and 0.3, and determine the slope and intercept of this line. This evaluation can be made graphically or by a linear least-squares fit of the data with Eq. (3) (see Chapter XX). From Eq. (4) calculate v_m and c for the sample studied.

Using Eqs. (5) and (6) and the known weight w of the adsorbent sample, calculate the *specific area* $\bar{A}$ in square meters per gram of solid. The cross-sectional area for an adsorbed N_2 molecule may be taken as $15.8\,\text{Å}^2$.

DISCUSSION

What factors can you think of that would tend to make the BET theory less reliable above $\theta = 0.3$, below $\theta = 0.05$?

APPARATUS

Adsorption apparatus (Fig. 3), containing high-area sample, attached to high-vacuum line equipped with McLeod gauge or other vacuum gauge, diffusion pump, and liquid-nitrogen trap (see Chapter XVII); high-purity helium and nitrogen gas; small electrical tube furnace; narrow-mouth taped Dewar flask and clamp for mounting; glass wool or clean towel; 0 to 30°C thermometer; stopcock grease.

REFERENCES

1. J. W. Williams, R. A. Alberty, and E. O. Kraemer, "The Colloidal State and Surface Chemistry," in H. S. Taylor and S. Glasstone (eds.), "A Treatise on Physical Chemistry," 3d ed., vol. II, pp. 594–611, chap. V, Van Nostrand, Princeton, N.J. (1951).
2. S. Brunauer, P. H. Emmett, and E. Teller, *J. Amer. Chem. Soc.* **60**, 309 (1938).
3. T. L. Hill, *J. Chem. Phys.* **14**, 263 (1946) and **17**, 772 (1949).
4. T. L. Hill, "Theory of Physical Adsorption," in "Advances in Catalysis," vol. IV, pp. 225–242, Academic Press, New York (1952).
5. *Ibid.,* pp. 242–255.
6. A. S. Joy, *Vacuum* **3**, 254 (1953).
7. W. E. Barr and V. J. Anhorn, *Instruments* **20**, 454, 542 (1947).

GENERAL READING

S. Brunauer, "Physical Adsorption," chaps. I–IV, Princeton University Press, Princeton, N.J. (1945).

CHAPTER
XI

MACROMOLECULES

EXPERIMENTS

28. Osmotic pressure
29. Intrinsic viscosity: chain linkage in polyvinyl alcohol
30. Helix–coil transition in polypeptides

EXPERIMENT 28
OSMOTIC PRESSURE

Macromolecules are very large molecules with molecular weights ranging in order of magnitude from 1000 to 1 000 000 g cm^{-1} or more. They are with a few exceptions organic molecules, and many examples such as cellulose and starch exist in nature. Macromolecules can also be made synthetically. These are often called polymers because structurally they are formed by the linking together of many monomer units to form chains of considerable length.

From measurements of the osmotic pressure of dilute solutions of a polymer, the number-average molecular weight of the polymer will be obtained.

THEORY

When a liquid solution of a solute B in a solvent A is separated from pure solvent A by a membrane that is permeable to A alone and both phases are at the same temperature and under the same pressure, the solvent molecules A will pass through the membrane into the solution. This can be prevented by applying a pressure to the solution which is greater, by a definite amount Π, than the pressure on the solvent. This *osmotic pressure* Π can be related to the vapor pressure of A, since at equilibrium the chemical potential of A in solution, μ_A, must equal the chemical potential of pure A, μ_A^0. If the partial molar volume of solvent $\bar{V}_A$ is taken as being independent of pressure, the result of equating chemical potentials is[1]

$$\Pi\bar{V}_A = RT \ln\frac{p_A^0}{p_A} = -RT \ln \gamma_A X_A \tag{1}$$

Approximating $\bar{V}_A$ by $\tilde{V}_A$, the molar volume of pure A, and assuming Raoult's law (i.e., $\gamma_A = 1$), we obtain

$$\Pi\tilde{V}_A \cong -RT \ln X_A = -RT \ln(1 - X_B) \tag{2}$$

where X_A and X_B are the mole fractions of A and B in the solution. For *dilute* solutions, $\ln(1 - X_B)$ can be replaced by $-X_B$ to give

$$\Pi = \frac{RT}{\tilde{V}_A} X_B \cong \frac{n_B RT}{V_A} \cong \frac{n_B RT}{V} \tag{3}$$

where V_A, the volume of solvent containing n_B moles of solute, is approximately equal to the total volume of the solution V.

In general, the observed osmotic pressures do not obey the ideal-solution law of Eq. (3). Owing to the large size of polymer molecules, interactions are important even in dilute solutions, but these can be taken into account fairly well by including an osmotic second virial coefficient:[2]

$$\Pi = \frac{n_B RT}{V}\left(1 + \frac{B n_B}{V}\right) \tag{4}$$

This equation can be written as

$$\frac{\Pi}{c_2} = \frac{RT}{M}(1 + \beta c_2) \tag{5}$$

where M is the molecular weight of the solute, c_2 is the concentration of solute in weight per unit volume of solution ($= Mn_B/V$), and β equals B/M. Thus by plotting Π/c_2 versus c_2, it is possible to determine both M and β.

Osmotic pressure is the one colligative property of solutions that is suitable for determining very high molecular weights, since Π is of the order of several Torr even for very dilute solutions. The other colligative properties—freezing-point depression, boiling-point elevation, and vapor-pressure lowering—show effects that are too small for accurate measurement; typical values for dilute polymer solutions would be 5×10^{-4} K, 10^{-4} K, and 10^{-4} Torr, respectively.

Number-average molecular weight. When a polymer sample is polydisperse (i.e., when there is a distribution of molecules with different weights), a determination of molecular weight must give some kind of average value. When a colligative property is involved, the average is a *number-average* molecular weight $\bar{M}_n$, since the effect depends only on the number of molecules per unit volume and is independent of their size or shape. This is easily seen for the case of osmotic pressure by comparing Eqs. (4) and (5). Molecular-weight determinations based on properties other than the colligative properties will give different kinds of average molecular weights. One common method involves light scattering,[3] which yields a *weight-average* molecular weight $\bar{M}_w$. Both $\bar{M}_n$ and $\bar{M}_w$ are defined and discussed in Exp. 29 along with the *viscosity-average* molecular weight $\bar{M}_v$. It should be noted that $\bar{M}_w$ will always be greater than $\bar{M}_n$ for a polydisperse polymer sample. (For a monodisperse polymer, $\bar{M}_n = \bar{M}_w = \bar{M}_v$.)

METHOD

Several methods of osmotic-pressure measurement have been described by several workers.[4] In the dynamic-equilibrium method, flow of solvent through the membrane can be prevented by applying an external gas pressure to the solution; the equilibrium pressure necessary to obtain zero flow will equal the osmotic pressure. Although this method is rapid, it involves a complex cell that must be leaktight. Two other methods that are frequently used are described below. In both techniques, the osmotic pressure is determined from the internal head of solution in a capillary tube on the solution side of the membrane (see Fig. 1).

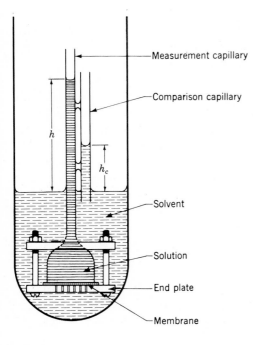

FIGURE 1
A simple osmometer; $\Delta h = h - h_c$ is the internal head corrected for capillary rise.

Static method. In the static method, solvent is allowed to diffuse through the membrane until there is no further change in the internal head. It is necessary to correct this head for the effect of surface tension (capillary rise) and to calculate the equilibrium concentration of the solution, which will differ from the initial concentration owing to passage of solvent through the membrane. The disadvantage of this method is that it is slow, equilibrium usually requiring several days. However, no attention is needed during this time.

Half-sum method. Fuoss and Mead[5] have proposed a rapid method that involves constant attention for about an hour. With the solution and solvent in thermal equilibrium, the internal head is adjusted initially to be close to the expected equilibrium value. To illustrate this method let us assume that the initial head is slightly above the equilibrium value. Frequent readings are made to follow the decrease in this head with time (curve *A* in Fig. 2). After sufficient time has elapsed to indicate an approximate asymptotic value, the head is adjusted to be roughly as far below this asymptote as it was initially above. Now readings are made to follow the increase with time (curve *B* in Fig. 2). By calculating one-half the sum of the ordinates of curves *A* and *B* at several values of the time, one can obtain curve *C* which converges rapidly to a constant value. Since only a little solvent flows out of the solution during measurements along curve *A* and approximately the same amount of solvent flows into the solution during measurements along the curve *B*, the equilibrium concentration may be assumed to be equal to the initial concentration. Correction for the effect of capillary rise is still necessary.

Osmometer. Of many designs that have been used, two are especially simple to construct and operate. The Schulz–Wagner osmometer[4] is of an excellent design but is limited to static measurements. A simplified version of this cell, suitable for student use, has been described elsewhere.[6]

The osmometer suggested for this experiment is based on the design of Zimm and Myerson[7] and is shown in Fig. 3. This osmometer consists of a small, heavy-wall cylindrical cell with a filling tube and a 0.5-mm i.d.

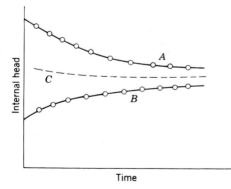

FIGURE 2
Fuoss-Mead method of half sums. Circles indicate the experimental points; the dashed line *C* indicates the average of curves *A* and *B*.

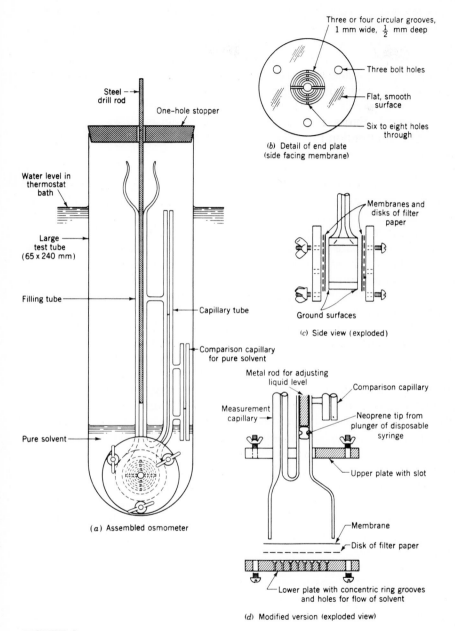

Three or four circular grooves, 1 mm wide, $\frac{1}{2}$ mm deep

Three bolt holes

Flat, smooth surface

Six to eight holes through

(b) Detail of end plate (side facing membrane)

Steel drill rod

One-hole stopper

Water level in thermostat bath

Large test tube (65 x 240 mm)

Filling tube

Capillary tube

Comparison capillary for pure solvent

Pure solvent

(a) Assembled osmometer

Membranes and disks of filter paper

Ground surfaces

(c) Side view (exploded)

Metal rod for adjusting liquid level

Comparison capillary

Measurement capillary

Neoprene tip from plunger of disposable syringe

Upper plate with slot

Membrane

Disk of filter paper

Lower plate with concentric ring grooves and holes for flow of solvent

(d) Modified version (exploded view)

FIGURE 3
Zimm-Meyerson osmometer: (a) the completely assembled apparatus; (b) detail view of one of the end plates; (c) side view of the cell; (d) detail view of modified version.

measuring capillary sealed into its side. The open ends of the cell are ground flat and smooth, so that the two membranes, which are held in place by perforated metal plates, also serve as gaskets. The end plates, machine screws, and wing nuts should be stainless steel (alternatively, brass is satisfactory if only organic liquids are to be used). The holes and grooves in the end plates may be replaced, for simplicity, by several drill holes if a disk of filter paper is used to support each membrane. Attached to the measuring capillary is a short length of capillary with the same inside diameter. This is immersed in the solvent to provide a correction for capillary rise.

A modified version of this osmometer is shown in Fig. 3(d). An advantage of this "thistle-tube" design is ease of assembly since only one membrane is used and it is mounted horizontally at the bottom of the cell. Two disadvantages are the somewhat more awkward clamping arrangement and the slower diffusion of solvent. The latter point is not serious if the half-sum method is used.

Once the cell is filled with solution, a drill rod is inserted into the filling tube. By raising or lowering this steel rod, one can adjust the position of the meniscus in the measuring capillary. Thus measurements can be made using the half-sum method or using the static method with the level set initially at the expected value in order to save time. Two designs are shown in Fig. 3 for the details of the steel rod–filling tube assembly. The simplest design, indicated in Fig. 3(a), is to use 2-mm i.d. capillary tubing and a length of steel drill rod chosen to give a very close fit to the capillary. An alternative design, shown in Fig. 3(d), utilizes a larger filling tube and a loosely fitting rod which has attached to it the neoprene tip from the plunger of a disposable syringe.

When volatile solvents are used, a small pool of mercury in the flared top of the filling tube will prevent evaporation around the metal rod. Also, a close-fitting one-hole stopper in the outer test tube which holds the pure solvent will retard evaporation of solvent and reduce absorption of water from the atmosphere if a water-soluble solvent is being used.

Membranes. Of the many materials used for membranes, the most convenient is cellulose in the form of nonwaterproofed cellophane sheet. Cellophane is a good membrane for methyl ethyl ketone, the solvent to be used in this experiment, and can be prepared in a range of porosities by treatment with strong basic solutions.[4,6] To condition a membrane in water for use with an organic solvent, it is necessary to wash it, successively, with 25, 50, 75, and 100 percent aqueous alcohol (or acetone) solutions and then displace the alcohol (or acetone) by similar washings with the desired organic solvent. Membranes should be stored in the solvent and should never be allowed to dry out.

EXPERIMENTAL

Polystyrene is a suitable polymer for investigation in methyl ethyl ketone (MEK) solvent. It is necessary to have a reasonably high-molecular-weight fraction to avoid diffusion of solute molecules through the membrane. If such solute diffusion occurs, the osmotic pressure will change with time.

An initial weight concentration of polymer that will produce about an 8-cm internal head is desirable. If an approximate value of the molecular weight is known, this concentration can be estimated from Eq. (3). Make up 50 mL of this solution in a volumetric flask. Pipette 25 mL of this solution into a clean 50-mL volumetric flask, and make up to the mark with solvent. Repeat this dilution procedure at least twice to obtain solutions that are one-quarter and one-eighth of the initial concentration. If these solutions are allowed to stand for a considerable time before the osmometer is filled, be sure to seal them carefully to avoid evaporation of solvent and consequent change of concentration.

The large test tube and the glass parts of the osmometer should be cleaned thoroughly with chromic acid cleaning solution, and rinsed with distilled water, alcohol, and MEK. The osmometer is then assembled by placing a disk of smooth, hard filter paper (moistened with MEK) and a round cellophane membrane (conditioned for and stored in MEK) on the metal end plates and bolting these plates carefully about the cell as shown in Fig. 3(c) or 3(d). The membranes should *never* be allowed to dry out. As soon as the osmometer is assembled, place it in the large test tube containing pure solvent.

It may be desirable to test the osmometer for leaks along the seal between the ground-glass edges of the cell and the metal end plates. This leak test can be accomplished by filling the osmometer with pure solvent and inserting the steel rod into the filling tube so as to create an internal head (liquid level in the measuring capillary above that in the comparison capillary). If the osmometer is leaktight, the level in the capillary should change only very slowly owing to diffusion through the membrane. If the level changes rapidly, a leak along the glass–metal seal is indicated and the cell should be taken apart, inspected for flaws, reassembled, and retested.

To fill and empty the cell it is convenient to use a large hypodermic syringe with a long stainless-steel needle that can be inserted down the filling tube all the way to the bottom of the cell. [If the osmometer has a large filling tube, it can be filled and emptied by means of a length of Teflon "spaghetti" tubing, which is inserted to the bottom of the cell through the filling tube. This spaghetti tubing is attached (by means of a tapered adapter, made by drawing down a piece of 8-mm glass tubing) to a water aspirator for emptying the osmometer or to a hypodermic syringe for filling it. When filling the cell, slowly withdraw the spaghetti tubing as the solution is introduced.]

After the osmometer has been leak-tested, rinse it several times with small quantities of solution and then fill it with the solution to about half-way up the filling tube. Be sure that there are no air bubbles in the cell; bubbles can be removed by gentle tapping or by removing some of the solution and refilling. Insert the steel drill rod into the filling tube so as to adjust the level in the capillary to approximately the expected level. *Immediately rinse* the syringe and needle with pure solvent (to prevent any deposit of polymer from evaporation of residual solution) and introduce the osmometer into its test tube, which should contain enough solvent to produce a meniscus in the reference capillary. [If a steel rod with rubber plunger tip is used, the osmometer is completely filled to the top of the filling tube. Then insert the plunger tip taking care not to produce a bubble. Slowly push the plunger in, about two-thirds of the way to the bottom, allowing the excess solution to exit through the measuring capillary. Then **very slowly** pull the the plunger back so as to bring the meniscus in the measuring capillary to approximately the desired position. Extreme slowness (with rotation of the plunger to avoid jerkiness) is necessary to allow the liquid in the measuring capillary to drain down the capillary walls without forming "plugs." **Caution:** Do not insert the plunger into the filling tube unless the tube contains liquid since the rubber tip may come off when the plunger is withdrawn from a dry tube.]

The osmometer should be mounted in a thermostat bath at 25°C that is regulated to within ±0.01°C or better. A glass-walled thermostat tank is best, since it will permit the test tube to be well immersed and enable the capillary levels to be observed through the glass wall. If such a bath is not available, immerse the test tube to the level of the solvent inside and read the capillary levels over the edge of the bath. The level of the liquid in the two capillaries should be read to within 0.1 mm using a cathetometer. This instrument consists of a telescope mounted on a vertical bar engraved with a millimeter scale. The telescope is moved up or down until the liquid level is in the center of the field and the position on the bar is recorded. A more complete description is given in Chapter XVIII. The internal head Δh is the difference $h - h_{comp}$ between the height of solution in the measuring capillary and the height of solvent in the comparison capillary. Determine the internal head for at least four concentrations of polymer.

Either the static method or the half-sum method described previously may be used. The method of half sums is recommended. Plot the height of the solution in the measuring capillary against time as soon as each data point is obtained. When the variation of h with time becomes quite slow, adjust the head and obtain the second set of readings. Do not wait for a constant asymptotic value of h on either set of readings. If the static method is employed, it is necessary to know the volume of the cell and the cross-sectional area of the measuring capillary in order to compute the equilibrium concentration. Since this correction is small, these dimensions need not be known to high precision. The volume of the cell can be obtained by carefully filling it with solvent using a calibrated syringe. To measure the area of the capillary, introduce a slug of mercury and measure its length. Then transfer the mercury to a weighing bottle and weigh. Using the known density of mercury, one can calculate the cross-sectional area.

When the measurements are completed, the osmometer and the test tube should be rinsed well and left filled with solvent if there is any possibility that it will be used again soon. Otherwise, the apparatus should be disassembled and its parts thoroughly rinsed and dried.

CALCULATIONS

For each solution, calculate the initial concentration in grams of solute per milliliter of solution. If the half-sum method was used, determine the equilibrium head h from the asymptotic value of the half sums as shown in Fig. 2. Then calculate the internal head Δh.

If the static method was used, correct the initial concentration for diffusion of solvent to obtain the equilibrium concentration c_2.[6] Calculate the osmotic pressure Π, in dyn cm^{-2} and Pa (the SI unit), for each solution from

$$\Pi = (\Delta h)\rho g \qquad (6)$$

where the solution density ρ is taken to be the same as the density of pure solvent and g is the acceleration due to gravity. The density of pure MEK is 0.803 g cm^{-3} at 25°C.

Tabulate your values for Δh, Π, c_2, and Π/c_2. Plot Π/c_2 versus c_2, and draw the best straight line through the experimental points. (It is possible that the experimental points may show a slight curvature; in this case, draw the straight line which is tangent to the curve at zero concentration.) From the intercept and slope of this line, calculate the number-average molecular weight $\bar{M}_n$ and the osmotic second virial coefficient β using Eq. (5).

APPARATUS

One or more osmometers (complete with end plates, bolts, large test tube, drill rod, and stopper); membranes (uncoated cellophane, designated as PUT grade, is available from Flexil Corp., Covington, IN 47932), conditioned for use in methyl ethyl ketone; four 50-mL glass-stoppered volumetric flasks; 25-mL pipette; pipetting bulb; several beakers; 25-mL hypodermic syringe with long needle; constant-temperature bath (preferably with glass walls) set at 25°C; large clamp for mounting the osmometer; cathetometer; stopwatch; mercury (optional).

High-molecular-weight polystyrene powder; reagent-grade methyl ethyl ketone (500 mL).

REFERENCES

1. R. S. Berry, S. A. Rice, and J. Ross, "Physical Chemistry," pp. 928–9, Wiley, New York (1980).
2. P. J. Flory, "Principles of Polymer Chemistry," pp. 279–282, 530–539, Cornell University Press, Ithaca, N.Y. (1953).
3. Ibid., pp. 283–303.
4. J. R. Overton, "Determination of Osmotic Pressure," in A. Weissberger and B. W. Rossiter (eds.), "Techniques of Chemistry: Vol. I. Physical Methods of Chemistry," part V, chap. 6, Wiley-Interscience, New York (1971).
5. R. M. Fuoss and D. J. Mead, J. Phys. Chem. **47**, 59 (1943).
6. F. Daniels, J. W. Williams, P. Bender, R. A. Alberty, C. D. Cornwell, and J. E. Harriman, "Experimental Physical Chemistry," 7th ed., pp. 123–132, McGraw-Hill, New York (1970).
7. B. H. Zimm and I. Myerson, J. Amer. Chem. Soc. **68**, 911 (1946).

GENERAL READING

G. W. Castellan, "Physical Chemistry," 3d ed., sec. 35.4, Addison-Wesley, Reading, Mass. (1983).

P. J. Flory, *op. cit.*

EXPERIMENT 29
INTRINSIC VISCOSITY: CHAIN LINKAGE IN POLYVINYL ALCOHOL

While the basic chemical structure of a synthetic polymer is usually well understood, many physical properties depend on such characteristics as chain length, degree of chain branching, and molecular weight, which are not easy to specify exactly in terms of a molecular formula. Moreover, the macro-molecules in a given sample are seldom uniform in chain length or molecular weight (which for a linear polymer is proportional to chain length); thus, the nature of the distribution of molecular weights is another important characteristic.

A polymer whose molecules are all of the same molecular weight is said to be *monodisperse*; a polymer in which the molecular weights vary from molecule to molecule is said to be *polydisperse*. Specimens that are approximately monodisperse can be prepared in some cases by fractionating a polydisperse polymer; this fractionation is frequently done on the basis of solubility in various solvent mixtures.

This experiment is concerned with the linear polymer polyvinyl alcohol (PVOH), $-(CH_2-CHOH-)_n$, which is prepared by hydrolysis of the polyvinyl acetate (PVAc) obtained from the direct polymerization of the monomer vinyl acetate, $CH_2=CH-OOCCH_3$. As ordinarily prepared, polyvinyl alcohol shows a negligible amount of branching of the chains. It is somewhat unusual among synthetic polymers in that it is soluble in water. This makes polyvinyl alcohol commercially important as a thickener, as a component in gums, and as a foaming agent in detergents.

A characteristic of interest in connection with PVOH and PVAc is the consistency of orientation of monomer units along the chain. In the formula given above it is assumed that all monomer units go together "head to tail." However, occasionally a monomer unit will join onto the chain in a "head-to-head" fashion, yielding a chain of the form

$$-(CH_2-CHX-)_n CH_2-CHX-CHX-CH_2-CH_2-CHX-(CH_2-CHX-)_n \cdots$$

head-to-head
linkage

reversed
monomer unit

where X is Ac or OH. The frequency of head-to-head linkage depends on the relative rates of the normal growth-step reaction α

$$\underset{\rightarrow}{R} \cdot + M \xrightarrow{k_\alpha} \underset{\rightarrow}{R} — \underset{\rightarrow}{M} \cdot \tag{1a}$$

and the abnormal reaction β

$$\underset{\rightarrow}{R} \cdot + M \xrightarrow{k_\beta} \underset{\rightarrow}{R} — \underset{\leftarrow}{M} \cdot \tag{1b}$$

(where $\underset{\rightarrow}{R} \cdot$ is the growing polymer radical, the arrow representing the predominant monomer orientation). The rates, in turn, must depend on the activation energies:

$$\frac{k_\beta}{k_\alpha} = \frac{A_\beta e^{-E_{a\beta}/RT}}{A_\alpha e^{-E_{a\alpha}/RT}} = S e^{-\Delta E_a/RT} \tag{2}$$

where $\Delta E_a = E_{a\beta} - E_{a\alpha}$ is the additional thermal activation energy needed to produce abnormal addition and S is the *steric factor* representing the ratio of the probabilities that the abnormally and normally approaching monomer will not be prevented by steric or geometric obstruction from being in a position to form a bond. Presumably the activation energy for the normal reaction is the lower one, in accord with the finding of Flory and Leutner[1,2] that the frequency of head-to-head linkages in PVAc increases with increasing polymerization temperature. The quantities S and ΔE_a can be determined from data obtained at two or more polymerization temperatures by a group of students working together. It should be of interest to see whether steric effects or thermal activation effects are the more important in determining the orientation of monomer units as they add to the polymer chain.

In this experiment, the method of Flory and Leutner will be used to determine the fraction of head-to-head attachments in a single sample of PVOH. The method depends on the fact that in PVOH a head-to-head linkage is a 1,2-glycol structure, and 1,2-glycols can be specifically and quantitatively cleaved by periodic acid or periodate ion. Treatment of PVOH with periodate should therefore break the chain into a number of fragments, bringing about a corresponding decrease in the effective molecular weight. All that is required is a measurement of the molecular weight of a specimen of PVOH before and after treatment with periodate.

METHOD[3]

For the determination of very high molecular weights, freezing-point depressions, boiling-point elevations, and vapor-pressure lowerings are too small for accurate measurement. Osmotic pressures are of a convenient order of magnitude, but measurements are time consuming. The technique to be used in this experiment depends on the determination of the intrinsic viscosity of the polymer. However, molecular-weight determinations from osmotic pressures are valuable in calibrating the viscosity method.

The coefficient of viscosity η of a fluid is defined in Exp. 4. It is conveniently measured, in the case of liquids, by determination of the time of flow of a given volume V of the liquid through a vertical capillary tube under the influence of gravity. For a virtually incompressible fluid such as a liquid, this flow is governed by Poiseuille's law in the form

$$\frac{dV}{dt} = \frac{\pi r^4 (p_1 - p_2)}{8\eta L}$$

where dV/dt is the rate of liquid flow through a cylindrical tube of radius r and length L and $(p_1 - p_2)$ is the difference in pressure between the two ends of the tube. In practice a viscosimeter of a type similar to that shown in Fig. 1 is used. Since $(p_1 - p_2)$ is proportional to the density ρ, it can be shown that for a given total volume of liquid

$$B = \eta/\rho t$$

$$\frac{\eta}{\rho} = Bt \qquad \eta = \rho Bt \qquad (3)$$

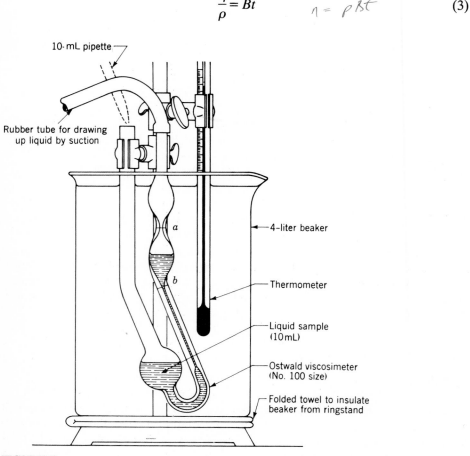

10-mL pipette

Rubber tube for drawing up liquid by suction

a

b

4-liter beaker

Thermometer

Liquid sample (10 mL)

Ostwald viscosimeter (No. 100 size)

Folded towel to insulate beaker from ringstand

FIGURE 1

Viscosimeter arrangement to be used if a glass-walled thermostat bath is not available.

where t is the time required for the upper meniscus to fall from the upper to the lower fiducial mark (a to b) and B is an apparatus constant that must be determined through calibration with a liquid of known viscosity (e.g., water). The derivation of this equation is similar to that given for Eq. (4-10). For high-precision work, it may be necessary to consider a kinetic-energy correction to Eq. (3). As in Eq. (4-16), we can write

$$\frac{\eta}{\rho} = BT - \frac{V}{8\pi Lt}$$

but the correction term is usually less than 1 percent of Bt and may be neglected in this experiment.

THEORY

Einstein showed that the viscosity η of a fluid in which small rigid spheres are dilutely and uniformly suspended is related to the viscosity η_0 of the pure fluid (solvent) by the expression[4]

$$\frac{\eta}{\eta_0} - 1 = \frac{5}{2}\frac{v}{V}$$

where v is the volume occupied by all the spheres and V the total volume. The quantity $(\eta/\eta_0) - 1$ is called the *specific viscosity* η_{sp}. For nonspherical particles the numerical coefficient of v/V is greater than $\frac{5}{2}$ but should be a constant for any given shape provided the rates of shear are sufficiently low to avoid preferential orientation of the particles.

The *intrinsic viscosity*, denoted by $[\eta]$, is defined as the ratio of the specific viscosity to the weight concentration of solute, in the limit of zero concentration:

$$[\eta] \equiv \lim_{c \to 0} \frac{\eta_{sp}}{c} = \lim_{c \to 0}\left(\frac{1}{c}\ln\frac{\eta}{\eta_0}\right) \tag{4}$$

where c is conventionally defined as the concentration in grams of solute per 100 mL of solution. In this case, the units for $[\eta]$ are $10^2\,\mathrm{cm^3\,g^{-1}}$. Both η_{sp}/c and $(1/c)(\ln \eta/\eta_0)$ show a reasonably linear concentration dependence at low concentrations. A plot of $(1/c)(\ln \eta/\eta_0)$ versus c usually has a small negative slope, while a plot of η_{sp}/c versus c has a positive and larger slope.[3] Careful work demands that either or both of these quantities be extrapolated to zero concentration, although $(1/c)(\ln \eta/\eta_0)$ for a single dilute solution will give a fair approximation to $[\eta]$.

If the internal density of the spherical particles (polymer molecules) is independent of their size (i.e., the volume of the molecule is proportional to its molecular weight), the intrinsic viscosity should be independent of the size of the particles, and hence of no value in indicating the molecular weight. This, however, is not the case; to see why, we must look into the nature of a polymer macromolecule as it exists in solution.

Statistically coiled molecules. A polymer such as PVOH contains many single bonds, around which rotation is possible. If the configurations around successive carbon atoms are independent and unrelated, it will be seen that two parts of the polymer chain more than a few carbon atoms apart are essentially uncorrelated in regard to direction in space. The molecule is then "statistically coiled" and resembles a loose tangle of yarn:

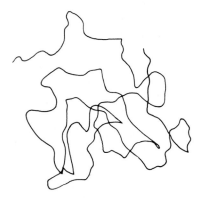

Simple statistical treatments[2] show that the mean distance between the two ends of the chain, and indeed also the effective mean diameter d of the coiled molecule regarded as a rough sphere, should be proportional to the square root of the chain length and thus to the square root of the molecular weight:

$$d \propto M^{1/2}$$

The volume v_m occupied by a molecule should then vary as $M^{3/2}$. The number of molecules in a given weight of polymer varies inversely with the molecular weight; hence the total volume of the spheres is

$$v \propto \frac{cV}{M} M^{3/2} = cVM^{1/2}$$

Therefore,

$$[\eta] = KM^{1/2} \tag{5}$$

where K is a constant. This treatment is much simplified; it ignores, among other things, the problem of "excluded volume"; that is, the fact that the chain cannot coil altogether randomly because it is subject to the restriction that no two parts of the chain may be at the same point in space at the same time. This restriction becomes more and more important the higher the molecular weight. Even more serious is the effect of solvent; the above treatment tacitly assumes a "poor solvent" which would barely get the polymer into solution. A "good solvent," by solvating the polymer, makes the size of the statistical coil increase faster with chain length than it otherwise would, owing to enhancement of the excluded volume effect. Accordingly, instead of Eq. (5) we might write

$$[\eta] = KM^a \tag{6}$$

where K and a are empirical parameters characteristic both of the polymer itself and of the solvent. The exponent a varies from about 0.5, for well-coiled polymer molecules in a poor solvent, to as much as 1.7, for a rigidly extended "rod-like" polymer molecule.

Flory and Leutner,[1] working with monodisperse specimens of PVOH differing from one another in molecular weight over a wide range (obtained by fractionating polydisperse commercial PVOH), established a correlation between the molecular weight, as determined from osmotic pressure measurements, and the intrinsic viscosity. They found that for PVOH in aqueous solution at 25°C,

$$[\eta] = 2.0 \times 10^{-4} M^{0.76}$$
$$M = 7.6 \times 10^4 [\eta]^{1.32} \tag{7}$$

Equation (7) also holds for a polydisperse sample of PVOH, but the molecular weight in this case is $\bar{M}_v$, the viscosity-average molecular weight defined below.

Number-average and viscosity-average molecular weight. For a polydisperse polymer, any determination of molecular weight must yield an average of some sort. When a colligative property such as osmotic pressure is used, the average is a number average:

$$\bar{M}_n = \frac{\displaystyle\int_0^\infty MP(M)\,dM}{\displaystyle\int_0^\infty P(M)\,dM} \tag{8}$$

where $P(M)$ is the molecular-weight distribution function; that is, $P(M)\,dM$ is proportional to the number of molecules with molecular weights between M and $M + dM$. However, the average obtained from intrinsic viscosity is not the same kind of average. It is called the viscosity average and is given by

$$(\bar{M}_v)^a = \frac{\displaystyle\int_0^\infty M^{1+a}P(M)\,dM}{\displaystyle\int_0^\infty MP(M)\,dM} \tag{9}$$

For a monodisperse polymer, $\bar{M}_v = \bar{M}_n = M$; for a polydisperse polymer the two kinds of averages are not equal but are related by a constant factor that depends on the distribution function $P(M)$ and the parameter a. A commonly encountered distribution function, and one that is likely to be valid for PVOH, is one that arises if the probability of a chain-termination reaction during the polymerization is constant with time and independent of the chain length already achieved. It is also the most likely function for the product resulting from cleavage with periodate if the head-to-head structures can be assumed to be randomly distributed along the PVOH chains. This distribution function is

$$P(M) = \frac{1}{\bar{M}_n} e^{-M/\bar{M}_n} \tag{10}$$

When this function is used as the weighting function in evaluating averages, it can be shown[3,5] that

$$\frac{\bar{M}_v}{\bar{M}_n} = [(1+a)\Gamma(1+a)]^{1/a}$$

where Γ is the gamma function. For $a = 0.76$ (the value given above for polyvinyl alcohol),

$$\frac{\bar{M}_v}{\bar{M}_n} = 1.89 \tag{11}$$

If a much exceeds 0.5, $\bar{M}_v$ is much closer to a weight-average molecular weight,

$$\bar{M}_w = \frac{\int_0^\infty M^2 P(M)\, dM}{\int_0^\infty M P(M)\, dM} \tag{12}$$

than it is to the number average $\bar{M}_n$. In fact, when $a = 1$, $\bar{M}_v$ and $\bar{M}_w$ are identical, and with the distribution assumed in Eq. (10), their ratio to $\bar{M}_n$ is 2.

Determination of frequency of head-to-head occurrences. We wish to calculate the fraction of linkages that are head to head (that is, the ratio of "backward" monomer units to total monomer units), on the assumption that degradation arises exclusively from cleavage of 1,2-glycol structures and that all such structures are cleaved. Let us denote this ratio by Δ. It is equal to the *increase* in the number of molecules present in the system, divided by the total number of monomer units represented by all molecules in the system. Since these numbers are in inverse proportion to the respective molecular weights,

$$\Delta = \frac{1/\bar{M}_n' - 1/\bar{M}_n}{1/M_0} \tag{13}$$

where $\bar{M}_n$ and $\bar{M}_n'$ are number-average molecular weights before and after degradation, respectively, and M_0 is the monomer weight, equal to 44 g cm^{-2}.

Thus,

$$\Delta = 44\left(\frac{1}{\bar{M}'_n} - \frac{1}{\bar{M}_n}\right) \tag{14}$$

Making use of Eq. (11) we can write

$$\Delta = 83\left(\frac{1}{\bar{M}'_v} - \frac{1}{\bar{M}_v}\right) \tag{15}$$

which permits viscosity averages to be used directly.

EXPERIMENTAL

Clean the viscosimeter thoroughly with chromic acid cleaning solution, rinse copiously with distilled water (use a water aspirator to draw large amounts of distilled water through the capillary), and dry with acetone and air. Immerse in a 25°C thermostat bath to equilibrate. Place a small flask of distilled water in a 25°C bath to equilibrate. Equilibration of water or solutions to bath temperature, in the amounts used here, should be complete in about 10 min.

If a stock solution of the polymer is not available, it should be prepared as follows. Weigh out accurately in a weighing bottle or on a watch glass 4.0 to 4.5 g of the dry polymer. Add it slowly, with stirring, to about 200 mL of hot distilled water in a beaker. To the greatest extent possible "sift" the powder onto the surface and stir gently so as not to entrain bubbles or produce foam. When all of the polymer has dissolved, let the solution cool and transfer it carefully and quantitatively into a 250-mL volumetric flask. Avoid foaming as much as possible by letting the solution run down the side of the flask. Make the solution up to the mark with distilled water and mix by *slowly* inverting a few times. If the solution appears contaminated with insoluble material that would possibly interfere with the viscosity measurements, filter it through Pyrex wool. (Since making up this solution may take considerable time, it is suggested that the calibration of the viscosimeter with water be carried out concurrently.)

In all parts of this experiment, avoid foaming as much as possible. This requires careful pouring when a solution is to be transferred from one vessel to another. It is also important to be meticulous about promptly and *very thoroughly* rinsing glassware that has been in contact with polymer solution, for once the polymer has dried on the glass surface it is quite difficult to remove. Solutions of the polymer should not be allowed to stand for many days before being used as they may become culture media for airborne bacteria.

Pipette 50 mL of the stock solution into a 100-mL volumetric flask, and make up to the mark with distilled water, observing the above precautions to prevent foaming. Mix, and place in the bath to equilibrate. In this and other dilutions, rinse the pipette *very thoroughly* with water and dry with acetone and air.

To cleave the polymer, pipette 50 mL of the stock solution into a 250-mL flask and add up to 25 mL of distilled water and 0.25 g of solid KIO_4. Warm the flask to about 70°C, and stir until all the salt is dissolved. Then clamp the flask in a thermostat bath and stir until the solution is at 25°C. Transfer quantitatively to a 100-mL volumetric flask, and make up to the mark with distilled water. Mix carefully, and place in the bath to equilibrate. (This operation can be carried out while viscosity measurements are being made on the uncleaved polymer.)

At this point, two "initial" solutions have been prepared: 100 mL of each of two aqueous polymer solutions of the same concentration (~0.9 g/100 mL), one cleaved with periodate and one uncleaved. To obtain a second concentration of each material, pipette 50 mL of the "initial" solution into a 100-mL volumetric flask and make up to the mark with distilled water. Place all solutions in the thermostat bath to equilibrate. The viscosity of both solutions of each polymer (cleaved and uncleaved) should be determined. If time permits, a third concentration (equal to one-quarter of the initial concentration) should also be prepared and measured.

The recommended procedure for measuring the viscosity is as follows:

1. The viscosimeter should be mounted vertically in a constant-temperature bath so that both fiducial marks are visible and below the water level. If a glass-walled thermostat bath which will allow readings to be made with the viscosimeter in place is not available, fill a large beaker or battery jar with water from the bath and set it on a dry towel for thermal insulation. The temperature should be maintained within ±0.1°C of 25°C during a run. This will require periodic small additions of hot water.

2. Pipette the required quantity of solution (or water) into the viscosimeter. Immediately rinse the pipette copiously with water, and dry it with acetone and air before using again.

3. By mouth suction through a rubber tube, draw the solution up to a point well above the upper fiducial mark. Release the suction and measure the flow time between the upper and lower marks with a stopwatch. Obtain two or more additional runs with the same filling of the viscosimeter. Three runs agreeing within about 1 percent should suffice.

4. Each time the viscosimeter is emptied, rinse it *very thoroughly* with distilled water, then dry with acetone and air. Be sure to remove *all* polymer with water before introducing acetone.

If time permits, the densities of the solutions should be measured with a Westphal balance. Otherwise, the densities may be taken equal to that of the pure solvent without introducing appreciable error.

CALCULATIONS

At 25°C, the density of water is 0.9970 g cm^{-3} and the coefficient of viscosity η_0 is 0.8937 cP. Using your time of flow for pure water, determine the apparatus constant B in Eq. (3).

For each of the polymer solutions studied, calculate the viscosity η and the concentration c in grams of polymer per 100 mL of solution. Then calculate η_{sp}/c and $(1/c)(\ln \eta/\eta_0)$. Plot both η_{sp}/c and $(1/c)(\ln \eta/\eta_0)$ versus c and extrapolate linearly to $c = 0$ to obtain $[\eta]$ for the original and for the degraded polymer.

Calculate $\bar{M}_v$ and $\bar{M}_n$ for both the original polymer and the degraded polymer, then obtain a value for Δ. Report these figures together with the polymerization temperature for the sample studied. Discuss the relationship between Δ and the rate constants k_α and k_β.

If a group of students have studied a variety of samples polymerized at different temperatures, it will be possible to obtain estimates of the difference in thermal activation energies ΔE_a and the steric factor S [see Eq. (2)]. It may be difficult to obtain a quantitative interpretation unless the details of the polymerization (amount of initiator, polydispersity, etc.) are well known. However, you should comment qualitatively on your results in the light of what is known about the theory of chain polymerization.[2,6]

DISCUSSION

Equations (14) and (15) were derived on the assumption that a 1,2-glycol structure will result from every abnormal monomer addition. What is the result of two successive abnormal additions? Can you derive a modified expression on the assumption that the probability of an abnormal addition is independent of all previous additions? Would such an assumption be reasonable? Are your experimental results (or, indeed, those of Flory and Leutner) precise enough to make such a modification significant in experimental terms?

APPARATUS

Ostwald viscosimeter; two 100- and two 250-mL volumetric flasks; a 10- and a 50-mL pipette; pipetting bulb; 250-mL glass-stoppered flask; one 100- and two 250-mL beakers; stirring rod; bunsen burner; tripod stand and wire gauze; 0 to 100°C thermometer; length of gum-rubber tubing. If polymer stock solution is to be prepared by student: weighing bottle; funnel; Pyrex wool.

Glass-walled thermostat bath at 25°C or substitute; polyvinyl alcohol (M.W. $\sim$60 000 to 80 000; 5 g of solid or 200 mL of solution containing 18 g L^{-1}); KIO$_4$ (1 g); chromic acid cleaning solution (50 mL); Westphal balance, if needed. The polyvinyl alcohol can be obtained from Polysciences Inc., 400 Valley Road, Warrington, PA 18976 and many other suppliers.

REFERENCES

1. P. J. Flory and F. S. Leutner, *J. Polym. Sci.* **3**, 880 (1948); **5**, 267 (1950).
2. P. J. Flory, "Principles of Polymer Chemistry," Cornell University Press, Ithaca, N.Y. (1953).
3. J. F. Johnson, J. R. Martin, and R. S. Porter, "Determination of Viscosity," in A. Weissberger and B. W. Rossiter (eds.), "Techniques of Chemistry: Vol. I. Physical Methods of Chemistry," part VI, chap. 2, Wiley-Interscience, New York (1977).
4. A. Einstein, "Investigations on the Theory of Brownian Movement," chap. III, Dover, New York (1956).
5. J. R. Schaefgen and P. J. Flory, *J. Amer. Chem. Soc.* **70**, 2709 (1948).
6. R. W. Lenz, "Organic Chemistry of Synthetic High Polymers," pp. 261–271, 305–369, Wiley-Interscience, New York (1967).

GENERAL READING

F. W. Billmeyer, Jr., "Textbook of Polymer Science," 3d ed., Wiley-Interscience, New York (1984).

EXPERIMENT 30
HELIX–COIL TRANSITION IN POLYPEPTIDES

Polymer molecules in solution can be found in many different geometric conformations, and there exist a variety of experimental methods (e.g., viscosity, light scattering, optical rotation) for obtaining information about these conformations.[1] In this experiment, the measurement of optical rotation will be used to study a special type of conformational change that occurs in many polypeptides.

The two important kinds of conformation of a polypeptide chain are the helix and the random coil. In the helical form the amide hydrogen of each "amide group"

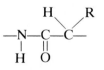

is internally hydrogen-bonded to the carbonyl oxygen of the third following amide group along the chain. Thus, this form involves a quite rigid, rod-like structure (see Fig. 1). Under different conditions, the polypeptide molecule may be in the form of a statistically random coil (see Exp. 29). The stable form of the polypeptide will depend on several factors—the nature of the peptide

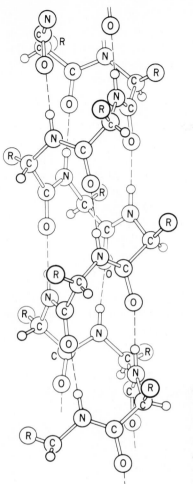

FIGURE 1
The Pauling-Corey alpha helix. In addition to the right-handed helix shown, a left-handed one is also possible (with the same *l*-amino acids). In proteins, the helix always seems to be right-handed. [*Reprinted by permission from L. Pauling, "The Nature of the Chemical Bond," p.* 500, *Cornell University Press, Ithaca, N.Y.* (1960).]

groups, the solvent, and the temperature. For example, poly-γ-benzyl-*l*-glutamate† (PBG)

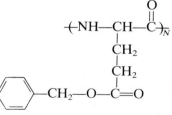

---

† Note that the usual chemical description of polypeptides is in terms of amino-acid residues, as in the PBG formula shown, rather than in terms of amide groups, which are more convenient for the present discussion.

has a helical conformation when dissolved in ethylene dichloride at 25°C, but it is in the random-coil form when dissolved in dichloroacetic acid at the same temperature. This difference is quite reasonable, since a hydrogen-bonding solvent like dichloroacetic acid can form strong hydrogen bonds with the amide groups and thus disrupt the internal hydrogen bonds that are necessary for the helical form. For a mixed solvent of dichloroacetic acid and ethylene dichloride, PBG can be made to transform from the random coil to the helix by raising the temperature over a fairly narrow range. This rapid reversible transition will be investigated here.

As we shall see below, the solvent plays a crucial role in determining which form is stable at low temperatures. When the helix is stable at low temperatures, the transition to the random coil at high temperatures is called a "normal" transition. For the case in which the random coil is the more stable form at low temperatures, the transition is called an "inverted" transition.

THEORY

We wish to present here a very simplified and approximate statistical-mechanical theory of the helix–coil transition. The treatment is closely related to that given by Davidson.[2] Let us consider the change

$$\text{Coil} \cdot NS = \text{helix} + NS \tag{1}$$

where the polymer molecules each consist of N segments (monomer units) and are dissolved in a solvent S that can hydrogen-bond to the amide groups when the chain is in the random-coil form. We shall (for convenience) artificially simplify the physical model of internal hydrogen bonding in the helix by assuming that the hydrogen bond formed by each amide group is with the *next* amide group along the chain (see Fig. 2) rather than the third following amide group as in the actual helix. We can then assume that the chain segments are independent of each other. Let q_1 be the molecular partition function of a *segment in the random-coil form* with a solvent molecule S hydrogen-bonded to it, and q_2 be the product of the partition function of a segment in the *helical form* times the partition function of a "free" solvent molecule S. We will then define a parameter s by

$$q_2 = sq_1 \tag{2}$$

Note that s is the ratio of partition functions and many of the contributions to q_1 and q_2 (e.g., vibrational terms) will cancel out. The principal contributions to s will involve differences between the helix and random-coil form, and we may guess that s can be represented in the general form[2]

$$s = s_0 e^{-\varepsilon/kT} \tag{3}$$

where $\varepsilon = \varepsilon_h - \varepsilon_c$ is the energy change per segment and s_0 is related to the entropy change per segment for the change in state of Eq. (1).

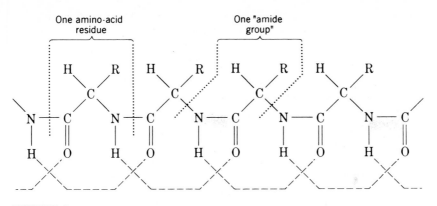

FIGURE 2
Hypothetical model of an internally hydrogen-bonded chain, with the simplification that *adjacent* "amide groups" are connected by hydrogen bonds. Note the distinction between the *amino-acid residue* and the *"amide group"*; the latter is the more convenient unit of structure for the present discussion.

First-order transition model. To simplify the model even further, let us assume for the time being that a given polymer molecule is either completely in the helical form or completely in the random-coil form. This assumption will lead to a first-order transition between these two forms. The partition function for a random-coil molecule q_c is then given by

$$q_c = q_1^N \tag{4}$$

while that for a helical molecule q_h will be

$$q_h = q_2^N = s^N q_1^N \tag{5}$$

if one neglects end effects (e.g., the fact that at the beginning of the chain the first carbonyl oxygen is not involved in an internal hydrogen bond).

At constant pressure, the two forms will be in equilibrium with each other at some temperature T^* at which $\Delta G = 0$ for the change in state (1). Since ΔV will be quite small for change (1), we can take $\Delta A \cong 0$ and $\bar{A}_c = \bar{A}_h$ at T^*, where $\bar{A}$ is the Helmholtz free energy per mole of polymer. For independent polymer molecules, we have[3]

$$\bar{A} = -kT \ln Q = -kT \ln q^{N_0} = -RT \ln q \tag{6}$$

where N_0 is Avogadro's number, q is the partition function for a single polymer molecule, and Q is the (canonical) partition function for a mole of molecules . Thus, one finds that $q_h = q_c$ (or $s = 1$) at T^*. With the use of Eq. (3), the value of T^* can be related to the parameters ε and s_0:

$$\frac{\varepsilon}{kT^*} = \ln s_0 \tag{7}$$

For an inert (non-hydrogen-bonding) solvent, one observes experimentally a "normal" transition: the helix, which is stable at low temperatures, is transformed at higher temperatures into the random coil. This case is represented by our model when $\varepsilon < 0$ and $s_0 < 1$. Since the solvent is inert, random-coil segments are essentially unbonded, whereas helical segments have internal hydrogen bonds and thus a lower energy $(\varepsilon_h - \varepsilon_c < 0)$. In the random-coil polymer molecule, considerable rotation can occur about the single bonds in the chain skeleton, but the helical form has a rather rigid structure. Therefore, there is a decrease in the entropy per segment on changing from the relatively free rotational configurations of the flexible random coil to the more rotationally restricted helix (and thus $s_0 < 1$). From Eqs. (3) to (6), it follows that $s > 1$ at low temperatures $(T < T^*)$ and thus $\tilde{A}_h < \tilde{A}_c$ and the helix is the more stable form. At high temperatures $(T > T^*)$, $s \cong s_0 < 1$ and the random coil is the more stable form.

For an active (hydrogen-bonding) solvent, one observes experimentally an "inverted" transition: the random coil is stable at low temperatures and changes into the helix when the solution is heated. This case is represented by our model when $\varepsilon > 0$ and $s_0 > 1$. Here the solvent plays a dominant role since a solvent molecule is strongly hydrogen-bonded to each random-coil segment. In terms of energy, there is little difference between polymers with random-coil segments and with helical segments since there is comparable hydrogen bonding in both forms. It is now almost impossible to predict the sign of ε, but it is certainly quite reasonable to find that the energy change associated with (1) may be positive $(\varepsilon_h - \varepsilon_c > 0)$. Although a helical segment will still have a lower entropy than a random-coil segment owing to the rigidity of the helix, the solvated random coil is now less flexible than previously. More importantly, the free solvent molecules will have a higher entropy than solvent molecules that are bonded to the random coil. Thus the entropy change associated with (1) is positive because of the release of S molecules (and therefore $s_0 > 1$). From Eqs. (3) to (6), we see that at low temperatures $s < 1$ and the random coil will be stable in this inverted case.

Recalling that ΔV is small so that $\Delta G \cong \Delta A$, we can make use of Eqs. (3) to (6) to obtain

$$\Delta \tilde{S} = -\left(\frac{\partial \Delta \tilde{G}}{\partial T}\right)_p \cong R \frac{\partial}{\partial T}\left(T \ln \frac{q_h}{q_c}\right) = NR \ln s_0 \tag{8}$$

The entropy change *per mole of monomer*, ΔS_m, is then given by

$$\Delta S_m \equiv \frac{\Delta \tilde{S}}{N} = R \ln s_0 \tag{9}$$

Since $\Delta G = 0$ for the first-order transition at p and T^*, we have for the enthalpy change *per mole of monomer*

$$\Delta H_m = T^* \, \Delta S_m = RT^* \ln s_0 = N_0 \varepsilon \tag{10}$$

The first-order theory presented above is related to the treatment of Baur and Nosanow,[4] who used a similar but perhaps physically less realistic model to derive the same results [that is, Eqs. (7), (9), and (10)].†

Cooperative transition model. It is an experimental fact that the helix–coil transition is *not* a first-order transition. In order to account for this, one must adopt a more realistic model by eliminating the assumption that an entire polymer molecule is all in a given form. Indeed, a polymer molecule can have some sections that are helical and others that are randomly coiled. The formation of a helical region will be a cooperative process. Since segment (i.e., amide group) n is hydrogen-bonded to segment $n + 3$, $n + 1$ to $n + 4$, $n + 2$ to $n + 5$, etc., it is difficult to initiate this ordered internal bonding; but once a single hydrogen-bond link is made, the next ones along the chain are much easier to achieve (a sort of "zipper" effect).

No attempt will be made here to develop the details of such a cooperative model. Readers with a sufficient background in statistical mechanics should refer to Davidson's presentation[2] or to the original papers in the literature.[5] The crucial idea is to introduce a "nucleation" parameter σ which is independent of temperature and very small in magnitude. It is then assumed that the partition function q_2 is changed to the value σq_2 for the *initial* segment of a helical section (i.e., the internally bonded segment directly adjacent to a solvent-bonded segment). Physically, this means that it is difficult to initiate a new helical section. For such a model, it can be shown[2,5] that X_h, the mole fraction of the segments in helical sections, will undergo a very rapid but *continuous* change from $X_h \cong 0$ when $s < 1$ to $X_h \cong 1$ when $s > 1$. For $\sigma \leqslant 10^{-2}$, X_h is very well represented by the approximation

$$X_h \begin{cases} \approx sF(s) & \text{for } s < 1 \\ = \frac{1}{2} & \text{for } s = 1 \\ \approx F(s) & \text{for } s > 1 \end{cases} \tag{11}$$

where the function $F(s)$ is given by

$$F(s) \equiv \frac{1}{2}\left\{1 + \frac{(s-1) + 2\sigma}{[(s-1)^2 + 4\sigma s]^{1/2}}\right\} \tag{12}$$

The resulting variation of X_h with the parameter s is shown in Fig. 3, which implies that a rapid cooperative transition occurs in the vicinity of a critical temperature T^* (the value of T for which $s = 1$). It is the fact that σ is very small that ensures a sharp transition. If σ were taken to be 1 (which

---

† In addition, Baur and Nosanow showed for the inverted case that the helical form will be stable only between T^* and another much higher transition temperature T', where the helix reverts to the random-coil form. (For PBG, it is estimated[5] that $T' \cong 700$ K.) This is physically reasonable in terms of our model: at very high temperatures, all hydrogen bonds (internal or with solvent) will be broken and the random coil will predominate owing to its higher entropy.

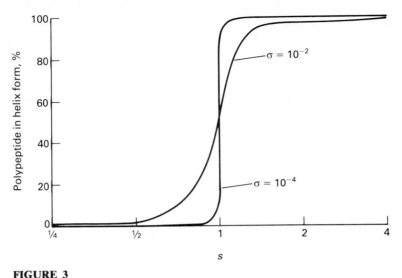

FIGURE 3

Percentage of polypeptide in helix form, as a function of the parameter s (on a log scale), for $\sigma = 10^{-4}$ and $\sigma = 10^{-2}$.

corresponds to ignoring the nucleation problem), this model would yield $X_h = s/(1 + s)$, which corresponds to a very gradual transition. For high-molecular-weight PGB, a value of $\sigma \simeq 2 \times 10^{-4}$ seems to be appropriate to obtain a fairly good quantitative fit to the experimental data.[2] It should be noted that the transition is sharpest, and the model is most successful, when polypeptides with $N \geqslant 1000$ are used.

METHOD

Since the helical form of a polypeptide is rigid and rod-like, its solutions are more viscous than corresponding solutions of the random-coil form, and the transition can be detected by measuring the intrinsic viscosity (see Exp. 29). However, the highly ordered structure of the helical form causes it to have an optical-rotatary power markedly different from that of a random coil. Indeed, use of optical rotation for following the transition gives better results than does intrinsic viscosity, since the optical rotation is directly related to the fraction X_h of segments in the helical form, whereas the hydrodynamic properties of the polymer molecule change substantially at the first appearance of random-coil regions in the helix (in effect, causing the otherwise straight rod to bend randomly in one or several places).

The polarimeter and its operation are described in Chapter XVIII. Although not stressed there, it is obviously necessary to use a polarimeter in which the temperature of the sample tube can be well controlled. The *specific rotation* $[\alpha]^t$ as determined with the sodium D doublet at $t°C$ is defined by Eq. (XVIII-9), and we note here that for a solution containing two solutes (helix and random coil) one can show that

$$[\alpha]^t = X_h[\alpha]_h^t + X_c[\alpha]_c^t \qquad (13)$$

where $[\alpha]_h^t$ and $[\alpha]_c^t$ are the specific rotations of helix and random-coil form. In a narrow range of temperature near T^*, $[\alpha]_h^t$ and $[\alpha]_c^t$ will be essentially constant. Thus measurements of $[\alpha]^t$ will determine the mole fraction X_h as a function of T.

EXPERIMENTAL

A solution of poly-γ-benzyl-l-glutamate (PBG) in a mixed solvent of dichloro-acetic acid (DCA) and ethylene dichloride, 76 volume percent in DCA, will be used. The dichloroacetic acid should be purified by vacuum distillation; the ethylene dichloride may be purified by conventional distillation in air. Mix the proper volumes of the two liquids to make about 50 mL of the desired solvent. In order to prepare approximately 25 mL of a 2.5 weight percent polymer solution, weigh accurately a sample of about 1 g of PBG and dissolve it in about 38 g of solvent (also accurately weighed). The resulting solution is very viscous; it must be mixed well and then allowed to stand for two or three days until any small undissolved particles have settled and the solution is clear. During this period, the solution should be placed in a polyethylene bottle and stored in a refrigerator. The remaining solvent should also be stored for future density determinations.

Since the solution is difficult to prepare, it is recommended that it be made available to the student. **Warning: The solvent is extremely corrosive.**

Detailed instructions for using the polarimeter and the temperature-regulating bath will be provided in the laboratory. After determining the path lengths of the cells to be used, fill a cell with pure solvent and zero-adjust the instrument. Next, rinse and fill a cell with the polymer solution.†

After mounting the solution cell in the polarimeter, it should not be necessary to move this cell during the rest of the experiment. Provisions must be made for measuring the temperature of the sample without obscuring the light path. Begin measurements with the temperature at about 10°C and read the optical rotation α at reasonably spaced temperature intervals until about 50°C is reached. The exact temperature intervals to be used are at your discretion, but should be close enough to follow the essential features of the variation of optical rotation; small intervals (perhaps 1 or 2°C) should be used when the optical rotation is changing rapidly with temperature (near 30°C), and larger intervals may be used when the changes with temperature are small. Be sure to allow sufficient time for thermal equilibration between readings.

---

† A difficulty in filling these cells is the problem of avoiding any air bubbles that might obscure part of the light path. Since the solution is viscous, this can be a tedious job and may require some time. If possible, it is wise to fill the cells about an hour before use and allow them to stand.

In order to calculate $[\alpha]^t$, the density of the solutions must be known at each temperature. The density of the solution is close enough to that of the solvent so that they can be assumed to be identical. The density of the solvent should be determined with a pycnometer for at least three different temperatures in the range 10 to 50°C. The volume of the pycnometer should first be determined by weighing it empty and also filled with distilled water at a known temperature. Since the density of water as a function of temperature is well known, the volume of the pycnometer can be determined. This procedure (see Exp. 9) is then repeated using the solvent, and since the volume of the pycnometer is now known the density can be calculated. The solvent density should be plotted against temperature and the best straight line drawn through the points obtained.

Record the composition of the solution and the molecular weight (or N value) of the polymer used.

CALCULATIONS

For each data point, calculate the specific rotation at temperature t from the expression

$$[\alpha]^t = 100 \frac{\alpha}{Lp\rho} \tag{14}$$

where α is the observed rotation in degrees, L is the path length of the cell in decimeters, p is the weight percent of solute in the solution, and ρ is the density of the solution in $g\,cm^{-3}$. These density values are interpolated from the ρ-versus-T plot for the solvent.

Construct a plot of $[\alpha]^t$ versus temperature, and determine the limiting (asymptotic) values of the rotation: $[\alpha]_h$ from the high-temperature plateau, $[\alpha]_c$ from the low-temperature plateau. If it is assumed that these represent *temperature-independent* specific rotations for the helix and random-coil forms, respectively, Eq. (13) gives for X_h:

$$X_h = \frac{[\alpha]^t - [\alpha]_c}{[\alpha]_h - [\alpha]_c} \tag{15}$$

Using this equation, you can add to your plot a scale of X_h values so that the plot will also represent X_h versus T. Determine and report the "transition" temperature T^* at which $X_h = 0.5$.

DISCUSSION

If one *arbitrarily* defined an equilibrium constant by $K = X_h/X_c$ and used the data in this experiment together with the Van't Hoff equation $d \ln K/d(1/T) = \Delta \bar{H}^0/R$ to determine the value of $\Delta \bar{H}^0$, to what (if anything) would such an enthalpy change refer?

APPARATUS

Polarimeter, with one or more optical cells (complete with windows, caps, and Teflon or neoprene washers); constant-temperature control for range 10 to 50°C; pycnometer; 25-mL pipette for filling the pycnometer and a small pipette for filling the cell; rubber bulb for pipetting; 0°C to 50°C thermometer; 250-mL beaker; gum-rubber tubing; lint-free tissues for wiping cell windows; 50-mL polyethylene bottles for PBG solution and excess solvent.

Poly-γ-benzyl-l-glutamate, M.W. 300 000 (1 g of solid or 25 mL of 2.5 weight percent solution); ethylene dichloride–dichloroacetic acid solvent (76 volume percent DCA) stored in polyethylene bottle; wash acetone. The polypeptide can be obtained from Sigma Chemical Co., P.O. Box 14508, St. Louis, MO 63178. Check the current "Chem. Sources" for other suppliers.

REFERENCES

1. M. A. Stahmann (ed), "Polyamino Acids, Polypeptides and Proteins," Wisconsin University Press, Madison, WI. (1962).
2. N. Davidson, "Statistical Mechanics," pp. 385–393, McGraw-Hill, New York (1962).
3. P. W. Atkins, "Physical Chemistry," 3d ed., p. 518, Freeman, New York (1986).
4. M. E. Baur and L. H. Nosanow, *J. Chem. Phys.* **38,** 578 (1963).
5. B. H. Zimm and J. K. Bragg, *J. Chem. Phys.* **31,** 526 (1959); B. H. Zimm, P. Doty, and K. Iso, *Proc. Natl. Acad. Sci.* **45,** 160A (1959).

GENERAL READING

P. Doty, J. H. Bradbury, and A. M. Holtzer, *J. Amer. Chem. Soc.* **78,** 947 (1956); P. Doty and J. T. Yang: *J. Amer. Chem. Soc.* **78,** 498 (1956).
C. H. Bamford, A. Elliott, and W. E. Handy, "Synthetic Polypeptides," Academic Press, New York (1956).

CHAPTER
XII

ELECTRIC AND MAGNETIC PROPERTIES

EXPERIMENTS

31. Dipole moment of polar molecules in solution
32. Dipole moment of HCl molecules in the gas phase
33. Magnetic susceptibility
34. NMR determination of paramagnetic susceptibility

EXPERIMENT 31
DIPOLE MOMENT OF POLAR MOLECULES IN SOLUTION

When a substance is placed in an electric field, such as exists between the plates of a charged condenser, it becomes to some extent electrically polarized. The polarization results at least in part from a displacement of electron clouds relative to atomic nuclei; polarization resulting from this cause is termed *electronic polarization*. For molecular substances atomic polarization may also be present owing to a distortion of the molecular skeleton. Taken together these two kinds of polarization are called *distortion polarization*. Finally, when molecules possessing permanent dipoles are present in a liquid or gas, application of an electric field produces a small preferential orientation of the dipoles in the field direction, leading to *orientation polarization*.

The permanent dipole moment μ of a polar molecule, as a solute molecule in a liquid solution in a nonpolar solvent or as a molecule in a gas, can be determined experimentally from measurements of the *dielectric constant* κ. This quantity is the ratio of the electric permittivity ε of the solution or gas to the electric permittivity ε_0 of a vacuum $(8.854 \times 10^{-12} \, \text{F m}^{-1})$:

$$\kappa = \frac{\varepsilon}{\varepsilon_0} \tag{1}$$

For that reason the dielectric constant is also known as the *relative permittivity* (with symbol ε_r). The dielectric constant is determined with a capacitance cell, incorporating a condenser of fixed dimensions in a suitable containment vessel. If C is the capacitance of the cell when the medium between the condenser plates is the solution or gas of interest, and C_0 is the capacitance when the medium is a vacuum, then[1]

$$\kappa = \frac{C}{C_0} \tag{2}$$

The present experiment deals with polar molecules in solution in a nonpolar solvent. Experiment 32 deals with polar molecules in a gas. The basic theoretical framework is the same for both.

Either of two systems can be studied in this experiment: (1) *o*- and *m*-dichlorobenzene, or (2) succinonitrile and propionitrile. In the first case, the results can be compared with the vector sum of C—Cl bond moments obtained from the known dipole moment of monochlorobenzene. In the second case, one obtains direct information about internal rotation in a simple 1,2-disubstituted ethane

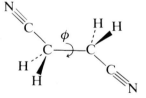

The angle ϕ is defined so that $\phi = 180°$ when the two C≡N groups eclipse each other (on viewing the molecule along the C—C axis). Thus, $\phi = 0$ corresponds to the most stable trans configuration, and $\phi = \pm 120°$ corresponds to two equivalent gauche configurations. Hindered rotation about the C—C sigma bond can be taken into account by defining temperature-dependent mole fractions in the trans, gauche plus, and gauche minus states. The dipole moment of each of these forms can be predicted from the C—C≡N bond moment measured in propionitrile (ethyl cyanide).

THEORY

An electric dipole[2] consists of two point charges, $-Q$ and $+Q$, with a separation represented by a vector $\mathbf{r}$, the positive sense of which is from $-Q$ to

$+Q$. The electric dipole moment is defined by

$$\mathbf{m} = Q\mathbf{r} \qquad (3)$$

A molecule possesses a dipole moment whenever the center of gravity of negative charge does not coincide with the center of gravity of positive charge. In an electric field, all molecules have an induced dipole moment (which is aligned approximately parallel to the field direction) owing to distortion polarization. In addition, *polar molecules* have a *permanent dipole moment* (i.e., a dipole moment that exists independently of any applied field) of constant magnitude μ and a direction that is fixed relative to the molecular skeleton.

The resultant (vector sum) electric moment of the medium per unit volume is known as the *polarization* $\mathbf{P}$.[3] For an isotropic medium, $\mathbf{P}$ is parallel to the electric field intensity $\mathbf{E}$, and to a first approximation it is proportional to $\mathbf{E}$ in magnitude. For a pure substance, $\mathbf{P}$ is given by

$$\mathbf{P} = \overline{\mathbf{m}}\,\frac{N_0}{\overline{V}} = \overline{\mathbf{m}}\,\frac{N_0\rho}{M} \qquad (4)$$

where $\overline{\mathbf{m}}$ is the average dipole moment of each molecule, N_0 is Avogadro's number, $\overline{V}$ is the molar volume, M is the molecular weight, and ρ is the density.

The moment of a polarized dielectric is equivalent to a moment that would result from electric charges of opposite sign on opposite surfaces of the dielectric. In a condenser these "polarization charges" induce equal and opposite charges in the metal plates that are in contact with them. These induced charges are in addition to the charges that would be present at the same applied potential for the condenser with a vacuum between the plates. Accordingly, the capacitance is increased by the presence of a polarizable medium. Thus the dielectric constant κ is greater than unity. By electrostatic theory it can be shown[3] that

$$\mathbf{D} = \varepsilon\mathbf{E} = \varepsilon_0\mathbf{E} + \mathbf{P} \qquad (5a)$$

or

$$\kappa\mathbf{E} = \mathbf{E} + \frac{1}{\varepsilon_0}\mathbf{P} \qquad (5b)$$

where $\mathbf{D}$ is the electric displacement and $\mathbf{E}$ is the electric field strength.

The average dipole moment $\overline{\mathbf{m}}$ for an atom or molecule in the medium is given by

$$\overline{\mathbf{m}} = \alpha\mathbf{F} \qquad (6)$$

where we denote by $\mathbf{F}$ the *local* electric field and α is the polarizability (which is independent of field intensity if the field is not so intense that saturation is incipient). To a good approximation $\mathbf{F}$ may be taken as the electric field intensity at the center of a spherical cavity in the dielectric, within which the atom or molecule is contained. Thus $\mathbf{F}$ is the resultant of $\mathbf{E}$ and an additional contribution due to the polarization charges on the surface of the spherical cavity; electrostatic theory[4] gives

$$\mathbf{F} = \mathbf{E} + \frac{1}{3\varepsilon_0}\mathbf{P} \tag{7}$$

Combining this with Eq. (5b) we obtain

$$\mathbf{F} = \frac{\kappa + 2}{\kappa - 1}\frac{1}{3\varepsilon_0}\mathbf{P} \tag{8}$$

and using Eqs. (4) and (6) we obtain for a pure substance

$$\frac{\kappa - 1}{\kappa + 2}\frac{M}{\rho} = \frac{1}{3\varepsilon_0}N_0\alpha \equiv P_M \tag{9}$$

This is the *Clausius–Mossotti* equation. The quantity P_M is called the *molar polarization* and has the dimensions of volume per mole.

If the molecules have no permanent dipole moment, only distortion polarization takes place. The corresponding polarizability is denoted by α_0. If each molecule has a permanent dipole moment of magnitude μ, there is a tendency for the moment to become oriented parallel to the field direction, but this tendency is almost completely counteracted by thermal motion which tends to make the orientation random. The component of the moment in the field direction is $\mu\cos\theta$, where θ is the angle between the dipole orientation and the field direction. The potential energy U of the dipole in a local field of intensity $\mathbf{F}$ is $-(\mu\cos\theta)F$, which is small in comparison with kT under ordinary experimental conditions. By use of the Boltzmann distribution, the average component of the permanent moment in the field direction is found to be

$$\overline{m}_\mu = [\mu\cos\theta\, e^{-U/kT}]_{av} = [\mu\cos\theta\, e^{\mu F\cos\theta/kT}]_{av}$$

$$\cong \mu\left[\cos\theta\left(1 + \frac{\mu F\cos\theta}{kT}\right)\right]_{av}$$

where the average is taken over all orientations in space. The average of $\cos\theta$ vanishes, but the average of its square is $\frac{1}{3}$; accordingly, as found by Debye,

$$\overline{m}_\mu = \frac{\mu^2}{3kT}F \tag{10}$$

Thus the total polarizability is given by

$$\alpha = \alpha_0 + \frac{\mu^2}{3kT} \tag{11}$$

and the molar polarization can be written in the form

$$P_M = \frac{\kappa - 1}{\kappa + 2} \frac{M}{\rho} = \frac{1}{3\varepsilon_0} N_0 \left(\alpha_0 + \frac{\mu^2}{3kT} \right)$$
$$= P_d + P_\mu \tag{12}$$

where P_d and P_μ are, respectively, the distortion and orientation contributions to the molar polarization:

$$P_d = \frac{1}{3\varepsilon_0} N_0 \alpha_0 \quad \text{and} \quad P_\mu = \frac{1}{3\varepsilon_0} N_0 \frac{\mu^2}{3kT} \tag{13}$$

Equation (12) is known as the Debye–Langevin equation.

Immediately it is clear that a plot of P_M versus $1/T$ utilizing measurements of κ as a function of temperature will yield both α_0 and μ. This technique is readily applicable to gases. In principle it is applicable to liquids and solutions also but it is seldom convenient owing largely to the small temperature range accessible between the melting point and boiling point.

In the foregoing derivation a static (dc) electric field was assumed. The equations apply also to alternating (ac) fields, provided the frequency is low enough to enable the molecules possessing permanent dipoles to orient themselves in response to the changing electric field. Above some frequency in the upper radio-frequency or far-infrared range, the permanent dipoles can no longer follow the field, and the orientation term in Eq. (12) disappears. At infrared and visible frequencies the dielectric constant cannot be measured by the use of a condenser. However, it is known from electromagnetic theory that in the absence of high magnetic polarizability (which does not exist at these frequencies for any ordinary materials)

$$\kappa = n^2$$

where n is the *index of refraction*. We then obtain from Eq. (12) the relation of Lorentz and Lorenz:

$$R_M \equiv \frac{n^2 - 1}{n^2 + 2} \frac{M}{\rho} = \frac{1}{3\varepsilon_0} N_0 \alpha_0 = P_d \tag{14}$$

where R_M is known as the *molar refraction*. Thus the distortion polarizability α_0 can in principle be obtained from a measurement of the refractive index at some wavelength in the far infrared where the distortion polarization is virtually complete. However, it is not experimentally convenient to measure the index of refraction in the infrared range. The index of refraction n_D measured with the visible sodium D line can usually be used instead. Although the atomic contribution to the distortion polarization is absent in the visible, and the electronic contribution is not in all cases at its dc value, these variations in the distortion polarization are small and usually negligible in comparison with the orientation polarization. The measurement of n_D can be made conveniently in an Abbe or other type of refractometer (see Chapter

XVIII). Thus, P_d can be measured independently from P_M, and P_μ can be determined by difference. By this means μ can be determined from measurements made at a single temperature.

Measurements in solution. We are here concerned with a dilute solution containing a polar solute 2 in a nonpolar solvent 1. The molar polarization can be written

$$P_M = X_1 P_{1M} + X_2 P_{2M} = \frac{\kappa - 1}{\kappa + 2} \frac{(M_1 X_1 + M_2 X_2)}{\rho} \tag{15}$$

where the X's are mole fractions, M's are molecular weights, and κ and ρ (without subscripts) pertain to the solution. Since a nonpolar solvent has only distortion polarization, which is not greatly affected by interactions between molecules, we can take P_{1M} to have the same value in solution as in the pure solvent:

$$P_{1M} = \frac{\kappa_1 - 1}{\kappa_1 + 2} \frac{M_1}{\rho_1} \tag{16}$$

We can then get P_{2M} from

$$P_{2M} = \frac{1}{X_2} (P_M - X_1 P_{1M}) \tag{17}$$

obtained by rearrangement of Eq. (15).

Values of P_{2M} calculated using Eq. (17) are found to vary with X_2, generally increasing as X_2 decreases. This effect arises from strong solute–solute interactions due to the permanent dipoles. This difficulty can be eliminated by extrapolating P_{2M} to infinite dilution ($X_2 = 0$) to obtain P_{2M}^0. Although this could be done by plotting P_{2M} for each of a series of solutions against X_2, it is easier and more accurate to follow the procedure of Hedestrand,[5] which is given below.

Let us assume a linear dependence of κ and ρ on the mole fraction X_2:

$$\kappa = \kappa_1 + a X_2 \tag{18}$$

$$\rho = \rho_1 + b X_2 \tag{19}$$

On writing out Eq. (17) explicitly and using Eqs. (16), (18), and (19), it is possible to rearrange terms and obtain the *limiting* expression

$$P_{2M}^0 = \frac{3 M_1 a}{(\kappa_1 + 2)^2 \rho_1} + \frac{\kappa_1 - 1}{(\kappa_1 + 2)\rho_1} \left(M_2 - \frac{M_1 b}{\rho_1} \right) \tag{20}$$

Thus measurements on solutions of the *slope* of κ versus X_2 and the *slope* of ρ versus X_2 enable us to calculate the limiting molar polarization of the solute in solution. If we assume that the molar distortion polarization in an infinitely dilute solution is equal to that in the pure solute, then

$$P_{2d}^0 = R_{2M} = \frac{n_2^2 - 1}{n_2^2 + 2} \frac{M_2}{\rho_2} \tag{21}$$

where n_2 is the index of refraction and ρ_2 is the density measured for the solute in the pure state. From these two expressions, we can obtain the molar orientation polarization of the solute at infinite dilution:

$$P_{2\mu}^0 = P_{2M}^0 - P_{2d}^0 = \frac{1}{3\varepsilon_0} N_0 \frac{\mu^2}{3kT} \tag{22}$$

On substituting the numerical values of the physical constants, we obtain

$$\mu = 42.7(P_{2\mu}^0 T)^{1/2} \times 10^{-30} \, \text{C m} \tag{23a}$$

$$= 12.8(P_{2\mu}^0 T)^{1/2} \, \text{debye} \tag{23b}$$

where $P_{2\mu}^0$ is given in $\text{m}^3 \, \text{mol}^{-1}$ units and T is in kelvin. The unit of molecular dipole moment universally used by chemists is the debye (D), historically defined as 10^{-18} esu cm where esu is the electrostatic unit of charge having the value 3.33564×10^{-10} C. Thus $1 \, \text{D} = 3.33564 \times 10^{-30} \, \text{C m}$.

An alternative expression for $P_{2\mu}^0$ has been given by Smith,[6] who improved a method of calculation first suggested by Guggenheim.[7] This expression presupposes knowledge of the index of refraction n of the *solution*, and assumes that

$$n^2 = n_1^2 + cX_2 \tag{24}$$

where c, like a and b, is a constant determined by experiment. It follows[6] that

$$P_{2\mu}^0 = \frac{3M_1}{\rho_1} \left[\frac{a}{(\kappa_1 + 2)^2} - \frac{c}{(n_1^2 + 2)^2} \right] \tag{25}$$

this expression is an approximate one which holds when $\kappa_1 - n_1^2$ is small (as in the case of benzene, where it is 0.03) and M_2 and b are not too large. It is a useful form when the index of refraction n_2 of the pure solute is inconvenient to determine; this is often the case when the pure solute is solid. Note also that Eq. (25) does not require a knowledge of the densities of the solutions.

Solvent effects. The values of P_{2M}^0 obtained from dilute solution measurements differ somewhat from P_M values obtained from the pure solute in the form of a gas. This effect is due to solvent–solute interactions in which polar solute molecules induce a local polarization in the nonpolar solvent. As a result, μ as determined in solution is often smaller than μ for the same substance in the form of a gas, although it may be larger in other cases. Usually the two values agree within about 10 percent. A discussion of solvent effects is given by LeFèvre.[4]

METHOD

A capacitance cell suitable for work with liquids or solutions is shown in Fig. 1; it is made with a small variable air condenser of the type formerly in common use in radios and electronic circuits. It should have a maximum capacitance of 50 to 200 pF. This device is more convenient than a fixed-plate capacitor, since with the latter device it is necessary to measure separately the stray capacitance due to electrical leads, etc. In the cell shown, the variable condenser is used in two positions: fully closed (maximum capacitance) and fully open (minimum capacitance); these positions are defined by mechanical stops for the pointer on the knob that rotates the condenser shaft.† The difference ΔC between the closed (b) and open (a) positions is independent of the stray capacitance. Thus, the dielectric constant of the liquid or solution is given by

$$\kappa = \frac{\Delta C(\text{liq})}{\Delta C(\text{air})} \tag{26}$$

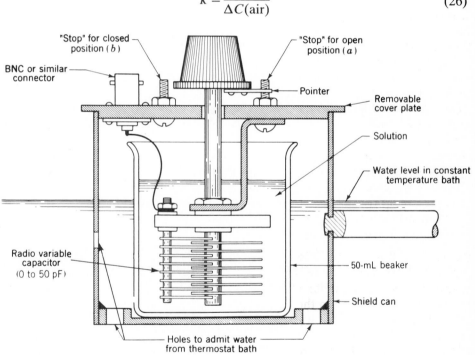

FIGURE 1
Variable-capacitance dielectric cell for solutions.

---

† For high-precision work with very stable electronics, a fixed-plate capacitor cell is preferable, since it eliminates any error due to lack of reproducibility in setting a variable capacitor. However, in this case one must determine the capacitance of the leads from measurements on a material of known dielectric constant.

where $\Delta C = C_b - C_a$. For liquids and solutions it is possible to use air instead of vacuum, since the dielectric constant of air is very close to unity.

We now turn our attention to methods of measuring the capacitance. A very precise method, and one that is particularly useful when the liquid or solution has a high conductance, involves the use of a capacitance bridge.[8] For the present system, in which the conductance is small, less expensive instruments can be used. Among these are digital capacitance meters made by a variety of manufacturers (see listings given by Tucker Electronics, Co., Garland, TX) with capacitance ranges down to 1–200 pF, readability of ±0.1 pF, and accuracy of 1 pF or better. These are very simple to operate, give capacitance readings directly, and are suitable for use in the present experiment, but their use gives little insight into the physical principles involved in capacitance measurement.

Several traditional methods are based on use of an electrical oscillator incorporating an inductance L and a capacitance C in parallel; this combination is known as a "tank" circuit. The frequency of such an oscillator is given by

$$f = \frac{1}{2\pi\sqrt{LC}} \tag{27}$$

The inductance L is fixed, while the capacitance C is the sum of several separate capacitances: that of the conductance cell, C_X; that of a separate variable tank capacitor incorporated in the oscillator, C_T (if present); and stray capacitances due to leads, etc., C_S. Also present in some methods is a precision air capacitor C_P graduated directly in capacitance units (generally picofarads), used when the equipment is operated in a null mode. In this mode the precision air capacitor is used to bring the frequency f to the same value in each measurement, and the change in reading of the precision air capacitor is then equal in magnitude to the change in capacitance of the cell.

An important and precise traditional method of measuring capacitance for dipole moment determinations is the *heterodyne beat* method, a particular form of the null method mentioned above. The output of an LC oscillator incorporating the capacitance cell and a precision variable air capacitor is mixed with the output of a fixed-frequency oscillator to produce a difference frequency in the audio range. This is typically applied to the vertical plates of an oscilloscope while the fixed-frequency output of a stable audio oscillator is applied to the horizontal plates. In each determination the precision air capacitor is adjusted so that the two audio frequencies are equal, as indicated by the appropriate Lissajous figure (circle, ellipse, or slant line) on the oscilloscope screen; see Fig. XVIII-6. An instrument that has been widely used for this purpose, incorporating the fixed oscillator, the variable oscillator, and the mixer, is the WTW Dipolmeter, Model DM01. This instrument, unfortunately, is no longer produced; however, such instruments are present in many physical chemistry laboratories. This instrument requires the minor modification of removing an existing capacitor from the tank circuit and introducing connections for leads for external capacitors, namely the capacitance cell and the precision air capacitor (such as General Radio 1304B).

With the advent of inexpensive, fast frequency counters, which count the individual cycles over a precisely fixed period (usually 1 s) and display the frequency digitally, it is more convenient to connect the r.f. output of the variable frequency oscillator directly to the frequency counter and determine the total capacitance with the aid of Eq. (27). This technique is highly suitable for the present experiment if a WTW Dipolmeter or another LC oscillator is available or can be constructed. (With the Dipolmeter only the variable frequency oscillator is used.) A simple LC oscillator circuit that can be constructed from inexpensive components has been described by Bonilla and Vassos;[9] this circuit, with a small modification to provide for one side of the tank to be grounded, is shown in Fig. 2. In this circuit, as in the WTW Dipolmeter circuit, all tank capacitances are in *parallel*. [This is *not* true of the circuit described in Ref. 4 of Exp. 32, as that circuit incorporates some *series* capacitance. If that circuit is employed, Eqs. (28)–(30) are not valid and Eqs. (32-3)–(32-5) must be used instead, unless the null mode is employed.]

It follows from Eq. (27) that

$$C = C_X + C_T + C_S = \frac{1}{4\pi^2 L f^2} \tag{28}$$

The capacitance of the cell can be determined from measurements of the frequency at the closed (b) and open (a) positions using the following equation, in which the capacitances C_T and C_S cancel out since they are held

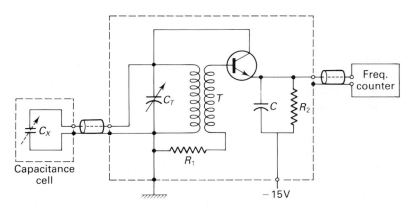

FIGURE 2
A simple oscillator circuit for capacitance measurements, adapted from Bonilla and Vassos.[9] The transistor is 1N3904 or a similar type. Suggested parameters: r.f. transformer T, 0.6 mH; R_1 and R_2, 12 kΩ; C, 300 pF; C_T, 0–100 pF. The circuit should be housed in a metal box for shielding, and connections with the capacitance cell and the frequency counter should be made with shielded coaxial cable with BNC or similar connectors.

constant:

$$C_X = \frac{1}{4\pi^2 L}\left[\frac{1}{f_b^2} - \frac{1}{f_a^2}\right] \tag{29}$$

To eliminate the apparatus constant $1/4\pi^2 L$ and determine the dielectric constant, we combine this with Eq. (2) and obtain

$$\kappa = \left[\frac{1}{f_b^2} - \frac{1}{f_a^2}\right]_{\text{sample}} \Big/ \left[\frac{1}{f_b^2} - \frac{1}{f_a^2}\right]_{\text{air}} \tag{30}$$

Another type of electronic circuit, the relaxation oscillator, can be used to measure capacitance; a simple apparatus for this purpose has been described by Kurtz *et al.*[10]

EXPERIMENTAL

Although several methods for making the capacitance measurements have been suggested, we will limit our discussion to the use of the *LC* oscillator and frequency counter shown in Fig. 2.

With the cell capacitor and cell beaker clean and dry, assemble the cell and mount it in the constant-temperature bath. Set the cell capacitor at the open position (*a*). Turn on the oscillator and the frequency counter, and adjust the tank capacitance C_T to yield a frequency in the range of 1.3–1.5 MHz. Wait a while to make sure that the apparatus is operating stably and not drifting in frequency. Determine the frequencies f_a and f_b in alternation at intervals of 30 s by alternating the position of the pointer knob between (*a*) and (*b*) until four to six frequency values have been recorded at each position. Make certain that you do not alter C_T or move the leads so as to affect C_S during either set of measurements.

Rinse the cell beaker and variable capacitor with the pure nonpolar solvent, fill the cell beaker to a level that will completely immerse the capacitor plates, and reassemble the cell. Make measurements in the open and closed positions as described above. Repeat this procedure with all the solutions of the polar solute in the nonpolar solvent.

Handle the capacitor with care; damage to the plates will affect the frequency values.

Average the frequency readings for each determination, and calculate the 95-percent confidence limit for each mean (see Chapter II).

Dichlorobenzene. Measurements should be made on pure benzene and on dilute solutions of *o*- and *m*-dichlorobenzene. Make up 50 or 100 mL of each solution as follows. Weigh a dry, clean volumetric flask; add an appropriate amount of solute; and weigh again. Now carefully make up to the mark with benzene and reweigh. Suggested concentrations are 1, 2, 3, and 4 mole percent of solute. The densities can be calculated from these weighings. If desired, the refractive index of each solution can also be measured with an Abbe refractometer so that Guggenheim's method of calculation can be used.

Succinonitrile. Measurements should be made on pure benzene and on dilute solutions of succinonitrile and propionitrile (or acetonitrile). Make up 50 or 100 mL of each solution in the same manner as described above. Densities can be calculated and/or refractive indexes can be measured.

CALCULATIONS

For each solution studied, calculate κ from Eq. (30) as well as ρ and X_2. Plot κ and ρ versus the mole fraction of solute X_2, and draw the best straight lines through your points; see Eqs. (18) and (19). Obtain the slopes a and b; the intercepts should agree with the pure solvent results. Using Eq. (20), calculate P_{2M}^0, the molar polarization at infinite dilution. Estimate P_{2d}^0 from Eq. (21) using the literature value of the refractive index of the solute n_2, and obtain $P_{2\mu}^0$.

Alternatively, Guggenheim's method can be used if the indexes of refraction were measured for each solution. Plot κ and n^2 versus X_2 and draw the best straight lines through your points; see Eqs. (18) and (24). Using Eq. (25) and the appropriate slopes a and c, calculate $P_{2\mu}^0$. It would be instructive to follow both of these procedures and see what difference (if any) the method of extrapolation to infinite dilution has on the $P_{2\mu}^0$ values.

Finally, calculate the dipole moment from Eq. (23) in units of 10^{-30} C m, and report it also in debye units (D).

DISCUSSION

Dichlorobenzene. Compare your experimental results with the values computed from a vector addition of carbon–chlorine bond moments obtained from the dipole moment of monochlorobenzene (5.17×10^{-30} C m, or 1.55 D). If they do not agree, suggest possible physical reasons for the disagreement.

Succinonitrile. An excellent discussion of this system has been presented by Braun, Stockmayer, and Orwoll,[11] who were the first to propose studying the dipole moment of succinonitrile. They show that the average square of the dipole moment is given by

$$\langle \mu^2 \rangle = \tfrac{8}{3}(1 - X_t)\mu_1^2 \tag{31}$$

where X_t is the mole fraction in the trans form and μ_1 is the C—C≡N bond moment as determined from the dipole moment of either propionitrile or acetonitrile. Thus a measurement of μ^2 for succinonitrile will determine the distribution of molecules among the three "rotational isomeric" states at the given temperature. [Note that $X_+ = X_- = \tfrac{1}{2}(1 - X_t)$ since the gauche plus and gauche minus states are equivalent.] It can also be shown from the appropriate Boltzmann populations that

$$X_t = (1 + 2e^{-\Delta E/RT})^{-1} \tag{32}$$

where $\Delta E = E(\text{gauche}) - E(\text{trans})$. Thus, the value of X_t obtained from Eq.

(31) will determine the value of ΔE and one can predict the temperature dependence of μ. Calculate a value of X_t and ΔE from your results and predict the value of μ at 280 and 350 K.

APPARATUS

Capacitance cell as shown in Fig. 1; oscillator as described in the text and Fig. 2; frequency counter (range at least 0.5–5 MHz); shielded coaxial cables with connectors; five 50- or 100-mL volumetric flasks; a 5-mL Mohr pipette; acetone wash bottle; rubber pipette bulb.

Benzene (analytical grade, 500 to 1000 mL; o- and p-dichlorobenzene (10 to 20 mL each), or succinonitrile and propionitrile (10 to 20 mL each); acetone for rinsing.

REFERENCES

1. P. W. Atkins, "Physical Chemistry," 3d ed., p. 577, Freeman, New York (1986).
2. Ibid., p. 576.
3. Ibid., pp. 577–581.
4. R. J. W. LeFèvre, "Dipole Moments," 3d ed., pp. 7–10 and chap. III, Methuen, London (1953).
5. G. Hedestrand, Z. Phys. Chem. **B2,** 428 (1929).
6. J. W. Smith, Trans. Faraday Soc. **46,** 394 (1950).
7. E. A. Guggenheim, Trans. Faraday Soc. **45,** 714 (1949).
8. C. P. Smyth, "Determination of Dipole Moments," in A. Weissberger and B. W. Rossiter (eds.), "Techniques of Chemistry: vol. I. Physical Methods of Chemistry," part IV, chap. 6, Wiley-Interscience, New York (1972).
9. A. Bonilla and B. Vassos, J. Chem. Educ. **54,** 130 (1977).
10. S. R. Kurtz, O. T. Anderson, and B. R. Willeford Jr., J. Chem. Educ. **54,** 181 (1977).
11. C. L. Braun, W. H. Stockmayer, and R. A. Orwoll, J. Chem. Educ. **47,** 287 (1970).

GENERAL READING

A. Chelkowski, "Dielectric Physics," Elsevier, Amsterdam (1980).
P. Debye, "Polar Molecules," Reinhold, New York (1929) [reprinted by Dover, New York (1945)].
R. J. W. LeFèvre, op. cit.

EXPERIMENT 32
DIPOLE MOMENT OF HCl MOLECULES
IN THE GAS PHASE

The permanent dipole moment μ of a polar molecule is determined in Exp. 31 from measurements of the dielectric constant of a solution containing such molecules as solute. In the present experiment the permanent dipole moment of a gas molecule is to be determined. The orientation polarization can be separated from the distortion polarization by means of measurements at more than one temperature, making use of the fact that the former is temperature dependent while the latter is not. An alternative method, which is recommended for this experiment, is to obtain the orientation polarization by subtracting from the molar polarization the distortion polarization as determined separately from the refractive index of the gas, which is determined by means of a laser interferometer.[1] Thus, the molar polarization needs to be determined at only one temperature.

The gas recommended is hydrogen chloride, HCl, because it has a large permanent dipole moment and because it has a boiling point low enough $(-83.7°C)$ to permit measurements of dielectric constant to be made down to Dry Ice temperature $(-78.5°C)$ if the option of measuring the dielectric constant at more than one temperature is chosen.

Thus, the experiment consists of two parts: the measurement of the dielectric constant of the gas by a method similar in principle to that of Exp. 31, and the measurement of the refractive index with an interferometer. The measurement of a dielectric constant at more than one temperature constitutes a complete stand-alone experiment, independent of the refractive index measurement; the measurement of refractive index can be used as a stand-alone experiment if a value for the low-frequency dielectric constant is supplied to permit the calculation of the dipole moment.

THEORY

The concepts of permanent dipole moment, induced dipole moment, and molar polarization are discussed in Exp. 31; this material should be reviewed. We assume further that deviations from the perfect-gas law are small in comparison with the experimental uncertainties.

Since the dielectric constant $\kappa \equiv \varepsilon/\varepsilon_0$ is only very slightly greater than unity and perfect-gas behavior is being assumed, we may write Eq. (31–12) in the form

$$\frac{(\kappa - 1)}{3} \frac{RT}{p} = \frac{N_0}{3\varepsilon_0} \left[\frac{\mu^2}{3kT} + \alpha_0 \right] \equiv P_M = P_\mu + P_d \tag{1}$$

where N_0 is Avogadro's number and ε_0 is the dielectric permittivity of a vacuum. A plot of P_M versus $1/T$ will yield $N_0\mu^2/9\varepsilon_0 k$ as the slope and $N_0\alpha_0/3\varepsilon_0$ as the intercept.

Properly speaking, the distortion polarization P_d is the sum of two parts, the atomic polarization P_a and the electronic polarization P_e. The atomic polarization results from distortion of the nuclear framework of the molecule in response to the electric field, while the electronic polarization results from distortion of the electron cloud on a time scale short compared to that in which the nuclei are able to move.

A separation of the molar polarization P_M into its three parts P_μ, P_a, and P_e can in principle be accomplished by measurements of the dielectric constant at three different frequencies, as shown in Fig. 1. At zero-frequency (dc) and low-frequency alternating electric field, all three parts are present: the molecules have time to reorient, the nuclei have time to move with respect to each other, and the electron cloud has time to distort. At frequencies corresponding to wavenumbers in the far-infrared range of about 20–300 cm^{-1} the molecules can absorb energy and change their rotational states, since their rotations are in or close to resonance with the alternating field. At frequencies well above this range the molecules can no longer reorient in response to the field. In principle, $P_d \equiv P_a + P_e$ can be measured directly with radiation in the infrared frequency range above 300 cm^{-1}. In practice, this is not done since this frequency range is too high for dielectric constant measurements and is too low for refractive index measurements with any convenient means (from the points of view of radiation sources and detectors, and the very long path-length required). At about 3000 cm^{-1} in the infrared range, the molecule can absorb vibrational energy from the alternating field and become promoted to the first excited vibrational state. Well above this frequency the nuclei can no longer respond to the alternating field; thus P_e can be measured in the visible range (with index of refraction measurements). In the ultraviolet and x-ray ranges electronic excitations can take place; with very hard x-rays even the electrons are unable to respond to the rapidly alternating field, and all polarization becomes negligible.

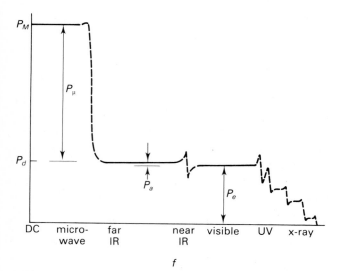

FIGURE 1
Schematic representation of the frequency dependence of the molar polarization of a gas of permanent dipoles. Dashed lines indicate ranges of complex behavior.

The dielectric constant becomes complex at frequencies at or near which the above-mentioned resonances and excitations take place. Measurements should be taken at frequencies that differ from such special frequencies by at least one order of magnitude.

If P_μ and P_d are separated through the temperature dependence of P_M, and P_e is determined at the frequency of visible light, it is possible in principle to determine P_a as $P_d - P_e$. Smyth[2] gives a value of $P_a = 0.5 \text{ cm}^3$ for HCl gas, determined in this way. However, analysis of literature data[1,3] indicates that this value, which is a small difference between large quantities, carries an uncertainty possibly as large as the value itself. Moreover, theoretical considerations (see Discussion) indicate that P_a should be very much smaller than this value. We will for this experiment assume that P_a is negligible in comparison to the experimental uncertainties in P_d and P_e, and reserve further consideration of P_a for the Discussion. Since the magnetic permeability of HCl gas is the same as that of a vacuum, $\kappa = n^2$ and we may use Eq. (31-14). Since the refractive index n of a gas at ordinary pressures is only very slightly greater than unity and since perfect gas behavior is being assumed, we may rewrite that equation in the form

$$P_d \simeq P_e = R_M = \frac{2(n-1)}{3} \frac{RT}{p} = \frac{N_0}{3\varepsilon_0} \alpha_0 \tag{2}$$

The interferometric determination of the index of refraction consists of determining the increase in the number of wavelengths in the light path of a Michelson interferometer as the gas is admitted to a previously evacuated cell that is placed in one of the two mutually perpendicular beam arms of the interferometer.

EXPERIMENTAL

Dielectric constant. The measurement of the dielectric constant of a gas is more difficult than the measurement of that of a solution. Because of the low density of a gas at ordinary pressures, its dielectric constant differs from that of a vacuum by a very small amount, of the order of a part per thousand. This requires a much higher order of experimental precision in the instrumentation used and in the procedure followed. To measure $(\kappa - 1)$ to one percent accuracy with a capacitance cell requires that the capacitance be measurable to a precision and reproducibility of one part in 10^5 or (preferably) better. This rules out the use of inexpensive capacitance meters for determining the small capacitance difference between the evacuated capacitance cell and the same cell containing a gas. The use of a high-resolution capacitance bridge is probably ruled out for this experiment unless one is lucky enough to be able to borrow one from a nearby research laboratory, as they are extremely expensive.† For the present experiment as here described, a dimensionally stable and well-shielded capacitance cell, a very stable LC oscillator, and an accurate frequency counter are required.

The stable LC oscillator contained in the WTW Dipolmeter mentioned in Exp. 31 can be used if that equipment is available. The oscillator circuit shown in Fig. 31-2 is probably not sufficiently stable; it lacks an amplifier stage to isolate the oscillator stage from the output load. A very stable solid-state LC oscillator for operation at about 1.5–2 MHz, constructed inexpensively from a published circuit diagram,[4] has been found to be satisfactory. This should be built in a metal box (which will serve as an electrical shield), with two BNC connectors—one for connecting with the cell and one for the output connection to the frequency counter. All such connections must be made with shielded cables. The capacitance cell replaces the capacitances shown in the LC tank circuit of the published diagram.‡ For best performance the temperature inside the oscillator box should be controlled with a small heating element, a temperature sensor, and a proportionating circuit (see Chapter XVI).

Suitable gas capacitance cells are not available commercially; it is necessary to construct one. A design for a capacitance cell comprising a cylindrical capacitor contained in a Pyrex glass jacket was given in a previous edition of this book.[5] A capacitance cell of superior stability, comprising a multi-plate capacitor in a metal jacket, is shown in Fig. 2. A satisfactory material for the cell is stainless steel. Although that material is not immune to corrosion by hydrogen chloride gas, our experience with a stainless-steel cell has been good. Monel or Inconel, which are superior in corrosion resistance, are much more expensive materials. The cell is demountable for cleaning (in case of corrosion) and for changing the number of plates if necessary to obtain the optimum capacitance (200–250 pF). The vacuum seal uses a gasket made from a wire-form hard solder such as Eutectoid 157; the wire solder is bent into a circle with the ends overlapping, and when the screws are uniformly tightened the ends of the wire cold weld together.

---

† The General Radio 161JA bridge, with six-figure resolution and very high accuracy, would be quite suitable but costs nearly $10 000; the Wayne–Kerr 4210 with somewhat lower resolution at about a quarter of that cost may be suitable. A used bridge might be obtained from a company that deals in second-hand electronics equipment (e.g., Tucker Electronics Co. of Garland, Texas).

‡ The tank inductor shown in that diagram may have to be replaced by one having a value or range of values of the inductance L which, with the tank capacitance C_{eff}, gives stable oscillatory behavior at a convenient frequency (preferably about 1.5–2 MHz) in accordance with Eq. (31-27). Suitable inductors are available from the J. W. Miller Division of Bell Industries, Compton, CA. Inductors of the 4400 and 4500 series from that source have powdered iron cores; the user-adjustable position of the core determines the inductance L. (The Miller 4508 or 4509 inductor is suitable with the capacitance cell here described.) The magnetic permeability of the core has a negative temperature coefficient, while the thermal expansion of the copper wire in the coils tends by itself to yield a positive temperature coefficient of inductance. Thus the magnitude and sign of the temperature coefficient of inductance are dependent on the core position.

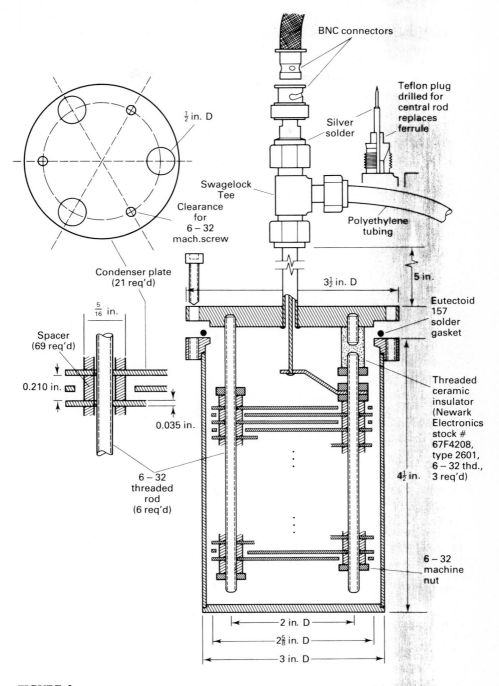

FIGURE 2
Suggested design for gas capacitance cell with a vacuum capacitance of about 250 pF. The
recommended material is stainless steel.

The cell and oscillator circuit should oscillate stably in the range 1–3 MHz and should have a frequency drift of less than 1 ppm per minute. The cell temperature should be held constant to 0.1°C by submerging it in a bath of water or trichloroethylene contained in a Dewar flask. If the frequency is sensitive to placement of nearby objects (including the experimenter's hands) look for defects in the electrostatic shielding; all visible parts of the apparatus should be at ground potential.

This cell has high thermal inertia. After transient changes in the internal temperature resulting from filling or evacuating the cell, thermal equilibrium is restored very slowly. This is particularly notable when a gas at room temperature is admitted to the cell at Dry-Ice temperature; in this case, effective restoration of equilibrium may require as much as 10 or 15 min. If the oscillator is stable with a small *but constant* drift rate, the approach to thermal equilibrium may be monitored by following the frequency. When the drift in oscillator frequency again becomes constant, thermal equilibrium may be considered to have been reached. Thus, the frequency should be plotted as a function of time, and when it achieves a constant slope it should be extrapolated to a time that is the same for the determinations being directly compared (i.e., gas and vacuum.) *It is important to take measurements for a long enough time to permit a valid extrapolation.*

The recommended procedure for determining the dielectric constant of an unknown gas with the cell here described depends on the relationship of changes in the capacitance of the cell to changes produced in the effective capacitance in the tank circuit. These are not the same because other capacitances are present besides that of the capacitor contained in the cell, C_{cell}. These always include the capacitance of the shielded cable connecting the cell to the oscillator and stray capacitances in the oscillator tank; these are *in parallel* with the cell, and lumped together may be called C_{par}. In addition, there is in the circuit described in Ref. 4 additional capacitance C_{ser} *in series* with the capacitances already mentioned. On following the rules for combining series and parallel capacitances, we obtain for the effective capacitance in the tank circuit:

$$C_{eff} = \frac{(C_{cell} + C_{par})C_{ser}}{C_{cell} + C_{par} + C_{ser}}$$

Only C_{cell} varies with the dielectric constant κ of the medium in the cell. On taking logarithms of both sides of this equation and differentiating with respect to C_{cell} we obtain:

$$\frac{1}{C_{eff}} \frac{dC_{eff}}{dC_{cell}} = \frac{1}{C_{cell} + C_{par}} - \frac{1}{C_{cell} + C_{par} + C_{ser}}$$

$$= \frac{C_{ser}}{(C_{cell} + C_{par})(C_{cell} + C_{par} + C_{ser})} \tag{3}$$

The dependence of the oscillator frequency f on the dielectric constant κ

of the gaseous medium in the cell is obtained by differentiating Eq. (31-27) and combining the result with Eq. (3):

$$\frac{df}{d\kappa} = -\frac{f}{2C_{\text{eff}}}\frac{dC_{\text{eff}}}{dC_{\text{cell}}}\frac{dC_{\text{cell}}}{d\kappa}$$

$$= -\frac{f}{2}\left[\frac{C_{\text{ser}}C_{\text{cell}}}{(C_{\text{cell}} + C_{\text{par}})(C_{\text{cell}} + C_{\text{par}} + C_{\text{ser}})}\right] \equiv -\frac{f}{2}K \qquad (4)$$

where K is the quantity in brackets (with C_{cell} set equal to its vacuum value). K is a dimensionless apparatus constant which, once carefully determined, need not be redetermined unless and until the apparatus is changed or readjusted. When C_{ser} approaches ∞ (equivalent to that capacitance being shorted out) and C_{par} approaches zero, K approaches unity.

On integrating over the very small change in dielectric constant we obtain (reversing signs for convenience)

$$f_{\text{vac}} - f_{\text{gas}} = \frac{f_{\text{vac}}}{2}K(\kappa_{\text{gas}} - 1) \qquad (5)$$

While in principle it is possible to measure all of the capacitances required for the calculation of K with Eq. (4), it is very difficult to do this in practice with sufficient accuracy. For our purposes it is very much preferable to determine K by measuring the frequency shift produced on introducing a reference gas of known dielectric constant into the previously evacuated cell. (Where students' time is limited, this may be done in advance by the teaching staff.) Dielectric constants of some convenient reference gases are given in Table 1.[6]

The molar polarization of these gases may be regarded as independent of temperature, and $\kappa - 1$ may be considered to be proportional to the density of the gas. Thus, at temperatures and pressures other than 293.2 K and 1 atm,

TABLE 1
Dielectric constants κ of some reference gases[a]

	$(\kappa - 1) \times 10^6$ at 20°C, 1 atm		
Gas	Radio frequency	Microwave	Optical
Ar	514.7	517.7	517.1
Air (dry, CO_2-free)	537.0	536.6	536.3[b]
N_2	547.2	547.8	548.3
CO_2	921.5	921.5	—

[a] From Ref. 6. Microwave and optical values are averages of two or more independent determinations. All values are stated to be uncertain in the final digit.

[b] Averaged from nine independent determinations; 95-percent confidence limits for the average are ±0.3.

assuming the perfect-gas law

$$(\kappa - 1)_{T,p} = (\kappa - 1)_{293.2,1} \frac{293.2}{T(\text{K})} p \text{ (atm)} \tag{6}$$

This equation may be considered as valid for the reference gases and also for HCl gas, within the precision of this experiment.

Given a value of κ_{gas} for the chosen reference gas from the above table, adjusted if necessary to the experimental temperature and pressure with Eq. (6), and given the value of the frequency shift obtained with this gas, Eq. (5) may be used to obtain the apparatus constant K. Once K is known, Eq. (5) may be used to determine the dielectric constant of the subject gas, say HCl, from the frequency shift obtained with that gas.

The overall experimental arrangement, including the gas-handling system, is shown in Fig. 3. A 0–1000 Torr mercury manometer is satisfactory for measuring the pressure of the gas in the cell; corrosion with HCl gas occurs but is very slow. Alternatively, a corrosion-resistant capacitance manometer (e.g., MKS—highly accurate but expensive) or variable reluctance manometer (e.g., Validyne—less accurate but also less expensive) may be used. A

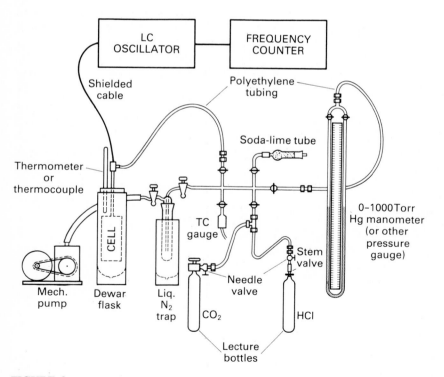

FIGURE 3
Gas-handling system for gas dielectric constant determination. Many connections are made with Swagelock and/or Ultra-Torr fittings.

thermocouple gauge is handy for leak testing but should not be exposed to HCl gas; it should be isolated with a stopcock. The HCl gas is provided from a lecture bottle fitted with a stem valve and a high-pressure needle valve. Connections between this and polyethylene tubing leading to the system are made with Swagelock fittings; Cajon Ultra-Torr fittings are preferable for all other connections made with polyethylene tubing.

Several hours before the experiment, the oscillator and the frequency counter should be turned on, and the capacitance cell, which should be in a Dewar flask containing water or trichloroethylene and a thermometer (or a thermocouple, taped to the cell), should be evacuated. Before any runs are made, the vacuum should be tested by closing off the cell and manometer from the vacuum pump and monitoring the pressure for at least 10 min. The frequency of the oscillator is then recorded, and if the oscillator is drifting it should be recorded at intervals of 30 s. When the drift is steady, HCl gas is slowly and carefully admitted to a pressure of approximately 1 atm. When equilibrium has been attained as indicated by restoration of a steady drift rate, the frequency should be recorded periodically over a long enough period to permit any needed linear extrapolation, and the pressure and the temperature should be recorded. The cell is then evacuated, and the frequency difference is determined in the same manner. This procedure should be repeated until at least four fillings have been made with HCl gas, yielding eight frequency differences. *During the experiment great care must be taken to avoid bumping the cell, cable, or oscillator.*

If the dielectric constant is also to be measured at Dry-Ice temperature ($-78.5°C$), the evacuated cell is immersed in a large Dewar flask containing barely enough trichloroethylene to submerge the main body of the cell, and crushed Dry Ice is added slowly until further addition of Dry Ice causes no increase in gas evolution; enough more is added to allow for further vaporization over the time of the experiment. The procedure described for determining the oscillator frequencies is repeated, with allowance for the additional time that may be required for the attainment of equilibrium at the lower temperature. It is possible also to make measurements above room temperature if means are available to obtain satisfactory temperature control for the duration of the measurements.

Since the capacitance of the cell, C_{cell}, varies somewhat with temperature owing to thermal expansion and contraction of the metal parts, it is advisable to determine the apparatus constant K for each temperature at which the cell is to be used. If this is to be done with CO_2 as the reference gas when the cell is cooled with Dry Ice, care must be taken to limit the filling pressure to significantly less than 1 atm to avoid condensation of CO_2 in the cell.

Refractive index. The Michelson interferometer divides an incoming beam of light into two beams that are exactly perpendicular to each other. This is done by a beam splitter, a half-silvered plane mirror inclined at 45° to the incoming beam. The two light beams are reflected back along the same light paths by adjustable plane mirrors at the ends of the two beam arms and recombine at the beam splitter where they undergo mutual interference. Depending on the relative phases of the two beams as they meet at the beam splitter, they produce an outgoing beam perpendicular to the incoming beam with an amplitude that is the sum of those of the two beams, or the difference, or some amplitude between these two values. If the transmission of the beam splitter is exactly 50 percent, the minimum amplitude is zero. The beam splitter and mirrors are mounted on a sturdy base for high dimensional stability. One (or both) of the mirrors has adjustment screws for making very small changes in the path length.

Figure 4 shows the interferometer set-up. The interferometer is of the type used in undergraduate physics demonstrations and laboratory courses. A small diverging lens (with a focal length of -20 to -50 mm) between the 0.5–2 mW He–Ne laser ($\lambda = 632.8$ nm) and the beam splitter of the interferometer diverges the beam so that a large area of the screen is illuminated, and (unless by accident the two beam-path lengths are exactly equal) a pattern of concentric interference fringes appears on the screen. The fringe pattern may be optimized by adjustment of one of the mirrors. (Preliminary adjustment of the mirrors *without* the diverging lens may be necessary to correct the alignment of the reflected beams if fringes are not apparent.) As the index of refraction changes, the fringes move outward or inward, appearing or disappearing at the center of the pattern. The direction of fringe movement with increasing index of refraction depends on the relative effective lengths of the two beam paths.

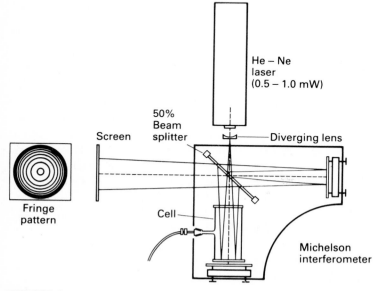

FIGURE 4
Laser interferometer set-up for gas refractive index determination.

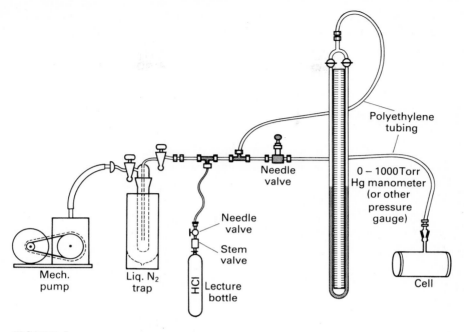

FIGURE 5
Gas-handling system for gas refractive index determination. Many connections are made with Swagelock and/or Ultra-Torr fittings.

The gas-handling system is shown in Fig. 5. The gas cell is similar to gas cells used in infrared spectroscopy (see Exp. 38), but ordinary Pyrex glass or quartz is more suitable for windows. Suitable cells with sealed-on windows are available commercially, or cells can be made by cementing windows to the ends of a cylindrical tube that will serve as the body of the cell with epoxy cement or Apiezon wax. In any case, the windows must be of accurately uniform thickness and must be accurately plane and parallel to each other, and of accurately known distance apart, preferably 10 cm. A suitable diameter is in the range of 2.5–5 cm. The commercially available cell normally has a taper joint; this can be attached by a Cajon Ultra-Torr fitting to a polyethylene tube through which the HCl gas is admitted to the cell. A thermometer, or preferably a thermocouple junction, should be taped or waxed to the cell.

As gas is *slowly* admitted to the evacuated cell through a needle valve, the fringes are counted as they appear or disappear at the center of the screen, or as they move past some designated point. (It is also possible to mount a photodiode on the screen and connect its amplified output to a strip-chart or X, Y recorder; this should give an oscillatory output, the cycles of which can be counted to a fraction of a fringe.) After the pressure in the cell has built up to approximately 1 atm and the gas supply has been shut off, counting of the fringes should continue until movement stops, indicating the attainment of equilibrium. At this point the pressure and the temperature should be recorded. The count of fringes made on filling the cell is now checked by slowly evacuating the cell and counting the fringes as they move in the opposite direction. If time permits, the entire procedure should be executed at least four times.

CALCULATIONS

Orientation polarization. If oscillator drift is negligible, the frequencies obtained under equilibrium conditions can be used directly in Eq. (5); otherwise the vacuum and HCl frequencies should be plotted against time and the readings at steady drift should be extrapolated back to a common time as mentioned earlier. The difference between the extrapolated values is $f_{vac} - f_{gas}$. The dielectric constant κ of the HCl gas at pressure p and temperature T is then calculated with Eq. (5), and corrected to 20°C and 1 atm with Eq. (6).

P_M is calculated with Eq. (1) from the measured values of $\kappa_{HCl} - 1$, p, and T for each run. Report the values of P_M together with their average in m^3 and in the conventional units of cm^3.

If values of P_M have been obtained at two (or more) temperatures, plot them against $1/T$. Calculate the permanent dipole moment μ from the slope; present it in units of 10^{-30} C m and also in debyes. (1 D = 3.33564 $\times$ 10^{-30} C m; see Exp. 31.) From the intercept determine P_d in m^3 and cm^3, and calculate α_0 in units of C^2 m^2 J^{-1}.

If P_M was measured at a single temperature, it is necessary to use data on the index of refraction n and Eq. (2) to estimate P_d. The orientation polarization P_μ, and the value of μ itself, can then be obtained with Eq. (1).

Distortion polarization. In the interferometer the light traverses the cell twice because of reflection in a mirror. If the cell is of length d, the number of wavelengths when the cell is evacuated is

$$N_{vac} = \frac{2d}{\lambda_0} = \frac{2dv}{c} \tag{7}$$

When the cell is filled with the gas to the desired pressure p at temperature T the number of wavelengths is

$$N_{gas} = \frac{2d}{\lambda} = \frac{2dv}{v} \tag{8}$$

Here λ_0 and λ are respectively the wavelengths of the light in the vacuum and in the gas, c and v are respectively the speeds of light in the vacuum and in the gas, and v is the frequency of the light:

$$v = \frac{c}{\lambda_0} = \frac{v}{\lambda} \tag{9}$$

The index of refraction is then

$$n \equiv \frac{c}{\upsilon} = \frac{\lambda_0}{\lambda} = \frac{N_{gas}}{N_{vac}} = 1 + \frac{\Delta N}{N_{vac}} \tag{10}$$

where $\Delta N = N_{gas} - N_{vac}$.

From the measured values of n, p, and T, use Eq. (2) to calculate the distortion polarization P_d and the distortion polarizability α_0 in units of $C^2 \, m^2 \, J^{-1}$.

If the low-frequency dielectric constant has been determined at more than one temperature *and* the refractive index has *also* been measured with visible light, calculate the atomic polarization from

$$P_a = P_M - P_\mu - P_e = P_d - P_e \tag{11}$$

Since the result is a small difference between two numbers having substantial uncertainties, it is particularly important to estimate the *uncertainty* in P_a to compare with the value of P_a itself, in order to establish whether the determined value of P_a is significantly different from zero.

DISCUSSION

It is instructive to compare the magnitude of the permanent dipole moment as determined in this experiment with the value that would result if net charges of $+e$ and $-e$ were centered at the H and Cl nuclei respectively. For this purpose we require the value of the internuclear distance[7]

$$r = 1.275 \times 10^{-10} \, m$$

as well as that of the magnitude of the electronic charge

$$e = 1.602 \times 10^{-19} \, C$$

Discuss this comparison in terms of mixed ionic and covalent character of the H—Cl bond. The dipole moment can also be discussed in terms of the difference in electronegativity between the H and Cl atoms.[8]

Let us now consider whether the neglect of the atomic polarization P_a is justified. A simple classical treatment of atomic polarization of a gas molecule has been given by Van Vleck,[9] Coop and Sutton,[10] and Smyth.[11] It is based on the approximation that the vibrations of the molecule are harmonic. Van Vleck argues that the result of this classical treatment is (to the same level of approximation) valid also under quantum mechanics.

We consider a diatomic molecule the axis of which is at an angle θ with the electric field direction. The electric field F has two effects: a turning moment on the molecule, which we ignore here because it is presumably taken account of in the orientation polarization, and a stretching or compression of the H—Cl bond due to interaction of the field with the charge distribution in the molecule. As the internuclear distance r changes from the equilibrium value r_e that it has in the absence of an applied electric field, the permanent dipole moment μ changes continuously from its equilibrium value μ_e. The energy of the molecule in the field may be written

$$U = -\mu F \cos \theta + \tfrac{1}{2}k(r - r_e)^2 \tag{12}$$

where k is the bond force constant. The molecule comes to a new equilibrium with a value of r for which U is a minimum:

$$\frac{dU}{dr} = -\frac{d\mu}{dr} F \cos \theta + k(r - r_e) = 0 \tag{13}$$

$$r - r_e = \frac{d\mu}{dr} \frac{F}{k} \cos \theta \tag{14}$$

The change in internuclear distance produces a change in permanent dipole moment, thereby producing a change in the component of polarization moment in the field direction (parallel to the z axis):

$$m_z = \frac{d\mu}{dr}(r - r_e)\cos\theta = \left(\frac{d\mu}{dr}\right)^2 \frac{F}{k}\cos^2\theta \tag{15}$$

When we average over all molecular orientations, $\cos^2\theta$ averages to $\tfrac{1}{3}$, and we obtain an *atomic polarizability*

$$\alpha_a = \frac{\langle m_z \rangle_{av}}{F} = \frac{1}{3k}\left(\frac{d\mu}{dr}\right)^2 \tag{16}$$

If we express the force constant in terms of the vibrational frequency v with the harmonic oscillator expression

$$k = 4\pi^2 \mu_{rm} v^2 \tag{17}$$

where in terms of the atomic masses m_H and m_{Cl}

$$\mu_{rm} = \frac{m_H m_{Cl}}{m_H + m_{Cl}} \tag{18}$$

is the reduced mass, then we have

$$\alpha_a = \frac{1}{12\pi^2 \mu_{rm} v^2}\left(\frac{d\mu}{dr}\right)^2 \tag{19}$$

The resulting expression for the atomic polarization is

$$P_a = \frac{N_0 \alpha_a}{3\varepsilon_0} = \frac{N_0}{36\pi^2 \varepsilon_0 \mu_{rm} v^2}\left(\frac{d\mu}{dr}\right)^2 \tag{20}$$

For evaluating this expression, we obtain from spectroscopic data[7]

$$\tilde{v} = \frac{v}{c} = 2991 \text{ cm}^{-1}$$

(where c is the velocity of light). The magnitude of the dipole derivative has been calculated theoretically by quantum mechanics[12] and determined experimentally from the transition moment from the ground state to the first excited vibrational state;[13] the values are

$$\left(\frac{d\mu}{dr}\right)_{\text{theor}} = 1.205 \text{ D}, 1.260 \text{ D}$$

$$\left(\frac{d\mu}{dr}\right)_{\text{exp}} = 1.206 \text{ D}$$

The student is encouraged to calculate α_a and P_a, and to compare their magnitudes with the experimental quantities α_0 and P_e respectively. Was the assumption that P_a is negligible justified?

The student should discuss the relative merits of the two methods described here for determining the permanent dipole moment of the HCl molecule.

APPARATUS

Gas dielectric constant. Gas dielectric cell (shown in Fig. 2); WTW Dipolmeter (modified as in Exp. 31) or very stable variable frequency LC oscillator with regulated 12-V power supply (see text); frequency counter (range at least 0.5–5 MHz); shielded coaxial cables with BNC connectors; gas handling system (shown in Fig. 3) incorporating mechanical vacuum pump, liquid nitrogen trap, closed-tube mercury manometer or other corrosion-resistant pressure gauge (0–1000 Torr), thermocouple gauge and associated electronics for vacuum testing, high-pressure stem valve for HCl lecture bottle, high-pressure needle valve(s), Swagelock and Cajon Ultra-Torr fittings, and polyethylene tubing; 0–100°C thermometer or thermocouple and readout meter; Dewar flask for liquid-nitrogen trap; large Dewar flask for constant-temperature bath for capacitance cell.

Lecture bottle of HCl gas; cylinder or lecture bottle of Ar gas, N_2 gas, or CO_2 gas (if used) or soda-lime tube for drying air (if used); trichloroethylene for constant-temperature bath; Dry Ice if low-temperature measurements are to be made; liquid nitrogen for trap.

Gas refractive index. Michelson interferometer [sources for such instruments are Sargent-Welch Scientific company (e.g., Cat. No. 3559), Central Scientific company (e.g., Cat. No 30666), and Ealing Scientific Company (e.g., Cat. No. 25–9069)]; 0.5–2 mW He–Ne laser and power supply; diverging lens (−20 to −50 mm focal length); small projection scrreen (a sheet of stiff white cardboard in a sturdy mount will do); gas cell (see text and Fig. 4); gas-handling system (see Fig. 5) incorporating mechanical vacuum pump, liquid-nitrogen trap, closed-tube mercury manometer or other corrosion-resistant pressure gauge (0–1000 Torr), high-pressure stem and needle valves for lecture bottle, Swagelock and Cajon Ultra-Torr fittings, and polyethylene tubing; 0–100°C thermometer or thermocouple and readout meter; Dewar flask for liquid-nitrogen trap.

Lecture bottle of HCl gas; liquid nitrogen for trap.

REFERENCES

1. D. A. Coe, and J. W. Nibler, *J. Chem. Educ.* **50,** 82 (1973).
2. C. P. Smyth, "Dielectric Behavior and Structure," p. 420, McGraw-Hill, New York (1955).
3. R. D. Nelson, Jr., D. R. Lide, Jr., and A. A. Maryott, "Selected Values of Electric Dipole Moments for Molecules in the Gas Phase," U.S. National Bureau of Standards NSRDS-NBS 10, p. 10, U.S. Government Printing Office, Washington, D.C. (1967).
4. "The ARRL Handbook for the Radio Amateur," chap. 10 (especially Fig. 14), American Radio Relay League, Newington, Conn. (1987) [the same circuit appears also in some earlier editions].
5. D. P. Shoemaker and C. W. Garland, "Experiments in Physical Chemistry," 2d ed., exp. 38 (pp. 303–309), McGraw-Hill, New York (1967).
6. "CRC Handbook of Chemistry and Physics," 65th ed., p. E-54, CRC Press, Inc., Boca Raton, Fla. (1984); A. A. Maryott and F. Buckley, "Table of Dielectric Constants and Electric Dipole Moments of Substances in the Gaseous State," National Bureau of Standards Circular 537, U.S. Government Printing Office, Washington, D.C. (1953).
7. K. P. Huber and G. Herzberg, "Molecular Spectra and Molecular Structure," p. 286, van Nostrand-Reinhold, New York (1978).
8. L. Pauling, "The Nature of the Chemical Bond," 3d ed., pp. 98–100, Cornell University Press, Ithaca, N.Y. (1960).
9. J. H. Van Vleck, "The Theory of Electric and Magnetic Susceptibilities," pp. 45–47, Clarendon Press, Oxford (1932).
10. I. E. Coop and L. E. Sutton: *J. Chem. Soc.,* 1269 (1932).
11. C. P. Smyth, *op. cit.,* pp. 416–422.
12. M. Kobayashi and I. Suzuki, *J. Mol. Spectrosc.* **116,** 422 (1986).
13. J. F. Ogilvie, W. R. Rodwell, and R. H. Tipping, *J. Chem. Phys.* **73,** 5221 (1980).

GENERAL READING

P. Debye, "Polar Molecules," Reinhold, New York (1929) [reprinted by Dover, New York (1945)].
C. P. Smyth, *op. cit.*
C. P. Smyth, "Determination of Dipole Moments," in A. Weissberger and B. W. Rossiter (eds.), "Techniques of Chemistry: Vol. I. Physical Methods of Chemistry," part IV, chap. 6, Wiley-Interscience, New York (1972).
J. H. Van Vleck, *op. cit.*
W. E. Vaughan, C. P. Smyth, and J. G. Powles, "Determination of Dielectric Constant and Loss," in A. Weissberger and B. W. Rossiter (eds.), *op. cit.,* part IV, chap. 5.

EXPERIMENT 33
MAGNETIC SUSCEPTIBILITY

When an object is placed in a magnetic field, in general a magnetic moment is induced in it. This phenomenon is analogous to the induction of an electric moment in an object by an electric field (see Exps. 31 and 32) but differs from it in that an induced magnetic moment may have either direction in relation to the applied field. If the induced moment is parallel to the external field (as in the electric case), the material is called *paramagnetic* or *ferromagnetic,* depending on whether the field due to the induced moment is small or large in comparison with the external field. If the moment is antiparallel to the external field, the material is called *diamagnetic*; the moment in this case is always small. This experiment will deal only with paramagnetic and diamagnetic substances in solution.

THEORY

If **M** is the magnetization (magnetic dipole moment per unit volume, analogous to the dielectric polarization **P**) induced by the field **H**, the volume magnetic susceptibility χ is defined by the equation

$$\mathbf{M} = \chi \mathbf{H} \tag{1}$$

For a paramagnetic substance, χ is positive; for a diamagnetic substance, it is negative. It is a dimensionless number, ordinarily very small in comparison with unity (except in the case of ferromagnetism) and essentially independent of **H** for fields readily available in the laboratory.

Magnetic succeptibilities are usually given in the literature on a mass or molar basis. Thus, while the volume susceptibility χ is induced moment per unit volume per unit applied field and is dimensionless, the mass susceptibility

$$\chi_{\text{mass}} = \frac{\chi}{\rho} \tag{2}$$

(where ρ is the density) is induced moment per unit mass per unit applied field and has *SI* units of $m^3 \, kg^{-1}$. The molar susceptibility

$$\chi_M = M\chi_{\text{mass}} = \frac{M}{\rho}\chi = \bar{V}\chi \tag{3}$$

(where M is the molecular mass and $\bar{V}$ is the molar volume) is induced moment per mole per unit applied field and has *SI* units of $m^3 \, mol^{-1}$.†

Diamagnetism. Nearly all known substances are diamagnetic. Diamagnetism results from the precession of the electronic orbits in atoms that occurs when a magnetic field is present. Volume diamagnetic susceptibilities are generally very small in magnitude compared with volume paramagnetic susceptibilities for pure substances. In paramagnetic substances the observed susceptibility is the net result of a paramagnetic contribution and a very much smaller

---

† The use of $cm^3 \, mol^{-1}$ units for χ_M was standard practice in the past (cgs system) and still persists. Such values are almost always cgs values for $\chi_M(ir)$, the irrational molar susceptibility given by $\chi_M(ir) = \chi_M/4\pi$.

diamagnetic contribution. In an estimation of the paramagnetism this diamagnetic contribution is often neglected, but in the case of aqueous solutions a correction should be made for the diamagnetic susceptibility of the water owing to the relatively large amount of it present.

Paramagnetism. The most important source of paramagnetism is the magnetic moment associated with the spin of the electron. The electron has two spin states, having spin magnetic quantum numbers $-\frac{1}{2}$ and $+\frac{1}{2}$, with the principal component of magnetic moment respectively parallel and antiparallel to the magnetic field direction. Spin paramagnetism (or in some cases ferromagnetism) exists in a substance if the atoms, molecules, or ions in it contain unequal numbers of electrons in the two possible spin states. This condition obviously exists when the atom, molecule, or ion contains an odd number of electrons [as in Fe^{3+}, Cu^{2+}, $(C_6H_5)_3C\cdot$ and other free radicals, etc.] It may also exist when the number of electrons is even, if a degenerate electronic level (such as a d or f atomic subshell) is only partially filled. For example, the free ferrous ion Fe^{2+} has six electrons outside the argon shell. The available orbitals of lowest energy are the five degenerate (i.e., equal-energy) $3d$ orbitals. Each of these may contain two electrons with their spins opposed (one with spin $+\frac{1}{2}$, the other with spin $-\frac{1}{2}$, in accord with the Pauli exclusion principle) or a single electron with either spin. No spin paramagnetism would occur if the six outer electrons of Fe^{2+} occupied three of the five $3d$ orbitals in pairs so that all spin magnetic moments cancelled. However, this would be contrary to a principle known as *Hund's first rule*, which states that, when several electronic orbitals of equal or very nearly equal energy are incompletely filled, the electrons tend to occupy as many as possible of the orbitals singly rather than in pairs, the electrons in singly occupied orbitals all having the same spin. In Fe^{2+} this rule predicts that one $3d$ orbital will contain a pair of electrons with spins opposed and the other four orbitals will each contain a single electron, the four spins being the same. Thus the free ferrous ion is paramagnetic. In molecular oxygen O_2, two molecular orbitals of equal energy each contain a single electron in accordance with Hund's first rule; consequently, oxygen gas is paramagnetic.

An atom, molecule, or ion containing one or more unpaired electrons with the same spin has a permanent magnetic dipole moment $\boldsymbol{\mu}$. In the absence of orbital contributions to the moment (see below), the magnitude μ of this moment is completely determined by the number of unpaired electrons n:[1]

$$\mu(\text{spin only}) = g_e\mu_e\sqrt{S(S+1)} = \sqrt{n(n+2)}\mu_e \tag{4}$$

where S is the spin quantum number (equal to the sum of the individual electron spin quantum numbers s_i), the electron g-factor g_e has been taken as 2 for simplicity rather than 2.0023, and μ_e is the *Bohr magneton* given by

$$\mu_e = \frac{eh}{4\pi m_e} = 9.274 \times 10^{-24}\,\text{J T}^{-1} \tag{5}$$

where T denotes tesla ($1\,T = 10^4$ gauss). Thus, for example, the "spin-only" magnetic moment of Fe^{2+} is

$$\mu = \sqrt{4 \times 6}\mu_e = 4.90 \text{ Bohr magneton}$$

Orbital magnetic moments may also contribute to paramagnetism. An electron in an orbital with one or more units of angular momentum behaves like an electric current in a circular loop of wire and produces a magnetic moment. When all orbitals in a subshell (e.g., all five $3d$ orbitals) are equally filled (with one electron each, as in Fe^{3+}, or two electrons each, as in Cu^+), the orbital moments cancel one another and there is no orbital contribution to the observed moment. In other cases (e.g., Fe^{2+}) an orbital contribution may arise, although usually it is "quenched" to a large extent by interactions with neighboring molecules or ions and does not contribute more than a few tenths of a Bohr magneton to the total moment. (Important exceptions are certain rare-earth ions, since quenching occurs to a much smaller extent for $4f$ orbitals than for $3d$ orbitals.) Atomic nuclei often possess spin magnetic moments, but these nuclear moments are so small as to have a negligible effect on magnetic susceptibility. They are important, however, in NMR spectroscopy.

In the absence of an applied field, the atomic moments in a paramagnetic substance orient themselves essentially at random owing to thermal motion and there is no net observable moment. In the presence of a magnetic field, the atomic moments tend to line up with the field, but the degree of net alignment is slight because of the disorienting effect of thermal motion. It is possible to show[1] that the paramagnetic contribution to the molal succeptibility is $N_0\mu^2/3kT$, where N_0 is Avogadro's number, k is the Boltzmann constant, and T is the absolute temperature. The total molal susceptibility can be written

$$\chi_M = N_0\alpha + \frac{N_0\mu^2}{3kT} \tag{6}$$

where α is the small (negative) diamagnetism per molecule. We can write this equation in the form

$$\chi_M = N_0\alpha + \frac{C}{T} \tag{7}$$

where C is called the *Curie constant* for the substance concerned. If C is determined by experiment, the magnetic dipole moment of the atom, molecule, or ion is obtained from it with the equation

$$\mu = \left(\frac{3kC}{N_0}\right)^{1/2} \tag{8}$$

Expressing this result in Bohr magneton units, we obtain

$$\mu = 2824\sqrt{C} \quad \text{Bohr magneton} \tag{9}$$

where C is in *SI* units ($m^3\,mol^{-1}\,K$).

Transition-metal complexes. The present experiment is largely concerned with complex ions of transition-group metals, such as hexahydrated or ammoniated ferrous or ferric ions and ferro- or ferricyanides. Here each metal ion is surrounded by a number of negative or neutral groups called *ligands*. This number is six in the cases cited and in other common cases may be four or eight.

Two distinct groups of complexes can be distinguished on the basis of experimental paramagnetic susceptibilities. *High-spin complexes* are those for which the effective magnetic moment is very close to the spin-only value for the free (gaseous) transition ion. *Low-spin complexes* have much lower moments than would be predicted for the free metal ion and can even be diamagnetic.

The early theory of transition-metal complexes, due largely to Pauling,[2] involved an explanation based on distinguishing between essentially ionic and essentially covalent bonding between the metal ion and its ligands. Where the number of unpaired electrons in the complex as deduced from the measured susceptibility is the same as that expected for the free metal ion (high-spin case), the bonding with the ligands was considered to be ionic (i.e., due to Coulomb attraction as in $[Fe(III)F_6]^{3-}$ or due to electrostatic polarization of neutral ligands by the cental ion as in $[Co(III)(H_2O)_6]^{3+})$. Where the number of unpaired electrons found in the complex is considerably less than the free-metal-ion value (low-spin case, as in most complexes with cyanides, ammonia, carbon monoxide, etc.), the bonding was considered to be covalent. It is assumed that the electrons are paired owing to the necessity of accommodating, in the atomic orbitals, some additional electrons donated by the ligands for forming electron-pair bonds. Thus, in $[Co(III)(NH_3)_6]^{3+}$ the 6 electrons outside the argon shell of Co^{3+} are augmented by 6 electron pairs from the ligands, giving 18 electrons. Of these, 12 electrons (or 6 pairs) are shared with the ligands to form 6 octahedral covalent bonds, using two $3d$, one $4s$, and three $4p$ orbitals of cobalt. The other 6 electrons are paired in the remaining three $3d$ orbitals. Since all electrons are paired with spins opposed, salts of this complex are diamagnetic.

The "crystal-field" or "ligand-field" theory of transition-metal complexes, first proposed by Van Vleck[3] and subsequently developed extensively,[4,5] has proven to be of great value in the interpretation of a wide range of properties. This theory explains transition-metal complexes in terms of the splitting of the five-fold-degenerate d level into two or more levels of different energy by perturbations due to the ligands. In the simplest version, this splitting is due purely to the electrostatic crystal field of the ligands. However, it is often necessary to consider also the metal–ligand orbital overlap (so-called "adjusted crystal-field theory"[5]). In any case, ligand-field theory is an adequate description as long as the d orbitals of the metal ion are well defined. If there is strong mixing between metal ion and ligand orbitals (as in metal carbonyls), a molecular-orbital theory is required.†

---

† As shown by Van Vleck,[3] Pauling's theory is a special case of the more general MO theory.

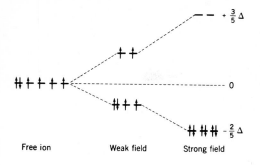

FIGURE 1
Weak field, e.g., $[Co(III)F_6]^{3-}$, and strong field, e.g., $[Co(III)(NH_3)_6]^{3+}$, splitting of the $3d$ level of Co(III) due to ligands with O_h symmetry. The splitting between the low-lying triplet and the upper doublet is defined as Δ. In the weak-field case, there are four unpaired spins; whereas all the spins are paired in the strong-field case.

We shall discuss ligand-field theory in the one-electron approximation, in which a one-electron d state is split by the ligand field and the individual states are then filled by the d electrons of the central ion (taking account of interactions). As an example, consider a d^6 complex with octahedral symmetry; Fig. 1 shows the situation for Co(III) complexes. For weak ligand fields, the splitting Δ is small and the electrons distribute themselves according to Hund's first rule. Thus a complex such as $[Co(III)F_6]^{3-}$ is a high-spin complex with four unpaired electrons. The energy of this configuration can be written as $(-\frac{2}{5}\Delta + P)$, where P is the average energy required to form an electron pair. When the ligand field is strong enough (Δ sufficiently large), the energy difference between the levels can no longer be overcome by the electron-unpairing tendency of Hund's rule. Thus a complex such as $[Co(III)(NH_3)_6]^{3+}$ is a low-spin complex: the six d electrons of the cobalt ion fill the lower triplet state with spins paired and the complex is diamagnetic. The energy of this configuration is $(-\frac{12}{5}\Delta + 3P)$. Analogous arguments can be applied to d^4, d^5, and d^7 complexes.

Thus ligand-field theory allows one to understand both high-spin and low-spin complexes. In particular, low-spin complexes can be explained without assuming a covalent electron-pair bond between the metal ion and the ligand. Indeed, there is no clear-cut distinction made between ionic and weak covalent character. According to the ligand-field theory, a low-spin complex simply means that the ligand field strength (and thus the splitting Δ) is greater than some critical value. It should be stressed that this critical value will be different for different complexes and one cannot equate increasing ligand field strength with increasing "covalency" in any simple way.

METHOD

Most methods for the determination of a magnetic susceptibility depend upon measuring the force resulting from the interaction between a magnetic field gradient and the magnetic moment induced in the sample by the magnetic field.[6] Assuming for simplicity that H varies only as a function of x, the x component of this force, per unit volume of the sample, is

$$f_x = \mu_0 M \frac{dH}{dx} = \chi \mu_0 H \frac{dH}{dx} \qquad (10)$$

where μ_0 is the vacuum permeability, equal to $4\pi \times 10^{-7}\,\mathrm{H\,m^{-1}}$. The henry (H) is the unit of magnetic inductance. This force is such as to tend to draw the sample into the strongest part of the field if the sample is paramagnetic (χ positive) or repel it into the weakest part if the sample is diamagnetic (χ negative). The work done on the system when a volume dV of the sample is carried from a point of field strength H_1 to a point of field strength H_2 is

$$dw = dV \int f_x\,dx = dV \chi \mu_0 \int H \frac{dH}{dx}\,dx = dV\,\chi\mu_0 \int_{H_1}^{H_2} H\,dH$$

$$= \tfrac{1}{2}\chi\mu_0 (H_2^2 - H_1^2)\,dV \tag{11}$$

The Gouy balance.[6,7] In the Gouy balance (Fig. 2) a long tube, divided into two regions by a septum, is suspended from one side of an analytical balance so as to hang vertically in a magnetic field. The septum is in the strongest part of the field and the two ends of the tube are in regions of essentially zero field

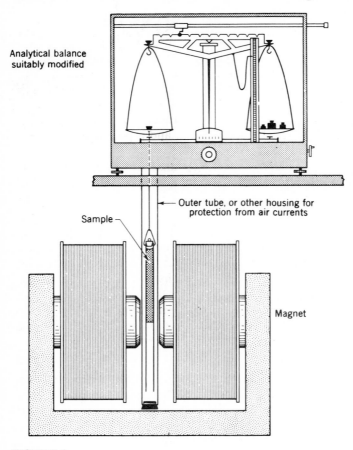

Analytical balance
suitably modified

Outer tube, or other housing for
protection from air currents

Sample

Magnet

FIGURE 2
Gouy balance.

strength. The part of the tube above the septum is filled with the sample to be investigated, and the part below is empty.

In addition to the downward gravitational force acting on the tube, a force is exerted by the magnetic field. Let us calculate the work done in *lowering* the tube, with cross-sectional area A, by an amount $|\Delta x|$. This is equivalent to bringing a volume $A\,|\Delta x|$ of the specimen from a region of zero field strength to a region of field strength H. Thus, ignoring gravitational work,

$$w = \tfrac{1}{2}\chi\mu_0 H^2 A\,|\Delta x| = f\,|\Delta x|$$

Therefore the *downward force* f on the tube due to its interaction with the magnetic field is[†]

$$f = \tfrac{1}{2}\chi\mu_0 H^2 A \qquad (12)$$

In making a measurement, the "apparent weight" W is determined in the absence of a field and then with a field present, and the difference is equated to f:

$$(W_{\text{field}} - W_{\text{no field}}) = f = \tfrac{1}{2}\chi\mu_0 H^2 A \qquad (13)$$

The weight W in each case is a *force* (in Newtons) obtained by multiplying the mass of the weights used (in kg) by the acceleration due to gravity (in m s^{-2}).

The magnet should provide a field of at least $0.4\,T(=4\,\text{kgauss})$ and preferably $0.6\,T$ or more. The field should be reasonably homogeneous over a region considerably larger than the diameter of the tube. For a sample tube up to 15 mm in diameter, a magnet with gap of 1 inch and a pole diameter of at least 3 inches is convenient.

For obtaining the weights in the presence and absence of a field it is most convenient to have an electromagnet, the field of which can easily be turned on and off. Such a magnet, with a regulated power supply, has the disadvantage of being rather expensive. A permanent magnet is usually less expensive, but special arrangements for making the no-field measurements are required. If the magnet is mounted on rails or on a pivot, it can be rolled or swung in and out of its normal position. If the magnet is stationary, the Gouy tube can be hung at two different levels (Fig. 3). If the latter method is to be used, the septum should be in the strongest part of the field in the lower position and in essentially zero field in the upper position. The sample tube should be long enough so that in either position the bottom end is in an essentially zero-field region *below* the strong part of the field.

A two-pan analytical balance is mounted on a sturdy table over the magnet as shown in Fig. 2. The Gouy tube is supported by a nonmagnetic wire

---

[†] In case the top of the tube is not at zero field strength, we should write

$$f = \tfrac{1}{2}\chi\mu_0(H_{\text{max}}^2 - H_{\text{min}}^2)A$$

Note that this force f as defined is a positive quantity for a paramagnetic sample, whereas f_x is negative when the positive x direction is taken to be upward.

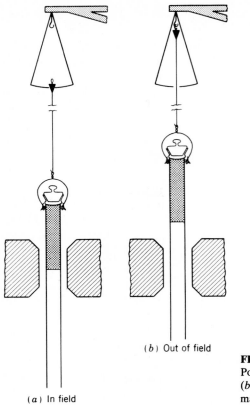

(b) Out of field

(a) In field

FIGURE 3
Positioning of Gouy tube for (a) in-field and (b) out-of-field weighings when a permanent magnetic is used.

or nylon thread which passes through a hole drilled in the base of the balance. If an electromagnet is used, the left-hand pan may be dispensed with and the wire attached directly to the stirrup. If, instead, a permanent magnet is used with the scheme shown in Fig. 3, a hole must be drilled in the pan so that the wire can pass through it to a hook which is attached to the stirrup for the no-field weighing or allowed to rest on the pan for the in-field weighing. The Gouy tube should be protected against air currents. Provision should also be made for mounting a thermometer near the Gouy tube.

EXPERIMENTAL

The procedure to be used in operating the Gouy balance will necessarily depend on the details of construction and cannot be given here with any completeness; a set of instructions should be compiled by the instructor and posted near the apparatus. The balance should be operated in the normal manner; the beam must be off the knife-edges when the Gouy tube is being mounted, demounted, or changed in position. It is important to **remove your**

watch when working near the magnet and to keep steel and iron tools or instruments out of the way.

In order to determine the "apparatus constant" $\mu_0 H^2 A/2$ that appears in Eq. (13), measurements are made on a material of known susceptibility. For this purpose, a common standard is an aqueous solution of nickel chloride, about 30 percent $NiCl_2$ by weight. Prepare 100 mL of such a solution with an accurately known weight fraction of $NiCl_2$ in air-free distilled water. When this solution is weighed on the Gouy balance, record the ambient temperature.

Solutions of the following salts† in air-free distilled water should be studied. In each case, the concentration ($\sim 0.5\ M$) must be precisely known. If possible, prepare the solutions in advance since some of these salts dissolve quite slowly.

$KMn(VII)O_4$, potassium permanganate (*Note:* this is soluble only to about $0.4\ M$)

$Mn(II)SO_4$, manganous sulfate

$[Fe(II)(H_2O)_6](NH_4)_2(SO_4)_2$, ferrous ammonium sulfate (*Note:* this is subject to slow air oxidation)

$K_4[Fe(II)(CN)_6]$, potassium ferrocyanide

$K_3[Fe(III)(CN)_6]$, potassium ferricyanide

It is important to know the densities of all solutions. A satisfactory procedure is to weigh the Gouy tube empty and then filled with water to a fiducial mark near the top. From the known density of water at this temperature, the volume is calculated. The tube is then filled with the solution to be studied up to the same fiducial mark and weighed. (All these weighings are in the absence of a magnetic field.) Alternatively, the density of each solution can be determined with a Westphal balance or a good hydrometer.

---

† Magnetic measurements may be made on powdered crystalline salts rather than on aqueous solutions. The volume susceptibilities are very much larger (an attractive feature if a strong magnet is not available), but the molar susceptibilities are not so easily interpretable in terms of atomic moments, owing to interaction effects in the crystalline state. In place of Eq. (7), we write (neglecting the diamagnetic term)

$$\chi_M = \frac{C}{T - \Delta}$$

This is called the *Curie–Weiss law.* The constant Δ may be positive or negative and for most compounds is less than 75 K in magnitude. To obtain a value of the Curie constant from which an atomic moment can be calculated, it is necessary to measure χ_M at more than one temperature. When the reciprocal of χ_M is plotted against absolute temperature, the slope is the reciprocal of the Curie constant C. The density ρ required in the calculations is not the crystal density, but an effective powder density determined in the manner described for solutions.

CALCULATIONS

Calibration. The mass susceptibility of an aqueous nickel chloride solution is given by[8]

$$\chi_{mass} = \left[\frac{10\,030p}{T} - 0.720(1-p)\right] \times 4\pi \times 10^{-9} \quad m^3\,kg^{-1} \qquad (14)$$

where p is the mass fraction of $NiCl_2$ and T is the absolute temperature. The second term in the brackets is the correction for the diamagnetism of the water used as solvent. This expression assumes that the solution is free of dissolved atmospheric oxygen.

From the χ_{mass} given by this expression and the density of the $NiCl_2$ solution, the volume susceptibility χ can be calculated from Eq. (2). With this and the measured weight difference, determine and report the apparatus constant $\mu_0 H^2 A/2$ appearing in Eq. (13).

Measurements on unknown solutions. The mass susceptibility of a solution is related to the molal susceptibility by

$$\chi_{mass} = \frac{\chi_M(\text{solute})}{M}p - 0.720 \times 4\pi \times 10^{-9}(1-p) \quad m^3\,kg^{-1} \qquad (15)$$

where p is again the mass fraction of solute. If the concentration c of the solute in units of $mol\,L^{-1}$ is known, we can write the volume susceptibility directly in the form

$$\chi = 1000c\chi_M - 0.720 \times 4\pi \times 10^{-9}(\rho - 1000cM) \qquad (16)$$

where χ_M, ρ, and M are all expressed in *SI* units ($m^3\,mol^{-1}$, $kg\,m^{-3}$, $kg\,mol^{-1}$, respectively). From the experimental weight difference, χ is determined by use of Eq. (13) and the known apparatus constant $\mu_0 H^2 A/2$. From Eq. (16) the solute molar susceptibility χ_M is obtained.

If the material is paramagnetic (χ_M positive), the small negative diamagnetic term $N_0\alpha$ can be neglected in Eq. (7) and the constant C can be determined. The atomic moment μ can then be calculated from Eq. (9). With neglect of any orbital contribution, the number of unpaired electrons can be found approximately with Eq. (4). Calculate the number of unpaired electrons for each paramagnetic substance studied.

DISCUSSION

In potassium permanganate, Mn(VII) has no unpaired electrons and thus no permanent magnetic moment. However, the magnetic field induces a small, temperature-independent paramagnetism because the field couples the ground state to paramagnetic excited states.[5] The other four compounds illustrate high- and low-spin cases of octahedral d^5 and d^6 complexes. (Neutral solutions of manganese sulfate are very pale pink and contain $[Mn(II)(H_2O)_6]^{2+}$ ions.) In the case of Fe(III) and the isoelectronic Mn(II), the ^6S ground state of the free ion has no orbital angular momentum and no effective coupling to excited states. Thus the magnetic moment of their high-spin complexes should be very close to the spin-only value. Comment on the spin type and ligand field strength of each complex studied.

APPARATUS

Gouy balance (comprising a magnet, power supply if needed, suitably modified analytical balance); glass-stoppered Gouy tube; several 100-mL volumetric flasks and glass-stoppered 200-mL flasks; 0°C to 30°C thermometer; Westphal balance or hydrometer (optional).

$NiCl_2$ (40 g); $MnSO_4$ (10 g); $KMnO_4$ (10 g); $K_4Fe(CN)_6$ (25 g); $K_3Fe(CN)_6$ (20 g); $Fe(NH_4)_2(SO_4)_2 \cdot 6H_2O$ (20 g); or a solution of each salt of an accurately known concentration (100 mL).

REFERENCES

1. P. W. Atkins, "Physical Chemistry," 3d ed., p. 597ff., Freeman, New York (1986).
2. L. Pauling, "The Nature of the Chemical Bond," 3d ed., pp. 161ff., Cornell University Press, Ithaca, N.Y. (1960).
3. J. H. Van Vleck, *J. Chem. Phys.* **3**, 807 (1935).
4. M. Gerloch, "Magnetism and Ligand-Field Analysis," Cambridge University Press, New York (1983).
5. F. A. Cotton and G. Wilkinson, "Advanced Inorganic Chemistry," 4th ed., pp. 535–542 and chap. 20, Wiley-Interscience, New York (1980).
6. L. M. Mulay, "Techniques for Measuring Magnetic Susceptibility," in A. Weissberger and B. W. Rossiter (eds.), "Techniques of Chemistry: Vol. I. Physical Methods of Chemistry," part IV, chap. 7, Wiley-Interscience, New York (1972).
7. M. M. Schieber, "Experimental Magnetochemistry," Wiley, New York (1967).
8. P. W. Selwood, "Magnetochemistry," 2d ed., p. 26, Interscience, New York (1956).

GENERAL READING

R. L. Carlin, "Magnetochemistry," Springer-Verlag, Berlin/New York (1986).
M. Gerloch, *op. cit.*
E. M. Purcell, "Electricity and Magnetism," 2d ed., McGraw-Hill, New York (1985).
H. L. Schläfer, "Basic Principles of Ligand Field Theory," chap. 2, Wiley-Interscience, New York (1969).
R. M. White, "Quantum Theory of Magnetism," Springer-Verlag, Berlin/New York (1983).
J. H. Van Vleck, "Electric and Magnetic Susceptibilities," Oxford University Press, New York (1932).

EXPERIMENT 34
NMR DETERMINATION OF PARAMAGNETIC
SUSCEPTIBILITY

The energy levels of a nucleus with a magnetic moment are changed in the presence of a magnetic field. Transitions between these levels can be induced by electromagnetic radiation in the radiofrequency region, and this resonance is useful in characterizing the chemical environment of the nucleus. For this reason, nuclear magnetic resonance (NMR) spectroscopy has developed as one of the most powerful structural methods used in organic and inorganic chemistry. In a slightly different application, in this experiment we will examine the effect of paramagnetic "impurities" on NMR solvent resonances and will use the predicted resonance shifts to deduce the paramagnetic susceptibility and electron spin of several transition-metal complexes.

THEORY

The magnetic moment of a nucleus with nuclear spin quantum number I is

$$\mu = g_N \mu_N \sqrt{I(I + 1)} \tag{1}$$

where g_N is the nuclear g-factor (5.5856 for a proton) and $\mu_N = eh/4\pi m_p$ is the nuclear magneton. Substitution of the charge e and mass m_p of a proton gives a value of $5.051 \times 10^{-27}\,\text{J}\,\text{T}^{-1}$ for μ_N. The symbol μ_N is the unit of nuclear magnetic moment and is smaller than the electronic Bohr magneton μ_e (defined in Exp. 33) by the electron-to-proton mass ratio.

The nuclear moment will interact with a *magnetic induction* (flux density)† B to cause an energy change (Zeeman effect)

$$E_N = -g_N \mu_N M_I B \tag{2}$$

Here M_I is the quantum number measuring the component of nuclear spin angular momentum (and magnetic moment) along the field direction, and it can have values $-I, -I + 1, \ldots, +I$. The effect of the field is thus to break the $2I + 1$ degeneracy and to produce energy levels whose spacing increases linearly with B (Fig. 1). Transitions among these levels can be produced by electromagnetic radiation provided that the selection rule $\Delta M_I = \pm 1$ is

---

† The vector quantity **B** is called either the magnetic induction or the magnetic flux density, although "magnetic field strength" would be a more appropriate name. Unfortunately, the latter name was given to **H** at the time when **H** was considered to be the fundamental magnetic-field vector. It is now known that **B** is the fundamental vector (analogous to **E**, the fundamental electric-field vector). To add to the confusion, **B** = **H** in a vacuum when Gaussian units are used. This book uses SI units, for which **B** = μ_0**H** in a vacuum and $\mu_0 \neq 1$. In SI units, **B** is expressed in tesla (1 T = 1 weber m^{-2} = 10^4 gauss).

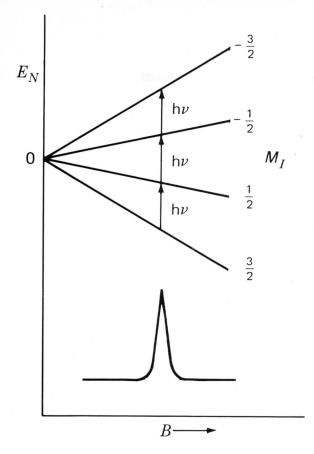

FIGURE 1
Energy levels, allowed transitions, and spectrum of a nucleus with $I = 3/2$ in a magnetic induction of magnitude B.

satisfied. In this case, the resonant frequency is given by

$$v = \frac{\Delta E_N}{h} = \frac{g_N \mu_N}{h} B \tag{3}$$

For protons, $v(\text{Hz}) = 4.26 \times 10^7 B$ (tesla) so that, for typical fields of 1–5 T, v falls in the radiofrequency region. In practice v is usually fixed at some convenient value (e.g., 60, 100, or 220 MHz) and the induction B is varied until resonance is achieved.

In general, the local induction B at the nucleus will differ from $\mu_0 H$, where H is the externally applied field, because of the magnetization M that is induced by H:

$$B = \mu_0 (H + M) = \mu_0 (1 + \chi) H \tag{4}$$

where μ_0 is the vacuum permeability, defined as $4\pi \times 10^{-7} \, \text{N A}^{-2} = 1.2566 \times 10^{-6} \, \text{N A}^{-2}$ (N = newton; A = ampere). The volume susceptibility χ (a dimensionless quantity) is the magnetic analog of the dielectric polarizability (Exp.

31), and it can be converted to a mass susceptibility χ_{mass} by dividing by the density ρ. Multiplication by the molar mass M gives the molar susceptibility

$$\chi_M = M\chi_{\text{mass}} = \frac{M\chi}{\rho} = \bar{V}\chi \tag{5}$$

where $\bar{V}$ is the molar volume. See Exp. 33 for a discussion of units. As noted in Exp. 33, χ_M has the form

$$\chi_M = N_0\alpha + \frac{N_0\mu^2}{3kT} = N_0\alpha + \chi_M^{\text{para}} \tag{6}$$

where $N_0\alpha$ is a diamagnetic contribution while the second term is a temperature-dependent paramagnetic contribution that is dominant if the magnetic moment μ is not zero. This will be true only if the electron spin or electronic orbital angular momentum is not zero, a case that is fairly common for transition-metal compounds.

Most organic compounds are not paramagnetic, and so it is the diamagnetic susceptibility that is important in determining the resonance condition. The diamagnetic contribution arises because the orbital motion of the electrons is altered by the presence of H so that there is a net orbiting of electrons about the field lines. This circulating charge in turn generates a magnetic induction B_d, that is proportional to the applied field and *opposed* to $\mu_0 H$,

$$B_d = \mu_0\chi H = \mu_0\frac{\rho}{M}N_0\alpha H = -\mu_0\sigma H \tag{7}$$

The resonant frequency is thus changed to

$$v = \frac{g_N\mu_N}{h}\mu_0 H(1 - \sigma) = v_0(1 - \sigma) \tag{8}$$

where v_0 is the resonance for a bare proton. The diamagnetic shielding constant σ is usually quite small ($\sim 10^{-5}$) and increases as the electron density about the nucleus is increased. Changes in B (and thus v) of a few parts per million (ppm) are typical when the chemical environment about a nucleus is changed. These chemical shifts relative to a convenient standard, such as tetramethylsilane, are easily measured with modern NMR instruments and hence serve to characterize the chemical bonding about a given nucleus. In addition, the relative intensities of, for example, proton resonances give a measure of the relative number of protons with different chemical environments (e.g., —CH$_3$ versus —CH$_2$— groups). Finally, the coupling of magnetic moments of nearby nuclei can produce spin–spin splitting patterns that are quite useful in identifying the functional groups present in the molecule. Some further discussion of these applications is presented in Exp. 44.

If a proton of a diamagnetic molecule is present in a solution containing a paramagnetic solute, the induction ("local field") B at the nucleus will be increased because of the alignment of the solute magnetic moments in the applied field. Evans[1] has shown that this increase in the local field is given by

$$\Delta B = \tfrac{1}{6}\mu_0(\chi_s - \chi_0)H \tag{9}$$

where χ_s and χ_0 are the susceptibilities of the solution with and without the paramagnetic solute, respectively. According to Eqs. (3) and (9), the proton resonance will shift by an amount

$$\frac{\Delta\nu}{\nu} = \frac{\Delta B}{B} \approx \frac{\Delta B}{\mu_0 H} = \tfrac{1}{6}(\chi_s - \chi_0) \tag{10}$$

with the approximation $B \approx \mu_0 H$ leading to negligible error.

To obtain the mass susceptibility χ_{mass} of the pure paramagnetic material, we assume that the volume susceptibility χ_s of the solution can be written as a sum of parts

$$\chi_s = \chi_{mass\,s}\rho_s = \chi_{mass}m + \chi_{mass\,0}(\rho_s - m) \tag{11}$$

Here ρ_s is the density of the solution containing m kilograms of paramagnetic solute per m^3 and $\chi_{mass\,0}$ is the mass susceptibility of the solution without the paramagnetic material. The density of the latter is ρ_0 so that $\chi_0 = \chi_{mass\,0}\rho_0$ and we obtain from Eqs. (10) and (11) the expression

$$\chi_{mass} = \frac{6}{m}\frac{\Delta\nu}{\nu} + \chi_{mass\,0} + \chi_{mass\,0}\frac{\rho_0 - \rho_s}{m} \tag{12}$$

The third term is a small correction which is unimportant for highly paramagnetic materials and is often neglected.[1] The value of χ_{mass} for a paramagnetic material can be determined, therefore, by measuring the difference in chemical shift of a proton in the solvent and in a solution containing a known weight of the paramagnetic solute. The value of $\chi_{mass\,0}$ can be obtained from tables such as those contained in Ref. 2 or by summing atomic susceptibilities χ_i according to Pascal's empirical relation

$$\chi_{mass\,0} = \frac{\chi_{M_0}}{M_0} = \frac{1}{M_0}\left[\sum_i t_i\chi_i + \sum_j \lambda_j\right] \tag{13}$$

where t_i is the number of atoms of type i in the molecule and the constitutive correction constants λ_j depend upon the nature of the multiple bonds. See Table 1 for some typical values of χ_i and λ_j.

The molar susceptibility χ_M for the solute is obtained by multiplying χ_{mass} by the molecular weight of the paramagnetic complex. As can be seen from Eq. (6), the paramagnetic contribution χ_M^{para}, and hence the paramagnetic moment μ, can be extracted by a temperature-dependence study (see also Exp. 33). For accurate results, a correction should be made to m to account for solvent-density changes with temperature.[3] Alternatively a measurement of χ_M at a single temperature can usually be combined with an adequate estimate of $N_0\alpha$ from Pascal's constants[2] to allow one to deduce the paramagnetic

TABLE 1
Pascal's constants χ_i for diamagnetic susceptibility[a] (units of $10^{-11}\,m^3\,mol^{-1}$)

Co^{2+}	-15	Br^-	-45	B	-8.8	N	open chain	-7.00
Co^{3+}	-13	CN^-	-23	Br	-38.5		ring	-5.79
Cr^{2+}	-19	Cl^-	-33	C	-7.5		monamides	-1.94
Cr^{3+}	-14	F^-	-14	Cl	-25.3		diamides, imides	-2.65
Cu^{2+}	-14	NO_3^-	-25	F	-7.9	O	alcohol, ether	-5.79
Fe^{2+}	-16	OH^-	-15	H	-3.68		aldehyde, ketone	$+2.17$
Fe^{3+}	-13	SO_4^{2-}	-50	I	-56.0		carboxylic $= O$	-4.22
K^+	-16			P	-33.0			
Ni^{2+}	-15			S	-18.8			

λ_j Corrections for bonds			
C=C	$+6.9$	C=N	$+10.3$
C≡C	$+1.0$	C≡N	$+1.0$
C=C—C=C	$+13.3$	C in aromatic ring	-0.30

[a] Adapted from Ref. 2.

contribution. If χ_M^{para} is obtained in this manner, the paramagnetic moment is given by

$$\mu = \sqrt{3kT\chi_M^{\text{para}}/N_0}$$
$$= 2824\sqrt{T\chi_M^{\text{para}}} \quad \text{Bohr magneton} \tag{14}$$

where the Bohr magneton (μ_e) is $9.274 \times 10^{-24}\,J\,T^{-1}$. In the absence of any orbital contributions, the value of μ is related to the number n of unpaired electrons in the d shell of the transition metal by

$$\mu(\text{spin only}) = \sqrt{n(n+2)} \quad \text{Bohr magneton} \tag{15}$$

As noted in Exp. 33, magnetic susceptibility measurements are thus quite useful in determining the electron configuration of a paramagnetic ion in a particular ligand field.

METHOD

The basic elements of an NMR spectrometer are outlined in Fig. 2. The principal magnetic field is provided by a permanent magnet ($\sim 1.5\,T$), an electromagnet (2.5 to 5.0 T), or a superconducting electromagnet (5.0 to 7.5 T). Since shifts of a few ppm are to be measured, the field must be quite stable and uniform. This is achieved by adding small adjustment coils to the magnet and by spinning the sample tube to average out residual in-homogeneities. An additional set of Helmholtz coils is wound around the pole faces to permit small linear variation in the field in recording spectra.

The sample tube is inserted into a probe region containing two

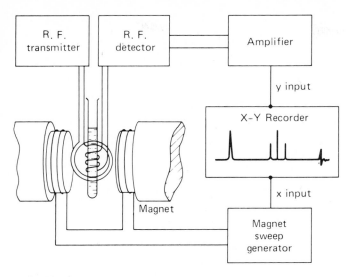

FIGURE 2
Schematic drawing of a nuclear magnetic resonance spectrometer.

radiofrequency coils wound at 90° to each other and to the magnetic-field coils. Radiation of some fixed frequency, usually 60 MHz or higher for protons, is sent through the sample by the transmitter coil. If the "local field" induction is such that Eq. (3) is satisfied, sample absorption and emission will occur. This re-radiated signal is detected by the receiver coil, which, being oriented at right angles to the transmitter, senses no signal in the absence of sample coupling. The signal from the receiver coil is amplified and displayed on a chart recorder to yield the NMR spectrum as a function of field. Since the absolute value of the magnetic field is not easily determined with high precision, field shifts are measured relative to some reference compound such as tetramethylsilane and are expressed as chemical shifts in ppm,

$$\delta_i \equiv \frac{H_r - H_i}{H_r} \times 10^6 \tag{16}$$

Here, H_r and H_i are field values for resonance by the reference nucleus and by nucleus i, respectively, for a fixed spectrometer frequency v. If H is fixed and v is varied, δ_i is given by

$$\delta_i \equiv \frac{v_i - v_r}{v_r} \times 10^6 \tag{17}$$

and many NMR chart displays are calibrated both in δ units and in frequency shift units appropriate to the fixed frequency source v_r (60 MHz for most small proton NMR instruments).

EXPERIMENTAL

Operation of the NMR instrument will be described by the instructor. Particular care should be taken in inserting and removing the NMR samples tubes to prevent damage to the probe. All tubes should be wiped clean prior to insertion to avoid contamination of the probe.

To provide an internal reference, the solvent is sealed in a capillary, which is placed at the bottom of the NMR tube used for the sample solution. For this purpose, a melting-point capillary is closed at one end and a syringe is used to add the reference solution. For aqueous studies, a 2% solution of t-butyl alcohol in water can be used as reference and as solvent for the paramagnetic solute. The shift with respect to the methyl resonance of the t-butyl group is then monitored. With organic ligands such as acetylacetonate (acac) groups, complexes such as $Cr(acac)_3$, $Fe(acac)_3$, and $Co(acac)_2$ are soluble in benzene and the proton resonance of the solvent is a convenient reference. The capillary is filled one-third full, the lower end of the capillary is cooled in ice, and the upper part is sealed off with a small hot flame. Alternatively, the reference solution can be placed in one compartment of a coaxial pair of cylindrical NMR tubes, which are commercially available. In either case, the spectral display should be expanded to permit an accurate measurement of the frequency shift.

The following complexes are suitable for study at room temperature. Either toluene or benzene can be used as the solvent, but it should be noted that toluene has *two* proton resonances.

d^3 $Cr(acac)_3$, chromium acetylacetonate (M.W. = 349.3 g mol^{-1})
d^5 $Fe(acac)_3$, ferric acetylacetonate (M.W. = 353.2 g mol^{-1})
d^8 $Co(acac)_2$, cobaltous acetylacetonate (M.W. = 257.2 g mol^{-1})

The concentrations are not critical (0.02 to 0.05 M) but they must be precisely known. These salts are available commercially or they can be readily synthesized by methods described in Ref. 4.

If the NMR spectrometer is equipped with a variable temperature probe, the $Fe(acac)_3$ shifts should be measured at four or five temperatures. A 0.025 M solution in toluene can be employed over the liquid range from $-95°C$ to $+110°C$. Care should be taken to seal and test the NMR tube if the high-temperature range is to be studied (i.e., heat the tube to $\sim135°C$ in a fume hood before mounting it in the spectrometer). Since the temperature variation of toluene density is appreciable,[3] the solute concentration should be corrected at each temperature T:

$$m_T \cong m_{rt}\left(\frac{\rho_T}{\rho_{rt}}\right) \tag{18}$$

where ρ_T is the density of *pure* solvent at temperature T and ρ_{rt} is that at room temperature. The density of toluene at various tempertures is given in Ref. 5. Approximately 15 min should be allowed for equilibrium to be reached after changing the temperature. An accurate measurement of the probe temperature should be made in the manner described in the spectrometer manual. This usually involves a measurement of the frequency separation between OH and CH resonances in methanol or ethylene glycol since this separation changes by about 0.5 Hz K^{-1} for a 60-MHz instrument.

Appropriate salts for alternative aqueous measurements are

d^5 Fe(NO$_3$)$_3$·9H$_2$O (0.015 M), ferric nitrate† (M.W. = 404.0 g mol^{-1})
d^5 K$_3$Fe(CN)$_6$ (0.06 M), potassium ferricyanide (M.W. = 329.3 g mol^{-1})
d^6 FeSO$_4$·7H$_2$O (0.02 M), ferrous sulfate (M.W. = 278.0 g mol^{-1})
d^8 NiCl$_2$ (0.08 M), nickel chloride (M.W. = 129.6 g mol^{-1})
d^9 CuSO$_4$ (0.08 M), cupric sulfate (M.W. = 159.6 g mol^{-1})

Three or more of these solutions should be made up in 10- or 25-mL volumetric flasks using as reference solvent ~2% (by volume) t-butyl alcohol in deoxygenated water. The Fe^{2+} and Fe^{3+} solutions illustrate the effect of oxidation state on the electron configuration of a single element. The Fe^{3+} and K$_3$Fe(CN)$_6$ solutions demonstrate the role of weak and strong ligand field splitting on the electron configuration (see discussion in Exp. 33).

The Fe(NO$_3$)$_3$ or NiCl$_2$ solutions can be studied from 0°C to 100°C in the same manner as described for the acetylacetonate complexes. In correcting for changes in m with temperature, the density variations of pure H$_2$O may be used.

CALCULATIONS

Neglecting the third term of Eq. (12), the mass susceptibility of a paramagnetic solute is readily determined from m and Δv. The diamagnetic mass suscep-tibilities $\chi_{\text{mass 0}}$ of the solvents, in SI units of m^3 kg^{-1}, are -8.8×10^{-9} (benzene), -9.0×10^{-9} (toluene), and -9.0×10^{-9} (t-butyl alcohol–water solvent). For temperature-dependence studies, correct m according to Eq. (18), calculate χ_M and plot χ_M versus $1/T$. The slope of the best straight line through these points is the Curie constant C (see Exp. 33) and the magnetic moment μ is given by

$$\mu = \left(\frac{3kC}{N_0}\right)^{1/2} = 2824\sqrt{C} \quad \text{Bohr magneton} \tag{19}$$

The value of $N_0\alpha$ is obtained from the intercept of your plot. Compare this value of the diamagnetic susceptibility of the solute with that estimated from Pascal's constants in Table 1. For single-temperature measurements, estimates of the latter type should be used to obtain χ_M^{para} for use in Eq. (14).

---

† This compound is hygroscopic so that the weighing should be done rapidly. It is also necessary to make the solution ~0.3 M in HNO$_3$ to prevent precipitation of Fe(OH)$_3$. The reference solvent used with this solution should have the same acid concentration.

Report your values of μ in Bohr magneton and calculate the number of unpaired electrons using the spin-only formula, Eq. (15).

DISCUSSION

In an aqueous solvent, the ions Fe^{2+}, Fe^{3+}, Ni^{2+}, and Cu^{2+} are all complexed to six H_2O molecules in an octahedral arrangement. The magnetic moments can be understood in terms of the weak-field splitting of the five d orbitals as described in Exp. 33. In $Fe(CN)_6^{3-}$, the CN^- ligands are strongly bound to the Fe^{3+} ion and the ligand-field splitting is larger. The structures of $Cr(acac)_3$ and $Fe(acac)_3$ are also pseudo octahedral with the two oxygen atoms of each $acac^-$ ion occupying adjacent metal ligand positions. For $Co(acac)_2$, a square planar or tetrahedral arrangement of the ligands might be expected, but the energetics favoring octahedral coordination are such that there is intermolecular association to fill the two vacant ligand sites. The result is a tetramer in which a pseudo octahedral arrangement is achieved by each cobalt atom.[6]

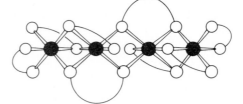

Discuss your observed magnetic moment values in terms of the number of unpaired electrons predicted for each transiton-metal complex assuming octahedral structures. Comment on any differences from the predicted spin-only moments.

APPARATUS

NMR spectrometer, with variable temperature capability if possible; several 10- or 25-mL volumetric flasks; NMR tubes; melting-point capillaries; syringe; torch for sealing capillaries; ethylene glycol and methanol in NMR tubes for temperature calibrations.

$Cr(acac)_3$, $Fe(acac)_3$, $Co(acac)_3$, where $acac$ = acetylacetonate (~ 1 g each); benzene or toluene (250 mL); $Fe(NO_3)_3 \cdot 9H_2O$, $K_3Fe(CN)_6$, $FeSO_4 \cdot 7H_2O$, $NiCl_2$, $CuSO_4$ (~ 1 g each); 2% by volume t-butyl alcohol in deoxygenated water (250 mL); 3 M HNO_3 for acidifying $Fe(NO_3)_3$ and corresponding reference solution.

REFERENCES

1. D. F. Evans, *J. Chem. Soc.* 2003 (1959).
2. L. N. Mulay and E. A. Boudreaux, "Theory and Applications of Molecular Diamagnetism," p. 303, Wiley-Interscience, New York (1976).

3. D. Ostfield and I. A. Cohen, *J. Chem. Educ.* **49,** 829 (1972).
4. T. H. Crawford and J. Swanson, *J. Chem. Educ.* **48,** 382 (1971).
5. E. W. Washburn, (ed.), "International Critical Tables," vol. III, p. 28, McGraw-Hill, New York (1929).
6. F. A. Cotton and G. Wilkinson, "Advanced Inorganic Chemistry," 4th ed., p. 877, Wiley-Interscience, New York (1980).

CHAPTER
XIII

SPECTRA AND MOLECULAR STRUCTURE

EXPERIMENTS

35. Absorption spectrum of a conjugated dye

36. Infrared spectroscopy: vibrational spectrum of SO_2

37. Raman spectroscopy: vibrational spectrum of CCl_4

38. Vibrational–rotational spectra of HCl and DCl

39. Vibrational–rotational spectrum of acetylene

40. Spectrum of the hydrogen atom

41. Band spectrum of nitrogen

42. Absorption and emission spectra of molecular iodine

43. Electron spin resonance spectroscopy

44. NMR determination of keto–enol equilibrium constants

EXPERIMENT 35
ABSORPTION SPECTRUM OF A CONJUGATED DYE

Absorption bands in the visible region of the spectrum correspond to transitions from the ground state of a molecule to an excited electronic state which is 170 to 300 kJ mol^{-1} above the ground state. In many substances, the lowest excited electronic state is more than 300 kJ mol^{-1} above the ground state and no visible spectrum is observed. Those compounds that are colored (i.e., absorb in the visible) generally have some weakly bound or delocalized electrons such as the odd electron in a free radical or the π electrons in a conjugated organic molecule. In this experiment we are concerned with the determination of the visible absorption spectrum of several symmetrical

440

polymethine dyes and with the interpretation of these spectra using the "free-electron" model.

THEORY

The visible bands for polymethine dyes arise from electronic transitions involving the π electrons along the polymethine chain. The wavelength of these bands depends on the spacing of the electronic energy levels. Bond-orbital and molecular-orbital calculations have been made for these dyes, but the predicted wavelengths are in poor agreement with those observed. We shall present here the simple free-electron model first proposed by Kuhn;[1,2] this model contains some drastic assumptions but has proved reasonably successful for molecules like conjugated dyes.

As an example, consider a dilute solution of 1,1'-diethyl-4,4'-carbocyanine iodide (cryptocyanine):

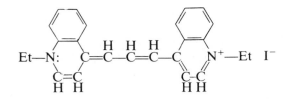

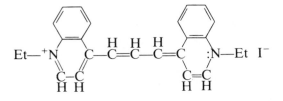

The cation can "resonate" between the two limiting structures above, which really means that the wavefunction for the ion has equal contributions from both states. Thus all the bonds along this chain can be considered equivalent, with bond order 1.5 (similar to the C—C bonds in benzene). Each carbon atom in the chain and each nitrogen at the end is involved in bonding with three atoms by three localized bonds (the so-called σ bonds). The extra valence electrons on the carbon atoms in the chain and the three remaining electrons on the two nitrogens form a mobile cloud of π electrons along the chain (above and below the plane of the chain). We shall assume that the potential energy is constant along the chain and that it rises sharply to infinity at the ends; i.e., the π electron system is replaced by free electrons moving in a one-dimensional box of length L. The quantum-mechanical solution for the energy levels of this model[3] is

$$E_n = \frac{h^2 n^2}{8mL^2} \qquad n = 1, 2, 3, \ldots \tag{1}$$

where m is the mass of an electron and h is Planck's constant.

Since the Pauli exclusion principle limits the number of electrons in any given energy level to two (these two have opposite spins: $+\frac{1}{2}$, $-\frac{1}{2}$), the ground state of a molecule with N π electrons will have the $N/2$ lowest levels filled (if N is even) and all higher levels empty. When the molecule (or ion in this case) absorbs light, this is associated with a one-electron jump from the highest filled level ($n_1 = N/2$) to the lowest empty level ($n_2 = N/2 + 1$). The energy change for this transition is

$$\Delta E = \frac{h^2}{8mL^2}(n_2^2 - n_1^2) = \frac{h^2}{8mL^2}(N + 1) \tag{2}$$

Since $\Delta E = h\nu = hc/\lambda$, where c is the speed of light and λ is the wavelength,

$$\lambda = \frac{8mc}{h}\frac{L^2}{N + 1} \tag{3}$$

Let us denote the number of carbon atoms in a polymethine chain by p; then $N = p + 3$. Kuhn assumed that L was the length of the chain between nitrogen atoms plus one bond distance on each side; thus, $L = (p + 3)l$, where l is the bond length between atoms along the chain. Therefore,

$$\lambda = \frac{8mcl^2}{h}\frac{(p + 3)^2}{p + 4} \tag{4}$$

Putting $l = 1.39$ Å $= 0.139$ nm (the bond length in benzene, a molecule with similar bonding) and expressing λ in nanometers, we find

$$\lambda \text{ (in nm)} = 63.7\frac{(p + 3)^2}{p + 4} \tag{5}$$

If there are easily polarizable groups at the ends of the chain (such as benzene rings), the potential energy of the π electrons in the chain does not rise so sharply at the ends. In effect this lengthens the path L, and we can write

$$\lambda \text{ (in nm)} = 63.7\frac{(p + 3 + \alpha)^2}{p + 4} \tag{6}$$

where α should be a constant for a series of dyes of a given type. If such a series is studied experimentally, this empirical parameter α may be adjusted to achieve the best fit to the data; in any event, α should lie between 0 and 1.[1]

In order to compare the results of this model with the more sophisticated bond- or molecular-orbital calculations let us use Eq. (5), which assumes that $\alpha = 0$, to calculate the wavelength λ for cryptocyanine (in which $p = 9$) and compare that value with those given by Herzfeld and Sklar:[4]

Free electron	$\lambda = 707$ nm
Bond orbital (case 1)	$\lambda = 3900$
Bond orbital (case 2)	$\lambda = 2900$
Molecular orbital	$\lambda = 2700$

Note that only the free-electron model predicts an absorption band in the visible in agreement with observation. While the orbital calculations are poor for polymethine dyes, they are in principle a superior approach and have given excellent results for unsaturated hydrocarbons.

METHOD

There are many commercial visible–UV spectrophotometers that are suitable for this experiment. These instruments range from simple single-beam devices such as the Spectronic model 20 to high-performance double-beam scanning spectrophotometers such as various Varian-Cary models. The components and operational principles of these instruments are discussed in Chapter XVIII and this material should be reviewed prior to undertaking the experiment.

It is necessary to define several terms commonly used in spectrophotometry. Absorption spectra are often characterized by the *transmittance T* at a given wavelength; this is defined by

$$T \equiv \frac{I}{I_0} \tag{7}$$

where I is the intensity of light transmitted by the sample and I_0 is the intensity of light incident on the sample. When the sample is in solution and a cell must be used, I is taken to be the intensity of light transmitted by the cell when it contains solution while I_0 is taken to be the intensity of light transmitted by the cell filled with pure solvent. Another way of describing spectra is in terms of the *absorbance A,* where

$$A \equiv \log \frac{I_0}{I} \tag{8}$$

A completely transparent sample would have $T = 1$ or $A = 0$, while a completely opaque sample would have $T = 0$ or $A = \infty$.

The absorbance A is related to the path length d of the sample and the concentration c of absorbing molecules by the Beer–Lambert law,[5]

$$A = \varepsilon c d \tag{9}$$

where ε is called the *molar absorption coefficient* when the concentration is expressed in moles per unit volume.† The quantity ε is an intrinsic property of

---

† The quantity ε was previously called the *extinction coefficient,* and this name is still frequently used in the scientific literature.

the absorbing material that varies with wavelength in a characteristic manner; its value depends only slightly on the solvent used or on the temperature. The SI unit for ε is $mol^{-1} m^2$, but a more practical and commonly used unit is $mol^{-1} L cm^{-1}$, which corresponds to using the concentration c in $mol L^{-1}$ and the path length d in cm.

For quantitative measurements it is important to calibrate the cells so that a correction can be made for any small difference in path length between the solution cell and the solvent cell. For analytical applications, one must check the validity of Beer's law, since slight deviations are often observed and a calibration curve of absorbance versus concentration is then required. Such quantitative techniques are described elsewhere[5] and will not be necessary in the present work.

EXPERIMENT

For the present experiment, a manually scanned spectrophotometer with a resolution of 10–20 nm is adequate although an autoscanning instrument is, of course, more convenient. Instructions for operating the spectrophotometer will be made available in the laboratory. Turn on the instrument as instructed, and allow it to warm up for a few minutes in order to achieve stable, drift-free performance.

Several polymethine dyes should be studied, preferably a series of dyes of a given type with varying chain length. In addition to the 1,1′-diethyl-4,4′-carbocyanine iodides mentioned previously, 1,1′-diethyl-2,2′-carbocyanine iodides and 3,3′-diethyl thiacarbocyanine iodides are suitable. Other possible compounds can be found in the literature.[6]

Choose any one of the available dyes, and prepare 10 mL of a solution using methyl alcohol as the solvent. The concentration should be approximately $10^{-3} M$. (If the solutions are to be kept for a long time, they should be stored in dark-glass bottles to prevent slow decomposition by daylight.)

Follow the spectrophotometer operating instructions carefully, and obtain the spectrum of this initial solution. Take absorbance readings at widely spaced intervals throughout the visible region (400 to 750 nm) until the absorption band is located. Then take readings at much closer intervals throughout the band. Dilute the initial solution and redetermine the spectrum. Repeat this procedure until a spectrum is obtained with an absorbance reading of about 1 at the peak (transmittance ~0.1). The band shape will change with concentration, since these dyes dimerize. At the final concentration used, the monomer band (higher wavelength band) will be much more prominent than the dimer band, which will disappear or remain as a shoulder on the low-wavelength side of the main peak.

Make up solutions (10 mL) of the other dyes at approximately the same molar concentration that gave the best results previously. Obtain their spectra in the same way.

CALCULATIONS

Present all the spectra obtained (plotting A versus λ if data were obtained manually); label each plot clearly with the name of the compound and the concentration used. Determine λ_{max}, the wavelength at the peak, for each dye studied.

Using Eq. (6), calculate λ from the free-electron model for each compound. If a series of dyes of a single type has been studied, choose α to give the best fit for this series. Alternatively, one could fix α by fitting any one of the series and use this value for all others in the series.

Report in a table your experimental and theoretical values of λ_{max}.

APPARATUS

Spectrophotometer, such as a Bausch & Lomb Spectronic or Beckman DU series; four Pyrex or polystyrene sample cells; lens tissue; several 10-mL volumetric flasks; wash bottle.

Reagent-grade methyl alcohol (150 mL); several polymethine dyes, preferably a series such as 1,1'-diethyl-4,4'-cyanine iodide, -carbocyanine iodide, and -dicarbocyanine iodide or 1,1'-diethyl-2,2'-cyanine iodide, -carbocyanine iodide, and -dicarbocyanine iodide (a few milligrams of each is sufficient). (Gallard-Schlesinger Chemical Mfg. Corp., 1001 Franklin St., Garden City, NY and K & K Laboratories, Inc., 121 Express St., Plainview, NY 11803, are possible but expensive suppliers of an entire series of dyes. Aldrich Chemical Co. and Kodak Laboratory & Research Products supply some of these dyes at lower cost. Check the current "Chem Sources" for other suppliers.)

REFERENCES

1. H. Kuhn, *J. Chem. Phys.* **17,** 1198 (1949); *Fortsch. Chem. Organ. Naturstoffe* **17,** 404 (1959).
2. I. N. Levine, "Quantum Chemistry," 3d ed., pp. 463–466, Allyn and Bacon, Boston, Mass. (1984).
3. P. W. Atkins, "Physical Chemistry," 3d ed., pp. 314–317, Freeman, New York (1986).
4. K. F. Herzfeld and A. L. Sklar, *Rev. Mod. Phys.* **14,** 294 (1942).
5. H. H. Bauer, G. D. Christian, and J. E. O'Reilly, "Instrumental Analysis," Allyn and Bacon, Boston, Mass. (1979); D. A. Skoog, "Principles of Instrumental Analysis," 3d ed., Saunders, New York (1985).
6. L. G. S. Brooker, *Rev. Mod. Phys.* **14,** 275 (1942); N. I. Fisher and F. M. Hamer, *Proc. Roy. Soc.,* Ser. A, **154,** 703 (1936).

GENERAL READING

H. H. Bauer, *et al., op. cit.*
G. A. Ewing, "Instrumental Methods of Chemical Analysis," 5th ed., McGraw-Hill, New York (1985).
D. A. Skoog, *op. cit.*

H. H. Willard, L. L. Merritt, Jr., J. A. Dean, and F. A. Settle: "Instrumental Methods of Analysis," 6th ed., Van Nostrand-Reinhold, New York (1981).

EXPERIMENT 36
INFRARED SPECTROSCOPY: VIBRATIONAL SPECTRUM OF SO_2

The infrared region of the spectrum extends from the long-wavelength end of the visible region at $1 \, \mu m$ out to the microwave region at about $1000 \, \mu m$. It is common practice to specify infrared frequencies in wavenumber units: $\tilde{\nu}(cm^{-1}) = 1/\lambda = \nu/c$, where c is the speed of light in $cm \, s^{-1}$ units. Thus this region extends from $10\,000 \, cm^{-1}$ down to $10 \, cm^{-1}$. Although considerable work is now being done in the far-infrared region below $400 \, cm^{-1}$, the spectral range from 4000 to $400 \, cm^{-1}$ has received the greatest attention because the vibrational frequencies of most molecules lie in this region.

THEORY

Almost all infrared work makes use of absorption techniques in which radiation from a source emitting all infrared frequencies is passed through a sample of the material to be studied. When the frequency of this radiation is the same as a vibrational frequency of the molecule, the molecule may be vibrationally excited; this results in loss of energy from the radiation and gives rise to an absorption band. The spectrum of a molecule generally consists of several such bands arising from different vibrational motions of the molecule.

Associated with each vibrational energy level there are also closely spaced rotational energy levels. In general, transitions occur from a particular rotational level (quantum number J'') in a given vibrational state (quantum number v'') to a different rotational level (quantum number J') in an excited vibrational state (quantum number v'). The appearance of this rotational fine structure will depend on the resolution of the spectrometer used, as shown schematically in Fig. 1. Vibrational–rotational spectra are discussed in Exps. 38 and 39 for the case of linear molecules and will not be considered here.

In this experiment we shall be concerned with the infrared bands of SO_2 gas at medium resolution, i.e., with only the vibrational spectrum. Electron diffraction studies show that SO_2 is a symmetrical, nonlinear molecule.[1] For a nonlinear molecule containing N atoms there are $3N - 6$ vibrational degrees of freedom. Thus, SO_2 has three basic patterns of vibration called "normal modes." These are shown in Fig. 2. All vibrations of the molecule may be expressed as linear combinations of these normal modes.

The frequencies of the normal modes of vibration are called *fundamentals*. For SO_2 all three fundamentals are infrared-active; they will

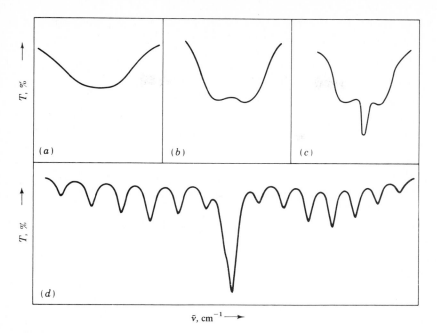

FIGURE 1
Schematic infrared band contours at (a) low, (b) and (c) medium, and (d) high resolution. For the band shown in (b) there is no Q branch ($\Delta J = 0$ transitions are forbidden); the bands shown in (c) and (d) have Q branches.

correspond to the three most intense bands in the spectrum. The fundamentals (v_1, v_2, v_3) may be assigned by analogy with the spectra of other molecules: the bending frequency v_2 should be that of the lowest frequency band, and the antisymmetric stretch v_3 should be that of the highest frequency band.

The structure of SO_2 is confirmed by the presence of all three fundamentals in the infrared spectrum. When this is true, the molecule cannot have a center of symmetry[1] as in O—S—O. Linear unsymmetric structures, such as O—O—S or O———S—O, can be ruled out by the shape of the rotational band envelopes in the infrared spectrum; discussion of this topic is beyond the scope of this book.

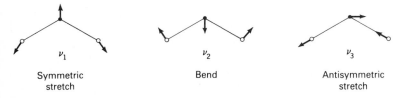

FIGURE 2
Normal modes of vibration for SO_2. The lengths of the arrows which indicate displacements from the equilibrium configuration have been exaggerated for clarity.

In addition to the fundamentals, other much weaker bands may be observed in the spectrum at high pressure. Such weak bands are either overtones $(2v_i, 3v_i, \ldots)$ or combination bands $(v_i \pm v_j, 2v_i \pm v_j, \ldots)$ that arise from the fact that small anharmonicities tend to couple the normal vibrations. On the basis of intensity, the transitions that involve binary overtones $(2v_i)$ and binary combinations $(v_i \pm v_j)$ are the most likely to be observed. Thus, any weak bands observed for SO_2 may be assigned as overtones or combinations of the strong fundamentals. Any such assignment should take into account rules that determine whether a given combination will be active in the infrared.[1] In the case of SO_2, none of the combinations are forbidden by symmetry. This is not generally true for molecules of higher symmetry; see the discussion in Exps. 37 and 39.

Valence-force model. The normal modes of vibration can be expressed in terms of the atomic masses (m_S and m_O), the O—S—O bond angle (2α), and several force constants that define the potential energy of the molecule. These force constants are analogous to the Hooke's law constant for a spring. If a force model is chosen involving fewer force constants than the number of normal modes, the force constants are overdetermined and a check on the validity of the model is then possible. The valence force model assumes that there is a large restoring force along the line of a chemical bond when the distance between the two atoms at the ends of this bond is changed from its equilibrium value. It also assumes a restoring force for any change in angle between two adjacent bonds. Adopting the notation used by Herzberg,[2] we can write the potential energy of SO_2 as

$$U = \tfrac{1}{2}[k_1(r_1^2 + r_2^2) + k_\delta \delta^2]$$

where r_1 and r_2 are changes in the S—O distances (whose equilibrium value is denoted by l), and δ is the change in the bond angle 2α. The quantities k_1 and k_δ/l^2 are force constants for stretching and bending motions, respectively. Using the classical mechanics of harmonic oscillators, one can obtain[2] for SO_2

$$4\pi^2 v_3^2 = \left(1 + \frac{2m_O}{m_S}\sin^2\alpha\right)\frac{k_1}{m_O} \tag{1}$$

$$4\pi^2(v_1^2 + v_2^2) = \left(1 + \frac{2m_O}{m_S}\cos^2\alpha\right)\frac{k_1}{m_O} + \frac{2}{m_O}\left(1 + \frac{2m_O}{m_S}\sin^2\alpha\right)\frac{k_\delta}{l^2} \tag{2}$$

$$16\pi^4 v_1^2 v_2^2 = 2\left(1 + \frac{2m_O}{m_S}\right)\frac{k_1}{m_O^2}\frac{k_\delta}{l^2} \tag{3}$$

If one expresses the frequencies in wavenumber units $\bar{v}(\text{cm}^{-1})$ and uses $m_O = 16$ and $m_S = 32$, $4\pi^2$ should be replaced by $4\pi^2 c^2/10^3 N_0 = 5.892 \times 10^{-5}$ to obtain k_1 and k_δ/l^2 in units of N m^{-1}. For SO_2, $2\alpha = 119.5°$ and $l = 1.432$ Å $= 0.1432$ nm.[3]

Vibrational partition function[4]. The thermodynamic quantities for an ideal gas can usually be expressed as a sum of translational, rotational, and vibrational contributions (see Exp. 3). We shall consider here the heat capacity at constant volume. At room temperature and above, the translational and rotational contributions to C_v are constants independent of temperature. For SO_2 (a nonlinear polyatomic molecule), the molar quantities are

$$\bar{C}_v(\text{trans}) = \tfrac{3}{2}R$$
$$\bar{C}_v(\text{rot}) = \tfrac{3}{2}R \tag{4}$$

The vibrational contribution to $\bar{C}_v$ varies with temperature and can be calculated from the vibrational partition function q_{vib} using

$$\bar{C}_v(\text{vib}) = R \frac{\partial}{\partial T}\left(T^2 \frac{\partial \ln q_{\text{vib}}}{\partial T}\right) \tag{5}$$

The partition function q_{vib} is well approximated by

$$q_{\text{vib}} = \prod_{i=1}^{3N-6} q_i^{\text{HO}} \tag{6}$$

where q_i^{HO} is the harmonic-oscillator partition function for the ith normal mode of frequency v_i. Since the energy levels of a harmonic oscillator are given by $(v + \tfrac{1}{2})hv$, one obtains[4]

$$q_i^{\text{HO}} = \sum_{v=0}^{\infty} \exp\left[\frac{-(v+\tfrac{1}{2})hv_i}{kT}\right] = \frac{e^{-hv_i/kT}}{1 - e^{-hv_i/kT}} \tag{7}$$

Combining Eqs. (5)–(7), we find

$$\bar{C}_v(\text{vib}) = R \sum_i \frac{u_i^2 e^{-u_i}}{(1 - e^{-u_i})^2} \tag{8}$$

where $u_i = hv_i/kT = hc\bar{v}_i/kT = 1.4388\bar{v}_i/T$ and the summation is over all of the normal modes.

EXPERIMENTAL

There are many different infrared spectrometers in current use, almost all of which are suitable for this experiment. A general description of the characteristics of infrared grating and Fourier transform infrared (FTIR) spectrometers is given in Chapter XVIII. Specific instructions should also be available for the operation of the instrument to be used in the laboratory. Before beginning the experiment, the student should review the pertinent material. **Use the spectrometer carefully**; if in doubt, ask the instructor.

The SO_2 gas cell is constructed from a short (usually 10 cm) length of large-diameter (4 to 5 cm) Pyrex tubing with a vacuum stopcock attached. Infrared-transparent windows are clamped against O-rings at the ends of the

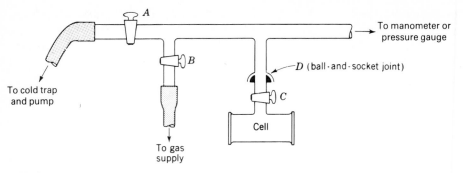

FIGURE 3
Gas-handling system for filling infrared cell.

cell or are sealed on the ends with Glyptal resin. Since SO_2 spectra should be recorded down to $400 \, cm^{-1}$, KBr windows are needed. For studies concentrating on the region 4000 to $700 \, cm^{-1}$, NaCl windows will suffice. Both types of salt windows become "foggy" on prolonged exposure to a moist atmosphere and should be protected (e.g., stored in a desiccator) when not in use.

Filling the cell and recording the spectra. An arrangement for filling the cell is given in Fig. 3. Attach the cell at D. With stopcock C open and B closed, open stopcock A and pump out the system. Make sure that the needle valve of the SO_2 cylinder is closed. Then open B and continue pumping.

Close A and **slowly** open the valve on the SO_2 cylinder. Fill the system to about 900 Torr (1.2 atm) with SO_2. Close the valve on the cylinder and then close B and C. Remove the cell from the vacuum line and take a spectrum.

Return the cell to the line. With B closed and C closed, open A and pump out the line for a few minutes. Now close A and open C. Open A **very slowly** and carefully to reduce the pressure to about 300 Torr. Close A and C, remove the cell, and record the spectrum. Repeat this process and record the specrum at pressures of 100, 20, and 5 Torr. Several spectra may be recorded on a single chart, using different colored inks to distinguish them. If it is not possible to observe v_2, the lowest frequency fundamental, owing to equipment limitations, a chart recording in the KBr region should be made available for reference.

CALCULATIONS AND DISCUSSION

Assign the fundamental bands of SO_2 and report their frequencies. Then assign any other bands, comparing the observed frequencies with those calculated from combinations of the fundamental values $\tilde{v}_1$, $\tilde{v}_2$, and $\tilde{v}_3$. If the $\tilde{v}_2$ fundamental frequency was not measured directly, estimate it from a combination band. Is there any indication of rotational structure observed?

Calculate k_1 and k_δ/l^2 from Eqs. (1) and (3) using your values of $\tilde{v}_1$, $\tilde{v}_2$, and $\bar{\nu}_3$. Check the valence-force model by calculating both sides of Eq. (2) and comparing them.

Using your values of $\bar{\nu}_1$, $\bar{\nu}_2$, and $\bar{\nu}_3$, calculate $\bar{C}_v(\text{vib})$ at 298 K and at 500 K from Eq. (8). Compare the spectroscopic value $\bar{C}_v = 3R + \bar{C}_v(\text{vib})$ with the experimental $\bar{C}_v$ value obtained from directly measured values[5] of $\bar{C}_p$ and the expression $\bar{C}_v = \bar{C}_p - R$: $\bar{C}_v = 30.5 \text{ J K}^{-1} \text{mol}^{-1}$ at 298 K and 37.7 J K^{-1} mol^{-1} at 500 K.

APPARATUS

Infrared spectrometer or Fourier transform (FTIR) instrument; gas cell with KBr windows; vacuum line for filling cell; cylinder of SO_2 gas with needle valve.

REFERENCES

1. G. Herzberg, "Molecular Spectra and Molecular Structure II. Infrared and Raman Spectra of Polyatomic Molecules," pp. 251–269, 285, Van Nostrand, Princeton, N.J. (1945).
2. *Ibid.,* pp. 168–172.
3. G. Herzberg, "Molecular Spectra and Molecular Structure III. Electronic Spectra and Electronic Structure of Polyatomic Molecules," p. 605, Van Nostrand, Princeton, N.J. (1966).
4. P. W. Atkins, "Physical Chemistry," 3d ed., chap. 22, Freeman, New York (1986); D. A. McQuarrie, "Statistical Thermodynamics," Harper & Row, New York (1976).
5. G. N. Lewis and M. Randall (revised by K. S. Pitzer and L. Brewer), "Thermodynamics," 2d ed., p. 419*ff.*, McGraw-Hill, New York (1961).

GENERAL READING

G. Herzberg, *op. cit.,* vol. II, chaps. II and III.
J. I. Steinfeld, "Molecules and Radiation," 2d ed., chap. 8, MIT Press, Cambridge, Mass. (1985).

EXPERIMENT 37
RAMAN SPECTROSCOPY: VIBRATIONAL SPECTRUM OF CCl₄

In Exp. 36, the vibrational frequencies of a molecule were determined by measuring the direct absorption of infrared radiation due to transitions between vibrational energy levels. Such changes in molecular state can also be caused by inelastic scattering of higher-energy visible photons. The energy (or frequency) *shifts* in the scattered radiation then give a direct measure of the vibrational frequencies of the molecule. Although the intensity of this scattered light is quite low, Raman was able to observe such scattering in 1928 and the effect was later named after him. With the advent of extremely intense monochromatic laser sources, Raman spectroscopy has become much faster and more sensitive, and instrumentation for such measurements is now reasonably common. Some of the advantages and unique features of Raman spectroscopy will be demonstrated in this experiment on liquid CCl_4.

THEORY

In Exp. 31, the effect of a static or low-frequency electric field on a molecule was considered and it was seen that a dipole moment is induced because of movement of the charged electrons and nuclei. At high optical frequencies ($\sim 10^{15}$ Hz) the nuclei cannot respond rapidly enough to follow the field, but polarization of the electron distribution can occur. For an isolated molecule, an oscillating radiation field of intensity $\mathbf{E}$ will induce a dipole moment of magnitude

$$\boldsymbol{\mu}_{ind} = \alpha \mathbf{E} \tag{1}$$

where α is the molecular polarizability. The field $\mathbf{E}$ oscillates at the frequency ν_0 of the light,

$$\mathbf{E} = \mathbf{E}_0 \cos 2\pi\nu_0 t \tag{2}$$

and hence the induced dipole will also change at this frequency. According to classical electromagnetic theory, any oscillating dipole will radiate energy; hence light of frequency ν_0 is emitted in all directions (except that parallel to the dipole). Such elastically scattered† light is termed Rayleigh scattering and, typically, one out of a million incident photons will be so scattered.

Inelastic or Raman scattering of light can be understood classically as arising from modulation of the electron distribution, and hence the molecular polarizability, because of vibrations of the nuclei. For example, for a diatomic molecule, α can be represented adequately by the first two terms of a power series in the vibrational coordinate Q:

$$\alpha = \alpha_0 + \left(\frac{d\alpha}{dQ}\right)_0 Q \tag{3}$$

Here the subscript zero indicates that the parameters α_0 and $(d\alpha/dQ)_0$ are evaluated at the equilibrium position. In the harmonic-oscillator model, Q for mode i oscillates at the vibrational frequency ν_i according to the relation

$$Q = A \cos 2\pi\nu_i t \tag{4}$$

where A is the maximum amplitude of vibration. Substitution of Eqs. (2)–(4) into Eq. (1) reveals the time dependence of $\boldsymbol{\mu}_{ind}$:

$$\boldsymbol{\mu}_{ind} = \alpha_0 \mathbf{E}_0 \cos 2\pi\nu_0 t + \left(\frac{d\alpha}{dQ}\right)_0 A\mathbf{E}_0 \cos 2\pi\nu_0 t \cos 2\pi\nu_i t$$

$$= \alpha_0 \mathbf{E}_0 \cos 2\pi\nu_0 t + \frac{1}{2}\left(\frac{d\alpha}{dQ}\right)_0 A\mathbf{E}_0[\cos 2\pi(\nu_0 + \nu_i)t + \cos 2\pi(\nu_0 - \nu_i)t] \tag{5}$$

---

†Elastic scattering means no energy change in the scattering process, hence no frequency shift.

The first term is responsible for Rayleigh scattering at ν_0, while the second and third terms produce inelastic Raman scattering at frequencies that are higher (anti-Stokes component) and lower (Stokes component), respectively, than the incident-light frequency. The frequency ν_0 usually corresponds to light in the visible region (typically the 514.5 nm line of the argon-ion laser), and the Raman-shifted light then occurs in the visible region but with an intensity that is 10^{-8} to 10^{-12} times that of the incident light.

The quantum mechanical treatment of light scattering[1] predicts the same general results but more explicitly involves the vibrational energy levels and wavefunctions of the molecule. Figure 1 shows the usual energy level representation of the scattering process. The virtual states are not real energy levels but serve as convenient intermediate levels for picturing the overall scattering process, which occurs in a time less than the period of a molecular vibration ($\sim 10^{-13}$ s). Should the incident radiation have a frequency that causes actual absorption to a real state n, the lifetime of the upper state is much longer ($\sim 10^{-8}$ s) and the resultant intense emission is termed fluorescence. In the absence of such a resonance, the ratio of the Raman intensity of the Raman anti-Stokes and Stokes lines is predicted[2] to be

$$\frac{I_A}{I_S} = \left(\frac{\nu_0 + \nu_i}{\nu_0 - \nu_i}\right)^4 \exp\left(\frac{-h\nu_i}{kT}\right) \tag{6}$$

Because of the ν^4 dependence, blue light is scattered more efficiently than red (a fact which accounts for the blue color of the sky). However, the Boltzmann exponential factor is dominant in Eq. (6), and the anti-Stokes features are always much weaker than the corresponding Stokes lines.

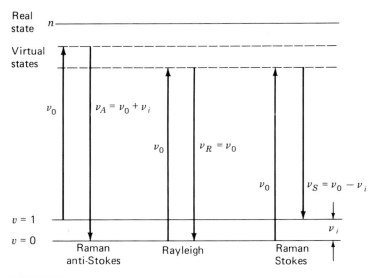

FIGURE 1
Energy levels and transition frequencies involved in Raman spectroscopy.

Raman selection rules. For polyatomic molecules, a number of Stokes Raman bands are observed, each corresponding to an allowed transition between two vibrational energy levels of the molecule. (An allowed transition is one for which the intensity is not uniquely zero owing to symmetry.) As in the case of infrared spectroscopy (see Exp. 36), only the fundamental transitions (corresponding to frequencies $v_1, v_2, v_3, \ldots$) are usually intense enough to be observed, although weak overtone and combination Raman bands are sometimes detected. For molecules with appreciable symmetry, some fundamental transitions may be absent in the Raman and/or infrared spectra. The essential requirement is that the transition moment P (whose square determines the intensity) be nonzero; i.e.,

$$P_{ind}(1 \leftarrow 0) = \int \psi_1 \mu_{ind} \psi_0 \, dQ = E \int \psi_1 \alpha \psi_0 \, dQ \neq 0 \tag{7}$$

for the Raman case, and

$$P(1 \leftarrow 0) = \int \psi_1 \mu \psi_0 \qquad dQ \neq 0 \tag{8}$$

for the infrared case. Here ψ_0 and ψ_1 are wavefunctions for states with vibrational quantum numbers $v = 0$ and $v = 1$, and the integration extends over the full range of the coordinate Q. Since these two constraints are not the same, infrared and Raman spectra generally differ in the number, frequencies, and relative intensities of observed bands. The prediction of allowed transitions for both cases represents an elegant example of the role of symmetry and group theory in chemistry. Although a discussion of this topic is beyond the scope of this book, the application of group theory to vibrational spectroscopy is presented in several standard texts (e.g., Refs. 2 to 4). A summary of the conclusions reached for a few simple molecular structures is provided in Table 1. Here, the general form of the vibrational coordinates is shown with the mode numbering and the infrared and Raman activity for the fundamental vibrations. Each molecular type has a unique set of symmetry elements (rotation axes, reflection planes, etc.). Associated with these elements are operations (such as rotation about symmetry axes, reflections, etc.) that transform the molecule from one configuration in space to another that is indistinguishable. This set of operations defines the point group, which is indicated in the table. The point group is associated with a unique set of symmetry species (denoted $\Sigma_g^+, A_1, B_1, E, F_2$, etc.), each of which behaves differently under the symmetry operations of the molecule. Every vibrational coordinate (and wavefunction) belongs to one of these species, as indicated in the table; it is knowledge of this symmetry that permits the prediction of the infrared and Raman activity of a given fundamental transition.

Raman depolarization ratio. A more complete discussion of vibrational selection rules must take into account the vector properties of the dipole

TABLE 1
Normal modes of vibration for several molecular types

Type	Point group	Vibrational modes,[a] symmetries, infrared-Raman activities, and Raman polarizations[b]			

B—A—B $D_{\infty h}$

ν_1 — Σ_g^+ (Rp)

ν_{2x} — ν_{2y} — Π_u (IR)

ν_3 — Σ_u^+ (IR)

C_{2v}

ν_1 — A_1 (IR, Rp)

ν_2 — A_1 (IR, Rp)

ν_3 — B_1 (IR, Rdp)

D_{3h}

ν_1 — A_1' (Rp)

ν_2 — A_2'' (IR)

ν_3 — E' (IR, Rdp)

ν_4 — E' (IR, Rdp)

C_{3v}

ν_1 — A_1 (IR, Rp)

ν_2 — A_1 (IR, Rp)

ν_3 — E (IR, Rdp)

ν_4 — E (IR, Rdp)

T_d

ν_1 — A_1 (Rp)

ν_2 — E (Rdp)

ν_3 — F_2 (IR, Rdp)

ν_4 — F_2 (IR, Rdp)

[a] Only one of two E or three F modes in the respective degenerate sets is shown for the AB_3 and AB_4 cases. Each mode of a degenerate set oscillates at the same frequency, and the description of the nuclear motion associated with each component of a degenerate mode is somewhat arbitrary.

[b] Rp indicates an allowed Raman band that is polarized; Rdp indicates an allowed depolarized Raman band.

moment and the tensor character of the polarizability α. Thus, Eq. (1) should be written

$$\begin{pmatrix} \mu_x \\ \mu_y \\ \mu_z \end{pmatrix}_{\text{ind}} = \begin{pmatrix} \alpha_{xx} & \alpha_{xy} & \alpha_{xz} \\ \alpha_{yx} & \alpha_{yy} & \alpha_{yz} \\ \alpha_{zx} & \alpha_{zy} & \alpha_{zz} \end{pmatrix} \begin{pmatrix} E_x \\ E_y \\ E_z \end{pmatrix} \tag{9}$$

Such considerations are of considerable importance in the study of oriented molecules, as in a crystal, but are less important for liquids and gases, in which the molecular orientation is random. However, in the case of Raman spectroscopy, not all information about the components of α is lost by orientational averaging. In particular, one finds that it is possible to distinguish totally symmetric from other molecular vibrations from a measurement of the depolarization ratio

$$\rho_l \equiv \frac{I_\perp}{I_\parallel} \tag{10}$$

of the Raman lines. Here, $I_\perp$ and $I_\parallel$ are the intensities of scattered light with polarization perpendicular ($\perp$) and parallel ($\parallel$) to the polarization of the linearly polarized exciting light. Theory shows that ρ_l is related to combinations of the tensor components of α and that $\rho_l = \frac{3}{4}$ for any vibration that is *not* totally symmetric and $0 \le \rho_l \le \frac{3}{4}$ for those that are totally symmetric.[5] Raman bands for which $\rho_l = \frac{3}{4}$ are called depolarized, and those with $\rho_l < \frac{3}{4}$ are called polarized. Totally symmetric vibrations are those in which the symmetry of the molecule does not change during the vibrational motion (e.g., the Σ_g^+, A_1', and A_1 vibrations of Table 1). For such vibrations, ρ_l values near zero are observed for highly symmetric molecules such as CO_2, SF_6, and CCl_4. Thus, the measurement of ρ_l will be used in this experiment to aid in assigning the frequency of the totally symmetric v_1 vibration of CCl_4.

Valence-force model. The simplest harmonic oscillator model for the potential energy U of a tetrahedral molecule such as CCl_4 can be written as

$$U = \tfrac{1}{2}k(r_1^2 + r_2^2 + r_3^2 + r_4^2) + \tfrac{1}{2}k_\delta(\delta_{12}^2 + \delta_{13}^2 + \delta_{14}^2 + \delta_{23}^2 + \delta_{24}^2 + \delta_{34}^2) \tag{11}$$

where r_i is the change in the length of C—Cl bond i from its equilibrium value l and δ_{ij} is the change in the angle between bonds i and j. From classical mechanics one obtains[6] the following relations between the vibrational frequencies and the force constants k and k_δ/l^2:

$$4\pi^2 v_1^2 = \frac{k}{m_{Cl}} \tag{12}$$

$$4\pi^2 v_2^2 = \left(\frac{3}{m_{Cl}}\right)\frac{k_\delta}{l^2} \tag{13}$$

$$4\pi^2(v_3^2 + v_4^2) = \left(1 + \frac{4m_{Cl}}{3m_C}\right)\frac{k}{m_{Cl}} + \left(1 + \frac{8m_{Cl}}{3m_C}\right)\left(\frac{2}{m_{Cl}}\right)\frac{k_\delta}{l^2} \tag{14}$$

$$16\pi^4 v_3^2 v_4^2 = \left(1 + \frac{4m_{Cl}}{m_C}\right)\left(\frac{2}{m_{Cl}}\right)\frac{k_\delta}{l^2}\frac{k}{m_{Cl}} \tag{15}$$

(See Exp. 36 for a discussion of the units to be used in these equations.) Since there are two force constants and four frequencies, one can test the quality of this force field by separately determining k and k_δ/l^2 from Eqs. (12) and (13) and from Eqs. (14) and (15).

METHOD

There are many different Raman spectrometers in current use, almost all of which are suitable for this experiment. A general description of Raman instruments with grating monochromators and laser sources is given in Chapter XVIII. Before beginning the experiment, you should review the pertinent material in Chapter XVIII and read carefully any available information about the actual instrument to be used in the laboratory. **Use this spectrometer with care**; if in doubt, ask the instructor for advice about proper procedures.

In order to measure depolarization ratios, a polaroid analyzer and a scrambler are placed in front of the entrance slit of the spectrometer. Since the laser source is polarized, ρ_l is obtained by measuring $I_\parallel$ and $I_\perp$ when the polaroid is oriented to pass only light whose polarization is parallel or perpendicular, respectively, to the polarization of the exciting light (see Fig. 2). The scrambler serves to completely depolarize the light passed by the

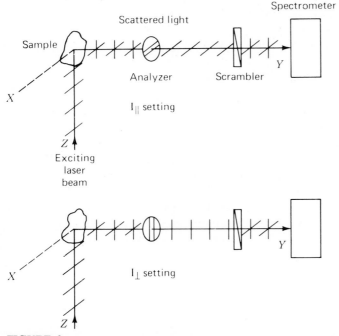

FIGURE 2
Polarization arrangements for Raman depolarization measurements.

polaroid; this is necessary since the reflection efficiency of the gratings in the spectrometer is polarization dependent. It should be mentioned that the theoretical value of $\frac{3}{4}$ for depolarized Raman bands applies strictly only for 90° scattering. Since the exciting source is focused and the collection angle is usually large, measured values of ρ_l can exceed $\frac{3}{4}$. If very precise values are desired, a collimated excitation beam and a smaller collection angle near 90° are used in conjunction with calibration by bands of known ρ_l value.

EXPERIMENTAL

Detailed instructions for the use of the Raman spectrometer will be provided in the laboratory. *It is particularly important to follow all safety instructions regarding the use of the excitation laser since serious eye damage can occur if the beam, or even a small reflection from it, should strike the eye.* **Safety goggles must be worn when inserting samples in the beam and when adjusting the optics for maximum signal.**

A convenient liquid cell for Raman spectroscopy is a 1-cm glass or quartz cuvette, with a clear bottom if the excitation beam enters from below. Alternatively, melting-point capillaries can serve as inexpensive cells; these have the added virtue when studying a scarce or expensive sample that very little sample volume is required. A 10 cm length of thin-walled capillary tubing is melted at the center and separated to form two 5 cm tubes. Any excess glass is removed and the end is sealed while twirling to form a hemispherical bottom. This tube can be used as is, or a flat bottom can be achieved by inserting a flat stainless steel or tungsten rod and pressing the heated bottom against a flat carbon plate. A hypodermic syringe can be used to add sample to the capillary up to a height of 1 to 2 cm. The top can then be sealed if the sample is sensitive to air or is volatile.

The scattering cylinder of the focused laser beam must be aligned vertically to be parallel to the entrance slit of the spectrometer. The collection lens is used to image this cylinder onto the slit. The sample is then inserted at the laser focal point with its long axis parallel to the beam and a quick survey scan is taken until a Raman band is found. The sample, the excitation beam, and the collection lens are then carefully adjusted for maximum signal at this spectrometer setting.

Scan the Stokes Raman spectrum of CCl_4 for shifts of 150 to 1000 cm^{-1} from the exciting line.† Record both parallel and perpendicular polarization scans so that you can determine the depolarization ratio of all bands. Indicate the spectrometer wavelength or wavenumber reading on the chart at several

---

† For reference, suitable Stokes and anti-Stokes Raman spectra of CCl_4 can be found in the review of Raman spectroscopy in Ref. 7. A high-resolution scan of the 460 cm^{-1} band of liquid and solid CCl_4 is available in Ref. 8. Raman spectra of several molecules are described in Refs. 9 and 10.

points in the scan to provide reference points for the determination of the Raman shifts for all bands.

Block the laser beam or spectrometer entrance slit and adjust the spectrometer to an anti-Stokes shift of 1000 cm^{-1}. **Caution:** *exposure of the sensitive phototube to the intense Rayleigh scattering line can seriously damage the detector.* Scan the anti-Stokes spectrum from -1000 to -150 cm^{-1} in the parallel polarization configuration and, using appropriate sensitivity expansion,† measure the ratio of anti-Stokes to Stokes peak heights for each band.

You will note some structure in the Stokes band near 460 cm^{-1}, which is due to chlorine isotopic frequency shifts. (See Exp. 38 for a discussion of isotope effects in diatomic molecules.) Re-scan this region at higher resolution (1 cm^{-1} or less) with an expansion of the chart display and measure the frequencies of each of the components.

CALCULATIONS AND DISCUSSION

Assign the fundamentals of CCl₄ using the fact that the v_1 and v_3 CCl₄ stretching motions are expected to occur at higher frequencies than the v_2 and v_4 CCl₄ bending motions. The depolarization ratios can be used to determine which of the two high-frequency bands should be assigned as v_1. To decide on the v_2 and v_4 assignments, use the fact that the only infrared band observed below 600 cm^{-1} is a feature near 300 cm^{-1}.

The v_3 vibration will be observed as a doublet due to Fermi-resonance splitting, a process that can occur if an overtone or combination energy level happens to fall near the energy level of a fundamental vibration and if both levels have the same symmetry. The v_3 frequency can be taken as the mean of the doublet peaks. From your values of v_1, v_2, and v_4, deduce which combination band is involved in the Fermi-resonance with v_3.

Tabulate your Stokes and anti-Stokes frequencies and also the intensity ratios I_A/I_S wherever possible. According to Eq. (6), the intensity ratio for each mode can be used to determine the vibrational temperature characterizing the Boltzmann distribution. In the present experiment, the vibrational temperature should agree with the ambient temperature of the laboratory. Does it? This technique of using information about spectral intensities to determine T is a very convenient method for finding the temperatue of flames, shock waves, and plasmas.

Calculate the valence force constants k and k_δ/l^2 from Eqs. (12) and (13), and then compare these values with those obtained from Eqs. (14) and

---

† Known scale expansions should be chosen that give peaks that are nearly full scale, so that the heights can be accurately measured. Some errors will arise in the measurement of Stokes and anti-Stokes intensities because of variation with frequency of spectrometer transmission and detector efficiency. These effects should be unimportant for bands involving small Raman shifts, and calibration corrections can be made if necessary for very accurate measurements.

(15). A more complete force field analysis[11] predicts that k in Eq. (12) should be replaced by $k_r + 3k_{rr}$ while k in Eqs. (14) and (15) should be replaced by $k_r - k_{rr}$, where k_r is the bond-stretching force constant and k_{rr} is a constant due to bond–bond interaction. Is k_{rr} important in CCl_4?

From the natural abundance of ^{35}Cl (75.5 percent) and ^{37}Cl (24.5 percent), calculate the relative intensities expected for the various components of the band near $460 \, \text{cm}^{-1}$ and compare these with your observations. Suggest a reasonable modification of one of Eqs. (12) to (15) that would permit you to predict the frequency spacing of these components.

Alternative or additional liquids suitable for Raman investigation include CS_2, PCl_3, $CHCl_3$, or benzene. Comparisons of similar molecules (e.g., CCl_4, $CHCl_3$, $CDCl_3$, or C_6H_6, C_6D_6, and substituted benzenes) can provide a basis for frequency assignments and can be used to test the transferability of force constants. Note that several of these samples are toxic and must be handled with care. Capillary samples can be prepared in a hood, frozen with liquid nitrogen, and then sealed with a flame (or simply sealed off with soft wax).

Aqueous solutions can also be studied, since the Raman spectrum of water is weak and quite broad. Possible samples would be $1 \, M$ to $3 \, M$ solutions of Na_2CO_3, $NaNO_3$, Na_2SO_3, or Na_2SO_4. Such samples are difficult to study using infrared spectroscopy, because of solvent absorption and the solubility of most infrared cell windows.

APPARATUS

Raman spectrometer with argon- or krypton-ion laser source; liquid-sample cell or melting-point capillaries; reagent-grade CCl_4; safety goggles (such as those available from Glendale Optical Co., Woodbury, NY 11797).

REFERENCES

1. D. A. Long, "Raman Spectroscopy," McGraw-Hill, New York (1977).
2. E. B. Wilson, Jr., J. C. Decius, and P. C. Cross, "Molecular Vibrations," p. 51, McGraw-Hill, New York (1955), reprinted in unabridged form by Dover, New York (1980).
3. D. C. Harris and M. D. Bertolucci, "Symmetry and Spectroscopy," Oxford, London (1978).
4. F. A. Cotton, "Chemical Applications of Group Theory," Wiley-Interscience, New York (1963).
5. E. B. Wilson, Jr., J. C. Decius, and P. C. Cross, *op. cit.*, p. 47.
6. G. Herzberg, "Molecular Spectra and Molecular Structure II. Infrared and Raman Spectra of Polyatomic Molecules," p. 182, Van Nostrand, Princeton, N.J. (1945).
7. R. S. Tobias, *J. Chem. Educ.* **44**, 2, 70 (1967).
8. H. J. Sloane, *Appl. Spectrosc.* **25**, 430 (1971).
9. F. P. DeHaan, J. C. Thibeault, and D. K. Ottesen, *J. Chem. Educ.* **51**, 263 (1974).
10. L. C. Hoskins, *J. Chem. Educ.* **52**, 568 (1975); **54**, 642 (1977).
11. E. B. Wilson, Jr., J. C. Decius, and P. C. Cross, *op. cit.*, p. 131.

GENERAL READING

D. A. Long, *op. cit.*
R. S. Tobias, *op. cit.*
G. Herzberg, *op. cit.*, chap. II.
F. A. Cotton, *op. cit.*, chaps. 1–5, 9.

EXPERIMENT 38
VIBRATIONAL–ROTATIONAL SPECTRA OF
HCl AND DCl

This experiment is concerned with the rotational fine structure of the infrared vibrational spectrum of a linear molecule such as HCl. From an interpretation of the details of this spectrum it is possible to obtain the moment of inertia of the molecule and thus the internuclear separation. In addition, the pure vibrational frequency determines a force constant that is a measure of the bond strength. By a study of DCl also, the isotope effect can be observed.

THEORY

The simplest model of a vibrating diatomic molecule is a harmonic oscillator, for which the potential energy depends quadratically on the change in internuclear distance. The allowed energy levels of a harmonic oscillator, as calculated from quantum mechanics,[1] are

$$E(v) = hv(v + \tfrac{1}{2}) \tag{1}$$

where v is the vibrational quantum number having integral values $0, 1, 2, \ldots$, v is the vibrational frequency, and h is Planck's constant.

The simplest model of a rotating diatomic molecule is a rigid rotor or "dumb-bell" model in which the two atoms of mass m_1 and m_2 are considered to be joined by a rigid, weightless rod. The allowed energy levels for a rigid rotor may be shown by quantum mechanics[1] to be

$$E(J) = \frac{h^2}{8\pi^2 I} J(J + 1) \tag{2}$$

where the rotational quantum number J may take integral values $0, 1, 2, \ldots$. The quantity I is the moment of inertia, which is related to the internuclear distance r and the reduced mass $\mu = m_1 m_2/(m_1 + m_2)$ by

$$I = \mu r^2 \tag{3}$$

Since a real molecule is undergoing both rotation and vibration simultaneously, a first approximation to its energy levels $E(v, J)$ would be the sum of expressions (1) and (2). A more complete expression for the energy levels of a diatomic molecule[2] is given below, with the levels expressed as *term values* T in cm^{-1} units rather than as energy values E in joules:

$$T(v, J) = \frac{E(v, J)}{hc} = \tilde{\nu}_e(v + \tfrac{1}{2}) - \tilde{\nu}_e x_e(v + \tfrac{1}{2})^2 + B_e J(J + 1)$$
$$- D_e J^2(J + 1)^2 - \alpha_e(v + \tfrac{1}{2})J(J + 1) \tag{4}$$

where c is the speed of light in cm s^{-1}, $\tilde{\nu}_e$ is the frequency in cm^{-1} for the molecule vibrating about its equilibrium internuclear separation r_e, and

$$B_e = \frac{h}{8\pi^2 I_e c} \tag{5}$$

The first and third terms on the right-hand side of Eq. (4) are the harmonic-oscillator and rigid-rotor terms with r equal r_e. The second term (involving the constant x_e) takes into account the effect of anharmonicity. Since the real potential $U(r)$ for a molecule differs from a harmonic potential U_{harm} (see Fig. 1), the real vibrational levels are not quite those given by Eq. (1) and a correction term is required. The fourth term (involving the constant D_e) takes into account the effect of centrifugal stretching. Since a chemical bond is not truly rigid but more like a stiff spring, it stretches somewhat when the molecule rotates. Such an effect is important only for high J values, since

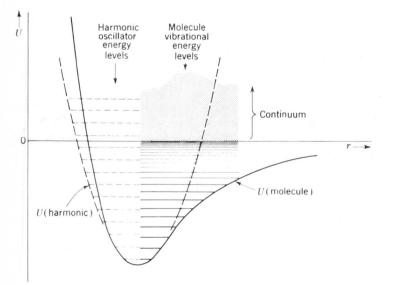

FIGURE 1

Schematic diagram showing potential energy U as a function of internuclear separation r for a diatomic molecule. The harmonic potential U_{harm} is indicated by the dashed curve. The vibrational levels are also shown.

the constant D_e is very small; we shall neglect this term. The last term in Eq. (4) accounts for interaction between vibration and rotation. During a vibration the internuclear distance r changes; this changes the moment of inertia and affects the rotation of the molecule. The constant α_e is also quite small, but this term should not be neglected.

Selection rules. The harmonic-oscillator, rigid-rotor selection rules[2] are $\Delta v = \pm 1$ and $\Delta J = \pm 1$; that is, infrared emission or absorption can occur only when these "allowed" transitions take place. For an anharmonic diatomic molecule, the $\Delta J = \pm 1$ selection rule is still valid but weak transitions corresponding to $\Delta v = \pm 2, \pm 3$, etc. (overtones) can now be observed.[2] Since we are interested in the most intense absorption band (the "fundamental"), we are concerned with transitions from various J'' levels of the vibrational ground state ($v'' = 0$) to J' levels in the first excited vibrational state ($v' = 1$). From the selection rule we know that the transition must be from J'' to $J' = J'' \pm 1$. Since $\Delta E = hv = hc\tilde{v}$, the frequency $\tilde{v}$ (in wavenumbers) for this transition will be just $T(v', J') - T(v'', J'')$. When $\Delta J = +1$ ($J' = J'' + 1$) and $\Delta J = -1$ ($J' = J'' - 1$), we find, respectively, from Eq. (4) that

$$\tilde{v}_R = \tilde{v}_0 + (2B_e - 3\alpha_e) + (2B_e - 4\alpha_e)J'' - \alpha_e J''^2 \qquad J'' = 0, 1, 2, \ldots \qquad (6)$$

$$\tilde{v}_P = \tilde{v}_0 - (2B_e - 2\alpha_e)J'' - \alpha_e J''^2 \qquad\qquad J'' = 1, 2, 3, \ldots \qquad (7)$$

where $\tilde{v}_0$, the frequency of the *forbidden* transition from $v'' = 0$, $J'' = 0$ to $v' = 1$, $J' = 0$, is

$$\tilde{v}_0 = \tilde{v}_e - 2\tilde{v}_e x_e \qquad (8)$$

The two series of lines given in Eqs. (6) and (7) are called R and P branches, respectively. These allowed transitions are indicated on the energy-level diagram given in Fig. 2. If α_e were negligible, Eqs. (6) and (7) would predict a series of equally spaced lines with separation $2B_e$ except for a missing line at $\tilde{v}_0$. The effect of interaction between rotation and vibration (nonzero α_e) is to draw the lines in the R branch closer together and spread the lines in the P branch farther apart as shown for a typical spectrum in Fig. 3. For convenience let us introduce a new quantity m, where $m = J'' + 1$ for the R branch and $m = -J''$ for the P branch as shown in Fig. 3. It is now possible to replace Eqs. (6) and (7) by a single equation

$$\tilde{v} = \tilde{v}_0 + (2B_e - 2\alpha_e)m - \alpha_e m^2 \qquad (9)$$

where m takes all integral values except zero. The separation between adjacent lines $\Delta\tilde{v}(m)$ will be

$$\Delta\tilde{v}(m) = \tilde{v}(m + 1) - \tilde{v}(m) = (2B_e - 3\alpha_e) - 2\alpha_e m \qquad (10)$$

Thus a plot of $\Delta\tilde{v}(m)$ versus m can be used to determine both B_e and α_e. With these known, $\tilde{v}_0$ can be obtained from any of the lines by using Eq. (9).

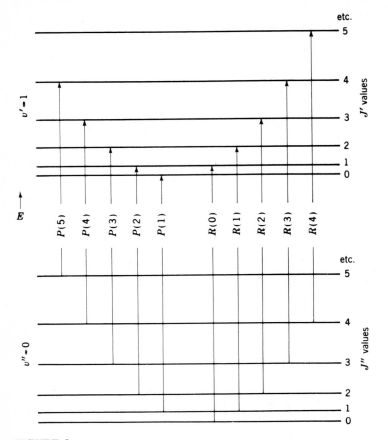

FIGURE 2

Rotational energy levels for the ground vibrational state ($v'' = 0$) and the first excited vibrational state ($v' = 1$) in a diatomic molecule. The vertical arrows indicate allowed transitions in the R and P branches; numbers in parentheses index the value J'' of the lower state. Transitions in the Q branch ($\Delta J = 0$) are not shown since they are not infrared active.

Isotope effect. When an isotopic substitution is made in a diatomic molecule, the equilibrium bond length r_e and the force constant k are unchanged, since they depend only on the behavior of the bonding electrons. However, the reduced mass μ does change, and this will affect the vibration and rotation of the molecule. For a harmonic oscillator model, the frequency ν_{harm} is given by

$$\nu_{\text{harm}} = \frac{1}{2\pi} \left(\frac{k}{\mu}\right)^{1/2} \tag{11}$$

If an asterisk is used to distinguish one isotopic molecule from another, Eq. (11) leads to the relation

$$\frac{\nu^*_{\text{harm}}}{\nu_{\text{harm}}} = \left(\frac{\mu}{\mu^*}\right)^{1/2} \tag{12}$$

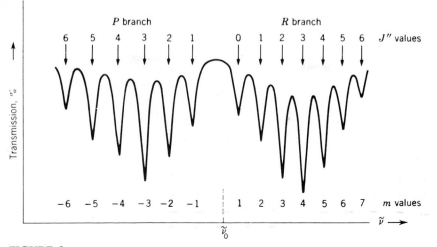

FIGURE 3
Schematic rotation-vibration infrared spectrum for a diatomic molecule.

For a real molecule the anharmonicity should be taken into account,[2] but since this is complicated and would require information from overtone vibrations ($\Delta v > 1$), we shall take Eq. (12) as an approximation and state that $\tilde{\nu}_0^*/\tilde{\nu}_0$ should be close to $(\mu/\mu^*)^{1/2}$. In the case of rotation, the isotope effect can be easily stated. From the definitions of B_e and I, we see that

$$\frac{B_e^*}{B_e} = \frac{\mu}{\mu^*} \tag{13}$$

Since HCl gas is a mixture of $H^{35}Cl$ and $H^{37}Cl$ molecules, an isotope effect should be present. However, HCl is predominantly $H^{35}Cl$ and the ratio of the reduced masses is only 1.0015; therefore quite high resolution is required to detect this effect. For this experiment, we shall assume that the HCl bands obtained are those of $H^{35}Cl$.

If deuterium is substituted for hydrogen, the ratio of the reduced masses, $\mu(D^{35}Cl)/\mu(H^{35}Cl)$, is 1.946. Thus there should be a very large isotope effect.

EXPERIMENTAL

A general description of infrared instruments is given in Chapter XVIII. Medium to high resolution is required for this experiment; a grating or FTIR instrument with at least $2\,\text{cm}^{-1}$ resolution is desirable. In addition, grating spectrometers require careful calibration. CO and CH_4 are suitable gases for calibration purposes. Detailed instructions for operating the spectrometer will be given in the laboratory. **Use this instrument carefully.**

The cell design and procedure for filling the cell are identical with those described in Exp. 36. Hydrogen chloride gas is available commercially in small

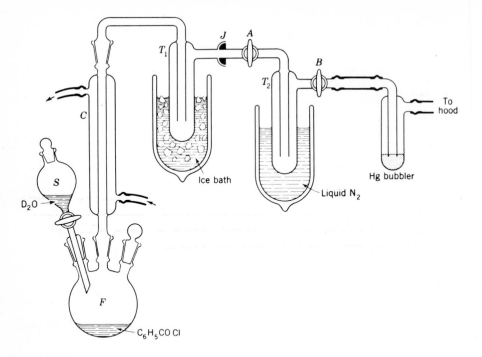

FIGURE 4
Apparatus for preparing deuterium chloride.

cylinders. Make several rapid scans of the HCl spectrum at various pressures to determine what pressure gives the best spectrum, and then record the spectrum at a very slow chart speed. Narrow slit widths are necessary to obtain the desired sharp bands.

Preparation of DCl. If DCl gas is not available in a commercial cylinder or bulb, it must be prepared in the laboratory. Deuterium chloride gas can be synthesized readily by the reaction between benzoyl chloride and heavy water:

$$C_6H_5COCl + D_2O \rightarrow C_6H_5COOD + DCl(g)$$
$$C_6H_5COOD + C_6H_5COCl \rightarrow (C_6H_5CO)_2O + DCl(g) \tag{14}$$

An arrangement for carrying out this reaction is shown in Fig. 4. Approximately 5 mL (0.3 mol) of D_2O in the separating funnel S is added slowly to 140 g (1 mol) of benzoyl chloride in flask F. With a water-cooled reflux condenser C attached and either ice or Dry Ice and trichloroethylene in trap T_1, the flask is gently heated. Stopcocks A and B are left open, and the system is swept out by allowing some DCl gas to escape through the mercury bubbler into a hood. Stopcock B is then closed, and trap T_2 is cooled with liquid nitrogen. After 20 min, remove the heat from flask F. Wait a few minutes, then close stopcock A and disconnect the ball-and-socket joint J. Keep trap T_2

in liquid nitrogen, and attach it to the system used to fill the cell (at stopcock B shown in Fig. 36-3). After the line is pumped out, the cell can be filled by allowing the DCl in the trap to warm up slowly until the desired pressure is achieved. (If a bulb of DCl gas is available, use the procedure given in Exp. 36.)

CALCULATIONS

Select your best HCl and DCl spectra, and index the lines with the appropriate m values as shown in Fig. 3. If $^{35}Cl/^{37}Cl$ splitting is seen, index the stronger ^{35}Cl lines. Make a table of these m values and the corresponding frequencies $\bar{v}(m)$. Express the frequencies in cm^{-1} units to tenths of a cm^{-1} if possible. Then list the differences between adjacent lines $\Delta\bar{v}(m)$, which will be roughly $2B_e$ but should vary with m. Plot $\Delta\bar{v}(m)$ against m, draw a straight line through the points, and check any points that seem out of line. Then carry out a linear least-squares fit to the data with Eq. (10), which will determine the intercept $(2B_e - 3\alpha_e)$ and the slope $(-2\alpha_e)$. Compute B_e and α_e. Using these values and Eq. (9), calculate $\bar{v}_0$ from several of the lines with low m values. For both HCl and DCl, present a table of $\bar{v}_0$, B_e, and α_e values, all in cm^{-1} units.

Calculate I_e, the moment of inertia, and r_e, the internuclear distance, for both HCl and DCl. Compute the ratios $\bar{v}_0^*/\bar{v}_0$ and B_e^*/B_e and compare them with the predicted values given by Eqs. (12) and (13). The masses (in atomic mass units) are $m_H = 1.007825$, $m_D = 2.014102$, $m_{^{35}Cl} = 34.968853$, and $m_{^{37}Cl} = 36.965903$.

DISCUSSION

Do your spectra show any indications of a $^{35}Cl-^{37}Cl$ isotope effect? Calculate the splitting expected for this effect in HCl and in DCl. Explain the band intensities in terms of the Boltzmann populations of the ground-state levels.

An interesting extension of this experiment is to use the spectroscopic data to calculate the equilibrium constant K for the gas-phase exchange reaction

$$DCl + HBr = HCl + DBr \qquad (15)$$

From statistical thermodynamics,[3] one can show in the ideal-gas approximation that K for isotopic exchange is related to the molecular partition functions q by the equations

$$K = \frac{p_{HCl}p_{DBr}}{p_{DCl}p_{HBr}} = \frac{q_{HCl}q_{DBr}}{q_{DCl}q_{HBr}} \qquad (16)$$

$$= \left[\frac{M_{HCl}M_{DBr}}{M_{DCl}M_{HBr}}\right]^{3/2}\left[\frac{B_{DCl}B_{HBr}}{B_{HCl}B_{DBr}}\right]\left[\frac{q_{HCl}^{vib}q_{DBr}^{vib}}{q_{DCl}^{vib}q_{HBr}^{vib}}\right] \qquad (17)$$

The term involving the molecular weights M_i comes from the translational contribution to K; the term involving B_i values represents the rotational contribution; and q_i^{vib} is the vibrational partition function for species i defined in Eq. (36-7). The expression given in Eq. (17) corresponds to the rigid-rotor, harmonic-oscillator approximation. If the values of B and v are determined experimentally for HBr gas (or taken from Ref. 2), one can use the isotopic mass relations given in Eqs. (12) and (13) to obtain B and v for DBr. Combining these data with those on HCl and DCl, a theoretical or "spectroscopic" value of K can be determined from Eq. (17).

An experiment has been described in which infrared intensities are used to obtain an experimental determination of K (Ref. 4), but this method is not straightforward. The accurate measurement of intensities is complicated by pressure-broadening effects and limitations in spectrometer resolution. A preferable and more accurate determination of K can be made with a mass spectrometer.

APPARATUS

Medium- or high-resolution infrared grating or FTIR spectrometer; gas cell with KBr windows; vacuum line for filling cell; cylinder of HCl gas with needle valve; three-neck round-bottom flask; reflux condenser; glass-stoppered dropping funnel; two traps, one with a stopcock on each arm; mercury bubbler; exhaust hood; bunsen burner or heating mantle; CO and CH_4 gas for calibration check (optional).

Heavy water (5 mL at least 70 percent D_2O); benzoyl chloride (140 g); ice or Dry Ice and isopropanol; liquid nitrogen; or a 5-liter flask of DCl gas (available from Cambridge Isotope Laboratories, 20 Commerce Way, Woburn, MA. 01801 and other suppliers of isotopically substituted compounds).

REFERENCES

1. P. W. Atkins, "Physical Chemistry," 3d ed., chap. 22, Freeman, New York (1986).
2. G. Herzberg, "Molecular Spectra and Molecular Structure I. Spectra of Diatomic Molecules," 2d ed., chap. III, Van Nostrand, Princeton, N.J. (1950); K. P. Huber and G. Herzberg, "Molecular Spectra and Molecular Structure IV. Constants of Diatomic Molecules," Van Nostrand–Reinhold, New York (1979).
3. D. A. McQuarrie, "Statistical Thermodynamics," Harper & Row, New York (1976).
4. J. E. Ruark, J. I. Ivers, and J. L. Roberts, *J. Chem. Educ.* **51,** 758 (1974).

GENERAL READING

G. Herzberg, *op. cit.*
J. M. Hollas, "Modern Spectroscopy," Wiley, New York (1987).
J. I. Steinfeld, "Molecules and Radiation," 2d ed., MIT Press, Cambridge, Mass. (1985).

EXPERIMENT 39
VIBRATIONAL-ROTATIONAL SPECTRUM
OF ACETYLENE

In this experiment, several vibrational–rotational infrared bands of C_2H_2 and C_2D_2 will be recorded at medium-to-high resolution (~ 1 cm^{-1}). These spectra will be analyzed to extract rotational constants for use in the calculation of accurate values for the C—H and C—C bond lengths. The role of symmetry and nuclear spin in determining the activities and intensity patterns of the spectral transitions is also examined. From such considerations, the infrared bands can be assigned to specific modes of vibration and values can be deduced for the fundamental vibrational frequencies of C_2H_2 and C_2D_2.[1,2]

THEORY

Vibrational levels and wavefunctions. Acetylene is known to be a symmetric linear molecule with $D_{\infty h}$ point group symmetry and $3N - 5 = 7$ vibrational *normal modes,* as depicted in Table 1. Symmetry is found to be an invaluable aid in understanding the motions in polyatomic molecules, as discussed in detail in Refs. 3–9. Group theory shows that each vibrational coordinate and each vibrational energy level, along with its associated wavefunction, must

TABLE 1
Fundamental vibrational modes of acetylene

Normal mode		Symmetry species	Description	Activity, band type[a]	Frequency (cm^{-1})[b]	
					C_2H_2	C_2D_2
H—C≡C—H	ν_1	Σ_g^+	sym. CH stretch	Rp, ‖	3372.8	2705.2
H—C≡C—H	ν_2	Σ_g^+	CC stretch	Rp, ‖	1974.3	1764.8
H—C≡C—H	ν_3	Σ_u^+	antisym. CH stretch	IR, ‖	3294.8	2439.2
H—C≡C—H $\atop$ H—C≡C—H	ν_4	Π_g	sym. bend	Rdp, ⊥	612.9	511.5
H—C≡C—H $\atop$ H—C≡C—H	ν_5	Π_u	antisym. bend	IR, ⊥	730.3	538.6

[a] The designations IR or R indicate that the fundamental transition for the mode is infrared or Raman active respectively, and the labels p and dp give the polarization of the Raman band (See Exp. 37 for a detailed discussion). Parallel bands (‖) have PR branches while perpendicular bands (⊥) show PQR branches.

[b] The frequencies are from compilations in Refs. 1 and 2.

have a symmetry corresponding to one of the symmetry species of the molecular point group. The $D_{\infty h}$ symmetry species corresponding to the different types of atomic motion in acetylene are indicated in the table. Motions that retain the center of inversion symmetry, such as the v_1, v_2, and v_4 modes of Table 1, are labeled g (*gerade*, German for even) while those for which the displacement vectors are reversed on inversion are labeled u (*ungerade*, odd). Modes involving motion along the molecular axis (z) are called parallel vibrations and labelled Σ while those involving perpendicular motion are labelled Π and are doubly degenerate since equivalent bending can occur in either x or y directions. From the appearance of the nuclear displacements, it can be seen that only the v_3 and v_5 modes produce an oscillating change in the zero dipole moment of the molecule and hence give rise to infrared absorption.

From the harmonic-oscillator model of quantum mechanics, the term value G for the vibrational energy levels for a linear polyatomic molecule can be written as[5]

$$G(v_1, v_2, \dots) = \sum_{i=1}^{3N-5} \bar{v}_i(v_i + \tfrac{1}{2}) \tag{1}$$

where $\bar{v}_i = v_i/c$ is the vibrational frequency of mode i in cm^{-1} when c is the speed of light in cm s^{-1}. Additional anharmonicity corrections, analogous to $\bar{v}_e x_e$ for diatomic molecule (see Exps. 38, 41, 42), can be added; but these are usually small (1–5 per cent of $\bar{v}_i$) and will be neglected in this discussion. The energy levels of some of the states of acetylene are shown in Fig. 1. Each level is characterized by a set of harmonic oscillator quantum numbers $v_1 v_2 v_3 v_4 v_5$, shown at the left of the figure. The *fundamental* transitions from the ground state are those in which only one of the five quantum numbers increases from 0 to 1; the two infrared active fundamentals v_3 and v_5 are indicated with bold arrows in the figure.

The set of quantum numbers of a level also serves to define the corresponding wavefunction, which, in the usual approximation, is written as a product of one-dimensional harmonic oscillator functions[3,5,7]

$$\psi_{v_1 v_2 v_3} \dots = \phi_{v_1}(Q_1)\phi_{v_2}(Q_2)\phi_{v_3}(Q_3) \cdots \tag{2}$$

The latter have the form

$$\phi_{0_i}(Q_i) = [\gamma_i/\pi]^{1/4} \exp[-\gamma_i Q_i^2/2] \tag{3a}$$

$$\phi_{1_i}(Q_i) = [4\gamma_i/\pi]^{1/4} \exp[-\gamma_i Q_i^2/2]\gamma_i^{1/2}Q_i \tag{3b}$$

$$\phi_{2_i}(Q_i) = [\gamma_i/4\pi]^{1/4} \exp[-\gamma_i Q_i^2/2][2\gamma_i Q_i^2 - 1] \quad \text{etc.} \tag{3c}$$

where $\gamma_i = k_i/hc\bar{v}_i$ and k_i is the quadratic force constant for the mode of frequency $\bar{v}_i$. The function ϕ is an even or odd polynomial in the *normal coordinate* Q_i, depending on the evenness or oddness of the quantum number v_i. In general, Q_i is a combination of bond-stretching and angle-bending coordinates that all oscillate in phase and at the characteristic frequency v_i.

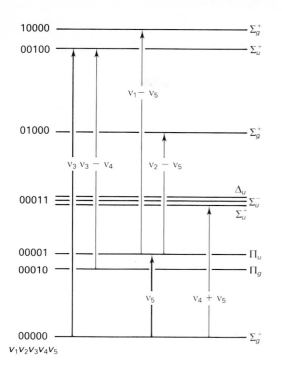

FIGURE 1
Vibrational energy levels and relevant infrared-active transitions for acetylene. The fundamental v_3 and v_5 transitions are indicated with bold arrows.

The precise combination is obtained by solution of Newton's equations as described in Refs. 4–9, but here we restrict our analysis to a consideration of the role of symmetry in limiting the possible mix of coordinates and the transitions between vibrational levels.

Symmetry relations. Each normal coordinate Q_i, and every wavefunction involving products of the normal coordinates, must transform under the symmetry operations of the molecule as one of the symmetry species of the molecular point group. The ground-state function in Eq. (3a) is a gaussian exponential function that is quadratic in Q, and examination shows that this is of Σ_g^+ symmetry for each normal coordinate, since it is unchanged by any of the $D_{\infty h}$ symmetry operations. From group theory, the symmetry of a product of two functions is deduced from the symmetry species for each function by a systematic procedure discussed in detail in Refs. 4, 5, 7, and 9. The results for the $D_{\infty h}$ point group applicable to acetylene can be summarized as follows:

$$g \times g = u \times u = g \qquad\qquad g \times u = u \times g = u$$
$$\Sigma^+ \times \Sigma^+ = \Sigma^- \times \Sigma^- = \Sigma^+ \qquad\qquad \Sigma^+ \times \Sigma^- = \Sigma^-$$
$$\Sigma^+ \times \Pi = \Sigma^- \times \Pi = \Pi \qquad\qquad \Sigma^+ \times \Delta = \Sigma^- \times \Delta = \Delta; \quad \text{etc.} \qquad (4)$$
$$\Pi \times \Pi = \Sigma^+ + \Sigma^- + \Delta; \qquad\qquad \Delta \times \Delta = \Sigma^+ + \Sigma^- + \Gamma$$
$$\Pi \times \Delta = \Pi + \Phi$$

Application of these rules shows that the product of two or more Σ_g^+ functions has symmetry Σ_g^+, hence the product function for the ground state level (00000) is of Σ_g^+ symmetry.

From Eq. (3b), it is apparent that the symmetry species of a level with $v_i = 1$ is the same as that of the coordinate Q_i. In the case of a degenerate level such as (00001), there are two wavefunctions involving the degenerate Q_{5x}, Q_{5y} pair of symmetry Π_u. The symmetry of combination levels involving two *different* degenerate modes is obtained according to the above rules and, for example, for the (00011) level, one obtains $\Sigma_g^+ \times \Sigma_g^+ \times \Sigma_g^+ \times \Pi_g \times \Pi_u = \Sigma_u^+ + \Sigma_u^- + \Delta_u$. Thus one sees that the product of two degenerate functions gives rise to multiplets of different symmetries. For *overtone* levels of *degenerate* modes, a more detailed analysis[7,9] is necessary in which it is found that levels such as (00020), (00003), and (00004) consist of multiplets of symmetry $\Sigma^+ + \Delta$, $\Pi + \Phi$, and $\Sigma^+ + \Delta + \Gamma$, respectively.

From such considerations, the symmetry species of each wavefunction associated with an energy level is determined, and these are indicated at the right in Fig. 1. It is important to realize that this symmetry label is the correct one for the true wavefunction, even though deduced from an approximate harmonic oscillator model. This is significant because transition selection rules based on symmetry are *exact* whereas, for example, the usual harmonic oscillator constraint that $\Delta v = \pm 1$ is only approximate for real molecules.

Selection rules. The probability of a transition between two levels i and j in the presence of infrared radiation is given by the transition moment P_{ij} [see Eq. (37-8)]

$$P_{ij} = \int \psi_i \mu \psi_j \, d\tau \tag{5}$$

For a given molecule, P_{ij} is a physical quantity with a unique numerical value that must remain unchanged by any molecular symmetry operation such as rotation or inversion. Hence to have a nonzero value, P_{ij} must be totally symmetric, i.e., $\Gamma(\psi_i) \times \Gamma(\mu) \times \Gamma(\psi_j) = \Sigma_g^+$ where $\Gamma(\psi_i)$ denotes the symmetry species of ψ_i, etc. The dipole moment component μ_z and the equivalent pair μ_x, μ_y are of symmetries Σ_u^+ and Π_u, respectively for the $D_{\infty h}$ point group and are usually indicated in point group (or character) tables.[3-9] From this and the rules of Eq. (4) it follows that, for a transition between two levels to be infrared-allowed, it is necessary that the symmetry species of the product of the two wavefunctions be the same as one of the dipole components. Thus from the Σ_g^+ ground state of acetylene, transition to the Σ_u^- or Δ_u members of the (00011) multiplet is forbidden while that to the Σ_u^+ level is allowed by the μ_z dipole component. Transitions involving μ_z are termed *parallel* bands while those involving μ_x, μ_y are called *perpendicular* bands.

In the case of a Raman transition, the same symmetry arguments apply, except that the dipole function μ must be replaced by the polarizability tensor elements α_{zz}, α_{xy}, etc. [see Eq. (37-7)]. For molecules of $D_{\infty h}$ symmetry, these elements belong to the symmetry species Σ_g^+, Π_g, and Δ_g so that the condition for a Raman-active transition is that the product $\Gamma(\psi_i) \times \Gamma(\psi_j)$ include one of these species. Thus from the Σ_g^+ ground state of acetylene, Raman transitions to the (10000) Σ_g^+, (01000) Σ_g^+, and (00010) Π_g levels are allowed and can be used to determine the ν_1, ν_2, and ν_4 fundamental frequencies respectively. As can be seen in Table 1, these three modes do not produce a dipole change as vibration occurs and thus these transitions are absent from the infrared spectrum. This is an example of the "rule of mutual exclusion," which applies for IR/Raman transitions of molecules with a center of symmetry.[4-9]

Although direct access to the (10000), (01000), and (00010) levels from the (00000) ground state level by infrared absorption is thus rigorously forbidden by symmetry, access from molecules in the (00010) or (00001) levels can be symmetry-allowed. For example, $\Gamma(00001) \times \Gamma(10000) = \Pi_u \times \Sigma_g^+ = \Pi_u = \Gamma(\mu_{x,y})$ and so the transition between these levels, termed a *difference band*, $\nu_1 - \nu_5$, is not formally forbidden. As can be seen in Fig. 1, the frequency $(\nu_1 - \nu_5)$ can be added to the fundamental frequency ν_5 to give the *exact* value of ν_1, the (10000)–(00000) spacing. Similarly the $\nu_2 - \nu_5$ and $\nu_3 - \nu_4$ difference bands are infrared-active and can be combined with ν_5 and ν_3 to deduce ν_2 and ν_4 respectively. Such difference bands are detectable for acetylene but will, of course, have low intensity because they originate in excited levels that have a small Boltzmann population at room temperature. The intensity of such bands increases with temperature, hence they are also termed "hot band" transitions.

Other non-fundamental bands often appear in infrared spectra and can be used to get an estimate of the fundamental frequencies. For example, from the ground state of acetylene, an infrared transition to the (00011) level is permitted and is termed the $\nu_4 + \nu_5$ *combination* band. The difference $(\nu_4 + \nu_5) - \nu_5$ can be used as an estimate of ν_4, but it should be noted that this is actually the separation between levels (00011) and (00001) and not the true ν_4 separation between (00010) and (00000). Because of anharmonicity effects, these two separations are *not identical* and hence the determination of fundamental frequencies from difference bands is to be preferred.

Force constants of acetylene. From the vibrational frequencies of the normal modes one can calculate the force constants for the different bond stretches and angle bends in the C_2H_2 molecule. In the most complete valence-bond, harmonic-oscillator approximation, the potential energy for C_2H_2 can be written as[5,8]

$$U = \tfrac{1}{2}k_r(r_1^2 + r_2^2) + \tfrac{1}{2}k_R R^2 + \tfrac{1}{2}k_\delta(\delta_1^2 + \delta_2^2) + k_{rr}r_1 r_2 + k_{rR}R(r_1 + r_2) + k_{\delta\delta}\delta_1\delta_2 \quad (6)$$

where r and R refer, respectively, to the stretching of the CH and CC bonds and δ represents bending of the H—C—C angle from its equilibrium value. The interaction constants k_{rr}, k_{rR}, and $k_{\delta\delta}$ characterize the coupling between the different vibrational coordinates and are usually small compared to the principal force constants k_r, k_R, and k_δ.

The normal modes are combinations of r, R, and δ coordinates that provide an accurate description of the atomic motions as vibration takes place. These combinations must be chosen to have a symmetry corresponding to the symmetry species of the vibration. Consequently, for example, there is no mixing between the orthogonal axial stretches and the perpendicular bending modes and U contains no cross terms such as $r\delta$ or $R\delta$. The process of finding the correct combination of coordinates, termed a normal coordinate analysis, basically involves the solution of Newton's equations of motion. This solution also give the vibrational frequencies in terms of the force constants, atomic masses, and geometry of the molecule.[4–9]

Such an analysis yields the following relations for acetylene[5,8]

$$4\pi^2 v_1^2 + 4\pi^2 v_2^2 = (k_r + k_{rr})(1/m_H + 1/m_C) + 2(k_R - 2k_{rR})/m_C \tag{7a}$$

$$4\pi^2 v_1^2 4\pi^2 v_2^2 = 2[(k_r + k_{rr})k_R - 2k_{rR}^2]/m_H m_C \tag{7b}$$

$$4\pi^2 v_3^2 = (k_r - k_{rr})(1/m_H + 1/m_C) \tag{7c}$$

$$4\pi^2 v_4^2 = (k_\delta - k_{\delta\delta})[1/R_{CH}^2 m_H + (1/R_{CH} + 2/R_{CC})^2/m_C] \tag{7d}$$

$$4\pi^2 v_5^2 = (k_\delta + k_{\delta\delta})(1/m_H + 1/m_C)/R_{CH}^2 \tag{7e}$$

When C_2D_2 frequencies are used, m_H should be replaced by m_D. The force constants for acetylene can be calculated from these relations using the measured vibrational frequencies and the bond lengths can be determined from the rotational analysis described below. If one expresses the frequencies in cm^{-1} units and the masses in atomic mass units, the factors $4\pi^2$ should be replaced by $4\pi^2 c^2 N_0^{-1} = 5.892 \times 10^{-5}$ (this includes a factor of 10^{-3} kg/g mass conversion). This substitution gives the force constants k_r, k_R, k_{rr}, and k_{rR} in $N\,m^{-1}$ units and the bending constants k_δ and $k_{\delta\delta}$ in units of $N\,m$.

Rotational levels and transitions. The vibrational–rotational energy levels for a linear molecule are similar to those for a diatomic molecule and, to a good approximation, are given in cm^{-1} units by the sum $G(v_1 v_2 \cdots) + F_v(J)$ where[5]

$$F_v(J) = B_v[J(J + 1) - l^2] - D_v[J(J + 1) - l^2]^2 \tag{8}$$

The general label v characterizes the set $v_1 v_2 v_3 \cdots$ and is added to F_v to account for the fact that the rotational constant B and centrifugal distortion constant D change slightly with vibrational level. B_v is related to the moment of inertia I_v by the equation

$$B_v = \frac{h}{8\pi^2 c I_v} \tag{9}$$

where

$$I_v = \sum_{i=1}^{N} m_i r_i^2 \tag{10}$$

and the sum is over all atoms in the molecule, having mass m_i and located a distance r_i from the center-of-mass of the molecule. The quantum number l characterizes the vibrational angular momentum about the linear axis and is $0, 1, 2, \ldots$ for levels of symmetries Σ, Π, $\Delta, \ldots$, respectively. This angular momentum derives from a rotary motion produced about the linear axis by a combination of the degenerate x and y bending motions. For acetylene, there are two bending modes, requiring l_4 and l_5 quantum numbers, which are sometimes shown as superscripts to the v_4 and v_5 labels.

The allowed changes in the rotational quantum number J are $\Delta J = \pm 1$ for parallel (Σ_u^+) transitions and $\Delta J = 0, \pm 1$ for perpendicular (Π_u) transitions.[3,5,7,8] Parallel transitions such as v_3 for acetylene thus have $P(\Delta J = -1)$ and $R(\Delta J = +1)$ branches with a characteristic minimum between them, as shown for diatomic molecules such as HCl in Fig. 38-3 and for the HCN v_3 mode in Fig. 2. However, perpendicular transitions such as v_5 for acetylene and v_2 for HCN (Fig. 2) have a strong central Q branch $(\Delta J = 0)$ along with P and R branches. This characteristic PQR versus PR band shape is quite obvious in the spectrum and is a useful aid in assigning the symmetries of the vibrational levels involved in the infrared transitions of a linear molecule.

The individual lines in a Q branch are resolved only under very high resolution, but the lines in the P and R branches are easily discerned at a resolution of $1 \, \text{cm}^{-1}$ or better. As discussed in Exp. 38, it is possible to represent both P and R transition frequencies with a single relation:

$$\tilde{v}_m = \tilde{v}_0 + B''l''^2 - B'l'^2 + (B' + B'')m + (B' - B'')m^2 \tag{11}$$

Here $\tilde{v}_0$ is the rotationless transition frequency corresponding to ΔG, the spacing between the two vibrational levels with $J = 0$. B' and B'' are the rotational constants of the upper and lower states, respectively, and the index $m = -J$ for P branch lines, $J + 1$ for R branch lines. The centrifugal distortion constants are neglected in this analysis since they are extremely small (typically $10^{-6} \, \text{cm}^{-1}$).

Intensities and statistical weights. The absolute absorption intensity of a vibrational–rotational transition is proportional to the square of the transition moment P_{ij} times the population in the lower state. P_{ij} varies only slightly for different rotational levels so that the principal factors determining the relative intensity are the degeneracy and the Boltzmann weight for the lower level

$$I_J \propto g_I g_J \exp[-hcBJ(J + 1)/kT] \tag{12}$$

The rotational degeneracy g_j is $2J + 1$, and the nuclear-spin degeneracy g_I varies with rotational level only when the molecule contains symmetrically equivalent nuclei.

A complete discussion of the factors that determine g_I is beyond the scope of this book but can be found in Refs. 5 and 7. Briefly, the total wavefunction Ψ_{tot} for molecules with equivalent nuclei must obey certain symmetry requirements upon exchange, as determined by the Pauli principle. Exchange of nuclei with half-integral spin, such as protons $(I = \frac{1}{2})$, must

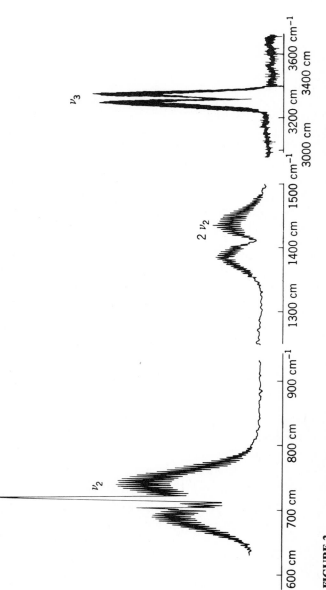

FIGURE 2

Portions of the infrared absorption spectrum of HCN. The ν_2 bending vibration is a perpendicular band and therefore has allowed P, Q, and R branches. The ν_3 [C—H stretching] vibration is a parallel band with P and R branches only.

produce a sign change in Ψ_{tot}. Such nuclei are termed fermions and are distributed among energy levels according to Fermi–Dirac statistics. Nuclei with integral nuclear spin, such as deuterium ($I = 1$), obey Bose–Einstein statistics and are called bosons; for these the sign of Ψ_{tot} is unchanged by interchange of the equivalent particles. The total wavefunction can be written, approximately, as a product function

$$\Psi_{tot} = \psi_{elec}\psi_{vib}\psi_{rot}\psi_{ns} \tag{13}$$

For the ground vibrational state of acetylene, $\Psi_{elec}\Psi_{vib}$ is symmetric with respect to nuclear exchange so that $\psi_{rot}\psi_{ns}$ must be antisymmetric for C_2H_2, symmetric for C_2D_2. For linear molecules the ψ_{rot} functions are spherical harmonics that are symmetric for even J, antisymmetric for odd J.[5,7] The ψ_{ns} spin product functions for two protons consist of three that are symmetric ($\alpha\alpha$, $\alpha\beta + \beta\alpha$, $\beta\beta$) and one that is antisymmetric ($\alpha\beta - \beta\alpha$) where α and β are the functions corresponding to M_1 values of $+\frac{1}{2}$ and $-\frac{1}{2}$ (see Exp. 34). Thus it follows that for C_2H_2, g_I is 1 for even J, 3 for odd J and the P and R branch lines will alternate in intensity. For C_2D_2, with spin functions α, β, γ representing the M_I values of $+1$, 0, -1, there are 6 symmetric nuclear spin combinations ($\alpha\alpha$, $\beta\beta$, $\gamma\gamma$, $\alpha\beta + \beta\alpha$, $\alpha\gamma + \gamma\alpha$, $\beta\gamma + \gamma\beta$) and three that are antisymmetric to exchange ($\alpha\beta - \beta\alpha$, $\alpha\gamma - \gamma\alpha$, $\beta\gamma - \gamma\beta$). Consequently the *even J* rotational lines are stronger in this case. The experimental observation of such intensity alternations confirms the $D_{\infty h}$ symmetry of acetylene, and in the present experiment serves as a useful check on the assignment of the J values for the P and R branch transitions.

EXPERIMENTAL

An infrared grating or Fourier transform (FTIR) spectrometer covering the spectral region from 600 to 4000 cm^{-1} is sufficient for this experiment, although extension to 400 cm^{-1} is desirable if the ν_5 band of C_2D_2 at about 540 cm^{-1} is to be studied. Table 2 indicates the spectral regions of interest and the approximate pressures that give satisfactory intensities. These pressures may require some adjustment depending upon the resolution capabilities of the instrument since the peak absorbance of a narrow line increases as the spectral resolution improves. For the survey scan, a resolution of 4 cm^{-1} is adequate to permit rapid recording at reasonable signal-to-noise ratio. The regions to be studied in detail should be scanned more slowly at an expanded scale to permit accurate frequency measurements. A resolution of at least 1.5 cm^{-1} is needed to resolve the rotational structure of the acetylene bands, and a value of 0.5 cm^{-1} or better is desirable. (The effect of resolution on the ν_5 mode of C_2H_2 can be seen in Fig. XVIII-30.) Detailed instructions for operating the spectrometer will be given in the laboratory.

TABLE 2
Infrared regions to be scanned for acetylenes

	C_2H_2	C_2D_2
	Scans at ~300 Torr	
Survey scan	400–4000 cm^{-1}	400–4000 cm^{-1}
$v_2 - v_5$	1235–1255 cm^{-1}	1220–1235 cm^{-1}
$v_1 - v_5$	2635–2650 cm^{-1}	2160–2175 cm^{-1}
$v_3 - v_4$	2675–2690 cm^{-1}	1920–1935 cm^{-1}
	Scans at ~25 Torr	
Survey scan	400–4000 cm^{-1}	400–4000 cm^{-1}
v_5	720– 740 cm^{-1}	530– 550 cm^{-1}
$v_4 + v_5$	1230–1410 cm^{-1}	980–1120 cm^{-1}
v_3	3275–3325 cm^{-1}	2425–2450 cm^{-1}

The C_2H_2 sample can be taken from a commercial gas cylinder† in the manner described in Exps. 36 and 38 or a sample can be synthesized as described below for C_2D_2. **Acetylene is flammable; there should be no flames in the filling or synthesis area** which, if feasible, should be in a hood. A 10-cm cell fitted with KBr windows should be filled to a pressure of about 300 Torr for a survey scan and for expanded traces of the weak difference band regions as indicated in Table 2. The cell pressure should then be reduced to about 25 Torr so that the strong v_3 and v_5 fundamentals and the $v_4 + v_5$ combination band have a more reasonable intensity. Expanded scans of the latter bands are recorded according to Table 2, preceded by a second survey scan.

C_2D_2 can be synthesized by addition of D_2O to calcium carbide using the apparatus shown in Fig. 3. About 2.5 g of calcium carbide is placed in flask F which is then evacuated to remove traces of H_2O; 0.5 mL D_2O is added with a syringe to flask F through the rubber septum and the entire system up to the vacuum stopcock V is allowed to "cure" at room temperature for about 5 to 10 min to allow deuterium exchange with H_2O adsorbed on the walls of the system. The pressure should be monitored and kept below 1 atm during this period, reducing it if necessary by opening and closing V.

The system is then evacuated, after which the cold traps are put in place

---

† Commercial acetylene is widely used for welding purposes and is shipped dissolved in acetone, in which it is extremely soluble. The acetone is retained by a porous filler material within the cylinder so that the discharged acetylene is typically >99%. If desired, residual traces of acetone can be eliminated by passage through a Dry Ice/isopropanol trap. In its free state, acetylene may decompose violently; the stability decreases at higher pressures. At pressures below 1 atm, the sampling conditions of this experiment, the gas can be handled safely but one should, of course, wear safety glasses and exercise reasonable judgement. Unalloyed copper, silver, and mercury should never be used in direct contact with acetylene, particularly when wet, owing to the possible formation of explosive acetylides.

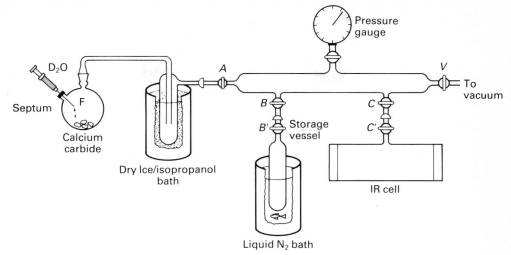

FIGURE 3
Apparatus for the synthesis of deuterated acetylene.

and 0.5 mL D_2O is added to flask F. The pressure will rise and then drop as the C_2D_2 is condensed in the storage vessel cooled by liquid nitrogen. When the pressure drops to a few Torr, another increment of D_2O is added and the procedure is repeated until a total of 4 mL has been added. When gas evolution has slowed or stopped, stopcock A is closed and the reaction flask and water trap are placed in the hood, open to allow any further reaction to occur harmlessly.

To fill the infrared cell, the system is evacuated with the liquid nitrogen trap still in place. Stopcock V is then closed and the nitrogen Dewar is lowered to allow the storage vessel to warm slowly until the pressure is about 300 Torr. The stopcocks C and C' leading to the infrared cell are then closed and, as warming continues, the system pressure is monitored and adjusted with stopcock V if necessary to keep it below 1 atm. When room temperature is reached, stopcock B' can be closed to save some residual C_2D_2 as "insurance" until all cell pressure adjustments and spectra are recorded. The spectral regions of C_2D_2 indicated in Table 2 should be scanned in the same manner as for C_2H_2.

Do not open stopcock B' to air with the storage vessel in liquid nitrogen since liquid oxygen will condense on top of the acetylene, forming a potentially explosive mixture. At the end of the experiment, the acetylene in the infrared cell and in the storage vessel can be disposed of by simply exhausting it through the roughing pump of the vacuum system.

CALCULATIONS AND DISCUSSION

Vibrational assignments and fundamental frequencies. Examine your survey spectra for C_2H_2 and C_2D_2 and note the striking difference between parallel and perpendicular bands. Determine the frequencies of as many of the transitions shown in Fig. 1 as your data allow. In doing this, take the Q branch maximum of each expanded spectrum of the v_5, $v_1 - v_5$, $v_2 - v_5$, and $v_3 - v_4$ perpendicular bands as a measure of these vibrational frequencies. The Q branch of the C_2D_2 $v_2 - v_5$ band may be difficult to detect since it overlaps some of the rotational structure of the $v_4 + v_5$ band of C_2HD, an inevitable impurity in C_2D_2.†

The parallel bands are expected to show a gap between the P and R branches at the position of the missing Q branch as in Fig. 38-3, but in fact such a gap is not seen for the acetylenes owing to overlap with combination and difference bands. For example, in the expanded trace of the v_3 region of C_2D_2, a weak feature seen at the Q branch position is *not* a consequence of a violation of selection rules but rather is due to overlapping R $(J'' = 2)$ branch lines of the difference bands $(v_3 + v_4) - (v_4)$ and $(v_3 + v_5) - (v_5)$.[10] This line happens to be the one of *minimum* intensity between the P and R branches and may be taken as a good approximate value of v_3 for C_2D_2. The corresponding v_3 region for C_2H_2 is more complicated[11] owing to additional overlapping absorption by a combination band $v_2 + v_4 + v_5$, and the v_3 value given in Table 1 can be used for subsequent calculations.

Use your data to obtain as many of the fundamental transition frequencies of acetylene as possible and compare the results with the literature values listed in Table 1. Use these values to assign other combination or difference bands that you observe in the C_2H_2 spectra, using band shapes and symmetry arguments as a guide. Draw a vibrational energy level diagram from 0 to 4000 cm^{-1} and show all the vibrational transitions you observe for C_2H_2. For C_2D_2, such an assignment task is more difficult because of C_2HD impurities, for which all the fundamental transitions are allowed because of the lower symmetry. One clue serving to identify the transitions of the latter species is the absence of intensity alternation in the P and R branches, since there is no longer exchange symmetry for the protons.

Rotational analysis. The $v_4 + v_5$ parallel combination bands of C_2H_2 and C_2D_2 should be analyzed to obtain the ground state B values for each species.‡ Note the alternation of line intensities and use the intensity predictions from the nuclear spin statistics as an aid in assigning an m value to each line in the P and

---

† A higher D/H ratio can be achieved by adding the acetylene to a storage bulb containing 5 mL of 99+% D_2O and about 1 g of basic alumina. The latter serves to promote the exchange of acidic protons on C_2H_2 with the D_2O and thereby improves the D content of the acetylene. For best results, the protons on the basic alumina should first be exchanged by adding a few mL of D_2O and evacuating the storage bulb prior to addition of more D_2O and acetylene. After exchange at room temperature for a few hours, the storage bulb is cooled in a Dry Ice/isopropanol bath and the enriched C_2D_2 is distilled into the infrared cell.

‡ The v_3 band of C_2D_2 and the v_5 perpendicular bands for both isotopic species are also suitable for analysis. In addition, the v_2, $2v_2$, and v_3 bands of HCN seen in Fig. 2, along with the DCN counterparts, can serve as alternatives for a similar vibrational-rotational study.

R branches. The feature that appears at the Q branch position is due to overlapping R lines of the difference bands $(2v_4 + v_5) - (v_4)$ and $(2v_5 + v_4) - (v_5)$.[11,12] These overlapping branches also cause the alternating intensity ratios to differ somewhat from the values of $3:1$ and $6:3$ predicted for C_2H_2 and C_2D_2.

Tabulate the transition frequencies and fit them to Eq. (11) using a least-squares method. Since this is a parallel transition between two Σ states, l' and l'' are zero in this equation. (If a perpendicular fundamental such as v_5 is to be analyzed, a value of $l' = l_5' = 1$ should be substituted.) Compare your rotational constants for the ground state with the literature values $B''(C_2H_2) = 1.176608 \text{ cm}^{-1}$, $B''(C_2D_2) = 0.847887 \text{ cm}^{-1}$ cited in Refs. 1 and 2. Assume that the structure is unchanged by deuteration and, using Eqs. (9) and (10), calculate the C—H and C—C bond lengths. Use the uncertainties from the least-squares analysis to calculate the uncertainty in these bond lengths and compare your results with values of $R_{CH} = 1.0625 \text{ Å}$, $R_{CC} = 1.2024 \text{ Å}$ that correspond to the *equilibrium* positions of the atoms on the potential energy surface.[2]

Force-constant determination. Calculate the force constants for acetylene using Eqs. (7) with the fundamental frequencies and the C—H and C—C bond lengths that you have determined. The bending force constants k_δ and $k_{\delta\delta}$ have units of energy, N m, when the angular displacements are in (dimensionless) radians. Compare values of k_δ and $k_{\delta\delta}$ obtained with C_2D_2 frequencies with those calculated for C_2H_2; how good is the assumption that the force constants are independent of isotopic substitution?

Compute the stretching force constants [k_r, k_R, k_{rr}, and k_{rR}] and discuss their magnitudes in terms of the strengths of the chemical bonds and the likely interactions among these. Two independent determinations of the quantity $k_r - k_{rr}$ are obtained using the isotopic data and Eq. (7c) but the calculation of $k_r + k_{rr}$, k_R, and k_{rR} requires the combined solution of Eqs. (7a) and (7b) for both isotopic species. In fact, Eq. (7b) for C_2D_2 is redundant and places no new constraint on the force constants since these factor out of the ratio of Eq. (7b) for the two isotopes:

$$\frac{v_1^2(D)v_2^2(D)}{v_1^2(H)v_2^2(H)} = \frac{1/m_D + 1/m_C}{1/m_H + 1/m_C} \tag{14}$$

Equation (14) is an example of a relation derived from a general *product rule*[5,8] that provides a useful method of checking frequency assignments without doing a detailed normal-coordinate analysis.

APPARATUS

Infrared grating or FTIR instrument with a resolution of 1.5 cm^{-1} or better; 10-cm gas cell with KBr windows; vacuum line with pressure gauge for synthesis and for filling cell, located in a hood if feasible; cylinder of acetylene.

Round-bottom flask (250 mL) with septum port; syringe; 5 mL D_2O (99+%); calcium carbide (3 g); D_2O trap and 1-liter storage flask with stopcocks; two Dewars; Dry Ice/isopropanol slurry; liquid nitrogen; basic alumina (optional).

REFERENCES

1. E. Kostyk and H. L. Welsh, *Can. J. Phys.* **58,** 534 (1980).
2. E. Kostyk and H. L. Welsh, *Can. J. Phys.* **58,** 912 (1980).
3. P. W. Atkins, "Physical Chemistry," 3d ed., chaps. 17–18, Freeman, New York (1986).
4. D. C. Harris and M. C. Bertolucci, "Symmetry and Spectroscopy," chaps. 1–3, Oxford University Press, New York (1978).
5. G. Herzberg, "Molecular Spectra and Molecular Structure II. Infrared and Raman Spectra of Polyatomic Molecules," chaps. II–III, Van Nostrand, Princeton, N.J. (1945).
6. J. M. Hollas, "Modern Spectroscopy," chaps. 6–7, Wiley, New York (1987).
7. I. N. Levine, "Molecular Spectroscopy," chaps. 1, 4–6, 9, Wiley Interscience, New York (1975).
8. J. I. Steinfeld, "Molecules and Radiation," chap. 6–8, MIT Press, Cambridge, Mass. (1978).
9. E. B. Wilson, J. C. Decius, and P. C. Cross, "Molecular Vibrations," chaps. 5–7, McGraw-Hill, New York (1955), reprinted in unabridged form by Dover, New York (1980).
10. S. Ghersetti and K. N. Rao, *J. Mol. Spectrosc.* **28,** 27 (1968).
11. K. F. Palmer, M. E. Mickelson, and K. N. Rao, *J. Mol. Spectrosc.* **44,** 131 (1972).
12. S. Ghersetti, J. Pliva, and K. N. Rao, *J. Mol. Spectrosc.* **38,** 53 (1971).

GENERAL READING

P. W. Atkins, *op. cit.*
G. Herzberg, *op. cit.*
J. M. Hollas, *op. cit.*
I. N. Levine, *op. cit.*
J. I. Steinfeld, *op. cit.*

EXPERIMENT 40
SPECTRUM OF THE HYDROGEN ATOM

Since the hydrogen atom has only one orbital electron surrounding a nucleus consisting of a single proton, it has a particularly simple spectrum. In this experiment, part of this spectrum will be determined. The observed frequencies of the lines can then be compared with the values predicted by quantum mechanics.

THEORY

When a high-voltage discharge takes place in H_2 gas at low pressures, many molecules are dissociated into atoms by electron impact. These atoms are often produced in excited electronic states. Such H atoms in excited electronic states spontaneously undergo transitions to lower energy electronic states with the emission of radiation. Quantum mechanics gives an expression[1,2] for the allowed electronic energy levels of an H atom in terms of a single quantum number† n;

$$E_n = -\frac{\mu e^4}{8h^2\varepsilon_0^2}\frac{1}{n^2} \qquad n = 1, 2, 3, \ldots \tag{1}$$

where e is the charge on the electron, h is Planck's constant, and ε_0^2 is the permittivity of vacuum. The reduced mass μ is given in terms of the mass of the electron m and the mass of the proton M by

$$\mu = \frac{mM}{m + M} \tag{2}$$

There is no selection rule for n; that is, transitions may occur between any of the levels given by Eq. (1). An energy-level diagram for the H atom is given in Fig. 1. Transitions from upper energy levels to lower levels are shown

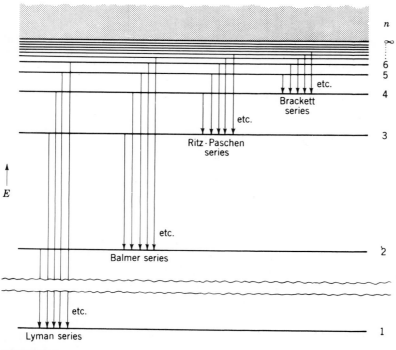

FIGURE 1
Electronic energy-level diagram for the hydrogen atom.

---

† Owing to spin–orbit coupling and quantum electrodynamical effects, the energy levels have a small dependence on another quantum number l, where l can assume any integral value less than n. This energy splitting is very small and can be neglected here.

by the vertical lines. These transitions form several series of lines depending on the quantum number of the lower state. Note that the energy levels converge toward a limit as $n \to \infty$. The shaded area above $n = \infty$ indicates a continuum of energy states corresponding to complete separation of the proton and the electron.

We shall be concerned with the Balmer series, the lines of which fall in the visible region of the spectrum. From Eq. (1), it is possible to predict the frequency of a transition from any upper state with quantum number n_1 to any lower state with quantum number n_2, since

$$\Delta E = E_{n_1} - E_{n_2} = h\nu = hc\bar{\nu} \tag{3}$$

where c is the speed of light in cm s^{-1} units and $\bar{\nu}$ is the frequency in wavenumber units of cm^{-1} (equal to the reciprocal of the wavelength λ in centimeters). Thus,

$$\bar{\nu} = \frac{\mu e^4}{8h^3 c \varepsilon_0^2} \left(\frac{1}{n_2^2} - \frac{1}{n_1^2} \right) = \mathscr{R} \left(\frac{1}{n_2^2} - \frac{1}{n_1^2} \right) \tag{4}$$

The quantity $\mathscr{R}$ is called the *Rydberg constant* and has the *calculated* value 109 677.5805 cm^{-1} for H. For the Balmer series, $n_2 = 2$ and $n_1 = 3, 4, 5, \ldots$. Equation (4) predicts a series of lines that converge to a high-frequency limit at $\mathscr{R}/n_2^2$.

METHOD

The frequency (or wavelength) of these Balmer lines can be determined experimentally by comparing the hydrogen spectrum with a reference spectrum for which the wavelengths are known. The iron or mercury spectra are common choices for reference spectra. (Such reference spectral lines must, of course, have been measured absolutely, as by an interferometric technique.) In this experiment, an Hg lamp will be used because the lines of the Hg spectrum can be identified quite easily. In other work the Fe arc is preferable, since there are many more lines (see Exp. 41).

A wide variety of spectrometers and spectrographs (instruments that record the spectra on a photographic plate) can be used to study spectra in the visible and near ultraviolet region. A description of such instruments is given in Chapter XVIII, and this material should be read as background for the present experiment. Figure 2 shows a schematic diagram for a typical prism spectrograph, and a brief description will be given here of this instrument. The sources shown in Fig. 2 are mounted in such a way that the hydrogen discharge tube will illuminate half of the entrance slit S and the mercury reference lamp will illuminate the other half. The Hg lamp is off the optical axis of the spectrograph, and a front-surface mirror M (or a 90° prism) is set at 45° to permit an image of this source to be focused on the slit by lens L_1. The lens L_2 renders the light from the slit parallel before it passes through the dispersing

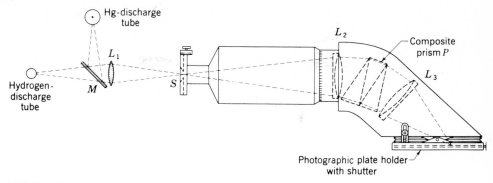

FIGURE 2
Spectrograph and source optics, viewed from above. (NOTE: the mirror M is just below the light path from the hydrogen discharge tube.)

prism P. A camera lens L_3 focuses an image of the slit on the photographic plate.†

The two common prism materials for use in the visible and near-ultraviolet regions are glass and quartz. Glass is preferable in the visible, since it gives a higher dispersion (and therefore better resolution). For wavelengths shorter than about 300 to 350 nm, most glasses absorb light strongly but quartz prisms may be used. Quartz also begins to absorb light below about 190 nm. Other prism materials (such as fluorite) may be used below 190 nm, but diffraction gratings are generally used in the far ultraviolet. Since the Balmer lines are more and more closely spaced and have decreasing intensity as the short-wavelength limit is approached, both high resolution and long exposures are needed to observe lines with high n_1 values.

Sources. The best Hg source is a low-pressure mercury lamp that operates from a small ballast transformer. **There is intense ultraviolet radiation from such a lamp, which is dangerous to the eyes.** Therefore a Pyrex glass jacket should be placed over the Hg arc to filter out this ultraviolet radiation.

Many designs of hydrogen discharge tubes are available.[3] A convenient commercial hydrogen source is a Geissler tube, since the pressure has been adjusted (0.1 to 0.5 Torr) to give strong Balmer lines. In addition, there is always a weak *band* spectrum due to molecular hydrogen; hydrogen atoms, formed by the electrode discharge, may recombine on the walls to give H_2 molecules in excited electronic states which will emit radiation. Such a band

---

† To align the sources properly, one may open the slit wide and look in at the camera end of the spectrograph. Adjust the source so that it appears in the center of the slit opening. Then close the slit to a narrow opening, and focus the instrument, using the Hg source until the yellow Hg line is well resolved into a doublet sharply focused in the plane of the photographic plate. A ground-glass plate may be placed in the film position and viewed with a small magnifier.

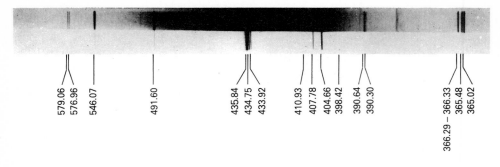

FIGURE 3
Mercury-arc reference spectrum. The upper half of this print has been strongly exposed to show the weak lines clearly. The wavelengths (in nm) of the Hg lines are given.

spectrum should be so weak that it will not interfere with determining the Balmer lines. The Geissler discharge tube operates from a neon-sign transformer at about 5000 V and 15 mA; it should glow steadily with a bright red color.

Hg reference spectrum. A print of the low-pressure Hg-arc spectrum is shown in Fig. 3. The upper half of the spectrum was strongly exposed in order to bring out the weak lines, and the lower half was exposed only a short time in order to show the strong lines clearly. The wavelengths of these lines in air, see Eq. (5), are given below the spectrum. The spectrum presented in Fig. 3 was made using Kodak 103a-O spectrographic plates; one can also use type 103a-F plates, which are much more sensitive in the long-wavelength end of the visible range.

EXPERIMENTAL

Instructions for the operation of the spectrograph or spectrometer to be used will be given in the laboratory. The procedure described below is appropriate for a spectrograph; obvious simplifying changes can be made if a scanning spectrometer with a photomultiplier detector is used.

   The placement of the sources and the focus and slit width settings of the spectrograph will be made in advance by the instructor. The sources should be arranged so that one half of the slit is illuminated by the Hg lamp and the other half by the hydrogen discharge tube. A shutter over the slit permits exposures to be made of each source independently without moving the photographic plate. This is important when wavelengths are to be determined by comparison. The photographic plate is placed in a holder that can be raised or lowered to permit several such pairs of spectra to be taken.

   The photographic plates are extremely light sensitive. The plate holder must be loaded and unloaded (and developing must be done) in total darkness. The plate should be placed in the holder in such a way that the emulsion side faces the shutter. The emulsion side can be detected in the dark by moistening a finger and touching the plate lightly near an edge. The emulsion side will feel sticky. Having loaded the holder, turn on the lights and place the holder in position on the spectrograph. Set the plate-holder position close to one end of its traverse. Check to see that the slit is properly illuminated by each source. Instructions for operating the sources will be given. The hydrogen discharge tube operates on **high voltage which is possibly lethal**; be very careful. Close the shutter over the slit.

Darken the room and open the shutter on the plate holder. Take an exposure with the hydrogen discharge tube and then with the Hg lamp. Each source should be turned off when not in use. Close the shutter over the slit, move the plate holder, and take another pair of exposures with exposure times different from the first set. Repeat this process several times.

After the final pair of exposures, *close the shutter on the plate holder* and remove the holder from the spectrograph. Turn on the room lights. Set up three developing trays: one with developer, one with water, one with fixer. Set a timer for 3 min. Darken the room and place the plate in developer, *emulsion side up*; start the timer. Agitate gently and remove after 3 min. Drain over the developer tray, rinse briefly in water, and place in the fixer. Allow to remain in the fixer at least 10 min. Lights may be turned on after about 5 min.

Examine the plate. If for some reason the plate is not usable, start over again but omit any long exposures if time does not permit them. If the plate is satisfactory, rinse it under running water for 20 min and then allow it to drain and dry for several hours (preferably overnight).

Measurement of spectral lines. If a comparator is available, mount the plate on the stage, focus on a suitable pair of exposures, and align the plate so that the spectral lines at both ends of the plate are parallel with the cross hair of the microscope. Starting at one end of the plate, move the stage or microscope until a line is under the cross hair and record the position. Continue this process for all lines, both Hg and H atom (disregard the H_2 molecular band spectrum). Always approach a line from the same direction to avoid errors due to backlash in the screw.

If a comparator is not available, a good enlarger may be used instead. Mount the plate in the enlarger, and project the spectra onto a large sheet of white paper mounted on a drafting board. Choose a suitable pair of exposures and mark the positions of all Hg lines and all H-atom lines with a sharp pencil (disregard H_2 molecular band spectrum). Avoid working near the edge of the image, where there may be distortion. Using a centimeter scale, measure the positions of all lines relative to some arbitrary zero position. Indicate which lines are Hg and which H.

Compare your Hg spectrum with that shown in Fig. 3, and assign wavelengths to all Hg lines. Using a least-squares routine and an empirical polynomial $\lambda = a_0 + a_1 x + a_2 x^2 + \cdots$, fit the variation of the wavelengths λ with the position x of the Hg lines on the plate. Once this calibration formula has been established, it is used for interpolation to convert the positions of the H-atom lines into their wavelengths.

CALCULATIONS

Obtain the frequencies of the Balmer lines $\bar{v}$ (in cm^{-1}) from the wavelength values λ (in nm). Present a table of both $\bar{v}$ and λ values together with the value of n_1 for each line observed.

Plot $\bar{v}$ against $1/n_1^2$ on a large sheet of graph paper. If the points fall on a straight line, this is a partial confirmation of Eq. (4). From the slope of this line, determine an experimental value of $\mathcal{R}$. In all this work, wavelengths in air have been used; these are related to vacuum wavelengths by

$$\lambda_{air} = \frac{\lambda_{vac}}{n_{air}} \qquad (5)$$

where n_{air} is the index of refraction of air, which equals 1.00027 at these wavelengths. To obtain a value of $\mathcal{R}$ referred to vacuum, one must divide the experimental value by n_{air}. Compare your vacuum value of $\mathcal{R}$ with the theoretical value.†

DISCUSSION

Show that a typical line in the Lyman series ($n_2 = 1$) lies in the ultraviolet and that a typical line in the Ritz–Paschen series ($n_2 = 3$) lies in the infrared.

Why do the Balmer lines become weaker toward shorter wavelengths? (The first Balmer line at about 650 nm may appear weak owing to poor film sensitivity in the red; it is actually the most intense of all the lines.) What are some experimental factors that influence the line width?

APPARATUS

Spectrograph or spectrometer; low-pressure mercury lamp (General Electric, Pen-Ray, or Hanovia Division of Engelhard Industries); hydrogen discharge tube (Geissler tube from Klinger Educational Products Corp., Jamaica, NY and others); mounts and power supplies for both sources; front-surface mirror; lens; spectroscopic plates (Kodak 103a-F); three developing trays; developer (Kodak D19); acid fixer; flashlight; timer; darkroom.

Comparator microscope (*or* enlarger, drawing board, T square and triangle, good centimeter scale).

---

† The best experimental value of $\mathcal{R}$ for hydrogen using vacuum wavelengths is known with even greater precision than the theoretical value, which is affected by uncertainties in e, h, c, and μ. The experimental value usually quoted is $\mathcal{R}_\infty$ (109 737.31534 cm^{-1}), the extrapolated value for an electron and a nucleus of infinite mass.

REFERENCES

1. I. N. Levine, "Physical Chemistry," 3d ed., p. 620–625, McGraw-Hill, New York (1988).
2. G. Herzberg, "Atomic Spectra and Atomic Structure," 2d ed., pp. 11–38, Dover, New York (1944).
3. G. R. Harrison, R. C. Lord, and J. R. Loofbourow: "Practical Spectroscopy," pp. 188–192, Prentice-Hall, Englewood Cliffs, N.J. (1948).

GENERAL READING

H. G. Kuhn, "Atomic Spectra," Longmans, London (1969).

EXPERIMENT 41
BAND SPECTRUM OF NITROGEN

Spectra that are observed in the visible and ultraviolet regions arise from transitions between electronic states. For atomic gases, such electronic spectra consist of individual sharp lines, as shown by the mercury specrum in Fig. 40-3. For molecular gases, the transitions take place between different vibrational–rotational levels of the upper and lower electronic states and a very large number of lines occur. Under low resolution, groups of lines very close together have the appearance of broad bands in the spectrum. Therefore, such molecular spectra are referred to as *band spectra*.

In this experiment, part of the emission spectrum of nitrogen is to be determined and its vibrational structure analyzed. Homonuclear diatomic molecules such as N_2 have no infrared spectrum,[1] and their ground electronic state is usually studied by Raman spectroscopy. However, electronic band spectra provide the most convenient source of the vibrational and rotational constants that characterize excited electronic states and even the higher vibrational levels of the ground electronic state. From a knowledge of the upper vibrational levels, the effect of anharmonicity is easily detected.

THEORY

For a visible emission spectrum, transitions occur from excited electronic states to lower-energy electronic states. There are many electronic states for N_2, and the energy-level diagram is quite complex[2] in contrast to the simple diagram for the hydrogen atom shown in Fig. 40-1. We shall discuss only the bands of the "second positive group," which are those to be observed in the present experiment. These bands arise from transitions from the $^3\Pi_u$ electronic state (denoted by C) to the $^3\Pi_g$ state (denoted by B), both of which are excited

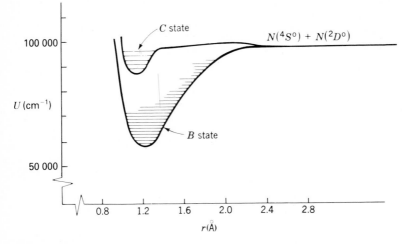

FIGURE 1
Potential energy curves for two excited electronic states of N_2.[2] B is the $^3\Pi_g$ state; C is the $^3\Pi_u$ state. (A few vibrational levels are shown.)

states.† However, the theory presented here is typical of most electronic transitions for diatomic molecules.

Associated with each electronic state there is a characteristic potential curve that represents the potential energy of the nuclei as a function of their internuclear separation r. The potential curves for states B and C of N_2 are given schematically in Fig. 1; allowed vibrational levels are shown by the horizontal lines, while rotational levels have been omitted for clarity. In general, there is different bonding involved in two different electronic states. As a result the two curves do not have their minima at the same value of r and do not have exactly the same shape.

The total energy of a diatomic molecule may be separated into translational energy and internal energy. We are concerned here with the internal energy E_{int} which can be expressed to a good approximation by

$$E_{\text{int}} = E_{el} + E_v + E_r \tag{1}$$

where E_{el} is the electronic energy, E_v is the vibrational energy, and E_r is the rotational energy. This electronic energy E_{el} refers to the minimum value of the potential curve for a given electronic state. The zero of energy is arbitrarily taken as the minimum in the potential curve for the lowest electronic state (ground state). It is convenient to divide Eq. (1) by the quantity hc, where c is expressed in units of cm s^{-1}, to get the so-called "term value" which has units

---

† It should be noted that this $C \rightarrow B$ transition is the one that produces spectacular lasing action in N_2.

of cm^{-1}. Thus

$$T_{int} = \frac{E_{int}}{hc} = T_{el} + G + F \tag{2}$$

where the vibrational and rotational term values E_v/hc and E_r/hc are given their conventional symbols G and F, respectively. The advantage of this change is that the frequency $\tilde{v}$ (expressed in cm^{-1}) for a transition between two electronic states can be simply expressed by

$$\tilde{v} \equiv T' - T'' = (T'_{el} - T''_{el}) + (G' - G'') + (F' - F'') \tag{3}$$

where the *single prime* refers to the *upper state* and the *double prime* refers to the *lower state*. Now let us denote $(T'_{el} - T''_{el})$ by $\tilde{v}_{el}$. For a given band system involving transitions from the same upper electronic state to the same lower electronic state, $\tilde{v}_{el}$ is a constant; in the case of the second positive group of N_2, $hc\tilde{v}_{el}$ corresponds to the energy separation between the minima of curves C and B in Fig. 1.

Vibrational structure. Since rotational energies are usually small compared with vibrational energies, it will be possible to neglect $(F' - F'')$ in Eq. (3) in order to discuss the vibrational structure of a band spectrum. In essence, this amounts to considering transitions among states without rotational energy. Corresponding to each spectral *line* expected on this basis, there is observed experimentally a *band* of very closely spaced lines resulting from the rotational structure that we are neglecting at this point. From quantum mechanics,[3] one obtains for an anharmonic oscillator the approximate expression

$$G = \tilde{v}_e(v + \tfrac{1}{2}) - \tilde{v}_e x_e(v + \tfrac{1}{2})^2 + \tilde{v}_e y_e(v + \tfrac{1}{2})^3 + \cdots \tag{4}$$

where the vibrational quantum number v has integral values $0, 1, 2, 3, \ldots$. (Note that $\tilde{v}_e$ is not to be confused with $\tilde{v}_{el}$ introduced earlier.) The terms in $\tilde{v}_e x_e$ and $\tilde{v}_e y_e$ take into account the effect of anharmonicity. In general, the cubic and higher terms in $(v + \tfrac{1}{2})$ are much smaller than the quadratic term, but we shall retain the cubic term explicitly. We can now write Eq. (3) as

$$v = \tilde{v}_{el} + (G' - G'') \tag{5a}$$

$$= \tilde{v}_{el} + \tilde{v}'_e(v' + \tfrac{1}{2}) - \tilde{v}'_e x'_e(v' + \tfrac{1}{2})^2 + \tilde{v}'_e y'_e(v' + \tfrac{1}{2})^3$$
$$- [\tilde{v}''_e(v'' + \tfrac{1}{2}) - \tilde{v}''_e x''_e(v'' + \tfrac{1}{2})^2 + \tilde{v}''_e y''_e(v'' + \tfrac{1}{2})^3] \tag{5b}$$

By writing

$$\tilde{v}_{00} = \tilde{v}_{el} + \tfrac{1}{2}(\tilde{v}'_e - \tilde{v}''_e) - \tfrac{1}{4}(\tilde{v}'_e x'_e - \tilde{v}''_e x''_e) + \tfrac{1}{8}(\tilde{v}'_e y'_e - \tilde{v}''_e y''_e) \tag{6a}$$

$$\tilde{v}_0 = \tilde{v}_e - \tilde{v}_e x_e + \tfrac{3}{4}\tilde{v}_e y_e \tag{6b}$$

$$\tilde{v}_0 x_0 = \tilde{v}_e x_e - \tfrac{3}{2}\tilde{v}_e y_e \tag{6c}$$

$$\tilde{v}_0 y_0 = \tilde{v}_e y_e \tag{6d}$$

Eq. (5) can be simplified to

$$\tilde{v} = \tilde{v}_{00} + \tilde{v}_0' v' - \tilde{v}_0' x_0' v'^2 + \tilde{v}_0' y_0' v'^3 - (\tilde{v}_0'' v'' - \tilde{v}_0'' x_0'' v''^2 + \tilde{v}_0'' y_0'' v''^3) \qquad (7)$$

Obviously, $\tilde{v}_{00}$ is the frequency of the transition from $v' = 0$ (in the upper state) to $v'' = 0$ (in the lower state) which is called the 0-0 band. Since there is no selction rule for the quantum number v, any v' value can be combined with any v'' value in Eq. (7) and we should expect a large number of "lines" (bands).

When the separation between vibrational levels in the upper electronic state and in the lower electronic state is not greatly different (that is, $\tilde{v}_0'$ is close to $\tilde{v}_0''$), the vibrational transitions form groups in the spectrum called sequences. For each sequence, Δv $(= v' - v'')$ is a constant. Such behavior occurs for N_2 and simplifies the analysis of the bands in the second positive group, as seen from Fig. 2.

Once the vibrational transitions in the spectrum have been assigned and the frequencies measured, it is convenient to organize the values of $\tilde{v}(v' \rightarrow v'')$ into a *Deslandres table*. As an example, a portion of such a table for PN vapor[4] is presented in Table 1. Note that the frequencies along diagonals (which correspond to bands with the same Δv) lie close together as shown by the schematic spectrum in Fig. 2. Now consider the first two rows in the table. Since the frequencies are for transitions from two adjacent upper vibrational states ($v' = 0$ and $v' = 1$) to common lower vibrational states ($v'' = 0$, 1, 2, or 3), there should be a constant separation between the rows—a separation corresponding to the upper state vibrational separation ($G_{v=1}' - G_{v=0}'$), as seen from Eq. (5a). The separation between the rows $v' = 1$ and $v' = 2$ will give ($G_{v=2}' - G_{v=1}'$), and so forth. The separation between successive rows should decrease as v' increases owing to the effect of anharmonicity; see Eq. (4). In exactly the same way, the separation between adjacent columns will give information about the vibrational levels of the lower electronic state. Agreement among the separations between corresponding frequencies of two rows (or two columns) is a definite check on the correctness of the entries in the Deslandres table. Inspection of the differences given in Table 1 shows a slight variation, which is, however, greater than the experimental error. This is caused by the use of band-head frequencies as discussed below.

Rotational structure. There are two deficiencies with the theory as presented so far: Eq. (7) predicts a spectrum of lines rather than the observed bands, and the differences in the Deslandres table are not quite constant as they should be. Both of these failures are due to the neglect of rotational transitions. For each electronic–vibrational transition discussed above, there are many rotational transitions, which give rise to a large number of closely spaced lines in the vicinity of each $\tilde{v}$ as given by Eq. (7). The theory of this rotational fine structure is too complex for presentation here, especially since it involves an interaction between rotational and electronic motions in the molecule. In addition, high resolution is required to observe this fine structure experimen-

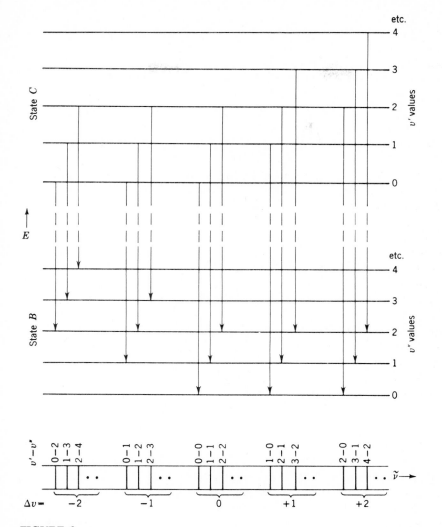

FIGURE 2
Vibrational energy levels for states B and C of N_2. The vertical arrows indicate transitions; transitions with the same Δv are grouped together to form sequences. The resulting vibrational-electronic "line" spectrum is shown at the bottom (for clarity only the first three lines in each sequence are shown).

tally. Under low resolution, the rotational lines will create the appearance of a band usually having at one end an abrupt change in intensity, called the *band head,* while the intensity falls off slowly at the other end.

This experiment will deal with these easily observed band heads in spite of the fact that Eq. (7) is valid only for the band origins $\tilde{\nu}_0$. (The origin is the position in the band that corresponds to a transition between vibrational states

TABLE 1
Deslandres table of PN bands[4] (Band-head frequencies in cm^{-1}. Differences between the entries in rows 0 and 1, 1 and 2 are in parentheses.)

v' \ v''	0	1	2	3	Average differences
0	39 698.8	38 376.5	37 068.7	—	
	(1 087.4)	(1 090.7)	(1 086.8)		(1 088)
1	40 786.2	39 467.2	38 155.5	36 861.3	
	(1 072.9)	(1 069 0)		(1 071.6)	(1 071)
2	41 859.1	40 536.2	—	37 932.9	
3	—	41 597.4	40 288.3	—	

without rotational energy.) Fortunately, the difference $\tilde{v}_{head} - \tilde{v}_0$ is small for N_2, varying from about -5 to -10 cm^{-1}. The use of band-head frequencies rather than band origins in Table 1 accounts for the slight lack of constancy in the separations, since $\tilde{v}_{head} - \tilde{v}_0$ varies somewhat.

Assignment of bands. Herzberg[5] discusses the vibrational assignment of a band system when clear sequences are observed on both sides of the 0-0 band. For this experiment, one may take one or two band heads as known from the literature and then assign the others from the sequence pattern shown in Fig. 2. Constancy of the differences in the Deslandres table will confirm the assignment.

EXPERIMENTAL

The method for determining the band spectrum of nitrogen is the same as that described in Exp. 40 for the study of the Balmer lines of hydrogen. It differs only in replacing the hydrogen discharge tube by a nitrogen discharge tube and using an iron arc for the reference spectrum instead of a mercury arc. For this experiment, the richer spectrum of lines from the iron arc is an advantage in obtaining accurate values of the wavelengths of the band heads for N_2.

Sources. An iron/neon hollow-cathode lamp or an iron arc can be used to provide the reference spectrum. A simple design for an iron arc is shown in Fig. 3. An iron oxide bead is used to stabilize the arc, which may be started by shorting the gap with a carbon rod. The nitrogen source is a Geissler discharge tube in which N_2 molecules are excited by an electrode discharge into upper electronic states. Such excited molecules will spontaneously undergo transitions to lower electronic states and in the process emit radiation. The discharge tube operates from a neon-sign transformer at about 5000 V and 15 mA; it should glow steadily with a pale greenish-blue color.

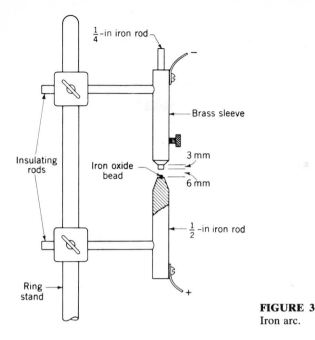

FIGURE 3
Iron arc.

Procedure. Follow the procedure given in the experimental section of Exp. 40 to obtain several pairs of N_2 and Fe-arc spectra. It may be convenient at the end to include one pair of exposures consisting of an Fe-arc spectrum and an Hg-arc spectrum. This will facilitate identification of the iron lines.

The lines should be measured using a comparator microscope as described in Exp. 40. Excellent reproductions of the Fe-arc spectrum together with the wavelength values for all the lines can be found in the literature.[6] Since the iron reference spectrum contains many closely spaced lines, it is not necessary to plot a calibration curve. Record the position of every N_2 band head together with the position of one Fe line on each side of the band head. Choose Fe lines that are as close to the band-head position as possible. Determine the wavelength of the band head by linear interpolation between these iron lines. A proper identification of the iron lines is crucial; checking them against a Hg or Ne spectrum may help. Make a list of the wavelengths (in nm) and the frequencies (in cm^{-1}) of all band heads measured.

CALCULATIONS

Several band-head wavelengths for the second positive group of N_2 can be obtained from Fig. 8(a) in Herzberg[7] as an aid to assignment. Assign the v' and v'' values for all band heads observed, and construct a Deslandres table of the frequencies. Indicate in this table the differences, and check their constancy to verify the assignment. Along the bottom and right-hand side of the table, record the average values of the differences between v' and v'' levels, respectively.

Draw a vibrational energy-level diagram for both electronic states like that shown in Fig. 2, and indicate on it the separation in cm^{-1} units between those levels for which the differences $G'_{v+1} - G'_v$ or $G''_{v+1} - G''_v$ are given in the Deslandres table. Use the average values of the differences as recorded in the table.

From Eq. (4) we see that

$$G_{v+1} - G_v = \bar{v}_e - 2\bar{v}_e x_e(v+1) + 3\bar{v}_e y_e(v^2 + 2v + \tfrac{13}{12}) + \cdots \tag{8}$$

In the case of state B, the $\bar{v}_e y_e$ value is very small and the last term in Eq. (8) can be neglected in fitting the experimental data. In the case of state C, the $\bar{v}_e y_e$ value is unusually large and this term must be retained. Thus the separations in the energy-level diagram should be used to determine the best values of the vibrational constants $\bar{v}''_e$ and $\bar{v}''_e x''_e$ for the lower electronic state B and the best values of $\bar{v}'_e$, $\bar{v}'_e x'_e$, and $\bar{v}'_e y'_e$ for the upper state C. From Eqs. (6b) and (7), calculate a value of $\bar{v}_{00}$ using several frequencies from the Deslandres table. Make a table of your values of $\bar{v}''_e$, $\bar{v}''_e x''_e$, $\bar{v}'_e$, $\bar{v}'_e x'_e$, $\bar{v}'_e y'_e$, and $\bar{v}_{00}$ and compare them with literature values.[8]

DISCUSSION

Is the rotational fine structure resolved in any of the bands studied? If so, compare it qualitatively with the high-resolution spectrum of the 0-2 band given in Herzberg[9] and indicate whether you feel rotational analysis would be possible from your spectrum.

APPARATUS

Medium-resolution spectrograph or spectrometer; nitrogen discharge tube (Geissler tube from Klinger Educational Products Corp., Jamacia, N.Y. and other sources); iron/neon hollow-cathode lamp, or iron arc and optional low-pressure mercury lamp; mounts and power supplies for all sources; front-surface mirror; lens; spectroscopic plates (Kodak 103a-F); three developing trays; developer (Kodak D19); acid fixer; flashlight; timer; darkroom.

Comparator microscope; reproduction of the iron-arc spectrum.[6]

REFERENCES

1. G. Herzberg, "Molecular Spectra and Molecular Structure I. Spectra of Diatomic Molecules," 2d ed., pp. 80, 131, Van Nostrand, Princeton, N.J. (1950).
2. F. R. Gilmore, RAND Corporation Memorandum R-4034-PR (June 1964). The potential curves are reproduced in K. E. Shuler, T. Carrington, and J. C. Light, *Appl. Optics Suppl.*, **2**, 85 (1965); and in J. I. Steinfeld, "Molecules and Radiation," 2d ed., MIT Press, Cambridge, Mass. (1985).
3. G. Herzberg, *op. cit.*, p. 151.

4. *Ibid.*, pp. 40–42.
5. *Ibid.*, p. 161.
6. A. Gatterer, "Grating Spectrum of Iron," Specola Vaticana, Vatican City (1951); C. E. Moore, "Atomic Energy Levels," vol. II, p. 49*ff*, Natl. Stand. Ref. Data Ser.-Natl. Bur. Stand. 35, U.S. Government Printing Office, Washington, D.C. (1971).
7. G. Herzberg, *op. cit.*, p. 32.
8. K. P. Huber and G. Herzberg, "Molecular Spectra and Molecular Structure IV. Constants of Diatomic Molecules," Van Nostrand-Reinhold, New York (1979); R. W. B. Pearse and A. G. Gaydon, "The Identification of Molecular Spectra," 4th ed., p. 217*ff*, Chapman and Hall, London (1976).
9. G. Herzberg, *op. cit.*, p. 46.

GENERAL READING

G. Herzberg, *op. cit.*, chap. IV.
J. M. Hollas, "Modern Spectroscopy," Wiley, New York (1987).
J. I. Steinfeld, *op. cit.*, chap. 5.

EXPERIMENT 42
ABSORPTION AND EMISSION SPECTRA OF MOLECULAR IODINE

Although the electronic spectra of condensed phases are typically quite broad and unstructured, the spectra of small molecules in the gas phase often reveal a wealth of resolved vibrational and rotational lines. Such spectra can be analyzed to give a great deal of information about the molecular structure and potential energy curves for ground and excited electronic states.[1,2] The visible absorption spectrum of molecular iodine vapor in the 490 to 650 nm region serves as an excellent example,[3-5] displaying discrete vibrational bands at moderate resolution and extensive rotational structure[6] at very high resolution. The latter structure is not seen at a resolution of ~0.2 nm, a common limit for commercial ultraviolet–visible spectrophotometers, but the vibrational features can be easily discerned in both absorption and emission measurements. In this experiment, the absorption spectrum of I_2 will be used to obtain vibrational frequencies, anharmonicities, bond energies, and other molecular parameters for the ground $X\,^1\Sigma_g$ and excited $B\,^3\Pi_{0u}^+$ states involved in this electronic transition. As an additional option, emission spectra[7,8] can be used to measure many more vibrational levels of the X state and hence to get improved values of the ground-state parameters.

THEORY

The relevant potential energy curves for I_2 are depicted in Fig. 1, which also shows some of the parameters to be determined from the spectra. The spacings

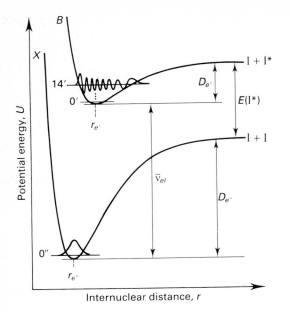

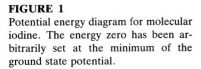

FIGURE 1

Potential energy diagram for molecular iodine. The energy zero has been arbitrarily set at the minimum of the ground state potential.

between levels in the two electronic states can be measured by either absorption or emission spectroscopy. Emission occurs following an absorption event if the upper state is not relaxed by a non-radiative collisional process (called *quenching*). The emission is termed *fluorescence* and the transition between two states is said to be spin allowed if the states have the same spin multiplicity (e.g., both are singlets or both are triplets). Fluorescence intensities are usually high, and the lifetime of the emitting state is short ($\sim 10^{-8}$ s). If the multiplicity changes in the transition, the emission is termed *phosphorescence*. In that case, the intensity is lower and the lifetime is longer ($\sim 10^{-3}$ s), since the transition is "forbidden" by the spin selection rules (which are only approximate owing to electron spin–orbit interactions). There is no strict selection rule for the change Δv in vibrational quantum number during an electronic transition; thus sequences of transitions are observed. Each band in the sequence contains rotational structure which, for I_2, is subject to the selection rule constraint that $\Delta J = \pm 1$.[9]

As discussed in Exp. 41, the frequency $\tilde{\nu}$ (in wavenumbers) for a transition between vibrational levels v'' and v' is given by,[9]

$$\tilde{\nu} = T' - T'' + G(v') - G(v'') + F(J') - F(J'') \tag{1a}$$

$$\approx \tilde{\nu}'_{el} + G(v') - G(v'') \tag{1b}$$

where $\nu_{el} = T' - T'' = T'$ since $T'' = 0$ for the ground electronic state. $G(v)$ is the vibrational term value which, for an anharmonic oscillator, is

$$G(v) = \tilde{\nu}_e(v + \tfrac{1}{2}) - \tilde{\nu}_e x_e(v + \tfrac{1}{2})^2 + \tilde{\nu}_e y_e(v + \tfrac{1}{2})^3 + \cdots \tag{2}$$

The rotational term difference $F(v', J') - F(v'', J'')$ will be ignored, since the rotational structure is not resolved in this experiment. The cubic term in $G(v)$ is also small and can be neglected in obtaining the transition frequency

$$\tilde{v}(v', v'') = \tilde{v}_{el} + \tilde{v}'_e(v' + \tfrac{1}{2}) - \tilde{v}'_e x'_e(v' + \tfrac{1}{2})^2 - \tilde{v}''_e(v'' + \tfrac{1}{2}) + \tilde{v}''_e x''_e(v'' + \tfrac{1}{2})^2 \quad (3)$$

If the quantum numbers v' and v'' are known, the measured frequencies in an absorption or emission spectrum can then be used with a multiple linear least-squares technique (see Chapter XX) to determine the parameters $\tilde{v}_{el}$, $\tilde{v}'_e$, $\tilde{v}'_e x'_e$, $\tilde{v}''_e$, and $\tilde{v}''_e x''_e$.

An alternative analysis procedure that is often used concentrates on the determination of $\tilde{v}_e$, $\tilde{v}_e x_e$ parameters within each electronic state. Differences between levels in the upper state are obtained from

$$\Delta\tilde{v}(v') \equiv \tilde{v}(v' + 1, v'') - \tilde{v}(v', v'') \simeq \tilde{v}'_e - 2\tilde{v}'_e x'_e(v' + 1) \quad (4)$$

A plot of $\Delta\tilde{v}(v')$ versus v', termed a Birge–Sponer plot, will thus have a slope of $-2\tilde{v}'_e x'_e$ and an intercept of $\tilde{v}'_e - 2\tilde{v}'_e x'_e$. The values of $\Delta\tilde{v}(v')$ for all v'' values are combined in this plot so that the two methods should give the same $\tilde{v}'_e$ and $\tilde{v}'_e x'_e$ parameters. A similar treatment can be used for lower-state differences $\Delta\tilde{v}(v'')$ to yield $\tilde{v}''_e$ and $\tilde{v}''_e x''_e$. The electronic spacing $\tilde{v}_{el}$ is then determined using these parameters and the observed frequencies in Eq. (3). This alternative procedure has the virtue of providing a visual representation of the data so that discordant points can be examined and the data can be fitted with a single least-squares treatment that is easily done on a hand-held calculator. The multiple linear regression technique is to be preferred, however, since it uses all the data with equal weighting and has minimum opportunity for calculational error in forming differences. Such regressions are only slightly more complicated on a hand-held calculator and on a computer they are particularly simple if a spreadsheet program (such as Lotus 1-2-3™ or Symphony™) or a fitting program is available.

Dissociation energies. Because of the anharmonicity term, the spacing between adjacent vibrational levels decreases at higher v values, going to zero at the point of dissociation of the molecule into atoms. From Eq. (4), the value of $v = v_{max}$ at which this occurs is $v_{max} = (1/2x_e) - 1$. Substitution of this into Eq. (2) gives an expression for the energy D_e required to dissociate the molecule into atoms:

$$D_e = G(v_{max}) = \tilde{v}_e(1/x_e - x_e)/4 \quad (5)$$

The energy D_0 to dissociate from the $v = 0$ level is smaller than D_e by the zero point energy $G(0) = \tilde{v}_e/2 - \tilde{v}_e x_e/4$ so that

$$D_0 = \tilde{v}_e(1/x_e - 2)/4 \quad (6)$$

The expressions used in Eqs. (3)–(6) assume that $\tilde{v}_e y_e$ and higher order anharmonicity terms can be neglected, an approximation that is good for the B state of I_2 but more typically leads to D_e values that are high by 10 to 30%.

The error for the X ground electronic state is particularly large if just the absorption data are used to deduce $\tilde{v}_e''$, $\tilde{v}_e''x_e''$, and D_e'' since only the $v'' = 0$, 1, 2 levels are appreciably populated at room temperature. Extension to higher levels, v'' up to ~30, is possible using the emission spectrum, so that improved values of $\tilde{v}_e''$ and $\tilde{v}_e''x_e''$ are obtained. The value of D_e'' remains poorly determined, however, since even the $v'' = 30$ level is less than half-way to the dissociation limit.

A more accurate value of D_e'' can be obtained by combining $\tilde{v}_{el}$ and D_e' values with $E(I^*)$, the difference in electronic energy of the iodine atoms produced by dissociation from the X and B states. The value of $E(I^*)$ is known to be 7603 cm^{-1} from atomic spectroscopy,[10] so that, as seen in Fig. 1,

$$D_e'' = \tilde{v}_{el} + D_e' - E(I^*) \tag{7}$$

Potential functions. Near the minimum in the potential energy curve of a diatomic molecule, the harmonic oscillator model is usually quite good. Therefore, the force constant k_e can be calculated from the relation

$$k_e = \left(\frac{\partial^2 U}{\partial r^2}\right)_{r_e} = \mu(2\pi c\tilde{v}_e)^2 \tag{8}$$

where μ is the reduced mass and c is the speed of light in cm s^{-1} units. The constant k_e is the curvature of the potential curve at the minimum distance r_e and, like the dissociation energy, serves as a measure of the bond strength.

At large displacements from the equilibrium position, the harmonic representation of the potential energy is invalid and a more realistic model is necessary. One simple function that is often employed is the Morse potential,

$$U(r - r_e) = D_e\{\exp[-\beta(r - r_e)] - 1\}^2 \tag{9}$$

which has the desired values of 0 at $r = r_e$ and D_e at $r = \infty$. The parameter β is determined by equating k_e to the curvature of the Morse potential at $r = r_e$, yielding

$$\beta = \left(\frac{k_e}{2hcD_e}\right)^{1/2} \tag{10}$$

This three-parameter function provides a very good approximation to the real potential energy curve at all distances except $r \ll r_e$, a region of no practical significance.

Rotational structure. Although rotational structure is not resolved in the present I_2 absorption experiment, each vibrational band consists of P ($\Delta J = -1$) and R ($\Delta J = +1$) branches as discussed in Exp. 38. For vibrational changes *within* a given electronic state, such as those measured for HCl in Exp. 38, the P and R branches are distinct, with a pronounced dip between them that characterizes the missing Q branch frequency for the "pure" vibrational transition (see Fig. 38-3). The spacing between lines in each branch is not constant, a slight asymmetry arising from a quadratic term, see Eqs. (38-9, 38-10; 39-11),

$$\tilde{v} = \tilde{v}_0 + (B' + B'')m + (B' - B'')m^2 \qquad (11)$$

This is a general equation for the transition frequencies in which $m = -J$ for the P lines and $m = J + 1$ for the R lines. For $B' < B''$ the m^2 term causes a decrease (increase) in line spacing in the $R(P)$ branch at high J values. The resultant asymmetry is small for HCl since $B' - B''$ is small.

If the upper and lower levels of a transition correspond to *different* electronic states, $B' - B''$ is generally much larger and the corresponding quadratic term in Eq. (11) will often cause a frequency maximum ($B' < B''$) in the R branch or frequency minimum ($B' > B''$) in the P branch. This reversal in the progression of lines at low values of J produces a sharp *band head*, which, in the case of I_2 occurs on the R branch edge at a J value as low as $J = 2$. The R branch thus folds back and merges with the P branch so that only a single band is seen for each transition to a vibrational level. A transition frequency measured at the intensity maximum of this band will be *lower* than the "pure" vibrational transition frequency assumed in Eq. (3). This error is not constant, varying from 20 to 50 cm^{-1} for I_2 as v' increases from 0 to the dissociation limit. For this reason, in the present experiment band-head frequencies, rather than band maxima, will be measured to obtain the best values of the transition frequencies and the vibrational spacings.

The *emission* bands of I_2 will also contain many rotational lines if the spectral width of the excitation source is broad enough to populate many upper state levels. However if the source is *monochromatic*, excitation to a single v', J' level can occur and the resultant spectral emission is greatly simplified. Assuming that there is no change to another level in the upper state owing to collisions, the emission to a given lower v'' level will consist of only the two transitions corresponding to $\Delta J = -1$ and $\Delta J = +1$. Since there is no restriction on Δv, one will observe sequences of doublets whose large spacings give the vibrational-level separations in the ground electronic state. The small spacing corresponds to $2B''(2J' + 1)$, the separation between the $J'' = J' + 1$ and $J'' = J' - 1$ levels in the lower v'' state. If a doublet of known J' value can be resolved, the splitting can be used to determine the rotational constant $B''(v'')$.

EXPERIMENTAL

Absorption spectrum. The absorption spectrum of I_2 vapor is easily obtained with any commercial visible spectrometer having a resolution of about 0.2 nm or better: see Fig. 2. A general description of such spectrometers is given in Chapter XVIII and the instrument manual of the instrument to be used should be consulted for specific operational details. Follow the guidelines provided by the instructor in recording the spectra at the highest resolution possible with the instrument. Calibration corrections to the wavelength readout should be

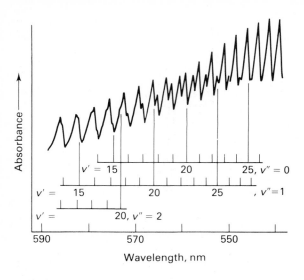

FIGURE 2

A portion of the medium-resolution spectrum of the visible $B \leftarrow X$ iodine absorption spectrum with assignments for the overlapping progressions for $v'' = 0, 1, 2$. The upper state v' values are indicated at the estimated band-head positions on the short wavelength side of each transition; the band maxima are at the *top* of the figure.

provided or made as described in Chapter XVIII. Unless these are quite variable over the 450 to 650 nm range, a single correction value is sufficient.

Crystals of I_2 can be placed in a conventional glass cell of 100-mm length which is then closed with a Teflon stopper. A usable spectrum can be obtained at room temperature (vapor pressure of I_2 ~0.2 Torr), although the absorption is much more intense if the cell is wrapped with heating tape to raise the temperature to ~40°C (vapor pressure ~1 Torr). In this case, to avoid condensation of I_2, the windows should be heated to a higher temperature by wrapping the ends of the cell with extra coils of the heating tape.

Emission spectrum. Several sources are suitable for exciting the emission spectrum of I_2. As shown in the high-resolution spectrum of Fig. 3, the "green line" at 546.074 nm emitted by a low-pressure mercury lamp is nearly monochromatic (the satellite lines are due to nuclear hyperfine structure and are weak enough to ignore). Below the mercury line is shown a portion of the iodine absorption spectrum, taken on a 10-m grating spectrograph, with some of the transitions identified. The narrow central component of the mercury emission line is seen to coincide precisely with a single absorption line of the I_2, spectrum, which has been identified[10] as a transition from $v'' = 0$, $J'' = 33$ in the ground state to $v' = 25$, $J' = 34$ in the excited state. [In conventional spectroscopic notation, this is the 25–0 $R(33)$ line.] Emission from this single state would yield doublet sequences, as discussed earlier.

For greater excitation efficiency, a more intense medium-pressure mercury source can be used in the present experiment. In this case, the exciting line is considerably broadened by the high temperature (Doppler effect) and relatively high pressure (Lorentz effect) existing in the lamp. This broadened emission line is absorbed by seven of the iodine lines, thus populating seven levels of the upper electronic state. Each of these excited states produces a

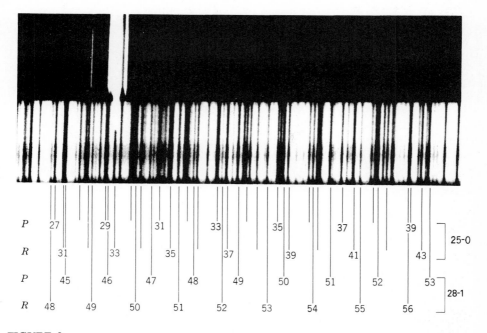

FIGURE 3

High resolution spectrum of the visible $B \leftarrow X$ iodine absorption spectrum. The mercury emission line at 546.074 nm is shown in the spectrum at the top. The total spectral width shown here is 0.4 nm.

doublet in returning to the ground state by photon emission. This closely spaced rotational fine structure is seen only under high-resolution conditions; at moderate resolution the emission multiplets appear as single features whose spacing is thus a measure of the vibrational energy levels in the ground state of the molecule.

If an argon-ion laser is available, the 514.527 nm line can be used as an alternative excitation source. Light of this wavelength causes two transitions to the B state from the X state: $v' = 43$, $J' = 12 \leftarrow v'' = 0$, $J'' = 13$ and $v' = 43$, $J' = 16 \leftarrow v'' = 0$, $J'' = 17$.[11] From these upper levels of the B state, an extended emission progression to many levels of the X state results. Under high resolution, triplet structure is observed owing to overlap of the P, R doublets expected for each originating J' level. A combination of the argon-ion laser with a Raman spectrometer is ideal for this experiment, since such instruments are designed to collect scattered light efficiently and to measure intensities that are much lower than the I_2 emission signals. A photomultiplier tube with "extended red" response is desirable for detection of the long-wavelength emission to high v'' levels.

For laser excitation, a 50-mm cylindrical glass cell with two flat end windows is used to contain the I_2; see Fig. 4(a). The focused laser beam enters

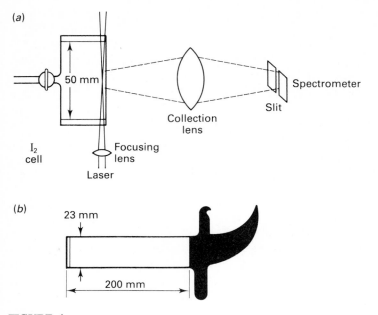

FIGURE 4
(*a*) Fluorescence cell for use with laser excitation. (*b*) Fluorescence cell with blackened Wood's horn and side arm for iodine solid, suitable for use with mercury lamp excitation.

and leaves through the windows, traversing the cell parallel with the entrance slit (to optimize the collection efficiency) and near the cell wall facing the spectrometer (to minimize reabsorption of the emitted light by I_2). The cell is prepared by adding several crystals of I_2, after which it is pumped down to 10^{-3} Torr or less residual air pressure and sealed off. No heating of the cell is necessary but it is essential that the cell should not leak, since the addition of air serves to quench the emission intensity very efficiently. The track of the laser beam through the cell should be quite visible to the eye. (One can also use a cell with a greaseless stopcock, which permits re-evacuation if necessary.)

A suitable sample cell for excitation using a mercury lamp is shown in Fig. 4(*b*). The front window should be an optical-quality Pyrex flat. The cell terminates in a blackened Wood's horn to cut down scattered excitation light. A suitable Hg excitation source is a 350-watt, medium-pressure quartz–mercury arc, which produces about 6 W at 546 nm. The lamp is 150 mm in length and, for best performance, should be mounted confocally with the cell in a polished elliptical reflector. A "didymium" glass filter interposed between the lamp and the cell cuts out the mercury yellow lines, which would excite a second emission series and make the analysis more complicated. The emission should be observable as a dull orange glow by looking down the end of the cell.

If the emission spectrum is to be recorded photographically, the procedure is essentially the same as that followed in studying the hydrogen atomic emission (Exp. 40) and the nitrogen band spectrum (Exp. 41). Since the emission is rather weak and in the red part of the spectrum, suitable high-speed, red-sensitive plates (Eastman Kodak Type 103a-F or 1-N) or film (Eastman Kodak Royal-X Pan Recording) should be used. Because these emulsions are very sensitive, all darkroom operations must be carried out without any safelight. A half-hour exposure should produce satisfactory spectra. The emission from a neon-filled lamp provides a suitable calibration spectrum.

CALCULATIONS AND DISCUSSION

Absorption spectrum. Assign the vibrational quantum numbers of the absorption bands using the numbering indicated in Fig. 2. This numbering choice is *not* obvious and was established, after some controversy, from considerations of intensity distributions[10] and isotopic frequency shifts.[12] Note that there are three overlapping progressions since there is appreciable population in the $v'' = 0$, 1, and 2 levels at room temperature. As a measure of the band-head position, take the *minimum* on the *short* wavelength side of each peak, estimating this as best you can for overlapping peaks from different progressions. Correct for any spectrometer calibration error and prepare a Deslandres table (see Table 41-1 and associated discussion in Exp. 41) as an aid to spotting any inconsistencies or poorly determined values.

Emission spectrum. The emission spectrum should form a regular progression to the red of the exciting line, with $v'' = 0$ corresponding to the exciting wavelength. Measure the wavelengths of as many emission bands as are observable, including any calibration correction if a scanning spectrometer is used. If multiplets are resolved, use the wavelength of the most intense member. For photographic recording, a dispersion curve for the spectral region between 500 and 700 nm is obtained by carefully measuring mercury and neon calibration lines with a comparator microscope. A compilation of prominent neon lines is given in Chapter XVIII.

Interpretation and discussion. Use a multiple linear least-squares technique and Eq. (3) to determine the parameters $\bar{v}_{el}$, $\bar{v}'_e$, $\bar{v}'_e x'_e$, $\bar{v}''_e$, $\bar{v}''_e x''_e$ from the absorption data. Calculate D_e and D_0 for both states using Eqs. (5) and (6) and compare the lower state D''_e value with the more accurate value obtained from Eq. (7). If emission data have been recorded, analyze these to get improved values for the X state parameters. If a Birge–Sponer plot of these data shows curvature, you might see whether inclusion of a $\bar{v}''_e y''_e$ term improves your fit.

Compare your constants with recent literature values.[5,13,14] Note that the $\bar{v}_e$ and $\bar{v}_e x_e$ values change if $\bar{v}_e y_e$ and higher anharmonicity terms are included in the analysis of the same data set,[5] so that close agreement may not be obtained with the literature results given in Refs. 13 and 14. Discuss other possible errors in the experiment or analysis that would add to the uncertainties obtained from the least-squares treatment.

Calculate the harmonic force constant k_e and the Morse parameter β for the two I_2 states. Using the known r''_e value (0.2666 nm)[13] plot the Morse curve for the ground electronic state of I_2. Compare this with the harmonic potential calculated from k_e.

To include the upper state potential curve on the same graph, it is necessary to know r'_e, which can be estimated from the observed intensities of the absorption spectrum in the following way. According to the Franck–Condon principle,[9,15] the intensity of an electronic transition is related to the overlap of the vibrational wavefunctions for the two states by

$$I(v', v'') \propto \left| \int \psi_{v'}(r)\psi_{v''}(r)\, dr \right|^2 \qquad (12)$$

This overlap will be the greatest when $\psi_{v'}(r)$ and $\psi_{v''}(r)$ have their maximum values at the same distance r. The maximum in $\psi_{0''}(r)$ occurs at $r''_0 \approx r''_e$ for a harmonic oscillator wavefunction but, for higher vibrational levels, this maximum approaches the classical turning point limits of the potential. Since r does not change during the transition (the heavy nuclei take time to move whereas the light electrons redistribute "instantly") the transition is said to be *vertical*. From Fig. 1, it can be seen that the $v'' = 0$ transition of greatest intensity, $\bar{v}(v^*{}')$, intersects the upper state potential curve at $r' = r''_e$ so that one can write

$$U'(r''_e - r'_e) = D'_e\{\exp[-\beta'(r''_e - r'_e)] - 1\}^2 + \bar{v}_{el}$$
$$= \bar{v}(v^*{}') + \bar{v}''_e/2 - \bar{v}''_e x''_e/4 \qquad (13)$$

Determine $\bar{v}(v^*{}')$ from your spectrum and then $r''_e - r'_e$ from this expression. Compare the resultant value of r'_e with the literature value of 0.3025 nm obtained by analysis of the rotational structure of the electronic spectrum.[13] Include the Morse potential curve for the upper state on your plot for the X state and comment on the differences in the various parameters determined for the two states.

As an optional computer exercise, the student might find it instructive to calculate and plot the wavefunctions for the first few levels, using harmonic oscillator and Morse potential models.[5]

APPARATUS

Medium-resolution absorption spectrometer; emission spectrometer with red-sensitive photomultiplier or emission spectrograph with red-sensitive plates or film (Kodak Type 103a-F or 1-N) and developing and fixing solutions; argon-ion laser or medium-pressure mercury arc (one suitable source is a 150-mm length 360 W UA-3 lamp. and power supply (420-UI), distributed by George W. Gates Co., Franklin Square, NY); didymium glass filters (Corning

Glass Co.); neon calibration lamp and power supply (available from, e.g., Oriel Corp. Stratford, CT); reagent-grade iodine; 100-mm glass cell with Teflon stoppers; heating tape with controlling variac; 50-mm cell for emission studies; vacuum system for pumping down emission cell; comparator microscope if photographic plates are used.

REFERENCES

1. G. Herzberg, "Molecular Spectra and Molecular Structure I. Spectra of Diatomic Molecules," 2d ed., Van Nostrand, Princeton, N.J. (1950).
2. G. Herzberg, "Molecular Spectra and Molecular Structure III. Electronic Spectra of Polyatomic Molecules," Van Nostrand, Princeton, N.J. (1966).
3. F. E. Stafford, *J. Chem. Educ.* **39,** 626, (1962).
4. R. D'alterio, R. Mattson, and R. Harris, *J. Chem. Educ.* **51,** 283 (1974).
5. I. J. McNaught, *J. Chem. Educ.* **57,** 101, (1980).
6. J. D. Simmons and J. T. Hougen, *J. Res. Natl. Bur. Stand.* **81A,** 25 (1977).
7. J. I. Steinfeld, *J. Chem. Educ.* **42,** 85 (1965).
8. J. Tellinghuisen, *J. Chem. Educ.* **58,** 438 (1981).
9. See Chap. 4 of ref. 1.
10. J. I. Steinfeld, R. N. Zare, J. M. Lesk, and W. Klemperer, *J. Chem. Phys.* **42,** 15 (1965).
11. R. B. Kurzel and J. I. Steinfeld, *J. Chem. Phys.* **53,** 3293 (1970); M. Rubinson, B. Garetz, and J. I. Steinfeld, *J. Chem. Phys.* **60,** 3082 (1974), see fn. 7.
12. R. I. Brown and T. C. James, *J. Chem. Phys.* **42,** 33 (1965).
13. K. P. Huber and G. Herzberg, "Molecular Spectra and Molecular Structure IV. Constants of Diatomic Molecules," p. 332, Van Nostrand Reinhold, New York (1979).
14. P. Luc, *J. Mol. Spectrosc.* **80,** 41 (1980).
15. E. U. Condon, *Amer. J. Phys.* **15,** 365 (1947).

GENERAL READING

G. Herzberg, "Molecular Spectra and Molecular Structure I. Spectra of Diatomic Molecules," 2d ed., chaps. 2–4, 8, Van Nostrand, Princeton, N.J. (1950).
J. M. Hollas, "Modern Spectroscopy," chaps. 6–7, Wiley, New York (1987).
J. I. Steinfeld, "Molecules and Radiation", 2d ed., chap. 5, MIT Press, Cambridge, Mass. (1985).

EXPERIMENT 43
ELECTRON SPIN RESONANCE SPECTROSCOPY

Electron spin resonance (ESR), also called electron paramagnetic resonance (EPR), is a form of magnetic resonance spectroscopy which is possible only for molecules with unpaired electrons. Despite this restriction, this sensitive technique has proved useful in the study of the electronic structures of many species, including organic free radicals, biradicals, triplet excited states, and most transition-metal and rare-earth species.[1-3] Important biological applications include the use of "spin labels" as probes of molecular environment in enzyme active sites and membranes. ESR has also been used to examine interior defects in solid-state chemistry, and to study reactive chemical species on catalytic surfaces. The present experiment, on several benzosemiquinone radical anions, provides an introduction to the theoretical principles and experimental techniques used in ESR investigations. The utility of simple Hückel molecular orbital theory in interpreting the experimental spectra is also demonstrated.[3-6]

THEORY

As indicated in Eqs. (33-4) and (33-5), a single unpaired electron has spin angular momentum characterized by the quantum number $S = \frac{1}{2}$ and a magnetic moment of magnitude

$$\mu(\text{electron}) = g_e \mu_e [S(S+1)]^{1/2} \tag{1}$$

where g_e, the electron g-factor, equals 2.0023 for a free electron and $\mu_e = eh/4\pi m_e c$, the Bohr magneton, equals 9.274×10^{-24} J T^{-1}. For atoms and molecules with many electrons, the total electron spin angular momentum can be as high as the sum of the spin angular momenta for each electron. Since spin is a vector quantity, the total spin is usually less because some of the individual spins are oriented so that they cancel each other. In fact, most of the electrons in molecular species are *paired*, with zero angular momentum per pair, because this arrangement gives a lower energy for the system. In practice the species of greatest interest are those in which the total spin is a small integral multiple of $\frac{1}{2}$: $S = \frac{1}{2}$ (radicals and some transition metal species), $S = 1$ (biradicals, triplets, and some transition metals), and $S > 1$ (some transition metals).

The energy of interaction of the electronic magnetic moment with a magnetic field (the Zeeman energy) is given by [1-3]

$$E = g_e \mu_e M_S B \tag{2}$$

where B is the magnetic induction and M_S is the quantum number that measures the component of the spin angular momentum along the field direction (z). For a single electron, $M_s = -\frac{1}{2}$ or $+\frac{1}{2}$, corresponding to two degenerate levels that split in the presence of the field, as depicted by the dashed lines in Fig. 1.

As for NMR spectroscopy, the selection rule for magnetic transitions is $\Delta M_s = \pm 1$ so that transition can take place between the $M_s = -\frac{1}{2}$ and $M_s = +\frac{1}{2}$ spin levels upon excitation by radiation of frequency

$$v = \frac{\Delta E}{h} = \frac{g_e \mu_e B}{h} \tag{3}$$

This relation is identical to Eq. (34-3) for a proton NMR transition except that, at a given field value, the splitting between the ESR levels is larger by the ratio $g_e \mu_e / g_N \mu_N = 658$. As a consequence, microwave radiation at a frequency of

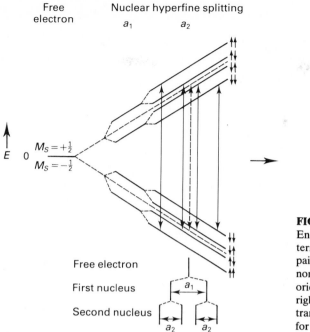

Free electron

Nuclear hyperfine splitting

a_1 a_2

E 0

$M_S = +\frac{1}{2}$
$M_S = -\frac{1}{2}$

Free electron

First nucleus a_1

Second nucleus a_2 a_2

FIGURE 1
Energy levels and spectral line pattern (stick spectrum) for an unpaired electron interacting with two nonequivalent protons whose spin orientations are indicated at the right. The dashed levels and dashed transition arrow indicate the case for an uncoupled free electron.

about 10 GHz is commonly employed for ESR spectroscopy, in contrast to the lower-energy radiofrequency radiation (about 100 MHz) used in NMR.

To chemists, the principal aspect of interest in electron spin resonance is that the electron spin energies are sensitive to the molecular environment, as in the NMR case. The ESR counterpart to the NMR chemical shift is a variation in the g_e factor for an electron in a compound. For systems with orbital angular momentum, such as transition metal atoms and ions, g_e can deviate substantially from 2.0023 if there is appreciable coupling between the electron spin and orbital motion. However for most organic radicals g_e changes only very slightly and its variation does not give nearly as much chemical information as does the NMR chemical shift for a nucleus.† Instead, the primary value of the ESR measurement comes from the fact that the unpaired electron is generally not localized on a given atom but samples the magnetic environment over much if not all of the molecule. If the molecule contains nuclei with magnetic moments, especially protons, the electron–nuclear interaction produces characteristic splitting patterns in the ESR spectrum that can be used to deduce the number of different types of nuclei and their geometrical symmetry.

---

† The g_e factor is actually a tensor quantity whose average reduces to a single value for isotropic media such as gases, liquids, or solutions such as those studied in this experiment. For anisotropic samples such as crystals and impurities in solids, measurement of the components of g_e can give added information about the local molecular environment.

More specifically, the total energy in a magnetic field can be written, to first order, as a sum of three contributions:

$$
\begin{array}{ccc}
\text{electron} & \text{nuclear} & \text{electron--} \\
E = \text{Zeeman} + \text{Zeeman} + \text{nuclear} \\
\text{energy} & \text{energy} & \text{coupling}
\end{array}
$$

$$E = g_e \mu_e M_s B - \sum_i g_{Ni} \mu_{Ni} M_{Ii} B + \sum_i a_i M_s M_{Ii} \tag{4}$$

The hyperfine splitting constant a_i has units of energy and is a characteristic parameter for the interaction of the unpaired electron with a nucleus of type i. It is the analog of the NMR J spin–spin constant describing the magnetic coupling between two nuclei.

The nuclear Zeeman term in Eq. (4) can be omitted in the following presentation since it does not change for levels involved in an ESR transition ($\Delta M_I = 0$ if $\Delta M_S = \pm 1$). Thus, for example, the energy levels for an unpaired electron interacting with two different protons can be written as

$$E = M_s [g_e \mu_e B + a_1 M_{I1} + a_2 M_{I2}] \tag{5}$$

giving rise to the pattern and transitions depicted in Fig. 1. Here the usual experimental procedure is assumed, in which radiation of fixed frequency impinges on the system as the magnetic field is varied. As indicated in the figure, the consequence of the electron–nuclear coupling in Eq. (5) is to split the free-electron levels by an amount $\pm \frac{1}{2} a_1 \pm \frac{1}{2} a_2$ and to produce a quartet of lines whose spacings yield directly the hyperfine splitting parameters. The latter are usually reported in gauss units (1 gauss $= 10^{-4}$ Tesla)

$$a_i(\text{gauss}) \equiv \frac{a_i(\text{Joule})}{g_e \mu_e} \times 10^{-4} \tag{6}$$

so they can be extracted directly from the spectrum. This statement is not changed by a more exact, second-order treatment[1-3] of the energy levels, which only produces a slight common shift in all transitions such that the hyperfine splittings remain the same.

If the two protons are equivalent ($a_1 = a_2$), the two central transitions in Fig. 1 merge and a triplet hyperfine pattern is produced with intensity ratios 1:2:1. In general, n equivalent protons give a spectrum of $n + 1$ hyperfine lines equally spaced by the hyperfine splitting constant a_H. The intensities are proportional to the degeneracy of the lower energy level involved in the transition; the relative values can be determined from the coefficients in a binomial expansion $(1 + x)^n$, which is equivalent to the relation

$$I_M = \frac{n!}{(\frac{1}{2}n + M)! \, (\frac{1}{2}n - M)!} \qquad M = -\tfrac{1}{2}n, \ -\tfrac{1}{2}n + 1, \ \ldots, \ \tfrac{1}{2}n - 1, \ \tfrac{1}{2}n \tag{7}$$

For the benzene radical anion, one thus expects a 7-line pattern with intensity

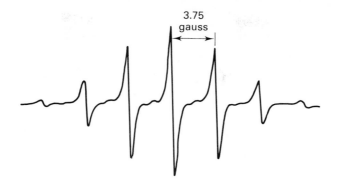

3.75
gauss

FIGURE 2
Proton hyperfine splitting pattern in the ESR spectrum of the benzene anion radical.

ratios of $1:6:15:20:15:6:1$, in good agreement with the ESR spectrum shown in Fig. 2.

In the present experiment, we are concerned with the hyperfine structure of the benzosemiquinone radical anions. The delocalized unpaired π electron

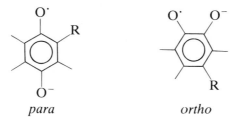

para   *ortho*

is of course distributed over the entire molecular frame of 6 C atoms and 2 O atoms. With $R = H$, by symmetry, it is clear that the four protons are all equivalent in the *para* species; hence, five hyperfine lines with relative intensities $1:4:6:4:1$, are expected in the ESR spectrum of this radical. By contrast, when R is not a proton, the three ring protons are not related by symmetry, and thus each may be expected to possess a different splitting constant. A hyperfine structure pattern of eight unequally spaced lines of equal intensity is expected. The line splittings and relative intensities in ESR spectra thus convey information about the geometric arrangement of the atoms.

The magnitude of the hyperfine splitting parameters also yield information about the electron distribution in the molecule. The theory of the electron–nuclear coupling interaction was first worked out by Fermi who showed that the constant a depended upon the electron density at the nucleus. For a free hydrogen atom, a is given by the Fermi contact interaction in the form[1-3]

$$a = (8\pi/3)g_e\mu_e g_N\mu_N\rho(0) \tag{8}$$

where $\rho(0) = |\psi(0)|^2$ is the unpaired electron density at the H nucleus. The wavefunction for the ground state of the H atom is well known, namely $\psi = (\pi a_0^3)^{-1/2} \exp(-r/a_0)$ with $a_0 = 0.0529$ nm equal to the Bohr radius. The resultant value for a in gauss is 507 for this "pure" s orbital while that for p, d, f and all other orbitals is zero since these have a node at the nucleus.

For the molecular case, the essential conclusion is that the orbital must have some s (or σ) character for the unpaired electron to interact with a magnetic nucleus. Consider, however, the case of the benzene radical anion, in which the electron is usually described as being in a π orbital with a node in the molecular plane. As a consequence, no coupling with the proton nuclei is expected, a prediction clearly in conflict with the hyperfine splitting of 3.75 gauss seen in the ESR spectrum of this species shown in Fig. 2. How then does the unpaired π electron density appear at the H nucleus?

The answer is that the electrons cannot be so neatly labeled as σ or π type, and part of the unpaired π electron density is transferred through the C—H sigma bonding electrons to the H nucleus through *exchange interactions*.[1-3] In the case of the π electron in the planar methyl radical, this process, termed spin polarization, results in a hyperfine constant of -23 gauss, about 5% of the limiting value of 507 gauss for an electron completely isolated on the H atom. The negative sign of a is not directly determined in the ESR experiment but is given by the theory. Thus one might say that the unpaired electron polarizes the CH bonding pair such that there is a net "negative spin excess of 5%" about the proton. For the benzene anion radical, the electron is equally distributed over six carbon atoms and one would expect a to be about $-23/6 = -3.83$ gauss, a value in good agreement with the experimental splitting of 3.75 gauss.

In the case of organic free radicals, McConnell[7] has shown that a simple empirical proportionality can be used to relate the observed hyperfine structure constant a_H and the unpaired electron spin density on the nearest carbon atom:

$$a_H = Q\rho_\pi \qquad (9)$$

The constant Q is of the order of -20 to -30 gauss for aromatic hydrocarbons, and the benzene anion value of $Q = -6 \times 3.75 = -22.5$ gauss is commonly used. This relation may also be applied to give the hyperfine constant a_H for splitting arising from protons on the first carbon of a substituent attached to a carbon in an aromatic system, e.g., each of three methyl hydrogens in the toluene radical cation. Again Q is in the range of -20 to -30 gauss, and a value of -28 is usually assumed[8] when an independent experimental determination cannot be made.

The hyperfine structure constant thus allows us to probe the electron distribution in radicals. Theoretically calculated values of the spin densities can then be compared with the experimental values obtained from Eq. (9). One of the simplest methods for calculating electron density in an aromatic hydrocarbon is to use Hückel molecular orbital (HMO) theory as discussed later.

The present experiment involves an investigation of the ESR spectra of the unsubstituted *ortho*- and *para*-benzosemiquinone radical anions, along with one or more of the methyl and *t*-butyl-derivatives. Aspects of interest include the elucidation of splitting constants from complex spectra, the examination of substituent effects in ESR spectra, and the interpretation of spectra using molecular-orbital calculations.

EXPERIMENTAL

A schematic of an ESR spectrometer is shown in Fig. 3; more detailed discussion of construction and operation of the instrument can be found in Ref. 1 and in citations therein as well as in the manuals accompanying the spectrometer to be used. The microwave source is a vacuum-tube Klystron or a solid-state Gunn diode which provides tunable monochromatic radiation at about 10 GHz at a power level adjustable with an attenuator from zero to a few tenths of a watt. Radiation of this frequency is conveniently directed to the sample cavity by rectangular tubing, termed X-band waveguide. The sample is placed in a quartz tube and inserted into a rectangular cavity assembly whose dimensions are such as to produce at the microwave frequency a standing wave with maximum magnetic field and minimum electric field at the sample position. This arrangement provides the most efficient coupling of radiation into the sample for the magnetic-dipole transitions of interest while at the same time minimizing energy absorption by nonresonant electric-dipole absorption (dielectric loss or "microwave cooking").

Typically, the source is tuned with the sample in place and then locked to match the cavity resonance frequency so as to achieve maximum energy storage and minimum reflected power. This reflected power is directed through

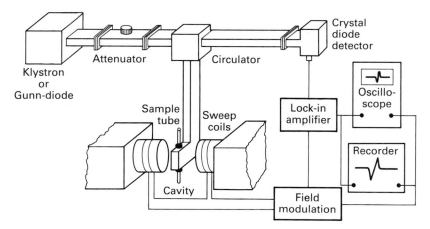

FIGURE 3
Schematic diagram of an ESR spectrometer.

a "one-way" coupler called a circulator to a crystal diode detector to convey information about sample absorption in the cavity. An iris opening to the cavity is adjusted to match the impedance of the cavity to that of the source so as to produce minimum reflection of radiation from the cavity. This condition gives maximum sensitivity for the impedance mismatch produced when sample absorption occurs in the cavity.

Resonance is achieved by varying the current in the electromagnet to sweep the field a few gauss about a typical center value of 3500 gauss. On resonance, energy is absorbed in the cavity and the amount of radiation reflected back to the detector is altered. To achieve high sensitivity, the magnetic field is modulated (typically at 30–100 kHz) by an amount chosen to be small compared to the transition line width. The resultant ac component from the detector is sent to a phase-sensitive lock-in amplifier, producing a derivative spectrum, the common mode of display in ESR spectroscopy.

Details of the operation of the spectrometer will be provided by the instructor. Familiarity with the instrument can be gained by some preliminary experimentation with a stable radical sample such as solid α,α'-diphenyl-β-picrylhydrazyl (DPPH), which is often used as an ESR calibration standard.

The semiquinone radicals are produced by base-induced oxidation of 1,4-dihydroxybenzene (hydroquinone) or 1,2-dihydroxybenzene (catechol) by molecular oxygen, present in dissolved form. Radical concentration will increase over a period of time as the oxidation reaction proceeds, and then decay as radical–radical reaction and other processes destroy the anions. Rates for these processes will depend on temperature, concentration of the dihydroxybenzene, and other parameters, so some experimentation may be necessary to obtain optimal spectra.

Prepare 5–10 mL of concentrated solution ($1\,M$ or greater) of the hydroquinones in methanol (ethanol or acetonitrile can also be used as solvents). A basic NaOH solution in methanol ($1\,M$ or greater) can be made by adding 1 g NaOH to 25 mL of alcohol. With a syringe, place 1–2 mL of the hydroquinone solution into a small beaker and then add a drop of basic methanol. Stir until the solution turns yellow, then transfer to a quartz ESR tube and record the spectrum. A scan range of 10–20 gauss and a modulation amplitude of 0.01 gauss would be suitable starting parameters. The spectra can be recorded in conventional 5-mm o.d. quartz ESR tubes although the high dielectric constant of methanol makes the cavity more difficult to tune. This problem can be reduced by inserting the tube so that the solution extends only partially into the cavity region or, better, by using a 2-mm o.d. ESR tube. Special ESR tubes for aqueous solutions are also available which have a flat rectangular sample section that can be oriented to maximize the sample volume at the central plane of the cavity where the electric field has a node, giving lower dielectric loss. **Special care should be taken to wipe all liquid off the outside of the tubes to avoid contamination of the cavity.** The cells should also be rinsed carefully with methanol or acetone prior to storage at the end of the experiment.

The parent *para*-benzosemiquinone radical anion may gradually increase in concentration and will last about two hours; the methyl- or *t*-butyl-substituted radicals form more quickly and have lifetimes of about 15 min. The color serves as some guide to the optimal concentration; a reddish color suggests that too much base has been added while a brown-black color indicates that radical–radical combination has occurred.

A satisfactory spectrum for the *ortho*-benzosemiquinone anion is more difficult to obtain by the above method because the radical–radical reaction is much faster. One simple way to promote the oxidation of the parent to provide a reasonable anion concentration is to increase the surface area of the solution to give greater access by oxygen in the air. This can be done by adding one drop each of catechol solution and base to a 5-mm o.d. EPR tube. Turn the tube to form on the tube wall a *film* of solution, which should be green; a brown color is the result of radical–radical reaction. Shake any excess solution from the tube and record the spectrum immediately. It may be necessary to experiment a bit to obtain optimal spectra.

Spectra for substituted catechols can be obtained in the same way as for the hydroquinones but these are much less stable and interference by other radical intermediates makes the interpretation of these spectra more difficult. Thus it is recommended that the student obtain spectra for the parent *ortho*- and *para*-benzosemiquinones and for one or more of the substituted methyl-, 2,3-dimethyl- or *t*-butyl-*para*-forms. All the expected lines in the parent and dimethyl anion spectra should be clearly resolved. For the methyl compound, it may be difficult to obtain a recording showing all 32 expected lines clearly separated, but at least 20 separate lines should be readily distinguishable. Resolution of all the hyperfine structure of the *t*-butyl species will depend upon the instrument and the sampling conditions. It may be possible to sharpen the lines somewhat by varying the concentration and by minimizing any excess dissolved oxygen gas which, since it is paramagnetic, can cause line broadening.

CALCULATIONS AND DISCUSSION

Analyze the spectrum of each of the benzosemiquinone radicals to obtain the hyperfine splitting constants, a_H. If you have at least 20 separate lines in the spectrum of the methyl derivatives it should be possible to deduce all the splitting constants for this system. Note that the separation between the two outermost lines in the spectrum is given by a simple linear combination of the constants. After you have a set of constants for a spectrum, refine the values to give the best fit to all the lines in the spectrum. Construct a predicted stick spectrum derived from your constants for comparison with your actual ESR spectrum.[9]

Molecular-orbital calculations. According to the McConnell relation, Eq. (9), the proton hyperfine splitting constants are proportional to ρ_π, the unpaired π

electron density of the carbon atom bonded to the proton. Using quantum mechanics, ρ_π can be calculated at different levels of approximation. The approach outlined here is one of the simplest, the Hückel molecular orbital (HMO) method.[2-6] This model assumes that a π molecular orbital, delocalized over n atoms, can be written as a linear combination of n atomic p_z orbitals (z is perpendicular to the molecular plane)

$$\psi = \sum_i c_i p_i \tag{10}$$

The corresponding one-electron π density at atom i is given by

$$\rho_{i\pi} = c_i^2 \tag{11}$$

The energy E and the coefficients c_i for the ground state are determined by use of the *variation principle*,[6] which says that one should choose the coefficients such that

$$E = \frac{\displaystyle\int \psi H \psi \, d\tau}{\displaystyle\int \psi^2 \, d\tau} = \text{minimum} \tag{12}$$

Here the Hamiltonian H is an effective one-electron energy operator whose explicit form need not be written in the HMO approximation. The variation principle ensures that the lowest-energy state is as close as possible to the true energy, and the coefficients are obtained from the minimization relations

$$\frac{\partial E}{\partial c_i} = 0 \qquad i = 1, \ldots, n \tag{13}$$

This results in n equations of the form

$$c_1(H_{11} - S_{11}E) + c_2(H_{12} - S_{12}E) + \cdots + c_n(H_{1n} - S_{1n}E) = 0$$
$$\vdots \qquad\qquad\qquad\qquad\qquad\qquad \vdots \tag{14}$$
$$c_1(H_{n1} - S_{n1}E) + c_2(H_{n2} - S_{n2}E) + \cdots + c_n(H_{nn} - S_{nn}E) = 0$$

In these equations, $H_{ii} = \int p_i H p_i \, d\tau \equiv \alpha_i$, called the *Coulomb integral,* is the energy of an electron in a $2p$ orbital, while $H_{ij} = \int p_i H p_j \, d_\tau \equiv \beta_{ij}$, the *resonance* or *bond integral,* represents the interaction energy of the two atomic orbitals. Both α_i and β_{ij} are negative energy quantities. $S_{ij} = \int p_i p_j \, d\tau$ is the overlap integral, which, in the simplest approximation, is taken to be 1 if $i = j$, and 0 otherwise. For a π system involving only carbon atoms, $\alpha_i \equiv \alpha$ and Eqs. (14) take the form

$$c_1(\alpha - E) + c_2\beta_{12} + \cdots + c_n\beta_{1n} = 0$$
$$\vdots \qquad\qquad\qquad\qquad \vdots \tag{15}$$
$$c_1\beta_{n1} + c_2\beta_{n2} + \cdots + c_n(\alpha - E) = 0$$

These equations have a nontrivial solution only if the corresponding *secular determinant* vanishes:

$$\begin{vmatrix} \alpha - E & \beta_{12} & \cdots & \beta_{1n} \\ & \vdots & & \vdots \\ \beta_{n1} & \beta_{n2} & \cdots & \alpha - E \end{vmatrix} = 0 \tag{16}$$

In the simple HMO model, it is assumed that, if atoms i and j are bonded, $\beta_{ij} \equiv \beta$ and that $\beta_{ij} = 0$ otherwise. Expansion of this determinant gives a polynomial equation with n roots of the form

$$E_j = \alpha - X_j \beta \qquad j = 1, \ldots, n \tag{17}$$

Since α and β are negative, negative values of the numerical quantities X_j correspond to lower-energy *bonding* MOs, and positive X_j values to *antibonding* MOs. The state of lowest X_j is thus the "best" ground electronic state and the other levels are approximations of the higher excited states.

To obtain the wavefunction for a given state of energy E_j, one substitutes $E = E_j$ into Eq. (15) and solves for the jth set of coefficients $c_{j1}, c_{j2}, \ldots, c_{jn}$. This process actually gives only ratios of coefficients; to determine numerical values we add the normalization condition

$$\int \psi_j^2 \, d\tau = \sum_i c_{ji}^2 = 1 \tag{18}$$

In the case of hetero-atom systems like the benzosemiquinones, the HMO approach is often adapted by making the assumption that $\alpha_O = \alpha_C + \beta_C \equiv \alpha + \beta$ and $\beta_O = \beta_C \equiv \beta$. This choice is made mainly for algebraic simplification since then, as in the homo-carbon case, the *relative* energies and the wavefunctions are not dependent upon a specific choice of α or β. This can be seen by dividing Eq. (15) by β to yield

$$\begin{array}{c} c_1 X + c_2 + \cdots + c_n = 0 \\ \vdots \qquad\qquad \vdots \\ c_1 + c_2 + \cdots + c_n X = 0 \end{array} \tag{19}$$

A more explicit form depends upon the molecular geometry, since it is assumed that $\beta_{ij} = 0$ for nonbonded atom pairs and hence many of the c_i coefficients can be set equal to zero. For *para*-benzosemiquinone the secular determinant takes the form shown in Table 1. The energies are obtained from the X_j solutions and Eq. (17) while the wavefunction for a given level j is obtained by substitution of X_j into Eq. (19).

In fact, the relative coefficients within a set of *symmetrically equivalent* atoms, such as C_2, C_3, C_5, and C_6 in *para*-benzosemiquinone can be determined by group theory alone. The appropriate set of symmetrized orbitals, also listed in Table 1, can be obtained by use of character tables and procedures described in Refs. 3–6. In matrix notation, the symmetrized combinations are $\phi = \mathbf{U}\mathbf{p}$ where ϕ and $\mathbf{p}$ are column vectors and $\mathbf{U}$ is the transformation matrix giving the relations shown in part (c) of Table 1. The

TABLE 1
Hückel molecular orbital calculations for *para*-benzosemiquinone

(a) *Secular determinant* — D_{2h} structure

$$
\begin{vmatrix}
X & 1 & 0 & 0 & 0 & 1 & 1 & 0 \\
1 & X & 1 & 0 & 0 & 0 & 0 & 0 \\
0 & 1 & X & 1 & 0 & 0 & 0 & 0 \\
0 & 0 & 1 & X & 1 & 0 & 0 & 1 \\
0 & 0 & 0 & 1 & X & 1 & 0 & 0 \\
1 & 0 & 0 & 0 & 1 & X & 0 & 0 \\
1 & 0 & 0 & 0 & 0 & 0 & X+1 & 0 \\
0 & 0 & 0 & 1 & 0 & 0 & 0 & X+1
\end{vmatrix} = 0
$$

(b) *Symmetrized determinant*

$$
\begin{vmatrix}
X & 1 & 2^{1/2} & & & & & \\
1 & X+1 & 0 & & & & 0 & \\
2^{1/2} & 0 & X+1 & & & & & \\
& & & X & 1 & 2^{1/2} & & \\
& & & 1 & X+1 & 0 & & \\
& & & 2^{1/2} & 0 & X-1 & & \\
& & & & & & X-1 & \\
& 0 & & & & & & X+1
\end{vmatrix} = 0
$$

(c) *Symmetrized orbitals*

$$
B_{1u} \begin{cases} \phi_1 = 2^{-1/2}[p_1 + p_4] \\ \phi_2 = 2^{-1/2}[p_7 + p_8] \\ \phi_3 = \tfrac{1}{2}[p_2 + p_3 + p_5 + p_6] \end{cases}
$$

$$
B_{2g} \begin{cases} \phi_4 = 2^{-1/2}[p_1 - p_4] \\ \phi_5 = 2^{-1/2}[p_7 - p_8] \\ \phi_6 = \tfrac{1}{2}[p_2 - p_3 - p_5 + p_6] \end{cases}
$$

$$
A_u \quad \{\phi_7 = \tfrac{1}{2}[p_2 - p_3 + p_5 - p_6]\}
$$

$$
B_{3g} \quad \{\phi_8 = \tfrac{1}{2}[p_2 + p_3 - p_5 - p_6]\}
$$

(d) Energies	Symmetry	Molecular orbitals
$X_8 = 2.115$	B_{2g}	$\psi_8 = 0.607\phi_4 - 0.195\phi_5 - 0.770\phi_6$
$X_7 = 1.303$	B_{1u}	$\psi_7 = 0.799\phi_1 - 0.347\phi_2 - 0.491\phi_3$
$X_6 = 1.000$	A_u	$\psi_6 = \phi_7$
$X_5 = -0.254$	B_{2g}	$\psi_5 = 0.496\phi_4 - 0.665\phi_5 + 0.559\phi_6$
$X_4 = -1.000$	B_{3g}	$\psi_4 = \phi_8$
$X_3 = -1.000$	B_{1u}	$\psi_3 = 0.817\phi_2 - 0.577\phi_3$
$X_2 = -1.861$	B_{2g}	$\psi_2 = 0.621\phi_4 + 0.721\phi_5 + 0.307\phi_6$
$X_1 = -2.303$	B_{1u}	$\psi_1 = 0.601\phi_1 + 0.461\phi_2 + 0.653\phi_3$

Notes: (a) Secular determinant corresponding to Eq. (16); (b) symmetry-factored determinant; (c) symmetrized orbitals in terms of p_z atomic orbitals; (d) molecular orbitals and their energies E_j given by $E_j = \alpha - X_j\beta$.

transformation $\mathbf{U}$ and its transpose $\mathbf{U}'$ can be used to simplify the solution of the secular matrix $\mathbf{X}$ since the matrix multiplication $\mathbf{UXU}'$ gives the block diagonal form shown in part (b) of Table 1.

Thus the 8×8 determinant of *para*-benzosemiquinone reduces to two 3×3 blocks of B_{1u} and B_{2g} symmetry along with two 1×1 determinants of A_u and B_{3g} symmetry. The student is encouraged to confirm that the roots of these determinants are those listed in the table and that, by use of Eqs. (18) and (19), the corresponding wavefunctions result. This algebraic exercise will solidify the understanding of this solution procedure, a common one for many problems in physical chemistry. It will also provide a deeper appreciation of the value of matrix diagonalization methods using computers. These yield both energies (eigenvalues) and coefficients (which give eigenvectors) quickly and accurately by finding a transformation matrix that diagonalizes the secular determinant. Specifically, the energies are calculated by diagonalizing the numerical matrix $\mathbf{A}$ obtained by setting $X = 0$ in the secular determinant. The wavefunction coefficients are given as row elements of the transformation matrix $\mathbf{T}$ which diagonalizes this matrix, i.e. $\mathbf{TAT'}$. Such a procedure is useful for the 4×4 symmetrized matrices displayed in Table 2 for *ortho*-benzosemiquinone and is almost essential for larger determinants.[10]

TABLE 2
Hückel molecular orbital calculations for *ortho*-benzosemiquinone

(a) *Secular determinant*

$$
\begin{vmatrix}
X & 1 & 0 & 0 & 0 & 1 & 1 & 0 \\
1 & X & 1 & 0 & 0 & 0 & 0 & 1 \\
0 & 1 & X & 1 & 0 & 0 & 0 & 0 \\
0 & 0 & 1 & X & 1 & 0 & 0 & 0 \\
0 & 0 & 0 & 1 & X & 1 & 0 & 0 \\
1 & 0 & 0 & 0 & 1 & X & 0 & 0 \\
1 & 0 & 0 & 0 & 0 & 0 & X+1 & 0 \\
0 & 1 & 0 & 0 & 0 & 0 & 0 & X+1
\end{vmatrix} = 0
$$

C_{2v} *structure*

(b) *Symmetrized determinant*

$$
\begin{vmatrix}
X+1 & 1 & 1 & 0 & & & & \\
1 & X+1 & 0 & 0 & & & 0 & \\
1 & 0 & X & 1 & & & & \\
0 & 0 & 1 & X+1 & & & & \\
 & & & & X-1 & 1 & -1 & 0 \\
 & 0 & & & 1 & X+1 & 0 & 0 \\
 & & & & -1 & 0 & X & 1 \\
 & & & & 0 & 0 & 1 & X-1
\end{vmatrix} = 0
$$

(c) *Symmetrized orbitals*

$$
B_2 \begin{cases} \phi_1 = 2^{-1/2}[p_1 + p_2] \\ \phi_2 = 2^{-1/2}[p_7 + p_8] \\ \phi_3 = 2^{-1/2}[p_3 + p_6] \\ \phi_4 = 2^{-1/2}[p_4 + p_5] \end{cases}
$$

$$
A_2 \begin{cases} \phi_5 = 2^{-1/2}[p_1 - p_2] \\ \phi_6 = 2^{-1/2}[p_7 - p_8] \\ \phi_7 = 2^{-1/2}[p_3 - p_6] \\ \phi_8 = 2^{-1/2}[p_4 - p_5] \end{cases}
$$

(d) *Energies*

Energies	Symmetry	Molecular orbitals
$X_8 = 2.127$	A_2	$\psi_8 = 0.664\phi_5 - 0.212\phi_6 + 0.536\phi_7 - 0.476\phi_8$
$X_7 = 1.197$	A_2	$\psi_7 = 0.579\phi_5 - 0.263\phi_6 - 0.149\phi_7 + 0.757\phi_8$
$X_6 = 1.095$	B_2	$\psi_6 = 0.475\phi_1 - 0.226\phi_2 - 0.768\phi_3 + 0.366\phi_4$
$X_5 = -0.262$	B_2	$\psi_5 = 0.505\phi_1 - 0.685\phi_2 + 0.312\phi_3 - 0.423\phi_4$
$X_4 = -0.748$	A_2	$\psi_4 = 0.129\phi_5 - 0.512\phi_6 - 0.737\phi_7 - 0.422\phi_8$
$X_3 = -1.477$	B_2	$\psi_3 = 0.226\phi_1 + 0.475\phi_2 - 0.366\phi_3 - 0.768\phi_4$
$X_2 = -1.576$	A_2	$\psi_2 = 0.455\phi_5 + 0.790\phi_6 - 0.383\phi_7 - 0.149\phi_8$
$X_1 = -2.356$	B_2	$\psi_1 = 0.685\phi_1 + 0.505\phi_2 + 0.423\phi_3 + 0.312\phi_4$

The molecular-orbital energies are indicated in the tables for the benzosemiquinones. Examination of the wavefunctions shows that the highest antibonding state has a node (sign change) between each pair of adjacent atoms. Examine the nodal surfaces for the other wavefunctions and comment on the correlation between energy and number of nodes.

For the benzosemiquinone radical anion, one has nine π electrons, with two each in the four lowest levels and the unpaired electron residing in ψ_5. Using the HMO results in the tables, calculate ρ_π for the carbons attached to the protons and compare with the values obtained from your experimental results and the McConnell relation [Eq. (9)] with $Q = -22.5$ gauss. Note that the calculations provide a basis for assignment of the hyperfine splitting constants to specific ring protons in *ortho*-benzosemiquinone. Draw out the possible valence bond resonance structures for both *ortho* and *para* compounds and discuss the relative importance of these.

Improvements in the simple HMO theoretical model are of course possible. For the *para*-benzosemiquinone anion, Vincow and Fraenkel[11] suggest that better parameters for oxygen are $\alpha_O = \alpha_C + h\beta$, $\beta_{CO} = k\beta_{CC} = k\beta$, with $h = 1.2$ and $k = 1.56$; values of $h = 2$ and $k = 2^{1/2}$ are also common choices for oxygen in C$=$O groups.[4,10] The student may find it interesting to repeat the calculations of Table 1 using these assumptions. For the *ortho* compound, higher-level self-consistent-field (SCF) calculations give very good agreement with experiment.[12]

Molecular-orbital calculations for the methyl- and *t*-butyl-substituted benzosemiquinones have also been done[4,13] and the resultant spin densities are listed in Table 3. In these cases, the transfer of unpaired spin density from the π system to the proton is explained in terms of hyperconjugation.[2,13] Use these theoretical results as an aid in assigning your experimental hyperfine splitting constants to specific protons. Can you rationalize the charge distributions in these species and the changes from the distributions in the parent benzosemiquinones?

TABLE 3
Calculated Hückel molecular orbital spin densities for substituted benzosemiquinone anions

	ρ_2	ρ_3	ρ_4	ρ_5	ρ_6
2-Methyl-*p*-[a]	0.120	0.0786	$\cdots$	0.111	0.112
2-*t*-Butyl-*p*-[a]	0.107	0.0704	$\cdots$	0.132	0.091
2,3-Dimethyl-*p*-[b]	—	—	$\cdots$	0.0884	0.0884
4-Methyl-*o*-[a]	$\cdots$	~0	0.175	0.158	0.0388
4-*t*-Butyl-*o*-[a]	$\cdots$	0.0110	0.171	0.163	0.045

[a] Ref. 12, [b] Ref. 4.

APPARATUS

ESR spectrometer (a relatively inexpensive teaching instrument is available from Micro-Now Instruments, Skokie, IL.); 2.5 mm o.d. ESR tube (or special ESR tube for aqueous samples) (Wilmad); 5-mL syringe; 5–10 mL beakers.

1,4-Dihydroxybenzene (hydroquinone), 2-methyl-, 2,3-dimethyl-, and 2-*t*-butyl-hydroquinones, 1,2-dihydroxybenzene (catechol), (~1 g each); methanol, NaOH pellets.

REFERENCES

1. J. E. Wertz and J. R. Bolton, "Electron Spin Resonance: Elementary Theory and Practical Applications," McGraw-Hill, New York (1972).
2. A. Carrington and A. D. McLachlan, "Introduction to Magnetic Resonance," pp. 1–23, 72–175, Harper and Row, New York (1967).
3. R. S. Drago, "Physical Methods in Chemistry," chaps. 1–3, 9, 13, Saunders, Pa. (1977).
4. A. Streitwieser, Jr., "Molecular Orbital Theory for Organic Chemists," Wiley, New York (1961).
5. K. Higasi, H. Baba, and A. Rembaum, "Quantum Organic Chemistry," Interscience, New York (1965).
6. P. W. Atkins, "Physical Chemistry," 3d ed., chaps. 16, 17, 20, Freeman, New York (1980).
7. H. M. McConnell, *J. Chem. Phys.* **24,** 764 (1956).
8. A. D. McLachlan, *Mol. Phys.* **1,** 233 (1958).
9. An IBM-PC program for simulating and plotting first derivative ESR spectra for simple radicals is "ESR" by Ronald D. McKelvey, *J. Chem. Educ.* **64,** 497 (1987). It can be obtained from Project SERAPHIM, NSF Science Education, Department of Chemistry, Eastern Michigan University, Ypsilanti, MI 48197. This program is especially useful in deducing coupling constants from experimental spectra by iterative comparison with spectra calculated for assumed hyperfine constants.
10. A useful program for Hückel MO calculations for up to 22 atom molecules, including N and O atoms, is "Hückel" by Ronald D. McKelvey. This program runs on the IBM-PC and is available from Project SERAPHIM, see Ref. 9.
11. G. Vincow and G. K. Fraenkel, *J. Chem. Phys.* **34,** 1333 (1961).
12. G. Vincow, *J. Chem. Phys.* **38,** 917 (1963).
13. C. Trapp, C. A. Tyson, and G. Giacometti, *J. Amer. Chem. Soc.* **90,** 1394 (1968).

GENERAL READING

N. J. Bunce, *J. Chem. Educ.* **64,** 907 (1987).
A. Carrington and A. D. McLachlan, *op. cit.*
R. S. Drago, *op. cit.*
A. Streitwieser, Jr., *op. cit.*
J. E. Wertz and J. R. Bolton, *op. cit.*

A useful compilation of EPR data is J. A. Pedersen, "Handbook of EPR Spectra from Quinones and Quinols," CRC Press, Florida (1985.

EXPERIMENT 44
NMR DETERMINATION OF KETO–ENOL EQUILIBRIUM CONSTANTS

In this experiment, proton NMR spectroscopy is used in evaluating the equilibrium composition of various keto–enol mixtures. Chemical shifts and spin–spin splitting patterns are employed to assign the spectral features to specific protons, and the integrated intensities to yield a quantitative measure of the relative amounts of the keto and enol forms. Solvent effects on the chemical shifts and on the equilibrium constant are investigated for one or more β-diketones and β-ketoesters.

THEORY

Chemical shifts. In Exp. 34, the Zeeman energy levels of a nucleus in an external applied field H were given as

$$E_N = -g_N\mu_N M_I B \tag{1}$$

where B is the magnetic induction ("local field") at the nucleus. As a result of the $\Delta M_I = \pm 1$ selection rule, a transition will occur at frequency

$$\nu_i = (g_N\mu_N/h)B_i \tag{2}$$

for a nucleus experiencing a local field B_i. The chemical shift in parts per million (ppm) of this nucleus relative to a reference nucleus r is defined by

$$\delta_i \equiv \frac{\nu_i - \nu_r}{\nu_r} \times 10^6 \equiv \frac{H_r - H_i}{H_r} \times 10^6 \tag{3}$$

Here the first definition is based on the resonant frequencies for a fixed external field H whereas the second is based on the more common experimental case where H is varied to achieve resonance at a fixed instrumental frequency ν. Tetramethylsilane (TMS) is usually used as the proton reference since it is chemically inert and its 12 equivalent protons give a single transition at a field H_r, higher than the field H_i found in most organic compounds. Thus δ is generally positive and increases when substituents are added that attract electrons and thereby reduce the shielding about the proton. This shielding arises because the electrons near the proton are induced to circulate by the applied field H (see Fig. 1(a)). This electron current produces a secondary field that *opposes* the external field and thus reduces the local field B at the proton. As a result, resonance at a fixed frequency such as 60 MHz requires a higher external field for protons with a larger shielding. This shielding effect is generally restricted to electrons localized on the nucleus of interest since random tumbling of the molecules causes the effect of secondary fields due to electrons associated with neighboring nuclei to average to zero. Nuclei such as ^{19}F, ^{13}C, and ^{11}B have more local electrons than hydrogen; hence their chemical shifts are much larger.

Long-range *deshielding* can occur in aromatic and other molecules with

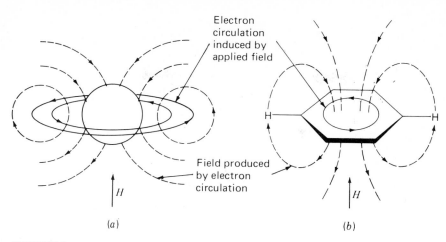

FIGURE 1
Shielding and deshielding of protons: (*a*) shielding of proton due to induced diamagnetic electron circulation; (*b*) Deshielding of protons in benzene due to aromatic ring currents.

delocalized π electrons. For example, when the plane of the benzene molecule is oriented perpendicular to *H*, circulation of the π electrons produces a ring current [see Fig. 1(b)]. This ring current induces a secondary field at the protons that is *aligned parallel* to *H* and thus increases the local field *B* at the protons. This induced field changes with benzene orientation but does not average to zero since it is not spherically symmetric. Because of this net deshielding effect, the resonance of the benzene protons occurs at a relatively low external field. The proton chemical shift δ for benzene is 7.27, greatly downfield from the value $\delta = 1.43$ that is observed for cyclohexane, in which ring currents do not occur. Similar deshielding occurs for olefinic and aldehydic protons because of the π electron movement. Typical values of δ for different functional groups are shown in Table 1 and additional values are available in Refs. 1 to 3. Although the resonances change somewhat for different compounds, the range for a given functional group is usually small and δ values are widely used for structural characterization in organic chemistry.

Spin–spin splitting. High-resolution NMR spectra of most organic compounds reveal more complicated spectra than those predicted by Eq. (2), with transitions often appearing as multiplets. Such *spin–spin splitting patterns* arise because the magnetic moment of one proton (A) can interact with that of a nearby nucleus (B), causing a small energy shift up or down depending upon the relative orientations of the two moments. The energy levels of nucleus A then have the form

$$E_A = -g_{N_A}\mu_N M_{I_A} B + h J_{AB} M_{I_A} M_{I_B} \tag{4}$$

and there is a similar expression for E_B. The spin–spin interaction is characterized by the coupling constant J_{AB}, and the effect is to split the energy

TABLE 1
Typical proton chemical shifts δ

CH$_3$ protons			Acetylenic protons	
(CH$_3$)$_4$Si	0.0		HOCH$_2$C≡CH	2.33
(CH$_3$)$_4$C	0.92		ClCH$_2$C≡CH	2.40
CH$_3$CH$_2$OH	1.17		CH$_3$COC≡CH	3.17
CH$_3$COCH$_3$	2.07		Olefinic protons	
CH$_3$OH	3.38		(CH$_3$)$_2$C=CH$_2$	4.6
CH$_3$F	4.30		cyclohexene	5.57
CH$_2$ protons			CH$_3$CH=CHCHO	6.05
cyclopropane	0.22		Cl$_2$C=CHCl	6.45
CH$_3$(CH$_2$)$_4$CH$_3$	1.25		Aromatic protons	
(CH$_3$CH$_2$)$_2$CO	2.39		benzene	7.27
CH$_3$COCH$_2$COOCH$_3$	3.48		C$_6$H$_5$CN	7.54
CH$_3$CH$_2$OH	3.59		naphthalene	7.73
CH protons			α-pyridine	8.50
bicyclo[2.2.1]heptane	2.19		Aldehydic protons	
chlorocyclopropane	2.95		CH$_3$OCHO	8.03
(CH$_3$)$_2$CHOH	3.95		CH$_3$CHO	9.72
(CH$_3$)$_2$CHBr	4.17		C$_6$H$_5$CHO	9.96

levels in the manner illustrated for acetaldehyde in Fig. 2. It is apparent from this diagram that the external field H does not effect the small spin–spin splitting that is characterized by the coupling constant J. The quantity J is a measure of the strength of the pairwise interaction of the proton spin with the spin of another nucleus. Since there are only proton–proton interactions in acetaldehyde, the same splitting occurs for both CH and CH$_3$ resonances.

The total integrated intensity of the CH and CH$_3$ multiplets follows the proton ratio of $1:3$. However, the intensity distribution within each multiplet is determined by the relative population of the lower level in each transition. Since the level spacing is much less than kT, the Boltzmann population factors are essentially identical for these levels. However, there is some degeneracy because rapid rotation of the CH$_3$ group around the C—C bond makes the three protons magnetically equivalent. The number of spin orientations of the CH$_3$ protons that produce equivalent fields at the CH proton determine the degeneracy. The eight permutations of the CH$_3$ spins shown in Fig. 2 thus lead to a predicted intensity ratio of $1:3:3:1$ for the CH multiplet. Similarly, the CH$_3$ doublet peaks will be of equal intensity, with a total integrated intensity three times that of the CH peaks. In a more general sense, it can be seen that n equivalent protons interacting with a different proton will split its resonance into $n + 1$ lines whose relative intensities are given by coefficients of the terms in the binomial expansion of the expression $(\alpha + \beta)^n$. Equivalent protons also interact and produce splitting in the energy levels. However, these splittings are symmetric for upper and lower energy states so that no new NMR resonances are produced.

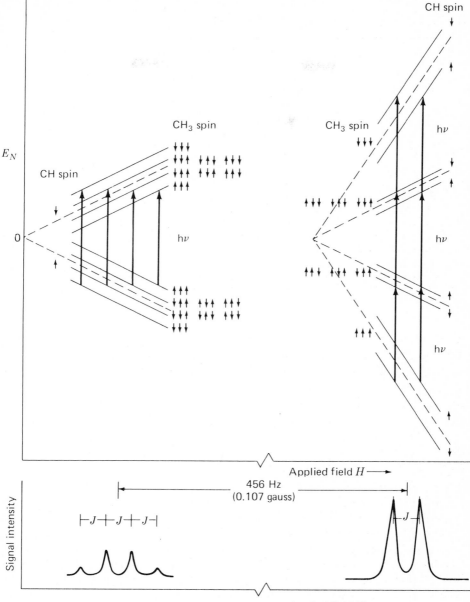

FIGURE 2
Energy levels, transitions, and 60-MHz NMR spectrum for acetaldehyde (CH₃CHO). The coupling constant $J = J_{CH_3} = J_{CH} = 2.2\,\text{Hz}$ ($= 5.2 \times 10^{-4}\,\text{gauss} = 5.2 \times 10^{-8}\,\text{T}$). For CH, the quantum number $M_I = -1/2$ or $1/2$. For the CH₃ group, $M_I = -3/2,\ -1/2,\ +1/2,\ +3/2$. The dashed lines represent the level spacing that would occur in the absence of the spin–spin interaction. The slopes of the energy levels are greatly exaggerated in the figure. Also, to be correct, all dashed lines should extrapolate to a common $E_N = 0$ at $H = 0$.

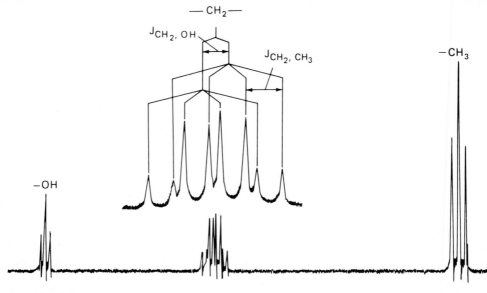

FIGURE 3
NMR spectrum of highly purified ethanol obtained at 100 MHz.

If a proton is coupled to more than one type of neighboring nucleus, the resultant multiplet pattern can often be understood as a simple stepwise coupling involving different J values. For example, the CH_2 octet that occurs for pure CH_3CH_2OH (Fig. 3) arises from OH doublet splitting ($J = 4.80$ Hz) of the quartet of lines caused by coupling ($J = 7.15$ Hz) with CH_3. It should be mentioned that such regular splitting and intensity patterns are only expected for two nuclei A and B if $|v_A - v_B| \gtrsim 10 \, J_{AB}$. The spectra for this weakly coupled case are termed *first order*. Since the difference $v_A - v_B$ (in Hz) increases with the field while J_{AB} does not, NMR spectra obtained with a high-field instrument (400 MHz) are often easier to interpret than those from a low-field spectrometer (60 MHz). However, even if the multiplets are not well separated, it is still possible to deduce accurate chemical shifts and J values using slightly more involved procedures, which are outlined in most texts on NMR spectroscopy.[1-5] Such an exercise can be done as an optional part of this experiment, although it will not be necessary for the determination of equilibrium constants.

The mechanism of spin–spin coupling is known to be indirect and to involve the electrons in the bonds between interacting nuclei. The spin of the first nucleus A is preferentially coupled antiparallel to the nearest bonding electron via the so-called Fermi contact interaction, which is significant only when the electron density is nonzero at the first nucleus. (Such is the case only for electrons in s orbitals since p, d, and f orbital wavefunctions have zero values at the nucleus.) This electron spin alignment information is transmitted

(a) low-energy case   (b) high-energy case

FIGURE 4
Illustration of nuclear spin–spin interaction transmitted via polarization of bonding electrons. The two electrons about each carbon will tend to be parallel since this arrangement minimizes the electronelectron repulsion (Hund's rule for electrons in degenerate orbitals).

by electron–electron interactions to the second nucleus B to produce a field which thus depends on the spin orientation of the first nucleus (Fig. 4). Since the strength of this interaction falls rapidly with separation, only neighboring groups produce significant splitting. A few typical spin–spin coupling constants are given in Table 2 and these, along with the chemical shifts, serve to identify proton functional groups. As mentioned above, the multiplet intensities also give useful information about neighboring groups. Thus NMR spectra can provide detailed structural information about large and complex molecules.

Keto–enol tautomerism. It is well known that ketones such as acetone have an isomeric structure, which results from proton movement, called the enol

TABLE 2
Typical proton spin–spin coupling constants

Coupling	J(Hz)	Coupling	J(Hz)
$\diagdown$C$\diagup$ with H, H	−20–+5	C=C with H (top), H (bottom)	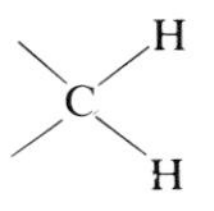 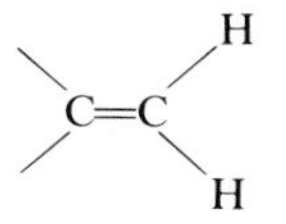 0–3.5
CH—CH	2–9	C=C with H, H	6–14
CH—(C)$_n$—CH	0		
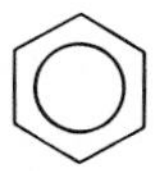 benzene ring: ortho-, meta-, para-	6–9, 1–3, 1	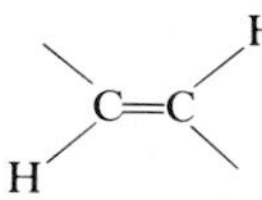 C=C with H, H	11–19

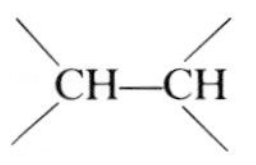

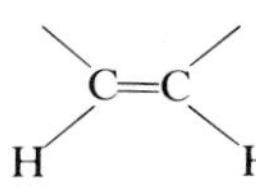

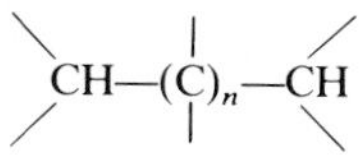

tautomer, an unsaturated alcohol:

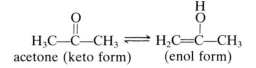

acetone (keto form)   (enol form)

For acetone, and the majority of cases in which this keto–enol tautomerism is possible, the keto form is far more stable and little if any enol can be detected. However, with β-diketones and β-ketoesters, such factors as intramolecular hydrogen bonding and conjugation increase the stability of the enol form and the equilibrium can be significantly shifted to the right.

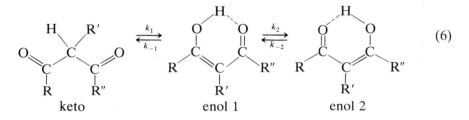

(6)

keto   enol 1   enol 2

The proton chemical environments are quite different for the keto and enol tautomers and the interconversion rate constants k_1 and k_{-1} between these forms are small enough that distinct NMR spectra are obtained for both forms. In principle, the two enols are also distinguishable when $R' \neq R''$. However, the intramolecular OH proton transfer is quite rapid at normal temperatures, so that a single (averaged) OH resonance is observed. In general, such averaging occurs when the conversion rates k_2 and k_{-2} (in Hz) exceed the frequency separation $\nu_1 - \nu_2$ (also in Hz) of the OH resonance for the two enol forms.[2] The magnetic field at the OH proton is thus averaged and resonance occurs at $(\nu_1 + \nu_2)/2$. Similarly, rapid rotation about the C—C bonds of the keto form explains why spectra due to different keto rotational conformers are not observed. Thus, distinct spectra are expected only for the two tautomers and these can be used to determine the equilibrium constant for keto-to-enol conversion:

$$K_c = \frac{(\text{enol})}{(\text{keto})} \tag{7}$$

where parentheses denote concentrations in any convenient units.

The keto arrangement shown in Eq. (6) is the configuration which is electrostatically most favorable, but the steric repulsions between R and R'' groups will be larger for this keto form than for the enol configuration. Indeed, experimental studies have confirmed that the enol concentration is larger when R and R'' are bulky.[4] This steric effect is less important in the β-ketoesters in which the $R \cdots R''$ separation is greater. For both β-ketoesters and β-diketones, α substitution of large R' groups results in steric hindrance between R' and R (or R'') groups, particularly for the enol tautomer, whose concentration is thereby reduced. Inductive effects have also been explored; in general, α substitution of electron-withdrawing groups such a —Cl or —CF$_3$ favor the enol form.[4]

The solvent plays an important role in determining K_c. This can occur through specific solute–solvent interactions such as hydrogen bonding or charge transfer. In addition, the solvent can reduce solute–solute interactions by dilution and thereby change the equilibrium if such interactions are different in enol–enol, enol–keto, or keto–keto dimers. Finally, the dielectric constant of the solution will depend on the solvent and one can expect the more polar tautomeric form to be favored by polar solvents. Some of these aspects are explored in this experiment.

EXPERIMENTAL

The general features of an NMR spectrometer were described briefly in Exp. 34, and more specific operating instructions will be provided by the instructor. Obtain several milliliters each of acetylacetone ($CH_3OCH_2COCH_3$, M.W. = 100.11, density = 0.98 g cm^{-3}) and ethyl acetoacetate ($CH_3CH_2OCOCH_2COCH_3$, M.W. = 130.45, density = 1.03 g cm^{-3}. Prepare small volumes of two solvents and three solutions.

Solvent A: Carbon tetrachloride, spectrochemical grade (M.W. = 153.83, density = 1.58 g cm^{-3} with 5% by volume tetramethylsilane (TMS) added. Prepare in a 10-mL volumetric flask.

Solvent B: Methanol, spectrochemical grade (M.W. = 32.04, density = 0.791 g cm^{-3}) with 5% by volume TMS added. Prepare in a 5- or 10-mL volumetric flask.

Solution 1: 0.20 mole fraction of acetylacetone in solvent A

Solvent 2: 0.20 mole fraction of acetylacetone in solvent B

Solution 3: 0.20 mole fraction of ethyl acetoacetate in solvent A

Use a 1-mL pipette graduated in 0.01-mL increments to measure out 0.010 mol of solute, and use a 2-mL graduated pipette to then add the correct amount (0.040 mol) of solvent. You may neglect the presence of the 5 percent TMS when determining the necessary volumes of solvent. **Warning:** All work with TMS should be carried out in a hood. All containers or samples containing TMS should be tightly sealed and stored at low temperatures because of its volatility.

Prepare an NMR tube containing about 1 inch of solvent A and another containing solvent B, and record both NMR spectra, setting the TMS signal at the chart zero. Repeat for solutions 1 to 3, taking care to scan above $\delta = 10$ ppm since the enol OH peak is shifted substantially downfield.

Determine which peaks are due to solute and measure chemical shifts for all solute features. Integrate the bands carefully at least three times, expanding the vertical scale by known factors as necessary in order to obtain accurate relative intensity measurements.

CALCULATIONS

Assign all spectral features using Table 1 and other NMR reference sources.[2,3,5] Tabulate your results and use your integrated intensities to calculate the percentage enol present in solutions 1 to 3. If possible, use the total integral corresponding to the sum of methyl (or ethyl), methylene, methyne, and enol protons. If this proves difficult because of overlap with solvent bands, indicate clearly how you used the intensities to calculate the percentage enol.

For both the enol and the keto form, compare experimental and theoretical ratios of the integrated intensities for different types of protons (e.g., methyl to methylene protons in the keto form).

Using Eq. (7), calculate K_c and the corresponding standard free energy difference $\Delta G°$ for the change in state keto$\rightarrow$enol in each solution.

DISCUSSION

Discuss briefly your assignments of chemical shifts and spin–spin splitting patterns of acetylacetone and ethyl acetoacetate. Which compound has a higher concentration of enol form and what reasons can you offer to explain this result? What changes would you expect in the NMR spectra of these two compounds if the interconversion rate between enol structures were much slower?

Compare the value of K_c for acetylacetone in CCl_4 with that in CH_3OH. What does your result suggest regarding the relative polarity of the enol and keto forms? Which form is favored by hydrogen bonding and why?

Compare your values of $\Delta G°$ with those for the gas phase ($\Delta G° = -9.2 \pm 2.1$ kJ mol^{-1} for acetylacetone, $\Delta G° = -0.4 \pm 2.5$ kJ mol^{-1} for ethyl acetoacetate).[6] What solvent properties might account for any differences you observe?

Additional compounds suitable for studies of steric effects on keto–enol equilibria include α-methylacetone ($CH_3COCHCH_3COCH_3$), diethylmalonate ($CH_3CH_2OCOCH_2COOCH_2CH_3$), ethyl benzoylacetate ($C_6H_6COCH_2COOCH_2CH_3$), and t-butyl acetoacetate (CH_3COCH_2COOt-Bu). Some other possible compounds are listed in Refs. 4 and 5. Further aspects of this equilibrium that could be studied include the effects of concentration, temperature, and solvent dielectric constants on K_c.[5]

APPARATUS

NMR spectrometer with peak integrating capability; several 5- and 10-mL volumetric flasks; precision 1-mL and 2-mL graduated pipettes; NMR tubes; spectrochemical grade CCl_4 and CH_3OH; tetramethylsilane, acetylacetone, and ethyl acetoacetate; fume hood.

REFERENCES

1. J. C. Davis, Jr., "Advanced Physical Chemistry: Molecules, Structure, and Spectra," Wiley-Interscience, New York (1965).
2. J. A. Pople, W. G. Schneider, and H. J. Bernstein, "High Resolution Nuclear Magnetic Resonance," McGraw-Hill, New York (1959); C. P. Slichter, "Principles of Magnetic Resonance," 2d ed., vol. I in P. Fulde (ed.), Solid State Sciences, Springer-Verlag, Berlin/New York (1980).
3. Compilations of NMR chemical shifts and coupling constants are given in F. A. Bovey, "NMR Data Tables for Organic Compounds," vol. I, Wiley-Interscience, New York (1967) and in L. F. Johnson and W. C. Jankowski, "C-13 NMR Spectra," Wiley-Interscience, New York (1972).
4. J. L. Burdett and M. T. Rogers, *J. Amer. Chem. Soc.* **86,** 2105 (1964).
5. M. T. Rogers and J. L. Burdett, *Can. J. Chem.* **43,** 1516 (1965).
6. M. M. Folkendt, B. E. Weiss-Lopez, J. P. Chauvel, Jr., and N. S. True, *J. Phys. Chem.* **89,** 3347 (1985).

GENERAL READING

C. P. Slichter, *op. cit.*
D. Shaw, "Fourier Transform NMR Spectroscopy," 2d., Elsevier, Amsterdam (1984).

CHAPTER

XIV

SOLIDS

EXPERIMENTS

45. Determination of crystal structure by x-ray diffraction
46. Single crystal x-ray diffraction with a precession camera
47. Lattice energy of solid argon
48. Statistical thermodynamics of iodine sublimation

EXPERIMENT 45
DETERMINATION OF CRYSTAL
STRUCTURE BY X-RAY DIFFRACTION

The object of this experiment is to determine the crystal structure of a solid substance from x-ray powder diffraction patterns. This involves determination of the symmetry classification (cubic, hexagonal, etc.), the type of crystal lattice (simple, body-centered, or face-centered), the dimensions of the unit cell, the number of atoms or ions of each kind in the unit cell, and the position of every atom or ion in the unit cell. Owing to inherent limitations of the powder method only substances in the cubic system should be chosen for study.

Knowledge of the crystal structure permits determination of the coordination number (the number of nearest neighbors) for each kind of atom or ion, calculation of interatomic distances, and elucidation of other structural features related to the nature of chemical bonding and the understanding of physical properties in the solid state.

532

THEORY

A perfect crystal constitutes the repetition of a single very small unit of structure, called the *unit cell,* in a regular way so as to fill the volume occupied by the crystal (see Fig. 1).

A crystal may for some purposes be described in terms of a set of three *crystal axes* **a**, **b**, and **c**, which may or may not be of equal length and/or at right angles, depending on the symmetry of the crystal. These axes form the basis for a coordinate system with which the crystal may be described. An important property of crystals, known at least a century before the discovery of x-rays, is that for any crystal the crystal axes can be so chosen that all crystal faces can be described by equations of the form

$$hx + ky + lz = \text{positive constant} \tag{1}$$

where x, y, and z are the coordinates of any point on a given crystal face, in a coordinate system with axes parallel to the assigned crystal axes and with units equal to the assigned axial lengths a, b, c; and where h, k, and l are *small integers,* positive, negative, or zero. This is known as the law of rational indices. The integers h, k, and l are known as the *Miller indices,* and when used to designate a crystal face are ordinarily taken relatively prime (i.e., with no common integral factor). Each crystal face may then be designated by three Miller indices *hkl,* as shown in Fig. 1. The law of rational indices historically formed the strongest part of the evidence supporting the conjecture that crystals are built up by repetition of a single unit of structure, as shown in Fig. 1.

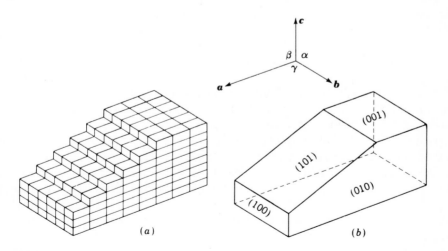

FIGURE 1
(*a*) Schematic diagram of a crystal, showing unit cells; (*b*) same crystal, showing axes and Miller indices.

The crystal axes for a given crystal may be chosen in many different ways; however, they are conventionally chosen to yield a coordinate system of the highest possible symmetry. It has been found that crystals can be divided into six possible systems on the basis of the highest possible symmetry that the coordinate system may possess as a result of the symmetry of the crystal. This symmetry is best described in terms of symmetry restrictions governing the values of the axial lengths a, b, and c and the interaxial angles α, β, and γ.

The crystal systems are as follows:[1]

Triclinic system. No restrictions.

Monoclinic system. No restriction on lengths a, b, c; however, $\alpha = \gamma = 90°$; $\beta \neq 90°$.

Orthorhombic system. No restrictions on a, b, c; however, $\alpha = \beta = \gamma = 90°$.

Tetragonal system. $a = b \neq c$; $\alpha = \beta = \gamma = 90°$.

Hexagonal system. *Hexagonal division*: $a = b = b' \neq c$; $\alpha = \beta = 90°$, $\gamma = \gamma' = 120°$ (there being three axes **a**, **b**, and **b'** in the basal plane, at 120° angular spacing; the **b'** axis is redundant). *Rhombohedral division*: $a = b = c$; $\alpha = \beta = \gamma \neq 90°$. (Although the coordinate systems in these two divisions are of different symmetry, hexagonal axes may be used in the description of rhombohedral crystals and vice versa.)

Cubic system. $a = b = c$; $\alpha = \beta = \gamma = 90°$ (cartesian coordinates).

The symmetry of a crystal is not completely specified, however, by naming the crystal system to which it belongs. Each crystal system is further subdivided into *crystal classes,* of which there are 32. On the basis of the detailed symmetry of the atomic structure of crystals, further subdivision is possible. Symmetry will not, however, be discussed in detail here.

Since a crystal structure constitutes a regular repetition of a unit of structure, the unit cell, we may say that a crystal structure is periodic in three dimensions. The periodicity of a crystal structure may be represented by a *point lattice* in three dimensions. This is an array of points that is invariant to all the translations that leave the crystal structure invariant and to no others. We shall find the lattice useful in deriving the conditions for x-ray diffraction.

To define a crystal lattice in another way, consider the crystal structure to be divided into unit cells (parallelepipeds in shape) in such a way as to obtain the smallest unit cells possible. These unit cells are then called *primitive.* Starting with a set of crystal axes that are parallel to three edges of a unit cell and equal to them in length, a point lattice may be defined as the infinite array of points the coordinates xyz of which assume all possible combinations mnp of integral values (positive, negative, and zero), and *only* integral values. There is then one lattice point per unit cell.

It is frequently found that it is not possible to find a primitive unit cell with edges parallel to crystal axes chosen on the basis of symmetry. In such a case the crystal axes, chosen on the basis of symmetry, are proportional to the edges of a unit of structure that is larger than a primitive unit cell. Such a unit is called a *nonprimitive* unit cell, and there is more than one lattice point per nonprimitive unit cell. If the nonprimitive unit cell is chosen as small as possible consistent with the symmetry desired, it is found that the extra lattice points (those other than the corner points) lie in the center of the unit cell or at the centers of some or all of the faces of the unit cell. The coordinates of the lattice points, in such a case, are therefore either integers or half-integers.

Within a given crystal system there are in some cases several different types of crystal lattice, depending upon the type of minimum-size unit cell that corresponds to a choice of axes appropriate to the given crystal system. This unit cell may be *primitive P* or in certain cases *body-centered I, face-centered F,* or *end-centered A, B,* or *C*, depending on which pair of end faces of the unit cell is centered. The lattices are designated as primitive, body-centered, face-centered, or end-centered depending on whether the smallest possible unit cell that corresponds to the appropriate type of axes is primitive, body-centered, face-centered, or end-centered. There are in all 14 types of lattice,[1] known as Bravais lattices. In the cubic system there are three: primitive, body-centered, and face-centered; these are shown in Fig. 2.

Bragg reflections. Let us now determine the geometrical conditions under which diffraction of x-rays by a crystal structure would take place. These are (except for intensity considerations) the same as the conditions for diffraction from the crystal lattice, if a "point-scattering center" is placed at each point of the lattice. We shall accordingly examine the geometry of diffraction by such a crystal lattice.

The principle by which x-rays would be diffracted by such a lattice is essentially the same as that by which light is diffracted by a ruled grating. When a plane wave impinges on a point-scattering center, the scattering center radiates a spherical wave. If there are two or more such scattering centers, the spherical waves will in certain regions tend to reinforce one another and in certain other regions tend to cancel one another out, that is, interfere.

If there are many scattering centers, it is possible for them to be so arranged that the individual spherical waves combine in a certain region a large

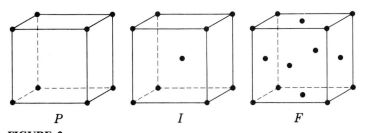

$$P \qquad\qquad I \qquad\qquad F$$

FIGURE 2
Cubic unit cells: *P*, primitive; *I*, body-centered; *F*, face-centered.

distance away to form a "reflected" plane wave with an amplitude that represents the sum of the amplitudes of the individual spherical waves at that distance. Such an arrangement is obtained when the scattering centers are confined to a plane surface, in which case the "reflected" plane wave is similar to that which would be obtained if the arrangement of scattering centers were replaced by a plane mirror at the same place. This is the situation represented in Fig. 3 by reflection of rays $\overline{AB}$ and $\overline{DE}$ from the plane Q. It is due to the fact that the path length, along a reflected ray of radiation, from a wave front of the incident radiation to a wave front of the reflected radiation is the same for every ray (for example, $\overline{AB} + \overline{BC} = \overline{DE} + \overline{EF}$) when the angle of incidence is equal to the angle of reflection.

It is not necessary, however, as a condition for maximum reinforcement, that all the path lengths be equal, provided that those which are not equal (for example, $\overline{AB} + \overline{BC}$ and $\overline{GH} + \overline{HJ}$ in Fig. 3) differ by a wavelength λ or by an integral number of wavelengths $n\lambda$. It can be seen from Fig. 3 that the condition for complete reinforcement is

$$n\lambda = 2D \sin \theta \tag{2}$$

(since $\overline{MH} = \overline{HN} = D \sin \theta$ and the path difference $\overline{MH} + \overline{HN}$ must be equal to $n\lambda$), where n is an integer, often called the order of the diffraction; D is the interplanar spacing; and θ is the angle which the incident ray and the reflected ray make with the reflecting planes. Equation (2) is called the *Bragg equation*, and θ is often called the *Bragg angle*. The angle through which the x-rays are deflected from their original direction by reflection from the planes is twice the Bragg angle, or 2θ.

If the angle θ deviates only slightly from that which satisfies the Bragg

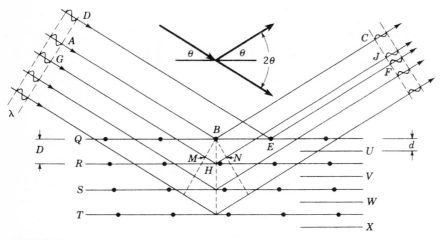

FIGURE 3
Condition for Bragg reflection from scattering centers confined to a set of equidistant, parallel planes. (Planes are perpendicular to the page; their traces are indicated by the horizontal lines.)

equation, reinforcement remains almost complete only if there is only a small number of reflecting planes. If, however, there is a very large (or "infinite") number of reflecting planes, the reflection from any given plane is cancelled by that from another plane a considerable distance away. Reflection should be observed, therefore, only for values of θ extremely close to those that satisfy the Bragg equation, and this is experimentally found to be the case.

We shall now modify the treatment slightly by taking note of the fact that a second-order reflection ($n = 2$) from the planes $Q, R, S, T, \ldots$ corresponds to a hypothetical first-order reflection from the planes $Q, U, R, V, S, W, T, X, \ldots$, only half of which contain lattice points. By inserting the required number of additional equidistant parallel planes containing no lattice points, we can dispense with the order n and write the Bragg equation in the following form, which is the form that will be used henceforth in this discussion:

$$\lambda = 2d \sin \theta \tag{3}$$

where d is now the interplanar spacing of the new set of planes. We shall now say that Eq. (3) is the necessary condition for reflection of x-rays of wavelength λ from a set of crystallographic planes with interplanar spacing d. A set of crystallographic planes is defined as an infinite array of equidistant and finitely spaced parallel planes so constructed that every point of the given crystal lattice lies on some plane of the set (though it is not necessarily true that all planes in the set contain lattice points). It is evident that sets of crystallographic planes may be constructed for a given crystal lattice in many ways (indeed, an infinite number of ways). The equations for such a set of planes are

$$\vdots$$
$$hx + ky + lz = -2$$
$$hx + ky + lz = -1$$
$$hx + ky + lz = 0$$
$$hx + ky + lz = 1 \tag{4}$$
$$hx + ky + lz = 2$$
$$\vdots$$
$$hx + ky + lz = N$$
$$\vdots$$

where x, y, and z are coordinates as previously defined and where, when the coordinate system corresponds to a primitive unit cell, the coefficients h, k, and l may take on any integral values, positive, negative, or zero. They may be called the Miller indices of the set of crystallographic planes and are related to Miller indices as applied to crystal faces. (However, they need not be taken relatively prime.) The different integral values of N, ranging from $-\infty$ to ∞, define different planes in the set. That any given lattice point must lie on one of the planes, in the case of a coordinate system corresponding to a primitive unit cell, is easily seen from the fact that its coordinates xyz must be integers mnp; since the Miller indices are integers, the quantity $hx + ky + lz$ must be an integer, and one of the equations in the set of Eqs. (4) is satisfied.

If the coordinate system is not chosen in correspondence to a primitive unit cell, not all the lattice points have coordinates that are integers and all lattice points will lie on planes in the set only if certain restrictions are placed on the combinations of values which the Miller indices hkl may assume.

For a body-centered lattice I, some of the lattice points have coordinates that are expressible as integers mnp and some have coordinates that must be expressed as half-integers $m + \frac{1}{2}$, $n + \frac{1}{2}$, $p + \frac{1}{2}$. For the latter,

$$hx + ky + lz = hm + kn + lp + \tfrac{1}{2}(h + k + l)$$

and for this quantity to be an integer, it is necessary that $h + k + l$ be even.

For a face-centered lattice F, some of the lattice points are at $xyz = mnp$; some at m, $n + \frac{1}{2}$, $p + \frac{1}{2}$; some at $m + \frac{1}{2}$, n, $p + \frac{1}{2}$; and some at $m + \frac{1}{2}$, $n + \frac{1}{2}$; p, where m, n, and p are in each case any three integers. In order that $hx + ky + lz$ be an integer, it is evidently necessary that $k + l$, $h + l$, and $h + k$ simultaneously be even. An equivalent restriction is easily seen to be that h, k, and l must be either all even or all odd.

When these restrictions are not obeyed, no reflections can be obtained from the set of crystallographic planes under consideration, for there will be lattice points lying between the planes and scattering out of phase with those in the planes, resulting in complete cancellation or interference. By observing experimentally what sets of planes reflect x-rays one can deduce what the restrictions are and thereby deduce the lattice type.

The interplanar distance d is determined by the Miller indices hkl. For the cubic system it is easy to show by analytic geometry that

$$d = \frac{a_0}{\sqrt{h^2 + k^2 + l^2}} = \frac{a_0}{M} \tag{5}$$

where a_0 is the length of the edge of the unit cube and

$$M^2 \equiv h^2 + k^2 + l^2 \tag{6}$$

Lattice type. From the angles at which x-rays are diffracted by a crystal it is possible to deduce the interplanar distances d, with Eq. (3). To determine the lattice type and compute the unit-cell dimensions, it is necessary to deduce the Miller indices of the planes that show these distances. In the case of a powder specimen (where all information concerning orientations of crystal axes has been lost) the only available information regarding Miller indices is that obtainable by application of Eqs. (5) and (6).

To find which of the three types of cubic lattice is the correct one, we make use of some interesting properties of integers. From Eq. (5) we see that

$$\left(\frac{1}{d}\right)^2 = \left(\frac{1}{a_0}\right)^2 M^2 = \left(\frac{1}{a_0}\right)^2 (h^2 + k^2 + l^2) \tag{7}$$

so that, if we square our reciprocal spacings, it should be possible to find a numerical factor which will convert them into a sequence of integers, which we shall find convenient to make *relatively prime*. We shall see that the type of lattice is determined by the character of the integer sequence obtained.

For a simple cubic (primitive) lattice, all integral values are independently possible for the Miller indices h, k, and l. Now it is possible to express most, but not all, integers as the sum of the squares of three integers. In Table 1 the various possible values of M^2 are listed in the column under P, and it is seen that there are gaps where the integers 7, 15, 23, 28, 31, 39, 47, and 55 are absent. (Other gaps occur at higher values of M^2.) For those values of M^2 that are possible, the first column of the table gives the Miller indices the sum of whose squares yield the M^2 values. In some cases it is seen that there is more than one possible choice.

For a simple (primitive) cubic lattice P there are no restrictions on the Miller indices and therefore none on M^2 except as noted above. In the case of a body-centered cubic lattice, only those values of M^2 can be allowed that arise from Miller indices whose sum is even. This has the effect of requiring M^2 to be even, as seen in the first of the two columns under I. We can then divide them by 2 and thereby reduce them to a *relatively prime* sequence, shown in the second column under I, for comparison with the sequence obtained from the $(1/d)^2$ values. We note immediately that the relatively prime sequence obtained differs from that for a primitive cubic lattice in having gaps at different places. By use of this fact it is almost always possible to distinguish between a primitive cubic lattice and a body-centered cubic lattice on the basis of a powder photograph. For the face-centered cubic lattice, application of the restriction that the indices must be all even or all odd produces the characteristic sequence 3, 4, 8, 11, 12, 16, ... given in the column under F. If most or all of the numbers in this sequence are present, and if none of the excluded numbers is present, the lattice is evidently face-centered cubic.

If no relatively prime sequence of integers can be found to within the experimental uncertainty of the measurements, the crystalline substance presumably does not belong to the cubic system.

When the cubic lattice type has been deduced, the unit-cell dimension a_0 can be calculated. From the unit-cell volume, the measured crystal density, and the formula weight, the number of formulas in one unit cell can be calculated.

Deduction of the structure. The arrangement of the atoms or ions in the unit cell is at least partly determined by symmetry considerations, but in most cases

TABLE 1
Possible values of M^2 for cubic lattices

hkl	M	P, M^2	M^2	$M^2/2$	F, M^2
			I		
100	1.0000	1			
110	1.4142	2	2	1	
111	1.7321	3			3
200	2.0000	4	4	2	4
210	2.2361	5			
211	2.4495	6	6	3	
220	2.8284	8	8	4	8
300, 221	3.0000	9			
310	3.1623	10	10	5	
311	3.3166	11			11
222	3.4641	12	12	6	12
320	3.6056	13			
321	3.7417	14	14	7	
400	4.0000	16	16	8	16
410, 322	4.1231	17			
411, 330	4.2426	18	18	9	
331	4.3589	19			19
420	4.4721	20	20	10	20
421	4.5826	21			
332	4.6904	22	22	11	
422	4.8990	24	24	12	24
500, 430	5.0000	25			
510, 431	5.0990	26	26	13	
511, 333	5.1962	27			27
520, 432	5.3852	29			
521	5.4772	30	30	15	
440	5.6569	32	32	16	32
522, 441	5.7446	33			
530, 433	5.8310	34	34	17	
531	5.9161	35			35
600, 442	6.0000	36	36	18	36
610	6.0828	37			
611, 532	6.1644	38	38	19	
620	6.3246	40	40	20	40
621, 540, 443	6.4031	41			
541	6.4807	42	42	21	
533	6.5574	43			43
622	6.6332	44	44	22	44
630, 542	6.7082	45			
631	6.7823	46	46	23	
444	6.9282	48	48	24	48
700, 632	7.0000	49			
710, 550, 543	7.0711	50	50	25	
711, 551	7.1414	51			51
640	7.2111	52	52	26	52
720, 641	7.2801	53			
721, 633, 552	7.3485	54	54	27	
642	7.4833	56	56	28	56

it is necessary to take account of the *intensities* of the Bragg reflections. The way this is done in present crystallographic practice is far too complicated to describe here.[2] We shall here only illustrate by a simple example how it is possible to use qualitative arguments based on intensity.

If the substance is a binary compound AB, and if its unit cell is simple cubic *P* with one formula (one atom of A and one of B) per cubic cell, the relative positions of the two atoms are fixed by symmetry. This is true of the salt cesium chloride, CsCl, the structure of which is shown in Fig. 4. One of the ions, Cs^+ say, may, without loss of generality, be placed at the origin. The other ion, Cl^-, must be at the center of the unit cell; if it is in any other position, the structure will lack the threefold rotational axes of symmetry which are always present along all four body diagonals of the unit cell in the cubic system.

In many cases, including even some with one formula of a binary compound for each lattice point, the positions of the atoms or ions are not necessarily given uniquely by symmetry, though the number of possible choices may not be large. The solution of the structure in some cases may lie in a simple clue from the intensities. Let it be supposed, for example, that a certain class of powder lines are relatively weak and that omitting these lines from the calculation leads to a "pseudo lattice" with a smaller number of atoms per lattice point than in the case of the true lattice. A possible hypothesis might then be that the weak lines owe their weakness to destructive interference between two kinds of atoms or ions and that, if the chemical difference between these two kinds of atoms or ions could somehow be removed, the interference would become complete and the pseudo lattice would become a true lattice. Knowledge of the pseudo lattice (and pseudo cell) may then show where the atoms or ions must be placed (irrespective of kind), and knowledge of the true lattice will then show which atoms are of each kind.

As an example of this kind of clue, let us deduce the relative positions of the ions in cesium chloride, bromide, and iodide (which all have the same structure) from intensity considerations, without appealing to any arguments based on symmetry. Putting the $(1/d)^2$ on a relatively prime basis, we obtain the integers 1, 2, 3, 4, 5, 6, $-$, 8, 9, 10, 11, 12, 13, 14, $-$, 16, . . . , which

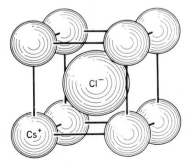

FIGURE 4
Unit cell of cesium chloride. Note that this structure is not body-centered cubic but would be if the two ions were identical.

clearly indicates a primitive lattice. From the density we find that there is one formula per lattice point. However, we see that all the odd-numbered lines are somewhat weak in CsCl, much weaker in CsBr, and very faint or absent in CsI. If we neglect them, the sequence of integers obtained (on dividing by 2 to obtain relative primes) indicates a body-centered cubic "pseudo lattice" with one-half formula, or one ion (irrespective of kind), at each pseudo-lattice point. In other words, if we were unable to distinguish between the two kinds of ions, the structure would look like one with a single atom of a single kind at each point of a body-centered cubic lattice. This clearly shows that in the real structure, an ion of one kind is located at (000) and an ion of the other kind is located at $(\frac{1}{2}\frac{1}{2}\frac{1}{2})$. It may be mentioned that the virtually complete obliteration of the odd lines in CsI is due to the fact that the Cs^+ ion and the I^- ion are *isoelectronic* (that is, have the same number of electrons, namely 54, which is the number in xenon); if any odd lines are observed at all, it is owing to the fact that, because of the different nuclear charges of the two ions, the sizes of the electron clouds of the two ions are slightly different.†

When there are more than two atoms per lattice point, the structure determination will be more complicated. One frequently used procedure is that of "trial and error," in which a number of "model" structures are successively proposed and tested by calculation of intensities and comparison with experiment, until a structure is found which yields satisfactory agreement. When the number of possible structures is too large for the practical application of trial-and-error methods, methods must be used that are too advanced for description here. With their use, crystal structures have been found in which there are thousands of atoms in a unit cell. In such cases and in most ordinary work, however, powder data would be inadequate for determining the structure, and diffraction patterns must be obtained from single crystals.

METHOD

There are several experimental techniques for realizing the diffraction conditions, the most powerful of which depends on having a single crystal of the substance to be studied;[2] see Exp. 46. In the present experiment we are concerned only with the Debye–Scherrer method (often called the powder method), which does not make use of a single crystal but rather of a powder obtained by grinding up crystalline or microcrystalline material.[3] This powder contains crystal particles of a few micrometers in size.

The diffraction pattern of a powdered material can be obtained by several different experimental methods. The one that we will illustrate in the present experiment is the *powder-photograph* method using the Straumanis technique.[4]

---

† The argument here depends upon the fact that it is the electrons, rather than the nuclei, that scatter x-rays.

Other specialized powder-photograph techniques exist[3] for extremely precise work. It is also possible to record powder diffraction directly in digital form or on a strip chart recorder with a *powder diffractometer.*[5] Such instruments are manufactured by the North American Philips Company, among others. In a powder diffractometer a radiation detector (a proportional counter or a crystal scintillation counter) with narrow entrance slits moves at a constant angular speed over the desired range of Bragg angles, picking up x-ray photons diffracted by the powder specimen. Each x-ray photon is converted by the detector into an electrical pulse. The stream of pulses is processed by an electronic circuit and is usually converted to a pulse rate (number of photons per unit time) that is plotted continuously on a strip chart, yielding a continuous curve having peaks that correspond to lines on a powder photograph. The present experiment can be done with such an instrument if it is available. Close supervision and instruction by qualified personnel is essential.

In the present method, the specimen for diffraction is obtained by sticking some of the powdered material onto a fine (0.1-mm) glass fiber with a trace of Vaseline or filling a very thin-walled glass capillary tube (0.2 mm diameter) with the powder. A narrow beam of parallel monochromatic x-rays, about 0.5 mm in diameter, impinges on this specimen at right angles to its axis. The source of x-rays is usually a Coolidge-type x-ray tube with a copper (or molybdenum) target, equipped with a filter (of nickel foil, in the case of a copper target) to remove all spectral components except the desired $K\alpha$ line. The narrow beam is formed by a *collimator,* which consists basically of a conical tube 5 or 6 cm long with a pinhole or slit at each end. On the opposite side of the specimen is a conical receptacle similar to the collimator but with only an entrance pinhole or slit and no exit; this is the *beam stop,* in which the undiffracted beam is trapped. The diffracted radiation is detected by a strip of photographic film bent into a cylinder and held firmly against the inside wall of a cylindrical camera, coaxial with the specimen. The arrangement of the collimator, specimen, and photographic film is shown schematically in Fig. 5. When the beam impinges on the randomly oriented particles in a stationary powder specimen, most of the particles will not diffract the x-rays at all. Only those particles will diffract x-rays which happen by chance to be so oriented that Eq. (3) holds for Bragg reflection from some set of crystallographic planes. The direction of the diffracted rays will then deviate from the direction of the incident beam by twice the Bragg angle. Since orientation is completely random with respect to the beam axis, the rays diffracted at a given scattering angle may lie with equal probability anywhere on a right circular cone with apex angle 4θ. In practice, the specimen is usually rotated about its own axis, so that during a single rotation all or nearly all the particles present will have an opportunity to reflect x-rays from any given set of crystallographic planes, the resulting diffracted radiation being distributed rather evenly over the cone. Where the cone intersects the photographic film, the latent image of a *powder line* is formed. When the film is removed from the camera and developed,

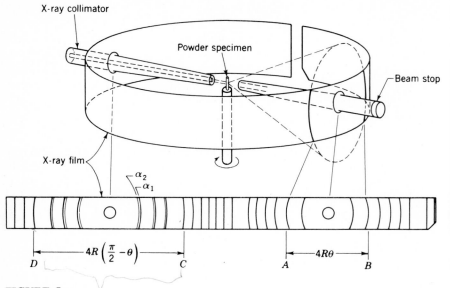

FIGURE 5
Schematic diagram illustrating x-ray powder method (Straumanis arrangement[4]). Note splitting of back-reflection lines.

fixed, washed, and dried, it is found to have on it a number of lines, each one of which is due to reflection from one or more sets of crystallographic planes.

The positions of the lines on the film can be measured with a comparator (see Exp. 40) or microphotometer, but adequate measurements can be made with a good millimeter scale. From measurements of the positions of the lines or distances between them, the Bragg angles can be calculated. The determinations of these angles is a purely geometric problem. The "effective radius" R of the film in the camera can be obtained by picking two sets of lines, A, B, and C, D as in Fig. 5, and measuring the four distances, $\overline{AC}$, $\overline{AD}$, $\overline{BC}$, and $\overline{BD}$:

$$4R = \frac{1}{\pi}(\overline{AC} + \overline{AD} + \overline{BC} + \overline{BD}) \tag{8a}$$

If it is desired to calculate angles in degrees rather than in radians, it is convenient to calculate

$$4R° = \frac{\pi}{180}(4R) = \frac{1}{180}(\overline{AC} + \overline{AD} + \overline{BC} + \overline{BD}) \tag{8b}$$

Then

$$\theta(\text{rad}) = \frac{\overline{AB}}{4R} \qquad \theta' = \frac{\pi}{2} - \theta(\text{rad}) = \frac{\overline{CD}}{4R} \tag{9a}$$

$$\theta(\text{deg}) = \frac{\overline{AB}}{4R°} \qquad \theta' = 90 - \theta(\text{deg}) = \frac{\overline{CD}}{4R°} \tag{9b}$$

One can then calculate $(1/d)$ from Eq. (3). In processing back-reflection data, it is unnecessary to convert from $(\pi/2 - \theta)$ to θ since $\sin \theta \equiv \cos(\pi/2 - \theta) \equiv \cos \theta'$. Thus

$$\frac{1}{d} = \frac{2}{\lambda} \sin \theta = \frac{2}{\lambda} \cos \theta' \tag{10}$$

When the measurements are made and $(1/d)$ is computed, account should be taken of the fact that the $K\alpha$ spectral line generally used as a monochromatic source of x-rays is not truly a single line but actually a closely spaced doublet. The doublet is ordinarily not resolved in the forward-reflection part of the film, but in the back-reflection part of the film the lines are usually observably split into two components, known as $K\alpha_1$ and $K\alpha_2$ in order of increasing wavelength (and therefore in order of increasing θ), in the intensity ratio 2:1. When the two components are not resolved, the position taken for the line should be an estimate of the position of the "center of gravity," and the value of the wavelength λ used in the calculation should be the weighted mean wavelength:

$$\lambda_{\text{mean}} = \tfrac{1}{3}(2\lambda_{\alpha_1} + \lambda_{\alpha_2}) \tag{11}$$

When the components are sufficiently well resolved to make possible the measurement of the positions of the two components separately, such measurements should be made and the actual values of the wavelengths should then be used in the calculations. Each component will then yield a separate value of $1/d$; the two values obtained should be in good agreement and may be averaged for the ensuing calculations.

For copper $(K\alpha)$ radiation,

$$\lambda_{\alpha_1} = 1.54050 \text{ Å}$$
$$\lambda_{\alpha_2} = 1.54434 \text{ Å} \tag{12}$$
$$\lambda_{\text{mean}} = 1.5418 \text{ Å}$$

In present-day x-ray crystallography, powder photography is only rarely used for complete structure determinations. It is, however, frequently used in the precise determination of unit-cell dimensions. Its most common technical use is as an analytical tool; powder patterns of thousands of crystalline substances are known.

EXPERIMENTAL

Preparation of powder specimen. A specimen of the material to be studied is finely pulverized in an agate mortar with an agate pestle. The powder is then loaded into a Pyrex or Lindemann glass capillary tube, 0.2 to 0.3 mm o.d., having a very thin wall (0.01 mm or less), so as to obtain a densely filled specimen about 1 cm in length. If the capillary tube is flared out at one end, it is easier to load; even without a flared-out end, however, the powder can be picked up by scraping one end of the capillary against the surface of the mortar

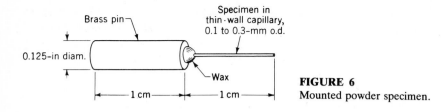

FIGURE 6
Mounted powder specimen.

and, with this end uppermost, agitating the powder by very gentle rasping with a file. One end of the capillary tube should be sealed off with a flame or with wax before filling, and the other end with wax after filling. The capillary tube is then affixed with wax to the end of a short ($\frac{3}{8}$ inch or less) length of $\frac{1}{8}$-inch brass rod; see Fig. 6.

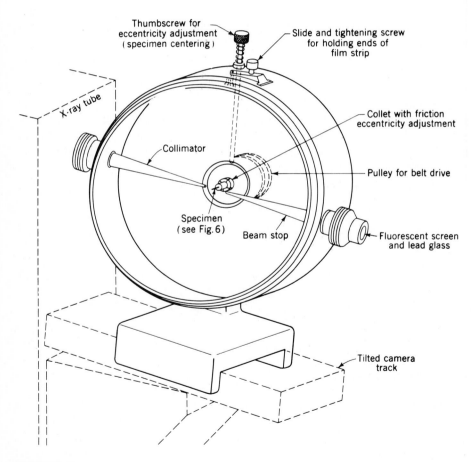

FIGURE 7
Philips powder camera (shown with cover not in place).

Installation of the specimen. A powder camera of the Straumanis design is shown in Fig. 7. It is a valuable instrument and must be handled carefully. The camera should not be used until the procedures for using it have been demonstrated to the student by an experienced user. To install and line up the specimen, proceed as follows.†

Remove the circular cover, and lay it down, inside surface upward, in a *safe place*. Remove the two slits (collimator and beam stop), and lay them down *inside the cover*.

By means of tweezers or long-nose pliers, insert the brass pin supporting the specimen into its receptacle, where it is held by friction. Carefully replace the collimator, being sure not to force it or to drop it.

Place the lens cap over the outside collimator opening, and set the camera down with the beam-stop hole facing a well-lighted surface (beam stop *not* in place). By hand, turn the pulley wheel on the outside of the camera. As it goes around, the silhouette of the specimen should appear to go up and down. Stop it in an "up" position, and by means of the screw on top of the camera push it down to the center of the visual field. Turn the screw back up, and rotate again to see if further adjustment is needed; if so, use the screw as before. Continue until the specimen appears to undergo no motion as the spindle is rotated.

Remove the lens cap, and replace the beam stop. As in the case of the collimator, do not force it or drop it. Replace the cover.

Loading the camera with film. Take the camera into the darkroom. Remove the cover and the collimator and beam stop. Turn off the room light, and under a safelight cut, punch, and insert the film in accordance with specific instructions given. (It is advisable to rehearse the procedure ahead of time with the light on, using a piece of exposed film.) Insert the collimator and beam stop and replace the cover.

Making the x-ray exposure. This should be done under the constant supervision of a qualified person; detailed instructions cannot be given here. **Be very careful to avoid any x-ray exposure to any part of the body**; make sure that all x-ray ports are covered when the unit is in operation. The optimum exposure time depends on the composition and size of the specimen, the tube current and operating potential, and the dimensions of the collimating system. Typical exposures are 4 to 20 hr. In the absence of a recommended exposure time, make a short trial exposure and a second longer exposure if necessary.

Developing the film. The camera is taken into the darkroom and the film is removed. It should be attached to a clip or other support and developed for 3 to 5 min, depending on the temperature of the developer. Agitate the film, but

---

† The procedure given is that applicable to the 114.6-mm camera manufactured by the North American Philips Co.[6]

do not allow its surface to rub against the walls or bottom of the tank or tray. Wash the film for 30 s and place it in the fixing bath. After the film has cleared, the room light may be turned on. The film should be in the fixing bath for at least twice as long as it takes to clear and should then be washed in running water for at least 30 min. It may then be carefully wiped with clean fingers, rinsed again with distilled water, and hung up to dry.

Measurement of crystal density. The density of the substance can be determined with a pycnometer or with a 5-mL (or smaller) volumetric flask used as a pycnometer. The liquid used should be one in which the substance is insoluble; for a water-soluble inorganic salt, medium-boiling petroleum ether is convenient. The procedure is described in Chapter I (Sample Report). Make two determinations.

CALCULATIONS

If it is not possible to take the x-ray powder pattern in the laboratory, a contact print of a powder photograph will be furnished and the x-ray equipment will be discussed and shown to the students. The instructions below should be followed with either the negative taken or the print provided.

Measure $\overline{AB}$ or $\overline{CD}$ (see Fig. 5) for as many lines as possible; estimate each distance to ± 0.1 mm. Enter the measurements, together with estimated intensity, into a table. Make a separate measurement for each component of a resolved doublet.

Pick a pair of sharp lines, A, B in the forward-reflection part of the film and another pair C, D in the back-reflection region. Measure the four distances $\overline{AC}, \overline{AD}, \overline{BC}, \overline{BD}$ as accurately as possible. Calculate $4R$ or $4R°$ from Eq. (8).

Calculate θ or θ' for each line or resolved component of a line [Eq. (9)] and then calculate $1/d$ from Eq. (10), averaging the values obtained for the two components of each resolved doublet. Calculate $(1/d)^2$.

Find a factor that will reduce the $(1/d)^2$ values to relatively prime integers within experimental error. Refer to Table 1 and identify the lattice type. Also, from Table 1, obtain the Miller indices hkl for each line.

The most precise values of a_0 are obtained from the lines in the back-reflection part of the film, where θ is close to 90°. This can be seen by combining Eqs. (3) and (5), solving for a_0, and differentiating:

$$da_0 = -\frac{M\lambda}{2\sin^2\theta}\cos\theta\,d\theta$$

$$= -a_0 \cot\theta\,d\theta$$

The experimental uncertainty in measuring θ is proportional to that of measuring the distances between pairs of lines and is approximately constant over the film unless lines in the back-reflection region are unduly faint or broad. However, even if the uncertainty of measurement is a little larger in the back-reflection region, the effect is ordinarily far outweighed by the $\cot \theta$ factor. In fact, lines very close to the collimator hole should always be used if at all possible, even if they are broad and diffuse. Another good reason for using only lines in the back-reflection region is that they are much more free of shifts due to absorption of x-rays by the specimen. From a few well-chosen lines in the back-reflection region, preferably resolved doublets, calculate a_0 from Eq. (5) and average the values obtained.

From the measured density and the a_0 value determined as above, calculate the number of atoms per unit cell and per lattice point. Report the lattice type, the value of a_0, and the number of atoms per unit cell.

DISCUSSION

Determine the crystal structure, if possible, by methods similar to those described for CsCl (see Theory). Draw a diagram of the cubic unit cell showing the positions of all atoms or ions.

APPARATUS

X-ray diffraction apparatus, complete with tube stand, x-ray tube, and power supply; Debye-Scherrer powder camera, preferably Straumanis type (such as North American Philips 114.6-mm camera); film cutter and punch; thin-wall Lindemann glass capillary tubes (may be purchased from Caine Scientific Sales Co., Chicago, IL.); agate mortar and pestle; small file; x-ray film (Eastman no-screen, 35-mm continuous strip); darkroom, equipped with x-ray developing tank or adequate trays; x-ray developer and fixer solutions; timer; thermometer; good millimeter scale; pycnometer or 5-mL volumetric flask; small pipette or eye dropper. Alternatively, the entire experiment can be done with an x-ray powder diffractometer such as those manufactured by North American Philips, used under the supervision and instruction of qualified personnel.

Small quantity of crystalline material of cubic structure for study (e.g., alkali halides; alkaline earth oxides; cuprous or silver chloride; simple metals such as aluminum or copper, finely powdered with a file); liquid (medium-boiling petroleum ether) for density work.

REFERENCES

1. M. J. Buerger, "Contemporary Crystallography," chap. 2, McGraw-Hill, New York (1970).
2. G. H. Stout and L. H. Jensen, "X-Ray Structure Determination," Macmillan, New York (1968).
3. B. D. Cullity, "Elements of X-Ray Diffraction," 2d ed., chaps. 6 and 7, Addison-Wesley, Reading, Mass. (1978).
4. M. J. Buerger, *Am. Miner.* **21**, 11 (1936); *J. Appl. Phys.* **16**, 501 (1945); M. Straumanis and A. Ievins, *Z. Phys.* **98**, 461 (1936).
5. B. D. Cullity, *op. cit.*, chap. 7.

GENERAL READING

M. J. Buerger, "Contemporary Crystallography," McGraw-Hill, New York (1970).
J. G. Glusker and K. N. Trueblood, "Crystal Structure Analysis: A Primer," Oxford University Press, New York (1972).
D. McKie and C. McKie, "Essentials of Crystallography," Blackwell, Oxford 1985.

EXPERIMENT 46
SINGLE-CRYSTAL X-RAY DIFFRACTION WITH THE BUERGER PRECESSION CAMERA

The introduction to x-ray crystallography given in the preceding experiment is mainly confined to the determination of a very simple cubic structure by the powder method. The present experiment is considerably more ambitious: it deals with a single crystal having a symmetry different from cubic. Analysis of this more general problem requires the use of the *reciprocal lattice,* a concept fundamental to crystallography and many aspects of solid-state physical chemistry. Another basic crystallographic concept—the *space group*—is also introduced. The specification of the space group of a crystal amounts to a complete statement of the symmetry of the infinitely periodic crystal structure.

Measurements of the reciprocal lattice by diffraction methods yield the lattice constants of the "reciprocal unit cell," from which one may calculate the lattice constants (unit-cell dimensions and angles) of the crystallographic unit cell in direct space. Once the lattice constants and the space group are established, the crystal structure can be determined (by methods beyond the scope of the present experiment) from the Bragg intensities associated with the reciprocal lattice points. A complete crystal structure is usually considered to include not only the positional coordinates of all of the atoms in the unit cell but also their average vibrational amplitudes in various directions. Structural chemistry information on molecular shapes and configurations, interatomic distances, bond angles, and intermolecular separation distances can then be easily determined.

The instrumentation used to explore the reciprocal lattice was once almost entirely photographic; nowadays crystal structure determinations are done with four-circle diffractometers under computer control. Nevertheless photographic methods are still valuable. One such photographic method, the *precession method,* is particularly effective in illustrating the concept of the reciprocal lattice. This method uses the Buerger precession camera.[1,2] As we shall see, this camera effectively "photographs the reciprocal lattice" to scale, layer by layer.

This experiment is devoted to the use of precession photographs for determining the lattice constants and space group of an orthorhombic crystal.

THEORY

Some elementary concepts of crystallography—the crystal lattice, the unit cell, lattice planes, Miller indices, and the Bragg equation—have been introduced in Exp. 45. These concepts are vital prerequisites for the study of the new concepts described here.

The reciprocal lattice. In Exp. 45 we dealt with the concept of a set of lattice planes—an infinite set of equally spaced parallel planes so defined that all points of the crystal lattice lie on planes of the set (although it is not necessary that all planes of the set contain lattice points). The infinite set of equations defining the set of lattice planes is indicated in Eqs. (45-4). The set of lattice planes is designated by the *Miller indices h, k, l,* each of which can independently take on all integral values from $-\infty$ to ∞, provided the coordinates *x, y, z* are defined with respect to a *primitive* unit cell. If the coordinates are defined with respect to a non-primitive cell, exemplified by the body-centered or face-centered cells discussed in Exp. 45, the Miller indices are subject to certain restrictions. These lattice planes are conceptually useful in the derivation of the Bragg equation [Fig. 45-3, Eq. (45-3)].

The crystal lattice, unit cell, and lattice planes exist in ordinary three-dimensional space, referred to in crystallography as *direct space* to distinguish it from a different kind of three-dimensional space called *reciprocal space* for reasons that will soon become apparent. The *reciprocal lattice,*[3] hereinafter abbreviated r.l., may be constructed in reciprocal space as follows.

First choose an arbitrary fixed point in reciprocal space and designate it as the origin of the r.l. ($hkl = 000$), a point indicated on the illustrations by the letter *O*. Now choose a set of lattice planes *hkl* in direct space. Construct in reciprocal space, extending from the origin *O,* a vector $\mathbf{q}_{hkl}$ with a direction perpendicular to the set of lattice planes in direct space and a length that is the *reciprocal* of the interplanar spacing d_{hkl}:

$$\mathbf{q}_{hkl} \perp (hkl) \qquad |\mathbf{q}_{hkl}| \equiv \frac{1}{d_{hkl}} \tag{1}$$

where (*hkl*) indicates the set of lattice planes corresponding to Miller indices *hkl*. The terminus of this vector is the *reciprocal lattice point hkl,* hereinafter abbreviated r.l.p. Since the set of lattice planes $\bar{h}\bar{k}\bar{l}$ (minus signs are customarily placed above the indices) is identical to the set *hkl*, there will be another vector in reciprocal space of the same length but in the opposite direction from the origin to r.l.p. $\bar{h}\bar{k}\bar{l}$. When this is done for all possible sets of lattice planes, the r.l. is completely defined. This procedure is perfectly general and does not depend on any particular assignment of crystal axes **a**, **b**, **c** to the lattice in direct space. In Fig. 1 this procedure is illustrated for the set of lattice planes (210). For the set (420), the direction of the r.l. vector **q** is the same but since *d* for (420) is half as large as that for (210), the vector $\mathbf{q}_{420}$ is twice as long as the vector $\mathbf{q}_{210}$. The student is urged to perform this construction for several

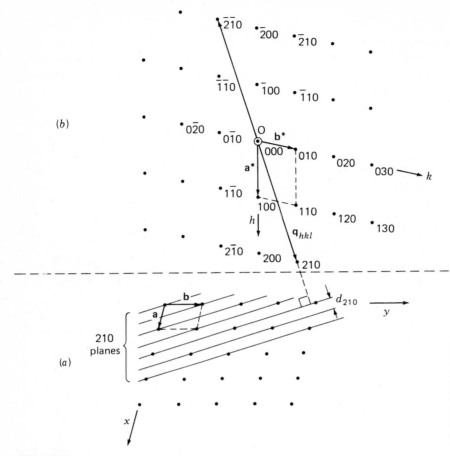

FIGURE 1
Construction of the reciprocal lattice (*b*) from the direct lattice (*a*). A two-dimensional section of each lattice is shown, corresponding to $z = 0$ in the direct lattice and $l = 0$ in the reciprocal lattice. The set of planes 210, of which the traces are shown in (*a*), are parallel to **c** (i.e., to the *z* axis) which comes out of the paper.

different sets of planes on graph paper, and verify that the points generated constitute a lattice.

Corresponding to the lattice basis vectors (axes) **a**, **b**, **c** in direct space, there are r.l. basis vectors **a***, **b***, **c*** in reciprocal space. The latter are defined in terms of the former in such a way that the following vectorial relations hold:

$$\mathbf{q}_{hkl} = h\mathbf{a}^* + k\mathbf{b}^* + l\mathbf{c}^*$$

$$\mathbf{a}^* \cdot \mathbf{a} = \mathbf{b}^* \cdot \mathbf{b} = \mathbf{c}^* \cdot \mathbf{c} = 1 \tag{2}$$

$$\mathbf{a}^* \cdot \mathbf{b} = \mathbf{b}^* \cdot \mathbf{a} = \mathbf{b}^* \cdot \mathbf{c} = \mathbf{c}^* \cdot \mathbf{b} = \mathbf{c}^* \cdot \mathbf{a} = \mathbf{a}^* \cdot \mathbf{c} = 0$$

Thus $\mathbf{a}^*$ is perpendicular to $\mathbf{b}$ and to $\mathbf{c}$, for example; it may or may not be parallel to $\mathbf{a}$ depending on the symmetry of the lattice.

For simplicity in the present experiment we shall confine ourselves to the orthorhombic system, which has orthogonal axes. In this system (as also in the tetragonal and cubic systems, where $\alpha = \beta = \gamma = 90°$)

$$a^* = \frac{1}{a} \qquad b^* = \frac{1}{b} \qquad c^* = \frac{1}{c} \tag{3}$$

It is very important to remember that these relations do *not* apply generally in the triclinic, monoclinic, and hexagonal systems, except for axes parallel to symmetry axes or normal to symmetry planes.

Any person working with diffraction will very soon appreciate the value of the r.l. in "mapping" the circumstances of diffraction. As a trivial example, consider the systematic extinctions corresponding to body-centered (I) and face-centered (F) cubic lattices in Exp. 45. The condition that $h + k + l$ for an I lattice must be even means that the r.l.p. for any set of Miller indices hkl for which the sum is odd simply does not exist, and the intensity for Bragg reflection from the corresponding set of planes must be zero.†

The Ewald construction. Let us rewrite the Bragg equation, Eq. (45-3), in the form

$$\frac{2}{\lambda} \sin \theta_{hkl} = \frac{1}{d_{hkl}} \tag{4}$$

We will now convert this equation to vectorial form. Let us define a vector of unit length $\mathbf{s}_0$ in the direction of the incident x-ray beam, and another of unit length $\mathbf{s}$ in the direction of the beam that is Bragg-reflected from the set of lattice planes hkl. These are shown in Fig. 2(a) together with a Bragg construction. From the vector diagram given it is clear that $|\mathbf{s} - \mathbf{s}_0| = 2 \sin \theta$. Because the reflection must be *specular* (i.e., like reflection from a mirror), the vectors $\mathbf{s} - \mathbf{s}_0$ is perpendicular to the set of lattice planes. But so also is the r.l. vector $\mathbf{q}_{hkl}$, which is of length $|\mathbf{q}_{hkl}| = 1/d_{hkl}$. Therefore,

$$\frac{\mathbf{s}}{\lambda} - \frac{\mathbf{s}_0}{\lambda} = \mathbf{q}_{hkl} \tag{5}$$

This is a general equation limiting the geometrical conditions of x-ray diffraction in the kinematic approximation. It is represented graphically in Fig. 2(b) in a diagram known as the *Ewald construction*. Let us construct a vector $\mathbf{s}_0/\lambda$ in reciprocal space (codirectional with $\mathbf{s}_0$ in direct space) with its terminus at the origin O of the r.l. Describe a sphere (called the Ewald sphere) of radius

---

† The student will find it instructive to show that as a result, the r.l. corresponding to an I direct-space lattice is actually an F lattice. It can likewise be shown that the r.l. corresponding to an F direct-space lattice is an I lattice.

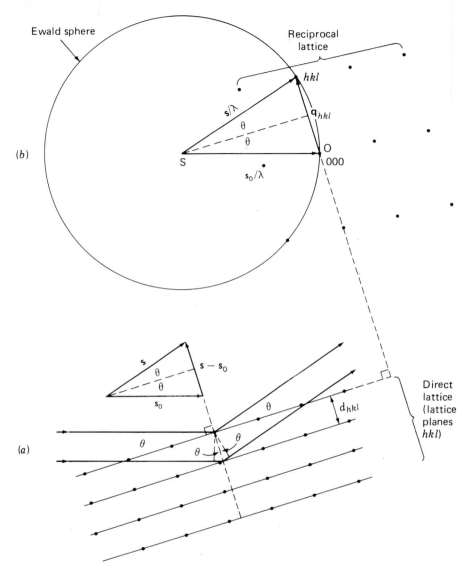

FIGURE 2
(*a*) Bragg construction [*cf.* Fig. (45-3)] with vector diagram; (*b*) Ewald construction.

$1/\lambda$ with its center S at the tail of the vector $\mathbf{s_0}/\lambda$. Now *if* the sphere happens to pass through any r.l.p. (other than the origin), say *hkl*, then the geometrical conditions for diffraction are satisfied and the direction of the diffracted beam is that of a vector $\mathbf{s}/\lambda$ drawn from the center S of the Ewald sphere to the r.l.p. *hkl*.

The orientation of the crystal in space not only determines the orientation of the crystal axes and the direct lattice, but also determines the orientation of the r.l. If the crystal is caused to rotate in direct space, the r.l. rotates correspondingly about its origin O in reciprocal space. For a beam of strictly parallel and absolutely monochromatic x-rays (so that s_0/λ is exactly defined) and a given arbitrary orientation of the crystal, the probability that the Ewald sphere passes through any r.l.p. (other than the origin) is infinitesimal. Therefore, in *diffraction cameras* the crystal is caused to undergo some kind of rotatory, oscillatory, or precessional motion. During the corresponding motion of the r.l., some of the r.l.p.'s pass through the stationary Ewald sphere. As each r.l.p. passes through the sphere, the geometrical conditions for diffraction are satisfied for an instant and a "flash" of reflected x-rays impinges upon the photographic film and exposes some grains in the emulsion. This process is repeated at this point on the film hundreds or thousands of times because of repetition of the mechanical cycle of the instrument. When the film is developed there appears a diffraction "spot" with a size and shape determined by the size and shape of the crystal specimen and by the angle of divergence ($\approx 1°$) of the incident x-ray beam. The intensity (blackness) of this spot depends on constructive or destructive interference of the x-ray wavelets scattered by the individual atoms in the unit cell. In the Buerger precession camera the film is moved in synchronization with the crystal motion so that the position of the spot on the developed film reveals the position (Miller indices) of the corresponding r.l.p. in reciprocal space.

The point group and crystal class. The theory of symmetry and the development of the point groups and space groups is a complex subject.[4] We will here dispense with formal group theory; indeed we will dispense with all kinds of symmetry that are not encountered in the one crystallographic system to which the present experiment is limited: the orthorhombic system.

We must first introduce the *point group,* a collection of symmetry elements that specify the symmetry of a finite object.† The only point-symmetry elements that occur in the orthorhombic system are:

• the identity element (trivial, possessed by every object),
• the two-fold rotation axis, designated by the symbol 2, representing invariance to a 180° rotation,
• the mirror plane, designated by the symbol *m,* representing invariance to reflection in a plane, and
• the center of inversion (often called *center of symmetry*), designated by the symbol *i,* representing invariance to inversion through a point.

---

† The point group can also be applied to an infinite object, such as a crystal lattice, a crystal structure, a reciprocal lattice, or an "intensity-weighted" reciprocal lattice.

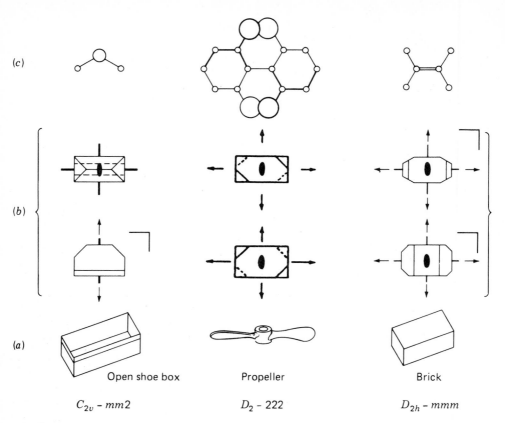

(c)

(b)

(a)

Open shoe box	Propeller	Brick
C_{2v} – $mm2$	D_2 – 222	D_{2h} – mmm

FIGURE 3

The three orthorhombic point groups with examples: (a) familiar objects, (b) hypothetical crystals showing the point group in their morphology, and (c) chemical molecules. In part (b), the conventional graphical symbols for the elements are superimposed: a right-angled line for a mirror plane parallel to the paper and a single straight line trace for one perpendicular to it; an arrow and a "lens" for twofold rotation axes, and a small circle for a center of inversion ("center of symmetry").

There are 32 crystallographic point groups, and the orthorhombic system contains three (see Fig. 3):

C_{2v} or $mm2$ = two mirror planes crossing at right angles, generating a two-fold axis at their line of intersection. Examples of objects conforming to this point group: an open shoe box; the molecules H_2O, SO_2, H_2CO (formaldehyde).

D_2 or 222 = three two-fold axes intersecting at right angles (actually, any two of the three suffice to generate the third). Examples: a two-bladed propeller without shaft; a molecule of 2,6,2',6'-tetraiodobiphenyl (twisted out of the planar configuration by steric repulsion of the iodine atoms).

D_{2h} or mmm = three mirror planes intersecting at right angles, generating three two-fold axes at their lines of intersection *and* (very importantly) a center of inversion. Examples: a brick; the molecules C_2H_4 (ethylene), and B_2H_6 (bridge structure of diborane). This point group contains all of the elements of $C_{2v} - mm2$ and all of the elements of $D_2 - 222$; those two point groups are said to be *subgroups* of $D_{2h} - mmm$.

   In each case, two symbols for each point group have been given. The first is the Schoenflies symbol, often used in spectroscopy. The second is the Hermann–Mauguin symbol, much more commonly used in crystallography. The three positions in the Hermann–Mauguin symbol for orthorhombic point groups correspond to the three assigned crystal axes **a**, **b**, **c** and each is occupied by a letter or numeral designating the symmetry element operating with respect to the corresponding direction. *The direction in the case of the mirror plane is that of its normal.* Clearly, with different axial assignments, a crystal may have point group symmetry $2mm$ or $m2m$; these are equivalent to $mm2$ and all three correspond to the Schoenflies symbol C_{2v}.
   Corresponding to each point group is a *crystal class*; all crystals having the same point-group symmetry are said to belong to the same crystal class. Thus there are three orthorhombic crystal classes, which are given the same identifying symbols as the corresponding point groups. The crystal class or point group of a given crystal is ultimately defined by its atomic structure, but the macroscopic properties of the crystal—its optical, elastic, piezoelectric, and pyroelectric properties, and also ideally its external morphology—must conform to the symmetry of the point group. However, the information required for the complete specification of the crystal point group is often not available from macroscopic properties alone.
   One might then hope that all of the required information can be obtained from x-ray diffraction experiments. Unfortunately, with ordinary photographic methods such as those employed in this experiment, it is generally not possible to detect any significant difference in Bragg reflection intensity between reflections hkl and $\bar{h}\bar{k}\bar{l}$. That is, the x-ray experiment by itself seems to confer upon the crystal an apparent center of inversion, and thus is incapable of showing whether or not the crystal really possesses a center of inversion. (This limitation is known as Friedel's law.) The intensity-weighted r.l., i.e., the r.l. weighted with the Bragg reflection intensities, therefore possesses for all orthorhombic crystals, the centrosymmetric $D_{2h} - mmm$ point-group symmetry. Accordingly, all orthorhombic crystals are said to have $D_{2h} - mmm$ "Laue symmetry." Without further information it is impossible to determine which of the three orthorhombic point groups is correct for a given crystal. As we shall see, the presence of *screw axes, glide planes,* or both can be inferred from certain systematic extinctions, and in some cases certain combinations of these can uniquely define the space group and point group. In many other cases, either physical methods must be used, or ambiguity will remain until the entire crystal structure has been successfully determined, possibly after trials with alternative space groups.

For crystals that do not conduct electricity, the final ambiguity may often be resolved by study of the piezoelectric or pyroelectric properties of the crystal. Crystals with $D_{2h} - mmm$ symmetry cannot show either the piezoelectric effect or the pyroelectric effect; those with $D_2 - 222$ can show the former but not the latter; those with $C_{2v} - mm2$ can show both. Details are given by Wooster and Breton.[5] Another method, novel and sensitive, applicable to transparent crystals (even if they have some electrical conductivity), involves determination of the capability of the crystal to generate second harmonics in a laser beam.[6]

The space group. The crystal structure, idealized, is perfectly "lattice periodic" and infinite in extent in all three dimensions. A complete statement of the symmetry of the structure must take this into account. Such a statement is the specification of the space group.

An excellent introduction to space groups is given by Buerger.[7] A complete tabulation of all 230 possible space groups in three dimensions, with extensive and detailed tables and diagrams, is given in vol. A, Symmetry Groups, of *International Tables for Crystallography*,[8] hereinafter abbreviated as *I.T.*

The same symmetry elements that we encountered in the point groups may exist in the space group, but they are repeated by the lattice and are present in an infinite number of positions. In addition, certain elements occur that do not occur in point groups because they cannot be possessed by a finite object. These include the elements of translation, corresponding to the periodic crystal lattice. In addition to these there are, in many space groups, elements that combine translations with rotations or reflections: in the orthorhombic system these are the two-fold screw axes and various glide planes. The two-fold screw axis corresponds to a 180° rotation *accompanied by a simultaneous translational shift of one-half a lattice translation* (lattice-point separation vector) in the direction of the rotation axis; it is designated by the symbol 2_1. The glide plane similarly corresponds to a reflection accompanied by a translational shift of one-half a lattice translation in a direction parallel to the plane of reflection; it is designated a, b, c, or n according to whether the translational shift is by $\mathbf{a}/2$, $\mathbf{b}/2$, $\mathbf{c}/2$, or a diagonal direction, e.g., $(\mathbf{b} + \mathbf{c})/2$ for an n glide plane normal to $\mathbf{a}$.† The two-fold screw axis and the glide plane are illustrated in Fig. 4.

As in the cubic system (Exp. 45), the lattice in the orthorhombic system may be primitive P, body-centered I, or face-centered F. In addition, a lattice in the orthorhombic system may be end-centered—i.e., the unit cell is centered on only one pair of opposing faces instead of all three. The lattice is then designated by a capital letter corresponding to the crystal axis *not* contained in the centered face: C if the centering point is at $(\frac{1}{2}, \frac{1}{2}, 0)$; A if it is at $(0, \frac{1}{2}, \frac{1}{2})$; B if

---

† An additional glide plane d, with a translational shift of one-quarter cell face diagonal, occurs in only two of the 59 orthorhombic space groups and will not be discussed here.

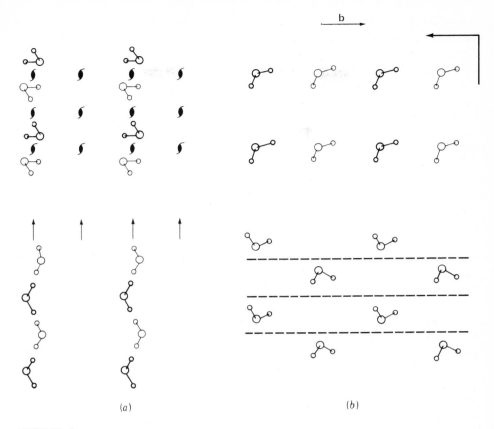

FIGURE 4

(a) 2_1 screw axis; (b) b glide plane. Symbols used for these elements are shown for the cases where (top) the 2_1 axis and the *normal* to the glide plane are perpendicular to the paper, and where (bottom) they lie in the paper. A dashed line is also used for the trace of an *a* glide plane; a dotted line is used for *c*, and dash-dot-dash for *n*.

it is at $(\frac{1}{2}, 0, \frac{1}{2})$; however, all three should be considered as examples of a single Bravais lattice type.

The structures of the space groups and their relation to the point groups may perhaps be best understood through the use of some examples. Consider the point group $C_{2v} - mm2$, which possesses a rotation axis 2 at the intersection of two perpendicular mirror planes *m*. We can immediately form five different *symmorphic* space groups by combining these elements of symmetry with the lattice translations of the different Bravais lattices: *P*, *C*, (*A*, *B*), *F*, *I*. These are

$$C_{2v}^1 - Pmm2, \quad C_{2v}^{11} - Cmm2, \quad C_{2v}^{14} - Amm2, \quad C_{2v}^{18} - Fmm2, \quad C_{2v}^{20} - Imm2$$

The Hermann–Mauguin symbol is formed by placing the capital letter symbolizing the Bravais lattice in front of the Hermann–Mauguin symbol of the point group.

It will be noted that *Cmm2* and *Amm2* are different space groups although *C* and *A* represent the same Bravais lattice. This is because in *Cmm2* the axis 2 is perpendicular to the centered unit-cell face, while in *Amm2* it is parallel to it. It is unnecessary to consider *Bmm2* as a separate space group; it is equivalent to *Amm2* by interchange of the **a** and **b** axis labels. Indeed, the Hermann–Mauguin symbols assigned above correspond only to the conventional assignments of axis labels; the space groups may also be written in terms of other axial assignments. Thus, for example,[8]

$$C_{2v}^{14}: Amm2 \doteq B2mm \doteq Cm2m \doteq Am2m \doteq Bmm2 \doteq C2mm$$

where $\doteq$ means "is equivalent to". Alternative settings such as these are used when it is necessary to assign orthorhombic axes in such a way that their lengths conform to the conventional relation

$$c \leq a \leq b \tag{6}$$

which is almost universally followed in the orthorhombic system.

The symmorphic space groups are only 73 (out of 230) in number, and are rarely encountered, particularly in molecular crystals. Molecules pack better when shifted by screw axes or glide planes which permits the "bumps" on one molecule to fit into the recesses of the next, or into intermolecular spaces, and permit oppositely charged groups to come into close proximity.

In addition to the symmorphic space groups, there are additional space groups in which rotation axes are replaced (or supplemented) by screw axes, and/or mirror planes are replaced (or supplemented) by glide planes. These nonsymmorphic space groups, together with the symmorphic ones, are said to be *isomorphic* with the point group and with each other, and all are said to belong to the crystal class of the point group. In the Hermann–Mauguin symbol, the character 2 for a rotation axis is replaced where necessary by 2_1 for the screw axis, and/or the character *m* for a mirror plane is replaced by *a*, *b*, *c*, *n*, or *d* for a glide plane. For example, there are 17 nonsymmorphic space groups isomorphic to point group $C_{2v} - mm2$, some of which are

$$C_{2v}^2 - Pmc2_1, \quad C_{2v}^8 - Pba2, \quad C_{2v}^{10} - Pnn2, \quad C_{2v}^{12} - Cmc2_1,$$
$$C_{2v}^{15} - Abm2, \quad C_{2v}^{22} - Ima2$$

These 17 plus the 5 symmorphic space groups constitute the 22 space groups in crystal class $C_{2v} - mm2$. There are 9 space groups in class $D_2 - 222$ and 28 in class $D_{2h} - mmm$, making a total of 59 space groups in the orthorhombic system.

To illustrate how a space group is presented diagrammatically in *I.T.*, the diagrams for space group $D_{2h}^{18} - Cmca$ are reproduced in Fig. 5. The "complete" Hermann–Mauguin designation for this space group is

$$C\frac{2}{m}\frac{2}{c}\frac{2_1}{a}$$

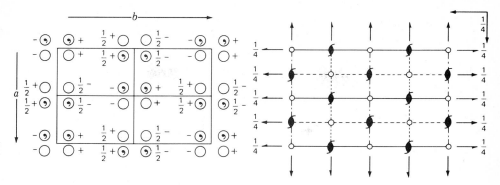

FIGURE 5

Diagrams for space group No. 64, $D_{2h}^{18} - Cmca$, reproduced by permission from I.T.[8] The right-hand diagram gives the symmetry elements and their positions in the unit cell. Each symmetry element parallel to the paper occurs at $z = 0$ and $z = \frac{1}{2}$ unless the fraction $\frac{1}{4}$ appears, in which case the element occurs at $z = \frac{1}{4}$ and $z = \frac{3}{4}$. Solid right-angled line with arrows: a and b glide planes parallel to the paper (with a single arrow this would indicate an a or a b glide depending on direction; without arrows, a mirror plane parallel to the paper). Solid, dashed, dotted, and dot-dashed lines: respectively m, b, c, and n planes perpendicular to the paper. Full arrow and half arrow: respectively 2 and 2_1 axes parallel to the paper. Lens with "tails": 2_1 axes perpendicular to the paper. Small circles: centers of inversion ("centers of symmetry"). The left-hand diagram shows a complete set of symmetry-equivalent points. The empty circles and the circles containing a comma represent points related enantiomorphically (i.e., left hand and right hand). z coordinates are indicated as follows: $+ = +z$, $- = -z$, $\frac{1}{2}^+ = \frac{1}{2} + z$, $\frac{1}{2}^- = \frac{1}{2} - z$.

where $2/m$ means that a two-fold rotation axis and the normal to a mirror plane are both parallel to **a**, etc. It will be observed in Fig. 5 that in addition to the elements explicitly given in the complete symbol, there are also b glides with normals parallel to **a** and to **c**, an n glide with its normal parallel to **b**, and a center of inversion. The b and n glides are generated by the interaction of other elements with the C-centering translation of the Bravais lattice.

Systematic extinctions. We now discuss the manner in which systematic extinctions are used to determine, completely or partially, the space group in the orthorhombic system.

As in the cubic case (Exp. 45) centered lattices impose certain conditions (called extinction rules) limiting possible reflections. We have already pointed out that a lattice extinction rule corresponds to a class of r.l.p.'s that have zero intensity in the reciprocal lattice. In addition to the rules given for I and F lattices in Exp. 45, the end-centered lattice gives rise to its own condition: for a C lattice it can easily be shown by use of Eq. (45-4) that reflection can only occur if $h + k$ is even. The screw axes and glide planes, in addition, impose their own conditions. For example, a 2_1 screw axis parallel to **c** permits $00l$ reflections to occur only if l is even. The extinction conditions for all lattices, screw axes, and glide planes (except d) applicable to the orthorhombic system are given in Table 1. It is important to emphasize that these are necessary, *but*

TABLE 1
Conditions limiting possible reflections ("extinction rules")

Type of reflection	Lattice type, or symmetry element and orientation	Condition(s) limiting reflection
hkl	Lattice I	$h + k + l = $ even
	Lattice F	$h + k = $ even, $k + l = $ even, $l + h = $ even (i.e., h, k, l are all even or all odd)
	Lattice C	$h + k = $ even
	Lattice A	$k + l = $ even
	Lattice B	$l + h = $ even
$00l$	$2_1 \parallel \mathbf{c}$	$l = $ even
$h00$	$2_1 \parallel \mathbf{a}$	$h = $ even
$0k0$	$2_1 \parallel \mathbf{b}$	$k = $ even
$hk0$	$a \perp \mathbf{c}$	$h = $ even
	$b \perp \mathbf{c}$	$k = $ even
	$n \perp \mathbf{c}$	$h + k = $ even
$0kl$	$b \perp \mathbf{a}$	$k = $ even
	$c \perp \mathbf{a}$	$l = $ even
	$n \perp \mathbf{a}$	$k + l = $ even
$h0l$	$c \perp \mathbf{b}$	$l = $ even
	$a \perp \mathbf{b}$	$h = $ even
	$n \perp \mathbf{b}$	$l + h = $ even

not sufficient, conditions for Bragg reflection; although a reflection may be permitted by all of the conditions in Table 1, it may be accidentally extinguished by destructive interference of wavelets scattered by different atoms located in various parts of the unit cell.

One must be careful about inferring from a single extinction rule that the symmetry element corresponding to it in Table 1 is necessarily present. Suppose, for example, that all $0k0$ reflections with k odd are absent. It would be a mistake to infer from that one fact that there is necessarily a 2_1 screw axis parallel to $\mathbf{b}$; one might only be observing a special case of $hk0$ reflections which are all absent when k is odd (as in space group $P2ab$, which does not have a 2_1 axis).

METHOD

The Buerger precession camera. A sheet of photographic film is of course only two-dimensional, and for ease in identifying the diffraction spots due to Bragg reflections it is convenient to record on it the information for only a two-dimensional section of the r.l. This can be accomplished with the precession camera,[1,2] which produces an undistorted photograph of one selected section of the intensity-weighted r.l. This camera is shown schematically in Fig. 6.

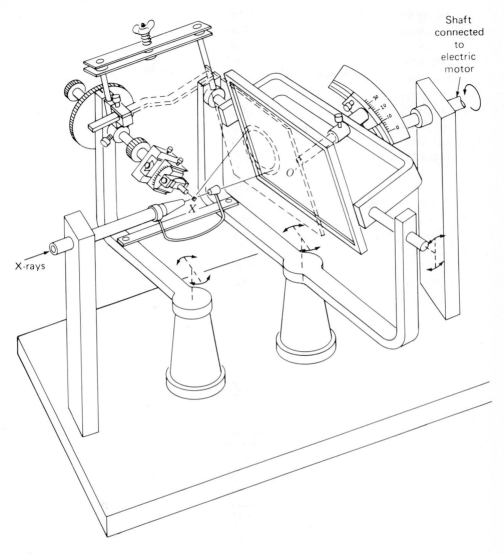

FIGURE 6
Schematic diagram of Buerger Precession Camera.

The crystal, its center in a fixed position X in the x-ray beam, is made to undergo a precessional motion such that the normal to the r.l. section being photographed rotates around the incident x-ray beam at a constant angle μ. *The photographic film is made to undergo exactly the same precessional motion* so that its plane is always parallel to the plane of the r.l. section being photographed. A "layer screen" consisting of a metal plate with an annular opening (shown with dashed lines in the figure) prevents reflections from reaching the film that do not come from the desired r.l. section. In the simplest form of operation, the r.l. section chosen is a so-called "equatorial" section, which includes the r.l. origin O, and the precessional motion of the film is such that the center point of the film (corresponding to the origin O of the r.l.) coincides with a point O' that remains stationary during the precession. For "nonequatorial" sections, which do not include the origin of the r.l., the film is displaced toward the crystal from this stationary point by an amount and in a direction corresponding to the displacement of the desired r.l. section from the origin.

The principles of operation are illustrated in Fig. 7. In direct space the

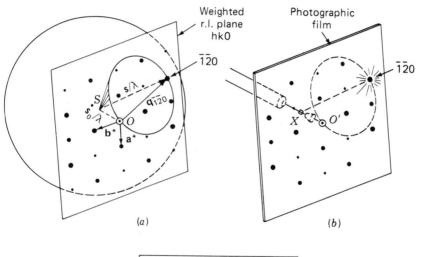

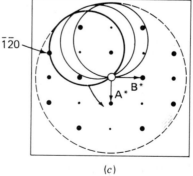

FIGURE 7

Operating principles of the Buerger precession camera. (a) Ewald sphere, with the $hk0$ section plane of the reciprocal lattice intersecting it at a circle; r.l.p. $\bar{1}\bar{2}0$ satisfies the conditions for diffraction. (b) Parallel situation in precession camera; the crystal occupies a position analogous to the center of the Ewald sphere, and the film occupies a position analogous to the $hk0$ section plane. Spot $\bar{1}\bar{2}0$ is getting an instantaneous exposure. (c) The film, showing the motion of the circle of intersection.

crystal and the film are in the same relation to one another as the center of the Ewald sphere and the r.l. layer being photographed. The possible vectors $\mathbf{s}/\lambda$ are confined to the surface of a cone; the corresponding diffracted x-rays are confined to a similar cone that intersects the film in a circle. As the crystal and the film precess, the circle rotates around the center of the film, sweeping out a circular area. In general, therefore, every "spot" within that area gets exposed twice per revolution. In Fig. 7 the r.l. points and the corresponding exposed spots on the film are represented by dots the sizes of which are proportional to the intensities of the reflection.

Figure 8(a) shows several sections or "layers" of the r.l. intersecting the Ewald spheres; the traces shown correspond to a set of lattice planes in the r.l. The spacing between planes is designated d^*. For illustration, we here assume that axes have been assigned so that different planes of the set correspond to successive values of the Miller index l; in this case

$$d^* = \frac{1}{c} \tag{7}$$

(This relation holds, incidentally, regardless of whether the crystal axes are orthogonal or not.) We denote by ζ the distance from the equatorial r.l. plane (i.e., the one passing through the origin O) to the r.l. plane being photographed. It is equal to 0 or some multiple of d^*; with the above assumed axial assignment

$$\zeta = ld^* = \frac{l}{c} \tag{8}$$

The distance from the center of the Ewald sphere to the r.l. origin O is $1/\lambda$; the corresponding distance from the crystal X to the point O' on the cassette mechanism which remains stationary during the precessional motion is D (usually 6 cm). Therefore the factor by which distances on the film must be multiplied to convert them to distances in reciprocal space is

$$f = \frac{1/\lambda}{D} = \frac{1}{\lambda D} \tag{9}$$

Thus, if on the film (Fig. 7(c)) we measure the reciprocal cell dimensions A^* and B^* in some convenient units (centimeters, say), we calculate the reciprocal cell dimensions by

$$a^* = fA^* = \frac{A^*}{\lambda D} \qquad b^* = fB^* = \frac{B^*}{\lambda D} \tag{10}$$

In the interest of brevity we will not give here the equations for the various instrumental parameters required for equatorial and nonequatorial photographs. These are given fully in Ref. 1; moreover, tables and graphs are usually supplied with the instrument.

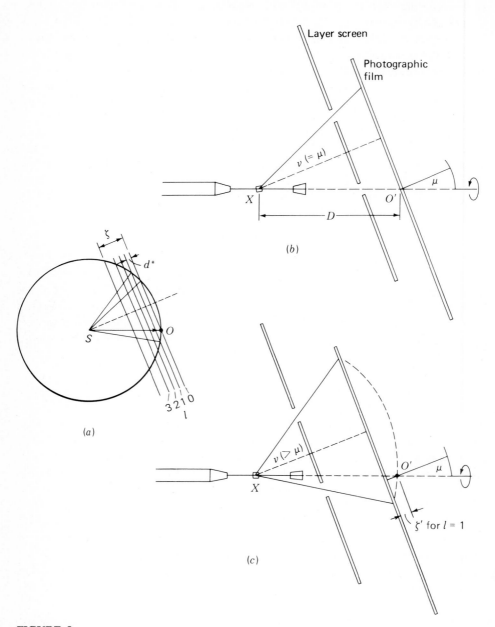

FIGURE 8

Details of geometry. Planes of reciprocal lattice sections, film, and layer screen are perpendicular to the paper; their traces are shown. (*a*) Ewald construction. (*b*) Parallel situation in the precession camera, set up to photograph $l = 0$. (*c*) Precession camera set up to photograph $l = 1$.

Determination of the lattice constants and space group. This process is best described in terms of an illustrative hypothetical example.

A crystal in the form of a stubby needle about 0.4 mm long and 0.16 mm in diameter was mounted with the needle axis parallel to the spindle. After satisfactory alignment was achieved, an equatorial photograph was taken at $\mu = 30°$ with CuKα radiation. This is shown schematically in Fig. 9(a); it

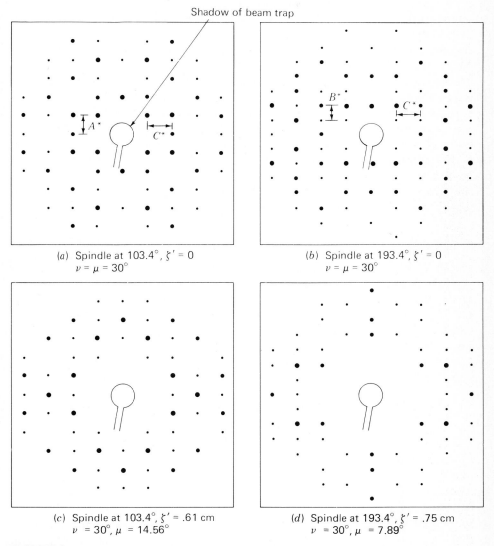

(a) Spindle at 103.4°, $\zeta' = 0$
     $\nu = \mu = 30°$

(b) Spindle at 193.4°, $\zeta' = 0$
     $\nu = \mu = 30°$

(c) Spindle at 103.4°, $\zeta' = .61$ cm
     $\nu = 30°, \mu = 14.56°$

(d) Spindle at 193.4°, $\zeta' = .75$ cm
     $\nu = 30°, \mu = 7.89°$

FIGURE 9
Hypothetical precession photographs (see text). These are idealized (i.e., show no idiosyncracies of spot shape, doubling, white adiation streaks, or tendency to higher intensity near the limiting circles).

showed clearly two "mirror lines" and accordingly has two-dimensional point symmetry *mm*. This did not yet prove that the crystal is orthorhombic since certain r.l. equatorial sections in monoclinic, tetragonal, hexagonal, and cubic crystals have the same apparent symmetry. The spindle was rotated 90° and another picture taken (which almost certainly would *not* show the two mirror lines if the crystal were monoclinic, but might easily show *more* symmetry if it were of higher symmetry than orthorhombic). The resulting photograph, also showing *mm* symmetry, is shown in Fig. 9(*b*). Additional photographs at other nonequivalent spindle settings showed only a single vertical mirror line; orthorhombic symmetry of the crystal (Laue symmetry $D_{2h} - mmm$) is therefore established. Following the convention of Eq. (6), the direction in which the minimum spacing between spots in a mirror-line direction was greatest in the two films was designated the **c*** axis, which happened to be parallel with the spindle axis. The next greatest minimum separation was the vertical direction on the first film; that direction was designated the **a*** axis. The shortest was **b***, in the vertical direction in the second film. The corresponding separations A^*, B^*, C^* were measured as accurately as possible with a millimeter scale. Distances between the ends of several parallel rows of spots were measured and divided by the number of intervening spaces to give the corresponding A^* (or B^* or C^*) value; the values for the different rows were then averaged. The values obtained were

$$A^* = 0.747(2) \text{ cm} \qquad B^* = 0.611(2) \text{ cm} \qquad C^* = 0.974(3) \text{ cm}$$

where, as is conventional in crystallographic literature, the digit in parentheses represents the estimated standard deviation (e.s.d.) in the final digit of the given figure. On multiplying by the scale factor $f = 1/\lambda D = 1/[1.5418 \times 6.00(2)] = 0.1081(4) \text{ Å}^{-1} \text{cm}^{-1}$, where 1.5418 Å is the mean wavelength of CuKα radiation (see p. 545), the following tentative reciprocal lattice constants were obtained:

$$a^* = 0.0808(4) \text{ Å}^{-1} \qquad b^* = 0.0660(3) \text{ Å}^{-1} \qquad c^* = 0.1053(4) \text{ Å}^{-1}$$

These yielded the following tentative lattice constants in direct space:

$$a = 12.38(6) \text{ Å} \qquad b = 15.14(7) \text{ Å} \qquad c = 9.50(4) \text{ Å}$$

To ascertain whether glide extinctions are doubling the reciprocal cell in one or more directions, two more pictures were taken of levels tentatively identified as $k = 1$ and $h = 1$. (See Figs. 9(*c*) and 9(*d*).) It will be noted that there are "holes" at the centers of these two patterns, in which no diffraction spots appear. This feature of nonequatorial photographs results from the fact that the inner r.l.p.'s always remain within the rotating diffraction circle. The separations between spots, A^*, B^*, C^*, were found to be the same as on the equatorial photographs, and the *mm* symmetry was again apparent in both photographs. Doubling of the reciprocal cell in any direction was therefore ruled out.†

It was observed on the layer $h0l$ (Fig. 9(a)) that all $h00$ reflections with h odd were missing, as were all $00l$ reflections with l odd; likewise on the layer $0kl$ (Fig. 9(b)) the odd $0k0$ and $00l$ reflections were uniformly missing. As no other systematic absences were apparent, including absences due to centering, the space group is established as

$$D_2^4 - P2_12_12_1$$

which is isomorphic with the point group $D_2 - 222$. This is the second most common space group encountered for organic crystals; the most common is the monoclinic space group $C_{2h}^5 - P2_1/c$ (with a 2_1 screw axis parallel to the b axis, perpendicular to a c glide plane).

It is to be noted that in the case of space group $P2_12_12_1$ the reciprocal lattice systematic extinctions uniquely determine the space group; no space group giving those extinctions exists in crystal classes $C_{2v}^{i} - mm2$ and $D_{2h} - mmm$. Therefore the absence of a center of inversion is established. Another example of a space group that can be uniquely determined by extinctions is $D_{2h}^{15} - Pbca$ (or the equivalent, $Pcab$); here the three glide planes, and therefore also a center of inversion, are established by extinctions. An important and frequently encountered space group, however, is $D_{2h}^{16} - Pnma$, which by extinctions alone cannot be distinguished from $C_{2v}^9 - Pn2_1a$. If in this case the crystal shows polar external morphology, or a piezoelectric or pyroelectric effect, or second harmonic generation in a laser beam, the lower symmetry $C_{2v}^9 - Pn2_1a$ is established.

EXPERIMENTAL

The object of the present experiment is to determine by means of precession photographs the lattice constants a, b, c and the space group (or set of possible alternative space groups) for an appropriate crystalline substance. Substances with orthorhombic symmetry suitable for this experiment include $KMnO_4$, Rochelle salt (sodium potassium D-tartrate tetrahydrate), nickel dimethylglyoxime, lithium acetate dihyrate, the amino acids DL-alanine and L-threonine, and the organic compounds cyanamide, acetamide, and acetanilide. Many others may be found in Ref. 9.

---

† It is possible in certain cases that even the above precautions will fail to save the experimenter from being deceived by glide extinctions. For example, if the crystal has an F lattice, the body-center r.l.p.'s (h, k, l, all odd) will not be found by a procedure limited to taking the four photographs described above. Such body-center points (if they exist) will be revealed by an hhl equatorial photograph, or by a nonequatorial photograph at $k_{tent} = \frac{1}{2}$ or at $h_{tent} = \frac{1}{2}$. If a photograph of either kind shows the body-centered points, then the tentative a^*, b^*, c^* must be halved, and the tentative a, b, c must be doubled.

Specimen preparation. If the substance chosen is soluble in water or some convenient organic solvent, diffraction specimens are usually best prepared by crystallization from the appropriate solvent. Inorganic salts and highly polar organic substances (acids, amides) are usually soluble enough in water; most others dissolve in alcohol, toluene, or heptane, or mixtures of these. For best results the solubility should not be too high (2 to 5 g per 100 mL of solvent at 25°C is convenient). Careful attention should be given to the possibility that the crystallized product may have a degree of solvation (e.g., hydration) different from that desired. A particular hydrate, for example, will differ from the anhydrous material and from other hydrates in crystal structure, unit cell, and probably symmetry. It will also differ in density, and so the measured density should be checked against the literature value for the desired material.

Usually a rather simple procedure will work for growing crystals. A saturated solution is prepared by shaking an excess of the solid material with the solvent at a temperature somewhat higher than room temperature (typically 40°C to 60°C). The solution is filtered or decanted into a beaker and warmed about 10°C higher than the previous temperature to ensure that all particles are dissolved. The beaker is then covered and set aside overnight or for a few days. Crystals may be withdrawn with tweezers or an eye dropper and placed on filter paper to dry.

The crystal specimens should be examined under a low-power binocular microscope for morphological clues to symmetry. It may be helpful to make sketches of typical crystals in the notebook. If the crystals are transparent, further clues to the symmetry may be obtained by examining the crystals under a polarizing microscope, between crossed polarizers. An orthorhombic crystal, transmitting through parallel prismatic faces, should show optical extinction when the apparent symmetry axes are parallel to either of the two polarization directions.

The density of organic crystals and the lighter inorganic ones can often be measured by the flotation method, using mixtures of two miscible liquids having densities that bracket that of the specimen. For polar organic crystals containing no elements heavier than oxygen, light hydrocarbons such as kerosene ($\rho = 0.79$ g cm^{-3} at 25°C) and methylene iodide ($\rho = 3.32$ g cm^{-3} at 25°C) are usually satisfactory. An initial mixture of the two liquids is made in proportions corresponding roughly to an initial estimate of the crystal density, and one or a few crystals dropped in. If the crystals tend to sink, some methylene iodide is added and the mixture stirred until again homogeneous; if the crystals tend to float, some kerosene is added. When the composition has been adjusted so that the crystals neither sink nor float perceptibly, the density of the mixture is measured with a pycnometer (see Exp. 9). The method obviously fails if the crystals dissolve in or react with the solvent, or are too dense for any convenient flotation medium. If no flotation procedure is satisfactory, the density can be determined by the procedure described in the Sample Report (Chapter I).

Taking the precession photographs. Manifestly, the necessary photographs can only be obtained in a laboratory possessing an x-ray generator, a Buerger precession camera, and personnel trained in the use of this equipment. Therefore, the myriad details of mounting the crystal, centering and aligning it on the precession camera, making the parameter settings, and taking and developing the photographs will not be given here. In the absence of the necessary equipment and qualified personnel, the exercise of determining the unit cell and space group may be performed with a set of photographs obtainable from some x-ray diffraction laboratory.

CALCULATIONS

The lattice constants and the space group (or set of alternative space groups) should be determined in the manner already described. The number Z of formula units (or molecules) of the compound in the unit cell should then be calculated from

$$Z = \frac{V_c}{V_f} \tag{11}$$

The volume V_f occupied in the crystal by one formula unit, in units of $\mathring{A}^3$, is given by

$$V_f = \frac{F}{0.6023\rho} \tag{12}$$

where F is the formula weight in g mol^{-1}, ρ is the crystal density in g cm^{-3}, and the orthorhombic unit cell volume, also in units of $\mathring{A}^3$, is given by

$$V_c = a \cdot b \cdot c \tag{13}$$

The number Z must be an integer within experimental error. If it is not, an error must be suspected, in the x-ray measurements, in the measured density (air bubbles, etc.), or in the calculations; or perhaps the composition of the crystal is different from that assumed owing to solvation or other factors.

DISCUSSION

The student should draw as many conclusions as possible from the space group and the number of formula weights or molecules, consulting *I.T.*[8] and publications on the structure in the literature for help when necessary.

REFERENCES

1. M. J. Buerger, "The Precession Method in X-ray Crystallography," Wiley, New York (1964).
2. M. J. Buerger, "Contemporary Crystallography," chap. 9, McGraw-Hill, New York (1970).
3. *Ibid.*, chap. 4.

4. *Ibid.*, chaps. 1 and 2.
5. W. A. Wooster and A. Breton, "Experiental Crystal Physics," 2d ed., pp. 82–89, Clarendon Press, Oxford (1970).
6. S. C. Abrahams, *J. Appl. Cryst.* **5**, 143 (1972); S. K. Kurtz and T. T. Perry, *J. Appl. Phys.* **39**, 3798 (1968).
7. M. J. Buerger, "Contemporary Crystallography," pp. 38–44, McGraw-Hill, New York (1970).
8. T. Hahn, (ed.), "International Tables for Crystallography," Volume A, "Space Group Symmetry," published for the International Union of Crystallography by D. Reidel, Dordrecht, Holland (1983).
9. J. D. H. Donnay and G. Donnay (eds.), "Crystal Data" (ACA Monograph No. 5), American Crystallographic Association, Washington, D.C. (1963).

GENERAL READING

M. J. Buerger, "Contemporary Crystallography," McGraw-Hill, New York (1970).
J. P. Glusker and K. N. Trueblood, "Crystal Structure Analysis: a Primer," Oxford University Press, New York (1972).
D. McKie and C. McKie, "Essentials of Crystallography," Blackwell, Oxford (1986).
G. H. Stout and L. H. Jensen, "X-Ray Structure Determination," Macmillan, New York (1968).

EXPERIMENT 47
LATTICE ENERGY OF SOLID ARGON

Accurate measurements of the sublimation pressures of solid argon are to be made as a function of temperature in the range 65 K to 78 K. Applying both a second- and third-law thermodynamic treatment to these equilibrium measurements, one obtains the heat of sublimation of argon. From this heat of sublimation and the Debye theory of lattice vibrations, the lattice energy of solid argon can be determined. Representing the Ar–Ar interaction by a Lennard–Jones 6,12 potential, the lattice energy will be calculated using potential parameters obtained from the room-temperature transport properties of argon gas. This "theoretical" value is then compared with the "experimental" thermodynamic value.

THEORY

The change in state involved here is simply

$$\text{Ar}(s) = \text{Ar}(g) \qquad (\text{const } p, T) \tag{1}$$

The molar enthalpy of sublimation $\Delta \bar{H}$ is rigorously related to the slope dp/dT of the sublimation pressure curve by the Clapeyron equation, which has been discussed in Exp. 13. The exact expression is

$$\Delta \bar{H} = T \frac{dp}{dT} (\bar{V}_g - \bar{V}_s) \tag{2}$$

If $\bar{V}_s$ is assumed to be negligible in comparison to $\bar{V}_g$ and if the vapor is assumed to be an ideal gas, an approximate form of the Clapeyron equation is obtained

$$\Delta \bar{H} \simeq -R \frac{d \ln p}{d(1/T)} \tag{3}$$

which will yield an approximate value of $\Delta \bar{H}$ directly from the slope of a plot of $\ln p$ versus $1/T$. This approximate Clapeyron equation is very nice for examination questions or the quick analysis of crude data, but it should never be used to represent accurate experimental measurements. Fortunately, the equation of state of $Ar(g)$ has been determined experimentally down to the triple point of argon. Table 1 gives values of B, the second virial coefficient, and dB/dT as a function of temperature. Also, the molar volume of the solid can be taken to have the constant value $25 \, cm^3 \, mol^{-1}$ over the investigated temperature range. Thus, one can use the exact Clapeyron equation to determine second-law values of $\Delta \bar{H}$.

On the basis of the third law, the thermodynamic properties of a pure substance [such as S^0, $(H^0 - H_0^0)/T$ and $(G^0 - H_0^0)/T$, where $S^0 = \int_0^T C_p^0 d \ln T$, $H^0 - H_0^0 = \int_0^T C_p^0 \, dT$ and $G^0 = H^0 - TS^0$] can be tabulated as functions of temperature. Table 2 gives these functions for $Ar(s, \text{cubic})$, $Ar(l)$ and $Ar(g)$. In the preparation of this table, experimental values of $\bar{C}_p^0$ for the solid and liquid phases and the experimental enthalpy of fusion were used. For the gas phase, $\bar{C}_p^0 = 5R/2$ (except at extremely low T) and $\bar{S}^0$ were calculated from the Sackur–Tetrode equation. We can now proceed to evaluate third-law values of the enthalpy of sublimation. For the change in state (1), $\Delta \bar{G}^0 = -RT \ln K$, where the equilibrium constant $K = f_g/a_s$. At low pressures, only the second virial coefficient need be considered and a satisfactory approximation for $\Delta \bar{G}^0$ is

$$\Delta \bar{G}^0 = -RT \ln p - Bp - \bar{V}_s(1-p) \tag{4}$$

Indeed, the last term in the above expression is required only for data of the highest accuracy. From experimental $\Delta \bar{G}^0$ values given by Eq. (4) and

TABLE 1
Second virial coefficient of argon[a] (B in units of $cm^3 \, mol^{-1}$)

T, K	B	dB/dT	T, K	B	dB/dT
66	−497	23.9	76	−335	11.2
68	−454	19.8	78	−314	9.9
70	−418	17.0	80	−295	8.8
72	−386	14.7	83.81	−265	7.1
74	−359	12.8	87.30	−242	5.9

[a] Extrapolated from measurements above 84 K; see M. A. Byrne, M. R. Jones and L. A. K. Staveley, *Trans. Faraday Soc.* **64**, 1747 (1968).

TABLE 2
Thermodynamic properties of argon[a]
(All quantities are in units of $J\,K^{-1}\,mol^{-1}$)

T, K	$\tilde{S}^0$	Ar(s, l)		$\tilde{S}^0$	Ar(g)	
		$\dfrac{\tilde{H}^0 - \tilde{H}_0^0}{T}$	$\dfrac{\tilde{G}^0 - \tilde{H}_0^0}{T}$		$\dfrac{\tilde{H}^0 - \tilde{H}_0^0}{T}$	$\dfrac{\tilde{G}^0 - \tilde{H}_0^0}{T}$
0	0	0	0	0	0	0
10	1.092	0.828	−0.264	84.275	20.786	−63.489
20	6.259	4.418	−1.841	98.684	20.786	−77.898
30	12.611	8.234	−4.377	107.111	20.786	−86.325
40	18.556	11.339	−7.217	113.094	20.786	−92.308
50	23.870	13.845	−10.025	117.730	20.786	−96.944
60	28.652	15.912	−12.740	121.521	20.786	−100.735
70	33.024	17.698	−15.326	124.726	20.786	−103.940
72	33.870	18.037	−15.833	125.312	20.786	−104.526
74	34.710	18.384	−16.326	125.880	20.786	−105.094
76	35.547	18.723	−16.824	126.433	20.786	−105.647
78	36.376	19.066	−17.310	126.972	20.786	−106.186
80	37.208	19.405	−17.803	127.500	20.786	−106.714
83.81(s)	38.798	20.075	−18.723	128.466	20.786	−107.680
83.81(l)	53.003	34.280	−18.723			
87.30	54.815	34.681	−20.134	129.316	20.786	−108.530

[a] The entries for Ar(s) and Ar(l) were calculated from data in Refs. 1 and 2; the entries for Ar(g) are statistical thermodynamic values for a monatomic ideal gas.

interpolated values of $(\tilde{G}^0 - \tilde{H}_0^0)/T$ from Table 2, we may now determine $\Delta \tilde{H}_0^0$, the enthalpy of sublimation at absolute zero:

$$\frac{\Delta \tilde{G}^0}{T} = \left(\frac{\tilde{G}^0 - \tilde{H}_0^0}{T}\right)_g - \left(\frac{\tilde{G}^0 - \tilde{H}_0^0}{T}\right)_s + \frac{\Delta \tilde{H}_0^0}{T} \tag{5}$$

A value of $\Delta \tilde{H}_0^0$ may be calculated from each experimental value of the sublimation pressure, and these $\Delta \tilde{H}_0^0$ values must be constant within the limits of experimental error. The value of $\Delta \tilde{H}^0$ at any temperature can be calculated from $\Delta \tilde{H}_0^0$ and the entries in Table 2. It is important to recognize that $\Delta \tilde{H}^0$ refers to the change in state

$$Ar(s) = Ar(\text{perfect gas}) \qquad (1\ bar,\ \text{const}\ T) \tag{6}$$

whereas $\Delta \tilde{H}$ refers to change in state (1) involving the real gas at pressure p. If the enthalpy of the solid phase is assumed to be independent of pressure for low pressures, then third-law values of $\Delta \tilde{H}$ can be calculated from

$$\Delta \tilde{H} = \Delta \tilde{H}^0 + \left(B - T\frac{dB}{dT}\right)p \tag{7}$$

If all the experimental data were perfect, the second-law value of $\Delta \tilde{H}$

calculated from Eq. (2) and the third-law value calculated from Eqs. (4), (5), and (7) would be identical. However, in this case, the third-law result should be the more reliable since it is difficult to determine dp/dT with very high accuracy.

An independent value for $\Delta\bar{H}$ has been reported by two different investigators from direct calorimetric measurements of the enthalpy of vaporization.[1-3] The calorimetric measurements, which are in good agreement with each other, give a $\Delta\bar{H}_0^0$ value which is about 63 J higher than the third-law value. This discrepancy, although small, is disturbing because it is well outside the limits of error claimed for the various experimental results. These studies of the thermodynamic properties of argon were made prior to the discovery in 1964 of a metastable hexagonal phase of argon, and the earlier investigators assumed without proof that their solid phase was the face-centered cubic phase. Until more is known concerning the hexagonal phase, it is impossible to say how the presence of this phase might have affected the earlier measurements.†

Lattice energy. If we measure the energy content of argon on an arbitrary scale which defines $\bar{E}^0(g)$ to be zero at 0 K, then the total energy of the crystal lattice at 0 K is $-\Delta\bar{E}_0^0$, which for all practical purposes is the same as $-\Delta\bar{H}_0^0$. There are two additive contributions to this absolute-zero energy content of solid argon: the *lattice energy* Φ_0, which is the potential energy of the argon atoms *at rest* in the lattice relative to a zero of energy for these atoms in the gas at infinite separation, and the zero-point vibrational energy, which arises from the quantum-mechanical vibrational motion of the atoms about their equilibrium positions.[4] In terms of the Debye characteristic temperature Θ_D, the zero-point vibrational energy equals $\frac{9}{8}R\Theta_D$. Thus, the lattice energy is given by

$$\Phi_0 = -\Delta\bar{H}_0^0 - \tfrac{9}{8}R\Theta_D \tag{8}$$

where $\Theta_D = 93$ K for argon.[2]

It is possible to calculate a theoretical value of the lattice energy for a molecular crystal if data are available on the potential energy between atoms as a function of their separation. A commonly used form for the interatomic potential (see Fig. 1) is due to Lennard–Jones:[4,5]

$$u(r) = 4\varepsilon\left[\left(\frac{\sigma}{r}\right)^{12} - \left(\frac{\sigma}{r}\right)^{6}\right] \tag{9}$$

where $u(r)$ is the potential energy for *two* atoms at a distance r. The r^{-12} term is an empirical function to describe the repulsion at short distances, and the r^{-6}

---

† Face-centered cubic argon is the stable solid phase of pure argon, but upon the addition of as little as 1 percent of N_2 to Ar, the solid phase crystallizes with the close-packed hexagonal structure. A study of the sublimation pressures of solid solutions of Ar with small amounts of N_2 would give valuable information concerning the hexagonal phase, and such a study could be made with the apparatus used in this experiment.

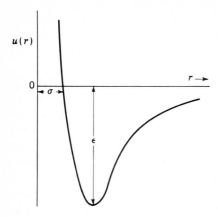

FIGURE 1

Lennard-Jones potential $u(r)$ as a function of inter-atomic distance r. The characteristic parameters ε and σ determine this potential curve; see Eq. (9).

term represents the r dependence of the potential energy found by London to describe the attraction at large distances due to induced dipole–dipole interaction. The Lennard–Jones potential for a given atom is characterized by the two constants ε and σ (which are shown in Fig. 1). These parameters can be evaluated from an analysis of gas data (second virial coefficient, Joule–Thomson effect, or gas viscosity); the best values for argon[6] are

$$\frac{\varepsilon}{k} = 119.5 \text{ K} \qquad \sigma = 3.405 \text{ Å} \tag{10}$$

For the solid it is assumed that the total potential energy (i.e., lattice energy) is the sum of all pair potentials $u_{ij}(r_{ij})$. The result of this summation for a face-centered cubic lattice (such as argon) is[7]

$$\Phi_0 = 2N_0\varepsilon\left[12.132\left(\frac{\sigma}{d}\right)^{12} - 14.454\left(\frac{\sigma}{d}\right)^{6}\right] \tag{11}$$

where d is the distance between nearest neighbors. Note that almost all the repulsion part of the potential comes from nearest-neighbor interactions (nearest neighbors alone would give 12 for both coefficients). Since r^{-6} falls off much more slowly than r^{-12}, the coefficient of the second term in Eq. (11) is considerably greater than 12. A knowledge of the lattice spacing for solid argon and the parameters of Eq. (10) will permit a calculation of the lattice energy for comparison with the experimental value obtained from Eq. (8).

EXPERIMENTAL

Temperatures below 77 K can be achieved with a liquid-nitrogen bath by reducing the pressure over the nitrogen by pumping, thus lowering its boiling point. It is possible to lower the bath temperature to 63 K before liquid nitrogen begins to solidify. Measurement of the temperature can be made by using a resistance thermometer or a thermocouple (see Chapter XVI) or by measuring the vapor pressure of nitrogen.

The low-temperature cell, shown in Fig. 2, consists of a copper block containing a small chamber A for solid argon and a large chamber B that can be used as a N_2 vapor-pressure bulb. If a thermocouple or a platinum resistance thermometer (PRT) is used, it should be mounted in the copper cell near the argon chamber. The cupronickel (or stainless steel) connecting tubes are soldered through a cap with a side-arm. This cap fits over the top of a tall glass Dewar flask and is attached to the Dewar with a rubber sleeve.† The side-arm is connected by heavy-wall vacuum hose to a high-capacity mechanical vacuum pump.

If the vapor pressure of liquid nitrogen is used for the temperature measurement, one allows N_2 gas to condense in chamber B and the pressures can be read directly on a closed-tube manometer. If a copper–constantan thermocouple or a platinum resistance thermometer is used, it must be well calibrated, since accurate absolute temperatures are needed. If chamber B is not used, all further instructions concerning it may be disregarded.

Procedure. The apparatus should be assembled by connecting the necessary manometers, filling lines, and pumping lines; see Fig. 3. Evacuate each chamber by closing clamp or stopcock C and stopcocks $A2$ and $B2$ and opening stopcocks $A1$ and $B1$. Then close stopcocks $A1$ and $B1$ and test for leaks in either chamber by waiting approximately 15 min to see if the manometers indicate any increase in pressure. The rate of increase in pressure should not exceed 5 Torr/h.

Filling chambers A and B with gas should be done with care to avoid breaking the manometers. If a simple reducing valve is used, the high-pressure valve on the storage cylinder should never be opened when gas is being admitted to the system. Allow the gas contained in the section between the two valves to expand into the chamber; then close the needle valve and refill this section. Repeat this procedure several times (if necessary) until the chamber is full. Flushing the chamber once or twice with the appropriate gas will reduce the possibility of contamination.

Fill chamber A and its manometer with argon at 760 Torr and close stopcock $A2$. **Slowly** raise the Dewar filled with liquid nitrogen until the copper cell is almost completely immersed (do *not* immerse the connecting tubes above the cell). The argon pressure should drop to about 200 Torr. **Slowly** allow N_2 gas to enter chamber B until the pressure is constant (about 760 Torr, and then close stopcock $B2$. Now raise the Dewar all the way up, and attach the cap to the top of the Dewar. The copper cell should be sufficiently far below the liquid-nitrogen level to ensure that the cell will remain submerged throughout the run. (If necessary, additional liquid nitrogen may be added through the filling port in the cap at this time.)

---

† The top of a rubber surgical glove makes an excellent sleeve.

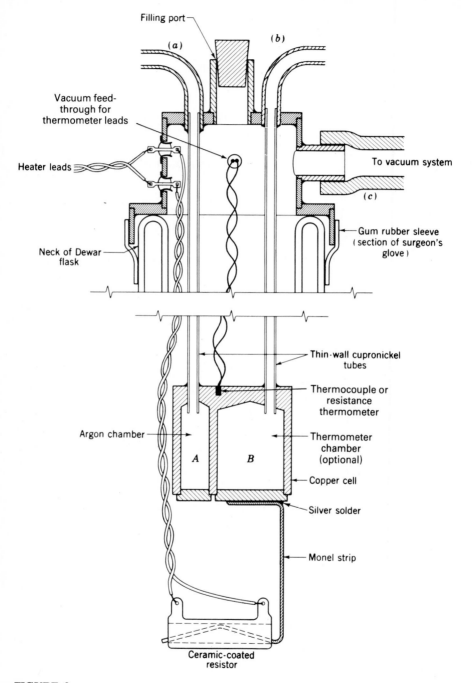

FIGURE 2

Low-temperature vapor-pressure apparatus. (Metal parts to be well tinned and soft-soldered, except where otherwise indicated.)

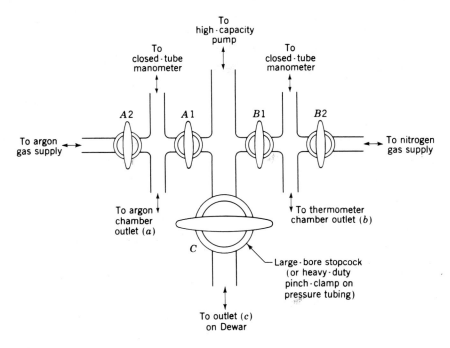

FIGURE 3
Vacuum and gas connections for the apparatus shown in Fig. 2.

More argon may now be added to chamber A, although this is not necessary if the measurements are to be made on warming. Experience has shown that the argon pressures will not be reproducible unless the argon is pumped very briefly prior to any measurement made after the cell has been cooled. To avoid the possibility of pumping off too much argon, it is convenient to make the measurements when the cell is warming rather than cooling. The careful student will want to confirm the fact that reproducible results can be obtained on both warming and cooling.

Close the liquid-nitrogen filling port with a tightly fitting rubber stopper and begin pumping on the refrigerant by opening clamp or stopcock C. After the cell has cooled to approximately 65 K, the Ar pressure should be measured at some arbitrary constant temperature. Then pump chamber A briefly and measure the Ar pressure again at the same constant temperature. If the two pressures differ, repeat the procedure. Usually the pressure will be reproducible after the second pumping. No further pumping should be necessary at higher temperatures, although it is advisable to check for reproducibility. At each of a series of constant temperatures, record the vapor pressure of solid argon and measure the temperature. If the vapor pressure of liquid nitrogen in chamber B is being used as the thermometer, the constancy of this pressure is a good indication that the temperature is stable. As an aid to rapidly

TABLE 3
Vapor pressure of liquid nitrogen[a]
(Equilibrium temperatures are given corresponding to the specified vapor pressure)

p, Torr	T, K	p, Torr	T, K	p, Torr	T, K
100	63.500	240	68.757	500	73.975
120	64.521	280	69.785	560	74.860
140	65.412	320	70.703	620	75.675
160	66.207	360	71.535	680	76.431
180	66.926	400	72.298	740	77.138
200	67.584	440	73.004	800	77.803

[a] Based on data in Ref. 8.

converting pressure readings into *approximate* bath temperatures, some nitrogen vapor pressures are listed in Table 3. The Ar and N_2 pressure readings should be estimated to within ±0.2 Torr (i.e., read mercury levels to within 0.1 mm). Proper light conditions are very important for an accurate measurement of the position of the meniscus in a manometer.

If only external heat leaks served to warm up the system, the rate of warmup would be inconveniently slow. Intermittent heating with an electric heater element immersed directly in the liquid nitrogen will provide a suitable rate of heating between readings. Emphasis should be placed on obtaining good equilibrium data rather than a large number of data points. Four or five good experimental points are sufficient to establish the sublimation curve.

Warning. A closed system containing a liquid with a normal boiling point well below room temperature can cause the destruction of a vacuum system and injuries to personnel. *Make certain that the apparatus is properly pumped out or vented before it is allowed to warm up.* As you should at all times in the laboratory, be sure to wear safety glasses.

CALCULATIONS

Convert all manometer readings to pressures in Torr. The correction for the density of mercury as a function of temperature must not be ignored (see p. 697). Compute the temperatures and tabulate them along with the corresponding argon sublimation pressures. If N_2 vapor pressures were used for thermometry, one can obtain T from the Antoine-type empirical equation given by Armstrong:[8]

$$\log p_{N_2} \text{(Torr)} = 6.49594 - \frac{255.821}{(T - 6.600)} \tag{12a}$$

or

$$T(\text{K}) = 6.600 + \frac{255.821}{6.49594 - \log p \text{ (Torr)}} \tag{12b}$$

If a thermocouple or *PRT* was used, convert their readings to temperatures with an appropriate calibration equation. It may be wise to check the calibration by measuring the bath temperature at atmospheric pressure. Calculate this bath temperature from Eq. (12) using the pressure determined with a barometer. If necessary, one can then apply a constant additive correction to all *PRT* resistance or thermocouple emf readings.

Represent the argon sublimation pressures by an equation of the form

$$\ln p \text{ (Torr)} = -\frac{A}{T} + b \tag{13}$$

Only two constants are used because the temperaure range of the measurements is quite limited. After the constants A and b are determined, calculate the deviation $p_{obs} - p_{calc}$ for each of the experimental points and add this to your table of p and T values. These deviations will serve as an estimate of the precision of the measurements.

Employing the second-law approach, calculate $\Delta \tilde{H}$ at 66 K, 70 K and 76 K from your experimental data, using both the exact and approximate Clapeyron equations. Estimate the experimental error in $\Delta \tilde{H}$ and comment on any differences between the values obtained with Eqs. (2) and (3).

Utilizing the third-law approach, calculate $\Delta \tilde{H}_0^0$ from each of your experimental sublimation pressures. Then calculate $\Delta \tilde{H}^0$ at 66 K, 70 K and 76 K, using what you consider to be the best value of $\Delta \tilde{H}_0^0$ (justify this choice). From Eq. (7), obtain $\Delta \tilde{H}$ values at the corresponding temperatures and compare these values with those obtained through the second-law treatment.

The report should show all numerical values of the various terms involved in the thermodynamic equations used and all appropriate graphs. Also compare your $\Delta \tilde{H}$ values with those reported in the literature.[2,9]

Finally, use Eq. (8) to determine the "experimental" value of the lattice energy of argon at 0 K. X-ray diffraction data[10] give 5.30 Å for the cubic unit-cell parameter of solid argon at 4 K. Find the nearest-neighbor distance d, and use Eqs. (10) and (11) to calculate a "theoretical" value of Φ_0. Compare the theoretical and experimental values.

APPARATUS

Tall Dewar flask (preferably with vertical unsilvered strip to permit viewing the cell and the liquid level); support for Dewar; copper cell and cap assembly as shown in figure; rubber stopper to fit filling port; rubber sleeve; eight lengths of heavy-wall rubber tubing; high-capacity mechanical vacuum pump; gas-handling assembly with four stopcocks (see figure); heavy-duty hose clamp; two closed-tube manometers; Variac for heater supply. If a copper–constantan thermocouple is to be used, a Dewar flask for ice-water reference junction and either a potentiometer circuit or a digital voltmeter are needed.

Supplies of argon and dry nitrogen gas at 1 atm; liquid nitrogen.

REFERENCES

1. P. Flubacher, J. Leadbetter, and J. A. Morrison, *Proc. Phys. Soc.* (*London*) **78,** 1449 (1961).
2. R. H. Beaumont, H. Chihara, and J. A. Morrison, *Proc. Phys. Soc.* (*London*) **78,** 1462 (1961).
3. K. Clusius, *Z. Phys. Chem.* **31B,** 459 (1936); A. Frank and K. Clusius, *Z. Phys. Chem.* **42B,** 395 (1939).
4. J. O. Hirschfelder, C. F. Curtiss, and R. B. Bird, "Molecular Theory of Gases and Liquids," 2d ed., pp. 1035–1044, Wiley, New York (1967).
5. J. A. Beattie and W. H. Stockmayer, "The Thermodynamics and Statistical Mechanics of Real Gases," in H. S. Taylor and S. Glasstone (eds.), "A Treatise on Physical Chemistry," Vol. *2,* "States of Matter," 3d ed., chap. 2, pp. 305–306, Van Nostrand, Princeton, N.J. (1951).
6. E. Whalley and W. G. Schneider, *J. Chem. Phys.* **23,** 1644 (1955).
7. J. A. Beattie and W. H. Stockmayer, *op. cit.,* p. 309.
8. G. T. Armstrong, *J. Res. Natl. Bur. Stand.* **53,** 263 (1954).
9. A. M. Clark, F. Dim, J. Robb, A. Michels, T. Wassenaar, and Th. Zwietering, *Physica* **17,** 876 (1951).
10. O. G. Peterson, D. N. Batchelder, and R. O. Simmons, *Phys. Rev.* **150,** 703 (1966).

EXPERIMENT 48
STATISTICAL THERMODYNAMICS OF
IODINE SUBLIMATION

This experiment is in some respects similar to two other experiments concerning enthalpy changes attending phase transformations, namely Exps. 13 and 47. However, it differs from them in that the experimental data, which are vapor pressures of solid iodine at several temperatures, are obtained from optical absorption measurements. As in the other experiments mentioned, the enthalpy change (here the heat of sublimation of solid iodine) can be calculated with the Clausius–Clapeyron equation, which requires the values of vapor pressures at two or more temperatures.

The system $I_2(s)$–$I_2(g)$ also provides an opportunity for the application of statistical mechanics to derive thermodynamic information from spectroscopic data. For the gas phase, the vibrational frequency of the I_2 molecule, needed in formulating the vibrational partition function, can be obtained from the absorption spectrum in the visible region (see Exp. 42); the rotational partition function in the gas phase will be calculated from the known internuclear distance in the iodine molecule. For the crystalline phase, published phonon dispersion curves, obtained by inelastic neutron scattering spectroscopy, will be used to determine the vibrational frequencies. With the above information and statistical mechanical theory, the molar energy difference $\Delta \bar{E}_0^0$ between the vibrational ground states of crystalline and gaseous iodine can be determined from a measurement of vapor pressure at *one* temperature. From the fully defined partition functions for both crystalline and gaseous iodine, the entropy and enthalpy changes attending sublimation of iodine can be calculated at any temperature T. The value of $\Delta \bar{H}_{sub}$ obtained in this way will then be compared with the value obtained with the Clausius–Clapeyron equation.

THEORY

This section will not be concerned with the Clausius–Clapeyron equation, which is discussed adequately in Exps. 13 and 47. The discussion here will focus on the application of statistical mechanics to the phase equilibrium

$$I_2(s) = I_2(g) \tag{1}$$

It is required for equilibrium that the chemical potential of I_2 be the same in the two phases:

$$\mu_s = \mu_g \tag{2}$$

The basic question is: *How can these chemical potentials be determined?*

Statistical-mechanical background. We will present a brief review of the basic statistical-mechanical concepts needed in this experiment because standard textbooks in physical chemistry vary widely in their approach.

Given the canonical partition function Q for a system,

$$Q = \sum_i e^{-E_i/kT} \tag{3}$$

where E_i is the energy of the ith quantum state of the entire system, the Helmholtz free energy is given by the equation

$$A = -kT \ln Q \tag{4}$$

For a one-component system the chemical potential per mole is given by[1]

$$\mu = \left(\frac{\partial A}{\partial n}\right)_{T,V} = -kT \left(\frac{\partial \ln Q}{\partial n}\right)_{T,V} = -RT \left(\frac{\partial \ln Q}{\partial N}\right)_{T,V} \tag{5}$$

where n is the number of moles in the system and N is the number of molecules in the system ($= N_0 n$, where N_0 is Avogadro's number).

In order to evaluate the canonical partition function Q for a gas we shall consider the system to be composed of an aggregate of essentially independent particles (molecules). As we shall see later, a crystal may be considered to a good approximation as an aggregate of independent harmonic oscillators. Each of these has its own microcanonical partition function:

$$q_i = \sum_j e^{-\varepsilon_j^{(i)}/kT} \tag{6}$$

where $\varepsilon_j^{(i)}$ is the energy of the ith oscillator in the jth quantum state of that oscillator. Since the oscillators are only very weakly interacting, the canonical partition function of the solid is a simple product of the microcanonical partition functions of the individual oscillators:

$$Q_s = \prod_i q_i \qquad \ln Q_s = \sum_i \ln q_i \tag{7}$$

Since many of these oscillators differ from each other in the values of their frequencies, energy levels, and partition functions, it is convenient to define a new quantity q_s which is the geometric mean of all of the q_i for the crystal:

$$q_s \equiv \left[\prod_{i=1}^{M} q_i \right]^{1/M} \qquad \ln q_s \equiv \frac{1}{M} \sum_{i=1}^{M} \ln q_i \tag{8}$$

where M is the number of oscillators. Then, for the crystal,

$$\ln Q_s = M \ln q_s = 3tN \ln q_s \tag{9}$$

where t is the number of atoms in a molecule. Since $\ln q_s$ can be shown to be independent of N, we find from Eq. (5)

$$\mu_s = -3tRT \ln q_s \tag{10}$$

For a one-component ideal gas, the microcanonical partition function for an individual molecule is q_g. Therefore, under all ordinary conditions, we may write for a gas

$$Q_g = \frac{q_g^N}{N!} \tag{11}$$

where the division by $N!$ takes into account the fact that the individual molecules are indistinguishable. With the aid of the Sterling approximation for $\ln(N!)$ we obtain

$$\ln Q_g = N \ln q_g - N \ln N + N \tag{12}$$

Using Eq. (5) again, we obtain

$$\mu_g = -RT \ln \frac{q_g}{N} \tag{13}$$

We will now develop expressions for the microcanonical partition functions q_s and q_g to substitute into Eqs. (10) and (13).

Gaseous I_2. The partition function q_g is very well approximated as a product of terms arising from translational, rotational, vibrational, and electronic degrees of freedom:

$$q_g = q_{trans} q_{rot} q_{vib} q_{el} \tag{14}$$

The translational partition function is given by[2]

$$q_{trans} = \left(\frac{2\pi m k T}{h^2} \right)^{3/2} V \tag{15}$$

where m is the molecular mass, k is the Boltzmann constant, T is the absolute temperature, h is Planck's constant, and V is the volume within which the molecule is constrained to move.

For a molecule as massive as I_2, the rotational energy levels are very closely spaced and the partition function has the simple form[3]

$$q_{rot} = \frac{kT}{\sigma hc\tilde{B}_0} = \frac{T}{\sigma\Theta_{rot}} \tag{16}$$

Here σ is the symmetry number of the molecule, c is the velocity of light, and $\tilde{B}_0$ is the rotational constant (conventionally expressed in units of cm^{-1} with c expressed in $cm\ s^{-1}$ units) defined by

$$\tilde{B}_0 \equiv \frac{h}{8\pi^2 Ic} \tag{17}$$

where I is the moment of inertia of the molecule

$$I = \mu r_0^2 \tag{18}$$

The reduced mass μ (not to be confused with chemical potential) is defined by

$$\mu = \frac{m_1 m_2}{m_1 + m_2} \tag{19}$$

where m_1 and m_2 are the respective atomic masses. In I_2 the interatomic distance r_0 is 0.2667 nm, and the rotational constant $\tilde{B}_0$ is 0.037315 cm^{-1}.[4] The quantity Θ_{rot} is the *rotational characteristic temperature,* given by

$$\Theta_{rot} = \frac{hc\tilde{B}_0}{k} \tag{20}$$

The factor hc/k has the value 1.43877 cm K. Since the I_2 molecule is end-for-end symmetric, $\sigma = 2$.

For the vibrational partition function the molecule is regarded as a quantum-mechanical harmonic oscillator, for which[5]

$$q = (1 - e^{-hv_0/kT})^{-1} = (1 - e^{-\Theta_{vib}/T})^{-1} \tag{21}$$

where v_0 is the molecular vibration frequency and Θ_{vib} is the *vibrational characteristic temperature,*

$$\Theta_{vib} = \frac{hv_0}{k} = \frac{hc\tilde{v}_0}{k} \tag{22}$$

For the I_2 molecule, $\tilde{v}_0$ has the value 213.3 cm^{-1}.[4]

Equation (21) as written applies to an oscillator for which the reference energy is the energy of the vibrational ground state ($v = 0$); i.e., the $v = 0$ state in a gas molecule has been assigned zero energy. For the present situation, in which I_2 molecules in the vapor phase are in equilibrium with crystalline iodine, it is more convenient to take the reference energy to be that of an I_2

molecule in the crystal when the *crystal* is in its ground vibrational state.†
Accordingly, the energy of the vibrational ground state of an I_2 molecule in the
ideal gas phase is taken to be $\Delta\varepsilon_0$, which is the energy required to remove a
molecule from the crystal at absolute zero temperature. Thus, we should write
for the I_2 molecule in the gas phase

$$q_{\text{vib}} = (1 - e^{-\Theta_{\text{vib}}/T})^{-1} e^{-\Delta\varepsilon_0/kT} \tag{23}$$

It remains to deal with q_{el}. The excited electronic states of I_2 are
separated from the ground electronic state by an energy difference that is very
large compared to kT. Therefore

$$q_{\text{el}} = 1 \tag{24}$$

Let us now introduce $\Delta\bar{E}_0^0 = N_0 \Delta\varepsilon_0$, the energy needed to sublime 1 mol
of crystalline I_2 into the ideal gas phase at the absolute zero, and replace V by
its ideal-gas equivalent NkT/p. We can then combine Eqs. (13) to (16), (23),
and (24) to obtain

$$\mu_g = \Delta\bar{E}_0^0 - RT \ln\left[\left(\frac{2\pi mkT}{h^2}\right)^{3/2} \frac{kT}{p} \frac{T}{\sigma\Theta_{\text{rot}}} (1 - e^{-\Theta_{\text{vib}}/T})^{-1}\right] \tag{25}$$

Crystalline I_2. The partition function for the crystalline state of I_2 consists
solely of a vibrational part; the crystal does not undergo any significant
translation or rotation, and the electronic partition function is unity for the
crystal as it is for the gas.

The geometric mean partition function for the crystal can be expressed as

$$q_s = \left[\prod_{i=1}^{M} (1 - e^{-\Theta_i/T})^{-1}\right]^{1/M} \tag{26}$$

where Θ_i is defined in terms of $\bar{\nu}_i$ in the same way as Θ_{vib} is defined in terms of
$\bar{\nu}_0$ in Eq. (22). Since the number of iodine atoms is $2N$ for a crystal containing
N molecules of I_2 and since each atom contributes three degrees of freedom,
the number of modes of vibration for the crystal is

$$M = 3 \times t \times N - 6 = 6N - 6 \cong 6N \tag{27}$$

The subtracted number 6 represents the 3 translational and 3 rotational
degrees of freedom of the crystal as a whole and will henceforth be ignored.

We now present a brief discussion of the vibrations occurring in a
crystal.[6,7] The crystal can be thought of as a gigantic molecule with a huge
number of normal modes, and the student may find it useful to review the
discussion of normal modes for small molecules given in Exps. 36, 37, and 39.
In the case of the I_2 crystal, each primitive (smallest) unit cell contains two

---

† It should be noted that the location of the energy zero is arbitrary; a different, but equally
reasonable, choice is made in Exp. 47. All that matters is a consistent choice for the two phases in
equilibrium—here gaseous and crystalline I_2.

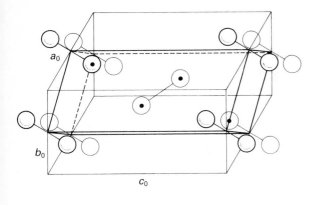

FIGURE 1

The crystal structure of $I_2(s)$.[8] The primitive unit cell, outlined in heavy lines, contains two molecules, identified by dots at the atomic centers (one half molecule each at the upper left and lower right corners, and one molecule in the body center). The light lines outline an orthorhombic non-primitive unit cell of dimensions $a_0 = 0.727$ nm, $b_0 = 0.479$ nm, $c_0 = 0.979$ nm. All molecules are in planes parallel to the **b** and **c** axes. (Not all molecules in the orthorhombic cell are shown.)

molecules.[8] Figure 1 shows that these two molecules are distinguished easily because their spatial orientations are different. As a consequence of this crystal structure, there are 3×4 atoms = 12 mechanical degrees of freedom associated with each unit cell. In the gas phase, there would be three translations, two rotations, and one vibration for each of the two I_2 molecules. In the crystal, however, only vibrations occur: six lattice modes, four librational modes, and two internal vibration (bond-stretching) modes.

Let us consider first the center-of-mass motions for each of the two I_2 molecules in a unit cell. These types of motion account for six degrees of freedom and give rise to two kinds of lattice vibration. When both I_2 molecules *in a given cell* move in phase with each other (say, for example, both are displaced in the $+x$ direction at the same time), there are three so-called *acoustic* vibrations. When the two I_2 molecules *in a given cell* move out of phase (say one is displaced in the $+x$ direction while the other is displaced in the $-x$ direction), there are three *optic* vibrations.†

The four librations (torsional oscillations or rocking motions) arise because the crystal-field potential prevents the I_2 molecule from rotating as it would in the gas phase. There are some special crystals, called plastic crystals, in which symmetrical molecules that interact weakly can still undergo hindered rotation in the solid phase, but $I_2(s)$ is not one of these. The librational motions for each I_2 occur about two axes (α, β) perpendicular to the I—I bond direction. The librations of the two I_2 molecules in the same unit cell are coupled—giving rise to SL_α, AL_α and SL_β, AL_β vibrations, where SL denotes symmetric libration (angle displacements in phase) and AL denotes anti-symmetric libration (angle displacements out of phase).

Finally, there are two I—I bond-stretching vibrations that are essentially the same as the gas-phase stretching mode. As expected, these vibrations are

---

† The name *optic mode* comes from the behavior of ionic crystals such as Na^+Cl^-. When Na^+ and Cl^- in a given cell move out of phase with each other, there is an oscillating electric dipole. Optical absorption will occur for light having frequency equal to that of the optic lattice mode.

coupled to produce a *SS* (symmetric stretch) in-phase vibration and an *AS* (antisymmetric stretch) out-of-phase vibration. In the latter case, one I_2 bond is stretching while the other is being compressed. As a result of interactions in the crystalline phase,[8,9] these *SS* and *AS* vibrations have lower frequencies than the gas-phase vibration at 213.3 cm^{-1}.

Now we must consider the fact that the motions of the I_2 molecules in any given unit cell are coupled to those of the molecules in other unit cells. An entire crystal of $N/2$ unit cells has $12 \times (N/2) = 6N$ degrees of freedom. Thus it would seem necessary to solve a $6N \times 6N$ secular determinant to obtain the normal-mode frequencies. However, symmetry and the periodicity of the lattice can be used to greatly simplify the problem,[6,7] and we can talk about 12 vibrational modes associated with each of $N/2$ discrete values of a *wave vector* **k**. This wave vector has a magnitude

$$k = \frac{2\pi}{\lambda} \tag{28}$$

and a direction that specifies the propagation direction of a *traveling wave* (*i.e.*, of the "crests and troughs" of the periodic displacements). The vibrational wave motion in the crystal can be represented by traveling-wave equations of the general form

$$A_j(\mathbf{r}, t) = A_{j0} \cos(2\pi v_j t - \mathbf{k} \cdot \mathbf{r}) \tag{29}$$

where A_j is the instantaneous amplitude of a displacement of type j ($j = 1$ to 12) in the cell at point **r**. Equation (29) describes the twelve normal modes associated with a given **k**, i.e., with a given wavelength and direction for the periodic displacements of molecules in *different cells*. All allowed **k** values lie inside a *Brillouin zone (BZ)*†, a region bounded by a polyhedron in reciprocal space that is centered around k_x, k_y, $k_z = 0, 0, 0$.[6,7] As $k \rightarrow 0$, adjacent cell displacements approach being in phase, and $\lambda \rightarrow \infty$; when $k \rightarrow k_{max}$ at the Brillouin zone boundary, $\lambda \rightarrow \lambda_{min}$, a minimum wavelength for the **k** direction.

The v versus k curves, called phonon dispersion curves,[6,7] are shown in Fig. 2 for the **a** axis direction in an I_2 crystal. These and similar dispersion curves in other directions were obtained by Smith *et al.*[9] using the technique of inelastic neutron scattering.[10-12] The frequencies of internal stretching and libration are not affected greatly by the coupling between unit cells; i.e., each v_j is roughly constant for all **k** values for these modes. In contrast, the center-of-mass motion is strongly affected, especially for the acoustic branches TA_1, TA_2, and LA. These lattice vibrations are three-dimensional analogs of the one-dimensional vibrations of a violin string or the air in an organ pipe and the two-dimensional vibrations of a drum head. In the continuum (long-wave) limit, they represent three-dimensional vibrations in a bowl of Jello. Such

---

† The *BZ* is the locus of all points in reciprocal space that are closer to 0, 0, 0 than to any other reciprocal lattice point; its volume is equal to that of the primitive unit cell in the reciprocal lattice. See Exp. 46 for a discussion of the reciprocal lattice.

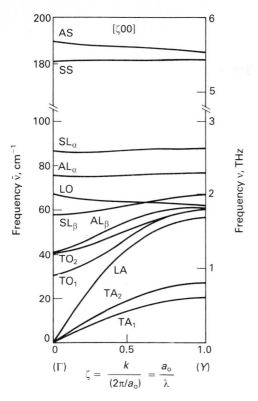

FIGURE 2
Phonon dispersion curves for $I_2(s)$ in the **a**-axis direction from the center of the *BZ* (Γ) to the boundary (*Y*) at 77 K. Adapted by permission from Smith *et al.* (Ref. 9).

acoustic frequencies range from 0 at the *BZ* center (point Γ) to ~1–2 THz at the *BZ* edge [1 terahertz(1 THz) $= 10^{12}$ Hz $= 33.3$ cm^{-1}]. The notation *TA* means transverse (shear) acoustic, and *LA* means longitudinal (compression–rarefaction) acoustic.

In order to assign frequency values $\tilde{\nu}_j$ to each of the 12 branches, we average the available values[9] over the Brillouin zone. The resulting values are given in Table 1, where limiting values at the zone center (point Γ) and zone edge (points *Y*, *T*, or *Z*) are also given. The choice of a single frequency for each mode corresponds to a version of the Einstein model for a solid.[1,6] This is quite reasonable for all branches except the three acoustic branches. For those three modes, the Debye model[1,6] would provide a better approximation. However, the simpler Einstein approximation for TA_1, TA_2, and LA is adequate for the present purposes.

We can now formulate the desired expressions for q_s and μ_s. Using Eqs. (8) to (10), together with the fact that $t = 2$ for I_2 and each unit cell contains two I_2 molecules, we find

$$\ln q_s = \frac{1}{6N} \sum_{i=1}^{6N} \ln q_i = \frac{1}{6N} \sum_{i=1}^{12(N/2)} \ln q_i \tag{30}$$

TABLE 1
Discrete (representative) phonon frequencies in I$_2$ crystals^a

Mode no. j	Type of mode	Representative frequency $\bar{\nu}_j$, cm^{-1}	Frequency range	
			cm^{-1}	Point(s) in BZ^b
1	TA_1	21.0	0–40.5	Γ, T
2	TA_2	26.5	0–53.1	Γ, T
3	LA	33.0	0–56.2	Γ, Y
4	TO_1	41.0	30.7–60.0	Γ, Y
5	TO_2	49.0	41.0–60.0	Γ, Y
6	AL_β	51.5	41.7–61.0	Γ, Y
7	SL_β	58.0	57.7–66.5	Γ, Y
8	LO	59.0	65.4–46.5	Γ, Z
9	AL_α	75.4	flat	Γ
10	SL_α	87.4	flat	Γ
11	SS	180.7	flat	Γ
12	AS	189.5	flat	Γ

^{a} Estimated from Ref. 9.

^{b} Points T and Z are not shown in Fig. 2; they are elsewhere on the surface of the BZ. See Ref. 9.

The number of discrete **k** values is $N/2$, the number of primitive unit cells in the crystal. Each of these is assumed to yield the *same* set of 12 branch frequencies ν_j. Thus we can simplify Eq. (30) to

$$\ln q_s = \frac{1}{6N} \frac{N}{2} \sum_{j=1}^{12} \ln q_j = \tfrac{1}{12} \sum_{j=1}^{12} \ln q_j \tag{31}$$

where 12 is the number of degrees of freedom per unit cell. Finally, we obtain for I$_2(s)$

$$\ln q_s = -\tfrac{1}{12} \sum_{j=1}^{12} \ln(1 - e^{-\Theta_j/T}) \tag{32}$$

and

$$\mu_s = -6RT \ln q_s = \frac{RT}{2} \sum_{j=1}^{12} \ln(1 - e^{-\Theta_j/T})$$

$$= \frac{RT}{2} \ln\left[\prod_{j=1}^{12} (1 - e^{-\Theta_j/T}) \right] \tag{33}$$

Equilibrium between crystal and gas. On substituting the expressions of Eqs. (25) and (33) into Eq. (2) and doing some rearranging and simplifying, we obtain

$$\ln p - \ln\left[\frac{T^{7/2} \prod_{j=1}^{12}(1 - e^{-\Theta_j/T})^{1/2}}{(1 - e^{-\Theta_{vib}/T})} \right] = \ln\left[\left(\frac{2\pi mk}{h^2}\right)^{3/2} \frac{k}{\sigma\Theta_{rot}} \right] - \frac{\Delta\bar{E}_0^0}{RT} \tag{34}$$

If the value of p is determined at *one* temperature, this equation can be solved for $\Delta \bar{E}_0^0$, the value of which is needed (along with Θ_{rot} and Θ_{vib}) to determine the chemical potential of gaseous I_2. Once $\mu_s(T)$ and $\mu_g(T)$ are both known, one can calculate $\Delta \bar{S}_{\text{sub}}$ and $\Delta \bar{H}_{\text{sub}}$. By contrast, the Clausius–Clapeyron equation, given by

$$\ln p = \text{constant} - \frac{\Delta \bar{H}_{\text{sub}}}{R} \frac{1}{T} \tag{35}$$

in its approximate integrated form, requires at least two values of p at different temperatures in order to obtain a value of $\Delta \bar{H}_{\text{sub}}$.

Equation (35) has obvious similarities to Eq. (34). This correspondence can be enhanced by replacing $\Delta \bar{H}_{\text{sub}}/RT$ with $\Delta \bar{E}_{\text{sub}}/RT + 1$, which is equivalent since $\Delta(p\bar{V}) \cong RT$ is an excellent approximation under the conditions of the present experiment. However, $\Delta \bar{E}_{\text{sub}}$ is temperature dependent and refers to the energy of sublimation at the temperature of the experiment rather than at absolute zero. This temperature dependence is reflected in the statistical treatment by the variation with T of the second term on the left-hand side (LHS) of Eq. (34).

If p values have been measured at several temperatures, the LHS of Eq. (34) can be plotted against $1/T$, and the value for $\Delta \bar{E}_0^0$ can be determined from the slope of a straight line fitted graphically or by least squares. In addition, the intercept can be compared with the predicted value of the constant term on the RHS of Eq. (34). Alternatively, it is possible to calculate a $\Delta \bar{E}_0^0$ value from each p, T data point and see how well these values agree.

Entropy and enthalpy of sublimation. Since we have a system of only one component, the chemical potentials for I_2 in crystalline and gaseous forms, given in Eqs. (33) and (25) respectively, are equivalent to the molar Gibbs free energies $\bar{G}_s$ and $\bar{G}_g$, aside from an additive constant. The entropies of the two phases can be obtained by differentiating with respect to temperature. The expressions obtained are

$$\bar{S}_s = -\left(\frac{\partial \bar{G}_s}{\partial T}\right)_p = -\left(\frac{\partial \mu_s}{\partial T}\right)_p$$

$$= \frac{R}{2} \sum_{j=1}^{12} \left[\frac{\Theta_j/T}{e^{\Theta_j/T} - 1} - \ln(1 - e^{-\Theta_j/T})\right] \tag{36}$$

$$\bar{S}_g = -\left(\frac{\partial \bar{G}_g}{\partial T}\right)_p = -\left(\frac{\partial \mu_g}{\partial T}\right)_p$$

$$= \frac{\Delta \bar{E}_0^0 - \mu_g}{T} + \frac{7}{2}R + R\frac{\Theta_{\text{vib}}/T}{e^{\Theta_{\text{vib}}/T} - 1} \tag{37}$$

The heat of sublimation at temperature T is

$$\Delta \bar{H}_{\text{sub}} = T \, \Delta \bar{S}_{\text{sub}} = T(\bar{S}_g - \bar{S}_s) \tag{38}$$

EXPERIMENTAL

The determination of the vapor pressure of solid iodine at temperatures from 20°C to 70°C in steps of about 10°C is accomplished through spectrophoto-metric measurements of the absorbance A of the iodine vapor in equilibrium with the solid at the absorption maximum (520 nm), and also at 700 nm where the molar absorption coefficient of iodine vapor is so small as to be negligible. On the short-wavelength side of the maximum there is no accessible wavelength at which the absorption is negligible; hence the baseline can only be drawn from the long-wavelength side. In most instruments the absorbance is indicated directly; at every wavelength it is related to the incident beam intensity I_0 and the transmitted beam intensity I by the equation

$$A_\lambda = \log\left(\frac{I_0}{I}\right)_\lambda \tag{39}$$

The absorption peaking at 520 nm is for the transition

$$X\,^1\Sigma_g^+ \to B\,^3\Pi_{0^+u} \tag{40}$$

(see Exp. 42). There is also a transition

$$X\,^1\Sigma_g^+ \to A\,^3\Pi_{1u} \tag{41}$$

with a broad absorption peaking at about 700 nm, but its extinction is quite small in comparison to the other band and can be neglected.[13]

The absorption cell containing iodine in the solid and vapor states must also contain air or nitrogen at about 1 atm to provide pressure broadening of the extremely sharp and intense absorption lines of the rotational fine structure (which can be individually resolved only by special techniques of laser spectroscopy). The reason lies in the logarithmic form of Eq. (39). Within the slit width or resolution width of the kinds of spectrophotometers that may be used in this experiment, low-pressure $I_2(g)$ exhibits many very sharp lines separated by very low background absorption. The instrument effectively averages transmitted intensity I, not absorbance A, over the sharp peaks and background within the resolution width, but the logarithm of an average is not the average of the logarithm. If the extremely sharp lines are so optically "black" that varying the concentration has little effect on the amount of light transmitted in them, the absorbance is mainly controlled by the background between the lines, and the contribution of the lines to the absorbance is largely lost. Increasing the concentration of gas molecules increases the number of molecular collisions and thus decreases the time between them. This can greatly broaden the lines and lower their peak absorbances, causing them to overlap and smooth out the spectrum over the resolution width so that the absorbance readings are meaningful averages over that range. This effect is readily demonstrated experimentally by comparison of spectra taken of I_2 vapor with and without air present.[14]

Figure 3 shows the absorption spectrum of I_2 vapor over the range of

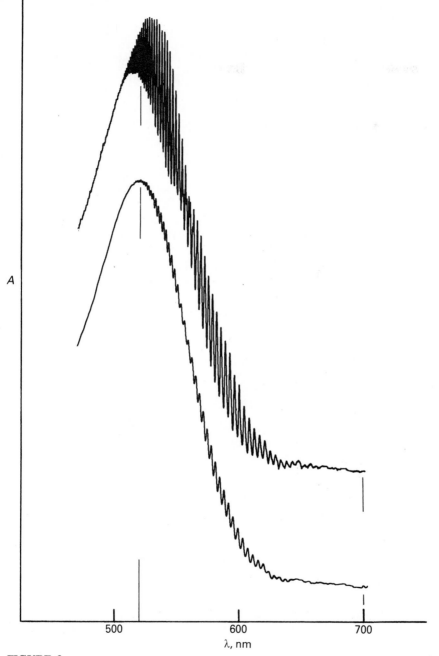

FIGURE 3

Absorbance of $I_2(g)$, in equilibrium with the solid at 26°C and in the presence of air at about one atmosphere, for wavelengths ranging from 470 to 700 nm. The top curve is at moderately high resolution. The bottom curve is at relatively low resolution, about the highest suitable for this experiment.

interest at the vapor pressure of iodine at 27°C, at moderate resolution and at low resolution. A low-resolution spectrum, obtained with wide slits, is preferable for this experiment, since the vibrational structure is averaged out, facilitating the determination of the absorbance at 520 nm. An even lower resolution may be entirely satisfactory for this experiment.

The spectrometer to be used need cover only the visible portion of the electromagnetic spectrum. Preferably it should be a double-beam instrument, to allow for compensation for absorption by the cell windows. However, a single-beam instrument may be used if the absorption spectrum of an equivalent set of cell windows is obtained separately so that the absorbances of the windows can be subtracted from those of the iodine-containing cell. The spectrophotometer must have a cell compartment large enough to contain the water-jacketed absorption cell that must be used for measurements at several temperatures. Further details concerning spectrophotometers are given in Chapter XVIII and in the operating manuals of the specific instrument to be used.

A suitable water-jacketed absorption cell is shown in Fig. 4. The central iodine-containing cell should be constructed of Pyrex or optical glass. The end

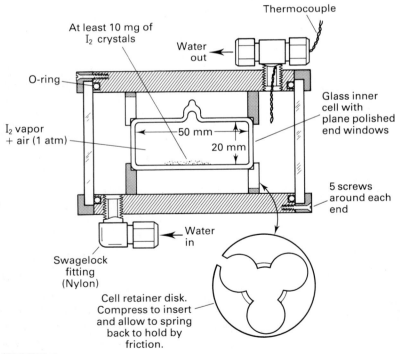

FIGURE 4

Suggested design for water-jacketed Pyrex spectrophotometer cell containing iodine. The recommended material for the main part of the jacket is Delrin, which can be easily machined. The outer windows may be either glass or clear plastic.

windows, which must be planar and polished, must be sealed (glassblown) to the body of the cell with a torch, not cemented; cements are unsatisfactory because of the corrosive action of iodine vapor. The cell can be fabricated with commercially available windows by an experienced glassblower, or made by modifying a commercially available spectrophotometer cell, or fabricated on special order by a commercial vendor (*e.g.*, NSG Precision Cells, Inc., Farmingdale, NY, or the Harrick Scientific Corporation, Ossining, NY). The optical path between the end windows should be about 4 or 5 cm and should be accurately known. At least 10 mg of solid iodine crystals (for a cell of about 15 cm^3 volume) must be introduced into the cell through a side-tube before the side-tube is sealed off close to the body of the cell with a torch, with about 1 atm of air inside. The details of the water jacket design are very flexible but must ensure that the end windows of the inner cell are bathed in the circulating water to prevent condensation of iodine crystals on the windows at the higher temperatures of the experiment. A calibrated thermocouple should be provided for measuring the temperature inside the water jacket. Plastic tubing is used to connect the water jacket to a water thermostat bath equipped with a variable temperature control and a circulating pump. The cell, with its water jacket constructed as shown, can be used year after year with little or no maintenance.

Note. A separate set of windows, both for the inner cell and for the water jacket, should be available for optical compensation as indicated above.

For each temperature, sufficient time must be allowed for the contents of the inner cell to come to equilibrium with the circulating water. The spectrophotometer can be set to 520 nm and used as a monitor to follow the approach to equilibrium. When the absorbance and the thermocouple reading are no longer changing, the temperature should be noted and the absorption spectrum should be recorded from 700 to 450 nm. It may be necessary to experiment with different settings of the absorbance scale, the slit and/or gain controls, the scanning speed, and the damping time to obtain optimum performance.

CALCULATIONS

For each temperature, determine the net absorbance as the difference between the absorbances at 520 and 700 nm:

$$A = A_{520} - A_{700} \tag{42}$$

The net absorbance is related to the concentration c and the pressure p of Ɪdine vapor in the cell as follows (assuming the perfect-gas law):

$$A = \varepsilon dc = \frac{\varepsilon d}{RT} p \tag{43}$$

where c is the concentration of I_2 vapor, p is the partial pressure of I_2 vapor, d

TABLE 2
Molar absorption coefficients of iodine vapor at $\lambda = 520$ nm[a]

t, °C	ε, L mol^{-1} cm^{-1}
20.0	691
30.0	682
40.0	672
50.0	663
60.0	654
70.0	646

[a] Based on an equation derived by Sulzer and Wieland (Ref. 13). The function "Tg" in their equation is the hyperbolic tangent. They tested this equation in the range 423–1323 K (150–1050°C). Their cells apparently contained no gas other than I_2 vapor, but at those temperatures the vapor pressures were apparently high enough to obtain the needed pressure broadening without the addition of another gas. Their equation satisfactorily fits lower-temperature data obtained with a cell containing about 1 atm of air.

is the optical path length inside the inner absorption cell, and ε is the molar absorption coefficient for I_2 vapor at 520 nm and temperature T.

The value of the molar absorption coefficient ε at each temperature can be found from Table 2 by interpolation. Then p may be calculated with Eq. (43). In the calculations of this experiment, considerable care must be taken with units. It is desirable to obtain p in pascals for the statistical-mechanical calculations; accordingly, ε should be converted into units of m^2 mol^{-1}, d should be in m, and R should be in units of J K^{-1} mol^{-1}.

To determine $\Delta \tilde{H}_{\text{sub}}$ from the approximate Clausius–Clapeyron equation [Eq. (35)] plot $\ln p$ against $1/T$ and determine the slope of the best straight-line fit to the data by graphical or least-squares methods.

The values of p and T can now be used for the statistical-mechanical calculations. In order to calculate the rotation characteristic temperature Θ_{rot} with Eq. (20), use the literature value[4] for the rotational constant $\tilde{B}_0 = 0.037315$ cm^{-1} [or calculate $\tilde{B}_0$ from the internuclear distance in the molecule, $r_0 = 0.2667$ nm, with Eqs. (17) to (19)]. From the literature value of the molecular vibrational frequency in the gas phase,[4] $\tilde{\nu}_0 = 213.3$ cm^{-1}, calculate the vibration characteristic temperature Θ_{vib} with Eq. (22). From the phonon dispersion data in Table 1, calculate the 12 vibration characteristic temperatures Θ_j.

Calculate $\Delta \tilde{E}_0^0$ with Eq. (34) at each temperature. Unless the calculations are well organized they can be exceedingly onerous with a hand-held calculator, particularly in the case of the crystal partition function. Be careful to save intermediate results for checking and for possible use in later calculations. A programmable calculator will be very helpful, or a personal computer may be programmed for the repetitive calculations. A particularly easy and rapid way to do these calculations on a computer is with a spreadsheet program, such as Symphony™ or Lotus 1-2-3™.

Do the values obtained for $\Delta \bar{E}_0^0$ agree satisfactorily? If not, check the calculations and/or consider possible systematic errors. Plot the LHS of Eq. (34) against $1/T$, and determine both $\Delta \bar{E}_0^0$ and the constant term graphically or by least squares. Does this value of $\Delta \bar{E}_0^0$ agree with the average of the values obtained by direct application of Eq. (34)? Does the constant term agree with the theoretical value?

Calculate the molar entropies $\bar{S}_s$ and $\bar{S}_g$ of the crystalline and vapor forms of I_2 at 320 K with Eqs. (36) and (37), and obtain the molar heat of sublimation $\Delta \bar{H}_{sub}$ with Eq. (38). Compare it with the value obtained by the Clausius–Clapeyron method and with any literature values that you can find.[15]

DISCUSSION

Of the two methods of determining $\Delta \bar{E}_0^0$ with Eq. (34), which do you judge gives the more *precise* value? Which gives the more *accurate* value? Which provides the better test of the overall statistical-mechanical approach? Compare this approach with the purely thermodynamic method using the integrated Clausius–Clapeyron equation, taking into account the approximations involved in the latter [see Exp. (13)]. State the average temperature corresponding to your Clapeyron value of $\Delta \bar{H}_{sub}$.

Comment on the choice of representative values of $\bar{\nu}_j$ for the 12 vibrational modes of the crystal. How much would reasonable changes (say 10–20 percent) in these values affect the results of the calculations? If possible, comment on the effect of using the Debye approximation for the acoustic lattice modes instead of the Einstein approximation.

Other experimental and theoretical methods have been developed for the determination of the heat of sublimation of solid iodine; these too are suitable for undergraduate laboratory experiments or variations on this experiment. Henderson and Robarts[16] have employed a photometer incorporating a He–Ne gas laser, the beam from which (attenuated by a $CuSO_4$ solution) has a wavelength of 632.8 nm, in a "hot band" near the long-wavelength toe of the absorption band shown in Fig. 3. Stafford[17] has proposed a thermodynamic treatment in which a free-energy function (*fef*), related to entropy, is used in calculations based on the third law of thermodynamics. In this method, either heat capacity data or spectroscopic data are used, and, as in the present statistical-mechanical treatment, the heat of sublimation can be obtained from a measurement of the vapor pressure at only one temperature.

APPARATUS

Visible spectrophotometer (preferably double-beam); water-jacketed absorption cell containing solid I_2 and air or N_2 (see text and Fig. 4); water thermostat with adjustable temperature regulator and circulating pump; plastic tubing to connect circulating pump with water jacket of absorption cell; calibrated thermocouple and associated instrumentation (see Chapter XVI).

REFERENCES

1. N. Levine, "Physical Chemistry," pp. 745–746, 854–858, McGraw-Hill, New York (1983).
2. P. W. Atkins, "Physical Chemistry," 2d ed., pp. 673–674, Freeman, New York (1982).
3. *Ibid.*, pp. 697–699.
4. Adapted from R. P. Huber and G. Herzberg, "Molecular Spectra and Molecular Structure. IV. Constants of Diatomic Molecules," p. 332, Van Nostrand-Reinhold, New York (1979).
5. P. W. Atkins, *op. cit.,* pp. 699–700.
6. C. Kittel, "Introduction to Solid State Physics," 5th. ed., Wiley, New York (1976).
7. N. W. Ashcroft and N. D. Mermin, "Solid State Physics," Saunders, Philadelphia (1976).
8. F. van Bolhuis, P. B. Koster, and T. Migghelsen, *Acta Crystallogr.* **23**, 90 (1967).
9. H. G. Smith, M. Nielsen, and C. B. Clark, *Chem. Phys. Lett.* **33**, 75–78 (1975); H. G. Smith, C. B. Clark, and M. Nielsen in J. Lascombe (ed.), "Dynamics of Molecular Crystals," pp. 4411–46, esp. fig. 2, Elsevier, Amsterdam (1987).
10. C. Kittel, *op. cit.,* pp. 120–121.
11. N. W. Ashcroft and N. D. Mermin, *op. cit.,* pp. 470–474.
12. G. E. Bacon, "Neutron Diffraction," 3d ed., chap. 9, Oxford University Press, Oxford (1975).
13. P. Sulzer and H. Wieland, *Helv. Phys. Acta* **25**, 653 (1952).
14. J. G. Calvert and J. N. Pitts, Jr., "Photochemistry," p. 184, ref. 424 in chap. 5, Wiley, New York (1966).
15. D. A. Shirley and W. F. Giauque, *J. Am. Chem. Soc.* **31**, 4778 (1959).
16. G. Henderson and R. A. Robarts, Jr., *Am. J. Phys.* **46**, 1139 (1978).
17. F. Stafford, *J. Chem. Educ.* **40**, 249 (1963).

GENERAL READING

N. Davidson, "Statistical Mechanics," chaps. 6–8, McGraw-Hill, New York (1962).
N. W. Ashcroft and N. D. Mermin, *op. cit.,* chaps. 4, 5, 7, 22–24.
C. Kittel, *op. cit.,* chap. 4.

CHAPTER
XV

ELECTRONIC DEVICES AND MEASUREMENTS

The rapid development of solid-state electronic devices in the last two decades has had a profound effect on measurement capabilities in chemistry and other scientific fields. In this chapter we consider some of the physical aspects of the construction and function of electronic components such as resistors, capacitors, inductors, diodes, and transistors. The integration of these into small operational amplifier circuits is discussed and various measurement applications are described. The use of these circuit elements in analog-to-digital converters and digital multimeters is emphasized in this chapter, but modern integrated circuits (ICs) have also greatly improved the capabilities of oscilloscopes, frequency counters, and other electronic instruments discussed in Chapter XVIII. Finally, the use of potentiometers and bridge circuits, employed in a number of experiments in this text, is covered in the present chapter.

CIRCUIT ELEMENTS

Figure 1 shows the symbols of common elements used in electronic circuits. These can be classed as either *passive* components, such as resistors, capacitors, inductors, and diodes, or *active* components, such as bipolar and field-effect transistors, silicon controlled rectifiers (SCRs), etc. Some of the key

599

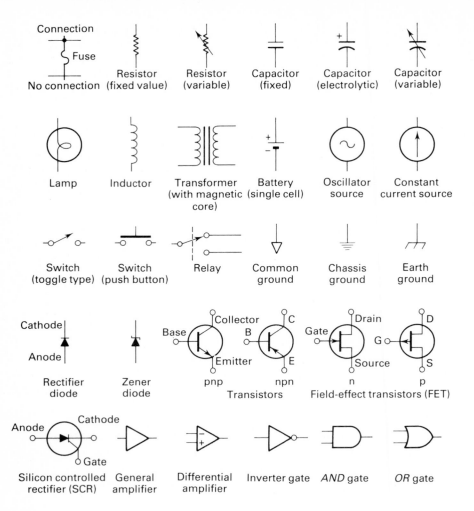

FIGURE 1
Common circuit component symbols.

features and physical characteristics of these devices are summarized in the first two sections of this chapter.

Resistors. Resistors are found in three commercial forms. The most common type, used for noncritical applications, is the graphite *composition resistor,* a mixture of graphite and silica in which their ratio determines the resistance. These resistors are generally molded with a binder in the shape of a cylinder, with two electrodes bonded at the ends. *Film resistors* are ceramic insulating rods coated with a thin conducting film of carbon, metal, metal oxide, or cermet, a mixture of glass and metal alloys. The film is then cut with a spiral groove to increase the effective length so as to produce the desired resistance. Electrodes are attached at each end and the unit is encased in a glass or ceramic material. *Wire-wound resistors* are made similarly but by wrapping thin wire of an appropriate length around a ceramic insulator, which is then coated. The wire is formed from alloys such as manganin or constantan whose resistance changes only slightly with temperature, about 10 p.p.m./°C versus 100 to 1000 p.p.m./°C for film or composition resistors. The color code for the labeling of fixed resistors is shown in Fig. 2.

Precision wire-wound resistors are used for critical applications in which low noise and resistance stability are important but, since they are coils, they have an inductive reactance that must be taken into account for ac circuits. By folding the insulated wire in the middle and carefully forming parallel windings, the inductive component can be reduced by about a factor of 100. Precision standard resistors with an accuracy of about 50 p.p.m. in the range of 1 Ω to 10 mΩ are available from Leeds and Northrop, Electro Science Industries, and other companies. These are often combined in the form of decade resistance boxes to provide a wide range of fixed values.

Resistors whose values can be varied are termed *potentiometers*. They are made of the same materials as fixed resistors but have a movable wiper to contact a coil of resistance wire or a strip of resistive film at any point along the resistor. Potentiometers used only occasionally to adjust a circuit are called *trimmers*, while those employed for high-wattage applications such as control of heating mantles and ovens are called *rheostats*. Precision potentiometers have played an important role in the measurement of electrical quantities and are considered in detail in a later section.

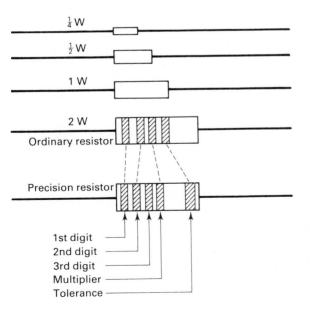

FIGURE 2

Resistor code for ordinary (2-digit) and precision (3-digit) resistors; several composition resistors with different power ratings are shown drawn to scale.

Capacitors. A capacitor consists of two conductors separated by an insulator. The capacitance C gives a measure of the charge separation Q produced when a potential V is applied across the electrodes and is given by $C = Q/V$. C is determined by the geometry and the dielectric component of a capacitor. For the simplest case of two plates of area a separated by a distance d the capacitance is

$$C = \varepsilon_0 \kappa_d \frac{a}{d} \tag{1}$$

where $\varepsilon_0 = 8.854 \times 10^{-12}\,\mathrm{F\,m^{-1}}$ is the permittivity of free space and κ_d is the dielectric constant of the insulating material between the plates. For a cylindrical capacitor, such as is sometimes used for dipole-moment determinations, the cell capacitance is

$$C = \frac{2\pi\varepsilon_0 \kappa_d l}{\ln(r_2/r_1)} \tag{2}$$

where l is the length of concentric cylinders of radii r_1 and $r_2 = r_1 + d$ with the dielectric material in between. This reduces to Eq. (1) when $d/r_1 \ll 1$, where d is the gap between the cylinders.

Capacitors are produced in several forms. *Film* capacitors are commonly made of two thin metal foils, separated by a dielectric material such as paper or plastic. The foils are rolled into a tubular form with conducting leads added at each end, yielding capacitance values in the range of 0.01 to 50 μF. Often the conducting film is coated on one side of the dielectric film to give a very compact structure. Plastic dielectrics include Mylar, Teflon, polystyrene, polycarbonate, and other materials. The best film capacitors are Teflon capacitors, which have a very low current leakage and temperature dependence and are not greatly affected by humidity.

Mica and *ceramic* disk capacitors consist of stacks of the rigid dielectric material, interleaved with metal foils, deposited metallic films, or conducting silver paste. These are encapsulated in epoxy or a fired glass to produce a very durable device that can withstand high temperatures and mechanical shock. The ceramic material is usually titanium dioxide (TiO_2) or barium titanate ($BaTiO_3$). Ceramic capacitors are also produced in tube form with plated electrodes on the inside and outside of the tube. Ceramic capacitors have a range of 100 pF to 1 μF, while mica capacitors cover the range of 5 pF to 10 μF.

For applications requiring high capacitance and where the polarity is fixed, as in dc power supplies, inexpensive *electrolytic* capacitors are commonly used. These employ an aluminium or tantalum film that is electrolyzed to form a thin oxidized layer that serves as the dielectric material. A low-conductance paste or solution is added between layers and, because the films are extremely thin, large capacitance values are possible. These range from 0.01 μF up to 100 000 μF, with the largest values only applicable in cases where the voltage does not exceed a few volts. Electrolytic capacitors are usually manufactured in tubular form with the metal electrode indicated as positive. This polarity must be maintained in use since a negative potential causes electrolytic destruction of the oxide film, leading to shorting and sometimes explosive rupture of the sealed capacitor.

Capacitor values are generally indicated on the body of the capacitor. If no units are shown and the value is less than 1 the unit is μF; if the value is greater than 1, the units are in pF. Some mica and ceramic capacitors are color-coded as indicated in Fig. 3.

Variable capacitors are available with values up to a few hundred picofarads. These are commonly formed from sets of interleaved plates, one fixed and the other attached to a shaft. Rotation changes the effective area and thereby the capacitance. Arrangements with sliding cylinders are also used and dielectrics include air, mica, and ceramic. *Varacter diodes, p–n* junction diodes in which the capacitance is determined by the reverse bias voltage, are now finding increasing use in circuits, because their capacitance can be actively controlled by other electronic circuit elements.

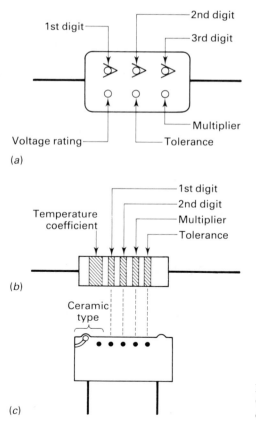

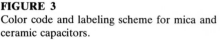

FIGURE 3
Color code and labeling scheme for mica and ceramic capacitors.

Inductors and transformers. Inductance is related to the fact that a moving charge has associated with it a magnetic field. Changes in this field arising from variations in the charge current result in the generation of a voltage in the circuit that is given by

$$V = L \frac{dI}{dt} \tag{3}$$

where the proportionality constant L is called the inductance. The induced voltage acts to oppose the current change and its magnitude increases with the frequency of oscillation of the current. The unit of inductance is the *henry* (H), after Joseph Henry, an early American investigator of inductive effects.

Inductors, sometimes termed *chokes,* are usually made of coils of wire wound adjacent to one another to concentrate the magnetic flux produced by each coil. Ten to one hundred loops about a hollow support serve to produce inductances in the μH to mH range, which is suitable for most high-frequency circuits. Large inductance values require hundreds of turns, usually about a core of ferromagnetic material such as iron. This core serves to greatly concentrate the magnetic flux in the region of the coil so as to yield inductance values up to several hundred henrys. Large inductors are useful principally in power applications and as filter elements.

Inductors are never pure, since there is always resistance in and capacitance between the coil windings. Because of this, and because of the relatively large size and cost of inductors, inductive reactance is not used as much as capacitive reactance for ac control in electronic circuits.

The combination of two or more inductors arranged so that the coils are inductively coupled to each other is a *transformer.* The coupling can be through air or a metallic core. Power transformers are widely used to provide various voltages from the standard 60 Hz 115 V power lines, the scaling being given simply by the turns ratio of the output (secondary) to input (primary) inductors. Smaller pulse transformers also find use in coupling high-frequency ac signals while isolating dc levels of primary and secondary circuits.

SEMICONDUCTOR DEVICES[1-6]

The *p–n* junction. The two most common semiconductor materials, the group IV elements silicon and germanium, have electrical resistivities intermediate between values of about 10^{-6} Ω cm for metals such as copper and 10^{11} to 10^{17} Ω cm for common insulators. This intermediate behavior is a consequence of the gap between the localized valence electron energy band and the delocalized conduction band which lies at higher energy. For Si this gap is 1.03 eV and for Ge it is 0.72 eV. For such pure (intrinsic) materials, thermal excitation is only occasionally sufficient to excite an electron into the conduction band. When excitation does occur, a positive hole is produced at the atom from which the electron is released and both the electron and the hole contribute to the electrical conductivity under an applied potential. The hole movement occurs in stepwise fashion by the hopping of a bound electron from a neighboring atom in the lattice. The charge mobility of a hole in the valence band is about half that of an electron in the conduction band.

The concentration of electrons or holes in a semiconductor can be greatly enhanced by adding electron-rich or electron-deficient elements to the lattice. When Group V elements such as phosphorus, arsenic or antimony replace silicon or germanium atoms in the lattice, the energy necessary for excitation of the extra valence electron of each Group V atom to the conduction band is quite small ($\sim$0.1 to 0.05 eV). This results in a large increase in the number of conducting electrons. An equal number of positive Group V ions (holes) are produced, which form four tetrahedral lattice bonds like the host Si and Ge atoms. Electron transfer from the host neighbors to positive holes is not favored, since the host electrons are tightly bound in forming lattice bonds. Thus the positive hole is not particularly mobile and the *majority* carriers of current are the electrons in the case of this *n*-type (negative type) semiconductor.

If the added impurity (dopant) is a Group III element such as aluminum or gallium, a positive hole is produced when a valence electron from an adjoining silicon or germanium atom jumps to the dopant atom. The energy for this process is also much smaller than the band gap energy and conduction in the valence band by the majority hole carriers is greatly increased in this *p*-type semiconductor. Materials whose enhanced conductivity is a consequence of added impurities are termed *extrinsic* semiconductors.

Diodes. A semiconductor *diode* is made by forming adjacent *n*- and *p*-type regions within a single crystal of silicon or germanium, usually by a carefully controlled diffusion process. The result is the production of an interface region, the *p–n junction*, in which there is an excess of holes on the left and an excess of electrons on the right. Diffusion of charge carriers across the junction results in electron–hole recombination and formation of a *depletion* region of greatly reduced conductivity, as depicted in Fig. 4(*a*). The effective separation of charge across the interface produces a *junction potential* of about 0.6 V for silicon, 0.3 V for germanium.

The utility of the diode comes from the fact that, when an external potential is applied, the junction acts as a unidirectional valve. If the biasing is in the forward direction, as shown in Fig. 4(*b*), the applied voltage reduces the junction potential, causing the majority carriers to drift across the junction to give a low effective resistance. If the biasing is reversed, as in Fig. 4(*c*), the depletion region is broadened and the effective resistance is increased by a factor of 10^3 to 10^8 over that for forward biasing. Since the width of the depletion region varies with applied voltage, the capacitance of the junction also varies, typically over a range of 1 to 15 pF. *Varacter* diodes are designed especially for this capacitance control feature.

The current–voltage *characteristic curves* for representative germanium

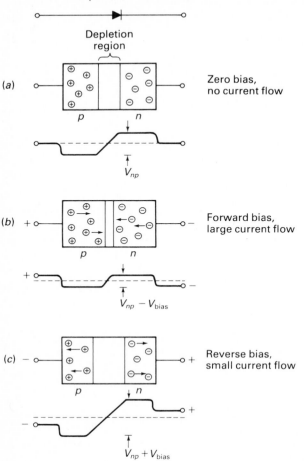

Direction of positive current flow

Depletion region

(a)   Zero bias, no current flow

p n

V_{np}

(b)   +   −   Forward bias, large current flow

p n

+   −

$V_{np} - V_{bias}$

(c)   −   +   Reverse bias, small current flow

p n

+

−   +

$V_{np} + V_{bias}$

FIGURE 4
Effect of bias voltage on the depletion region of a p-n junction diode and on the potential for positive current flow.

and silicon p–n junctions are shown in Fig. 5. For a perfect diode the resistance would change from infinity to zero as the bias goes from negative to positive. Actual diodes provide a close approximation to this ideal behavior. The nearly constant value of current at negative bias and the exponential rise for positive voltage is adequately described by the relation

$$I = I_i \left[\exp\left(\frac{eV}{kT}\right) - 1 \right] \tag{4}$$

Here I_i is the intrinsic current at zero bias voltage, e is the electron charge, and V is the bias voltage. I_i itself also has an exponential dependence on temperature since at zero bias the charge carriers are of thermal origin. The resultant temperature dependence of the one-way characteristic of a diode is large and places an upper operational temperature limit for diodes of about

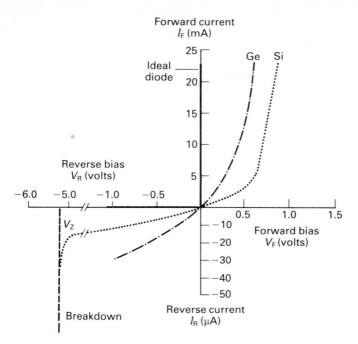

FIGURE 5
Current-voltage characteristics of germanium and silicon diodes. The heavy lines represent ideal behavior. Note the scale change for the region of large reverse bias near the breakdown voltage V_Z.

150°C for Si and 85°C for Ge. Despite this temperature dependence, diodes are widely used as rectifiers in power applications where substantial currents pass through the junction. To avoid excessive internal heating, a rectifier diode is generally provided with fins for radiative cooling or is mounted in good thermal contact with a substrate of large thermal mass.

A typical diode bridge rectifier arrangement is shown in Fig. 6. The diodes are arranged in such a fashion that positive current flows as indicated with solid arrows when the ac voltage is positive and as indicated with dashed arrows on the negative portion of the ac cycle. The current flow through the load resistor is thus always in the same direction and the frequency of the positive waveform oscillation is twice that of the input frequency. A capacitor filter is commonly used to dampen this oscillation and to provide a positive dc source with some ac ripple. The latter can be further reduced by use of a Zener diode, a $p-n$ junction operated at negative bias in the breakdown region of Fig. 5.

The reverse-bias breakdown voltage V_Z of a diode is a strong function of the dopant concentration and depletion layer thickness of the $p-n$ junction. As the concentrations increase from about 10^{15} to 10^{17} atoms/cm^3, the depletion layer thickness varies from 100 to 0.1 μm at typical breakdown voltages of five

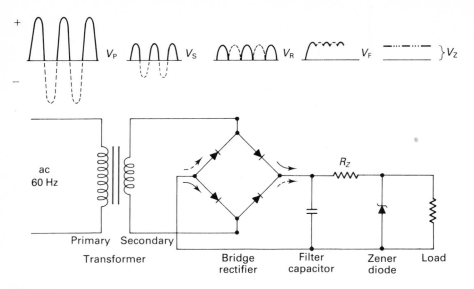

FIGURE 6

Waveforms produced by a bridge rectifier circuit with capacitor filter and Zener diode voltage regulation. The solid arrows show the direction of current flow on the positive cycle of the ac input, the dashed arrows that for the negative cycle. The Zener diode holds the voltage constant as long as $V_F > V_Z$. R_Z serves to hold the Zener diode current below its rated maximum value.

volts to several hundred volts.[1] This breakdown occurs because electrons produced in the depletion region by thermal excitation are accelerated by the applied potential. If this potential is large enough, collisions of these electrons with other atoms can generate additional electrons that are similarly accelerated to produce an *avalanche* current that increases very quickly at the critical Zener voltage V_Z. The diode resistance varies very sharply for a small voltage deviation from V_Z and thus a Zener diode provides a convenient way to hold a dc voltage constant, as shown in Fig. 6. If the source voltage increases, the diode resistance decreases so that more current passes through the diode and the IR drop across the load is held constant. Each Zener diode has an upper current limit of course, so that a current-limiting resistor, R_Z in Fig. 6, is added to avoid destruction of the diode.

Diodes are available with Zener voltages down to about 4 V, a regime in which electron tunneling through a very thin depletion region is the primary mechanism for conduction. The avalanche mechanism is predominant above about 6 V. Two or more Zener diodes can be placed in series to obtain almost any control voltage above four volts and a voltage precision of 0.01 percent or better is possible from carefully designed Zener reference sources. For best performance, a temperature-stabilized housing is used along with special temperature-compensated diodes to give a temperature coefficient with voltage of 0.001% °C^{-1} compared to a value of about 0.1% °C^{-1} for an ordinary Zener diode.

Bipolar junction transistor. The $p-n$ junction diode finds its most important application when two junctions are combined back-to-back to form the *bipolar junction transistor* (BJT). This remarkable device consists of two n- or p-type semiconductor regions separated by a narrow region of semiconductor of the opposite type. Figure 7 shows the general features of *npn* and *pnp* planar transistors, along with their conventional symbols. In a typical fabrication, an *npn* transistor is made from a silicon wafer which is first n-doped and then oxidized to provide a protective SiO_2 coating. A circular region of the oxide layer is etched away with HF and a p-type impurity such as boron (provided by decomposition of gaseous B_2H_6 on the hot wafer surface) is allowed to diffuse into the silicon to form a *base* region of 1 to 50 μm thickness. The wafer is then reoxidized and the process is repeated with n-doping with phosphorous (from gaseous PH_3) over a smaller central *emitter* region. A *pnp* transistor is made in an analogous manner. Since the gaseous diffusion doping process can be accurately controlled and hundreds of transistors can be made at the same time from single wafer, transistors with well-defined characteristics can be fabricated at very low cost.

There are three distinct circuit configurations for the junction transistor, as depicted in Fig. 8. These differ in the choice of the transistor terminal that is chosen to be common to the input and the output. An operational circuit for biasing an *npn* transistor in the *ON* or conducting state is shown in Fig. 8(*a*) for the *common-emitter* (CE) configuration. One junction, the emitter, is biased in the forward direction and the other, the *collector*, is reverse biased.

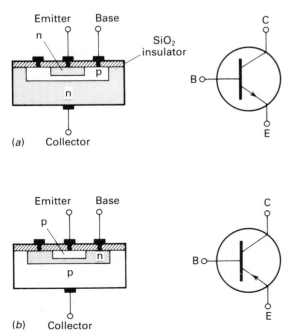

(a)

(b)

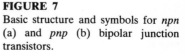

FIGURE 7
Basic structure and symbols for *npn* (a) and *pnp* (b) bipolar junction transistors.

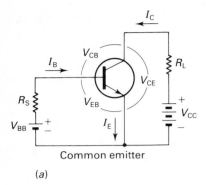

Common emitter

(a)

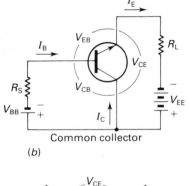

Common collector

(b)

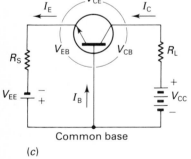

Common base

(c)

FIGURE 8

Circuit arrangements for three wiring configurations of a *npn* bipolar junction transistor.

Resistance to electron flow from the emitter to the base is thus small and the current I_E is consequently large. The doping level and thickness of the base are kept small so that only a few (1–5 percent) of the emitter electrons combine with positive holes in the base region to produce the small base current I_B. The majority of the electrons reach the collector–base junction and cross it owing to the positive potential (V_{CE}) of the collector.

The important control feature of a transistor comes from the fact that a small change in V_{BE}, and hence I_B, has a large effect on the number of charge carriers in the emitter–base junction. Control of the effective resistance from collector to emitter is thereby achieved and the transistor can be employed as a current amplifier. The current gain of the transistor is defined as $\beta = I_C/I_B$ and ranges from about 5 to 1000 for typical transistors. If the V_{BE} bias is reversed, both junctions have high resistance and the transistor is turned off. Transistors thus can serve as convenient on–off switches as well as fine-control "valves" for current flow.

The common-emitter transistor configuration is frequently used for amplification, because it provides both current and voltage gain. The disadvantage of the CE configuration is its relatively low input resistance owing to the forward biasing of the base-to-emitter junction. In contrast, the *common-collecter* (CC) arrangement provides a high input impedance (≈ 1 MΩ) and low output impedance (30–50 Ω). Thus it is common in circuits to follow a CC buffering input stage with a CE amplifier to provide high-input-impedance and voltage gain for the overall circuit.

The current–voltage *characteristic curves* for a typical common-emitter *npn* transistor are shown in Fig. 9(a). These consist of a family of curves, one

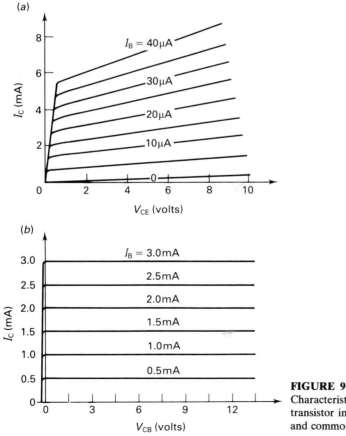

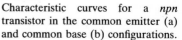

FIGURE 9
Characteristic curves for a *npn* transistor in the common emitter (a) and common base (b) configurations.

for each value of base current I_B, in which the collector current I_C is plotted as a function of collector–emitter voltage V_{CE}. It is apparent that for V_{CE} greater than 1 V, the collector current is approximately constant and is proportional to the base current. The transistor gain, β, at a particular V_{CE} value is easily estimated from such curves by determining the collector current for a given base current.

The characteristic curves for a *common-base* (CB) configuration are shown in Fig. 9(b). The gain for this circuit is very nearly equal to unity and the curves show that the collector current I_C, for a given value of I_E, is virtually independent of V_{CB}. This CB configuration thus serves as a useful constant-current source for applications in which the load resistance, and hence V_{CB}, varies. The input resistance for this configuration is low ($\approx 30\ \Omega$) and the output impedance is typically a few megohms.

Field-effect transistors. The junction field-effect transistor (JFET) is a simple semiconductor device that is superior to the bipolar junction transistor (BJT) as a high-input-impedance amplifier and for many switching applications. As depicted in Fig. 10, it consists of a narrow *n*- or *p*-type conductive channel embedded between two connected regions of *p*- or *n*-type silicon that serves as a controlling *gate*. Conductivity between the input to the channel (the *source*) and the output (the *drain*) is determined by the charge-carrier density in the channel. This in turn is controlled by the gate bias voltage. If the gate is

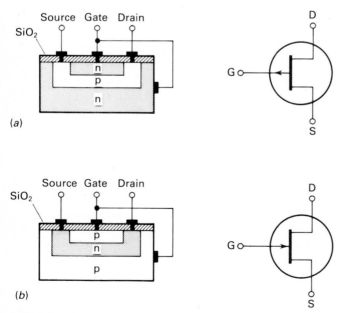

FIGURE 10
Basic structure and symbols for *p*-channel (a) and *n*-channel (b) junction field effect transistors (JFETs).

reverse biased with respect to the source, the electric field of the gate creates a depletion region in the channel, effectively narrowing the conductive region and decreasing the drain current I_D.

The current–voltage characteristic curves for a typical n-channel JFET are shown in Fig. 11. As the gate bias V_{GS} becomes more negative, the effective resistance of the channel increases and the drain current drops. This is reflected in the change of slope of the different V_{GS} curves at low drain-source voltages V_{DS}, an "ohmic-behavior" region in which a JFET acts as a variable voltage-controlled resistor. As V_{DS} is increased, the depletion region between the gate and the drain expands and eventually "pinches off" the channel at a voltage V_P, indicated as a dashed line in the figure. Further increases in V_{DS} have no effect on the drain current until eventually ionization breakdown occurs and the current rises sharply.

The JFET serves as a useful switch since, when the gate bias V_{GS} is zero, the resistance is low (25 to 100 Ω). In contrast to this ON condition, application of a large negative V_{GS} bias increases the resistance to $10^9\,\Omega$ or greater for the OFF condition. As an amplifier, the JFET has the advantage that an input voltage applied to the gate sees a very high input impedance ($\sim 10^9\,\Omega$) because the gate channel junction is reverse biased. In contrast, the BJT has a low base impedance since the base–emitter junction is forward biased. The simple unipolar structure of the JFET conduction channel also results in a low noise level and JFETs are often used as initial amplifiers for low-level signals. The JFET is somewhat slower than the BJT device, however.

A modification of the JFET that provides an even higher input impedance is the metal oxide semiconductor (MOS) field-effect transistor

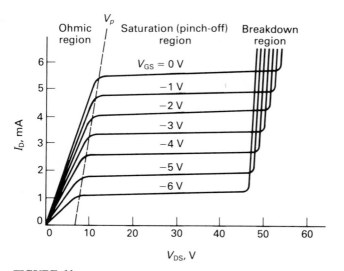

FIGURE 11
Current-voltage characteristics of an n-channel JFET.

(MOSFET). In this device an insulating layer of silicon dioxide is sandwiched between the gate electrode and the extremely thin channel material as shown in Fig. 12. The gate field is able to extend through the insulator into the channel to control the drain current but the insulator provides an extremely high resistance to current flow through the gate $(10^{12}$ to $10^{15}\,\Omega)$. The current–voltage characteristic curves of the MOSFET are similar to those for the JFET. Because of the high gate resistance of the MOSFET, static charge build-up can produce voltages high enough to cause discharge through the thin insulating layer. To avoid this, some MOSFET gates are protected by Zener diodes. Loose components are usually packaged in conducting foam and care to avoid static charge is necessary in handling these devices.

The MOSFET can be operated in either a *depletion* mode, in which an effective reverse bias appears at the surface of the channel, or in an *enhancement* mode, in which there is a forward bias. In the latter case no channel need be actually diffused between source and drain regions; the gate field is sufficient to create a conducting channel of opposite type on the surface of the substrate that separates source and drain regions. Frequently, *n*- and *p*-type MOSFET elements are produced on the same substrate and are connected in series to form *complementary* MOSFETs (CMOS). This series

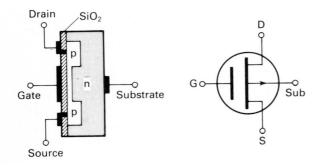

(a)

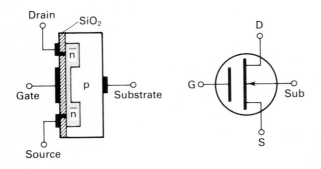

(b)

FIGURE 12
Structure of and symbol for *p*-channel (a) and *n*-channel (b) depletion mode MOSFETs.

combination of transistors has found widespread use in modern integrated logic circuits, since the pair acts as two switches in series that always have opposite states. The result is that the circuit consumes almost no power when it is not being switched. The large-scale integrated-circuit production of many CMOS transistors on a single chip has fueled today's microcomputer revolution. Since the first production of the single-chip transistor in 1959, component densities have increased exponentially, reaching more than a million transistors per chip in 1987 with predictions of a one-billion-component chip by the year 2000. The electronic control capability and computing power of such devices will have a profound effect on measurement instruments and on all aspects of science in the coming decades.

OPERATIONAL AMPLIFIERS

The use of semiconductor devices has been greatly facilitated by the incorporation of many discrete elements on a single microchip of specific purpose, such as amplification, memory storage, switching, time delay, and so on. An example is the operational amplifier (op amp), a device whose function is easy to understand and whose internal construction has been carefully designed to provide nearly ideal characteristics of high input impedance, linear gain over a wide frequency range, and low output impedance. While it is not necessary to know the full details of the inner operation of the simple triangular symbol used to represent the operational amplifier in order to make effective use of it, some knowledge of its basic circuit elements is helpful in appreciating its specifications and limitations. A more extensive discussion of these aspects can be found in Refs. 1–6.

Figure 13 shows a simplified view of the three basic stages of an LF351 operational amplifier. The input stage consists of two matched p-channel JFETs used in a differential mode. This arrangement provides a very high input impedance and efficient cancellation of any noise voltages common to both inputs. The second stage consists of several BJT amplifiers to provide an overall current gain of about 10^6. The final output stage serves to provide current with an output impedance of a few ohms or less. The device actually has many more transistors, not indicated in the figure, which serve to provide appropriate biasing levels and current sources.

Typical specifications of several "industry standard" operational amplifiers are provided in Table 1.[5] The 741 BJT amplifier consists of 20 transistors incorporated in a single chip that is produced in high volume at a cost of about $0.25. The 741 was the first general-purpose op amp and is still widely used because it is inexpensive, robust, and adequate for most routine applications at frequencies below about 10 kHz. The 301A and LM–116 transistors are more recent bipolar versions with improved specifications.

Operational amplifiers that contain both bipolar and MOS or FET transistors on the same chip are termed BiMOS or BiFET op amps. These families have much higher input impedances and a frequency response

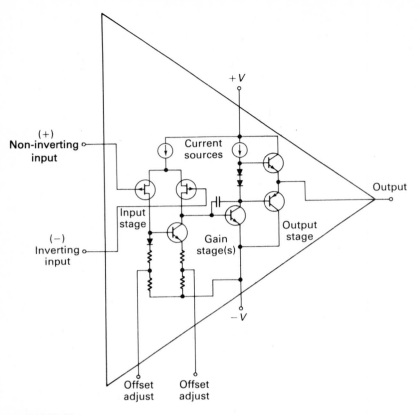

FIGURE 13
Simplified schematic of an LF351 BiFET integrated circuit operational amplifier with FET input stage and BJT amplification stages. From Ref. 1 with permission.

extending above 1 MHz. The 3140 op amp is pin-compatible with the 741. In addition to these standard units, a number of manufacturers produce premium op amps with significantly improved performance at a cost that ranges from $5 to $200. For laboratory and research applications, where the component cost is usually small compared to design and construction costs, it is almost always best to choose these premium devices. Full specifications and helpful applications literature can be obtained from Analog Devices, Burr-Brown, Com-Linear, Avantek, and other companies.

In its overall function, an op amp is basically an amplifier of voltage differences, and it is usually represented by a triangular symbol as shown in Fig. 14. The plus and minus signs on the input terminals refer to the polarity of the output voltage produced by a positive voltage at that input with respect to the other. All voltages are referenced to the ground line and this is often omitted from the symbol. The output voltage V_0 is related to the input difference

$$V_0 = A(V_+ - V_- + \Delta V) \approx A(V_+ - V_-) \tag{5}$$

TABLE 1
Comparative specifications of industry-standard operational amplifiers

Device	Bipolar			BiMOS		BiFET	
	741C	301A	LM-11C	3140	3130	LF351	LFT356
Input offset voltage V_{os}, mV (max)	6	7	0.6	15	15	10	0.5
Input voltage drift $(dV_{os}/dT)_{max}$, $\mu V\,°C^{-1}$					5	10	3
Input bias current I_b, nA (max)					0.05	0.2	0.05
Input offset current I_{os}, nA (max)					0.03	0.1	0.01
Input noise voltage V_n, nV $(Hz)^{-1/2}$						16	12
Input impedance R_i, MΩ					5×10^6	10^6	10^6
Open-loop voltage gain a, V(out), mV(min)					50	25	50
Common mode rejection ratio CMRR, dB (typ)	90	90	110	90	90	100	100
Unity-gain bandwidth f_u, MHz	1	>4	5	4.5	15	4	4
Slew rate $(dV_o/dt)_{max}$, V μs^{-1}	0.5	1.5	0.3	0.9	30	13	12
Output swing using ± 15 V V_{max}, V (typ)	10	12	12	13	13	12	12
Output current $(I_o)_{max}$, mA	25	20	15	20	20	20	25
Power dissipation P_{max}, mW (typ)	500	500	500	600	600	500	570
Compensation	int.	ext.	ext.	int.	ext.	int.	int.
Approximate price, dollars	0.30	0.50	3.00	1.00	1.00	0.50	1.00

where A is the amplifier open-loop gain (10^4 to 10^8), which is ideally the same for both inputs. The quantity ΔV is a small voltage offset ($\sim$1 mV or less) that is characteristic of the amplifier and can be canceled by applying a small compensating voltage to the inverting ($-$) input.

For stable operation, the operational amplifier is generally used with a

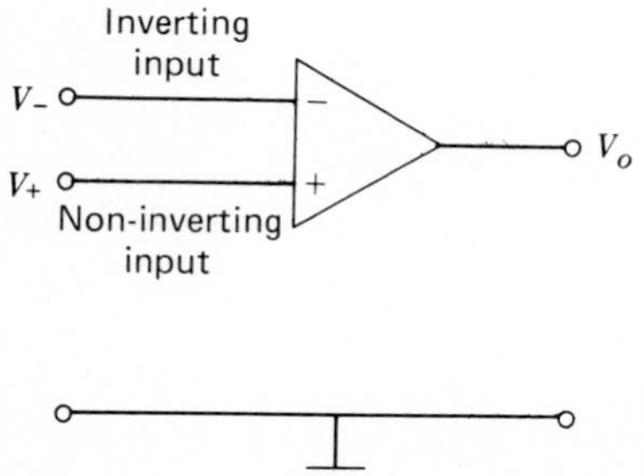

FIGURE 14
Representation of an operational amplifier.

feedback loop that can involve a resistor, a capacitor, or a diode. The essential function of the amplifier is to produce whatever voltage is required at the output to hold, by means of the feedback loop, the inverting input at a voltage close (within 1 mV or so) to that of the noninverting input. When the feedback is achieved with a simple connector as shown in Fig. 15(a), the output acts as a *voltage follower* with $V_0 = V_i$. Such a circuit is commonly used when a voltage source cannot supply sufficient current to drive a voltage-measuring device or

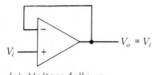

(a) Voltage follower

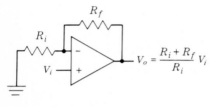

(b) Non-inverting voltage amplifier

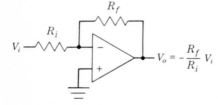

(c) Inverting voltage amplifier

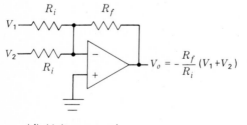

(d) Voltage summing amplifier

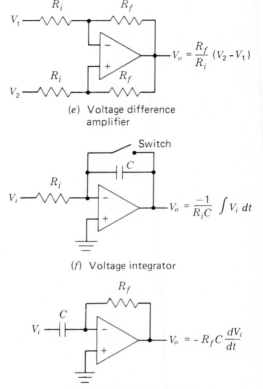

(e) Voltage difference amplifier

(f) Voltage integrator

(g) Voltage differentiator

(h) Current-to-voltage amplifier

FIGURE 15
Some circuit configurations with operational amplifiers.

other load. The input impedance is that of the amplifier (ranging from 10^5 to $10^{14}\ \Omega$) and the output impedance is small (10^{-1} to $10^3\ \Omega$).

If amplification of a small input voltage is desired, either of the circuits shown in Figs. 15(b) and 15(c) can be used. In the latter case, the noninverting input is grounded and the amplifier acts to hold the inverting ($-$) terminal at a *virtual ground,* i.e., $V_- \cong 0$, but there is no actual connection to ground. (Actually, V_- approaches zero as the gain A approaches infinity.) The source resistance relative to ground potential is determined by the input resistor R_i, and a current of

$$I = \frac{V_i - V_-}{R_i} \cong \frac{V_i}{R_i} \tag{6}$$

flows into the inverting input terminal. Since the input resistance to the amplifier is very high, essentially all of this input current must flow out through the feedback resistor R_f, corresponding to a potential at the output of

$$V_0 = V_- - IR_f \cong -\frac{R_f V_i}{R_i} \tag{7}$$

The polarity is thus reversed and the output voltage is amplified by the factor R_f/R_i. This gain must be less than the open-loop gain A, and the feedback resistance R_f must be small compared to the amplifier input resistance. If several voltage inputs are applied to the inverting input, as in Fig. 15(d), the amplifier can be used to perform a summing operation with simultaneous amplification. Feedback arrangements to perform voltage subtraction, integration, and differentiation are shown in Figs. 15(e) to 15(g); other operations such as multiplication, division, and logarithmic manipulations are also possible.

For many applications in spectroscopy, electrochemistry, chromatography, and other areas, a current, instead of a voltage, is generated that is directly proportional to a property of interest. For example, electron emission from a photocathode is linearly related to variations in the incident-light intensity. In a photomultiplier, this cathode current is amplified and the anode current I (typically 10^{-9} to 10^{-3} A) can be converted to a voltage $V_0 = IR_f$ by passage through a load resistor R_f. However, if R_f is large, the potential at the anode varies significantly with anode current. As a result, the gain of the tube changes with current so that the anode current (and the voltage across R_f) is no longer linearly related to light intensity. A current-to-voltage amplifier, such as that depicted in Fig. 15(h) can be used to solve this problem. This arrangement offers no impedance to current flow into the virtual ground at the inverting input and the anode potential is constant. Of course, no current actually flows into the amplifier; instead the output level drops to a negative potential to ensure that all of the photomultiplier current flows through the feedback resistor R_f. The same magnitude of voltage signal is produced as with a simple load resistor, but with opposite polarity. The amplifier in effect acts as a zero-impedance device while generating a large voltage across R_f, which can then be read with a meter or recorder of relatively low input impedance. Summation, integration, and other operations involving current sources can be done with the circuits shown in Figs 15(a) to 15(f) if the input resistance R_i is simply eliminated.

In principle, an ideal operational amplifier produces an output that is proportional only to the differential voltage $V_+ - V_-$ and is independent of the *common mode voltage* [CMV = $\frac{1}{2}(V_+ + V_-)$]. The extent to which this is true of a real amplifier can be judged by the *common mode rejection ratio:*

$$\text{CMRR} = \frac{\text{gain for } (V_+ - V_-)}{\text{gain for CMV}}$$

Typical values of CMRR for ordinary operational amplifiers are 10^3 to 10^5. Frequently gain and CMRR factors are given in decibels, which is 20 times the base-ten log of the numerical factor. A CMRR factor of 10^5 corresponds to 100 dB. High values of CMRR are particularly important for applications where common noise signals (such as 60-Hz pickup from power lines) occur on both inputs to the amplifier. Examples include signal amplification of outputs from transducers such as thermocouples, thermistors, strain gauge bridges, and other devices. For such applications, instrumentation or isolation amplifiers are generally used.

An instrumentation amplifier consists of three or more operational amplifiers and is carefully designed to achieve gains in the range 1 to 10^4 with very high values of input impedance (10^8 to $10^{10}\ \Omega$) and CMRR (10^4 to 10^6). Figure 16 shows a simplified circuit diagram for an instrumentation amplifier used to amplify a thermocouple output. With such an arrangement, the two thermocouple leads can have common voltages as large as ± 10 V. For common

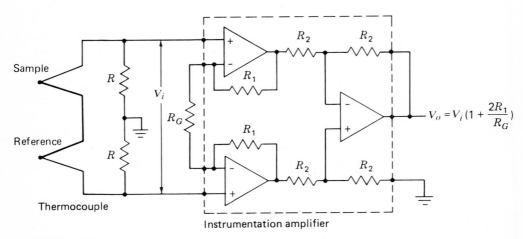

Instrumentation amplifier

FIGURE 16
Thermocouple amplification using an instrumentation amplifier.

mode voltages in excess of 10 V, an isolation amplifier is used in which the input is electrically isolated from the output and from the power supply. With such devices, CMRR values of 10^8 or more are possible for common mode voltages that exceed 1000 V.

ANALOG-TO-DIGITAL CONVERSION

Most voltage or current measurements in physical chemistry involve conversion from an analog form to a digital number at some recording stage, so that subsequent numerical analysis can be performed. For example, the student acts as an analog-to-digital converter in nulling a potentiometer and then recording a Chromel–Alumel thermocouple output as, say, 10.24 mV. This digital number then might be used with a reference table to deduce that the temperature at the sensing point is 252°C. This same analog-to-digital voltage conversion and even the conversion to temperature) can also be done with integrated circuit (IC) chips consisting of combinations of operational amplifiers acting in a linear mode for amplification and in a nonlinear fashion for switching purposes and interfacing to a display or a computer. These elements form the heart of digital voltmeters (DVM), which are readily available with three to six display digits at accuracies of 0.1 to 0.003 percent of full scale. With the development of large-scale ICs, the cost of such meters has dropped dramatically (ranging from $50 to $3000 in 1987, depending upon the accuracy and number of digits), and it is certain that they will find increasing use in instrumentation and in instructional laboratories.

A variety of methods are used for the digitizing operation but only the two most common techniques will be discussed here. The dual-slope integration technique (Fig. 17) is used with most display meters when millisecond conversion times are adequate. The conversion involves two steps. In the first step the signal is applied to an operational amplifier, which provides extremely high input impedance plus gain for low-level inputs. At the start of a measurement cycle, control circuits simultaneously switch the amplifier output to the integrator input, remove the short from the integrating capacitor, and open a gate to pass clock pulses to a counter. The integrator output, which begins from an initial value of zero, produces a linear ramp with slope and direction proportional to the instantaneous amplitude and polarity of the input voltage (cf., Fig. 15(f)). The integration continues until a fixed number of counts have accumulated (10^4 for a 4-digit meter). At this time, in the second step of the cycle, the integrator input is switched to a reference voltage whose polarity is opposite to that of the input signal. The counter is reset to zero and again begins to count while the reference voltage drives the integrator back to zero. The slope of this second ramp is proportional to the reference voltage, and the time to reach zero is exactly proportional to the input voltage. When the integrator reaches zero, the counter gate is closed and the count displayed is numerically equal to the unknown input voltage.

The dual-slope method has a number of important features. Conversion

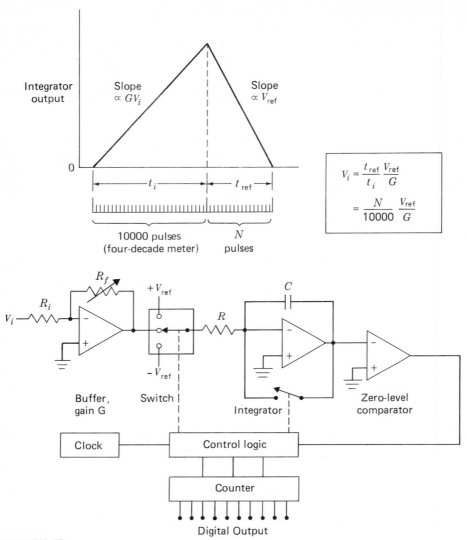

FIGURE 17

Dual-slope type A/D converter. For calibration, either V_{ref} or the buffer gain G can be adjusted to give the correct display for a known input voltage.

accuracy is independent of the effect of temperature or aging on the clock frequency and integrating capacitor value since they only need remain stable during a conversion cycle. Accuracy is thus determined by the accuracy and stability of the reference source. Resolution is limited primarily by the analog resolution of the converter. Because of the integration operation the converter gives excellent high-frequency noise rejection, and acts to average the signal over the integration period. Moreover, by choosing the clock frequency to be a multiple of 60 Hz, such as 600 kHz, it is possible to achieve nearly complete rejection of 60-Hz ac fluctuations on the input signal since this ac noise averages to zero when one (or more) full 60-Hz cycle equals the 10^4 count period. The minimum period for this process is 16.67 ms so that this advantage of the dual-slope method of integration is not obtained at high digitizing rates.

For applications requiring high resolution and high speed, the successive approximation A/D conversion method is used. This method is illustrated in Fig. 18 and consists of comparing the unknown input against a precisely

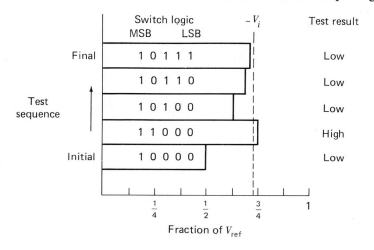

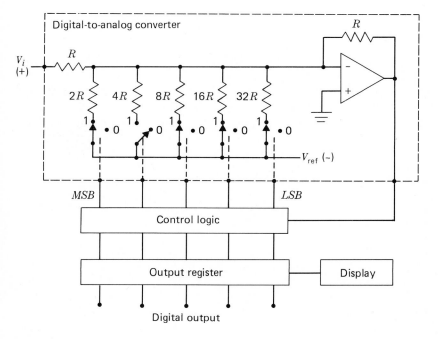

FIGURE 18
Successive approximation type A/D converter. A 5-bit converter is shown for simplicity of illustration, but commercial instruments typically use 10 to 16 bits.

generated internal voltage from a digital-to-analog (D/A) converter. The D/A converter operates by switching in various binary fractions of a reference voltage (V_{ref} = full scale, typically 1 to 10 V) to the summing point (−) of an operational amplifier. The polarities of V_{ref} and V_i are opposite so that subtraction occurs when V_i is also introduced into the summing point. The output polarity of the operational amplifier thus indicates the relative magnitudes of V_i and the total fraction of the reference voltage passed by the control logic. At the start of a conversion cycle, the D/A converter's most-significant binary bit (MSB), which is one-half full scale, is compared with the input. (For a discussion of binary numbers, see Chapter XXI). If smaller than the input, the MSB is left on and the next bit (which is one-fourth full scale) is added. If the MSB is larger than the input, it is turned off when the next bit is turned on. The procedure is analogous to that used in placing weights on one side of a two-pan analytical balance. The process of comparison is continued down to the least-significant bit (LSB) which is 2^{-n} of full scale for a converter of n bits. At this point the output register contains the complete output digital number, which can be sent to a computer or other processing device. The bit-comparison operation can be done in as little as 100 ns, so that a 10-bit conversion (with an accuracy of 1 part in $2^{10} = 1024$) can be done in 1 μs. In practice, conversion times of 10 to 100 μs are typical for high-resolution converters, which generally have a 10-bit to 16-bit digitizing capability.

DIGITAL MULTIMETERS (DMM)

An A/D converter is basically a dc voltage measuring device but, with the appropriate signal pre-processing, it can also serve to measure resistance and dc current as well as ac voltage and current. A block diagram of a digital multimeter (DMM) is shown in Fig. 19. The A/D unit, commonly based on the dual-slope A/D conversion method, will have a fixed input range

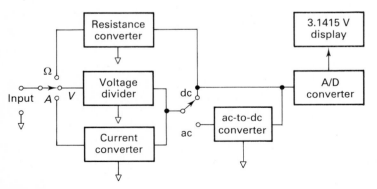

FIGURE 19
Block diagram of a digital multimeter (DMM).

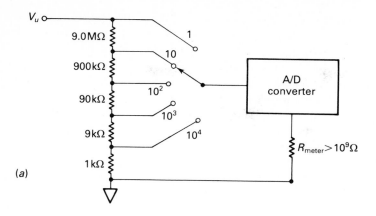

(a)

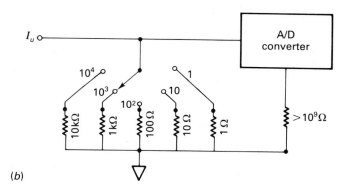

(b)

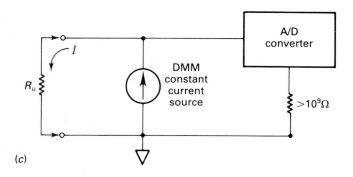

(c)

FIGURE 20

(a) Voltage divider network for DMM. For accurate measurement of an unknown voltage V_u, the input impedance of the meter should be much larger than the resistor network sum. (b) Shunt network to convert current to voltage for a DMM measurement of an unknown current I_u. (c) Circuit for measurement of an unknown resistance R_u with a DMM.

corresponding to the most sensitive scale. The voltage attenuator is just a resistance divider network (Fig. 20(a)) that allows one to measure larger voltages. On some DMMs with autoranging capability, the full-range setting is changed automatically to ensure the highest resolution without over-ranging. The circuitry also senses the polarity of the input, which is displayed along with the voltage, the correct decimal position, and the appropriate voltage unit.

For current measurements, the input current can be routed through precision resistors to produce a proportionate dc voltage (Fig. 20(b)). A separate input is sometimes provided for high currents to avoid damage to sensitive elements. For ac measurement of voltage and current, the attenuated signal is usually rectified and filtered to present a dc voltage to the A/D converter. The ac converter may be either a true RMS (root-mean-square) or an averaging type. Both types of meters display the same RMS result for sine-wave inputs but, for example, the averaging meter gives a value which is high by 11.1% for a square wave and low by 3.8% for a triangular wave.

To determine resistance, the DMM is equipped with an operational amplifier constant-current source that provides a known current I in the unknown resistor R_u (Fig. 20(c)). The resultant voltage drop is then measured by the A/D circuit. Some meters provide more accurate resistance measurement by providing leads for the current source which are separate from those used for measurement of the IR drop across the unknown resistor. Since very little current passes through the latter leads, this eliminates any error due to the voltage drop across the measuring leads and connections, which can be significant in the simpler two-lead meter.

POTENTIOMETER CIRCUITS

Although a precision DMM provides fast and accurate measurement of most electrical quantities, the traditional use of potentiometers circuits is still common in many physical chemistry laboratories. In part this is probably due to "economic inertia" (potentiometers do not easily wear out), but it is also true that the potentiometer enables one to obtain very accurate measurements over a wide voltage range. For example, a 5-decade DVM with a most-sensitive scale of 20 mV provides a resolution of 1 μV, but on the 2 V scale the voltage can only be read to 100 μV. Using a potentiometer to accurately cancel out most of the voltage permits the use of the more sensitive scale. This nulling principle is a common technique in improving the accuracy of physical measurements and its clear illustration in voltage applications is of distinct pedagogical value. Potentiometers also serve as key elements in bridge circuits used in a number of the experiments presented in this text.

Basic circuit. As the name implies, the potentiometer is a device for measuring electrical potential (voltage) difference. The principle of the potentiometer can be described in terms of the simple "slide-wire" potentiometer shown schematically in Fig. 21. With a constant direct current flowing

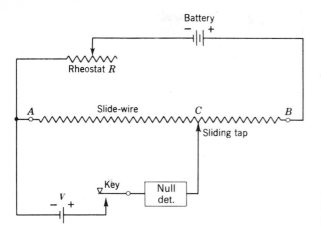

FIGURE 21
Schematic diagram of a simple slide-wire potentiometer.

through the uniform-resistance slide-wire *AB,* the potential difference between *A* and *C* should be accurately proportional to the length *AC.* Thus the potentiometer is a device for selecting with high precision any desired voltage that is less than a certain maximum value. This maximum value depends on the construction of the slide-wire and the voltage of the battery. The voltage across part of the slide-wire (*AC*) may be compared with another voltage by connecting the negative pole of the external voltage *V* with the negative end of the slide-wire (at *A*) and connecting its positive pole with the positive slide-wire contact *C* through a null detector and a tapping key. When this key is depressed, a nonzero indication on the null detector will be observed if the voltage of *V* and that between *A* and *C* are not equal. The sign of the imbalance will obviously depend on whether the potential at *C* is higher or lower than that at the positive pole of *V*. If the position of the sliding contact is adjusted until a zero indication is observed on depressing the tapping key, the voltage of *V* must be the same as that between *A* and *C*. Indeed, a linear scale marked off directly in volts can be placed alongside the slide-wire, and the voltage between *A* and *C* can be read directly from this scale, provided that the current through the slide-wire has been properly adjusted by manipulation of the rheostat *R*.

The adjustment of the current through the slide-wire is accomplished by connecting a source of known voltage, such as a *standard cell,* in place of the potential *V*. The slider *C* is then placed at the setting on the slide-wire scale corresponding to the known potential, and the rheostat *R* is adjusted until a null indication is observed on depressing the tapping key. This standardization of the potentiometer against a standard cell should be repeated frequently during the experimental work because the battery voltage usually shows a tendency to drift slightly.

The accuracy of a voltage measurement depends most of all on the design and the quality of the potentiometer unit used. For a precision potentiometer, the relative accuracy of measurement (ability to measure very small changes in voltage) will depend a great deal on the null detector sensitivity, while the absolute accuracy will depend more on the accuracy of the standard-cell voltage value.

Null detectors. For many years, the standard null detector for voltage comparisons was the D'Arsonval, or moving-coil, galvanometer. This detector senses current and consists of a rectangular coil of many turns of insulated copper wire mounted in the field of a magnet, usually a horseshoe-shaped permanent magnet. This coil is suspended from a fine wire or very thin metal strip, which serves as an electrical lead and also as a torsion fiber. Attached to the bottom of the coil is a loosely coiled wire, which serves as the other lead. Current flowing through the coil moves perpendicular to the magnetic lines of force and produces a torque on the coil. Therefore, the coil will rotate in the field until the magnetic torque is balanced by an opposite mechanical torque introduced by twisting the suspension. Motion of the coil can be detected in two ways: by the position of an indicating pointer attached to the coil or by the position of a light beam reflected from a small mirror attached to the suspension. The pointer-type galvanometer is cheaper, more rugged, and less sensitive than the reflecting type. For any type of D'Arsonval galvanometer the amount of deflection per unit of current through the coil will depend on certain design parameters (the size of the coil, the number of turns of the wire in the coil, the magnetic field intensity, and the stiffness of the suspension.) Typical current sensitivities are 0.5 to $1000 \, \mathrm{nA \, mm^{-1}}$ or, in terms of voltage sensitivity, 0.5 to $1000 \, \mu\mathrm{V \, mm^{-1}}$.

More recently, galvometers have generally been replaced by stable, high-gain operational amplifiers of high sensitivity. These permit the use of ordinary meters, microphones, or other devices to indicate the null condition. Two microvoltmeter–null detectors that are often used for this purpose are available from Keithley (Model 155) and Fluke (Model 845AB). Both instruments have an input impedance of $1 \, \mathrm{M\Omega}$ and provide an analog display with $\pm 20 \, \mathrm{nV}$ accuracy on the most sensitive scale of $1 \, \mu\mathrm{V}$.

For most applications involved in the experiments of this text, a precision digital multimeter, operated on its lowest voltage range, serves as a suitable null detector. For example, the Fluke model 8842A provides a resolution of $0.1 \, \mu\mathrm{V}$ with an accuracy of $2 \, \mu\mathrm{V}$ on its lowest $20 \, \mathrm{mV}$ scale. Similar performance is obtained with a Keithley model 196 and, at higher cost, the Keithley nanovoltmeter model 181 gives $10 \, \mathrm{nV}$ resolution and $50 \, \mathrm{nV}$ accuracy on a $2 \, \mathrm{mV}$ scale. Such multimeters also serve as very flexible instruments for the measurement of resistance as well as ac–dc voltage and current.

Standard cell. The electrochemical cell used as a voltage standard is the Weston cell, shown in Fig. 22. The voltage of this cell changes only slightly with temperature and is given in absolute volts by

$$V_{\mathrm{W}} = 1.01865 - 4.1 \times 10^{-5}(T - 20) - 9.5 \times 10^{-7}(T - 20)^2 \quad \text{volts} \qquad (8)$$

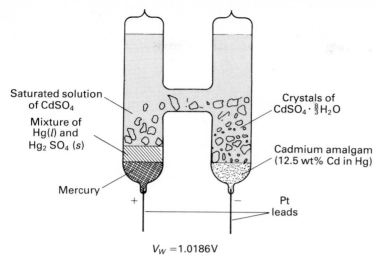

Saturated solution of CdSO$_4$

Mixture of Hg(l) and Hg$_2$SO$_4$ (s)

Mercury

Crystals of CdSO$_4 \cdot \frac{8}{3}$H$_2$O

Cadmium amalgam (12.5 wt% Cd in Hg)

Pt leads

$+$ $-$

$V_W = 1.0186$V

FIGURE 22
The saturated standard Weston cell.

where the temperature T is in Celsius degrees.[7] In actual practice an unsaturated cell is often used in which the CdSO$_4$ solution is made saturated at 4°C and the solution is slightly unsaturated at the operating temperature. This cell has the advantage of a lower temperature coefficient (1×10^{-5} V °C^{-1}) but the voltage at 20°C can vary from 1.0185 to 1.0195 for different cells. However, the calibrated value for a given cell is reproducible to about 10 μV. For highest accuracy, standard cells should be calibrated against a reliable reference every few years and, in use, one should *never* pass more than 10^{-4} A through the cell. A simple voltage-follower circuit (Fig. 15(a)), with the Weston cell in the feedback loop, can be used to deliver current at the standard voltage without significant loading of the standard cell.

Leeds and Northrup Student Potentiometer. The L & N Student Potentiometer, Model 7651, is a medium-precision unit (accuracy of ±0.5 mV on the 1.6 V scale and ±0.01 mV on the 0.016 V scale) which is still widely used in instructional laboratories for measuring voltages in the range 0.001 to 1.6 V. Indeed, a number of the experiments in this book assume the availability of this instrument (or some other of comparable quality and similar construction). Therefore, we shall discuss its operation in some detail. Figure 23 shows a complete potentiometer circuit based on the Student Potentiometer unit; note that several auxiliary components are required in addition to the potentiometer unit.

In the Student Potentiometer the potential difference is selected, not between a fixed end and a variable point, but rather between two variable points, one of which is moved in steps (by means of a step switch) and the other of which is moved continuously (a dial "slide-wire"). The stator of the

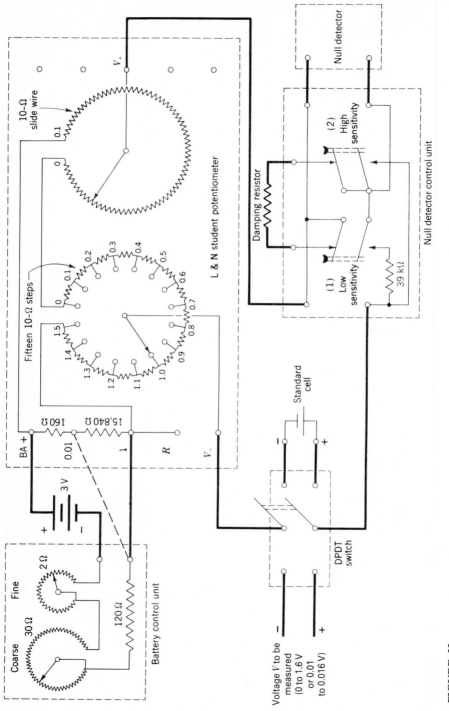

FIGURE 23

Complete wiring diagram for a potentiometer circuit based on the L & N Student Potentiometer. The internal-wiring diagram of the potentiometer is somewhat simplified.

step switch has 16 contact points, between each adjacent pair of which is a noninductively wound fixed resistor of 10-Ω resistance, all the resistances being precisely equal to one another. The slide-wire is in series with the 15 fixed resistances, and its resistance is precisely equal to the resistance of any one of the fixed resistors. The 16 points of the step switch are numbered in tenths of a volt from 0 to 1.5 V, and the scale of the slide-wire is so graduated (the smallest graduations being 0.0005 V) as to cover the range 0 to 0.1000 V. A potential can be estimated to 0.0001 V from the slide-wire dial reading. The desired potential difference is obtained between binding post V^-, which is connected to the rotor of the step switch, and binding post V^+, which is connected to the sliding-contact rotor of the dial slide-wire. Thus the desired potential difference or emf is the sum of a step-switch reading and a slide-wire reading. The proper current through the 15 resistances and the slide-wire is obtained from batteries through rheostats, and the current supply is normally connected to the binding posts marked BA^+ and 1. The potentiometer is standardized against a standard cell in the same way as described previously in the case of the simple slide-wire potentiometer.

If, *after standardization,* the negative lead of the battery circuit is switched from binding post 1 to that marked 0.01, the range of the Student Potentiometer changes from 0–1.6 to 0–0.016 V; that is, all readings must be multiplied by 0.01. This change of scale is accomplished by means of an internal resistive network so designed (see Fig. 23) that the current flowing through the step-switch resistors and the slide-wire is reduced to precisely 1 percent of the value established during the standardization. The total resistance in the battery circuit is, however, not affected on switching from post 1 to 0.01. This low-scale arrangement permits one to make a more precise measurement of very small emf values, such as thermocouple emfs below 16 mV or the potential drop across a 1-Ω standard resistor when less than 16 mA of direct current are flowing through it.

The battery circuit shown in Fig. 23 consists of two 1.5-V dry cells connected in series with a battery control unit that permits both coarse and fine adjustment of the current through the potentiometer unit. It is possible to achieve better current stability by using a 2-V low-discharge wet cell. In that case, the 120-Ω fixed resistor in the control unit should be replaced by a 20-Ω resistor so that a current of 10 mA can be obtained.

If a high-impedance microvoltmeter is used to monitor the voltage imbalance, the inputs to the null detector control unit shown in Fig. 23 can be attached directly to the microvoltmeter. This should be set initially to a high scale and then be gradually reduced to the most sensitive range as the null position is approached.

For a galvanometer detector, more care is required to limit the current and the null detector control unit shown in Fig. 23 is desirable. This consists of two spring-loaded push buttons either of which, when depressed, will close the circuit through the galvanometer. When the low-sensitivity button (1) is depressed, a protective resistance of 39 kΩ is in the galvanometer circuit to limit the current flow when the potentiometer setting is far from the proper value. When the high-sensitivity button (2) is depressed, this protective resistance is *not* in the circuit. Therefore, button 1 should be used during all preliminary adjustments, and button 2 should be depressed only when the deflections are too small to detect using button 1. Either button should be depressed only briefly and not held down. There is also a position on the control unit for mounting an external damping resistor so that it will be across the galvanometer whenever *neither* button is depressed. This resistor is chosen to match a critical dampening value (CRDX) which is a characteristic of a particular galvanometer model—typical values range from about $30\,\Omega$ to $10\,k\Omega$. This resistance serves to damp oscillations of the galvanometer needle and, ideally, should be the total resistance in series with the galvanometer when button 2 is pushed.

To summarize, a detailed outline is given below of the procedure for using the circuit shown in Fig. 23.

1. Connect the components, taking care that the batteries, standard cell, and unknown emf are connected with the correct polarities. (Until the student has become thoroughly familiar with the potentiometer circuit, it is advisable to omit the connections to the standard cell until the wiring has been checked by an instructor.)

2. Set the step switch and slide-wire dial to the value of the voltage of the standard cell. By means of the DPDT switch, place the standard cell in the circuit.

3. Standardize the potentiometer against the standard cell by adjusting the coarse battery control for a voltage null. If using a galvanometer, adjust until there is no deflection when galvanometer button 1 is depressed. Then adjust the fine battery control until there is no deflection on depressing button 2. Do not hold button 2 down for a prolonged period.

4. Switch from the standard cell to the unknown potential V by reversing the DPDT switch. **Do not disturb the battery current setting.** Vary the potentiometer step switch and slide-wire dial setting for a null as in step 3. *Note:* If it is impossible to achieve balance at any setting of the potentiometer, either the unknown voltage V is greater than $1.6\,V$ or V is connected into the circuit with the incorrect polarity.

5. Record the voltage value for V, then restandardize and repeat the measurement.

WHEATSTONE-BRIDGE CIRCUITS

Direct-current Wheatstone bridge. The dc Wheatstone-bridge circuit provides a simple means of accurately determining an unknown resistance. As shown in Fig. 24(*a*), an arbitrary dc potential drop is established across the bridge from A to C and a null detector with tapping key serves as a detector of current flow

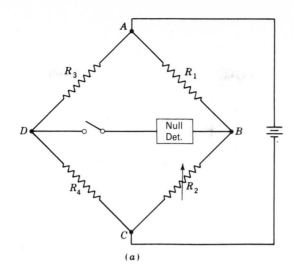

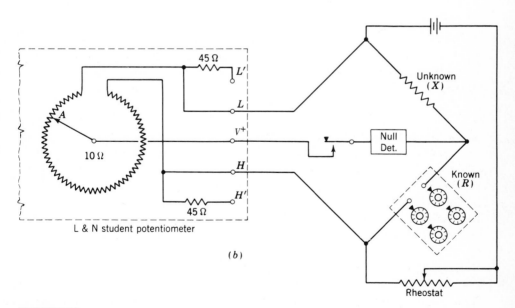

FIGURE 24

The dc Wheatstone bridge: (a) general schematic diagram of the bridge; (b) wiring diagram, based on the use of the slide-wire in an L & N Student Potentiometer. Slide-wire internal circuit slightly idealized.

from B to D. Since direct current is involved, all arms of the bridge are treated as purely resistive elements. When the bridge is balanced (i.e., zero current when the tapping key is closed), the potential at B must be the same as that at D and it follows that

$$\frac{R_1}{R_2} = \frac{R_3}{R_4} \tag{9}$$

A form of this bridge incorporating the slide-wire in an L & N Student Potentiometer is shown in Fig. 24(b). The slide-wire, which has a scale A reading from 0 to 1000, constitutes two of the four arms of the bridge. When terminals L and H are used, one obtains from Eq. (9) the condition of balance

$$X = \frac{A}{1000 - A} R \tag{10}$$

Often the known resistance R is variable, for example, a decade resistance box. In this case, R can be varied until A is somewhere in the range 450 to 550. A considerably greater sensitivity can then be achieved by shifting the L lead to L' and the H lead to H', obtaining the so-called "long bridge." The two left arms of the bridge now each contain an additional resistance 4.5 times that of the slide-wire itself, and the balance condition then becomes

$$X = \frac{4500 + A}{5500 - A} R \tag{11}$$

In general, use of the long bridge is preferable because of the greater sensitivity provided. Commercial Wheatstone bridges are available in a form in which R_3/R_4 can be set by a step switch to accurate decimal ratios from 10^{-3} to 10^3, and R_2 is a precision four- or five-decade resistance box. The use of the Wheatstone bridge with platinum resistance thermometers is discussed in Chapter XVI.

Alternating-current Wheatstone Bridge.[8–10] A Wheatstone bridge can be operated with an alternating- as well as a direct-current source. This is advantageous even when measuring an unknown element that is a pure resistance, since high-sensitivity lock-in-detection techniques can be used for the null detection (see Chapter XVIII). For measurements such as those in Exp. 17 on the conductance of electrolyte solutions, the use of alternating current is necessary to prevent polarization of the electrodes in the conductance cell. The basic circuit for an ac bridge is the same as that shown in Fig. 24(a) except that the voltage source is an oscillator, operating usually at 1 kHz (or any convenient frequency in the 0.5–10 kHz range) and the detector is an oscilloscope or lock-in detector.

In general, one must treat each arm of the bridge as a complex impedance Z:

$$Z = |Z| \exp(i\phi) = |Z|(\cos \phi + i \sin \phi) \tag{12}$$

where ϕ is the phase angle by which the voltage vector is advanced with respect to the current vector. One can also express Z in the form

$$Z = R + iX \tag{13}$$

where R is the ac resistance and X is the reactance. It then follows that

$$|Z| = (R^2 + X^2)^{1/2} \quad \text{and} \quad \tan \phi = \frac{X}{R} \tag{14}$$

The general bridge balance condition is

$$Z_1 Z_4 = Z_2 Z_3 \quad \text{or} \quad |Z_1||Z_4| \exp[i(\phi_1 + \phi_4)] = |Z_2||Z_3| \exp[i(\phi_2 + \phi_3)] \tag{15}$$

which requires that

$$|Z_1||Z_4| = |Z_2||Z_3| \tag{16}$$

$$\phi_1 + \phi_4 = \phi_2 + \phi_3 \tag{17}$$

Equations (16) and (17) represent the necessity to balance *both* the real (resistive) and the imaginary (reactive) elements. If one starts from the Z expression given in Eq. (13) instead of Eq. (12), the resulting balance conditions can be written in the form

$$(R_1 R_4 - R_2 R_3) = (X_1 X_4 - X_2 X_3) \tag{18}$$

$$(X_1 R_4 - X_2 R_3) = (X_3 R_2 - X_4 R_1) \tag{19}$$

In general, two adjustable elements are sufficient to satisfy relations (18) and (19). In practice, it is desirable to have the adjustments in the resistances independent of those in the reactances. This can be achieved in a *ratio bridge*.[8] Let us put the unknown element in arm 1 of the bridge, allow arm 3 to have a variable impedance Z_3, and choose the ratio arms 2 and 4 so that $Z_2/Z_4 = K$ where K is some real positive number (i.e., $R_2/R_4 = X_2/X_4 = K$). This is equivalent to having the same phase shift in legs 2 and 4, i.e., $\phi_2 = \phi_4$. It follows from Eq. (15) that the balance condition for this ratio bridge is

$$(R_1 + iX_1) = (R_3 + iX_3)K \tag{20}$$

Equating real and imaginary parts yields

$$R_1 = R_3 K = R_3 R_2/R_4 \tag{21}$$

$$X_1 = X_3 K = X_3 X_2/X_4 \tag{22}$$

and, from Eq. (19),

$$\frac{X_1}{R_1} = \frac{X_3}{R_3}; \quad \frac{X_2}{R_2} = \frac{X_4}{R_4} \tag{23}$$

$$\phi_1 = \phi_3 \quad \phi_2 = \phi_4$$

Equation (21) is the familiar Wheatstone resistance balance condition and Eq. (22) is a new condition that is needed when phase shifts can occur. Together,

Eqs. (20) and (21) ensure that the alternating potential at points B and D in Fig. 24(a) will be equal amplitude and exactly in phase at all times.

Since the reactance can be inductive or capacitive, and can be parallel or in series with the resistance, there are many variations of the ac bridge. An arrangement suitable for conductive measurements is discussed in Exp. 17 and an ac bridge used for capacitance determinations is shown in the following section. A more extensive treatment of ac bridges can be found in Refs. 8–10.

CAPACITANCE MEASUREMENTS

The measurement of capacitance is of importance in physical chemistry primarily for the study of dielectric properties. As discussed in Exps. 31 and 32, dipole moments can be determined from the dielectric constants of gases or dilute solutions. There is also considerable interest in the dielectric properties of liquids and solids. When there are strong dipolar interactions between molecules, as in a polar liquid, there is an appreciable relaxation time for dielectric polarization. In that case, *dielectric loss* (dissipation of electrical energy as heat) occurs and the dielectric constant is frequency dependent.[11,12] In the following discussion we shall be concerned mostly with methods of measuring the capacitance of samples with very low dielectric loss (gases or dilute solutions in nonpolar solvents).[13] For such samples, the resistance of the dielectric cell is very high and its capacitance is independent of frequency.

Heterodyne-beat method. The heterodyne-beat method is capable of extremely high precision, and it is a method commonly used for dipole-moment measurements. However, this method is limited to the investigation of gases or of solutions and liquids with very low conductance (cell resistance greater than $10^4\,\Omega$). If a substance with appreciable conductance is placed in the dielectric cell, which is part of the LC tank circuit of a high-frequency oscillator, too much energy will be dissipated in the tank circuit and there will not be enough feedback to maintain proper oscillation. The heterodyne-beat method is described in Exp. 31, and considerable details on the circuitry are available in the literature.[13]

Resonance method. While the resonance method is not capable of the very high accuracy of the heterodyne-beat method, it is quite satisfactory for many investigations and has the advantages of simple and low-cost circuitry. This method can best be explained in terms of the schematic diagram shown in Fig. 25. The primary circuit consists of an oscillator with the inductance coil L_1 in the plate circuit of the oscillator tube. The secondary circuit is loosely coupled via the inductances to the primary circuit. When the oscillator is in operation, a constant high-frequency alternating current flows through L_1 and a small amount of current also flows in the secondary circuit. The amount of current flowing in the secondary will depend both on the value of L_2 and on the total capacity $C = C_T + C_P + C_X$. By varying C one can achieve the *resonance*

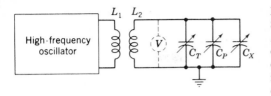

FIGURE 25
Schematic circuit for a resonance apparatus: L_1 and L_2 are fixed inductances; C_T is a coarse tuning capacitor; C_P is a precision air capacitor; X_C is the dielectric cell. A high impedance voltmeter V can be used as the detector (see text).

condition: $f = 1/2\pi\sqrt{L_2 C}$, where f is the frequency of the oscillator. When resonance occurs, the current flow in the secondary will be at its maximum value. This resonance point is detected by observing either a minimum in the dc plate current of the oscillator tube or a maximum in the ac voltage across the secondary. (The latter method requires the use of a high-impedance voltmeter.) A variable dielectric cell can be used (as in Exp. 31), or a fixed-plate cell can be used by employing the substitution method. In either case, C_X is obtained from the difference in the settings of C_P necessary to achieve resonance. The coarse-tuning condenser is used only to adjust the total capacity C to an initial value near the resonance value and should *not* be disturbed during the actual measurements.

In the above discussion, the secondary circuit was treated as a pure LC network, and that is usually not true in practice. For dielectric-constant measurements this is of no concern, since there is always a unique value of C that will make the secondary circuit resonant. We can also see that the resonance method has one great advantage over the heterodyne-beat method: solutions or liquids with appreciable conductance can be studied. This is due to the fact that the dielectric cell is in the secondary circuit, which is only loosely coupled to the oscillator, and cannot affect the feedback necessary for proper oscillation. Indeed, it is possible to determine dielectric loss with this method, since the sharpness of resonance (Q of the secondary circuit) is related to the loss.[11,12] The lower the loss in the dielectric, the sharper the resonance and the larger the Q value.

Bridge method. An ac impedance bridge used for dielectric measurements is very similar in principle to the ac Wheatstone bridge described previously; however, the reactance X is now large, since the dielectric cell has a large capacitance. Both the theory and the proper experimental techniques of ac bridge measurements are extensively discussed by Hartshorn[9] and by Hague[8]. Precision commercial bridges are available from Boonton Electronics, Electro Scientific Industries (ESI), and General Radio (Gen Rad Corp.). Only a very brief treatment of the "capacity bridge" will be given here.

The type of bridge often used for dipole moment work is a simple capacity bridge with two resistance arms (the "ratio arms") and two capacitance arms. Figure 26 shows a schematic diagram of a bridge with equal ratio arms of fixed resistance R and with one fixed capacitance arm C. The measuring capacitance arm consists of a parallel combination of a precision air capacitor C_P and the dielectric cell C_X. Large ($\sim 10^4$-Ω) noninductively wound

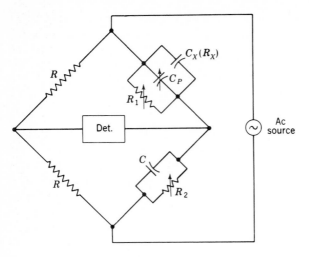

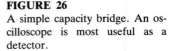

FIGURE 26
A simple capacity bridge. An oscilloscope is most useful as a detector.

resistors R_1 and R_2 are attached in parallel with the capacitances. These resistors help to balance out any small differences between the resistances of the capacitance arms and are very necessary to achieve balance if the sample has appreciable conductance (i.e., if cell C_X has a finite resistance R_X). Measurements can be made with a fixed-plate dielectric cell by using the substitution method, where a balance is first achieved with C_X in the bridge and then a second balance is achieved with C_X removed. Alternatively, a variable-capacity cell can be used as in Exp. 31; this has the advantage of eliminating any effects due to the capacity or resistance of the leads. If the sample has negligible conductance, only an adjustment of the C_P setting is required to achieve balance. For samples having appreciable conductance (finite R_X), it is necessary to adjust both C_P and the resistors R_1 and R_2, alternately, until the balance point is reached. (The cell resistance R_X is then given by $1/R_X = 1/R_2 - 1/R_1$, and the dielectric loss can also be calculated.) In either case, C_X is equal to the difference in the two settings of C_P required to balance the bridge.

REFERENCES

1. H. V. Malmstadt, C. E. Enke, and S. R. Crouch, "Electronics and Instrumentation for Scientists," Benjamin, Menlo Park, Calif. (1981).
2. J. H. Moore, C. C. Davis, and M. A. Coplan, "Building Scientific Apparatus," chap. 4, Addison-Wesley, Reading, Mass. (1983).
3. J. J. Brophy, "Basic Electronics for Scientists," 4th ed., McGraw-Hill, New York, (1983).
4. A. J. Diefenderfer, "Principles of Electronic Instrumentation," Saunders, Pa. (1979).
5. R. J. Higgins, "Electronics with Digital and Analog Integrated Circuits," Prentice-Hall, Englewood Cliffs, N.J. (1983).
6. B. H. Vassos and G. W. Ewing, "Analog and Digital Electronics for Scientists," 3d ed., Wiley, New York (1985).
7. J. J. Lingane, "Electroanalytical Chemistry," 2d ed., Interscience Pub., New York (1958).

8. B. Hague, "Alternating Current Bridge Methods," 5th ed., Pitman, London (1943).
9. L. Hartshorn, "Radio-frequency Measurements by Bridge and Resonance Methods," Wiley, New York (1941).
10. B. M. Oliver and J. M. Cage, "Electronic Measurements and Instrumentation," McGraw-Hill, New York (1971).
11. W. E. Vaughan, C. P. Smyth, and J. P. Powles, "Determination of Dielectric Constant and Loss," in A. Weissberger (ed.), "Techniques of Chemistry," vol. I, part IV, chap. V, Interscience, New York (1972).
12. C. P. Smyth, "Dielectric Behavior and Structure," McGraw-Hill, New York (1955).
13. C. P. Smyth, "Determination of Dipole Moments," in A. Weissberger (ed.), *op. cit.*, chap VI.

GENERAL READING

References 1–6, *op. cit.*

CHAPTER XVI

TEMPERATURE

Temperature is one of the most important variables in thermodynamics and physical chemistry. In this chapter we are concerned with the methods and instruments that are used in measuring and controlling temperature.

TEMPERATURE SCALES

The nature of temperature scales has been briefly discussed in Exp. 1. The *thermodynamic temperature scale,* based on the second law of thermo-dynamics, embraces the Kelvin (absolute) scale and the Celsius scale, the latter being defined by the equation

$$t(\text{Celsius}) \equiv T(\text{Kelvin}) - 273.15 \qquad (1)$$

The size of the Kelvin degree is defined by the statement that the triple point of pure water is exactly 273.16 K. The practical usefulness of the thermo-dynamic scale suffers from the lack of convenient instruments with which to measure absolute temperatures routinely to high precision. Absolute tempera-tures can be measured over a wide range with the helium gas thermometer (appropriate corrections being made for gas imperfections), but the apparatus is much too complex and the procedure much too cumbersome to be practical for routine use.

The *International Practical Temperature Scale* of 1968 (IPTS-68)[1,2] is basically arbitrary in its definition but is intended to approximate closely the thermodynamic temperature scale. It is based on assigned values of the temperatures of a number of defining fixed points and on interpolation formulas for standard instruments (practical thermometers) that have been calibrated at those fixed points. The fixed points of IPTS-68 are given in Table 1, and a number of secondary reference points recommended for use with IPTS-68 are given in Table 2. A new international standard (IPTS-88), based on work by the International Electrotechnical Commission, is scheduled for adoption in 1988. The temperature values for the fixed points and reference points are unlikely to change very much, but the interpolation procedures will probably be simplified.

Between the triple point of hydrogen ($-259.34°C$) and the antimony point ($630.74°C$), the standard instrument is the platinum resistance thermometer. Unfortunately, the IPTS-68 interpolation formulas are quite complicated in the range below $0°C$. In this range, the basic formula is of the form

$$W = W_t^*(T) + \Delta W(T) \tag{2}$$

where $W = R_t/R_0$ represents the ratio of the resistance of the platinum sensing element at a given temperature t to its resistance at $0°C$. W_t^* is the resistance ratio of an ideal platinum thermometer (a single-valued reference function given numerically by a 20-term series),[1] and ΔW is a deviation function involving up to four constants that are determined by the measured deviations at various fixed points. Four different deviation functions are used in four distinct temperature ranges below $0°C$.[1]

TABLE 1
Fixed points of the international practical temperature scale of 1968[a]

		Assigned value	
Fixed point	**Phases in equilibrium[b]**	$T_{68}(K)$	$t_{68}(°C)$
Hydrogen triple point[c]	Solid, liquid, vapor	13.81	−259.34
Hydrogen at $\frac{25}{76}$ atm[c]	Liquid, vapor	17.042	−256.108
Hydrogen boiling point[c]	Liquid, vapor	20.28	−252.87
Neon boiling point	Liquid, vapor	27.102	−246.048
Oxygen triple point	Solid, liquid, vapor	54.361	−218.789
Oxygen boiling point	Liquid, vapor	90.188	−182.962
Water triple point	Solid, liquid, vapor	273.16	0.01
Water boiling point	Liquid, vapor	373.15	100
Zinc freezing point	Solid, liquid	692.73	419.58
Silver freezing point	Solid, liquid	1235.08	961.93
Gold freezing point	Solid, liquid	1337.58	1064.43

[a] *Metrologia*, **5**, 35 (1969).

[b] Except for the triple points and the hydrogen point at 17.042 K, all temperature values are for equilibrium states at one standard atmosphere (101 325 Pa). Corrections can be made for small pressure deviations.

[c] Equilibrium mixture of ortho- and para-hydrogen.

TABLE 2
Partial list of secondary reference points[a]

Fixed point	Phases in equilibrium[b]	T_{68}(K)	t_{68}(°C)
Hydrogen[c]	Liquid, vapor	20.397	−252.753
Nitrogen triple point	Solid, liquid, vapor	63.148	−210.002
Nitrogen	Liquid, vapor	77.348	−195.802
Carbon dioxide	Solid, vapor	194.674	−78.476
Mercury	Solid, liquid	234.288	−38.862
Ice point	Ice, air-saturated water	273.15	0.
Phenoxybenzene triple point	Solid, liquid, vapor	300.02	26.87
Indium	Solid, liquid	429.784	156.634
Tin	Solid, liquid	505.118	231.968
Lead	Solid, liquid	600.652	327.502
Mercury	Liquid, vapor	629.81	356.66
Sulfur	Liquid, vapor	717.824	444.674
Antimony	Solid, liquid	903.89	630.74
Copper	Solid, liquid	1357.6	1084.5
Platinum	Solid, liquid	2045.	1772.
Tungsten	Melting	3660.	3387.

[a] *Metrologia*, **5**, 35 (1969).

[b] See footnote *b* to Table 1.

[c] Normal hydrogen.

In the range from 0°C to 630.74°C, the situation is simpler. The basic equation is

$$\frac{R_t}{R_0} = 1 + At' + Bt'^2 \tag{3}$$

where t' is related to the temperature t by the empirical formula

$$t = t' + 0.045\left(\frac{t'}{100}\right)\left(\frac{t'}{100} - 1\right)\left(\frac{t'}{419.58} - 1\right)\left(\frac{t'}{630.74} - 1\right) \tag{4}$$

The constants R_0, A, and B can be determined by calibration at the ice point, the boiling point of water, and the freezing point of zinc (419.58°C). A typical set of values would be $R_0 = 100\,\Omega$, $A = 3.9077 \times 10^{-3}\,\text{K}^{-1}$, $B = -5.765 \times 10^{-7}\,\text{K}^{-2}$. Note that $t' = t$ at the three fixed points specified above. If one wishes to use calibration points with temperatures t that differ from the standard set, it is awkward to determine the corresponding t' values from Eq. (4). Fortunately, t' is always close to t and one can get an adequate approximation to t' by using a modified version of Eq. (4) with t replacing t' at all four places in the correction term. The temperature error arising from this approximation is less than 0.1 mK.[3] Once calibration is complete, a platinum resistance reading R_t can be converted into the corresponding Celsius temperature t by solving the quadratic equation (3) to obtain the value of t' and then substituting this value into Eq. (4) to get t.

Between the antimony point (630.74°C) and the gold point (1064.43°C), the standard instrument is a type S thermocouple consisting of a platinum wire and a wire that is a 10% rhodium–90% platinum alloy.[4] The voltage of this thermocouple defines the temperature t according to the formula

$$v(t) = a + bt + ct^2 \qquad (5)$$

where $v(t)$ is the voltage obtained when one junction is at t and the other is at 0°C. The three constants are evaluated from the values of v at the freezing points of antimony, silver, and gold.

Above the gold point, an optical pyrometer is used and IPTS-68 is defined by the Planck radiation law in the form

$$\frac{J_\lambda(T)}{J_\lambda(T_{Au})} = \frac{\exp(c_2/\lambda T_{Au}) - 1}{\exp(c_2/\lambda T) - 1} \qquad (6)$$

where J_λ is the radiant flux per unit wavelength interval at wavelength λ. If λ is expressed in nanometer units, one should use $c_2 = 1.43877 \times 10^7$ nm K.

Although IPTS-68 conforms to the thermodynamic temperature scale much more closely than its predecessor (IPTS-48), the distinction between the two scales is not of great importance except for precise work.[5] In the earlier scale, the relatively simple Callendar–van Dusen interpolation formula was used between the oxygen point and 0°C; and this formula is still useful for many practical purposes (see the discussion on p. 657).

TRIPLE-POINT AND ICE-POINT CELL

Since the triple point of water is defined to be exactly 273.16 K on the thermodynamic temperature scale, this is an especially important fixed point. It is also a point which can be reproduced with exceptionally high accuracy. If the procedure of inner melting (described below) is used, the temperature of the triple point is reproducible within the accuracy of current techniques (about ±0.00008 K). This precision is achieved by using the NBS-designed *triple-point cell* shown in Fig. 1. This cell, which is about 7.5 cm in outer diameter and 40 cm in overall length, has a well of sufficient size to hold all thermometers that are likely to be calibrated.

The procedure for using a triple-point cell is as follows. Cool the cell by placing it in a bath of crushed ice and distilled water. Fill the thermometer well with powdered dry ice in order to form a clear mantle of ice around the well. When this mantle is about 0.5 cm thick, remove the dry ice and place a warm tube in the well just long enough to free the mantle and provide a thin layer of liquid next to the wall. If the mantle is free, it will rotate about the well when the cell is given a quick twist about the vertical axis. This "inner melting" provides an additonal step of purification. As the water freezes to form the mantle, impurities are left behind in the liquid near the outer wall. When the

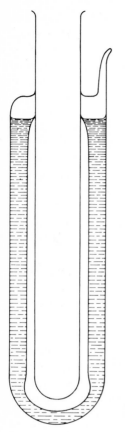

FIGURE 1
Triple-point cell for water (NBS design). Note the clear mantle of ice formed around the inner well.

mantle is partly melted, the water near the inner well is exceptionally pure and this water does not mix rapidly with the water outside the mantle. The equilibrium between the inner water layer, the ice mantle, and the water vapor in the top of the cell occurs at the defined triple-point temperature. In order to ensure good thermal contact between the wall of the well and the thermometer being calibrated, the well is filled with ice water. It is also wise to precool the thermometer in an ice bath before it is placed in the well.

Although the triple point of water is the most fundamental and stable of all fixed points, the ice point also serves as a convenient fixed point for less demanding calibrations of practical thermometers. The ice point refers to an ice and pure water mixture in equilibrium with air saturated with water vapor at a total pressure of 1 atm. Procedures for preparing an ice bath are given in the section on temperature control. Commercial ice-point devices that eliminate the need for frequent attention are also available.[6] Such devices consist of a sealed chamber containing distilled water; the outer walls are cooled with a thermoelectric element to create a shell of ice. The increase in volume that accompanies the conversion of liquid into solid is detected by the expansion of a bellows, which operates a microswitch controlling the cooling element. The alternating cycle of freezing and thawing maintains a constant temperature and eliminates the slow drift in temperature that will occur in an unattended ice bath as melting proceeds. An ice-point device will maintain a reference temperature that is constant to within ± 0.04 K, but the absolute value may differ from 273.15 K $\equiv 0°C$ by a somewhat larger amount.

THERMOMETERS

In this section, the design and operation of familiar liquid thermometers, thermocouples, platinum resistance thermometers, thermistors, and optical pyrometers are discussed in detail. Briefer descriptions are also given of a variety of special thermometric devices such as quartz thermometers, germanium resistance thermometers, and silicon-diode thermometers.

Liquid-in-glass thermometers. Major emphasis is given to the mercury thermometer, although any liquid with a thermal expansion coefficient significantly greater than that of glass can be used. The familiar household thermometer uses an organic liquid such as toluene or alcohol containing a little red dye to enhance visual contrast.

Mercury thermometers.[7] By far the most common type of laboratory thermometer is the mercury thermometer, based on the differential volume thermal expansion of liquid mercury (about 1.8×10^{-4} K^{-1}) and glass (about 0.2×10^{-4} K^{-1}). In the manufacture of such thermometers, the stem, a capillary of uniform bore, is usually marked at two points (say 0°C and 100°C) and then graduated uniformly in between, on the tacit assumption that the volume of a fixed mass of mercury in glass is a linear function of the temperature. The error resulting from this assumption (about +0.12°C at 50°C for a 0°C to 100°C mercury thermometer made with Corning normal thermometer glass) is ordinarily smaller than that due to variations in the bore of the capillary.

Laboratory thermometers are commonly available in two types: solid stem and enclosed scale. The former has a stem of solid glass, with a scale engraved on the outside surface; the latter has a slender capillary and a separate engraved scale, both enclosed in an outer glass shell. The enclosed-scale type is preferable for thermometers with extremely fine threads largely because it suffers less from parallax in reading.

Since the mercury in the thread, as well as that in the bulb, is susceptible to thermal expansion, it is important in precise work to take account of the temperature of the thermometer stem. Most thermometer calibrations, especially those for enclosed-stem types, are for *total immersion*—it is assumed that the thread is at the same temperature as the bulb. Other thermometers are meant to be used with partial immersion, often to a ring engraved on the stem, and the remainder of the stem is assumed to be at room temperature (say

25°C). For precise work stem corrections should be made if the stem temperatures differ significantly from those assumed in the calibration. The correction that should be *added* to the thermometer reading is given by the equation

$$\Delta t_{\text{corr}} = -0.00016(\Delta t_{\text{stem}})(\Delta L_{\text{stem}}) \tag{7}$$

where Δt_{stem} is the amount by which the temperature of the stem *exceeds* that for which the calibration applies (or the amount by which it exceeds the thermometer reading itself, if the calibration is for total immersion) and ΔL_{stem} is the length of mercury thread, expressed in degrees Celsius, for which the temperature is different from that assumed in the calibration. To obtain Δt_{stem} a second thermometer may be positioned near the first, with its bulb near the midpoint of ΔL_{stem}.

For most purposes a partial-immersion thermometer need not be stem-corrected because of a few degrees variation in room temperature or few degrees error in the immersion level. On the other hand, it is usually worthwhile to apply stem corrections to readings of a total-immersion thermometer when it is used in partial immersion, particularly when reading temperatures well removed from room temperature.

Other important sources of error in mercury thermometers are parallax and sticking of the mercury meniscus. The former can largely be avoided by careful positioning of the eye when reading or by use of a properly designed attached magnifier. It can be eliminated entirely by use of a cathetometer (see Chapter XVIII). Sticking of the mercury meniscus is due to the fact that the contact angle of mercury to glass (see Fig. 26-5) varies depending on whether the mercury surface is advancing or receding, and thus the capillarity pressure due to surface tension is variable. This combines with the small but finite compressibility of the mercury and the elasticity of the glass to yield a small variability in meniscus position, especially for very sensitive thermometers. Gentle tapping of the stem before taking readings usually leads to reproducible results. Readings of sensitive mercury thermometers are also slightly pressure dependent; pressure coefficients may be as high as 0.1 K bar^{-1}.

When precision of better than about 1 percent of full scale is required, it is necessary to use a calibration chart which gives corrections to be added to or subtracted from the readings. Detailed procedures for calibrating liquid-in-glass thermometers are given by Swindells.[8] The simplest method involves comparison of the thermometer to a standard thermometer, which is itself well calibrated. Such calibrated standard thermometers can be obtained commercially. The two thermometers are mounted next to each other in a well-stirred thermostat bath. The standard thermometer is immersed as specified in its calibration (usually to the top of the thread) and the test thermometer is immersed to the depth at which it will be used subsequently. The readings should be compared at several temperatures within the range of interest. Prior to this calibration procedure, the standard thermometer should be checked. Although it is best to use a triple-point cell (see p. 643), an ice bath made from finely divided ice and distilled water (see p. 666) can also be used. If necessary, a constant additive correction can be made in the calibration table for the standard thermometer. The best calibration method is to calibrate the test thermometer against a platinum resistance thermometer, since such thermometers have exceptional reproducibility and long-term stability. The calibrations of mercury thermometers should be checked at a single point from time to time, as significant irreversible changes may take place in the glass, particularly if the thermometer is used above 150°C.

Special thermometers are made for calorimetric and cryoscopic work, where it is desired to measure very accurately (to 0.01 K or even 0.001 K) a temperature *difference* of the order of a few kelvin. For these thermometers the fineness of scale graduation has little to do with the accuracy with which the thermometer measures a single temperature. The scale may be in error by several tenths of a kelvin, but this error cancels out in taking differences. A typical thermometer for bomb calorimetry has a range of 19°C to 35°C, with graduations of 0.02°C. For measuring freezing-point depressions with water or benzene as solvent, a range of −2°C to +6°C with graduations of 0.01° is convenient. Such thermometers require careful handling. Not only are they relatively fragile, but they are susceptible to certain malfunctions (separation of the mercury column, bubbles in the bulb) arising principally from the extreme fineness of the thread. Whenever possible keep these thermometers upright; *avoid overly rapid heating or cooling.* If the mercury thread separates (which often happens when thermometers are shipped), cool the bulb in an ice–salt mixture to bring the mercury entirely into the bulb and tap if necessary to bring any bubbles to the top of the bulb. Then allow the thermometer to warm to room temperature in an upright position.

Beckmann thermometer. The Beckmann thermometer is a differential mercury thermometer with a range of about 6 K and graduations of 0.01 K or 0.02 K. It differs from an ordinary mercury thermometer in having a provision for changing the position of this 6 K interval on the temperature scale by adjusting the amount of mercury in the bulb and thread. For this purpose a special reservoir is provided at the top of the thermometer (see Fig. 2). To raise the setting, the bulb is warmed so as to discharge droplets of excess mercury into this reservoir. To lower the setting, the bulb is warmed to bring the mercury thread into the bottom of the reservoir, where it is then united with a column of excess mercury which is brought into position by inverting the thermometer. The bulb is cooled to draw mercury into the thread. At a few degrees above the desired range the excess mercury is broken away from the thread by a sharp tap with the fingers. The thermometer is then brought to an upright position in such a way as to bring the excess mercury to its normal position in the reservoir.

The Beckmann thermometer is fragile and often subject to difficulties in use. Fixed-scale thermometers are ordinarily preferable for calorimetric and cryoscopic work in the usual temperature ranges if they provide sufficient resolution. Thermocouples, platinum resistance thermometers, and thermistors

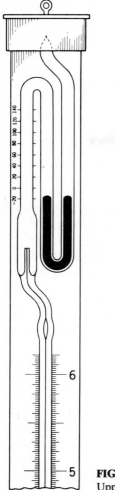

FIGURE 2
Upper end of a Beckmann thermometer.

are all capable of even higher resolution measurements of temperature differences than are Beckmann thermometers; but such instruments are more expensive when a resolution of 5 mK or less is desired.

Special liquid thermometers. For temperatures below the freezing point of mercury (−39°C), one can use several kinds of liquid-in-glass thermometers. Toluene thermometers may be used down to −95°C, and pentane thermometers will operate as low as −130°C. Mercury thermometers constructed from special glasses may be used far above the normal boiling point of mercury (357°C). However, outside the ordinary mercury range it is usually more convenient, as well as more accurate, to use thermometric devices of other types, especially thermocouples or resistance thermometers.

Thermocouples.[2,9] Thermocouples provide one of the most convenient and versatile means of measuring temperatures over a wide range from very low ($-250°C$) to very high ($2300°C$) values. All that is required is two kinds of fine wire made from appropriate metals or alloys, a suitable precision voltage-measuring instrument (a good potentiometer or very good digital voltmeter), and a constant-temperature bath (almost always an ice-water bath) for the reference junction.

When two dissimilar metals are placed in contact, a transfer of electrons from one to the other takes place and a double charge layer forms at the junction surface. As in the case of a junction between a metal electrode and an electrolyte solution, the resulting electrical potential must be measured in the presence of a reference junction, which in the present instance is a junction of the same two metals at a known temperature. As shown in Fig. 3, the thermoelectric potential difference ΔV is measured between the ends of two wires of the same kind. One wire of metal A leads to the junction at the reference temperature T_1, and the other wire of metal A leads to the junction at the unknown temperature T_2; these two junctions are connected directly by a wire of the second metal B, which completes the circuit. The potential drop ΔV is a measure of the temperature difference $\Delta T = T_2 - T_1$. In cases where ΔT is not too large, ΔV is roughly proportional to the difference between the unknown and the reference temperature:

$$\Delta V = \alpha \Delta T \tag{8}$$

where α is called the Seebeck coefficient in honor of Thomas Seebeck, who discovered the thermoelectric effect. In fact, ΔV is not exactly a linear function of ΔT since the sensitivity α varies slowly with temperature. Thus, a well-known reference temperature T_1 and calibration tables are needed to obtain values of T_2 from measured ΔV values. It is common practice to use the ice point ($0°C$) as the reference temperature; in this case, ΔT equals the value of T_2 in degrees Celsius.

The presence of an isothermal junction block in Fig. 3 should also be noted. Almost all high quality voltage-measuring instruments have input terminals made of copper. Unless metal A of the thermocouple pair is also copper, connection of the thermocouple to the voltage instrument will create two new bimetallic junctions J_3 and J_4. The presence of any temperature

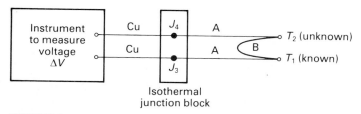

FIGURE 3
Schematic diagram of a two-junction thermocouple setup.

difference between the input terminals would then generate an unwanted extra thermoelectric potential drop and cause an error in the determination of T_2. In practice, this error voltage is usually very small and can often be neglected. However, for high-precision measurements this source of error can be eliminated by using an isothermal junction block. This block is made from an electrical insulator with high thermal conductivity, which ensures that J_3 and J_4 are at the same temperature, T_{block}. The value of T_{block} has no effect on the measured ΔV values.

The isothermal junction block is also used in commercial thermocouple devices that utilize only a single junction. Since T_{block} can have any value, it is made equal to the reference temperature T_1, as shown in Fig. 4(a). Another simplification is the elimination of wire A in the junction block, so that two junctions (Cu–A and A–B) at the same temperature T_1 are replaced by one (a Cu–B junction) as shown in Fig. 4(b). It can be shown empirically that this modification has very little effect on the measured value of ΔV. The isothermal junction block can be mounted in an ice–water bath so that $T_1 = 0°C$. Many commercial thermocouple systems (voltage-measuring instruments plus thermocouple assembly) do not utilize an ice bath but employ software or hardware compensation schemes so that the reference temperature T_1 can have an unregulated value equal to the ambient room temperature.[6] For high-precision work, a well-regulated ice bath should always be used.

The most useful thermocouples for general work are listed in Table 3.[10] Both pure metals and various alloys are used. Many of these alloys are nickel alloys. *Constantan* (sold under the trade names Advance and Cupron) is available in two different compositions: 60%Cu + 40%Ni is designed for use with type J thermocouples, and 55%Cu + 45%Ni is used for both type T and

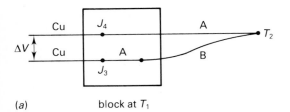

(a)　　　block at T_1

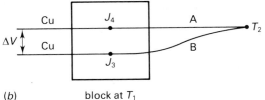

(b)　　　block at T_1

FIGURE 4
Configurations for a "one-junction" thermocouple arrangement utilizing an isothermal junction block held at a reference temperature T_1; see text for details.

TABLE 3
Thermocouple characteristics

ANSI Type	Metals	Useful temperature range (°C)	Sensitivity (μV K^{-1})			Notes
			−200°C	25°C	300°C	
T	Copper Constantan	−200 to 370	15.8	40.7	58.1	R, I, V, O (mild)
J	Iron Constantan	0 to 750	21.8^a	51.7	55.4	R, I, V, O (moderate T)
E	Chromel Constantan	−200 to 900	25.1	60.9	77.9	I, O
K	Chromel Alumel	−200 to 1250	15.2	40.5	41.5	I, O, L
N^b	Nicrosil Nisil	−200 to 1250	9.9	26.8	35.4	I, O, L
S	Platinum Pt + 10%Rh	0 to 1450	—	6.0	9.1	I, O

Notes. Atmospheres in which bare-wire thermocouples can be used: R = reducing, O = oxidizing, I = inert, V = vacuum, L = limited use in vacuum or reducing atmospheres.
^{a} Not recommended for low temperatures.
^{b} Proposed letter code not yet adopted by ANSI.

type *E* thermocouples.† *Chromel* is the trade name of a 90%Ni + 10%Cr alloy, and *Alumel* is the trade name of a 95%Ni + 2.5%Mn + 2%Al + 0.5%Co alloy. Two other nickel alloys with trade names are Nicrosil (84.4%Ni + 14.2%Cr + 1.4%Si) and *Nisil* (95.5%Ni + 4.4%Si + 0.1%Mg).

The *copper–constantan* thermocouple, called type *T,* is widely used. This has the important advantage that one of the metals (metal A in Fig. 3) is copper. Thus an isothermal junction block is not needed, and no stray thermoelectric potentials can occur at the input terminals of the voltage instrument even if a temperature difference does occur there. Furthermore, if a thermocouple is to be used without calibration, the possible systematic errors are smaller for type *T* than for other types of thermocouples. *Iron–constantan* (type *J*) thermocouples are popular because of their high sensitivity and the wide range of atmospheres in which they can be used. Individual thermo-couples may show variations from the standard voltage–temperature conver-sion table values owing to trace impurities in the iron. Type *J* couples should never be used above 760°C, since a magnetic phase transition at the temperature can shift the calibration at lower temperatures after cycling above

† The letter codes used for thermocouples have been established by the American National Standards Institute (ANSI).

760°C. A *Chromel–constantan* (type *E*) thermocouple is attractive for low-temperature use owing to its high sensitivity and low thermal conductivity. The *Chromel–Alumel* (type *K*) couple is convenient because it can be used over a very wide temperature range and the sensitivity is almost constant above 25°C (e.g., $\alpha = 36 \, \mu V \, K^{-1}$ at 1250°C). There is some calibration instability due to an order–disorder phase transition at ~500°C. The *Nicrosil–Nisil* (type *N*) thermocouple is a more recently developed couple for use at high temperatures. It is similar to type *K* with changes in the alloy composition to reduce problems associated with the order–disorder transition and to improve the resistance to oxidation at high temperatures. However, there can be inconsistencies in the thermoelectric voltages that depend on the wire gauge. *Platinum–platinum + 10% rhodium* (type *S*) thermocouples have low sensitivity but can be used at temperatures as high as 1700°C. The calibration stability is excellent, and the type *S* thermocouple is the standard device specified by IPTS-68 to measure temperatures in the 630.74°C to 1064.43°C range. However, it may well be replaced in this role by the platinum resistance thermometer when the new IPTS-88 scale is agreed upon.

Other thermocouples not listed in Table 3 include the chromel–gold couple for cryogenic use below −200°C (useful range 4 K to above 100 K) and several tungsten–tungsten + rhenium couples that can be used up to 2300°C.

Although extensive tables are given in various handbooks for converting measured thermoelectric voltages into temperatures, the best tables are those published by the National Bureau of Standards.[6,11] No matter what tables are used, thermocouple wires should be selected carefully and one or more specimens of each production lot (the spools are marked with this number) should be calibrated at a number of widely-spaced temperatures. For very precise work, each individual thermocouple should be calibrated. Deviations between the reading of a given thermocouple and the entries in standard conversion tables are due to strains and small compositional variations that can occur during fabrication of the wires. Over the temperature range from −200°C to 300°C, these systematic errors can vary from ±1°C to ±4°C for different thermocouples. As an example, the uncertainty in the absolute accuracy of an uncalibrated type *T* thermocouple is the larger of ±0.8°C *or* ±0.75 percent of the Celsius temperature above 0°C (±1.5% of the Celsius temperature below 0°C). For type *K,* the analogous values are ±2.2°C or ±0.75 percent above 0°C (±2.0% below 0°C).

Junctions to be used only at low or moderate temperatures may be joined with soft solder (be sure to use a noncorrosive flux such as rosin) or silver solder, but it is better to weld the two metals together with an electric arc or an oxygen–gas flame. The two wires are stripped of any insulating material and then twisted together tightly over a distance of ~0.5 cm. If a glass-blowing torch is used, the end of the junction should be placed in the reducing region at the tip of the blue part of the flame. Remove the wires as soon as the metals melt and form a small bead at the end. This bead should be pinched with a pair of long-nosed pliers and examined carefully to make sure it is hard and metallic rather than a bead of oxide (which crumbles easily). The two wires must be electrically insulated from each other, and some consideration must be given to the temperature characteristics of the insulating materials. In particular, at high temperatures nothing should come into contact with the wires, particularly Chromel-P and Alumel, that will form a liquid flux (low-melting eutectic) with the protective oxide film.

Thermocouple wire is available commercially either as bare wire or as wire protected with a variety of insulators. The best choice of insulation for use up to 280°C is Teflon. Its resistance to abrasion, chemical reaction, solvents, and humidity is excellent; and its flexibility is good. Braided fiberglass is very good for use up to 480°C, but it has poor abrasion resistance and quite high porosity. Special braided ceramic fibers are available for use as high as 1425°C. Bare wires and Alundum tubes can also be used above 450°C. The junctions of any thermocouple, and the wire itself when bare wire is used, must always be protected carefully from corrosion or mechanical damage. For use in liquids and solutions at moderate temperatures, a thermocouple is usually placed in a closed glass tube (~6 mm in diameter) with a small amount of nonvolatile oil or silicone grease at the bottom to improve the thermal contact. It is also possible to bond thermocouple junctions to a wide range of solid surfaces (metals, ceramics, glass, plastics) with high thermal conductivity epoxy adhesives. Fully assembled thermocouple probes are also available commercially with both wires mounted in a stainless-steel or Inconel protective sheath and insulated from each other with a compacted ceramic such as MgO, which is good up to 1650°C. The junctions can be obtained exposed as a bare butt weld or bare bead, or enclosed in the sheath metal (grounded or ungrounded options are available).

A satisfactory environment for the 0°C reference junction is provided by a slushy mixture of ice and distilled water in a Dewar flask, with a ring stirrer and a monitoring mercury thermometer. Elaborate thermoelectric ice–water chambers are also available; these are convenient for prolonged periods of use but rather expensive. As mentioned previously, many commercial thermocouple systems eliminate the ice bath by placing the cold junction on an isothermal block that is at room temperature and compensating for the resulting error. This is a convenient but less accurate procedure.

For ordinary work with thermocouple wires of large or medium diameter, the thermoelectric potential can be measured with a good digital voltmeter (DVM) or a dc potentiometer. The L&N Student Potentiometer, set on the 0.01 scale as described in Chapter XV, can be used; but a L&N model K-3 Potentiometer is much more satisfactory. If a continuous record of the temperature is required, the thermocouple can be connected to a millivolt strip-chart recorder; or a DVM with an interface to a computer can be used to store frequent periodic readings. For precision work, a potentiometer capable of measuring voltage changes of $\sim 0.1\,\mu\text{V}$ or a high-resolution ($6\frac{1}{2}$ digit) high-impedance DVM are required. Such equipment used with a calibrated copper–constantan thermocouple having one measuring junction and one reference junction maintained carefully at the ice point permits one to measure the absolute temperature around room temperature with an uncertainty of about ±0.01 K. However, the precision is much better than this, and temperature differences can be measured to within ±0.002 K or better.

There are several precautions that should be observed in order to enhance the long-term stability of a thermocouple.

1. Use the largest diameter wire that is feasible without conducting too much heat away from the area of the measurement junction.

2. If very thin wires are required, use larger-diameter extension wires for regions where there is almost no temperature gradient (typically, the room temperature leads that go to the voltage-measuring device).

3. Avoid mechanical stresses and vibrations that might cause strain distortions in the wires.

4. Avoid very large temperature gradients along the wire.

5. Operate the thermocouple only at temperatures below its design limit.

6. If operating in a hostile atmosphere, use an adequate external sheath to protect the wires.

Platinum resistance thermometers.[2,12] The platinum resistance thermometer (PRT) is capable of extremely high accuracy, owing to the high purity attainable for platinum and the high reproducibility of its temperature coefficient of resistivity. For a standard PRT used to determine absolute temperature values with the highest possible accuracy, the resistance element is a coil of pure platinum wire, carefully annealed both before and after winding, and enclosed in a sealed sheath (glass or thin-walled metal tube) containing dry air or helium as a heat-transfer gas. The design of standard PRTs is aimed at minimizing errors due to heat conduction and radiation and maximizing stability and calibration reproducibility. Such PRTs are rather fragile and sensitive to mechanical vibration. They also have low-resistance elements, typically an ice-point resistance R_0 of 25 Ω, which means a sensitivity of only 0.1 Ω K^{-1}.

In most experimental applications, one uses "practical" PRTs that are less expensive, more compact, more rugged, and more sensitive (owing to the choice of R_0 values of 100 Ω or 200 Ω) than standard PRTs.† There are, of course, some trade-offs since practical PRTs are slightly less stable and have poorer interpolation accuracy than standard PRTs. In practical PRTs, the platinum wire is wound on a glass or ceramic bobbin, and the assembly is hermetically sealed in a glass or ceramic capsule. Considerable care is taken to match the thermal expansion coefficients of platinum and the bobbin in order

---

† Standard PRTs can be obtained from Leeds & Northrup, Philadelphia, PA. Practical PRTs are available in a wide variety of formats from several companies such as Rosemount Inc., Minneapolis, MN, and Omega Engineering Inc., Stamford, CT.

to minimize strain-induced resistance changes. Since practical PRTs are quite small (typically 25 mm long and 1.5–4.5 mm in diameter), they have moderately fast responses to temperature changes. The 90% response time in stirred or flowing liquids are typically 1 to 3 s. A recent development is the thin-film PRT, in which a metal slurry is deposited on a ceramic substrate. These devices are very small and have 90% response times of 0.3 s or less, but their stability is significantly less than that of wire-wound PRTs.

The basic principle underlying the operation of any resistance thermometer[2] is a finite temperature coefficient of resistivity

$$\alpha_t = \frac{1}{R}\left(\frac{\partial R}{\partial T}\right)_p \cong \frac{1}{R_0}\frac{dR}{dt} \tag{9}$$

where t is the Celsius temperature and R_0 is the resistance at 0°C. For platinum, α_t is roughly constant over a wide temperature range and is very reproducible for different wire samples. Since $\alpha_t \cong 0.0039\,\mathrm{K}^{-1}$ and R_0 is typically 100 Ω in a practical PRT, a change of 1 K will cause a resistance change of only 0.39 Ω. In many applications, especially when measuring low or high temperatures, rather long electrical leads are needed. Such leads are usually made of fine wire to reduce heat conduction and thus have appreciable resistance (say 5–10 Ω). Clearly, any accurate determination of temperature requires a careful measurement of R_t for the PRT at temperature t free from any errors due to lead resistance. This can be accomplished by using a four-lead PRT, which has a pair of identical leads connected to each end of the resistance element (i.e., separate current leads and voltage leads).

Measurement of R_t can be made with a Wheatstone bridge or with a potentiometric method. A three-terminal bridge configuration is shown in Fig. 5, where leads a, b, A, B have resistance R_a, R_b, R_A, R_B. Since the two fixed

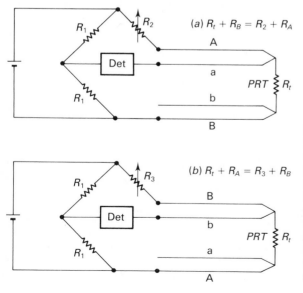

FIGURE 5
Measurement of a four-lead PRT with a three-terminal Wheatstone bridge. The resistance $R_t = (R_2 + R_3)/2$; see text for details. If the lead resistances R_A and R_B are equal and there are no disturbing thermal emfs at the lead junctions, R_t can be determined from either of the single bridge balance conditions since R_2 and R_3 are then the same.

resistance arms have been chosen to have the same values R_1, the balance condition is simply

$$R_t + R_B = R_2 + R_A \tag{10}$$

for the configuration in Fig. 5(a), and

$$R_t + R_A = R_3 + R_B \tag{11}$$

for the configuration in Fig. 5(b). There are special three-pole double-throw switches to allow one to switch easily from configuration a to b. Differences in lead resistances and also any thermal voltages (thermocouple effects at junctions between platinum and other metals like copper) are canceled out by balancing the bridge in both configurations and calculating R_t from

$$R_t = \tfrac{1}{2}(R_2 + R_3) \tag{12}$$

Note that use of an ac bridge will also eliminate any errors due to thermal voltages, which are dc effects.

A potentiometric set-up for measuring R_t is shown in Fig. 6. The detector can be a precision dc potentiometer or a high-resolution, high-impedance digital voltmeter. Voltage drops are measured across R_t and across a series standard resistance R_s that has a stable constant value, usually $100\,\Omega$. The value of R_t is then given by

$$R_t = R_s(V_t/V_s) \tag{13}$$

In order to avoid self-heating (electrical I^2R_t heating caused by the measuring current I), one must use a DVM that operates at $1\,\text{mA}$ or less on the appropriate resistance scale. The same limitation applies to the current flowing through R_t in the Wheatstone bridge. A current of $1\,\text{mA}$ through a PRT of $100\,\Omega$ corresponds to a power dissipation of $0.1\,\text{mW}$, which will heat a typical PRT by $\sim50\,\text{mK}$ in still air, by $\sim5\,\text{mK}$ in flowing air, and by less than $1\,\text{mK}$ in a stirred liquid.

For a PRT with $R_0 = 100\,\Omega$, a temperature resolution of $\pm1\,\text{mK}$ requires that R_t be measured with a *precision* of $\pm4 \times 10^{-4}\,\Omega$; i.e., a random error

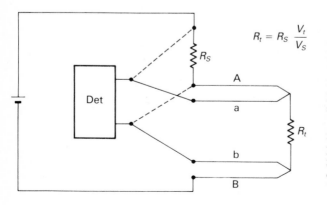

$$R_t = R_S \frac{V_t}{V_s}$$

FIGURE 6

Potentiometric method of determining the resistance R_t of a four-lead PRT. The measured potential drops across R_t and a series standard resistance R_s are V_t and V_s, respectively. If thermal emfs at lead junctions are a problem, the PRT leads can be reversed as in Fig. 5b; in this case, $R_t = R_s(V_t + V_{t,r})/(V_s + V_{s,r})$ is used to get R_t.

$\leq 4 \times 10^{-4}$ percent. Thus the measurement bridge, potentiometer, or digital voltmeter must be a high-precision instrument. In the case of a DVM, one needs at least 6-digit resolution. It should be noted that owing to decreasing sensitivity as T approaches $0\,\mathrm{K}$, PRTs do not make attractive practical thermometers at very low temperatures.

The absolute accuracy of a PRT temperature value depends not only on the *accuracy* of the resistance ratio R_t/R_0 but also on the quality of the calibration. A calibration can be carried out by using fixed points or by making a comparison with a standard PRT.[13] The fixed-point method is more accurate if carefully done but is much more time consuming. In the comparison method, both thermometers are placed in good thermal contact with an "isothermal" metal block, which can then be placed in an ice bath and in convenient stirred liquid baths over the range $-150°C$ to $400°C$. For a calibration point below $-150°C$, liquid nitrogen at its normal boiling point is the best choice. The freezing points of several metals can be used for calibration above $400°C$.

The best calibration procedure is to use the IPTS-68 standard formulas; see Eqs. (2) to (4). However, the interpolation scheme below $0°C$ is unfortunately complicated and is likely to be changed when the IPTS-88 standard is adopted. For practical use, the Callendar–van Dusen equation, which was the basis for IPTS-48, is still a very convenient form. This is especially true if one wishes to determine R_t digitally under computer control and convert it into a temperature with a reasonably simple algorithm. The general form of the Callendar–van Dusen equation is

$$R_t = R_0 \left\{ 1 + \alpha \left[t - \delta \left(\frac{t}{100} \right) \left(\frac{t}{100} - 1 \right) - \beta \left(\frac{t}{100} - 1 \right) \left(\frac{t}{100} \right)^3 \right] \right\} \tag{14}$$

where t is in degrees Celsius and α, β, δ are calibration constants. The constant $\beta = 0$ by definition above $0°C$. It is obvious from Eq. (14) that

$$\alpha = \frac{R_{100} - R_0}{100 R_0} \tag{15}$$

Thus α equals the average value of the resistivity coefficient α_t over the range $0°C$ to $100°C$. A more convenient form of Eq. (14) is

$$W \equiv \frac{R_t}{R_0} = 1 + At + Bt^2 + Ct^3(t - 100) \tag{16}$$

where $A = \alpha(1 + \delta/100)$, $B = -\alpha\delta/10^4$, $C = -\alpha\beta/10^8$ and $C = 0$ above $0°C$. Note that the constant A equals the value of α_t at $t = 0°C$. A calibration at $0°C$ and at two points above $0°C$ will yield R_0, A, and B (or α and δ); one more calibration point below $0°C$ will then determine C (or β).

The typical uncertainties in temperature values based on a four-point Callendar–van Dusen calibration over the range $-180°C$ to $260°C$ vary from $\pm(10–20)\,\mathrm{mK}$ over this range but rise to $\pm 80\,\mathrm{mK}$ at $350°C$ because an extrapolation beyond the fitted calibration range is involved. A crude calibration can be achieved on the basis of only two points at 0°C and 100°C. These two points will determine R_0 and α; see Eq. (15). One can then *assume* that $\alpha\delta = 0.005\,8755$ and $\beta = 0.11$ for $t < 0°C$.[13] The resulting calibration formula can be used to interpolate reasonably well between 0°C and 100°C (errors of about ±50 mK at 50°C) and to extrapolate roughly down to $-50°C$ and up to 150°C (errors grow to about ±200 mK).

Once the constants in the calibration formula are known, there are two ways to proceed in converting R_t readings into temperatures. One can prepare a conversion table of closely spaced R_t–T entries at even intervals of R_t and linearly interpolate between entries. Alternatively, one can directly calculate t from Eq. (16). An analytic formula is used above 0°C since W is quadratic in t; an iterative numerical method must be used below 0°C.

The International Electrotechnical Commission has recommended that IPTS-88 utilize the Callendar–van Dusen interpolation formula or some variant of it in order to avoid the awkward IPTS-68 equations for temperatures below 0°C. The proposed constants for use in Eqs. (14) and (16) are $R_0 = 100\ \Omega$, $A = 3.90802 \times 10^{-3}$, $B = -5.802 \times 10^{-7}$, $C = -4.27350 \times 10^{-12}$ (or $\alpha = 0.003\,85$, $\delta = 1.507\,01$, $\beta = 0.111$). The platinum wire used in PRTs that conform to this proposed standard is a platinum alloy containing small amounts of several different elements (mostly noble metals) adjusted so as to have the required $\alpha = 0.003\,85$. This is now widely used in Europe and by some American companies; other American firms use a wire for which $\alpha = 0.003\,92$.

Thermistors.[2,14] Thermistors (a portmanteau contraction of "thermally sensitive resistors") are semiconductor devices consisting of sintered mixtures of metallic oxides such as NiO, Mn_2O_3, Co_2O_3. They are usually extrinsic p-type materials containing excess oxygen above the stoichiometric amount. These ceramic-like resistors have large *negative* temperature coefficients of resistivity. The variation of the resistance R can be roughly approximated by

$$R = R_\infty \exp\left(\frac{\Delta E}{2kT}\right) \tag{17}$$

where ΔE is the electron energy gap. For typical thermistors, $\Delta E/2k \approx 3500$ K, which means a 4 percent change in resistance per kelvin at room temperature. Thus, the magnitude of the temperature coefficient of resistivity $\alpha = R^{-1}(dR/dT)$ is about 10 times that of metallic resistors, and high resolution as a resistance thermometer can be attained with a less sensitive bridge or DVM than is required for a platinum thermometer. In addition, the resistivity is so high that very small thermistor elements can still have high resistances. By using a thermistor with a high resistance in the temperature range of interest (say 10 kΩ to 100 kΩ), complications due to lead and contact resistances are eliminated and two-wire measurements are adequately precise.

Figure 7 shows a comparison of the temperature dependence of R_t/R_{25} for a platinum resistance thermometer and three typical thermistors. The

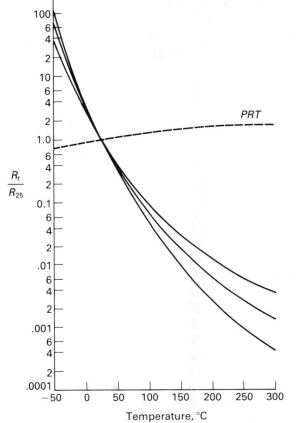

FIGURE 7
Resistance versus temperature characteristics for several typical thermistors. The behavior of a platinum resistance thermometer (dashed line) is shown for comparison.

greater thermistor sensitivity is obvious, but one should also note that the practical operating range for a given thermistor is relatively narrow. Let us assume that a resistance range from 2 kΩ to 20 kΩ is ideal for the measuring device to be used. Then a thermistor with $R_{25} = 20$ kΩ will operate over a range from 25°C to about 100°C, while one with $R_{25} = 200$ kΩ is useful from 100°C to about 200°C. Figure 7 also clearly shows that practical thermistors deviate from Eq. (17), the simple exponential expression valid for many intrinsic semiconductors. The standard empirical curve-fitting technique used to represent the thermistor $R(T)$ variation is to consider $\ln R$ to be a polynomial in $1/T$ or *vice versa*. These equations are usually truncated at the cubic term, yielding

$$\ln R = A_0 + A_1/T + A_2/T^2 + A_3/T^3 \tag{18}$$

$$\frac{1}{T} = a_0 + a_1(\ln R) + a_2(\ln R)^2 + a_3(\ln R)^3 \tag{19}$$

It is common practice to drop the quadratic terms for the temperature range above 0°C; however, these quadratic terms must be retained in calibrating thermistors for use at low temperatures. In any event, neither Eq. (18) nor (19) provide a good representation of the $R(T)$ variation over a wide temperature range. Any given calibration is accurate to within ±0.1 K over a range of about 80 K. Two other disadvantages of thermistors in comparison with platinum resistance thermometers are their lower reproducibility and poorer long-term stability. A comparison of many performance characteristics of thermocouples, PRTs, and thermistors is given in Table 4.

The most stable and most useful type of thermistor is the glass-coated bead. In this design, a sintered oxide bead (0.075 to 1 mm in diameter) is sealed in glass, resulting in a probe bead with a diameter ranging from 0.125 to 1.5 mm. Important advantages of this small thermistor size are low probe heat capacity and rapid temperature response (0.1 to 1 s in still air and 5 to 15 ms in water). Such bead or microbead thermistors can be obtained commercially† with room-temperature resistances ranging from $10^3\ \Omega$ to $5 \times 10^6\ \Omega$.

TABLE 4
Comparison of performance characteristics for thermocouples, platinum resistance thermometers (PRTs), and thermistors

Characteristic	Thermocouple	PRT	Thermistor
Sensitivity	Low to moderate	Low	High (+)
Temperature range[a]	Very wide (+)	Wide (+)	Narrow (−)
Lowest temperature	4 K	15 K	77 K
Highest temperature[b]	2600 K	1000 K	575 K
Linearity	Fair	Good	Poor
Calibration stability	Very good (+)	Excellent (+)	Fair
Ease of measurement	Fair	Fair	Good
Measurement problems	Cold junction, Thermal emfs	Lead and contact resistance	Self heating
Mechanical stability	Excellent (+)	Very good	Fair
Size	Tiny (+)	Moderate	Very small (+)
Time response	Very fast (+)	Moderate	Fast (+)
Heat capacity[a]	Very low (+)	Moderate	Low (+)
Interchangeability	Good	Very good	Poor
Cost	Very low (+)	Moderate	Low

Note. A distinct advantage is indicated by (+) and a disadvantage by (−)

[a] For a given unit.

[b] For continuous operation.

---

† Microbead thermistors are available from Thermometrics, Edison, NJ; Fenwal Electronics, Inc., Framingham, MA; Victory Engineering Corp., Springfield, MA, and many other firms.

The measurement of thermistor resistance can be carried out with two-lead bridge circuits or high-quality digital multimeters, as described in the preceding section on platinum resistance thermometers. The major concern is to limit the electrical power dissipation in the thermistor during the measurement. Small bead thermistors are sensitive to self heating; typical values range from 0.5 K to 4 K per milliwatt of power dissipated. To avoid serious errors, this heating effect must be negligible compared with the desired accuracy of the temperature measurement. For a self-heating temperature rise of less than 1 mK, the power dissipation must be less than $\sim 1 \mu$W, which implies, for example, a current of less than 7 μA through a 20 kΩ thermistor.

The calibration of a thermistor with either Eq. (18) or (19) requires at least three calibration points even if a_2 and A_2 are set equal to zero. Preferably, one should calibrate at 5 or 6 points uniformly distributed over the temperature range of interest. These equations should be considered as interpolation formulas; they do not provide a trustworthy extrapolation for temperatures outside the calibration range. One of the problems associated with calibrating thermistors is their fast response and the possibility of a thermal lag between the thermistor and the standard thermometer. If a thermistor and a mercury thermometer or a large PRT are positioned independently in a regulated constant-temperature bath, the thermistor will show larger and more rapid temperature fluctuations than the standard thermometer. The proper procedure is to mount both thermistor and standard thermometer in good thermal contact with a large metal block, which will damp out short-term fluctuations in the bath temperature.

Unfortunately, the stability of a thermistor resistance is not as good as that achieved with PRTs. The principal causes of instability in hermetically-sealed bead thermistors are changes in lattice defect concentrations and changes in the electrical contact with the leads. Both of these changes can be caused by operating at high temperatures or by a transient electrical power surge. Contact resistance can also change during rapid cooling. As a result, thermistors "age" (i.e., exhibit a drift in resistance at constant temperature over a long time). This aging occurs even at room temperature and more rapidly at elevated temperatures. Typical drift rates for bead thermistors with $R_{25} = 10$ kΩ are only ~ 3 mK per 100 days at 30°C to 60°C but can reach 2 mK per day at 200°C.[14,15]

Optical pyrometers.[16,17] The optical pyrometer can be used for the determination of temperatures above ~ 900 K, where blackbody radiation in the visible part of the spectrum is of sufficient intensity to be measured accurately. The blackbody emitted radiation intensity at a given wavelength λ in equilibrium with matter at temperature T is given by the Planck radiation law,

$$J_\lambda = \frac{c_1}{\lambda^5} \frac{1}{\exp(c_2/\lambda T) - 1} \tag{20}$$

where J_λ is the blackbody radiant flux per unit wavelength interval and c_1 and

c_2 are universal constants $(c_1 = 2\pi c^2 h = 3.74177 \times 10^{-16}$ W m^2; $c_2 = hc/k = 0.0143877$ m K). Figure 8 shows the blackbody radiant flux per unit wavelength interval in units of W m^{-2} per μm.

The optical pyrometer is essentially a photometer that measures the radiation intensity in a given wavelength interval. We shall here consider only the most common type of optical pyrometer, the disappearing-filament type. The object whose temperature is to be measured is viewed through a telescope containing, at an image plane, a lamp with a pure tungsten filament enclosed in an evacuated glass tube. The temperature of this filament can be varied by adjustment of the current passing through it. A red filter in the eyepiece and the insensitivity of the human eye to long-wavelength radiation create a narrow wavelength range (centered at 655 nm) for visual observation. When the current has been adjusted so that the filament becomes invisible against the blackbody radiation, the temperature is obtained from the measured current with a calibration curve or chart.

Fixed points for the calibration of the optical pyrometers are the silver and gold points and higher secondary fixed points such as those given in Table

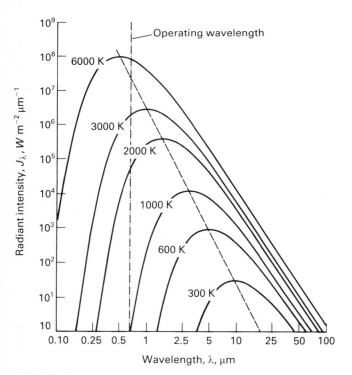

FIGURE 8

Blackbody radiative intensity per unit wavelength as a function of wavelength and temperature. The effective wavelength of 655 nm for optical pyrometers is indicated by the vertical broken line. The position of maximum intensity $\lambda_{\max}$ at temperature T is given by the simple expression $\lambda_{\max} T = 2898\ \mu$m K, and this is shown by the dashed line.

2. Calibration at other temperatures can be accomplished by use of a rotating sector or a filter of accurately known transmission factor between the fixed-point source and the pyrometer, in order to simulate a source of lower temperature in accordance with the Planck equation. Such sectors or filters are used also to permit the optical pyrometer to be used for the measurement of temperatures much above 2000 K.

Optical-pyrometer measurements are most reliable when the object being examined is the interior of a furnace or cavity of uniform temperature viewed through a small opening. Readings for an exposed surface are dependent upon the emissivity ε of the substance concerned, which for an ideal "blackbody" is unity and for actual materials is less than unity. The emissivity in the visible range is near unity for carbon ($\varepsilon \approx 0.85$) and oxidized metals, but it is considerably less for platinum ($\varepsilon = 0.3$) and other unoxidized metals, especially when they are polished. The difference between the brightness temperature T_B obtained from the optical pyrometer and the actual temperature can be approximated by

$$\frac{1}{T} - \frac{1}{T_B} \cong \frac{\lambda}{c_2} \ln \varepsilon \qquad (21)$$

and correction tables are available that take the emissivity and any nonideal transmission of the viewing device into account. Under the best conditions, the optical pyrometer is accurate to about 0.2 percent of the absolute temperature (e.g., ± 4 K at 2000 K). It is also a convenient instrument for less precise measurements at high temperatures, such as routine measurement of furnace temperatures, etc.

Cryogenic thermometers.[18] The measurement of temperatures in the low-temperature region, below 77 K (liquid nitrogen boiling point) and especially below ~30 K, requires special devices. Although the platinum resistance thermometer is the standard IPTS-68 device down to 13.81 K, its sensitivity decreases rapidly at low temperatures and approaches zero as T approaches 0 K. As practical thermometers, PRTs are limited to use above about 35 K. Special gold–Chromel thermocouples are available for low-temperature use. The alloy Au + 0.07%Fe yields the best sensitivity ($\geq 15 \, \mu$V K^{-1} above 10 K) and quite good stability, but the reproducibility of gold–chromel couples is poor compared with that of the thermocouples listed in Table 3 for use at 77 K and above.

There are several special types of resistance thermometers for use in the range 1 K to 30 K: germanium resistors, carbon resistors, and carbon-glass resistors. Bulk germanium thermometers, which are widely used, have good stability (± 0.5 mK long term) and excellent sensitivity at low T values ($d \ln R / dT \cong -2/T$). More recently, germanium wafers have been developed with smaller heat capacities and better "high-temperature" sensitivities that allow them to be used up to 100 K. Problems with bulk and wafer germanium thermometers include high cost and difficulties with contact resistance. Carbon resistors are inexpensive and sensitive, but they are subject to poor long-term stability and require frequent recalibration. Carbon–glass resistance thermometers are significantly more reliable and easier to use than carbon and have characteristics comparable to those of germanium thermometers.

There are several semiconductor devices (p-type GaAs and Si) used at low temperatures; these diode thermometers provide an output voltage that increases as T is decreased. The Si-diode thermometer is more sensitive (typically -50 mV K^{-1} in the 2–30 K range and -3 mV K^{-1} from 30 K to 80 K for a diode giving 1 V at 30 K) and more linear than germanium or carbon-glass resistance thermometers. It is probably the practical thermometer of greatest utility in the range 4.2 K to 80 K.

Another attractive and widely used special thermometer is the quartz frequency thermometer, which is described in the next subsection since it can be used over the very wide range from 4 K to 500 K.

Other thermometric devices. The most basic measurement of temperature, yielding the thermodynamic temperature T_{therm}, involves the use of gas thermometers. These are complex devices that are now used almost exclusively to obtain highly accurate absolute values of the fixed-point temperatures. However, simple gas thermometers resembling that described in Exp. 1 can be used for moderately precise measurements, especially at high or low temperatures.

The vapor pressure of a pure liquid or solid is a physical property sensitive to temperature and thus suitable for use as a thermometer. The use of a liquid-nitrogen vapor-pressure thermometer is suggested for the range 64–78 K in Exp. 48. At very low temperatures (1–4.2 K), the vapor pressure of liquid helium can be used.

A recent innovation is the integrated-circuit temperature transducer, which is available in both current- and voltage-output versions.[6] This device provides an output linearly proportional to the absolute temperature (typically 1 μA K^{-1} or 10 mV K^{-1}). As semiconductor devices, they resemble thermistors in having limited range, self-heating problems, and mechanical fragility. However, they do provide a convenient way of obtaining an analog voltage proportional to T for various automated control devices.

Finally, the quartz thermometer should be emphasized as a versatile and high-precision device.[19] This thermometer makes use of the temperature dependence of the resonance frequency f_r of a quartz single crystal. This frequency variation is small (35 p.p.m. K^{-1}) but very reproducible, and it can be measured with excellent precision. The temperature dependence of f_r depends strongly on the orientation of the quartz crystal slab. The frequency of a temperature-sensitive crystal beats against that of another crystal whose frequency is essentially independent of temperature. By counting the beat frequency, a resolution of $\pm 10^{-4}$ K can be achieved over the range 200–500 K and better than ± 0.02 K over the range 4–200 K. However, the sampling time is 10 s for $\pm 10^{-4}$ K resolution, which is inconveniently long; happily, it is only

0.1 s for ±0.01 K resolution. The output is digital, and there are no lead-wire problems. Hysteresis and aging do occur, so that calibrations are not better than +0.1 mK and a calibration drift of ~10 mK per month occurs. However, the quartz thermometer is generally a very convenient device for high precision, as opposed to high absolute accuracy.

TEMPERATURE CONTROL

In addition to the measurement of temperature, it is often necessary to maintain a constant temperature. The importance of this type of control in experimental physical chemistry is illustrated by the fact that 27 of the 49 experiments described in this book require temperature control of some kind. Many physical quantities such as rate constants, equilibrium constants, vapor pressures, and magnetic susceptibilities are sensitive functions of temperature and must be measured at a known temperature that is held constant to within ±0.1 K or better. Certain physical techniques are even more demanding; for example, the use of the dilatometer in Exp. 21 requires that the temperature be controlled to within ±0.002 K.

There are two basic methods of achieving a constant temperature.

1. Phase equilibrium is maintained at constant pressure between two phases of a pure substance or three phases of a two-component system.
2. A temperature sensor provides a feedback signal to control the input of heat (or refrigeration cooling in some cases) in order to maintain the temperature close to any arbitrary desired value.

Method 1 is the simplest approach; method 2 is more flexible and generally more useful.

1   Phase-Equilibrium Baths

The greatest disadvantage of this method is the impossibility of attaining a desired *arbitrary* temperature unless a liquid–vapor system is used with a complicated manostat to maintain an arbitrary boiling pressure. In addition, it is often difficult to maintain temperature control for very long periods of time with this method. There are, however, several advantages: economy and simplicity in operation, excellent temperature stability, and potentially high precision in the absolute temperature of the bath. Several commonly used systems of this type are listed below, together with brief comments on their use.

Liquid-nitrogen bath. Liquid nitrogen at its normal boiling point (77.35 K = −195.80°C) provides a very convenient low-temperature bath. Since O_2 dissolved in liquid N_2 will raise the temperature of the bath, the mouth of the Dewar flask should be plugged loosely with glass wool or cotton to retard the slow condensation of atmospheric oxygen. The temperature of a liquid-nitrogen bath can be calculated on the assumption that the nitrogen vapor pressure is equal to the atmospheric pressure. Clearly, temperature stability will depend on the absence of large changes in pressure; fortunately, the boiling point of nitrogen changes by only $0.011 \, \text{K Torr}^{-1}$ near 77 K.

"Dry Ice" bath. Solid carbon dioxide in equilibrium with CO_2 vapor at 1 atm will provide a temperature of $194.67 \, \text{K} = -78.48°\text{C}$. Thus, Dry Ice, which is inexpensive and readily available, would seem to be very suitable for a constant-temperature bath at moderately low temperatures. Unfortunately, the use of Dry Ice alone is complicated by two difficulties: the problem of obtaining and maintaining the proper pressure of CO_2 gas and the problem of achieving good thermal contact between the Dry Ice and the object to be cooled. Although these difficulties can be overcome by careful bath design and the use of a heater to cause a constant evolution of CO_2 gas, it is much easier to use a bath consisting of Dry Ice and a liquid that does not freeze at $-78.5°\text{C}$. This liquid provides good thermal contact throughout the bath and prevents air from diluting the CO_2 gas at the surface of the Dry Ice as it would at an exposed Dry Ice surface. The traditional choice of liquid was acetone, but acetone is flammable and thus not safe for general use. Recommended nonflammable liquids are isopropanol and trichloroethylene. In making up a Dry Ice bath, it is necessary to minimize the foaming which occurs owing to rapid evolution of gas when Dry Ice is placed in contact with liquid initially at room temperature. The Dry Ice should be pulverized. If a special grinder is not available, one can wrap chunks of Dry Ice in a towel and pound them with a mallet. Be careful in handling pieces of Dry Ice; it can cause painful "burns" if held in the bare hand for even a few seconds. The use of tongs or insulated gloves is strongly recommended. A Dewar flask is first filled about two-thirds full with isopropanol or trichloroethylene, and then small quantities of very finely powdered Dry Ice are added **slowly** with a spatula. This Dry Ice will evaporate almost immediately, and there will be considerable foaming at the surface. Add more Dry Ice only after the foaming has subsided. After a while the liquid will have cooled to the point where the evaporation of Dry Ice is much slower, and some Dry Ice will begin to accumulate on the bottom of the Dewar. At this point, small lumps of Dry Ice can be added without causing serious foaming. Good temperature control with this bath is ensured only if it is well stirred and there is a slow but steady stream of CO_2 bubbles rising from the bottom. **Caution:** Trichloroethylene is suspected of being a very weak carcinogen; avoid ingestion and contact with the skin.

Ice bath. The ice–water equilibrium at 0°C provides an excellent constant-temperature bath. The ice should be washed, and distilled or deionized water must be used. To avoid thermal gradients between water at 4°C (maximum density) at the bottom of the Dewar and a 0°C liquid surface in which ice is floating, it is usually necessary to stir this bath. As a bath for the reference junction of a thermocouple, gradients can be eliminated by completely filling the Dewar with ice and adding only a small amount of cold distilled water; the weight of ice above the liquid will force ice down to the very bottom of the Dewar.

Vapor baths. Boiling water (100.0°C), naphthalene (218.0°C), and possibly sulfur (444.7°C) can be used as vapor baths. In each case, the object whose temperature is to be controlled is immersed in the refluxing or condensing vapor. See Exp. 1 for details of a steam bath.

2 Systems Utilizing Temperature Controllers

The most important method of achieving temperature control is to use a sensitive thermometer that generates an electrical signal, such as a resistance thermometer, thermistor, or thermocouple. Comparison of this signal with a reference signal that establishes the *set point* provides an error signal to a feedback circuit that controls the power to the heater. Thus any temperature deviation from the desired value detected by the sensor will be corrected automatically.

As in any system employing feedback to maintain a steady-state condition, certain design criteria must be met in order to obtain reasonably rapid response to environmental changes while avoiding excessive "hunting" or even uncontrolled oscillations. The performance of the system will depend upon such factors as the sensitivity and speed of response of the thermosensing element, the conduction or circulation of heat in the system (stirring, convection), and the fraction of the total energy input (heat input plus stirring work, etc.) that is being controlled by the regulator.

To provide damping of oscillations, a proportionating circuit is usually employed; in the vicinity of the desired temperature the current or power to the heater is made roughly proportional to the difference between the actual temperature and a temperature setting that is slightly above the desired value. Such a circuit usually has an adjustment for providing an optimum range of proportionation, since a proportionating system provides stability at the expense of some precision of temperature control.

Design of controllers. A brief review of the operating principles for temperature-control systems will be given below; more details are available in the literature.[20,21] The simplest type of controller is an on–off device, in which the heater is either fully on or fully off. This inexpensive and easily implemented control is suitable for moderate-quality control of large, sluggish baths and furnaces. A serious disadvantage of such a controller is its inherent cycling about the set point (i.e., the desired constant temperature value T_s). Figure 9 shows the time response after a step increase in the set point of an on–off controlled system. Not only does the temperature overshoot the desired value T_s when the set point is changed from a previous value T_s', but there is a

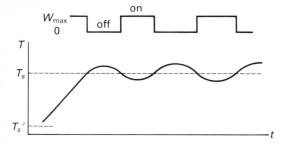

FIGURE 9

ON-OFF controller behavior: the temperature variation observed after a step increase in the set-point temperature from T_s' to T_s. Note the long-term oscillation that persists about T_s. The heater power W can have only two values: 0 or W_{max}.

persistent oscillation about T_s that never disappears. This long-term cycling can be eliminated by the use of proportional control.

With a proportional controller, the heater power supplied to the system is changed by an amount proportional to an error signal corresponding to the temperature difference $(T_s - T)$; see Fig. 10. The controller output can be either time proportionating or current proportionating;[21] only the latter will be considered here, as it is the more commonly used mode. Proportional controllers provide a fast response and show very little cycling about the set point, which means better temperature control. Two disadvantages are a steady-state *offset* from the set-point temperature and the presence of damped oscillations associated with a change in the set point, both of which are illustrated in Fig. 11. The offset can be reduced by decreasing the proportional band PB (i.e., increasing the controller gain), but there is a limit to this approach since instability in the form of undamped oscillations will occur at too high a gain.

The overshoot oscillations occurring after a change in set point can be reduced greatly by adding to the simple proportional control a second control signal proportional to the time derivative of the temperature. This feature "anticipates" the magnitude of future error signals and smooths out overshoot

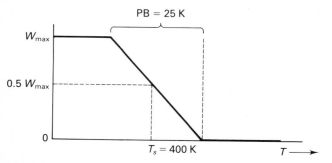

FIGURE 10

Heater power output W from a proportional controller as a function of sensor temperature. The steeply sloping linear ramp extends over a temperature range called the proportional band PB. The controller gain is inversely proportional to the width of this band: $(1/W_{max}) \, dW/dT = \text{PB}^{-1}$. Note that an on-off controller is a limiting case of proportional control with a proportional band of zero width.

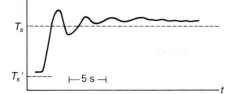

FIGURE 11
Response of a proportionally controlled system to a step increase in the set point from T_s' to T_s. In this example, the gain is roughly one-half the critical gain and the transient oscillations are well damped. Note the offset from the set-point temperature.

oscillations. The offset problem could, of course, be eliminated by a manual reset of the set-point value. However, this can also be done automatically by adding an integrating control feature, for which the control signal is proportional to the integral of the error signal. As long as there is any difference between the control temperature and the set-point temperature, a small corrective change in the heater power will occur to slowly reduce this difference to zero. Figure 12 illustrates the behavior of systems with proportional + derivative (PD) or proportional + integral + derivative (PID) controllers.

The proper design of constant-temperature systems requires judicious choice of sensor and controller and also careful attention to mechanical/ thermal layout of the bath or block that serves as the heat bath. Since control problems are more acute in maintaining constant temperature at low temperatures, much of the pertinent literature is in cryogenic journals. However, the same general principles are equally valid for control at low, moderate, or high temperatures. A very clear mathematical analysis of the coupling between electrical and thermal responses has been given by Forgan,[22] and the conclusions of his analysis will summarized here. A typical temperature-control system is shown schematically in Fig. 13. Frequency-domain responses are indicated; e.g., the Fourier transform of the temperature T as a function of time t gives an amplitude $T(\omega)$. The symbols $h(\omega)$, $b(\omega)$, and $\theta(\omega)$ denote the thermal responses of the heater to electrical power input, of the block to energy input from the heater, and of the thermometer to temperature

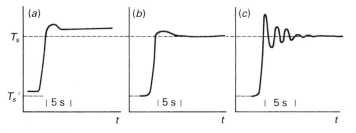

FIGURE 12
Response of a system with (a) proportional + derivative (PD) control and (b) proportional + integral + derivative (PID) control when the gain is 15% of the critical gain. Part (c) shows the response with PID control when the gain is 50% of the critical gain, a practical limit for good regulation. Compare these responses to that shown in Fig. 11 for simple proportional (P) control.

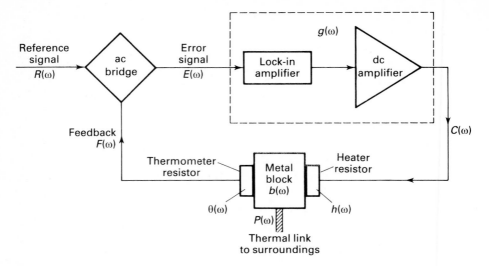

FIGURE 13

Schematic diagram of a temperature-controlled block at a set-point temperature above the ambient temperature of the surroundings. The notation is that of Forgan.[22] The dc voltage V applied to the heater is modified by the small correction voltage $C(\omega)$; the resulting heater power output is $W(\omega) = V^2/R_h + 2VC(\omega)/R_h$, where R_h is the electrical resistance of the heater. See text for other details.

variations of the block. The electrical gain $g(\omega)$ represents the response of the lock-in amplifier and the dc operational amplifier. The temperature $T(\omega)$ of the block is related to the control signal $C(\omega)$ and the perturbation $P(\omega)$, which is mostly a heat leakage from the block to colder surroundings.

There are four time constants τ associated with the three thermal responses $h(\omega)$, $b(\omega)$, $\theta(\omega)$ and the electrical gain $g(\omega)$. The effect of a coupled set of four time constants on the behavior and stability of the system is a complex problem with rather messy solutions. However, a reasonably simple result is obtained for the limiting case in which

$$\tau_{max} = \tau_a > \tau_b > \tau_c > \tau_d = \tau_{min} \tag{22}$$

and all four τ values are well separated. In this case, the best arrangement is to have the time constants of the block and of the lock-in amplifier (τ_a and τ_b) be long and the time constants of the heater and of the thermometer (τ_c and τ_d) be very short. This increases the critical open-loop dc gain and allows operation at high amplifier gain, which leads to good regulation control. Indeed, one wants the time constants of the block, which characterizes how rapidly the block cools when the heater is off, to be as long as possible, say 10 s or more. The time constant of the thermometer must be very short (1–10 ms is desirable), since the controller is actually controlling the temperature of the thermometer, which must therefore be closely coupled to the block. The heater should have a small thermal mass and be in very good contact with the block so that τ_c will be short. The absence of temperature gradients within the block (or bath) is also important; the heater should be distributed as uniformly as possible over the block (or the bath should be well stirred).

Stirred water baths. The water bath, equipped with stirrer, temperature sensor, control circuit, and heater provides the most commonly used means of temperature control in the 15°C–80°C range. There are several commercial sources of regulated water baths with fluid volumes of 6–50 liters.† A good design for a water bath for general student laboratory use consists of a large rectangular tank, perhaps 18 by 36 inches in horizontal area and 18 inches deep, with a water capacity of about 170 liters. The tank should be constructed of welded stainless steel; the inclusion of glass windows in two or more sides is a convenience but complicates the design. The thermometer, temperature controller, and stirrer are best mounted in the middle; this arrangement leaves the ends free for experimental work. There should be adequate provision for mounting rods and clamps to support flasks, dielectric cells, etc. Depending on the experiment being done, two or four sets of apparatus can be operated in a single bath of this size. The bath should be provided with the following items.

Stirrer. The bath fluid must be well stirred in order to avoid temperature gradients within the bath. A large centrifugal water circulator, driven by a $\frac{1}{20}$-hp motor, provides adequate stirring. It should be positioned carefully so that the effluent stream will cause efficient circulation throughout the entire tank.

Heater. A single copper-sheathed immersion heater ($\frac{1}{4}$–$\frac{3}{8}$ inch diameter) of 250–W capacity is adequate for temperature control up to about 30°C. It is recommended that the heater be positioned low in the bath and in the effluent stream from the stirrer. At higher temperatures, a second heater may be needed. In this case, one can be maintained at a constant power and the other can be regulated by the temperature controller, or both can be controlled, depending on their power ratings, the temperature desired, and other design factors. Blade heaters sheathed in stainless steel have longer time constants than copper-sheathed tube types and will not give as good temperature control, but they can be used as unregulated constant-power heaters.

Thermometers. The sensor probe should be placed in the bath in a central location removed from the heater and stirrer units. If a single cell or sample holder immersed in the bath is the only object of interest, the probe should be placed near this cell but not touching it. The measuring thermometer should be placed in good thermal contact with the cell or sample holder if there is only one, or positioned close to the location of multiple cells.

---

† Grant constant-temperature baths are available from Science/Electronics, Inc., P.O. Box 986, Dayton, OH 45401; Lauda baths are available from Brinkmann Instruments Co., Cantiague Rd., Westbury, NY 11590.

Temperature controller. There are many commercial units providing low (±1 K) or moderate (±0.05 K to ±0.1 K) resolution control. Controllers capable of ±5 mK to ±10 mK regulation are available from Grant and from Lauda. Even better constancy (±2 mK or less) can be achieved with temperature controllers from Tronac Inc., 1804 S. Columbia Lane, Orem, UT 84058. These instruments permit high-resolution (±1 mK) selection of the bath temperature, adjustment of the gain to match the rate of heat loss, and adjustment of the time constant for the integrating circuit.

Water-level and cooling control. When a water thermostat bath is operated for long periods, it is subject to loss of water by evaporation to a point that will interfere with its normal operation unless the water is replaced. Therefore the tank should have an overflow drain and should be provided with a continuous supply of cold water that can be controlled from a few drops per minute to a small, steady stream. This supply of water also performs an important control function. A $\frac{1}{20}$-hp motor delivers constantly about 37 W of mechanical energy to the bath, and, on days when the room temperature is not far below the desired bath temperature, this amount of energy is itself sufficient to maintain the bath temperature above the desired level. A stream of cold water may compensate largely or entirely for the stirring heat. The adjustment of the water flow is often critical and must be checked frequently.

Miscellaneous: To reduce rusting of iron hardware in the bath, a zinc or aluminum electrode with an applied anodic potential of a few volts with respect to the tank as ground may be provided. This should be protected from accidental contact by a perforated plastic tube.

Low-temperature modifications.[23] For operation at temperatures in the range 5°C to ~15°C (~20°C in the summer), one needs a supply of chilled cooling water. For work down to about −15°C, one needs a sealed refrigeration unit that can cool the bath sufficiently. The water in the bath must be replaced by water + ethylene glycol (antifreeze), and better insulation may be required.

Stirred oil baths. Open water baths are subject to substantial evaporation loss above 50°C, and even closed water baths present problems above ~80°C. In the past, "cylinder oil" was frequently used in the range 70°C to 250°C, but there are safety concerns connected with fire hazards and toxicity. Commercial peanut oil is an inexpensive bath fluid that can be used up to about 225°C. It is nontoxic and even smells good, but fire hazards still remain. The best modern bath fluids for high-temperature use are high-molecular-weight polyethylene glycol or polypropylene glycol (both of which are essentially nonflammable with "flash points" of 260°C) and Dow-Corning 210H or 550 silicone fluids (nonflammable polydimethylsiloxanes with flash points of 310°C). Both types of materials are nontoxic and have low vapor pressures even at 200°C; the glycols can be used in closed baths up to 230°C and the silicones can be used up to 300°C. Unfortunately, none of these materials is inexpensive ($20–

$50 per liter), but they have long service lifetimes at high temperatures, especially the silicone fluids. For even higher temperatures, up to about 425°C, a eutectic mixture of sodium, potassium, and lithium nitrates (14, 56, and 30 wt%, respectively) can be used as the bath fluid.

Stirred fluid baths designed to operate below −15°C require low-viscosity, low-freezing-point fluids such as Univis (a nonflammable oil) or several of the Dow-Corning silicone fluids. Large water-cooled compression refrigeration units and heavy-duty fluid stirrers are also needed. Many of the low-temperature units available commercially use circulator pumps instead of stirrers, as described below.

Circulating liquid baths. These devices consist of a fluid reservoir, a heater, a temperature sensor, a controller, and a circulating pump. The most commonly used general-purpose unit is made by Haake, and this system will be described here. A large variety of excellent circulating units can also be obtained from Lauda. Some of these are designed for operation at low temperatures (down to −70°C) as well as above room temperature.

Since the Haake constant-temperature circulating bath makes use of plastic components, the maximum temperature of these pumps is 150°C and the useful range is 25°C to 150°C. Distilled water may be used as the circulating liquid in the 25°C to 90°C range, and Nujol, glycerine, or ethylene glycol may be used in the 90°C to 150°C range. Approximately 1.5 liters of liquid is required to keep the fluid in the operating range $\frac{1}{2}$ inch to 2 inches below the top of the fluid reservoir. The fluid may be pumped through a closed external system or pumped into an open bath located *above* the circulating pump with an overflow line to return fluid to the thermostat reservoir.

The electrical connections on top of the Haake unit provide circuits to the pump, the heater, and the controller. A switch allows both the pump and temperature-regulating portions of the device to be turned on or off. A neon light indicates when the heater is on. The temperature of the thermostat is set by turning a rotating magnet after loosening the locking screw on the side of the thermoregulator. Rotation of the knurled screw on the top of the magnet housing rotates the magnet which moves the temperature indicator bar up or down. The temperature *at the top* of the temperature indicator bar is the approximate temperature which the reservoir will maintain. The exact temperature of the fluid is read on a separate thermometer immersed in the reservoir and held in place with a stopper. The temperature of the circulating liquid will remain constant to within ±0.02°C. With the addition of special cooling accessories, this unit may also be used to provide constant-temperature liquid at temperatures below 0°C (e.g., methanol in the temperature range −70°C to +30°C). The manufacturers' instructions should be consulted for operating details at low temperatures.

Constant-temperature sample blocks. The design principles discussed at the beginning of this section apply just as well to the control of a metal block containing the sample of interest as to a fluid bath. This "dry design" usually involves a three-stage vacuum thermostat: an outer vacuum jacket that is immersed in a bath at roughly constant temperature, an inner vacuum can that serves as an adiabatic/radiation shield, and the sample block (usually copper). For low-temperature operation (4 K–300 K), the outer bath is a cryogenic liquid such as nitrogen or helium, and excellent commercial controllers are available from Lake Shore Cryotronics.[24] For operation in the moderate temperature range from −40°C to +70°C, the outer jacket is maintained at a temperature below the desired control temperature with a commercial circulating device, and microcomputer digital control is capable of maintaining a long-term temperature constancy of better than ±1 mK.[25] For operation at high temperatures (300–1300 K), a water jacket can serve as the outer bath, and digital control yields temperature stability of ±0.1 K at 1000 K.[26] Less elegant high-temperature control can be achieved with electric furnaces; see the next subsection.

3   Special Systems

A few constant-temperature systems of lower precision involving "air baths" are described below.

Air thermostats. Complex gas-handling systems are not well accommodated in liquid thermostat baths. A double-wall air thermostat can be constructed with suitable insulation (air, fiberglass, vermiculite, foamed plastic, etc.) and with double glass windows and doors where needed. Air is circulated by a blower or fan that has its motor mounted externally in order to reduce uncontrolled heating. Electrical heating elements with large exposed surface areas, such as helical wire coils, should be used instead of the immersion-type heaters used in liquid baths. Temperature control is accomplished in much the same way as in a water bath, but the time constants may be very different.

Ovens and furnaces. Commercially available laboratory ovens can be used up to temperatures as high as 200°C or 300°C (~575 K). They are usually regulated to within ±(1–2) K with a bimetallic thermoswitch. A bimetallic strip, which consists of two dissimilar metal strips bonded together, undergoes differential thermal expansion. At some arbitrary temperature the two metals are of equal length and the strip is flat. As the temperature changes, one metal will expand or contract more than the other, thus causing the strip to bend or curl up. Although such thermoswitches are cheap, mechanically rugged, and capable of operating over a wide range, their accuracy is only about 1 percent and the contacts sometimes stick.

Air circulation in typical ovens is by convection, and the temperature inside the oven is usually nonuniform. For obtaining better temperature uniformity and improved control, a lining made of thick sheet aluminum or copper can be installed.

Electric furnaces are used for higher temperatures. These are obtainable commercially, but satisfactory ones can be made in the laboratory (see Fig. 24-1). A pregrooved Alundum core is wound with suitable heating wire (Nichrome, Chromel A, Kanthal) of 12 to 20 gauge, and covered with a thick coating of Alundum cement. This is surrounded by several inches of powdered magnesia for thermal insulation, in a large metal container. After assembly, the furnace is slowly heated to slightly above 1000°C to set the cement. A cover plug may be cut from firebrick. Internal metal radiation shields (platinum is best but very expensive; molybdenum may be used if a neutral or reducing atmosphere is maintained) are needed if a high degree of temperature uniformity is required. Temperature is controlled with a Variac or similar transformer operated manually or by a proportionating control circuit with a thermocouple potentiometer or resistance thermometer bridge.

For temperatures above 1000°C special high-temperature furnaces, including induction furnaces, vacuum furnaces, and arc furnaces, are used, but these are beyond the scope of this book.

REFERENCES

1. The International Practical Temperature Scale of 1968, *Metrologia* **5,** 35 (1969); **12,** 7 (1976).
2. J. F. Schooley, "Thermometry," CRC Press, Boca Raton, FL. (1986).
3. R. E. Bedford and C. G. M. Kirby, *Metrologia* **5,** 83 (1969).
4. This standard may change in the near future. See L. A. Guildner, H. J. Kostkowski and J. P. Evans, *Metrologia* **15,** 1 (1979).
5. T. B. Douglas, *J. Res. U.S. Natl. Bur. Stand.* **73A,** 451 (1969); L. A. Guildner, *PTB-Mitteilunger* **90,** 41 (1980).
6. "Temperature Measurement Handbook and Encyclopedia," Omega Engineering, Inc., Stamford, Conn. (1974).
7. J. Busse, in "Temperature, Its Measurement and Control," vol. I, p. 228, Reinhold, New York (1941).
8. J. F. Swindells, "Calibration of Liquid-in-Glass Thermometers," Natl. Bur. Stand. Monogr. 90, U.S. Government Printing Office, Washington, D.C. (1965).
9. R. P. Benedict and H. Hoersch (eds.), "Manual on the Use of Thermocouples in Temperature Measurement," ASTM Special Publ. 470A, Omega Press, Stamford, Conn. (1981).
10. F. R. Caldwell, "Thermocouple Materials," Natl. Bur. Stand. Monogr. 40, U.S. Government Printing Office, Washington, D.C. (1962).
11. "Thermocouple Reference Tables," Natl. Bur. Stand. Monogr. 125, U.S. Government Printing Office, Washington, D.C. (1979) [also available from Omega Press, Stamford, Conn.].
12. "Platinum Resistance Temperature Sensors," Bull. 9612, Rosemount Inc., Minneapolis, Minn. (1972).
13. "Temperature Calibration and Interpolation Methods for Platinum Resistance Thermometers," Report 68023F, Rosemount, Inc., Minneapolis, Minn. (1980).
14. M. Sapoff, "Thermistors," a five-part series appearing in *Measurements and Control,* issues 80–84 (Apr.–Dec. 1980) [available as a reprint from Thermometrics, 808 U.S. Highway 1, Edison, N.J. 08817].
15. S. D. Wood, B. W. Mangum, J. J. Filliben, and S. B. Tillett, *J. Res. Natl. Bur. Stand.* **83,** 247 (1978).
16. H. J. Kostkowski and R. D. Lee, "Theory and Methods of Optical Pyrometry," Natl. Bur. Stand. Monogr. 41, U.S. Government Printing Office, Washington, D.C. (1962).

17. P. H. Dike, W. T. Gray, and F. K. Schroyer, "Optical Pyrometry," L&N Techn. Publ. A1.4000, Leeds & Northrup, Inc., North Wales, PA.
18. L. G. Rubin, B. L. Brandt, and H. H. Sample, "Cryogenic Thermometry: A Review of Recent Progress," in J. F. Schooley (ed.), "Temperature: Its Measurement and Control in Science and Industry," vol. 5, part II, p. 1333ff, American Institute of Physics, New York (1982).
19. A. Benjaminson and F. Rowland, "The Development of the Quartz Resonator as a Digital Temperature Sensor with a Precision of 10^{-4} K," in H. H. Plumb (ed.), "Temperature: Its Measurement and Control in Science and Industry," vol. 4, part IV, p. 701ff, Instrument Society of America, Pittsburgh, PA (1972).
20. J. Sandford, "Understanding Temperature Control," in "Instruments and Control systems," June 1977, pp. 37–42.
21. C. de Silva and M. A. Aronson, "Process Control," a series appearing in "Measurements and Control," esp. part 3 (issue 106, Sept. 1984, pp. 165–170) and part 4 (issue 107, Oct. 1984, pp. 133–145) much of this material has also appeared in "Principles of Temperature Control," available from Gulton Industries Inc., Servonic Division, Costa Mesa, Calif. 92627.
22. E. M. Forgan, *Cryogenics* **14,** 207 (1974).
23. G. I. Williams, and W. A. House, *J. Phys. E: Sci. Instrum.* **14,** 755 (1981).
24. J. M. Swartz and L. G. Rubin, "Fundamentals for Usage of Cryogenic Temperature Controllers," (1985) available from Lake Shore Cryotronics, Inc., 64 East Walnut St., Westerville, Ohio 43081.
25. R. B. Strem, B. K. Das, and S. C. Greer, *Rev. Sci. Instrum.* **52,** 1705 (1981).
26. W. M. Cash, E. E. Stansbury, C. F. Moore, and C. R. Brooks, *Rev. Sci. Instrum.* **52,** 895 (1981).

GENERAL READING

R. P. Benedict, "Fundamentals of Temperature, Pressure, and Flow Measurements," 3rd ed., Wiley-Interscience, New York (1984).

T. J. Quinn, "Temperature," Academic Press, New York (1983).

J. F. Schooley, "Thermometry," CRC Press, Boca Raton, FL. (1986). "Temperature, Its Measurement and Control in Science and Industry," vol. 3 (C. M. Herzfeld, ed., Reinhold (Krieger reprint), New York (1962)], vol. 4 [H. H. Plumb, ed., Instrument Society of America, Pittsburgh, PA (1972)], vol. 5 [J. F. Schooley, ed., American Institute of Physics, New York (1982)].

CHAPTER XVII

VACUUM TECHNIQUES

This chapter contains a discussion of various topics of importance in the theory, design, and practice of handling gases at low pressures. Major emphasis is given to high-vacuum techniques, since they play an important role in many physical chemistry research problems. For vacuum applications in this book, see Exps. 5, 24, 27, and 47. Some of the material presented is also pertinent to the problem of pumping on refrigerant baths; for an application of this technique, see Exp. 47. This chapter is not intended as a comprehensive treatment; more detailed information is available in the monographs cited in the General Reading list.

INTRODUCTION

The term vacuum refers to the condition of an enclosed space that is devoid of all gases or other material content. It is not experimentally feasible in a terrestrial environment to achieve a "perfect" vacuum, although one can approach this condition extremely closely. It is possible routinely to obtain a vacuum of 10^{-6} Torr and with more sophisticated techniques 10^{-10} Torr $(1.3 \times 10^{-13}$ bar); it is even possible by special techniques to obtain a vacuum of 10^{-15} Torr, or about 30 molecules per cm^3. One *Torr,* the conventional unit of pressure in vacuum work, is the pressure equivalent of a manometer reading of one millimeter of liquid mercury, the temperature of which is 0°C, at standard gravity of 9.806 65 m s^{-2}; 1 Torr $\equiv \frac{1}{760}$ atmosphere.

For Dewar flasks, metal evaporation apparatus, and most research apparatus, a vacuum of 10^{-5} to 10^{-6} Torr is sufficient; this is in the "high vacuum" range, while 10^{-10} Torr would be termed "ultrahigh vacuum." However, for many routine purposes a "utility vacuum" or "forepump vacuum" of about 10^{-3} Torr will suffice, and for vacuum distillations only a "partial vacuum" of the order of 1 to 50 Torr is needed.

THEORETICAL BACKGROUND

The basic theory needed here is the kinetic theory of gases,[1] which is discussed in Chapter IV. It will be assumed that gases always obey the perfect-gas law and have an essentially Maxwellian distribution of molecular velocities. The most important properties for our present purpose are the mean free path λ, the coefficient of viscosity η, and the coefficient of thermal conductivity K; in particular, we are concerned with the dependence of these three properties on pressure at constant temperature. According to Eqs. (IV-3), (IV-13), and (IV-18), λ varies inversely with p while η and K are independent of p. The expression given by Eq. (IV-3) for λ is correct over the entire pressure range, but Eq. (IV-13) is valid only at pressures high enough so that λ is small compared with the dimensions of the container. At lower pressures, where λ becomes comparable to the apparatus dimensions, both η and K decrease with decreasing pressure. When λ is considerably larger than the apparatus dimensions (i.e., at pressures less than 0.01 Torr for most apparatus), η and K will vary linearly with the pressure. This limiting behavior is a consequence of molecular-flow conditions where the mechanism of momentum or kinetic-energy transport depends on collisions between gas molecules and the walls rather than on intermolecular collisions.[1] In this limiting region, the pressure dependence of viscosity or thermal conductivity serves as the basis for the operation of several types of vacuum gauges.

PUMPING SPEED FOR A SYSTEM

Detailed calculations of pumping speed are seldom necessary for designing a system to obtain a static vacuum in a small, closed line. However, certain qualitative principles of vacuum-line design are important. When one must pump out a large volume or pump on a system where there is a continuous evolution of gas inside the system (as when pumping on a refrigerant bath, for example), a careful quantitative design is essential. Overall pumping speed will depend on the characteristics of the pump (or pumps) used and also on the impedance to gas flow through the connecting tubing.

Speed of the pump. The pumping speed S, usually expressed in liter s^{-1}, of a given pump operating at a pressure p is defined as the volume of gas measured at a constant temperature T and the given pressure p that is removed from the system per unit time. This quantity is closely related to the flux Q, which is the quantity of gas (in pressure–volume units) flowing through a given plane in unit time. For an ideal gas, the flux is given by

$$Q \equiv RT \frac{\delta N}{\delta t} = \frac{\delta(pV)}{\delta t} = p \frac{\delta V}{\delta t} \tag{1}$$

where $\delta V/\delta t$ is the volume of gas at temperature T flowing in unit time past a plane in the system where the steady-state pressure is p, $\delta N/\delta t$ is the number of moles flowing through that plane in unit time, and the gas constant R is conveniently expressed as $62.37 \, \text{Torr liter} \, \text{K}^{-1} \, \text{mol}^{-1}$. Thus we see that $S = Q/p$.

The pumping speed is approximately constant over a wide pressure range since it depends on the length of time required for exhausting the gas contained at a given instant within the pumping element itself. This time is called the "cycle" of the pump. For a rotary pump, it is the time for a shaft rotation; for a diffusion pump or an ion pump, it is the time required for certain molecular processes to take place that are independent of pressure. The pumping speed of a given pump begins to decrease when the pressure is sufficiently low and becomes zero when the pressure reaches some ultimate attainable pressure p^*. The variation of the pumping speed S with the pressure p can be roughly approximated by

$$S = S^* \left(1 - \frac{p^*}{p} \right) \tag{2}$$

where S^* is the intrinsic pumping speed. It is not possible to make an accurate calculation of pump speeds, but there are reliable experimental methods for measuring S^*. Both dynamic and static methods have been used. In the dynamic method, the change in the steady-state pressure is measured when a metered leak is closed; in the static method, the rate of pressure change is measured for the evacuation of a bulb of fixed volume. Details of these methods are given by Dushman.[2] Once S^* is known, Eq. (2) permits one to calculate S, which is the quantity of practical interest.

Conductance. In pumping down a system, gas must be conducted through orifices and tubes, and frequently through valves and cold traps as well. At ordinary pressures the speed of flow is limited by the viscosity of the gas, and at very low pressures it is limited by the finite speed of molecular motion and by diffusive effects of molecular collisions with the walls. In either case, a given segment i of the apparatus conducting the gas may be said to have conductance C_i which is the flux of gas through it divided by the pressure difference:

$$C_i = \frac{Q}{(p_1^i - p_2^i)} \tag{3}$$

where p_1^i and p_2^i are taken as fixed pressures at the inlet and outlet of the segment, respectively. Since conductance has the same dimensions as pumping speed, the same units (liter s^{-1}) will be used for both.

The overall conductance of two or more segments is given by

$$C = \sum_i C_i \tag{4}$$

if they are connected in parallel, and by

$$Z \equiv \frac{1}{C} = \sum_i \frac{1}{C_i} \equiv \sum_i Z_i \tag{5}$$

if they are connected in series. Here Z, the reciprocal conductance, is often called the impedance. There is an obvious analogy in behavior between Q, p, and Z on the one hand and the corresponding direct-current electrical quantities I, E, and R on the other. An important consequence of Eq. (5) is that *the conductance of a series sequence of segments is always less than the conductance of any one segment.*

The conductance of a given segment is a function of its geometry and a function of the pressure, temperature, and properties of the gas being pumped. At pressures of about 10^{-2} Torr and above, gas flow in most vacuum apparatus takes place by viscous flow.† At pressures below 10^{-3} Torr, molecular flow is the principal mechanism for gas transport. We shall now cite appropriate formulas for calculating C in a variety of simple cases.

(a) Viscous flow in cylindrical tubing. Combining Eq. (1) with Eq. (4-8) for the laminar flow of a gas (and noting that $\delta N/\delta t$ has the same meaning as ϕ_N) we obtain

$$Q = \frac{\pi d^4}{128\eta L} \frac{p_1 + p_2}{2} (p_1 - p_2) \tag{6}$$

where d is the diameter of the tube, L is its length, η is the viscosity and p_1 and p_2 are the inlet and outlet pressures. Comparing Eqs. (3) and (6) we see that

$$C_\eta = \frac{\pi d^4}{128\eta L} \bar{p} \tag{7}$$

where $\bar{p} = (p_1 + p_2)/2$ is the average pressure in the tube. The subscript η on C_η is a reminder that this expression is valid only for the viscous flow that occurs at high pressures. For air at 298 K, Eq. (7) reduces to

$$C_\eta(\text{air}) = \frac{180 d^4}{L} \bar{p}(\text{Torr}) \quad \text{liter s}^{-1} \tag{8}$$

---

† Turbulent flow will only occur if the Reynolds number (see Exp. 4) is very high. This is uncommon in vacuum work, and we shall neglect turbulence completely.

where $\bar{p}$ is now expressed in Torr and d and L are in cm.† When the tube is short enough that end effects perturb the steady-state laminar velocity distribution over an appreciable fraction of the length of the tube, Eq. (8) must be modified:[2]

$$C(\text{air}) = 180 \frac{d^4}{L} \frac{\bar{p}(\text{Torr})}{[1 + 0.38(Q/L)]} \tag{9}$$

where Q is the flux in Torr liter s^{-1}.

For viscous flow in a tube 100 cm long and 1 cm in diameter, the conductance decreases from 1350 liter s^{-1} at 760 Torr (1 atm) to 1.78 liter s^{-1} at 1 Torr. As we shall see below, the conductance for molecular flow in the same tube has the constant value 0.122 liter s^{-1}. In partial-vacuum applications where viscous flow is dominant, conductance limitations are seldom a problem. Accordingly, most of our attention will be focused on the case of molecular flow, which is important in high-vacuum work.

(b) Molecular flow. At very low pressures the mean free path λ of a gas molecule is larger than the critical dimensions of the apparatus, and collisions of a molecule with the walls are more frequent than collisions with other molecules. [At 25°C, $\lambda = (a/1000p)$; when λ is expressed in cm and p in Torr, $a = 5.1$ for air, 3.4 for water, and 2.7 for mercury.] When the pressure is sufficiently low that intermolecular collisions are negligible, the conductance of a long uniform tube of length L, cross-sectional area A, and perimeter H was shown by Knudsen[3] to be

$$C = \frac{4}{3} \frac{A^2}{HL} \bar{c} \tag{10}$$

where $\bar{c} = (8RT/\pi M)^{1/2}$ is the mean molecular speed and M is the molecular weight. For a cylindrical tube with $L \gg d$,

$$C = \frac{\pi d^3}{12L} \bar{c} \tag{11}$$

Note that this expression for the conductance is independent of pressure. For air at 298 K, Eq. (11) becomes

$$C(\text{air}) = \frac{12.2d^3}{L} \quad \text{liter s}^{-1} \tag{12}$$

Equations (10) to (12) do not include the effect of the conductance of the inlet aperture where the tube is connected to another region of larger cross section (to a bulb, for example). Unless L is very much larger than d, this aperture conductance will have a significant effect on the total conductance of

---

† While SI units are generally used in this book, units of cm, liter, and Torr are used here since they are more appropriate in describing practical vacuum systems and pumping speeds.

the tube. From Eqs. (3) and (IV-7) we see that for molecular flow through an orifice of area A,

$$C = \tfrac{1}{4}A\bar{c} \tag{13}$$

(For air at 298 K and a circular opening of diameter d, $C = 9.15d^2$ liter s^{-1}.) The ratio of the aperture conductance to that of a long cylindrical tube of the same diameter is $3L/4d$. A short tube may be regarded as two circuit elements—the entrance of the tube regarded as an orifice and the tube proper—in series, with their impedances additive. The entrance of the tube is then equivalent to an addition of $\Delta L = 4d/3$ to the length of the tube. The exit is not regarded as having any impedance *per se*, although it may have impedance by virtue of being the entrance to a much smaller tube in the sequence. This approximate treatment gives for air in a cylindrical tube at 298 K,

$$C(\text{air}) = \frac{12.2d^3}{L} \frac{1}{1 + 4d/3L} \quad \text{liter s}^{-1} \tag{14}$$

Equation (14) might be applied, for example, to the bore of a vacuum stopcock placed in a line of considerably larger diameter. A somewhat more exact treatment[2] gives, for a wide-bore stopcock with $d = 1$ cm, $L = 3.2$ cm, a conductance for air at 25°C of 2.6 liter s^{-1}; a narrower-bore stopcock with $d = 0.40$ cm, $L = 2.0$ cm will have a conductance of only 0.29 liter s^{-1}.

A tube with a bend in it will have a lower conductance than a straight tube. A crude way to take this into account is to consider each bend as an orifice having a diameter equal to the tube diameter. This will add $4d/3$ to the length (measured along the centerline of the tube) for each bend. To make more accurate calculations would require taking into account the shape of the bend.

The conductance of a *cold trap*, chilled with liquid nitrogen to prevent mercury or pump-oil vapors from entering the system or volatile substances in the system from contaminating the pump, can also be of considerable practical importance. Such a trap is diagrammed schematically in Fig. 1. Dushman[2] has investigated the effects of varying the trap dimensions and has shown that the optimum conductance for a trap of outer diameter d_2 is obtained when $d_1/d_2 = 0.62$. With this condition maintained, the conductance for air at 25°C is approximately given by

$$C(\text{air}) \simeq \frac{1.4d_2^3}{L} \quad \text{liter s}^{-1} \tag{15}$$

However, if the entire trap is cooled with liquid nitrogen the conductance will be reduced by a factor of $(\tfrac{77}{298})^{1/2} \simeq 0.5$. More detailed calculations[2] on a liquid-nitrogen trap with $L = 21.6$ cm and $d_2 = 4.76$ cm indicate that the conductance should be in the range 3 to 3.5 liter s^{-1}, which is in fair agreement with the above approximation.

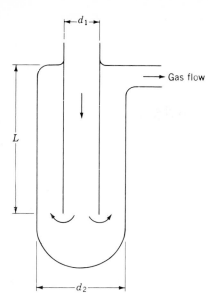

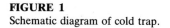

FIGURE 1
Schematic diagram of cold trap.

(c) Intermediate pressures. When the flow is neither purely viscous nor purely molecular, one can use an empirical expression of Knudsen[3]

$$C = C_\eta + FC_{\text{molecular}} \tag{16}$$

The factor F is a complicated function of M, T, d, η, and $\bar{p}$; it has values for air which vary between ~0.8 at high pressures and 1.0 at low pressures.

Overall pumping speed. Consider, as an example, a large bulb connected via a tube of conductance C to a pump of speed S. We wish to find S_T, the overall pumping speed for this composite system. The rate of gas removal from the bulb Q equals $S_T p_1$ where p_1 is the pressure in the bulb. Also, Q must equal $C(p_1 - p_2)$ for the flow through the tube where p_2 is the outlet pressure at the pump end of the tube. Finally, Q equals Sp_2 at the pump itself. Since

$$\frac{Q}{C} = p_1 - p_2 = \frac{Q}{S_T} - \frac{Q}{S}$$

we can write

$$\frac{1}{S_T} = \frac{1}{C} + \frac{1}{S} \tag{17}$$

Although the speed of a pump differs from the conductance of a tube in being defined in terms of a single pressure rather than a pressure difference, Eq. (17) shows that, for the kind of system being considered, S can be treated in the same way as C in calculations involving series or parallel combinations of pumps and pumping lines.

Thus the reciprocal of the overall pumping speed S_T of a pump *plus* vacuum line is the sum of the impedances of all of the segments from the orifice representing the entrance to the vacuum line to the pump itself; typically,

$$\frac{1}{S_T} = \frac{1}{C_{\text{entrance}}} + \frac{1}{C_{\text{tube}}} + \frac{1}{C_{\text{stopcock}}} + \frac{1}{C_{\text{tube}}} + \frac{1}{C_{\text{bend}}} + \frac{1}{C_{\text{tube}}}$$
$$+ \frac{1}{C_{\text{trap}}} + \frac{1}{C_{\text{tube}}} + \frac{1}{S} \tag{18}$$

Up to this point, pumping speeds have been discussed for the case of fixed pressures. For the case of evacuating a fixed volume V, not including the space in the pump or vacuum line itself, the flux Q becomes $-V \, dp/dt$. (The negative sign is required since the pressure p in the volume V is decreasing.) Therefore, the pressure will obey the differential equation

$$S_T = \frac{Q}{p} = -\frac{V}{p}\frac{dp}{dt} \tag{19}$$

A general solution of Eq. (19) is complicated since S_T varies with the pressure.[2] For the simple case where S_T is considered constant over a range of pressures, one finds

$$p = p_0 e^{-(S_T/V)t} \tag{20}$$

where p_0 is the pressure at $t = 0$. The length of time for the pressure in volume V to fall by one decade (power of 10) is then

$$\Delta t_{0.1} = \frac{2.303}{S_T} V = \sum_i (2.303 \, Z_i) V \tag{21}$$

One frequently sees in books and catalogs the quantity $2.303Z_i \equiv 2.303/S_i$ quoted for a vacuum component rather than the impedance or conductance. The quantity $2.303Z_i$ represents the amount of additional time, per liter of space to be evacuated, that insertion of this component will add to the time for a decade fall in pressure. Thus, for the large-bore stopcock already described, $2.303Z_{\text{stopcock}} = 0.88$ s liter^{-1} for air; including it in a line evacuating a 50-liter enclosure for which $\Delta t_{0.1}$ was previously 100 s will increase the decade time to 144 s.

In practice, Eq. (21) is valid only during the preliminary stages of pumping, i.e., until the pressure drops to a point where the pumping speed S begins to decrease and/or an appreciable fraction of the gas being pumped out is entering the system through leaks, desorption from the walls, or backflow from the pump, etc. Very long times are often required in the later stages to attain a vacuum close to the ultimate vacuum for the system.

DESORPTIONS AND VIRTUAL LEAKS

Under high or ultrahigh vacuum conditions, much or virtually all of the gas being pumped is being desorbed from the walls of the system. If we have a surface of area A that is covered with a layer of adsorbed molecules each of which occupies an average area σ, the number of moles of gas that will be produced by complete desorption is $A/\sigma N_0$, where N_0 is Avogadro's number. For A in cm^2, σ in Å^2 (1 Å $= 10^{-8}$ cm), the quantity of gas in Torr-liter is, at room temperature,

$$\int Q\, dt = 3.1 \times 10^{-4} \frac{A}{\sigma} \tag{22}$$

A 1-liter glass bulb has an inside surface area of 485 cm^2; if each adsorbed molecule occupies 10 Å^2 on the surface, the content of adsorbed gas is about 0.015 Torr liter. At 10^{-6} Torr, 15 000 liters would have to be pumped; at 10^{-10} Torr, 150 000 000 liters. Fortunately the greater part of this gas can be pumped off at higher pressures where the volumes to be pumped are smaller.

Adsorption of gases on surfaces is of two kinds: physical and chemical. Physical adsorption is generally reversible and is characterized by heats of adsorption of only a few kilojoule per mole. Physical adsorption is ordinarily negligible at more than 50 or 100°C above the boiling point of the gas. Chemical adsorption (called chemisorption) is often irreversible, comes to equilibrium very slowly except at elevated temperature (often 1000°C or higher), and is characterized by heats of adsorption of the order of 80 to 800 kJ mol^{-1}. Examples are oxygen or carbon monoxide chemisorbed on metals. Water is of special importance because it is usually present at room temperature on glass or metal surfaces in both physically and chemically adsorbed forms at the same time, with its strength of attachment to the substrate varying over a wide range. Other substances that may be adsorbed on the walls of a vacuum system are N_2, CO_2, NH_3, oxides of nitrogen, hydrocarbons, and other organic substances.

Most physically adsorbed gases are readily desorbed in the early stages of pumping, when the pressure is above 10^{-3} Torr. The approach to 10^{-6} or 10^{-7} Torr is slow because of the slow desorption of water at room temperature. For practical purposes 10^{-6} or 10^{-7} Torr may be the lowest pressure that can be achieved because the time required to desorb all of the water is so long at this temperature.

A remedy for this situation is to "bake out" the system, in order to hasten the desorption of water and other sorbates and to increase the amount desorbed, so that these substances can be pumped off quickly at higher pressures. For most purposes 250°C is a high enough temperature, and one which is not high enough to damage most high-vacuum components; bakeout up to 400°C is occasionally practiced. A bakeout to 150°C is better than none at all, as is a selective bakeout of most components of a system, leaving others (such as greased stopcocks) at room temperature. A glass system is often selectively baked out by careful flaming with a hand torch while the system is being pumped, or baked out with heating tapes or infrared lamps.

The most common source of gas in a vacuum system, excluding the gases (air) present originally and those produced by desorption, is *leaks*. The prevention of leaks—pinholes in blown glass, faulty welded seams, faulty gaskets, scratched flanges, channels in greased stopcocks and over O rings, diffusion through slightly porous materials—is a matter of careful choice of materials, apparatus design, construction, and operation; the detection and repair of leaks is a technology in itself that will be discussed later. A leak that admits only 1 mm^3 of air (at 1 atm) per minute is putting 1.3×10^{-5} Torr liter s^{-1} into the pumping system; to keep up with this at 10^{-6} Torr would require a system with a pumping speed of 13 liter s^{-1}.

The *virtual leak* is also a common problem. This is leakage into the vacuum, not of air from outside, but rather of air or vapor from some source inside the system: a drop of a slightly volatile substance slowly vaporizing, or a small pocket of air trapped in a screw thread and slowly leaking out. A cubic millimeter of air at 1 atm trapped under a screw (7.6×10^{-4} Torr liter) is 7 600 000 liter at 10^{-10} Torr; if that amount of air took a week to leak out at that pressure, an average of 12 liter s^{-1} of the pumping capacity would be taken up for that time; the same would be true of 0.01 mg of a slightly volatile substance (M.W. 250) that required a week to volatilize completely.

A very small amount of a thermally decomposable organic substance in the system during bakeout, or in a part of the system that is operated above 200°C to 300°C, can also give vacuum contamination that will take a long time to get rid of. The higher the vacuum, the more important it is that the system be clean. The presence of a greasy film on inside metal surfaces should be precluded by a "degreasing" with trichloroethylene, particularly if a vacuum better than 10^{-7} Torr is desired.

In a glass vacuum system stopcock grease is commonly used for lubricating stopcocks, ground joints, O rings, seals for rotating shafts, etc; vacuum pump oil is also sometimes used for lubricating purposes. These materials are specially prepared and refined to be as free as possible from volatile components. Stopcock greases may have stated "vapor pressures" as low as 10^{-6} Torr; the values may be even lower for diffusion pump oils. However, they can pick up air, water, and especially organic vapors and later release them slowly from solution into the vacuum. This nuisance should be minimized by keeping the area of these materials that is exposed to the vacuum as small as possible, by use of careful technique in greasing stopcocks, etc. Vacuum pump oil has a tendency to "creep" on surfaces; one of the functions of a cold trap is to prevent pump oil from creeping into the clean part of the vacuum system.

PUMPS

The principal types of vacuum pumps will be discussed in this section. The very simple Toepler pump is quite useful for the quantitative transfer of small amounts of gas from one part of a system to another. Rotary oil pumps are used for pumping on refrigerant baths and as the forepump for "backing" low-pressure pumps. For most high-vacuum work, diffusion pumps are utilized to achieve pressures of about 10^{-6} Torr; however, there are now special types of pumps that can be used to achieve ultrahigh vacuums of 10^{-9} Torr or less. The water aspirator is a very useful pump for many routine operations.

Water aspirator. The water aspirator is probably the vacuum "pump" most frequently used by chemists. It produces only a partial vacuum but one that is entirely adequate for a large number of chemical operations, such as filtration by suction, distillation under reduced pressure, and evacuation of desiccators.

This device operates through Bernoulli's principle. When a flowing fluid encounters a constriction that forces it to flow at a higher velocity, the hydrostatic pressure exerted by the fluid decreases by an amount corresponding to the increase in kinetic energy density plus any energy losses due to turbulence or viscous resistance. If the diameter of the tube increases following the constriction, the pressure may rise again if the difference in kinetic energy density is not completely lost to turbulence and viscosity. In the aspirator, water enters from a pipe at mains pressure (about 4 atm) and leaves against 1 atm external pressure from a nozzle of lesser diameter. In an internal constriction of still smaller diameter the pressure drops nearly to zero. The air from the system being evacuated enters at this constriction and is carried along with the water.

The ultimate vacuum achievable with a water aspirator is limited by the vapor pressure of the water itself; any significant drop in the pressure much below this will be reversed by the vaporization of water. The vapor pressure of water is 12.8 Torr at 15°C, 17.5 Torr at 20°C, and 23.8 Torr at 25°C. A lesser vacuum (higher pressure) can be achieved by reducing the flow of water into the aspirator and/or bleeding air into the system through a controlled leak.

When the water supply to an aspirator is shut off or greatly reduced while the system is under vacuum, the water in the aspirator can be sucked into the system. Modern aspirators often contain a check valve to prevent this, but this should not be relied upon; a trap bottle should always be provided between the aspirator and the system.

A word of **warning**: The flow rate of water through an aspirator is subject to change, often sudden change, as the water main pressure fluctuates. This may produce changes in the pressure in the system being evacuated, or even cause "suckback" of water into the trap bottle. If a constant pressure (especially for reduced-pressure distillation) is required, one must be alert for such changes, and be ready to make required adjustments.

Toepler pump.[2] The Toepler pump permits one to transfer gas from a bulb A into another bulb B without loss or contamination. This is accomplished by alternately lowering and raising the mercury level in a reservoir connected with both bulbs. When the mercury level is lowered, gas expands from A into the reservoir; when it is raised, this gas is forced into B. Many cycles of operation are necessary to effect an almost complete transfer of the gas. Obviously, this transfer could be accomplished much more easily for a condensable gas by merely cooling bulb *B* with liquid nitrogen. The Toepler pump is most useful in handling those few substances (such as He, H_2, CO, N_2, Ar, CH_4) which either condense below 77 K or have an appreciable vapor pressure at that temperature. One of the most common applications of this device is to remove gas from a reaction bulb for quantitative analysis at the end of a kinetics run; for example, in Exp. 24 a Toepler pump would be necessary to remove quantitatively the hydrogen gas from the furnace.

Mechanical pumps. Perhaps the commonest form of vacuum pump is the mechanical pump that operates with some sort of rotary action, with moving parts immersed in oil to seal them against back streaming of exhaust as well as to provide lubrication. These pumps are used as forepumps for diffusion pumps. Other common laboratory applications are the evacuation of desiccators and transfer lines, and distillation under reduced pressure. These pumps have ultimate pressures ranging from 10^{-4} to 0.05 Torr, and pumping speeds from 0.16 to 150 liter s^{-1} or more, depending on type and intended application. The Cenco Hyvac 2 or the Welch Duo-Seal 1400B are relatively inexpensive laboratory pumps capable of good ultimate vacuum (10^{-4} Torr) but with relatively low pumping speed (0.42 liter s^{-1} at 1 atm; about half that at 1 mTorr). The Cenco Pressovac or the Welch Duo-Seal 1399B are inexpensive all-purpose pumps giving a somewhat higher pumping capacity (0.58 liter s^{-1} at 1 atm) in exchange for a less good ultimate vacuum (0.015 Torr). The Pressovac can also be used as a source of positive pressure above 1 atm. Other pumps with different capabilities are manufactured by Cenco, Welch, Edwards, Kinney, and other companies.

One common design of the rotary pump is that of the Welch Duo-Seal pump, as shown in Fig. 2. In this type of pump, a rotating drum, tangent to the inside of the cylinder (with a clearance of only 0.003 mm), has deep slots that hold two (or more) vanes that slide along the cylinder wall, compressing the gas and carrying it to the exhaust port. Some pumps tend to be noisy (hammering sound) when pumping a good vacuum; the hammer noises tend to be intermittent or muffled when the vacuum is less good. By contrast the Duo-Seal and Edwards pumps are most quiet when the vacuum is best; this is a definite advantage for continuous operation in a room where people are working.

There are considerable variations on this basic design. Many commercial types are compound pumps in which two single stages are mounted on the same shaft and are connected in series as a means of increasing the pumping speed and improving the ultimate vacuum. The pumping speed at high pressures depends on both the size and design of the pump.

An important limitation on the ultimate vacuum achievable with a rotary oil pump is the characteristics of the oil used. It must of course have a negligible vapor pressure and contain no significant quantity of dissolved

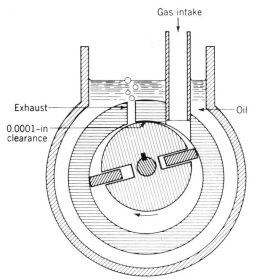

Gas intake

Exhaust

0.0001-in
clearance

Oil

FIGURE 2

Schematic diagram of a rotary oil pump similar to the Welch Duo-Seal; based on the Gaede oil pump design.

volatile components. Accordingly, *care should be taken to prevent vapors, especially those of organic solvents, from entering the pump.* A Dry Ice–trichloroethylene trap is adequate to prevent this. The oil should be viscous enough to serve as a lubricant, but fluid enough to permit air bubbles from the exhaust to rise rapidly to the oil surface so as not to re-enter the pump with the oil. *The oil must be clean*; dirt or metal particles suspended in the oil will cause abrasion and wear at points where tolerances are critical, and chemical fumes (e.g., hydrogen halides, sulfur dioxide, oxides of nitrogen) will cause corrosion. The oil level and cleanliness should be inspected at regular intervals, and the oil should be replaced whenever it is suspect. In the event of any serious contamination (e.g., corrosive fumes, water, or mercury) the pump must be disassembled by a competent mechanic, cleaned with solvents, dried, carefully inspected, reassembled (with due care for the extremely critical mechanical tolerances), and filled with fresh oil.

Although direct-drive pumps are manufactured, many rotary oil pumps are belt-driven by an electric motor. The belt should be tight enough to prevent slippage but not so tight that it will produce wear on the motor and pump bearings and on the belt itself. Since the belt drive of a pump presents a mechanical hazard for fingers, loose clothing, and long hair, it should always be covered by a safety guard. With care and periodic oil changes, rotary pumps can be left on continuously and will operate well for years. When a rotary pump must be used under conditions in which significant amounts of condensable vapors—generally water vapor—pass into it, a pump with a *vented exhaust* should be used. In this pump a quantity of room air is introduced into the exhaust as a diluent to prevent condensation and resultant contamination of the oil. It is inadvisable to permit organic vapors to enter even a vented exhaust pump, because of their high solubility in the oil; it is much better to freeze them out with a cold trap.

A mechanical pump giving very high pumping speeds, a few hundred to a few thousand liters per second in the 10 to 10^{-2} Torr range, is the *Roots pump.* As illustrated in Fig. 3, this pump, also known as a Roots blower, consists of a pair of two-lobed rotors which rotate in opposite directions. The rotors have a clearance of a few tenths of a millimeter and are synchronized by a gear drive so that they "push" the gas on the inlet side through to the exhaust outlet. There is no oil seal between the rotors so that compression ratios of only about 10:1 to 100:1 are achieved at typical rotational speeds of 3000 rpm. Because of this compression however, a rotary pump of lower pumping speed in series with the blower is adequate to prevent back-streaming at high throughput levels. Two-stage Roots pumps are also produced that can reduce the pressure to the 10^{-3} to 10^{-5} Torr range. Roots pumps are widely used for pumping large chambers used in molecular-beam studies.

A mechanical pump providing even lower vacuum levels is the *turbomolecular pump,* in which one or more balanced rotors (turbine blades) spin at 20 000 to 50 000 rpm. At these rotation rates, the periphery moves at a speed that exceeds the mean molecular speeds of most molecules and gas–rotor collisions impart a momentum component to the gas in the direction of the exhaust. Compression ratios up to $10^6:1$ can be achieved as long as the outlet pressure is kept below about 0.1 Torr by a forepump. The 10 cm Welch 3134 turbomolecular pump has an ultimate vacuum limit of 10^{-9} Torr and a pumping speed of 280 liter s^{-1} from 10^{-2} to 10^{-8} Torr. Other pumps with speeds up to 10 000 liter s^{-1} are available, although these rates are reduced somewhat for light gases such as H_2 and He, which have high molecular velocities. The turbomolecular pump is a very clean pump, requires no trap between itself and the system, and is not sensitive to contamination (although in some cases the forepump should be protected by a cold trap.) It is not widely used in routine laboratory applications, however, as it is quite expensive and demands careful maintenance.

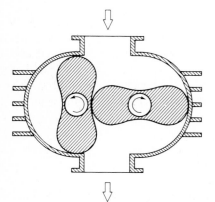

FIGURE 3
Roots blower.

Diffusion pumps. The diffusion pump is the standard means of achieving high vacuum (about 10^{-6} Torr); with special care, ultrahigh vacuum (10^{-10} Torr) can be achieved. The operation of the diffusion pump is superficially similar to that of an aspirator, although the principles are quite different. The molecules of the gas being pumped diffuse into a jet of hot vapor, that carries them along by essentially viscous flow inside a condenser in which the vapor is gradually condensed to a liquid of low vapor pressure. The liquid falls back into a boiler where it is again vaporized and brought to the jet nozzle. The exhaust is essentially isolated from the intake by a sort of "plug" of viscous-flowing vapor, which contains the gas being pumped as a minority component; the very slight pressure difference against which the pump must operate (normally less than 10^{-1} Torr) is overcome by the momentum possessed by the vapor as it emerges from the jet nozzle. The gas being pumped is carried away by a forepump, which is usually a rotary oil pump.

Two very different kinds of pump fluids have been employed in diffusion pumps. For many years inexpensive *mercury diffusion pumps,* similar to that shown in Fig. 4, were used in small laboratory-bench glass vacuum systems. Mercury pumps are now somewhat in disfavor owing to the health hazards associated with mercury and the high probability of contamination of the vacuum system with mercury (the vapor pressure of mercury at room temperature is ~1.5 mTorr). The *oil diffusion pump* eliminates the safety hazard and can serve for both small glass and larger metal vacuum systems.

The essential features of an all-glass single-stage mercury diffusion pump are shown in Fig. 4. Glass one- to three-stage oil pumps of slightly more complex construction are available from Ace Glass, Kontes Scientific Glassware, Pope Scientific, Inc., and other manufacturers. These are either air- or water-cooled and have pumping speeds ranging from 8 to 30 liter s^{-1}. Before the heater of a water-cooled diffusion pump is turned on, it is important to make sure that the forepump has reduced the pressure below the limiting head pressure (about 10^{-1} Torr), and that a reliable flow of cooling water is flowing through the condenser. If water is turned on after the condenser has become hot, the glass will very likely crack, even if it is Pyrex. A reliable (although expensive) precaution against damage is to power the heater through the contacts of a holding relay, with the current through the hold coil passing through a flow-operated switch in the outlet side of the condenser, so that cessation of flow through the condenser—whether by water-main failure, blockage of the line, or a detached hose—will cut off the current to the heater. To turn the heater on again requires the positive act of resetting the relay. For protection against floods from a detached hose, a solenoid valve can be placed in the water pipe line. A switch opeated by a thermocouple gauge in the vacuum foreline can also be placed in the holding circuit to cut off the diffusion pump in the event of forepump failure. A suggested plumbing and electrical layout is shown in Fig. 5 (needless to say, a similar arrangement can protect a distillation setup from water failures, and the occupants of the floor below from floods.)

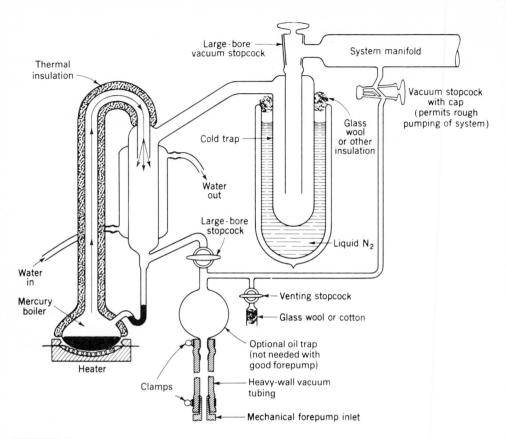

FIGURE 4

Typical all-glass mercury diffusion pump, shown with appropriate connections to the vacuum manifold and to the mechanical forepump. A glass oil diffusion pump can be substituted if mercury is to be avoided.

The fluids used in diffusion pumps are usually hydrocarbon esters (e.g., di-*n*-octylphthalate) that are heated to 130 to 160°C in operation or silicone or polyphenyl ether fluids that operate at higher temperatures (180 to 280°C). The latter fluids are quite expensive but give lower ultimate vacuum levels (10^{-7} to 10^{-11} Torr) and greater resistance to oxidation by excessive amounts of air. **Warning:** When hot, a diffusion pump should never be exposed to vacuum levels above about 0.1 Torr; *never vent an oil diffusion pump to air when it is hot.*

Oil diffusion pumps of metal construction have the general characteristics shown in Fig. 6. The pump consists of a chamber with oil at the bottom that is heated to boiling at about 0.5 Torr. The streaming vapor is redirected to provide the pumping action by the control jet assembly. The limiting forepressure at which a single-jet diffusion pump will operate increases with decreasing annular spacing between the exit rim of the nozzles and the pump

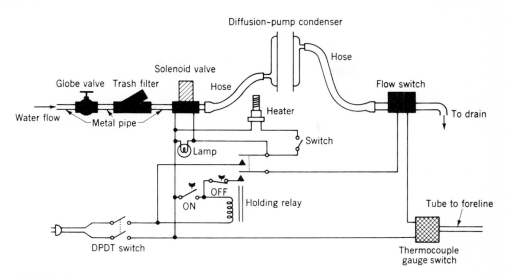

FIGURE 5

Suggested plumbing and electrical circuitry for protection from water-supply failure and floods, for use with diffusion pumps (and other things!).

wall. Therefore, high-speed diffusion pumps usually have two or more stages of jet nozzles and are referred to as multistage pumps. By the proper design of the size and spacing of these jets, the pump acts like several separate diffusion pumps connected in series. The smallest, fastest stage is positioned nearest to the system, and the largest, slowest stage (which operates at the highest forepressure) is nearest to the forepump, as shown in Fig. 6. Oil diffusion pumps have inlet diameters that range from 5 cm, with a pumping speed of 100 to 300 liter s^{-1}, to an extreme of 1.2 m, with a speed of about 10^5 liter s^{-1}. The latter enormous pump finds use in aerospace test chambers and other specialized cases; 5- to 15-cm pumps are more typical for laboratory applications.

An oil diffusion pump can be used without a trap if it is sufficiently well provided with baffles to stop back-streaming molecules, for the oils used usually have room-temperature vapor pressures less than 10^{-7} Torr when pure. However, pump oils generally undergo some decomposition in service, and some decomposition products may find their way into the system. Liquid nitrogen cold traps are usually added when backstreaming and oil creep along the walls of the system must be prevented.

Ion pump. A pump now much used in ultrahigh vacuum systems is the ion pump, of which the Varian Vac-Ion pump is an example. In this pump a "gas discharge" such as is commonly observed in air and other gases at about 10^{-2} Torr (e.g. neon signs, mercury and sodium lamps) is generated by a strong electric field and maintained all the way down to 10^{-10} Torr by

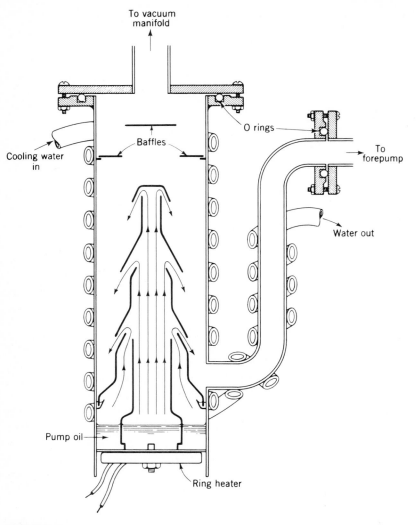

FIGURE 6
Schematic drawing of a multistage, all-metal oil diffusion pump. The arrows indicate the flow of oil vapor.

application of a strong magnetic field from an Alnico magnet. This field constraints the ions to move in circular or helical paths thereby greatly increasing their path lengths between wall collisions and giving them the opportunity of colliding with neutral gas molecules and ionizing them so as to continue the discharge. The ions eventually strike certain parts of the specially designed titanium electrodes, eroding them by a process known as sputtering. The titanium atoms come to rest on other titanium pump surfaces, to maintain a fresh and very active titanium surface where the real pumping action takes place by surface adsorption. As the pumped molecules are adsorbed they are continually buried by additional titanium atoms which refresh the surface.

The ion pump needs no forepump, but it is necessary to reduce the pressure initially to 10^{-2} or 10^{-3} Torr before turning the pump on. With care, this may be accomplished with a rotary oil pump, but it is often done instead with a "sorption pump" such as the Varian Vac-Sorb pump in which the air in the system is adsorbed on a molecular sieve (synthetic zeolite, Linde type 5A) chilled with liquid nitrogen. This type of pump has the advantage of presenting no danger of contaminating the system with oil.

The ion pump has an ultimate vacuum capability of 10^{-10} Torr or better, and may have a pumping speed for air of several hundred liters per second depending on its size. The components of air are chemisorbed on the fresh titanium and are thus permanently fixed by chemical bonds. Noble gases (helium, argon, etc.) are pumped at a much slower rate, as they are only physically adsorbed on the surface and are held principally by being "plastered over" by titanium atoms. An ion pump that has pumped a considerable quantity of argon is subject to a condition called argon instability, in which bursts of argon are released at intervals into the system; other noble gases show a similar effect.

The ion pump is widely used in all-metal systems, which are usually fabricated of stainless steel with copper gaskets between machined stainless-steel flanges. These systems are baked out for several hours at 250°C (or even as high as 400°C) to desorb surface gases (mainly water). The ion pump is baked out at the same time while it is in operation; the temperature should not exceed 250°C, to protect the magnet).

An important application of systems of this kind is to fundamental studies of ultraclean surfaces, and of chemisorbed molecular monolayers on such surfaces, by such techniques as low-energy electron diffraction (LEED) and electron energy-loss spectroscopy (EELS). Ultrahigh vacuum is necessary to provide ample time for the preparation and study of such surfaces. Assuming a molecular cross section for N_2 or O_2 of 10 Å^2 ($= 0.1 \text{ nm}^2$) and a sticking probability of unity, it would take only about 5 s for the surface to be covered with a monolayer at 10^{-6} Torr; this time is extended to many hours at 10^{-10} Torr.

Cryopump. As the name cryopump or cryogenic pump implies, this pump operates at very low temperatures; in its most effective form it is a liquid-helium-chilled "cold trap." The pumping surface is at 4.2 K if the liquid helium is at 1 atm, or lower if the pressure over the coolant is reduced by pumping. The liquid helium must be protected from too rapid evaporation by cold walls and by baffles chilled with liquid nitrogen (77 K), which greatly reduce the input of energy in the form of blackbody radiation. Commercial cryopumps are available that utilize a closed-cycle helium refrigerator to cool condensing surfaces in a turnkey fashion. These pumps, which cool to 10–20 K, often include a cold cryosorbent material such as charcoal that absorbs noncondensable gases such as H_2, He, and Ne. The cryopump would normally be used in a system already pumped to a high or ultrahigh vacuum by some other pump. It would be useful, for example, for rapidly pumping down argon following argon-ion bombardment of a surface in a LEED apparatus having an ion pump. The liquid helium cryopump is effective under the best conditions down to 10^{-13} to 10^{-15} Torr for all gases except helium, hydrogen, and possibly neon.

VACUUM GAUGES

A large variety of gauges is available for the measurement of low pressures, and the range of useful operation depends a great deal on the type used (see Table 1). This section contains a discussion of the principles of operation for the most commonly used vacuum gauges. More specific details of construction and operation are given in Refs. 2, 4–6, and in the appropriate manufacturers' catalogs. Pressure transducers based on diaphragms are discussed in the section on gas-handling procedures in Chapter XIX.

Manometers. A U-tube manometer filled with mercury is simple to construct, requires no calibration, and operates over a wide pressure range. With each arm connected to a separate region, the manometer can be used to obtain differential pressures or can be used as a null device. Most commonly, one arm is evacuated and the manometer directly indicates the total pressure. A temperature correction is necessary to obtain the absolute pressure. This

TABLE 1
Operational range of various vacuum gauges (given in Torr)

Bourdon gauge	1–1000
Mercury manometer	1–1000
Oil manometer	0.03–10
Diaphragm gauges	
Capsule	1–1000
Piezoresistive	1–1000
Magnetic reluctance	0.1–1000
Capacitance	10^{-4}–1000
Thermocouple gauge	10^{-3}–3
Pirani gauge	10^{-4}–0.3
McLeod gauge	10^{-5}–1
Ionization gauges	
Alphatron	10^{-4}–100
Penning (cold cathode)	10^{-7}–10^{-2}
Thermionic	10^{-8}–10^{-3}
Bayard-Alpert	10^{-11}–10^{-3}
Redhead (magnetron)	10^{-12}–10^{-3}
Residual gas analyzer	10^{-14}–10^{-4}

correction allows one to convert the observed reading $p(\text{mm})$ in millimeters of mercury to Torr (1 Torr $\equiv 1/760$ atm $= 1$ mm Hg at 0°C and standard gravity):

$$p(\text{Torr}) = p(\text{mm})\left[1 - \frac{3\alpha t - \beta(t - t_s)}{1 + 3\alpha t}\right]$$

$$\approx p(\text{mm})\,(1 - 3\alpha t) = p(\text{mm})\,(1 - 1.8 \times 10^{-4}t) \qquad (23)$$

where 3α is the volume coefficient of expansion for mercury (average value of 3α between 0 and 40°C is $18.2 \times 10^{-5}\,\text{K}^{-1}$), β is the linear expansion coefficient of the scale, t is the Celsius temperature of the manometer, and t_s is the temperature at which the scale was graduated. Equation (23) does not contain any correction for the small effect due to the difference between local and standard values of gravity. For precise work, large-bore tubing ($\sim$25 mm o.d.) is used to avoid distortion of the meniscus, and the mercury levels are measured with a cathetometer (see Chapter XVIII). A suitable manometer for medium-precision (± 0.5 mm) work is shown in Fig. 7. In this design the scale is constructed by carefully covering a smooth hardwood board with good-quality millimeter paper and the U tube is rigidly attached to this board. Before being mounted and filled with vacuum-distilled mercury, the manometer should be thoroughly cleaned inside (with nitric acid) and dried, evacuated with a good forepump, and preferably flamed with a torch to release adsorbed water.

For pressures between 0.03 Torr and about 10 Torr, oil manometers are more accurate than mercury manometers, since oil has a much lower density. For example, dibutyl phthalate, which is often used, has a density of 1.046 g cm^{-3} at 20°C. The absolute pressure is given by

$$p(\text{Torr}) = \frac{\rho_{\text{oil}}}{13.59}\,h \qquad (24)$$

where h is the reading of the oil manometer in millimeters. However, oil-filled manometers have two drawbacks. Most gases are quite soluble in oil, and this may cause frothing or erratic behavior when the pressure changes suddenly. Also, the wetting of the walls of the manometer by a viscous oil causes sluggish response.

McLeod gauge. This gauge is very widely used because of its simplicity of operation and its wide pressure range. In addition, the readings depend only on the geometry of the gauge and not on the properties of the gas whose pressure is being measured. Its operation involves the compression of a known large volume V of gas at an unknown pressure p into a known small volume v where the final pressure p' can be measured. Both p' and p are quite low, and the perfect-gas law can be used in the form $pV = p'v$, since the compression is carried out at constant temperature.

A typical McLeod gauge design is shown in Fig. 8. A closed capillary (0.5 to 2 mm i.d., length 12 to 20 cm) of uniform cross section is sealed to the top of

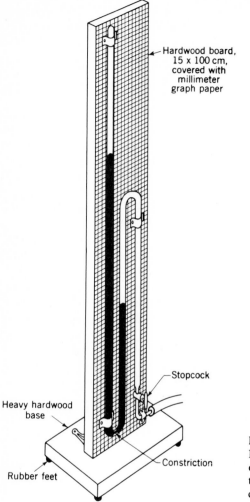

Hardwood board,
15 x 100 cm,
covered with
millimeter
graph paper

Stopcock

Heavy hardwood
base

Constriction

Rubber feet

FIGURE 7

Design for a closed-tube laboratory manometer. The capillary constriction acts to retard the mercury flow and thus prevents damage in case of sudden changes in pressure.

a large bulb (volume 200 to 700 ml). A reference capillary of the same bore is mounted parallel to the closed capillary to minimize the effects of capillary depression on the readings of the differences in mercury level. The cross-sectional area A of the closed capillary is determined before the gauge is assembled by measuring the length and weight of a slug of mercury placed in the capillary. The volume V is measured from the cutoff plane P and includes the volume of the bulb and that of the capillary tubing. This is determined by measuring the weight of water required to fill the gauge prior to sealing off the closed capillary but after the side-arm is attached.

To make a reading with the McLeod gauge, open stopcock A *slowly* and, if necessary, reduce the pressure on the mercury reservoir to prevent mercury

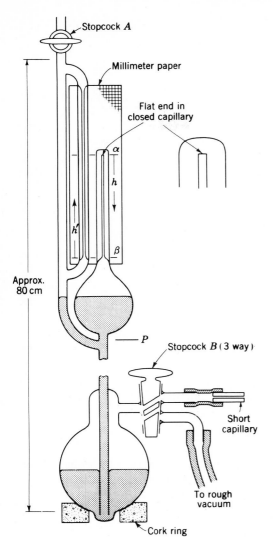

Stopcock *A*

Millimeter paper

Flat end in
closed capillary

α

h

h′

β

Approx.
80 cm

P

Stopcock *B* (3 way)

Short
capillary

To rough
vacuum

Cork ring

FIGURE 8
A typical McLeod gauge.

from rising above the cutoff plane *P*. After waiting about a minute for pressure equilibrium, *carefully* open stopcock *B* to the atmosphere to allow the mercury level to rise slowly in the gauge. As the mercury rises above the plane *P*, the gas contained in the bulb is compressed into the closed capillary, where its pressure can be measured.

There are two methods of reading a McLeod gauge. For low-pressure measurements the mercury is raised until the level in the reference capillary is at point α (even with the top of the closed capillary) and the level in the closed capillary *h* is then read. Applying the perfect-gas law we find that $pV =$

$(h + p)(Ah)$; neglecting p in comparison with h, one obtains

$$p = \frac{A}{V}h^2 = kh^2 \tag{25}$$

where the apparatus constant k is completely determined by the calibration data. For a typical gauge with $A = 1 \text{ mm}^2$ and $V = 200 \text{ cm}^3$, $k = 5 \times 10^{-6}$ Torr (mm Hg)$^{-2}$. With such a gauge, pressures down to 10^{-5} Torr can be measured easily and pressures of 10^{-6} Torr (within a factor of 2 or 3) can be measured with difficulty. Especially when measuring such low pressures, be sure to tap the capillary tube *gently* with finger or pencil while adjusting the meniscus position.

A "stick vacuum" is one in which the gas bubble at the top of the closed capillary is so small that when the mercury level in the comparison tube is lowered a few centimeters the column in the closed capillary "sticks" as a result of surface tension. Further lowering causes the column to break loose from the top. Usually, in determining whether there is a "stick vacuum," the level in the comparison capillary is initially raised one or two centimeters above the top of the closed capillary before it is lowered. With a gauge of the dimensions given above a stick vacuum would be less than 10^{-6} Torr.

The second method of reading a McLeod gauge is useful at higher pressures (above ~0.05 Torr). In this technique the mercury is raised until the level in the closed capillary is at a fixed point β (near the bottom of the capillary) and the level in the reference capillary h' is then read. Again, by using the perfect-gas law and neglecting p in comparison with h', we obtain

$$p = \frac{v}{V}h' \tag{26}$$

where v, the volume between α and β, equals $A(\alpha - \beta)$. The range of the gauge when used in this way can be extended to higher pressures by having a small bulb just above β, so that v is larger.

There are several disadvantages to the McLeod gauge that somewhat offset its attractive features. Since it contains mercury, a cold trap (containing "dry ice" or liquid nitrogen) may be necessary between the gauge and the system to avoid unwanted mercury vapor. It is slow in operation, which is a handicap when used for leak detection. Finally, it cannot measure the pressure of a vapor that will condense when compressed into the capillary. Such condensation will occur as soon as h (or h') exceeds the room-temperature vapor pressure of the substance.

Pirani gauge. The operation of this gauge depends on the pressure variation of the thermal conductivity of a gas at low pressures. A fine wire or ribbon filament, often in the form of a coil for greater sensitivity, is sealed into a glass envelope which is connected to the vacuum system. The filament is made of a metal with a high temperature coefficient of resistance (such as platinum, nickel, or tungsten) and is heated electrically. For a fixed energy input the temperature of the filament, and therefore its resistance, will depend on the gas pressure (the resistance will decrease as the pressure increases). The filament resistance can be measured by making the gauge one arm of a Wheatstone bridge. Since the performance of the gauge depends on its geometry and design and on the nature of the gas, it is necessary to calibrate a Pirani gauge against a McLeod gauge. This can be done by maintaining a constant filament current and determining the resistance as a function of pressure or by determining the current necessary to maintain a constant resistance at each pressure. Ambient-temperature fluctuations can cause erratic readings, since thermal conduction depends on the temperature of the walls of the envelope as well as that of the filament. For accurate measurements, one can thermostat the gauge or use a compensating dummy gauge in an opposite arm of the Wheatstone bridge. This dummy gauge is made as nearly like the measuring gauge as possible except that it is evacuated and sealed off.

Thermocouple gauge. This type of gauge, which also depends on the pressure variation of gaseous thermal conductivity, is almost identical with the Pirani gauge. It differs from the Pirani gauge in that the temperature (rather than the resistance) of the filament is measured by a very fine wire thermocouple that is welded to the midpoint of the filament. The thermocouple output is usually measured with a low-resistance microammeter ($\sim 70\,\Omega$) connected in series with the thermocouple. The thermocouple gauge must also be calibrated against a McLeod gauge; as with the Pirani gauge there are two ways to perform this calibration. However, the standard method of calibration and operation is to maintain constant filament current and obtain the thermocouple emf as a function of pressure. The thermocouple gauge is a rugged, inexpensive instrument that is well suited for leak detection, since it is direct reading and has a rapid response to pressure changes.

Thermocouple and Pirani gauges may be electrically coupled to a switch which will shut down the system in the event that a bad leak or pump failure causes the pressure to rise above a preset level. Alternatively, such a switch may be used to control a solenoid-operated throttle valve between system and pump so as to maintain the vacuum at a specified level; see Manostats in Chapter XIX.

Ionization gauges. Ionization gauges are of several types, having in common the production of positive ions from the molecules present and the measurement of the ion current to a cathode. In the *Alphatron gauge* the ions are produced by the action of alpha rays emitted by a small amount of radioactive material. In the *Penning gauge* a cold-cathode discharge produces the ions; the *Redhead* or *magnetron gauge* uses a magnetic field to constrain the ions to long spiral paths so that the cold-cathode "discharge" can be maintained as low as 10^{-13} Torr.

However, most ionization gauges are *thermionic gauges,* that obtain positive ions through bombardment of the gas molecules with electrons thermionically produced from a heated filament and accelerated by a grid biased 100 to 200 V positive with respect to the filament. The ions are collected by a cathode typically about 20 to 50 V negative with respect to the filament, and the ion current in the cathode circuit is measured with a sensitive electrometer.

Thermionic ionization gauges are usually constructed like electron tubes (with a central filament, outside cathode plate, and anode grid in between), but with their glass envelopes provided with a stem of large-diameter tubing for connection to the system. In large metal systems, however, it is common to use the gauge in the form of a so-called "nude gauge"; that is, the gauge structure is supported from electrical feedthroughs in its mounting flange so that it protrudes, without any envelope, into the vacuum chamber itself. This has the advantage of substantially eliminating errors due to the limited conductance of the connecting tube, in view of the fact that the gauge itself often acts either as an ion pump or as a virtual leak (owing to outgassing of the filament). The filament of the gauge is usually tungsten, either bare or coated with a thoria activator which permits it to be operated at a lower temperature. Tungsten is notorious for emitting large amounts of CO until it has been thoroughly outgassed by many hours of incandescent heating under high vacuum. Iridium wire is some times used because it gives less gas and does not oxidize easily. Another satisfactory filament is rhenium wire activated with a coating of lanthanum boride; this operates at a much lower temperature than a tungsten filament. In any case, when a thermionic ionization gauge is employed, attention must be given to thoroughly outgassing the filament under vacuum before pressure readings are taken. The pumping or "clean-up" characteristics of the ionization gauge, which give rise to some uncertainties in measuring the system pressure, can be reduced to tolerable proportions by restricting application of high voltages to grid and collector to the very short interval needed to measure the electron and ion currents.

Thermionic ionization gauges cannot measure pressures lower than $\sim 10^{-8}$ Torr since the minimum cathode current is produced by a phenomenon altogether independent of the presence of gas molecules, namely the photo-electric ejection of electrons from the cathode plate by soft x-rays produced by the impact of the electrons on the anode grid. This led Bayard and Alpert to invent an "inverted" ionization gauge in which the cathode attracting the positive ions is a slender wire *inside* the anode grid, and the thermionic filament is outside. The *Bayard–Alpert gauge* has an "x-ray limit" which is of the order of 10^{-10} or 10^{-11} Torr, and is widely used in ultrahigh vacuum work.

The ionization gauge has to be calibrated (e.g., against a McLeod gauge at relatively high pressures) and the calibration is in general different for every gas. The ion current for a given gas is, however, linear with the gas pressure over a very wide range, extending almost down to the x-ray limit.

The ionization gauge has a number of disadvantages, besides those implicit in the foregoing paragraphs. Considerable auxiliary electrical equipment is needed. The hot filament may decompose organic vapors that may be present; it may be "poisoned" by some gases and thereby lose its emissivity; it can easily burn out if exposed to air while hot; and it gradually erodes in use through evaporation, slow oxidation, and ion bombardment. Many gauges (especially nude gauges) are so constructed that a burned-out or poisoned filament can easily be replaced in the laboratory. The cold-cathode gauge is preferred in cases where the pressure is generally greater than 10^{-5} Torr, such as in vacuum lines primarily devoted to synthesis or the handling of gases.

Residual gas analyzer. The residual gas or partial-pressure analyzer is a form of ionization gauge in which the positive ions are mass-analyzed either by a conventional sector-mass-spectrometer arrangement (employing a permanent magnet) or by a quadrupole-mass-spectrometer arrangement (which requires no magnet). The ion current is amplified either inside by an electron multiplier arrangement or outside by a vacuum-tube electrometer, or by both. Devices of this kind are sold by Balzers, Varian, and by other companies and typically are capable of unit mass resolution for singly ionized molecules up to masses of a few hundred atomic units. The sensitivity is of the order of 10^{-10} Torr over most of the mass range. The partial-pressure analyzer is very useful in determining the actual composition of the residual gases inside a vacuum system, and in hunting for leaks (with a helium "torch" outside the vacuum); it is virtually indispensable for surface adsorption and desorption studies under ultrahigh vacuum.

Other gauges. The *capacitance manometer,* although expensive, possesses nearly all of the advantages of a McLeod gauge without the disadvantageous presence of mercury vapor. The pressure is sensed by the deflection of a membrane that is one element of a capacitor attached to a capacitance bridge. This gauge is an absolute instrument, is independent of the kind of gas present, and is available in many ranges and sensitivities. For example, the model 310 gauge from MKS Instruments covers the 10^{-2} to 1000 Torr range with an accuracy of 0.05 percent, and similar gauges are manufactured by Datametrics and Setra Systems. In its most sensitive form a capacitance gauge can measure pressures in the neighborhood of 10^{-4} Torr to within a few percent. It can also be used to measure differences in pressure between two vessels.

MISCELLANEOUS VACUUM COMPONENTS

Cold traps. Some discussion of cold traps has already been given; see Figs. 1 and 4 and the accompanying text. A trap should be designed so that a molecule must make two or more collisions with a cold surface in order to get through, but must be designed so that its conductance does not unduly limit the overall pumping speed of the system. The trap should also be designed with enough space below the inner tube to accommodate all condensate, once it has liquefied and run down the sides, without blocking the passage for air or other gases.

For a mechanical-pump vacuum, a mixture of Dry Ice and trichloro-ethylene (−79°C) is satisfactory; the latter serves simply as a medium for heat transfer from the wall of the trap to the "dry ice." Acetone is often used instead but is not recommended because of its high volatility and flammability. Add Dry Ice *slowly* to the liquid, waiting for gas evolution to cease before adding the next portion; when further addition causes no evolution, as much may be added as is needed to maintain the low temperature for some time.

For high vacuum, the trap should be chilled with liquid nitrogen (77 K) in order to reduce sufficiently the vapor pressure of volatile substances that may be present, especially water. **Caution:** In order to avoid condensation of liquid air, the trap should not be chilled with liquid nitrogen until the pressure in the system has been reduced to a few Torr.

The efficiency of a trap is greatly increased by the addition of an adsorbent, such as activated charcoal or a molecular sieve (e.g., Linde 5A). However, it is very important in such a case to allow for desorption of gases as the trap warms up, especially if any significant quantity of gas may have leaked into the system and become adsorbed.

Joints and seals. For many purposes it is convenient to fabricate a vacuum system in one piece. In an all-glass system the parts are fused together with a glass-blower's torch; in an all-metal system parts are soft-soldered, silver-soldered, brazed, or welded. It should be noted that soft-soldered joints, especially large ones, are relatively unreliable for vacuum purposes. (Never hard-solder or braze a joint that has been soft-soldered, unless every trace of soft solder has been removed by machining.) Metal and glass parts also can be permanently sealed together in a vacuum-tight way. This is most easily done with commercially available Kovar seals. Kovar, an Fe–Ni–Co alloy, can be soldered or brazed to other metals, and the glass part, usually Pyrex with a graded seal to the special glass that is sealed to the Kovar, is easily glassblown to other Pyrex glass. Glass can also be soldered to most metals by use of indium (m.p. 155°C) or an indium alloy such as In95%–Ag5% (m.p. 145°C), provided both surfaces are first wetted ("tinned") with the indium solder.

Other vacuum-tight glass-to-metal seals are platinum wire in soft glass and tungsten wire in Pyrex glass. These two seals, useful for electrical feedthroughs, are fairly easy to make, although the tungsten requires a sodium-nitrite flux. With skill it is possible to make a vacuum-tight seal of "five nines" (99.999 percent pure) platinum wire or ribbon in Pyrex, despite the disparity in linear expansion coefficients. For bringing electrical contacts into metal vacuum systems it is convenient to use commercially available glass- or ceramic-insulated lead-throughs, which may be soft-soldered into holes in the metal.

Often it is desirable to make a system demountable. Swagelok and Cajon

Ultra-Torr compression fittings, described in Chapter XIX, are particularly convenient for joining glass, metal, and plastic tubes together. For an all-glass system, standard-taper ground glass joints or ground ball joints lubricated with a good stopcock grease (Apiezon type L or N, Dow-Corning Silicone High-Vacuum) may be used. Glass joints are also available with O rings of neoprene (usually lubricated with a trace of stopcock grease) or of Vitron (a plastic with extremely low vapor pressure, usually needing no grease to make a good seal). The two similar glass parts are clamped together with the O ring between them. This joint is more flexible than the tapered ground joint although much less flexible than the ball joint.

Metal systems are frequently assembled with flanges and O rings (circular gaskets with a circular cross section) or other gaskets. Metal gaskets (OFHC copper, gold, aluminum) between specially designed flanges are commonly used in all-metal ultrahigh vacuum systems that must be baked out. Neoprene O rings are usually adequate for high vacuum (10^{-6} Torr), and Viton O rings are sometimes used at very much lower pressures. When an O ring is used with metal flanges, a smooth groove (rectangular or trapezoidal in cross section) is machined into one flange and the facing surface is machined flat and as smooth as possible (see Fig. 9). Scratches across an O-ring groove or its mating surface are potential sources of leaks and must be carefully avoided. The O-ring groove should have a depth equal to about 80 percent of the cross-sectional diameter of the O ring and a width about 135 percent larger than that of the O ring. In use, the O ring is very lightly greased with stopcock grease and placed in the groove; the mating flange is then bolted on so that the metal surfaces are everywhere in direct and firm contact. O rings that are left for long periods

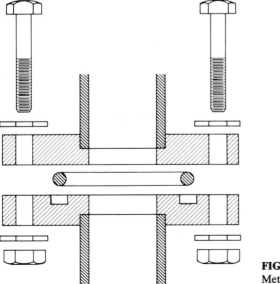

FIGURE 9
Metal O-ring flange joint.

bolted between flanges tend to lose their elasticity and may leak, especially if they have been exposed to heat.

For permanently closing off an evacuated glass system, a constriction in the line (with previously thickened walls) is first flamed for several minutes to outgas it as much as possible and is then heated with the torch until it melts, collapses, and can be pulled off. For a metal system, a tubulation of OFHC copper may be pinched off with a special tool. This is usually sufficient by itself, but a coat of soft solder carefully applied over the pinch-off will render the vacuum more secure.

Stopcocks and valves. It is almost always essential that various parts of a vacuum system can be isolated from each other. This is accomplished by stopcocks and valves. In an all-glass system, glass stopcocks lubricated with a low-vapor-pressure stopcock grease are generally used. The bore of a stopcock should be large enough that its conductance, calculated with Eq. (14), is in considerable excess of that of other more expensive components in series with it. However, large-bore stopcocks are themselves expensive, and compromises may be necessary.

Stopcocks for high-vacuum use are ground to a very good fit with small clearances, so that little grease is necessary and air streaks do not easily arise on turning the stopcock. The better grades are individually ground and are numbered on both the plug handle and on the barrel; these should not be mismatched. A diagonal-bore stopcock is less susceptible to leaks due to streaking than a straight-bore stopcock, since the "off" position of the former is 180° from the "on" position, rather than 90° as in the latter.

The technique of greasing a stopcock with enough, but not too much, grease and no air bubbles or streaks is important in vacuum work. Both parts of the stopcock are first thoroughly cleaned with solvents (a pipe-cleaner or Q-tips swab is useful). With a spatula or wooden applicator the grease is applied to the stopcock plug in four streaks running the length of the plug along each side, preferably avoiding the bore orifices. The optimum amount of grease in each streak has to be learned by experience; it depends on the size of the stopcock, but as a rough guide one may say that each streak should be 2 to 3 mm wide and about $\frac{1}{4}$ mm thick; bubbles should be avoided. The plug is then inserted into the barrel in the open position, keeping the plug centered as well as possible. The plug is pressed firmly into the barrel so that the grease spreads; to facilitate this, wiggle the plug back and forth a few degrees. *After* the streaks have joined and the grease has spread over the entire surface, rotate the plug a few times. Inspect the stopcock for striations from air bubbles, which may indicate that too little grease was used, or accumulations of excess grease in the bore, which indicates that too much was used. If either is excessive, clean and regrease the stopcock.

Good-quality vacuum greases have a high viscosity and a very low vapor pressure ($\sim10^{-6}$ Torr). It is important to choose a grease of the proper viscosity for the ambient temperature of the stopcock. If the grease is too firm, the stopcock will be difficult to turn and striations may appear; if too soft, the grease may flow out and leave too thin a film to make a vacuum seal. Apiezon (type L or N) and Dow-Corning Silicone High-Vacuum greases are recommended.

Glass and Teflon needle valves, sealed with O rings are available as substitutes for stopcocks. These are advantageous where fine control of gas flow is important and for grease-free systems. However, careful use and maintenance are necessary to prevent leaks past the O-rings seals and seats; ordinary vacuum stopcocks are adequate for most purposes.

For metal systems a great variety of high-conductance valves have been developed, many of which are operated through flexible metal bellows to obviate the possibility of atmospheric leaks. Some of these are sealed with neoprene or Vitron O rings. For ultrahigh vacuum purposes, valves have been developed in which stainless steel seats against a copper or gold gasket. Such valves may be baked out in the open position; a few can be baked out in the closed position without damage.

DESIGN OF VACUUM LINES[8]

Most vacuum problems encountered in physical chemistry require small static vacuum lines (i.e., systems of low pumping speed) capable of achieving ultimate pressures of 10^{-6} Torr or below. For such purposes an all-glass system is generally used. It has the advantage of ease of construction, even for complex designs, without the use of couplings or gaskets; it can be easily degassed after assembly; and the detection and repair of leaks are usually straightforward.

It may be helpful to make some qualitative generalizations about design parameters on the basis of the equations presented at the beginning of this chapter. In the high-vacuum region the conductance is determined by molecular flow and the overall pumping speed is given by Eq. (18). To achieve a desired pumping speed the conductances of all series components must be in excess of this value. For expensive components, such as the pump and vacuum stopcocks or valves, the speed or conductances should not be greatly in excess of those needed; for inexpensive components, such as glass tubes, the conductance should be made high enough that their impedances make up only a small part of the total. Thus, the tubes in high-vacuum lines (and especially in ultrahigh-vacuum lines) should be as large in diameter as practicable, should be as short as possible, and should have as few bends as possible. Careful thought should be given to the reduction of valves and stopcocks to the minimum number really needed, as valves and stopcocks are expensive when of high conductance, often cannot be baked out, and often provide unwanted sources of gas (dissolved in stopcock grease) and opportunities for leaks. In the pressure region above 10^{-1} Torr, viscous flow becomes important and the conductance for a given tube becomes considerably larger than for pure molecular flow. Thus, for pumping at higher pressures (as in the line between a diffusion pump and forepump) smaller-diameter tubes can be used.

In order to build and repair a glass vacuum system one must be familiar with certain techniques of glassblowing. An adequate description of the method for making even simple seals is beyond the scope of this chapter; the student should refer to one of the several detailed books on this subject.[9] Indeed, it is our opinion that even extensive reading about the proper techniques is not sufficient for the beginning glassblower. The advice and demonstration of a skilled worker and considerable personal practice are required to achieve the proper manual dexterity. Certain general principles, can, however, be mentioned briefly. Fortunately, glass has a quite high mechanical strength under compression; to take advantage of this, vacuum lines are usually constructed with cylindrical or spherical sections. In addition to strength, a low coefficient of thermal expansion is needed to avoid cracking during sudden or local temperature change while blowing the glass. Although fused silica is excellent in this regard ($\alpha = 6 \times 10^{-7}\,\mathrm{K}^{-1}$), it is impractical for general use owing to very high cost and a very high softening point of $\sim 1200°C$. (However, silica bulbs are used for high-temperature applications.) Pyrex glass is the proper type to use for general-purpose vacuum work. It has a linear thermal expansion coefficient of $3 \times 10^{-6}\,\mathrm{K}^{-1}$, which is about one-third that of soda-line glass ("soft" glass).

Since the annealing temperature† for Pyrex is 560°C and the proper working temperature (the point at which glass is fluid enough to flow readily) is about 800°C, an oxygen-gas torch is required to blow this glass. In general, it is easiest to work with tubing between 8 and 15 mm in diameter. The danger with small tubing is that the walls will collapse and form a solid plug; the difficulty with large tubing is that uneven heating over the surface will cause strains and subsequent cracking. Standard practice involves slowly heating a large area of the glass with a large, gas-rich flame until it begins to soften. Then a smaller, hotter flame is used to work the glass. An inexperienced glass blower may have better success by working with the hot flame on only part of the joint at a time, but it is vital to avoid appreciable cooling anywhere until the joint has been completed and annealed. After the seal is completed, the entire area must be heated enough to relieve any internal strains and then cooled slowly through the annealing temperature range. This is commonly done by lowering the temperature of the flame until a deposit of carbon soot appears on the glass surface.

Oil or mercury diffusion pumps made of glass are inexpensive and convenient for a general-purpose vacuum system. The diffusion pump is connected to the forepump by a short length of *heavy-wall*, large-diameter vacuum tubing. It is often wise to place a ballast bulb or oil trap in this section to avoid contamination in case oil accidentally backs up from the forepump.

---

† At the annealing temperature, strains in the glass will disappear in a few minutes and an evacuated bulb will collapse.

Provision for isolating the diffusion pump and for rough pumping on the system is also advantageous (see Fig. 4). The Hg diffusion pump is connected to the manifold via a cold trap and large stopcock. This manifold is a length of large-diameter (~35 mm) tubing mounted horizontally at a height of at least 1 m from the base of the vacuum bench. A vacuum gauge (McLeod gauge or thermocouple and ionization gauge) and the desired experimental appratus are attached directly to this manifold via stopcocks. Special vacuum stopcocks (preferably with an evacuatable cap on one end of the barrel, as shown in Fig. 4) must be used. The vacuum system should be carefully designed so as to minimize the number of stopcocks required, and the procedure for operating a vacuum system should be thought out so as to minimize the number of stopcock operations. In some cases it is necessary to make part of the apparatus demountable from the vacuum line. Long standard-taper joints, ball-and-socket joints, or O-ring joints can be used if a certain amount of flexibility is needed. The entire vacuum line should be clamped securely to $\frac{1}{2}$-inch rods that are mounted on a sturdy vacuum bench. The base of this bench should be at least a foot off the floor in order to provide space for locating the forepump.

In order to obtain a high vacuum of 10^{-6} Torr or less, it is necessary to eliminate "virtual leaks" due to the desorption of gases and vapors from the glass walls and the slow evolution of dissolved gas from the stopcock grease. These can be reduced by thorough cleaning of the glass prior to assembling the line and by the use of scrupulously clean grease. However, the main cause of these virtual leaks is adsorbed water vapor on the glass. This water can be driven off by degassing (heating the walls to about 200°C under vacuum). After the vacuum line is completely assembled and pumped down, the walls are carefully heated with a soft flame from a hand torch while pumping is continued. Do not heat the stopcocks or any greased joints; instead, slowly rotate each of these several times to speed up the removal of dissolved gas. It will normally require protracted pumping to achieve a good vacuum on a new line. After this initial degassing, avoid venting the line to the atmosphere if possible.

For "dynamic" vacuum lines, where there is a steady influx of gas from some unavoidable source or when pumping on refrigerant baths, high pumping speed is needed. This requires large oil diffusion or other pumps and large-diameter tubing; therefore, all-metal systems are normally used. The construction of such lines requires considerable knowledge of "vacuum plumbing" and of special equipment such as bellows devices, cold traps, and vacuum gaskets.

A general-purpose vacuum system employing an oil diffusion pump and a conventional mechanical forepump is shown in Fig. 10. This system is designed to provide a high vacuum in the left-hand manifold (HVM) and a forepump vacuum in the right-hand one (LVM). The forepump manifold pressure is monitored with a mercury manometer (M) and a thermocouple gauge (TC). The high-vacuum manifold is monitored with a conventional McLeod gauge

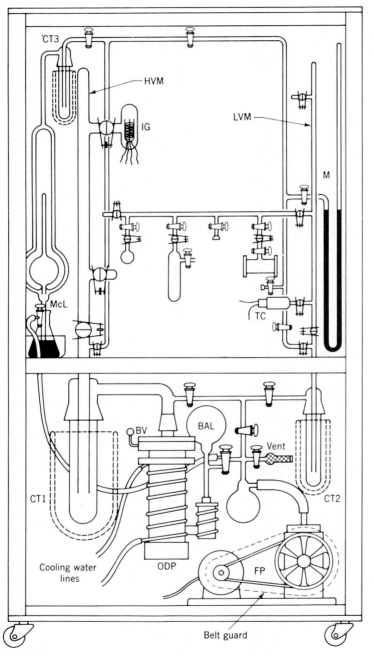

FIGURE 10
General-purpose vacuum system.

(McL) or an optional Bayard–Alpert ionization gauge (IG). Aside from the thermocouple and ionization gauges (which may be attached to the manifolds wherever desired), all pumps, traps, and gauges are fixed in position, while the system manifolds and their appendages are detachable (with O-ring joints) to provide easy adaptability to the needs of any particular experiment. The oil diffusion pump (ODP) can be isolated from the system by valves (butterfly valve BV and 3-way stopcock) and bypassed so that the high-vacuum manifold can be operated with the forepump (FP) alone.

After all joints have been checked for tightness, the traps checked for cleanliness, and the vent stopcock closed, the forepump may be turned on. If only a forepump vacuum is needed, all that remains is to chill the right-hand trap (CT 2) with "dry ice" and trichloroethylene. The manometer (M) and thermocouple gauge (TC) should be closed off with their stopcocks except when readings are being taken, as a precaution against contamination. If the oil diffusion pump is to be used, first wait until the forepump has reduced the pressure in the entire system (including the McLeod gauge) to less than 1 Torr; then chill the left-hand trap (CT 1) with liquid nitrogen. Do not do this earlier, or liquid oxygen may condense in the trap. When the pressure is down to 0.1 Torr, the water cooling of the oil pump and then the electric heater of the pump can be turned on.

The McLeod gauge (McL) is operated by carefully venting the mercury supply bulb to the air so that the mercury rises (not too fast!) into the bulb. Slow the mercury down to a crawl as the mercury in the reference capillary approaches the level of the top of the closed one, and tap gently with a finger or pencil. Cut off the air when that level is reached, and read the depression of the mercury in the closed capillary. To "pull down" the mercury requires that the forepump be isolated temporarily from the diffusion pump and manifolds, and opened to the mercury supply bulb long enough to reduce the pressure there to a few Torr. A ballast bulb (BAL) stores gas from the diffusion pump in the meantime. After closing the stopcock to the mercury supply bulb, wait one or two minutes before reopening the stopcock that connects the forepump to the oil diffusion pump.

When the diffusion pump is no longer needed, it is isolated at both inlet and outlet from the rest of the system, the power is turned off, and the refrigerant is taken away from the trap(s). The system may now be vented to air (preferably through a tube of drying agent). When the pump is cold it may be vented to air and the water cooling can be turned off. *Never vent an oil diffusion pump when it is hot.*

LEAK DETECTION

The design and construction of a vacuum system is only half the battle. To make it work properly, one must known how to detect and repair leaks. Leak trouble can be best avoided by very careful work in building the system; extra effort in constructing the line will be more than repaid in saving time in leak hunting. The most important general rule is to keep the system as simple as possible—every extra stopcock, joint, or appendage is a potential source of trouble. The entire design should be carefully thought out before any construction is started. Wherever possible use permanent joints and avoid hose connections, tapered joints, and flanges. It is best to build complicated systems in sections and vacuum-test each section on an auxiliary line before attaching it to the manifold.

In a glass system, very small pinholes (usually at a seal) can be detected by a spark discharge if the internal pressure is reduced to below about 0.1 Torr. Leak detectors utilizing a Tesla coil are available commercially; it is convenient to solder a short length of copper wire to the electrode tip of these leak testers in order to obtain a small and flexible probe. When this probe is passed slowly over the surface of the glass, a wide, low-intensity discharge occurs (which will excite a pale violet glow in the gas inside the system) unless the probe is close to a pinhole leak. Near a pinhole a concentrated, intense arc will form through the hole and excite a much brighter glow inside. The voltage on the coil should be no higher than necessary, or else the spark may puncture the glass at a weak point and create a leak. Once leaks have been detected, the system should be vented and the leaks fixed by a glassblower. Obviously, this method cannot be used near a metal electrode or clamp; a leak in such a region can often be detected by applying acetone and watching for a change in the color of the discharge produced by a Tesla coil at some point between this region and the pump. The same trick can be used for an all-metal system by connecting the system to a cold trap and mechanical pump through a length of glass tubing.

When a McLeod gauge is used to hunt for leaks in a assembled system, the best procedure is to pump down the entire system, then close off as many sections as possible (including the manifold), and let the system stand for several hours. The pressure is then measured in the gauge itself, gauge plus manifold, and so on, until a section with a high pressure is located. This section can then be carefully rechecked with a Tesla coil, and the stopcocks in it can be regreased. Direct-reading gauges, such as the thermocouple or ionization gauge, are more suitable for leak hunting, since they can be read continuously. They also show a rapid response to pressure changes and a selective response to certain gases or vapors. One standard technique is to swab or flood the suspected area with liquid acetone or methyl alcohol. When a small leak is covered with the liquid, there is a rapid decrease in the leak rate probably caused by the increase in viscosity. Thus, a gauge will indicate an abrupt drop in pressure. Another technique involves covering the suspected area with a plastic bag and filling this bag with hydrogen or helium gas. Since both the thermal conductivity and the ionizability of H_2 and He differ greatly from those of air, a change in gauge reading should occur if the covered area contains a leak. Once a leak is definitely established, a small jet of helium gas from a glass-blowing torch can be used as a probe to locate the exact position. Naturally, it will be difficult to find a leak by this method if there are several of them of about the same size.

It is often helpful to make a temporary repair of known leaks in order to hunt for others in the same leak-hunting operation. This can often be done with Apiezon Q, or "Q putty" as it is commonly called. This is a black puttylike material with very low vapor pressure. To be effective, however, this must be worked very patiently against the surface with the fingers to ensure good contact.

In order to find those very small and frustrating leaks that defy all other methods, one can use a mass-spectrometer leak detector. The so-called *helium leak detector* is a very useful device, incorporating a vacuum pump for evacuating the system to be tested and a mass spectrometer, adjusted to ^{4}He, for detecting helium entering through leaks. The helium gas is sprayed over the various parts of the system, and changes in the mass spectrometer response are indicated on a meter and often audibly by changes in a whistle tone. Although helium leak detectors are complex and expensive instruments, they have the great advantages of speed in operation, extreme sensitivity, and an unambiguous response. They are of special value for metal systems.

When leaks are found, they should be repaired in a permanent way. For glass systems, this involves venting the line and reblowing the glass at the site of the leak. Leaks in metal systems should be repaired by mechanical means: resoldering, rewelding, or replacement of a component. This is often impractical, and recourse must be had to sealants such as Apiezon W or picein (black wax) or Glyptal (an enamel that is most satisfactory when baked on). A leaky silver-soldered joint may often be fixed by cleaning and tinning with soft solder.

When a poor vacuum persists though no leak can be found, one must suspect the possibility of virtual leaks. The system should be thoroughly cleaned and degreased inside, and retested. It may also be necessary to look for gas pockets trapped by screw threads, etc., or to replace decomposable plastic or porous ceramic components with ones giving less gas evolution.

SAFETY CONSIDERATIONS

Vacuum apparatus probably presents fewer accident hazards than almost any other kind of laboratory apparatus of comparable complexity. However, these hazards are by no means entirely negligible, and we summarize the principal ones.

Implosion. This hazard is most important with glass apparatus, and is ever-present when large glass bulbs (over 1 liter in size) or flat-bottomed vessels (of any size) are evacuated. The force of atmospheric pressure makes dangerous missiles of glass fragments from imploding vessels. If at all possible, avoid evacuating Erlenmeyer flasks or other flat-bottomed vessels. Reduce the hazard of flying glass by placing strips of plastic electrician's tape on all large glass vessels that are to be evacuated. Wear safety glasses.

Explosion. If significant quantities of a gas have been liquefied or taken up by an adsorbent at low temperature, an explosion can result when the system warms up if adequate vents or safety valves have not been provided. An explosion of a different kind can take place if an oil diffusion pump (particularly a glass one) is vented to air while hot.

Liquid air. Do not refrigerate a trap with liquid nitrogen until the system pressure is reduced below a few Torr. (Liquid air boils at higher temperature than liquid nitrogen and can easily be condensed at 77 K.) Liquid air present in a trap may react explosively with any organic substances that condense there. The liquid air can also produce excess pressure and likely damage to the vacuum system should the trap warm without venting; this is a common mishap with vacuum systems.

Mercury poisoning. Mercury diffusion pumps, McLeod gauges, and mercury manometers all pose hazards of mercury contamination in the event of fracture of glass apparatus. See Appendix E.

REFERENCES

1. R. D. Present, "Kinetic Theory of Gases," McGraw-Hill, New York (1958); W. Kauzmann, "Kinetic Theory of Gases," Benjamin, New York (1966).
2. S. Dushman, "Scientific Foundations of Vacuum Technique," 2d ed. (J. M. Lafferty, ed.), Wiley-Interscience, New York (1962).
3. M. Knudsen, *Ann. Physik* **28,** 75 (1909).
4. H. Adam, "Foundations of Vacuum Technology", in B. W. Rossiter and J. F. Hamilton (eds.), "Physical Methods of Chemistry," 2d ed., vol. I, chap. 3, Wiley-Interscience, New York (1986).
5. R. P. Benedict, "Fundamentals of Temperature, Pressure, and Flow Measurements," 3d ed., Wiley, New York (1984).
6. J. F. O'Hanlon, "A Users Guide to Vacuum Technology," Wiley, New York (1980).
7. J. H. Moore, C. C. Davis, and M. A. Coplan, "Building Scientific Apparatus," chap. 3, Addison-Wesley, London (1983).
8. G. W. Green, "The Design and Construction of Small Vacuum Systems," Chapman Hall, London (1968); R. P. LaPelle, "Practical Vacuum Systems," McGraw-Hill, New York (1977).
9. "Laboratory Glass Blowing with Corning Glasses," Bulletin B-72, Corning Glass Works, Corning, New York (1961); *op. cit.,* Ref. 7, chap 2; W. E. Barr and V. J. Anhorn, "Scientific Glassblowing," Instruments Publishing, Pittsburgh (1959).

GENERAL READING

Refs. 2, 4-7, *op. cit.*
A. Guthrie, "Vacuum Technology," Wiley, New York (1963).
P. A. Redhead, J. P. Hobson, and E. V. Kornalsen, "The Physical Basis of Ultrahigh Vacuum," Chapman & Hall, London (1968).
A. Roth, "Vacuum Technology," 2d ed., North-Holland, Amsterdam (1982).
D. F. Shriver and M. A. Drezdzon, "The Manipulation of Air-sensitive Compounds," 2d ed., Wiley-Interscience, New York (1986).

CHAPTER
XVIII

INSTRUMENTS

This chapter consists of brief descriptions and discussions of certain devices and instruments that are commonly used in experimental physical chemistry. Considerably greater detail can be found in the references cited. It should also be noted that most commercial scientific instruments are furnished with a detailed instruction manual. This should be reviewed and thoroughly understood *before* the instrument is used.

BALANCES

The determination of mass is one of the most common and important measurements in experimental chemistry. It is common practice to use the terms *mass* and *weight* as interchangeable, but of course they have quite different meanings. Whereas the mass m of an object in kilograms is a measure of the amount of matter in that object, the weight w represents the gravitational force exerted on the object by the earth and should properly be expressed in newtons. Since $w = mg$ and g varies with geographical location, the weight of an object of a given mass will depend on where it is measured. The usage of such expressions as "a 10-gram weight" to mean a mass whose weight equals that of a 10-g mass arises naturally from the common method of comparison weighing. Unless otherwise specified (as in Exp. 33), the term weight as used in this book actually means the mass in grams; this should be clear from the context and from dimensional analysis.

We shall assume that the reader has some prior experience with the use of an equal-arm analytical balance, with which an object is weighted by determining the "weights" which must be added to the right-hand side of the beam in order to made the rest point of the loaded balance the same as the zero point (rest point of unloaded balance).

The design, construction, and operation of such two-pan balances are described in detail by textbooks on quantitative chemical analysis,[1,2] and the essential features of a magnetically damped "chain" balance of this type are given in the second edition of this book. In recent years, the use of single-pan "automatic" balances has grown rapidly because of their convenience and speed of operation. Given below is a brief description of the operation of two balances of this type.

Single-pan mechanical analytical balances. Figure 1 is a schematic drawing showing the key features of mechanical balances still in use in many university laboratories.[1-3] The weighing pan hangs on a hard knife edge of agate or sapphire on one end of the beam. The beam is supported at the center of gravity on another knife edge and a counterweight is permanantly fixed at the other end of the beam. The fixed weights are hung above the weighing pan and may be removed or replaced by operation of the weight-control knobs. The balance is damped by the motion of a piston in a cylinder (air damping) to prevent the beam from acting as a pendulum and oscillating for an inconvenient length of time. A well-adjusted balance usually swings just to equilibrium and stops or it reaches equilibrium after one or two brief

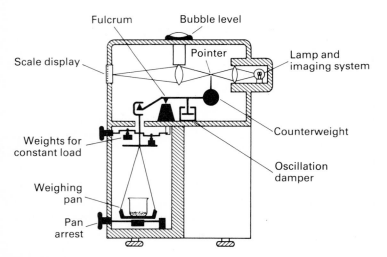

FIGURE 1
A single-pan mechanical analytical balance.

excursions. The rest position of the beam is displaced by putting an object on the pan. Sufficient weights are then removed with the weight-control knobs to bring the beam approximately back to its original position, and the residual displacement from the normal rest point is read on an optical scale that is directly calibrated in milligrams.

This "constant-load" condition gives optimal reproducibility over the entire mass range of the balance. Because of the delicacy of the knife edges and the sensitivity of the balance, the beam is always arrested and the knife edges lifted out of contact with the bearing surfaces whenever anything is being put on or removed from the pan, or, for that matter, whenever a weighing is not actually in progress. Failure to observe this precaution results in rapid wear of the knife edges with resulting decrease in sensitivity of the balance and erratic behavior.

To carry out a weighing, one first checks that the balance is level and steady and that the pan is empty and clean. Chemicals must never be weighed directly on the balance pan—corrosion or contamination of the balance pan can contribute serious errors! With the balance case closed and the pan empty the beam arrest is *gently* released and the rest point of the balance observed. If the empty pan does not give a reading of precisely zero on the optical scale, the optical scale should be adjusted to read zero. If a zero reading cannot be achieved with the adjustment knob, a more serious adjustment is required and the instructor should be notified. When the zero has been adjusted, the object to be weighed is picked up with tongs or a paper loop and placed carefully and gently in the center of the pan. The balance case is closed and the *partial beam release* is operated. The weight knobs are then manipulated until the approximate weight of the object is determined as shown by the optical scale indicating insufficient weight with one setting and overweight with the next individual weight added. With this last weight removed, the beam is then completely released, and after equilibrium has been established the fractional weight is read on the optical scale. The beam is then arrested, the sample is removed, the weights are replaced, and the zero on the balance is checked.

Single-pan electronic analytical balances. Modern analytical balances commonly employ an electromagnetic force principle to achieve a reproducible balance point.[3,4] Figure 2 shows the basic electromagnetic servo system in which the current necessary to hold a pointer at the null position is proportional to the mass placed on the coil platform. Such a servo system is incorporated into classical enclosures as depicted in Fig. 3. Knife edges are replaced by a parallelogram arrangement with flexure points to provide a limited movement of the weighting pan. The null point is chosen to be the relaxed position of the flexure pivots and the current necessary to maintain this position on addition of a mass object is measured and converted to a digital readout. Calibration is achieved by use of a standard weight and adjustment of the electronic circuitry to give the proper readout. Typical capacities are 30–200 g with a precision of 0.01 to 0.1 mg.

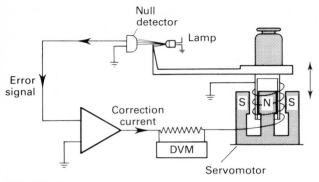

FIGURE 2
Simplified electromagnetic servo system for an electronic balance.

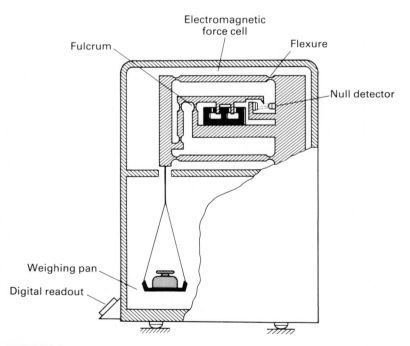

FIGURE 3
An electronic balance based on a parallelogram flexure support and an electromagnetic force cell. The flexure points allow limited movement about the null position and provide resistance to torsional distortion by off-center loading of the weighing pan. Drawing taken from Ref. 3 with permission.

In top-loading versions of the electronic balance, the weighing pan is supported on an upward extension of the moving bar of the parallelogram. This type of balance is very convenient for moderate precision weighing operations, and such balances have capacities of 0.1 to 10 kg with a precision of 1 mg to 1 g. Hybrid versions (electronic-mechanical balances) have also been produced, particularly for low-capacity (<10 g) micro and ultra-micro balances; these have a precision of 0.1 to 1 μg.

Operation of balances. Any type of precision analytical balance should be mounted level on a study bench, free from vibrations, in a room with a fairly stable temperature (no drafts or direct sunlight near the balance). Detailed instructions for the operation of the particular type to be used should be made available by the instructor. Indeed, if the student is not already familiar with the use of that particular design, demonstration by an experienced user and a practice weighing are strongly recommended. The most important general principle is the need for a sense of personal responsibility. Analytical balances are delicate instruments which are capable of excellent precision if they are used with respect and care. It is especially important to use beam and pan arrest controls properly, since they protect the components of the balance (especially the knife edges) from damage. If any difficulties arise, consult an instructor; **do not attempt repairs or adjustments.** After use, leave the balance clean and restored to the zero settings as a courtesy to other users.

Corrections and errors. Buoyancy will affect the results of a weighing, since air will exert a buoyant effect both on the object and on the weights; in general, these two effects will not cancel. The weight in vacuo W_v of an object can be obtained from the weight in air W_a by adding the weight of air displaced by the object and subtracting the weight of air displaced by the weights. Thus,

$$W_v = W_a + V\rho - v\rho \tag{1}$$

where V is the volume of the object, v is the volume of the weights, and ρ is the density of air. This relation applies to electronic balances, since these are calibrated with standard metal weights. Analytical weights are usually made of stainless steel (density 7.8 to 8.0 g cm^{-3}) or brass that is lacquered or plated with gold, nickel or chromium (density 8.4 g cm^{-3}). Taking 8.0 as a reasonable average density, we can replace v in Eq. (1) by $W_a/8.0$ to obtain

$$W_v = W_a\left(1 - \frac{\rho}{8.0}\right) + V\rho \tag{2}$$

Equation (2) is useful if V is known, but often it is not. However, if d_0 (the density of the object) is known, V can be replaced by W_v/d_0 to give

$$W_v = W_a\frac{1 - (\rho/8.0)}{1 - (\rho/d_0)} \cong W_a\left[1 + \left(\frac{1}{d_0} - \frac{1}{8.0}\right)\rho\right] \tag{3}$$

where the final approximation is excellent as long as $\rho/d_0 \ll 1$. Although the density of air varies with temperature, pressure, and moisture content, ρ can be taken to be about $0.0012\,\mathrm{g\,cm}^{-3}$ at any relative humidity over the range 15 to 30°C and 730 to 780 Torr. Buoyancy corrections are often neglected in the weighing of solids, but they are essential in weighing gases and are quite important in weighing large volumes of liquids (as when calibrating volumetric apparatus). For example, W_v for water is 0.1 percent higher than W_a, and the buoyancy correction for 100 g of water would be about 100 mg (much greater than any other source of weighing error). Further discussion of buoyancy corrections can be found in Refs. 5 and 6.

The most common source of error in weighings, with conventional two-pan balances particularly, is due to inaccurate weights. Even in a good set, the weights are often in error by as much as 1 mg. It is recommended that a given set of weights be used only with a single balance and that these weights be calibrated on that balance against a good secondary standard set meeting the tolerance limits set by the National Bureau of Standards. The calibration procedure is given in detail elsewhere.[2,4,7] With single-pan mechanical balances the weights are inside the case, effectively protected from handling and dust. When delivered new or reconditioned, or after periodic servicing (which should include a check of the weights), the weights should be within the manufacturer's specifications. For the most precise work it may be advisable for the user to make a prior check of the balance with a good secondary standard set of weights, and to look carefully for any evidence of dulled knife edges (hysteresis, poor reproducibility).

Finally, there are several other weighing errors that may occur as a result of poor technique but can usually be avoided. Volatile, hygroscopic, or efflorescent samples and samples that adsorb gases (e.g., CO_2 or O_2) should be kept in closed weighing bottles. An object should never be weighed while warm, since convective air currents will occur, causing the weighing to be in error. Weighings may also be in error because of the condensation of moisture on dry glass walls or the force produced by static charge caused by vigorous wiping of a glass surface.

BAROMETER

The Fortin barometer is simply a single-arm, closed-tube mercury manometer equipped with a precise metal scale (usually brass). The bottom of the measuring arm of the barometer dips into a mercury reservoir that is in contact with the atmosphere. The mercury level in this reservoir can be adjusted by means of a knurled screw that presses against a movable plate (see Fig. 4). When the meniscus in the reservoir just touches the tip of a pointed indicator, the zero level is properly established and the pressure can be determined from the position of the meniscus in the measuring arm. Both the front and back reference levels on a sliding vernier are simultaneously lined up with the top of this meniscus in order to eliminate parallax error, and the height of the arm

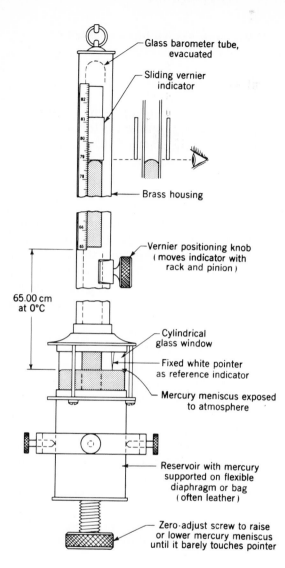

Glass barometer tube, evacuated

Sliding vernier indicator

Brass housing

Vernier positioning knob (moves indicator with rack and pinion)

65.00 cm at 0°C

Cylindrical glass window

Fixed white pointer as reference indicator

Mercury meniscus exposed to atmosphere

Reservoir with mercury supported on flexible diaphragm or bag (often leather)

Zero-adjust screw to raise or lower mercury meniscus until it barely touches pointer

FIGURE 4

Detailed sketch of a Fortin barometer. Uncorrected reading shown is 79.02 cm.

can be read to the nearest tenth of a millimeter using the vernier scale. A thermometer should be mounted on or near the barometer, since the temperature must be known in order to make a correction for thermal expansion. This correction is discussed in Chapter XVII and the appropriate formula is given by Eq. (XVII-23). The metal scale on most barometers is made of brass (linear coefficient of thermal expansion $1.84 \times 10^{-5} \, K^{-1}$) and is usually graduated so as to read correctly at 0°C. A table of barometer corrections over the range 16 to 30°C and 720 to 800 mm is given in Appendix D. High-precision work also requires corrections for the effect of gravity, residual gas pressure in the closed arm, and errors in the zero position of the scale.[8,9] Since these usually amount to only a few tenths of a millimeter, they will not be discussed here.

CATHETOMETER

A cathetometer is used for the accurate measurement of vertical distances, such as the height of the menisci in a wide-bore manometer. It consists of a heavy steel bar (or rod) mounted on a sturdy tripod stand. This steel bar is graduated in millimeters and supports a traveling telescope, which can be moved vertically through about 100 cm and can also be rotated in a horizontal plane. Leveling screws in the base of the stand are used to obtain an accurate vertical alignment of this steel scale, and the telescope mounting is equipped with a fine-adjustment screw and a spirit level to ensure accurate horizontal positioning of the telescope. The meniscus to be measured is brought into focus and aligned with respect to a cross hair in the eyepiece; the position of the telescope on the scale is then read to the nearest 0.1 or 0.05 mm with a vernier scale. A cathetometer is especially convenient for reading levels on an apparatus that must be immersed in a constant-temperature bath (e.g., an osmometer).

OSCILLOSCOPE

The basic principles of dc and ac voltage measurements are discussed in Chapter XV. In many applications these measurements are carried out by complex electronic instruments designed to produce a visual record of the detected signal, either as a trace on a luminescent screen, an inked record on chart paper, or a numerical (digital) readout of the voltage measured. Oscilloscopes are described briefly in this section. Other instruments for accomplishing these functions are described in the section on recorders and plotters. More detailed information may be found in standard textbooks on electronics.[10,11]

In a cathode-ray tube, electrons are emitted from a cathode and are then accelerated and focused by a series of special anodes to form a beam that impinges on the face of the tube. This face is coated with fluorescent material, and the beam produces a sharp visible spot. Displacement of this spot from the center of the screen can be achieved by passing the electron beam through the electrostatic field between a pair of charged plates. There are two independent sets of these deflection plates, which control, respectively, the vertical and the horizontal position of the spot. Since high voltages are required across these plates to produce a suitable displacement of the beam, oscilloscopes have wide-band amplifiers to provide voltage amplification for the input signals.

The primary use of an oscilloscope is to display the shape of a voltage waveform (i.e., to plot out voltage vertically against a horizontal time scale). To accomplish this, there must be applied to the horizontal deflection plates a sweep voltage that will cause the beam to move from left to right at a uniform rate and then return very rapidly to the starting point. All oscilloscopes have an internal sawtooth waveform generator to produce this linear timebase sweep.

The operation of an oscilloscope can best be described by reference to Fig. 5, which shows the typical layout of the controls of a commercial instrument. The face of the cathode-ray tube, in the upper left-hand corner, is covered by a removable plastic *graticule* on which is ruled a pair of centimeter scales at right angles. The graticule can be illuminated from the side for visibility; the level of illumination is controlled by the scale illumination knob, generally coupled to the on/off switch. A Polaroid camera can be mounted over the tube face if it is desired to make permanent photographic records of the oscilloscope display. The sharpness of the electron-beam image is optimized by the focus and intensity controls. It is good practice to turn the beam intensity down to quite a low level; in particular, when the beam is resting at one spot on the tube face, the intensity *must* be lowered until the spot is barely visible. Failure to do so can result in literally burning a hole in the tube phosphor at that point.

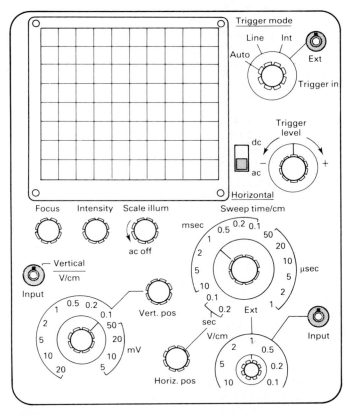

FIGURE 5
Simplified controls of a typical cathode-ray oscilloscope.

The signal to be measured is applied to the input connector of the vertical deflection amplifier, and centered on the tube face by means of the vertical position control. Voltages in excess of the deflection factor at the least sensitive position of the V/cm control must be attenuated by a suitable calibrated probe before being applied to the input connector.

The horizontal display is controlled by the horizontal amplifier and triggering sections of the oscilloscope. The sweep is initiated whenever a preset voltage level, furnished by the triggering section, is reached. If the trigger mode control is set to external, then this voltage must be supplied from some external source to the trigger input. At the internal setting, the triggering signal is the same as the vertical input signal, so that the wave form to be observed triggers itself. At the line setting, this voltage is taken from the 60-Hz line current which powers the oscilloscope. This mode is very useful in detecting sources of unwanted electronic noise that arise from the power-line voltage itself. In the auto mode, the sawtooth simply repeats itself at the end of each sweep, so that the horizontal deflection is not synchronized with any external source.

In all but the last triggering mode, the voltage at which the sweep is initiated is set by the triggering level control. Provision is also made for selecting this voltage with a positive or negative slope, and from the ac or dc portion of the input signal.

The horizontal sweep rate is determined by the sweep time/cm control. The leading edge of the waveform should be set to the left-hand end of the graticule ruling by use of the horizontal position control. The $y(x)$ relationship between two voltages, $x(t)$ and $y(t)$, can be obtained by setting the horizontal sweep control to external and applying $x(t)$ to the horizontal input and $y(t)$ to the vertical input. Thus the sawtooth generator is bypassed, and the amplified $x(t)$ signal is applied directly to the horizontal deflection plates. In this mode, the oscilloscope can be used as a very sensitive device for comparing the frequencies of two different sinusoidal waveforms. When the ratio of the two frequencies is a rational fraction, a symmetric closed pattern called a Lissajous figure will appear on the screen. The frequency ratio can be obtained from the form of the pattern by using the formula

$$\frac{f_H}{f_V} = \frac{n_V}{n_H} \tag{4}$$

where f_H and f_V are the frequencies applied at the horizontal and vertical inputs, n_H is the number of points at which the figure is tangent to a horizontal line, and n_V is the number of points of tangency between the figure and a vertical line. Several simple types of Lissajous figures are shown in Fig. 6(a).

Shown in Fig. 6(b) is the Lissajous figure for several different values of the phase angle between two signals of the same frequency and amplitude; note that the pattern is a circle when the two sine waves are 90° out of phase and is a straight line tilted at 45° when the phase angle is 0 or 180°. This fact

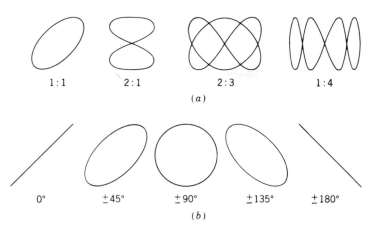

FIGURE 6
Lissajous figures: (*a*) several simple figures, with the ratios $f_H:f_V$ indicated; (*b*) the 1:1 figure for various values of the phase angle between signals of equal amplitude.

provides a convenient means of determining the phase shift in a circuit. For balancing an ac Wheatstone bridge, a signal from the ac power source is applied to the horizontal input, and the unbalance signal across the bridge is applied to the vertical input. The bridge is first balanced capacitively by obtaining a straight-line pattern. After this is done, the resistive balance is indicated by obtaining a horizontal line (zero vertical amplitude).

pH METER

A pH meter is a special type of millivolt potentiometer designed to measure the emf of a cell in which the electrolyte contains hydrogen ions.[12] A "glass electrode" is employed as the measuring electrode, and a calomel electrode is used as the reference electrode.

Calomel electrode.[13] In addition to pH measurements, there are many other emf cell measurements for which it is convenient to use the calomel electrode as a reference electrode against which a measuring electrode is compared. Actually, this "electrode" is really a half-cell that is connected via a KCl salt bridge to another half-cell containing the solution of interest. The *saturated* calomel electrode can be written as

$$Hg(l) + Hg_2Cl_2(s), \; K^+Cl^-(aq, \text{ sat.}), \text{ aq. electrolyte} \tag{5}$$

There are two other common versions of this half-cell: the *normal* and *tenth-normal* calomel electrodes, in which the KCl concentration is either 1.0 or 0.1 N. The saturated electrode is the easiest to prepare and the most convenient to use but has the largest temperature coefficient. The half-cell potential for each of the calomel electrodes has a different value relative to the

TABLE 1
**Half-cell potentials of calomel
reference electrodes**

KCl conc.	Potential at 25°C, V
0.1 N	0.3338
1.0 N	0.2800
Saturated	0.2415

standard hydrogen electrode; these emf values are given in Table 1. Calomel
electrodes can be easily prepared in the laboratory and are also available
commercially. Two typical calomel electrode designs are shown in Fig. 7.

Glass electrode.[12] This electrode is usually a silver–silver chloride electrode,
surrounded by a thin membrane of a special glass that is permeable to
hydrogen ions. The glass membrane is essentially a special type of salt
bridge—one in which the anions are immobile (have zero transference
number), since they are part of the porous glass framework through which the

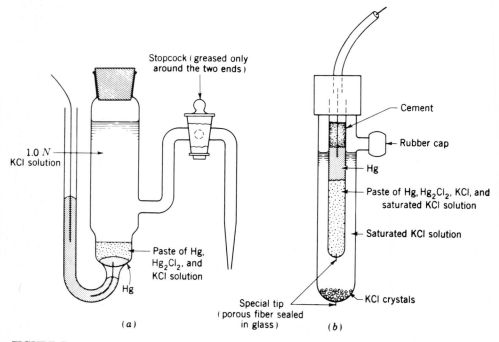

FIGURE 7
Two typical calomel electrode designs: (*a*) laboratory type, shown unsaturated; (*b*) commercial
type, shown saturated.

H^+ cations can move. The glass electrode may be formulated as

$$\text{aq. solution,} \quad H^+(\text{glass})^-, H^+Cl^- \ (aq, a_0) \mid AgCl(s) + Ag(s) \qquad (6)$$
containing H^+
at activity a_{H^+}

The change in state per faraday for this half-cell is thus

$$AgCl(s) + H^+(a_{H^+}) + e^- = Ag(s) + Cl^-(a_0) + H^+(a_0) \qquad (7)$$

The activity a_0 has a definite constant value, usually obtained by using $0.1\,M$ HCl as the solution inside the membrane. An important advantage of the glass electrode is that it can be used under many conditions for which the hydrogen electrode is subject to serious error.[14]

pH measurement.[12,14] The overall emf for a cell with a calomel and a glass electrode dipping into an aqueous electrolyte solution is

$$\mathscr{E} = \mathscr{E}' - \frac{RT}{\mathscr{F}} \ln a_{H^+} = \mathscr{E}' + \frac{2.303RT}{\mathscr{F}} (\text{pH}) \qquad (8)$$

where $\text{pH} \equiv -\log a_{H^+}$ and $\mathscr{E}'$ is the difference between the half-cell potential of the calomel reference electrode and the "standard potential" of the glass electrode. Obviously, $\mathscr{E}'$ will depend on the type of calomel electrode used and on the activity a_0 of hydrogen chloride in the inner solution of the glass electrode. Since these factors are kept constant, changes in emf are a direct indication of variations in pH.

Because of the high resistance of the glass membrane (10 to 100 MΩ) it is not practical to measure the emf directly. Instead, pH meters either use a direct-reading electronic voltmeter or electronically amplify the small current that flows through the cell and potentiometrically detect the voltage drop across a standard resistor. Both battery-operated and ac line-operated pH meters are available commercially from such firms as Beckman Instruments, Inc., Orion Research, Inc., and Corning Glass Works. Such pH meters are calibrated to read directly in pH units, have internal compensation for the temperature coefficient of emf, and have provision for scale adjustments.

Since the operation of a pH meter is very simple but slightly different for each model, no detailed operational procedure will be given here. However, a few general remarks are necessary. If a glass electrode and a silver–silver chloride electrode were placed in an HCl solution for which a_{H^+} equals a_0, the emf of this cell would ideally be zero (i.e., there should be no potential difference across the glass membrane). However, there is always some small emf (1 or 2 mV) across the membrane under these conditions. This so-called *asymmetry potential* is presumably due to strains in the membrane and may change slowly with time or be temporarily changed by exposure of the electrode to very strong acid or base. Therefore, it is necessary to compensate for this asymmetry potential by calibrating the pH meter frequently against a buffer solution of known pH. Also, pH readings on solutions of pH greater than 12 may be in error owing to a significant contribution from sodium-ion transference in the glass at these low hydrogen-ion concentrations. This difficulty can be avoided by the use of special high-resistance glass membranes.

POLARIMETER

The polarimeter (Fig. 8) is an instrument for measuring the optical rotation produced by a liquid or solution.[15] The *specific rotation* $[\alpha]_\lambda^t$ of a solute in solution at a given wavelength λ and Celsius temperature t is given by

$$[\alpha]_\lambda^t = \frac{100\alpha}{Lc} = \frac{100\alpha}{Lp\rho} \qquad (9)$$

where α is the angle in degrees through which the electric vector is rotated, L is the path length in *decimeters,* c is the concentration of solute in grams/100 cm^3 solution, p is the weight percent of solute in the solution, and ρ is the density of the solution in g cm^{-3}. The angle α is considered positive if the rotation of the electric vector as the light proceeds through the solution is in the sense of a left-hand screw or negative if in the sense of a right-hand screw.

Usually the optical rotation is measured with the sodium D yellow line (a doublet, 589.0 and 589.6 nm). For more precise work the 546.1 nm green mercury line may be used.

Light from the source (a sodium-vapor arc lamp or a mercury-vapor lamp with appropriate filter) is polarized by a Nicol prism, termed the polarizer, which consists of two prisms of calcite cemented together with canada balsam so that one of the two rays produced in double refraction (the "ordinary ray") is totally reflected at the interface and lost while the other (the "extraordinary

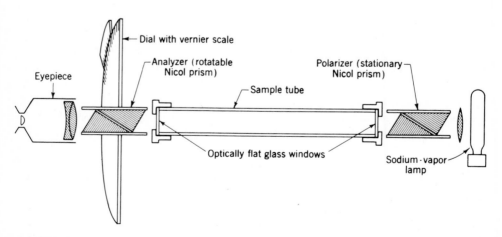

FIGURE 8
Schematic drawing of a polarimeter.

ray") is transmitted. The polarized light passes through the solution and then through a second Nicol prism, termed the analyzer, which can be rotated around the instrument axis. The normal position (zero rotation) is one in which the two Nicol prisms are at 90° to each other, and no light passes through. When an optically rotating medium is introduced between the two Nicol prisms, light is transmitted; the observed rotation α is the angle in degrees through which the dial must be turned (clockwise with respect to the observer if α is positive, counterclockwise if negative) in order to restore the field to complete darkness.

As the dial is turned, the intensity of emergent light is proportional to $\sin^2(\alpha - \alpha_d)$, where α_d is the setting for complete darkness. Since this quantity behaves approximately quadratically near the dark position, the setting is of limited sensitivity if only a single polarizer and a single analyzer are used. In many instruments the field of view is divided into two equal parts with polarization angles differing by a few degrees. This is done either by use of a composite polarizer (two Nicol prisms cemented together side by side with their planes of polarization at a small angle) or by use of an added Nicol prism covering half of the field of the polarizer. In use, the analyzer is adjusted so that the two fields appear equally bright.

Although laboratory polarimeters generally use Nicol prisms as polarizers and analyzers, dichroic crystals (such as tourmaline) or dichroic sheet polarizers (such as Polaroid) may be used in the construction of special apparatus.†

The polarimeter is commonly used in organic and analytical chemistry as an aid in identification of optically active compounds (especially natural products) and in estimation of their purity and freedom from contamination by their optical enantiomers. The polarimeter has occasional application to chemical kinetics as a means of following the course of a chemical reaction in which optically active species are involved. Since the rotation α is a linear function of concentration, the polarimeter can be used (in the same way that a dilatometer might be used) in studying the acid-catalyzed hydrolysis of an optically active ester, acetal, glycocide, etc.

In modern organic chemistry optical-rotatory dispersion,[16] the variation with wavelength of optical-rotatory power (and certain related properties), is used in molecular-structure investigation as a means of identifying and characterizing chromophore groups. Automatic polarimetric spectrophotometers of high complexity have been developed for this purpose.

---

† In the phenomenon known as dichroism, the optical absorption depends strongly on the orientation of the plane of polarization with respect to the crystallographic axes (or axis of preferred orientation). Commercial sheet polarizers are made from acicular dichroic crystals, herapathite (iodoquinine) in the case of Polaroid, suspended in a viscous or plastic medium and aligned by extrusion or stretching. Dichroic polarizers and analyzers are inferior to Nicol prisms for use in polarimeters because the transmission is considerably less than unity when the planes of polarization are parallel and the absorption is not quite complete when they are perpendicular.

RADIATION COUNTERS

Electromagnetic radiations with wavelengths below about 1 nm are called x-rays or gamma rays. These have a penetrating power that increases strongly as the wavelength decreases and is dependent on the atomic numbers of the atoms present rather than their state of chemical combination.

A 0.1-nm photon has about 8000 eV of energy, while ultraviolet radiation has only 3 or 4 eV. This much energy, absorbed in matter, may result in the virtually simultaneous production of many ions—enough so that under optimum conditions the absorption of a single x-ray or gamma-ray photon can be detected with an efficiency approaching 100 percent. The measurement of x-ray and gamma-ray intensities thus amounts to a *counting* of discrete events wherein individual photons are detected.[17]

The best known detector is the *Geiger–Müller counter*. This usually consists of a cylindrical container (the cathode) filled with an absorbing gas such as argon or krypton, with an insulated central wire (the anode) to serve as a collector for electrons. When an x-ray photon is absorbed, producing ions and electrons, acceleration of the ions and electrons to the electrodes results in collisions resulting in more ionization; thus a cascade process develops that results in a general gas discharge. This continues until the fall of potential between the electrodes is sufficient to quench the discharge and allow the potential to be restored. The Geiger–Müller counter with its associated circuitry is relatively simple but for many radiation-counting purposes suffers from a long "dead time" (0.1 to 1 ms) resulting from the complete discharge and required recharge. This dead time results in coincident counts and nonlinear response at counting rates higher than about 100 counts per second. Another limitation of the Geiger–Müller counter is that the size of the pulse generated is independent of the energy of the incoming particle.

To overcome these limitations the self-quenching *proportional counter* has been developed. This counter is very much like the Geiger–Müller counter in construction. The electrons and positive ions from the primary ionizing event go to their respective electrodes, but the production of additional electrons through positive-ion bombardment of the wall is prevented by molecules of some organic compound (i.e., ethanol) that is present as a "quench gas." Thus the tube does not discharge completely. The pulse is of very short duration, of the order of 1 μs. Therefore, counting rates of up to 10 000 counts per second are essentially linear with intensity. Since the pulse is very small, the proportional counter requires an exceedingly sensitive (high-gain) preamplifier, well shielded from electrical disturbances. Most important for many purposes is the fact that the pulse height depends upon the energy of the incident photon or other particle. The output of the preamplifier may be fed to an electronic pulse-height discriminator circuit, connected to two or more scaling and counting circuits, among which the pulses are distributed according to the height ranges in which they fall. The self-quenching proportional counter does not last indefinitely; after about 10^{11} counts the quench gas is entirely consumed, and the tube thereafter behaves like a Geiger–Müller counter.

Another commonly used detector is the *scintillation detector*. This makes use of a crystal that produces a scintillation (pulse of visible light) on absorption of an x-ray photon. The visible light is detected by a photomultiplier tube and associated amplifier circuit, which is sensitive enough to detect nearly every scintillation. The scintillating crystal is usually sodium iodide doped with an activator such as thallous iodide.

RECORDERS AND PLOTTERS

A conceptually simple modification of the basic potentiometer circuit described in Chapter XV permits one to make a permanent recording of voltage variations as a function of time or another parameter. The movable tap that contacts the slide-wire is controlled by a servomotor, and the null detector is a sensitive operational amplifier. The output of the amplifier is connected to the servomotor, which drives the tap up or down the slide-wire according to the sign of the off-balance voltage. A pen is attached to the sliding tap, which writes a trace on a paper chart. Strip-chart recorders have a roll of chart paper that is sprocket-driven by a low-speed motor or stepping motor, so that the pen produces an X, t record of voltage variations with time. In recording spectrophotometers, the chart is mechanically linked to the rotation of a grating to produce a record of voltage (related to light intensity) against wavelength. The position of the pen on a *stationary* chart paper can be controlled by two independent null detectors and servomotor mechanisms to record the simultaneous variation of two external voltages; such a device is called an X, Y recorder.

Other types of recording devices exist. In one form, the slide-wire potentiometer is replaced by a longer-lasting, noncontact capacitance transducer that produces a reference voltage which, on comparison with the input signal, drives a motor to reposition the pen for a zero difference. In other versions, the analog input signal is simply digitized and the resultant information is used with a stepping motor to position the pen on the chart by an amount proportional to the analog input.

In a completely digital recorder, termed a *plotter,* all voltage conversions are done externally and only the digital information is sent to the recorder, typically by a microprocessor or computer. Digital plotters have positioning accuracies of 0.02 to 0.2 mm, a resolution sufficient for printing of alphanumeric characters so that the record can include important information about the experimental conditions. If the data are to be used subsequently for analysis or for figures for reports or publications, the digital form is most convenient since it is easily stored on magnetic media such as disks and tapes for later recall and manipulation.

REFRACTOMETERS

The term *refractometer* is principally applied to instruments for determining the index of refraction of a liquid, although instruments also exist for determining the indexes of refraction of a solid.[18] The index of refraction n for a liquid or an isotropic solid is the ratio of the phase velocity of light in a vacuum to that in the medium. It can be defined relative to a plane surface of the medium exposed to vacuum as shown in Fig. 9(a); it is the ratio of the sine of the angle ϕ_v a ray of light makes with a normal to the surface in vacuum to the sine of the corresponding angle ϕ_m in the medium:

$$n = \frac{c_v}{c_m} = \frac{\sin \phi_v}{\sin \phi_m} \tag{10}$$

It is common practice to refer the index of refraction to air (at 1 atm) rather than to vacuum, for reasons of convenience; the index referred to vacuum can be obtained from that referred to air by multiplying the latter by the index of refraction of air referred to vacuum, which is 1.00027.

The index of refraction is a function of both wavelength and temperature. Usually the temperature is specified to be 20°C or 25°C. The former is more in accord with past practice, but the latter is somewhat easier to maintain with a constant-temperature bath under ordinary laboratory conditions. The wavelength is usually specified to be that of the yellow sodium D line (a doublet, 589.0–589.6 nm), and the index is given the symbol n_D.

Most refractometers operate on the concept of the *critical angle* ϕ_{crit}; this is the angle ϕ_m for which ϕ_v (or ϕ_{air}) is exactly 90° (see Fig. 9(b)). A ray in the medium with any greater angle ϕ_{m_1} will be totally reflected at an equal angle ϕ_{m_2} as shown in Fig. 9(c). The index of refraction is given in terms of the critical angle by

$$n = \frac{c_v}{c_l} = \frac{1}{\sin \phi_{crit}} \tag{11}$$

where c_l is the phase velocity of light in the liquid. In a refractometer the critical

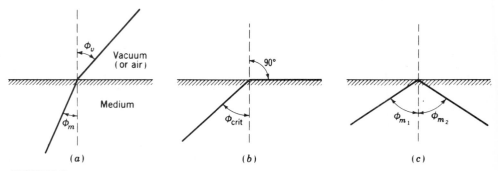

FIGURE 9
Reflection and refraction at an interface: (a) $\phi_m < \phi_{crit}$, (b) $\phi_m = \phi_{crit}$, (c) $\phi_{m_1} > \phi_{crit}$.

angle to be measured is that inside a glass prism in contact with the liquid, since the index of refraction of the glass is higher than that of the liquid. Therefore

$$\frac{c_l}{c_g} = \frac{c_l}{c_v}\frac{c_v}{c_g} = \frac{n_g}{n} = \frac{1}{\sin \phi_g}$$

where n_g is the index of refraction of the prism glass and ϕ_g is the critical angle in the glass. By trigonometry it can be shown that the index of refraction of the liquid is given by

$$n = \sin \delta \cos \gamma + \sin \gamma \sqrt{n_g^2 - \sin^2 \delta} \qquad (12)$$

where γ is the prism angle (angle between the two transmitting faces) and δ is the angle of the critical ray in air with respect to the normal to the glass-air prism face (see Fig. 10).

The most precise type of refractometer is the *immersion refractometer*. It contains a prism fixed at the end of an optical tube containing an objective lens, an engraved scale reticule, and an eyepiece. It also contains an Amici compensating prism (see below). In use, the instrument is dipped into a beaker of the liquid clamped in a water bath for temperature control. A mirror in the bath or below it reflects light into the bottom of the beaker at the requisite angle and with some angular divergence. The field of view is divided into an illuminated area and a dark area, as shown in Fig. 10; the scale reading that corresponds to the boundary-line (critical-ray) position is read and referred to a table to obtain the refractive index. This instrument is capable of measuring

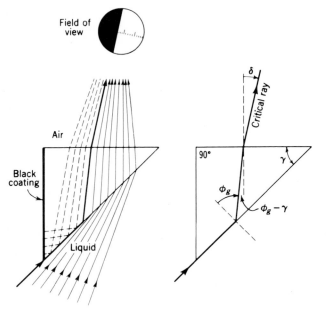

FIGURE 10
Essential features of an immersion refractometer. The behavior of the critical ray is shown in detail, since this represents the basic principle of almost all refractometers.

the refractive index to ±0.000 03. Its scale normally covers only a small range; a set containing several refractometers or detachable prisms is required to cover the ordinary range of refractive indexes for liquids (1.3 to 1.8).

The most commonly used form of refractometer is the *Abbe refractometer*, shown schematically in Fig. 11. This differs from the immersion refractometer in two important respects. First, instead of dipping into the liquid, the refractometer contains only a few drops of the liquid held by capillary action in a thin space between the refracting prism and an illuminating prism. Second, instead of reading the position of the critical-ray boundary on a scale, one adjusts this boundary so that it is at the intersection of a pair of cross hairs by rotating the refracting prism until the telescope axis

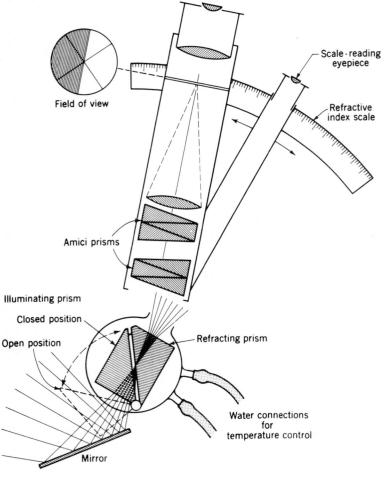

FIGURE 11
Schematic diagram of an Abbe refractometer.

makes the required angle δ with the normal to the air interface of the prism. The index of refraction is then read directly from a scale associated with the prism rotation.

The Abbe refractometer commonly contains two Amici compensating prisms, geared so as to rotate in opposite directions. An Amici prism is a composite prism of two different kinds of glass, designed to produce a considerable amount of dispersion but to produce no angular deviation of light corresponding to the sodium D line. By use of two counter-rotating Amici prisms the net dispersion can be varied from zero to some maximum value in either direction. The purpose of incorporating the Amici prisms is to compensate for the dispersion of the sample so as to produce the same result with white light that would be obtained if a sodium arc were used for illumination. This is achieved by rotating the prisms until the colored fringe disappears from the field of view and the boundary between light and dark fields becomes sharp. It should be borne in mind that the dispersion of a sample is not always exactly compensated, for dispersion is not exactly defined by a single parameter for all substances. The most precise results are obtained with illumination from a sodium arc, the Amici prisms being set at zero dispersion.

The Abbe refractometer is less precise (± 0.0001) than the immersion refractometer and requires somewhat less exact temperature control ($\pm 0.2°C$). For this purpose, water from a thermostat bath is circulated through the prism housings by means of a circulating pump. Alternatively, tap water is brought to the temperature of a thermostat bath by flow through a long coil of copper tubing immersed in the bath, and is then passed once through the refractometer and down the drain.

The procedure for the use of an Abbe refractometer is as follows.

1. If a sodium arc is being used, check to see that it is operating properly. The sodium arc should be treated carefully and should be turned on and off as infrequently as possible. It should be turned on at least 30 min before use.

2. Check to see that the temperature is at the required value by reading the thermometer attached to the prism housing.

3. Open the prism (rotate the illuminating prism with respect to the stationary refracting prism). Wipe both prism surfaces *gently* with a *fresh* swab of cotton wool dampened with acetone or benzene. **Caution:** the prisms must be treated with great care since scratches will decrease the sharpness of the boundary and permanently reduce the accuracy of the instrument.

4. When the prism surfaces are clean and dry, introduce the sample. In some models the refracting prism surface is horizontal and faces upward. In this case place a few drops of the liquid sample on the refracting prism and move the illuminating prism to the closed position. Other models have a prism configuration like that shown in Fig. 11. In that case bring the prisms into the closed position first and then squirt a small amount of sample into the filling hole.

5. Rotate the prism until the boundary between light and dark fields appears in the field of view. If necessary, adjust the light source or the mirror to obtain the best illumination.

6. If necessary, rotate the Amici prisms to eliminate the color fringe and sharpen the boundary.

7. Make any necessary fine adjustment to bring the boundary between light and dark fields into coincidence with the intersection of the cross hairs.

8. Turn on the lamp (if any) that illuminates the scale, and read off the value of the refractive index.

9. Open the prism and wipe it *gently* with a clean swab of cotton wool, dampened with acetone or benzene. When dry, close the prism.

If the sample is very volatile, it may evaporate before the procedure is completed. In this case or in the event of drift, add more sample.

One of the worst enemies of the refractometer is *dust*. A gritty particle may scratch the prisms badly enough to require their replacement. The cotton wool used for wiping the prisms should be kept in a covered jar. Each swab of cotton wool should be used only once and then discarded. *Do not rub* the prisms with cotton wool, and do not attempt to wipe them dry: if streaks are left when the acetone or benzene evaporates, wipe again with a fresh swab dampened with fresh solvent. Do not use lens tissue on the prism surfaces. Finally, the instrument should be protected with its dust cover when not in use, and the table on which the instrument is used should be kept scrupulously clean.

For adjustment of the scale a small "test piece" (rectangular block of glass of accurately known index of refraction) is usually provided with the refractometer. The illuminating prism is swung out and the surface of both the refracting prism and the test piece are carefully cleaned. They are then carefully brushed with a clean camel's-hair brush (which is normally kept in a stoppered container) and inspected at grazing incidence to detect particles of dust or grit. A very small drop (ca. 1 mm^3) of a liquid (such as 1-bromonaphthalene or methylene iodide) that has a higher refractive index than the refracting prism is placed on the test piece, and the latter is then carefully pressed against the refracting prism and carefully moved around to spread the liquid. The reading of refractive index is made in the usual way. If it is not in agreement with the true value of the test piece, an adjustment of the instrument scale is made or a correction is calculated.

The procedure for determining the index of refraction of an isotropic solid sample is similar; like the test piece it must have at least one highly polished plane face.

The refractometer is essentially an analytical instrument, used to determine the composition of binary mixtures (as in Exp. 14) or to check the purity of compounds. Its most common industrial application is in the food and confectionary industries, where it is used in "saccharimetry"—the determination of the concentration of sugar in syrup. Many commercially available refractometers have two scales: one calibrated directly in refractive index, the other in percent sucrose at 20°C.

The refractive index of a compound is a property of some significance in regard to molecular constitution. The *molar refraction,* defined by Eq. (31-14), is a constitutive and additive property; for a given compound it may be approximated by the sum of contributions of individual atoms, double bonds, aromatic rings, and other structural features.

SIGNAL-AVERAGING DEVICES

An important class of electronic measuring instruments is that designed to retrieve weak voltage signals from accompanying noise.[10,11] Since the frequency spectrum of "white" noise is very wide, typically from 0.1 Hz to several MHz with a $1/f$ intensity distribution, much of it can be eliminated with the use of a *frequency-selective amplifier* that passes only a narrow bandwidth of the input at a specified frequency f_0. By modulating the dc signal to be measured at this same frequency, the signal-to-noise ratio is greatly improved. This modulation can be achieved in a number of ways: mechanically chopping a light beam; applying a modulated electric field to the sample in Stark-modulated microwave spectroscopy; or applying a modulated magnetic field in nuclear magnetic or electron paramagnetic resonance spectroscopy are a few examples. The combination of a frequency-selective amplifier with a *phase-sensitive detector,* which locks the amplifier input to a reference voltage, is known as a *lock-in detector.* The schematic operation of one such detector is shown in Fig. 12.

If the signal event is of short duration compared to the repetition time, a *gated integrator* provides better signal averaging than a lock-in amplifier. In this device, current from the signal source is allowed to pass through a fast, gated transistor switch to charge an integrating capacitor on a low-noise op amp [see Fig. XV-9(f)]. By making the gate overlap only the signal, noise outside this time period is excluded and the voltage across the capacitor reaches an average value representative of the input signal. Alternatively, the voltage can be measured on each cycle, passed to a computer, and the capacitor discharged by a shorting transistor prior to the next input pulse; in this case signal averaging can be done by direct addition with the computer. A typical application of a gated integrator is in the study of transient signals produced by light sources or lasers with pulse durations of nanoseconds to microseconds but a duty cycle of only 1 to 100 Hz.

When time resolution of a signal is required, the sampling period of a gated integrator can be made small and stepped or swept slowly across a repetitive waveform. A device with this capability, termed a *boxcar integrator,* is available from several manufacturers with time resolutions down to the nanosecond and even picosecond range. Alternatively, one can use a *transient digitizer* or *digital oscilloscope* to capture a single-shot event in time by

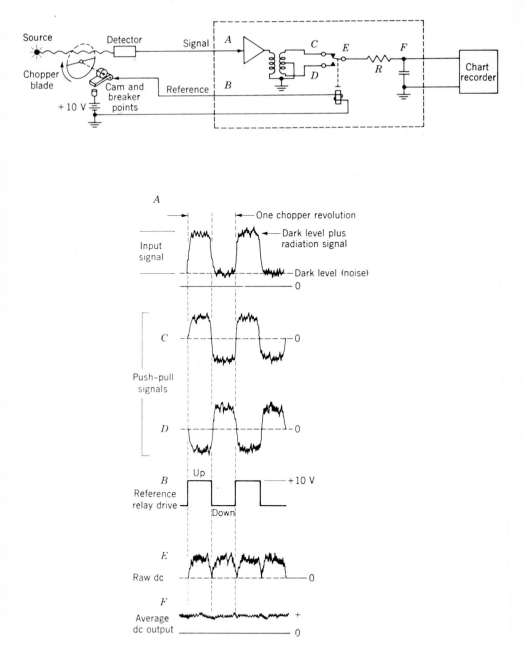

FIGURE 12

Schematic diagram of a very simple lock-in amplifier used to measure low-intensity light signals: block diagram of the components and voltage levels at various points indicated in the diagram. The signal-plus-noise coming in at A is amplified at C and D, and referenced to the signal coming in at B. The resulting dc signal at E is smoothed by the output capacitor to the average signal at F.

essentially doing many analog-to-digital (A/D) conversions in rapid succession. Such measurements can be done at rates up to about 100 kHz with relatively inexpensive A/D boards that plug into microcomputers, and at MHz and even GHz rates with more expensive commercial instruments. To handle the high data flow of such devices, signal averaging of repetitive sources is usually done by direct accumulation of the digitized signal in computer memory addresses assigned to each sampling interval.

SPECTROSCOPIC COMPONENTS

The interaction of electromagnetic radiation with matter serves as one of the most useful methods for studying the structures, energy levels, and dynamics of chemical systems. Figure 13 shows the energy transitions associated with different portions of the electromagnetic spectrum, and various experiments in this book utilize most of the types of spectroscopy associated with these changes. Some experimental details on magnetic resonance techniques are given in Exps. 34, 43, and 44, and the discussion here will center on the physical elements (sources, dispersion devices, and detectors) used for spectroscopy in the visible–ultraviolet (VIS–UV) and infrared (IR) regions.

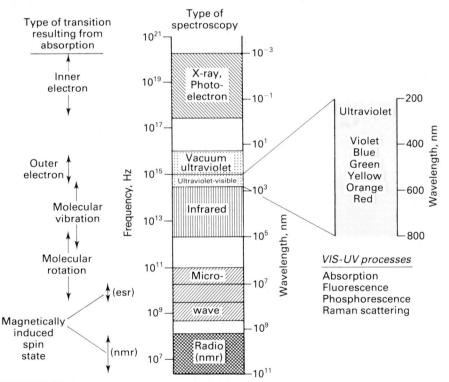

FIGURE 13
The electromagnetic spectrum and its applications in spectroscopy.

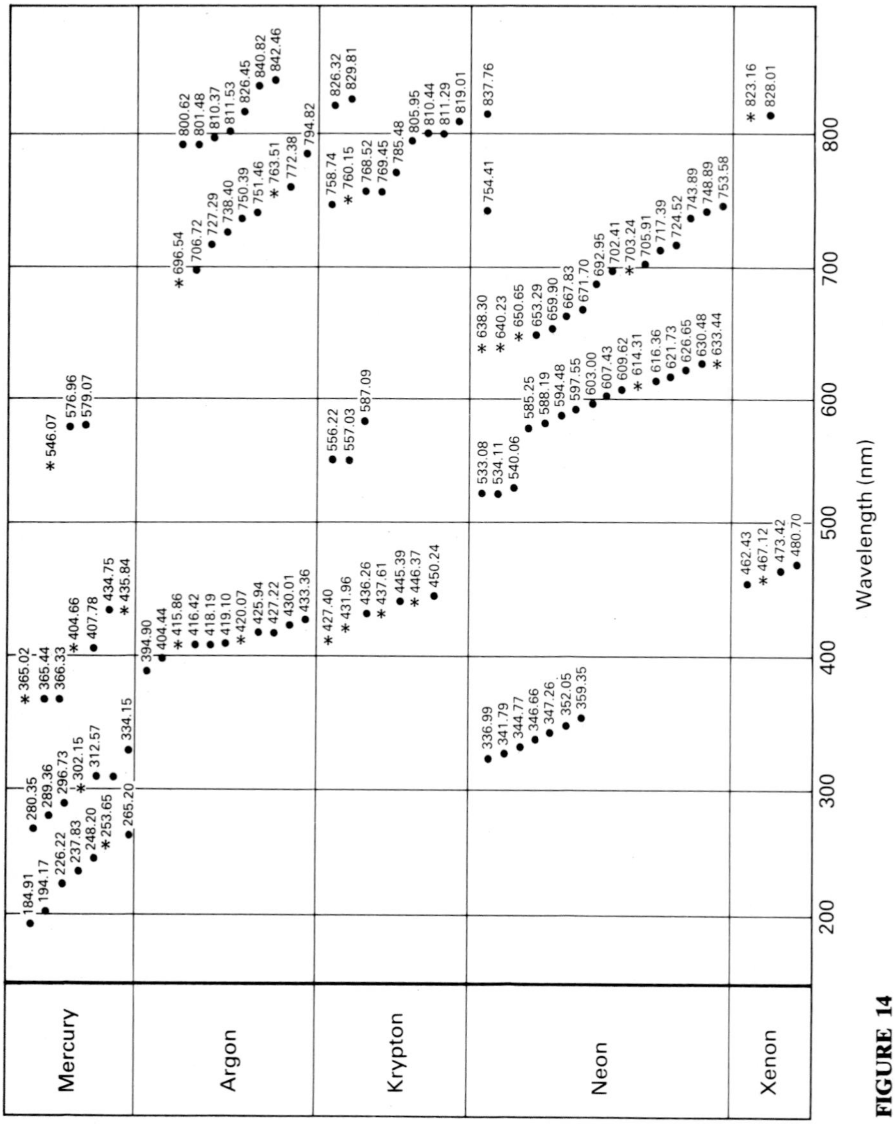

FIGURE 14
Calibration lines from low-pressure discharge lamps. The wavelengths in nm are given in air and the stronger lines are indicated with an asterisk.

Sources. In emission spectroscopy, the molecule or atom itself serves as the source of light with discrete frequencies to be analyzed. For example, in Exps. 40 and 41 dealing with emission from H atoms and N_2 molecules, electron excitation of hydrogen or nitrogen in a discharge tube provides an intense source whose spectrum is analyzed to extract information about the electronic and, for nitrogen, vibrational levels. Such low-pressure ($p < 10$ Torr) line sources are available with many elements, and lamps containing Hg, Ne, Ar, Kr, and Xe are often used for calibration purposes. The Pen-Ray pencil-type lamp is especially convenient for the visible and ultraviolet region, and some calibration wavelengths are given in Fig. 14. More complete compilations are to be found in Refs. 19–21.†

A continuum source is needed for absorption spectroscopy, and this is provided by discharge lamps filled to higher densities, such that pressures can exceed 100 bar at operational temperatures. The result is a broad continuum emission with superimposed line spectra, as shown for several lamps in Fig. 15. In commercial spectrometers, the deuterium lamp is commonly used for the UV region below 350 nm while the tungsten–halogen lamp is convenient for the 350–900 nm range. The latter is an example of a thermal source whose *spectral emittance* distribution $J(\lambda)$ closely approximates that predicted by the Planck formula for a blackbody radiator[24,25]

$$J(\lambda) = \frac{2\pi h c^2}{\lambda^5 [\exp(hc/\lambda kT) - 1]} \tag{13}$$

$J(\lambda)$ has units of W m^{-3} and its variation with temperature is shown in Fig. XVI-8. The wavelength of maximum emittance, λ_m, at temperature T obeys Wien's displacement law

$$\lambda_m T = 2.8978 \times 10^6 \text{ nm K} \tag{14}$$

while the total energy emitted per unit time per unit area is given by the Stefan–Boltzmann equation

$$J = \int_0^\infty J(\lambda)\, d\lambda = \frac{2\pi k^4}{15 c^2 h^3} T^4 = \sigma T^4 \tag{15}$$

where $\sigma = 5.6697 \times 10^{-8} \text{ W m}^{-2} \text{ K}^{-4}$.

---

† The wavelengths shown in Fig. 14 are those measured in air. Conversion to a vacuum wavelength or to a vacuum wavenumber value, $\bar{\nu}_{vac} = 1/\lambda_{vac} = 1/n_{air}\lambda_{air}$, can be done using the index of refraction for dry air, CO_2-free, at 1 bar, 15°C:

$$(n_{air} - 1) \times 10^8 = 6431.8 + 2,949,330(146 - \bar{\nu}^2)^{-1} + 25,536(41 - \bar{\nu}^2)^{-1}$$

where $\bar{\nu}$ must be inserted in units of μm^{-1}.[22,23] The correction $-\Delta\lambda/\lambda = \Delta\bar{\nu}/\bar{\nu} = \Delta n/n$ is about 0.028 percent over the visible–UV region; for the mercury green line, for example, $\lambda_{air} = 546.075$ nm corresponds to $\lambda_{vac} = 546.227$ nm. For more precise values, one must include a correction to the index of refraction n_{air} for the temperature, humidity, and CO_2 content of the air.[22,23] (The temperature correction can be obtained to adequate accuracy with the ideal gas law.)

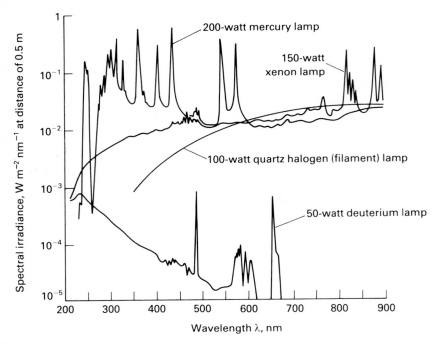

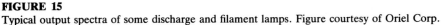

FIGURE 15
Typical output spectra of some discharge and filament lamps. Figure courtesy of Oriel Corp.

A source whose spectral emittance differs from that of a blackbody only by a multiplicative constant is termed a *graybody*. The proportionality constant is called the emissivity and denoted as ε. A tungsten source at 3000 K has an emissivity of 0.43 ± 0.03 in the 350–800 nm spectral range of typical use. The lamp emission can also be characterized by specification of a *color temperature,* which is the temperature at which a black body would have a spectral emittance closest in shape to that of the lamp. Tungsten filaments in ordinary light bulbs are operated at color temperatures of 2500–2600 K, while a xenon arc lamp has plasma emission that corresponds to a color temperature of 6000 K. To obtain higher intensities, a tungsten source can be operated at higher temperatures, albeit at a cost of shorter filament life. In practice, iodine is often added in quartz–halogen tungsten lamps to extend the filament life through the formation of volatile WI_2 near the quartz wall of the lamp. This compound then diffuses to the hot filament where it decomposes and redeposits the tungsten. A coiled-filament lamp of this type serves as an NBS standard of spectral irradiance and is available from commercial sources such as EG&G Inc.

A thermal source commonly used for infrared spectroscopy is the *globar,* a 6-mm-diameter rod of silicon carbide, typically operated at 1000–1500 K with an emissivity of 0.86 ± 0.04 in the 4000 to 600 cm^{-1} spectral region. Another resistively heated source is the *Nernst glower,* a smaller-diameter (1–3 mm) rod of semiconducting rare-earth materials such as ZrO_2, Y_2O_3, ThO_2, or CeO_2 which is operated at 1200–2000 K. Because of a negative temperature coefficient of resistance, it must be preheated to achieve conduction and must be used in series with a ballast resistor to prevent excessive currents and burnout. For the far-infrared region below $200\ cm^{-1}$, a mercury lamp with a fused quartz envelope gives higher intensities. The hot envelope serves as the source down to about $35\ cm^{-1}$; below this the mercury plasma contributes most of the emission.

Wavelength selection. Although most modern spectrophotometers now use a grating as a dispersion device, many of the older visible–ultraviolet or infrared spectrophotometers still in use employ prisms of quartz or flint glass or, for the infrared, salts such as LiF, NaCl, KBr, or CsI. A typical prism arrangement is shown in Fig. 16. The angular dispersion, D_a, of a polychromatic beam of radiation as it travels through a prism is due to the wavelength dispersion, $dn/d\lambda$, of the index of refraction

$$D_a = \frac{d\theta}{d\lambda} = \frac{d\theta}{dn}\frac{dn}{d\lambda} \tag{16}$$

The geometric factor $d\theta/dn$ is a function of the prism apex angle and, at a constant angle of incidence, changes only slightly with wavelength. Thus the resolving power of a prism is determined primarily by $dn/d\lambda$ as well as by the size (base length) of the prism. The refractive index variation for various materials is shown in Fig. 17.[26] It should be noted that the dispersion (slope) changes significantly with wavelength, with the greatest value occurring near the high- and low-wavelength limits of transmission determined by lattice absorption or by electronic absorption, respectively.

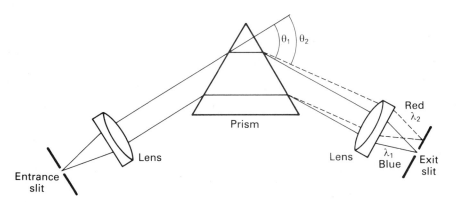

FIGURE 16
Schematic diagram of a simple prism spectrograph. Short wavelengths are refracted more than long wavelengths. The collimating lenses are multi-element achromats to provide a sharp image at the exit slit for a broad wavelength range.

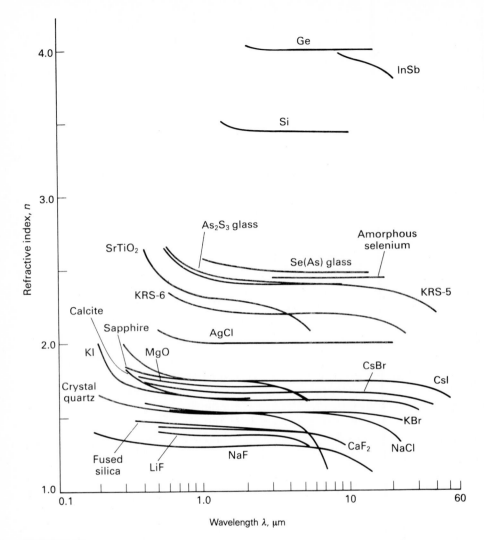

FIGURE 17
The refractive index for various optical materials. Figure from Ref. 26 with permission.

The dispersion of a grating is generally much higher than that of a prism and is more nearly constant with wavelength. The diffraction from a reflection grating is governed by the grating law

$$m\lambda = d(\sin \alpha - \sin \beta) = 2d \sin \theta \cos \phi \tag{17}$$

where d is the spacing between grating grooves and the angles are defined in Fig. 18. The order of diffraction, m, is an integer equal to ± 1, ± 2, ... for the dispersed orders of light; a value of $m = 0$ corresponds to undispersed specular reflection as for a plane mirror. For a fixed angle of incidence, the angular

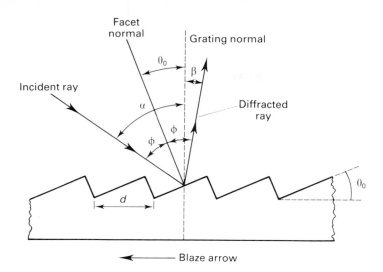

FIGURE 18
Cross section of a reflection grating with groove spacing d and blaze angle θ_0. Here α is the angle of incidence, β that of diffraction, and ϕ that of reflection. The blaze arrow, usually scribed on the back or top of a grating, indicates the direction (left or right) of enhanced diffraction of a beam incident along the grating normal.

dispersion of the diffracted beam is obtained by differentiation

$$\frac{d\beta}{d\lambda} = \frac{-m}{(d \cos \beta)} \qquad (18)$$

The resolving power of a grating thus is better in higher orders and for small groove spacings. It is also linear in the total number of grooves illuminated; i.e., it is proportional to the width of the grating.

At a particular angle of diffraction β, wavelengths for several orders will satisfy Eq. (17) but, since these are widely separated in magnitude, a simple bandpass filter of colored glass or other materials can be used to block undesired orders. Normally a grating is used in first order, since the reflection efficiency is highest there.

Ruled plane gratings are made by cutting regular grooves into an aluminum film on glass using a diamond tool. Line densities range from 20 grooves mm^{-1} for the far IR to 3600 or more grooves mm^{-1} for the VIS and UV ranges. Special interferometrically controlled ruling machines are required to produce master gratings over distances as great as 10 to 25 cm. Master gratings are expensive and most instruments use replica gratings that are of comparable quality. These are made by application of a parting agent such as silicone oil to a master, followed by evaporation of a film of aluminum. Epoxy resin is then poured onto the film, topped by a glass plate that bonds to the epoxy and provides a rigid support. After the epoxy is hardened, the replica is parted from the master, the oil is removed and an additional thin coat of aluminum is evaporated onto the grating. In the vacuum UV, the reflectivity of aluminum decreases and the grating efficiency is often improved by adding a thin layer ($\sim$30 nm) of MgF_2 to increase the reflectivity to about 70%.

To increase the reflection efficiency at a particular wavelength, the grating groove is usually cut at a *blaze angle* θ_0 whose value is chosen to be $\theta_0 = (\alpha - \beta)/2$, so that specular reflection from the groove facet and first order diffraction coincide. This concentrates most of the diffracted light at the corresponding blaze wavelength λ_B and provides high efficiency over a range from about $\frac{2}{3}\lambda_B$ to $2\lambda_B$. The reflectivity of a typical grating for the visible region is shown in Fig. 19 and it is seen that there is substantial variation depending upon the electric polarization of the incident light. To eliminate this dependence, the incident polarized light, such as occurs in Raman scattering, can be depolarized ("scrambled") by passing the radiation through a wedged quartz plate.

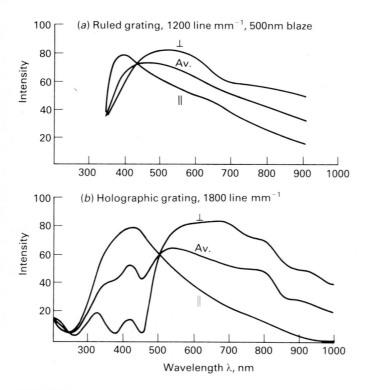

FIGURE 19
Diffraction efficiencies of ruled (*a*) and holographic (*b*) gratings for use in the visible region. The symbols $\parallel$ and $\perp$ correspond to light whose polarization is parallel (*P* type) or perpendicular (*S* type, *S* for senkricht) to the grating grooves. Also shown is the average which is characteristic of unpolarized light. Figure courtesy of Spex Industries, Inc.

Because of the difficulty of controlling the groove spacing during ruling, extraneous lines (grating "ghosts") and other defects sometimes occur in mechanically-ruled gratings. Since the late 1960s, an alternative method of production has involved the use of two collimated laser beams arranged to cause a pattern of interference lines on a substrate coated with a positive photoresist emulsion. After a few minutes' exposure, the emulsion is developed to produce a periodic groove pattern that is then overcoated with aluminum. Groove densities up to 6000 lines mm^{-1} are commercially available. The production of a well-defined blaze angle for the groove is more difficult with these *holographic gratings,* so the overall reflection efficiency tends to be somewhat lower than that of a ruled grating (see Fig. 19). However, since all grooves are formed at the same time, ruling errors are nonexistent and these gratings exhibit very low scattered light levels as well. Another special advantage of holographic gratings is that the grooves do not have to be straight lines. By interfering spherical wavefronts, groove patterns can be placed on concave substrates so as to eliminate the need for collimating and focusing optics in a spectrometer. In addition, the groove pattern can be designed to correct for spectrometer aberrations so that improved resolution with fewer optical elements becomes possible.

Radiation detectors.[26] Devices used for the detection of UV–VIS–IR electromagnetic radiation can be divided into two groups sensitive respectively to the *number* of incident photons (quantum detectors) or to the net *energy* of the beam (thermal detectors). In quantum detectors, photons interact directly with the electrons in the detector material and, if the energy of the photon is sufficient, cause a chemical reaction (photographic emulsion), a voltage or current change in the material (photovoltaic or photoconductive detectors), or actual ejection of an electron from the surface of the material (vacuum photocell or photomultiplier). Thermal detectors reveal the absorption of radiant energy through a change in a temperature-dependent property such as voltage, resistance, capacitance, or pressure and are generally slower and less sensitive than quantum detectors.

The use of photography in the 100–1000 nm spectral region has a long tradition in spectroscopy but is less common in modern instruments because of its low sensitivity and slow "readout" time with respect to most other photon detectors. We shall give no discussion of photography, as most of the techniques required for routine work are well known or given in instructions supplied with commercial photographic materials.[27] The principles of the photographic method are adequately described elsewhere.[28]

Photomultiplier tubes are the most common detectors used in VIS–UV spectrometers. These consist of a transparent evacuated tube containing a photoemissive material, the *photocathode,* plus a series of secondary electron emitters called *dynodes* that provide current gain. A simple tube with no dynodes is called a *phototube* or *photocell*; often gas is added to the tube to provide secondary electrons from collisions with the photoelectron as it is accelerated toward the anode by the tube potential. The photocathode is usually made of one or more alkali metals, a group V element such as P, As, Sb or Bi, and sometimes silver and/or oxygen. The formulation of the cathode is something of an art, the objective being to produce a material that absorbs all incident photons and that has the least resistance to photoemission. A very sensitive photocathode material of recent development employs GaAs and InP semiconductor substrates, coated with a thin film of Cs and CsO to lower the electron affinity below that of the substrate and hence to allow the electron to escape more easily. Figure 20 shows the spectral response curves of some typical photocathode materials.[29]

The first dynode in a photomultiplier is usually biased at a positive potential of about 75 to 150 V with respect to the photocathode, which has a potential of -1000 to -2000 V. An electron emitted from the photocathode is thus accelerated and, on impact with the dynode, generates additional

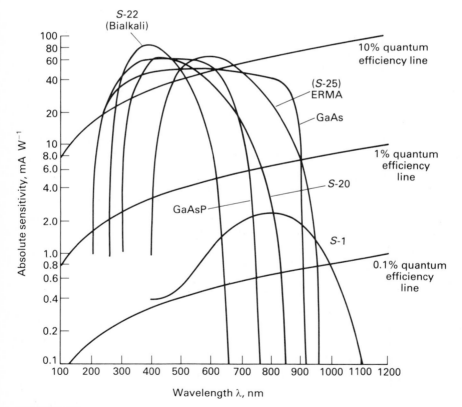

FIGURE 20

Wavelength dependence of the radiant sensitivity of various photocathode materials. S-1 = AgOCs, S-20 = Na_2KCsSb, S-22 = $KCsSb$ (bialkali), S-25 = $NaKCsSb$, also denoted as ERMA (extended red multi-alkali).

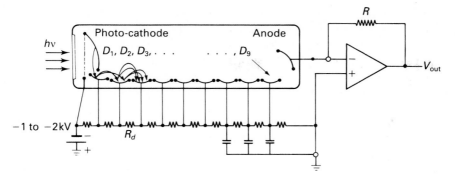

FIGURE 21
Typical connections for a photomultiplier tube. Secondary emission of electrons from 5 to 15 dynodes provides gains of up to 10^8. The dynode resistors are usually all the same and are about 50 kΩ. The capacitors on the last few dynodes store charge for improved operation with high current pulses. The output of the operational amplifier E_{out} is equal to the anode current i_a times the feedback resistance R.

electrons that are then accelerated toward the second dynode for further current amplification as depicted in Fig. 21. The actual dynode shape and positioning are designed to give the most efficient collection and redirection of electrons along the chain to the anode that delivers the current pulse to the measuring electronics. Common dynode materials are Be–Cu and Cs–Sb alloys, which generate 2 to 6 electrons per incoming electron. Dynode materials such as GaP can give gains of up to 40 at higher biasing voltages of 800 V.

Typically 5 to 15 dynode stages are used in photomultipliers to obtain gains that can be as high as 10^8. Each photon produces a current pulse of 5 to 20 ns duration because of the path differences traveled by the secondary dynode electrons. At pulse rates higher than 10^6 Hz, pulse coincidences become more frequent and this introduces counting errors. Thus direct-current measurement is preferred at high light levels. A photomultiplier should never be exposed to ambient or other high light levels with voltage on the tube, since destruction of the final dynode stages and even the photocathode can occur.

Because of their extreme sensitivity, photomultipliers are critical elements in applications involving low light levels, such as astronomy, emission spectroscopy (Exps. 40 to 42), and Raman spectroscopy (Exp. 37). Single-photon detection is possible and photon counting is the usual mode of operation under these signal-limited conditions. Even in complete darkness, some background thermal emission of electrons from the photocathode and the dynodes is always present, which can produce a dark count of 50 to 10^4 pulses per second. This emission decreases exponentially as the temperature is lowered and dark counts of 1 to 10 counts per second can be obtained for some tubes by cooling the photomultiplier to about −40°C thermoelectrically or to lower temperatures with solid CO_2 or liquid nitrogen. Some residual noise comes from cosmic rays and from nearby radioactive materials such as ^{40}K in the glass housing.

The short-wavelength limit for a photomultiplier is determined by the transmission of the window material covering the photocathode. Quartz can be used for tubes to be operated down to about 170 nm and tubes with LiF windows are available to permit operation down to 105 nm. In the 30 to 300 nm region, an ordinary photomultiplier can be used by coating the quartz or pyrex entrance window with a film of sodium salicylate which, upon absorption of UV light, fluoresces with nearly unit quantum efficiency at about 400 nm.

Although significantly less sensitive than photomultipliers, photodiodes also serve as useful detectors throughout the UV–VIS–near-IR regions. The most common photodiode consists of a $p-n$ junction formed on a silicon chip, as depicted in Fig. 22. The absorption of light with an energy in excess of the

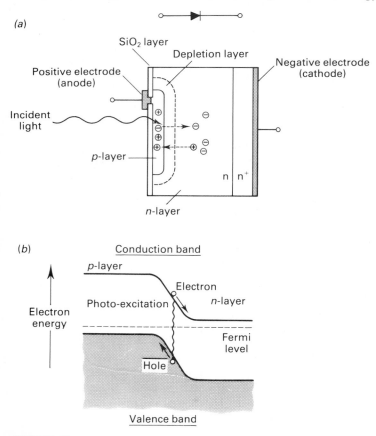

FIGURE 22

Cross section of a p-n photodiode (a) and potential energy diagram for the junction region (b). Absorption of a photon with energy in excess of the band gap produces charge separation of the resultant electron-hole pair to produce a voltage across the depletion layer.

band gap creates electron–hole pairs, which, owing to the internal energy barrier of the junction, separate to produce a potential difference across the junction. This produces a voltage proportional to the incident light intensity that can be measured directly without the need for a power supply. Typical response curves for some common photovoltaic detectors are shown in Fig. 23. The quantum efficiency of silicon can be as high as 90 percent, but the dark noise and noise from any subsequent amplification generally exceeds that of a photomultiplier. Silicon photovoltaic cells can deliver relatively large currents and are also used as panels for solar energy conversion. Photovoltaic cells made of selenium have found use in camera exposure meters.

Photovoltaic detectors are employed for low-noise, low-frequency applications. For faster and more accurate intensity measurements, it is usually best to operate the photodiode in a reverse-biased photoconductive mode. Here a positive voltage is applied to the n junction and the p junction is negative, creating a charge-depletion layer that reduces the conductance of the junction to a very low value. The creation of electron–hole pairs by light

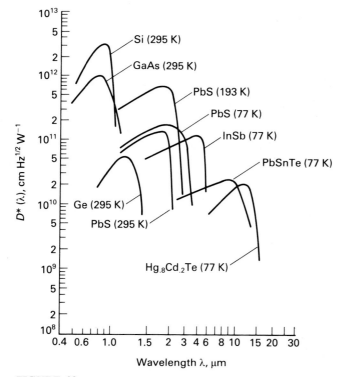

FIGURE 23
Typical detectivity D^* values as a function of wavelength for PbS photoconductive and various photovoltaic detectors. D^* is a figure of merit defined as $A^{1/2}/NEP$ where A is the detector area and NEP is the *noise-equivalent power*, the rms radiant power in watts of a sinusoidally modulated input incident on the detector that gives rise to an rms signal equal to the rms dark noise in a 1-Hz bandwidth. Data from Hughes Aircraft Company.

absorption then causes a current flow that is very linear in light intensity and that gives the device improved detectivity. The time response of a photodiode is very fast, less than 1 ns for a *pin* type diode which has a very thin insulating layer *i* between the *p* and *n* type semiconductor materials to permit operation at higher bias voltages.

Detector arrays of up to 2048 silicon diodes are commercially available and are used in some modern UV–VIS spectrometers to give an entire spectrum in a few milliseconds. Often these include an intensifier stage that consists of a sensitive photocathode whose surface is imaged to the array through a *microchannel plate,* a bundle of hollow glass fibers subject to a large voltage drop from one end to the other. Each fiber has an interior resistive coating for secondary emission and produces a gain of 10^3 to 10^4 in the electron pulse, which strikes a phosphor to produce a strong visible pulse of light detected by one of the diode elements. This signal is then much larger than the inherently high dark noise found in the silicon detector. Further discussion of silicon and other multichannel detectors can be found in Ref. 30.

A number of other solid-state materials, both pure (intrinsic) and doped with impurities to reduce the effective bandgap (extrinsic), have been developed for use in the near- to far-infrared regions. These generally require cooling to minimize noise from thermally generated charge carriers, especially for use in the far-IR. Lead sulfide is the most sensitive photoconductive detector for the near-IR region from 1 to 3 μm and can be operated at room temperature. InSb, operated in a photovoltaic mode at 77 K, is useful for the 3 to 5.6 μm range, while mercury–cadmium–telluride (MCT) detectors at 77 K are often used in commercial infrared instruments covering the 2.5 to 25 μm range. Figure 23 shows the spectral response curves for these and a few other detector materials. Most of these detectors are used for specialized application where speed and sensitivity are critical. Examples include pulsed laser measurements and remote sensing of surface temperatures by orbital satellites.

The most common sensing devices found in commercial infrared spectrometers are thermal detectors. Although generally less sensitive than quantum detectors, these devices have broad, flat spectral response curves (see Fig. 24) and most can be operated at room temperature. One of the earliest types of IR detectors, the thermopile, is still widely used. This consists of a multijunction thermocouple, each junction being composed of semiconductor elements or of metal pairs such as copper–constantan, bismuth–silver, or antimony–bismuth. An equal number of junctions is kept in the dark in the detector housing to provide a reference to compensate for changes in the ambient temperature. The irradiated junctions are blackened to increase the absorption and have small leads in an evacuated housing to minimize cooling by heat conduction through the leads and residual gas. The time response is consequently slow, in the milliseconds range; but a temperature difference of 10^{-6} K can be detected. The far-IR limit of the detector is determined by the window material and is about 400 cm^{-1} for KBr and 200 cm^{-1} for CsI. Polyethylene is often used as a window for the 10 to 600 cm^{-1} far-IR region.

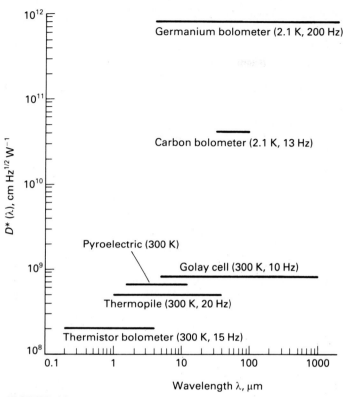

FIGURE 24

Typical detectivity D^* values for various thermal detectors assuming total absorption of incident radiation. See the legend of Fig. 23 for a definition of D^*. The operating temperature, modulation frequency and typical useful wavelength range are shown for each detector. Figure from Ref. 26 with permission.

A more recent type of detector offering greater sensitivity and much higher frequency response is the pyroelectric detector. This is essentially a small capacitor made of a ferroelectric material such as triglycine sulfate (TGS), its deuterated form (DTGS), or lithium tantalate. When placed in an electric field, a surface charge results from alignment of the internal electric dipoles of these materials. Heating caused by absorption of radiant energy causes reorientation of the dipoles and a consequent change in surface charge that produces a capacitance change, sensed as a current pulse. The detector is inherently an ac device and can only be used with chopped or pulsed sources. The output is a linear function of the input light intensity, and the time response can be as fast as a few picoseconds, since only charge reorientation is involved. In the infrared, the detector is blackened so the time constant is longer due to the time required for heat transfer into the bulk material. The DTGS pyroelectric detector is the most common room-temperature detector used in modern Fourier-transform infrared instruments.

The bolometer is another type of thermal detector that can offer extreme sensitivity for specialized applications.† This is essentially a resistance thermometer, usually with a platinum, nickel, carbon, or germanium element, although a semiconductor thermistor can also be used. Typically, two elements are used in a bridge circuit with one exposed to radiation and the other kept dark as a reference. The germanium bolometer provides exceptional sensitivity and is especially useful in the far-IR (see Fig. 24). Bolometer detectors have also found recent use in low-density molecular-beam experiments in which vibrational energy deposited in molecules by infrared laser absorption is detected as a temperature rise when the molecules strike the detector.

SPECTROSCOPIC INSTRUMENTS

Most spectroscopic measurements involve the use of an appropriate combination of source, dispersive device, and detector to analyze the absorption or emission spectrum of a sample. If only the wavelength or frequency of the radiation is measured, the resultant instrument is called a *spectrometer.* If the instrument provides a measure of the relative intensity associated with each wavelength, it is called a *spectrophotometer,* but this fine distinction is often ignored. Absorption spectra are often characterized by the *transmittance T* at a given wavelength; this is defined by

$$T \equiv \frac{I}{I_0} \tag{19}$$

where I is the intensity of light transmitted by the sample and I_0 is the intensity of light incident on the sample. When the sample is in solution and a cell must be used, I_0 is taken to be the intensity of light transmitted by the cell filled with pure solvent.

Another way of describing spectra is in terms of the *absorbance A,* where

$$A \equiv \log\left(\frac{I_0}{I}\right) = -\log T \tag{20}$$

The absorbance is related to the path length d of the sample and the concentration c of absorbing molecules by the Beer–Lambert law,

$$A = \varepsilon c d \tag{21}$$

where the proportionality constant ε is called the *absorption coefficient* (or, in older literature, the *extinction coefficient*). When the concentration is ex-

---

† The reputed sensitivity of the bolometer has inspired the following anonymous limerick:
  Simon Langley invented the bolometer,
  Which is really a kind of thermometer
  That can measure the heat
    of a polar bear's seat
  At a distance of half a kilometer.

pressed in moles per liter, ε is called the molar absorption coefficient. The quantity ε is a property of the absorbing material that varies with wavelength in a characteristic manner; its value depends only slightly on the solvent used or on the temperature. Quantitative absorption measurements are widely used in chemistry, and accurate determinations require careful calibration of instruments and cells to confirm the validity of the Beer–Lambert law over the concentration range of interest.[4,30]

Visible–ultraviolet spectrophotometers. There are many commercial spectrophotometers that are suitable for the experiments in this text. Often, for instructional purposes, greater insight and more flexibility can be obtained by assembly of an instrument from modular components, available from such sources as Oriel, PTR Optics, Schoeffel, Spex, or other companies. More commonly, an integrated instrument is used to provide greater reliability under use by relatively large numbers of students of varying backgrounds.

The simplest VIS–UV absorption instruments are relatively inexpensive ($1000–2000) single-beam devices such as the Bausch and Lomb Spectronic 20 and Turner 350 models shown in Fig. 25. In a simple single-beam instrument, a reference detector provides feedback to hold the source intensity constant. The light from a tungsten source is focused through a slit and is dispersed by a grating (or by a prism in older instruments) whose rotation by an external manual knob governs the spectral wavelengths that are reimaged through an exit slit. The Spectronic 20 has a particularly simple arrangement in which the beam is slightly convergent on a 600-line-per-mm grating, an economical compromise that provides a spectral bandwidth of 20 nm at the exit slit. The Turner 350 monochromator uses a single concave mirror as a collimating and imaging element, in a symmetrical optical arrangement called a *Fastie–Ebert* mounting, to achieve a resolution of 8 nm. Other designs add additional lenses or mirrors to produce a collimated beam for diffraction and provide resolution of 2 to 10 nm is monochromators of comparable size. The selected light then passes through a cylindrical or rectangular sample cell made of pyrex or polystyrene for the visible region, quartz for the UV range. The transmitted light intensity is measured with a gas-filled phototube or, in some instruments, with a solid-state silicon photodiode or a photomultiplier. The amplified signal is displayed on an analog or digital meter to give a linear representation of the percentage transmission. Scanning versions of a single-beam instrument provide this signal for display on a monitor or strip-chart recorder as the grating rotation angle is varied to generate a spectrum.

The Varian-Cary 219, 2200, and 2300 instruments, whose optical schematic is shown in Fig. 26, are examples of a much more sophisticated double-pass absorption spectrophotometer with multiple sources and detectors to cover a 185–3150 nm scan range. A swing mirror is used to select between tungsten and deuterium source lamps. A filter wheel serves to eliminate overlapping spectral orders from the diffraction grating. The latter is two-sided and can be rotated 180° so as to provide better diffraction efficiency over the broad

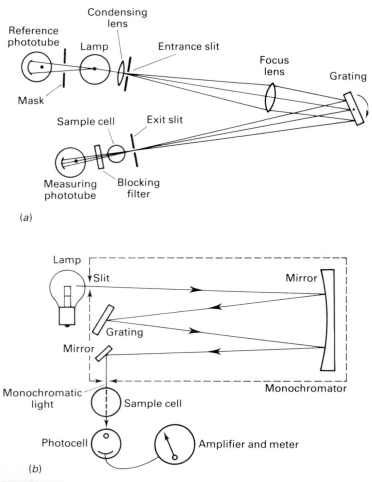

(a)

(b)

FIGURE 25

Two examples of simple single-beam spectrophotometers. (a) Spectronic 20 from Bausch and Lomb; (b) Turner 350 from Sequoia-Turner Co.

spectral range. All slits are curved for highest resolution (0.07 nm) in this *Czerny–Turner* monochromator configuration and special optics match the source image to this shape. The exiting beam is sent alternately through sample and reference cells by a rotating half-sector mirror and is then detected by a phototube or, for the 900–3150 nm region, by a lead-sulfide detector. Simpler versions of such a double-beam instrument employ a single-pass monochromator with 1–2 nm resolution. A spectrum is acquired by recording the detector signal as the grating is rotated or, in some recent instruments, by holding the grating fixed and positioning a multi-element diode array detector in the focal plane containing the dispersed radiation.

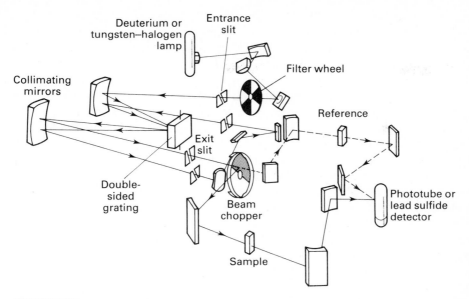

FIGURE 26
Optical diagram of Varian-Cary models 219, 2200, 2300 double-beam spectrophotometers from the Instrument Division of Varian Associates.

The Czerny–Turner configuration is also used in many emission or Raman instruments. A typical double monochromator arrangement for Raman spectroscopy is shown in Fig. 27. The excitation source is usually the 514.5 nm or the 488.0 nm line of an argon-ion laser at a typical power of 0.1 to 2 W in a

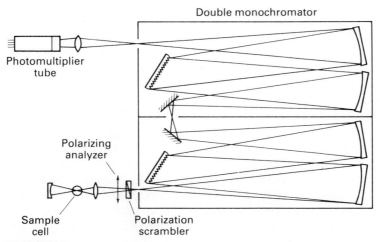

FIGURE 27
Schematic diagram of a Raman spectrometer. The exciting laser beam enters from below ($\perp$ to the plane of the drawing).

spectral width of $0.25 \, \text{cm}^{-1}$. The laser beam is focused into the sample by a lens (not shown in Fig. 27) to form a scattering cylinder approximately 0.1 mm in diameter and 10 mm in length. The scattered radiation is usually collected at an angle of 90° to the direction of the source beam by a large lens and is dispersed by the spectrometer. Since the Raman light intensity is quite low, stray Rayleigh light scattered off the slits, gratings and mirrors of the spectrometer is minimized by the use of holographic gratings and a double (or triple) monochromator. The gratings have 1800 to 2400 lines per mm for high dispersion and provide resolution down to $0.2 \, \text{cm}^{-1}$. A red-sensitive photo-multiplier tube is used as a detector in a photon-counting mode. The tube is cooled to −40°C for reduced dark count and a discriminator circuit serves to reject noise pulses whose amplitude differs significantly from that produced by a electron ejected from the photocathode by a Raman-scattered photon. Count rates of 10^4 to 10^6 pulses per second are easily obtained with solids or liquid samples such as benzene or carbon tetrachloride; for gases at 1 bar, count rates of 10^2 to 10^4 are more typical. In recording a spectrum, the count rate is converted to an analog signal that is displayed on a monitor or strip-chart recorder as the gratings in the spectrometer are rotated in a synchronous fashion. Vibrational and/or rotational frequencies are deduced from the Raman spectrum by measuring the wavenumber values of each peak and finding the shift from the known exciting frequency. The spectrometer is easily calibrated[31] by using some of the emission lines shown in Fig. 14. Additional details of Raman sampling techniques and depolarization measurements can be found in Exp. 37.

Infrared spectrophotometers.[32] Infrared spectroscopy is one of the most powerful tools for the quantitative and qualitative identification of molecules, and a large variety of infrared instruments are in current use. Most are double-beam spectrometers and the optical arrangement shown in Fig. 28 for a Perkin–Elmer model 1400 is representative. The radiation from the thermal source traverses sample and reference paths and, by means of a rotating-sector mirror, is imaged in an alternating fashion onto the entrance slit of the monochromator. First-order dispersion is provided by one or more gratings on a rotatable carousel. The light exits through a slit and higher orders of radiation are rejected by broadband filters. The desired radiation is imaged onto a thermopile detector and the resultant ac signal is detected with phase-sensitive electronics so that a separate measure of sample and reference beam intensities is obtained. A chopping frequency of about 13 Hz is common because it gives a good match to the time constant of the thermopile detector.

In older instruments, an optical nulling scheme is used in which a wedged combs is inserted into the reference beam until the example and reference beam energies are identical. A potentiometer attached to the wedge drive system provides a voltage that gives a measure of the beam transmittance. In more recent instruments, the energy is simply measured and ratioed directly,

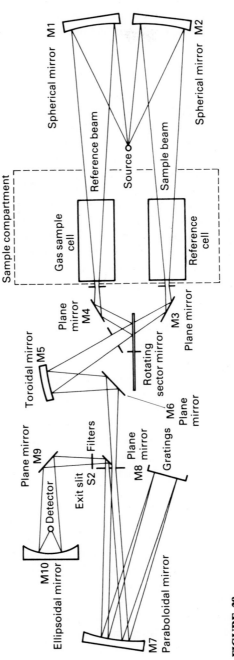

FIGURE 28

Optical diagram of the Perkin-Elmer model 1400 infrared spectrophotometer. Courtesy of Perkin Elmer Corporation.

giving better performance for strongly absorbing samples. Most instruments give a display that is linear in wavenumber $\tilde{\nu}$ rather than wavelength λ by using a cam to provide an appropriate rotation of the grating. Calibration to about $\pm 1\,\mathrm{cm}^{-1}$ can be checked with polystyrene films; for more critical measurements, vibrational–rotational transitions of H_2O, CO_2, CH_4, NH_3, and other standard gases are used.[33]

Routine instruments covering the 400 to $4000\,\mathrm{cm}^{-1}$ vibrational range with a resolution of 2 to $4\,\mathrm{cm}^{-1}$ are relatively inexpensive and are suitable for qualitative identification of materials and measurement of vibrational frequencies of liquids and solids and solutions of these in various non-aqueous solvents. Solids are often sampled as a dispersions in an oil such as Nujol (a mull) or as a suspension in KBr, prepared by pressing a powered mixture to form a pellet. Gases are examined in a simple cell with salt windows, such as that described in Exp. 36, where the vibrational spectrum of SO_2 is studied. If vibrational–rotational information is desired, as in Exp. 38 on HCl–DCl and in Exp. 39 on acetylene, an instrument of higher resolution is desirable. Grating spectrometers with a resolution of $0.5\,\mathrm{cm}^{-1}$ or better have been manufactured but these are now rapidly being displaced by Fourier-transform infrared (FTIR) instruments, which provide superior high-resolution performance at lower cost. The latter instruments also give much better results in the far-IR region from 10 to $600\,\mathrm{cm}^{-1}$.

Fourier-transform infrared spectrometer.[34,35] Most FTIR instruments are based on the Michelson interferometer configuration depicted in Fig. 29. A collimated beam of light from an infrared source is divided into two halves by a beam splitter, typically KBr coated with germanium to give 50 percent reflectance. Reflections from a fixed mirror $M1$ and a moving mirror $M2$ are recombined and imaged onto a detector element that gives the net intensity. As one of the mirrors is moved, interference between the two beams occurs, which, for a monochromatic source of wavelength λ, produces a periodic signal as depicted in the figure. The detector goes through one cycle for a mirror movement of $\lambda/2$ and hence the frequency f of this oscillation is given by $f = 2v/\lambda = 2v\tilde{\nu}$, where v is the mirror velocity. For infrared light in the 400 to $4000\,\mathrm{cm}^{-1}$ range (1.2 to $12 \times 10^{13}\,\mathrm{Hz}$), a typical mirror velocity of $0.05\,\mathrm{cm\,s}^{-1}$ gives $f = 40$ to $400\,\mathrm{Hz}$, well within the response time of the pyroelectric and other infrared detectors commonly used in FTIR instruments.

For a polychromatic source, the signal is a sum of such cosine waves, with all adding constructively at the zero-path-difference point where the two mirrors are equidistant from the beam splitter. At other distances, the waves interfere and the detector signal, the *interferogram,* drops rapidly as shown in the figure. The interferogram is thus a sum of cosine waves, each of which has an amplitude and frequency proportional to the source intensity at a particular infrared frequency. Recovery of this desired information is achieved by performing an Fourier transform, a process greatly aided by the use of a computer and a fast-Fourier-transform algorithm developed in recent years

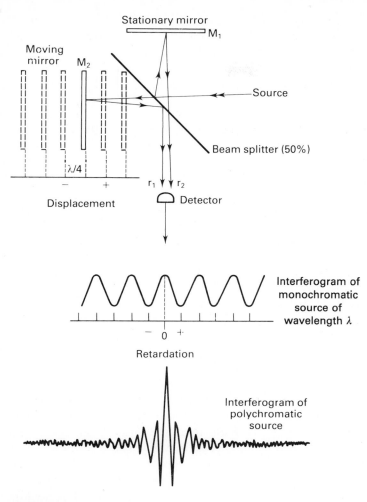

FIGURE 29
Schematic diagram of a Michelson interferometer. The detector signal variation as a result of mirror motion is displayed for the cases of monochromatic and polychromatic sources.

(Cooley–Tukey procedure).[34] The resolution is determined by the total mirror travel L as illustrated in Fig. 30 and is essentially $\Delta\bar{\nu} = 1/2L$; one commercial instrument (Bomem) provides a resolution of 0.0025 cm^{-1}.

Figure 31 shows the optical configuration of a commercial FTIR instrument, the Mattson Sirius 100, in which corner-cube mirror reflectors are used in place of plane mirrors. These "cat's eye" reflectors serve to reduce the sensitivity of the mirror alignment and make it possible to use a simpler mirror drive. The position of the mirror is determined precisely by using the fringes produced by interference of a helium–neon laser reference beam. Since the

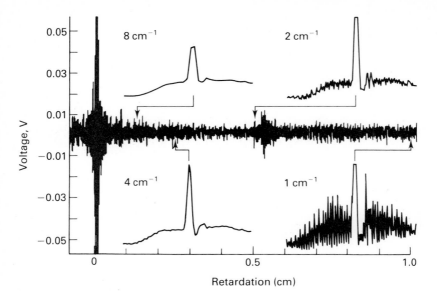

FIGURE 30

Interferogram and transformed spectra for acetylene in the 670–800 cm^{-1} v_5 bending region. The transformed spectra show the effect of retardation distance on the spectral resolution; the 1 cm^{-1} case corresponds to a total mirror travel of 0.5 cm. The vertical scale of the interferogram is greatly expanded; the peak-to-peak voltage at zero retardation is ~4 V.

frequency of the laser is accurately known, the infrared frequencies are determined to a few hundredths of a wavenumber and routine calibration with gas standards is unnecessary. This high precision also facilitates signal averaging over many scans and comparisons between spectra. Most FTIR instruments are single-beam devices, requiring separate measurements of the source background intensity and the signal when the sample is inserted. In some commercial instruments, a double-beam capability is provided by mirrors that oscillate to redirect the beams between sample and reference cells.

Many FTIR spectrometers offer options to extend the coverage into the far-IR and, more recently, into the near-IR and even VIS–UV regions. In general the signal-to-noise ratio is much higher for FTIR instruments than for grating spectrometers, and this allows the use of shorter scan times. This improvement is largely due to two major advantages of an FTIR over a grating instrument: higher beam intensities due to the elimination of slits and a multiplex advantage arising from the fact that all source frequencies are monitored simultaneously. The gain in signal-to-noise from the latter advantage is $N^{1/2}$, where N is the number of resolution elements to be examined. This is particularly helpful for high-resolution studies; for a scan of 4000 cm^{-1} at 0.1 cm^{-1} resolution, the advantage is 400. More detail on these instrumental aspects of FTIR, on the transform process itself, and on sampling techniques can be found in Refs. 34 and 35.

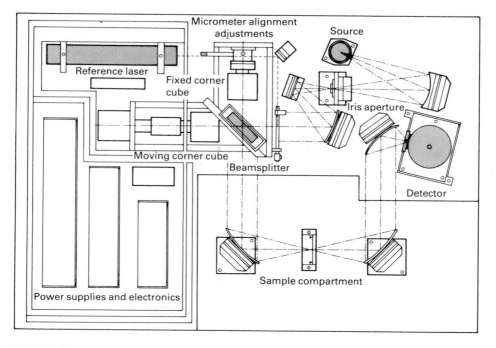

FIGURE 31
Optical diagram of the Mattson Sirius 100 FTIR instrument. Courtesy of Mattson Instruments, Inc.

TIMING DEVICES

Inexpensive quartz-crystal digital stopwatches are convenient devices for time interval measurements of durations ranging from ten seconds to several hours, and they offer accuracies of 0.0001 to 0.01 percent. The principal error in the use of these is the reaction time of the experimenter who is manually operating the start and stop buttons. Reaction times vary greatly from one individual to another, but a reasonable estimate of the error in a time interval for a (sober) operator would be ±0.2 s.

More accurate time-interval measurements require electronic or mechanical triggering of the start–stop points. For example, the heating period in a calorimetric experiment can be timed automatically by using a fast double-pole switch to control both the timer and heater circuit simultaneously. Most electronic counters have a time-interval mode of operation in which successive electronic trigger pulses can turn a high-frequency (10–100 MHz) counter on and off with an accuracy of ±1 count. For the most precise timing, a high-speed electronic frequency counter can be used to count the oscillations of an ultrastable crystal-controlled oscillator. Hewlett-Packard offers several instruments capable of making time measurements accurate to within a few parts in 10^8. The international standard of time at present is the cesium atomic-beam magnetic-resonance "atomic clock," which is stable to a few parts in 10^{12}.

WESTPHAL BALANCE

The Westphal balance is an instrument for measuring the density or specific gravity of a liquid by application of the principle of Archimedes. Although it is not usually capable of the very high accuracy obtainable with a pycnometer (Exp. 9), it is easier and more rapid to use. It is far more accurate than a hydrometer.

The Westphal balance measures gravimetrically the buoyancy exerted on a glass-enclosed body of definite volume immersed in the liquid. This body is an elongated glass bulb weighted with mercury and containing a thermometer. It is suspended by means of a slender wire from one arm of a special balance. The volume of the test body is carefully adjusted to some definite value, say 5 mL, by grinding the glass at the bottom. A detailed description of the operation of this device is given elsewhere.[36]

REFERENCES

1. D. A. Skoog and D. M. West, "Fundamentals of Analytical Chemistry", 4th ed., Saunders, Philadelphia (1982).
2. I. M. Kolthoff, E. B. Sandell, E. J. Meehan, and S. Bruckenstein, "Quantitative Chemical Analysis," 4th ed., chap. 19, Macmillan, New York (1969).
3. R. M. Schoonover, *Anal. Chem.* **54,** 976A (1982).
4. D. A. Skoog and D. M. West, "Analytical Chemistry," 4th ed., Saunders, Philadelphia (1986).
5. R. M. Schoonover and F. E. Jones, *Anal. Chem.* **53,** 900 (1981).
6. R. Batlino and A. G. Williamson, *J. Chem. Educ.* **61,** 51 (1984).
7. T. W. Lashof and L. B. Macurdy, "Precision Laboratory Standard of Mass and Laboratory Weights," Natl. Bur. Stand. Circ. 547, U.S. Government Printing Office, Washington, D.C. (1954).
8. G. W. Thomson and D. R. Douslin, "Determination of Pressure and Volume," in A. Weissberger and B. W. Rossiter (eds.), "Techniques of Chemistry: Vol. I. Physical Methods of Chemistry," part V, chap. 2, Wiley-Interscience, New York (1971).
9. W. G. Brombacher, D. P. Johnson, and J. L. Cross, "Mercury Barometers and Manometers," Natl. Bur. Stand. Monogr. **8,** U.S. Government Printing Office, Washington, D.C. (1960).
10. H. V. Malmstadt, C. G. Enke, and S. R. Crouch, "Electronics and Instrumentation for Scientists," Benjamin-Cummins, Reading, Mass. (1981).
11. J. Diefenderfer, "Principles of Electronic Instrumentation," 2d ed., Saunders, Philadelphia (1979).
12. C. C. Westcott, "pH Measurements," Academic Press, New York (1977).
13. D. J. G. Ives and G. J. Janz (eds.), "Reference Electrodes," Academic Press, New York (1969).
14. H. H. Willard, L. L. Merritt, Jr., J. A. Dean, and F. A. Settle, Jr., "Instrumental Methods of Analysis," 6th ed., chaps. 21, 22, Van Nostrand, Princeton, N.J. (1981).
15. W. Heller and H. G. Curme, "Optical Rotation—Experimental Techniques and Physical

Optics," in A. Weissberger and B. W. Rossiter (eds.), "Techniques of Chemistry: Vol. I. Physical Methods of Chemistry," part IIIC, chap. 2, Wiley-Interscience, New York (1972).

16. C. Djerassi, "Optical Rotatory Dispersion," McGraw-Hill, New York (1960); K. P. Wong, *J. Chem. Educ.* **51,** A573 (1974); **52,** A9, A89 (1975).

17. G. Friedlander, J. W. Kennedy, W. S. Macias, and J. M. Miller, "Nuclear and Radiochemistry," 3d ed., Wiley, New York (1981).

18. S. Z. Lewin and N. Bauer, in I. M. Kolthoff and P. J. Elving (eds.), "Treatise on Analytical Chemistry," vol. 6, part 1, chap. 70, Wiley-Interscience, New York (1965).

19. C. W. Mathews and K. N. Rao, "Tables of Standard Data," in K. N. Rao and C. W. Mathews (eds.), "Molecular Spectroscopy: Modern Research," chap. 8, Academic Press, New York (1972).

20. G. R. Harrison, "M.I.T. Wavelength Tables," M.I.T. Press, Cambridge, Mass. (1969).

21. A. R. Striganov and N. S. Sventitskii, "Tables of Spectral Lines of Neutral and Ionized Atoms," IFI/Plenum, New York (1968).

22. B. Edlin, *J. Opt. Soc. Am.* **43,** 339 (1953); H. Barrell and J. E. Sears, *Trans. Roy. Soc.* **A238,** 1 (1939).

23. G. Strey, *Spectrochim. Acta* **25A,** 163 (1969); K. Burns, K. B. Adams, and J. Longwell, *J. Opt. Soc. Am.* **40,** 339 (1950).

24. R. A. Alberty, "Physical Chemistry," 7th ed., chap. 11, Wiley, New York (1987).

25. P. W. Atkins, "Physical Chemistry," 3rd ed., chap. 13, Freeman, New York (1986).

26. J. H. Moore, C. C. Davis, and M. A. Coplan, "Building Scientific Apparatus," chap. 4, Addison-Wesley, Reading, Mass. (1983).

27. See, for example, Kodak publications CIS 51, "Kodak Materials for Emission Spectrography" (1982), and P-140, "Characteristics of Kodak Plates for Scientific and Technical Applications" (1984).

28. T. H. James, "The Theory of the Photographic Process," 4th ed., Macmillan, New York (1977).

29. R. J. Pressley, (ed.), "Handbook of Lasers," CRC Press, Boca Raton, Fla. (1971).

30. J. D. Ingle and S. R. Crouch, "Spectrochemical Analysis," Prentice-Hall, New York (1988).

31. S. B. Kim, R. M. Hammaker, and W. G. Fateley, *Appl. Spectrosc.* **40,** 412, (1986).

32. J. E. Stewart, "Infrared Spectroscopy," Dekker, New York (1970).

33. A. R. H. Cole, "Tables of Wavenumbers for the Calibration of Infrared Spectrometers," 2d ed., Pergamon, Oxford (1977).

34. P. R. Griffiths and J. A. de Haseth, "Fourier Transform Infrared Spectrometry," Wiley-Interscience, New York (1986).

35. R. J. Bell, "Introductory Fourier Transform Spectroscopy," Academic Press, New York (1972).

36. N. Bauer and S. Z. Lewin, "Determination of Density," in A. Weissberger and B. W. Rossiter (eds.), "Techniques of Chemistry: Vol. I. Physical Methods of Chemistry," part IV, chap. 2, Wiley-Interscience, New York (1972).

CHAPTER
XIX

MISCELLANEOUS
PROCEDURES

Physical chemistry laboratory work involves many manual arts, techniques, and procedures in addition to those described in earlier chapters. In this chapter we shall deal with some of the more important of these.

VOLUMETRIC PROCEDURES

Several experiments in this book require either the preparation of solutions of precisely known concentrations or successive dilutions of a solution to obtain a series of solutions with known concentration ratios. For the benefit of those without previous knowledge of the necessary volumetric techniques, we shall present a summary of the more important aspects of volumetric methods. Considerably greater detail is available in many standard textbooks of quantitative chemical analysis.[1]

Volumetric apparatus. Volumetric glassware is of three principal kinds: (1) volumetric flasks, (2) pipettes, and (3) burettes; see Fig. 1.

The *volumetric flask* has a ring engraved around its neck; when the bottom of the liquid meniscus is level with this ring, at the designated temperature, the volume of liquid contained is that indicated by the engraved label. Frequently the letters "TC," meaning "to contain," are present on the

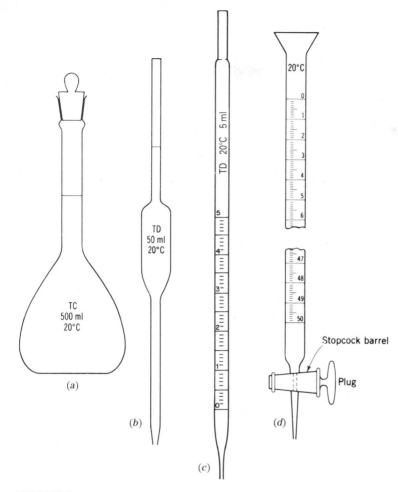

FIGURE 1
Typical volumetric glassware: (*a*) volumetric flask, (*b*) transfer pipette, (*c*) measuring pipette, (*d*) burette.

label. When such a flask is used "to deliver," it should be allowed to drain for at least 1 min; even so, its accuracy is not so high when used to deliver as when used to contain.

The technique for using a volumetric flask will be illustrated by a description of the proper way to make up a standard solution from a weighed amount of a solid. Once the solid has been introduced into the clean flask with the aid of a funnel, solvent is added to rinse out the funnel and rinse down the neck of the flask. A sufficient amount of the solvent is then added to dissolve the solid; shaking and swirling is most efficient when the flask is only about half full. When the solid has dissolved, additional solvent is added until the liquid level is about 2 cm below the calibration mark. The flask is stoppered, and the solution is mixed well by repeated inversion and shaking. The flask is then placed upright and the stopper is removed and rinsed down with a small amount of solvent. The flask is allowed to drain for a minute or two, and the neck is inspected to make sure that there are no clinging drops above the calibration mark. Finally the liquid level is raised until the bottom of the meniscus is even with the calibration mark by addition of solvent dropwise from a pipette or a dropper. The flask is then stoppered, and the solution is mixed thoroughly.

It should also be noted that graduated cylinders are often used for rough measurements of liquid volume. When used to deliver a given volume, the smaller sizes are not at all accurate. When used to contain, the larger graduated cylinders may be accurate to within ~3 percent. Graduated cylinders are much more convenient than volumetric flasks when solutions of approximate concentration are made up by dilution of concentrated reagents.

The *transfer pipette* has a single ring engraved on its upper stem, and its label frequently contains the letters "TD," meaning "to deliver." The pipette is filled by drawing liquid well above the ring using a pipetting bulb. **As a safety precaution, never draw liquid into a pipette by mouth**. Slip off the pipetting bulb and close the top of the upper stem with a finger tip. By careful manipulation of the finger tip the meniscus is allowed to fall to the ring. On delivering the contents into a vessel, the pipette is held at an angle of about 35° from the vertical and the tip is held against the wall of the vessel to prevent spattering. When the meniscus stops falling, the pipette should be held for about 10 seconds to allow drainage and then withdrawn after again touching the tip to the wall of the receiving vessel but *without* blowing out the liquid held in the tip by capillary action.

Measuring pipettes are essentially small burettes without stopcocks. As with burettes, the need for adequate drainage time cannot be ignored. Because of the double reading involved in the use of a measuring pipette and the additional manipulation required they are neither as precise nor as accurate as transfer pipettes, but they are quite satisfactory for semiquantitative operations.

Micropipettes are made in sizes from 0.1 μL to 500 μL (0.5 mL) in several designs. Some are simply capillaries that fill completely by capillary action and are blown or rinsed out. Others are capillaries with a single calibration mark that are filled beyond the mark by capillary action and then drawn down to the mark by touching the tip carefully to absorbent paper. Very often micropipettes are essentially miniature measuring pipettes fitted with a "pipette control," usually an uncalibrated hypodermic syringe or a threaded piston in a cylinder, which permits suction or pressure to be applied to adjust the level of the liquid with high precision. Micropipettes can be used with considerable precision if sufficient care is exercised, but the importance of droplets left on the outside of the tip, internal cleanliness, and uniformity of technique is even greater than with pipettes of ordinary size.

The *burette* is a cylindrical tube of uniform cross section, graduated in volume units along its length. A stopcock and drawn-out tip are attached at the bottom. A burette with bad chips in the tip should be replaced or repaired. Burette stopcocks are made in a variety styles; those having a smooth glass barrel and a fitted Teflon plug are the most desirable. The set nut that holds the Teflon stopcock in position must be tightened sufficiently to avoid leakage but not overtightened lest the plug be deformed or the threads damaged. The traditional ground-glass stopcock requires careful lubrication with an appropriate stopcock grease.

In use, the burette should be mounted vertically and firmly by means of a clamp that allows the entire burette scale to be visible from the front. With the burette graduations facing the operator, the stopcock handle is generally at his right. A right-handed individual should learn to operate the stopcock with his left hand in order to leave his right hand free for swirling, manipulating a wash bottle, etc. The crook between the thumb and forefinger fits around the burette at the top of the barrel, and the handle is grasped with the thumb and first two fingers. With practice very fine control can be obtained this way, and a fraction of a drop can be delivered if necessary.

In reading the burette, avoid parallax error. The line of sight should be accurately perpendicular to the burette axis; the rings engraved all the way around the burette will help. Read the *bottom* of the meniscus to about a tenth of the smallest division. Some help in seeing the meniscus can be obtained from a white card containing a very broad horizontal black line, held behind the burette so that the line is reflected in the meniscus. Move the card upward until the top edge of the line and the bottom edge of its reflection become tangent in your field of view; read the position of the point of tangency.

The burette is filled with the required solution (after two or three rinsings with small portions of the same solution; see below) to a point above the zero graduation. Solution is run out into a beaker until the meniscus has dropped to zero or some position on scale. Be sure that no air bubbles are present, especially in the tip, and that no drop is hanging when the initial reading is made. The burette is now ready to deliver solution. After running out any significant volume, allowed at least 30 seconds for drainage from the burette walls before reading. Also, before reading, remove any hanging drop by touching the tip against the wall of the receiving vessel. The volume delivered is the difference between initial and final readings, subject to calibration corrections and temperature corrections if needed.

For the delivery of small volumes, particularly of highly volatile materials or materials susceptible to the atmosphere, the use of a *hypodermic syringe* is recommended. Good-quality syringes may be read with reasonable accuracy or the amount of liquid discharged may be found by weighing. Pyrex syringes are reasonably resistant to most reagents, but some care should be exercised with the choice of needles. The use of ordinary stainless-steel hypodermic needles as delivery tips with syringe burettes has a number of hazards. Although stainless steel tends to be quite resistant to many reagents, the needles are brazed into the shanks with an acid-susceptible alloy. Teflon needles are available commercially and are to be preferred.

Accuracy and calibration. The National Bureau of Standards has established specifications[2] concerning the shape, dimensions, graduations, and capacity tolerances of precision volumetric glassware. Various capacity tolerances are given in Table 1. Volumetric flasks that meet these specifications are designated "Class A" and are usually so marked. Precision pipettes, with a serial number and certificate, can also be obtained. The tolerances of most commercially available glassware are greater than those listed in Table 1 by approximately a factor of 2 or 3.

Whenever very reliable volumetric measurements are required, the glassware should be calibrated, preferably at 20°C. Volumetric flasks (TC) are best calibrated by weighing them empty and then filled with water. Pipettes are calibrated by filling with water, delivering their contents into a weighed glass-stoppered weighing bottle, and then reweighing. A burette is filled with water, and 10 to 20 percent at a time is run out into a weighed glass-stoppered weighing bottle or flask, which is stoppered and weighed after each addition (without being emptied in between); a calibration curve (required correction plotted against burette reading) is then prepared, analogous to a thermometer calibration curve. In each case the correct density of water at that temperature $(0.99823 \text{ g cm}^{-3}$ at 20°C) must be used in the calculations, and air-buoyancy corrections (see Chapter XVIII) are applied. Further details on calibration procedures and calculations are given elsewhere.[1,2]

TABLE 1
National bureau of standards specifications for capacity tolerances of precision volumetric glassware.

Capacity in mL less than and including	Volumetric flasks		Pipettes	
	Calibrated to contain (TC) ± mL	Calibrated to deliver (TD) ± mL	Transfer (TD) ± mL	Measuring (TD) ± mL
1	0.01	—	0.006	0.01
2	0.015	—	0.006	0.01
5	0.02	—	0.01	0.02
10	0.02	0.04	0.02	0.03
25	0.03	0.05	0.03	0.05
50	0.05	0.10	0.05	
100	0.08	0.15	0.08	
200	0.10	0.20	0.10	
300	0.12	0.25		
500	0.20	0.40		
1000	0.30	0.60		
2000	0.50	1.00		
Above 2000	1 part in 4000	1 part in 2000		

The NBS has specified 20°C as the *normal temperature* for volumetric work. The cubical coefficient of expansion of Pyrex is about $0.9 \times 10^{-5} \, K^{-1}$; that of water at 20°C is about $2.1 \times 10^{-4} \, K^{-1}$. The expansion of glass will be of importance only in very precise work; that of water, however, will affect molar concentrations and will be of significance if the actual temperature is more than about 5°C removed from 20°C.

Cleaning of glassware. Volumetric work of any quality depends upon glassware with a surface clean enough to be wetted uniformly by water and aqueous solutions; if the meniscus pulls away from areas of the glass leaving dry spots, the glass requires cleaning. If cleaning with an ordinary laboratory detergent (such as trisodium phosphate or an organic sulfonate) is not sufficient, a chromic-acid cleaning solution may be needed. About 10 g of $Na_2Cr_2O_7$ is dissolved in the minimum quantity of hot water, and after cooling, about 200 mL of concentrated sulfuric acid is slowly added with stirring. The cleaning solution must be kept in a glass-stoppered bottle. If after much use it appears greenish, it should be replaced. **Contact of chromic acid cleaning solution with any organic material** (e.g., wood, cloth) **or with the skin must be carefully avoided**. Glassware to be cleaned is ordinarily filled with this solution, which may be moderately warm but should not be hot. Stopcocks should be dismantled and degreased with solvent beforehand; glassware should be reasonably dry to avoid dilution of the solution with water. The solution will attack most fillers and pigments used to fill graduations of burettes and other volumetric glassware; confine the solution to the inside surfaces as much as possible. After 15- to 30-minutes of contact, this solution should be poured back into the storage bottle and the glassware should be thoroughly rinsed with distilled water. The clean glassware is usually allowed to drain if it not to be used immediately; burettes, however, are often filled with distilled water and covered by an inverted beaker pending use.

Rinsing. Volumetric glassware need not be dry before filling with the appropriate aqueous solution if the vessel is first rinsed two or three times with the solution. Several *small* portions are more effective than the same total volume of solution in a single portion.

Titration. In titration with a standard solution, the amount of solution required to react quantitatively with a given sample is measured with a burette. A description of a burette and its proper manipulation has been given on p. 769. In order to determine the equivalence point (the point at which exactly stoichiometric quantities of sample and titrant have been brought together), it is necessary to find a chemical or physical property that changes very rapidly at this point. Many properties have been used successfully, but the most common method is the visual observation of a color change in a chemical indicator present in very small concentration. This observable change takes place at the so-called "end-point," which must lie very close to the equivalence point. The technique of titration is principally concerned with approaching the end point with reasonable speed without "running over"; it is best learned by practice but there are descriptions in the literature[1] that may be helpful.

Preparation and storage of solutions in large quantities. In a physical chemistry laboratory experiment that is performed by many students, many liters of each solution are normally required. For the preparation and storage of such solutions, 18-liter carboys (such as those used for commercial handling of acids) are convenient. Reagents are weighed out on an open-pan balance or measured out with a graduated cylinder and introduced into the carboy, water is added to fill the carboy, the contents are well mixed, and samples are withdrawn by pipette for standardization by titration. The carboy is clearly labeled and placed in the laboratory for student use. A convenient arrangement is shown in Fig. 2.

If the solution must be made up to a precisely specified concentration, a 2000-mL volumetric flask can be used one or more times. If a very large quantity is required, the concentration of the solution in a carboy can be adjusted by two or three progressively smaller additions of reagent or of water each followed by a titration.

A special word should be said about *mixing,* particularly in the case of large volumes. It will not suffice merely to swirl the carboy for a few minutes. When the carboy is filled, enough air space (1 or 2 liters) should be left to permit effective mixing. Most convenient for thorough mixing is an electric stirrer, comprising a motor with a long shaft and a swivel-mounted propeller that will go through the mouth of the carboy. This will produce adequate

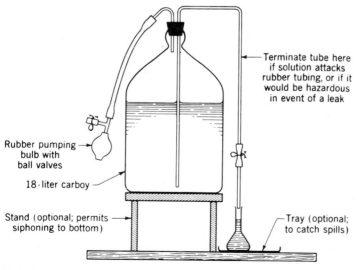

FIGURE 2
Carboy configuration for dispensing large quantities of solution.

mixing in 15 to 20 min. If such a device is not available, the carboy should be tightly stoppered and turned onto its side on a table covered with a towel. It should then be *vigorously shaken* by a rapid back-and-forth rolling motion sufficient to distribute air bubbles throughout the volume. This should be continued for at least 1 or 2 min. After the carboy has been allowed to stand upright for a few minutes, samples should be withdrawn both from the bottom and from the top for titration.

PURIFICATION METHODS

Mercury. Reagent or triple-distilled mercury is available commercially and can be used without further treatment for virtually all purposes in a physical chemistry laboratory. Through ordinary use mercury becomes dirty and should be returned to the stockroom in exchange for new reagent mercury. If the contamination is merely a surface accumulation of dust and oxides, sufficient cleaning can be accomplished in the laboratory by "filtering" the mercury through a piece of filter paper in which a small pinhole has been punched. The small quantity of mercury that does not flow through the pinhole will support all the scum and can be poured off into a flask for proper disposal. *Mercury vapor is toxic.* Mercury spills should be meticulously cleaned up. Where inaccessible, mercury droplets may be covered with a fine dusting of sulfur to retard evaporation.

Water. Ordinary distilled water is pure enough for most purposes. In many situations, elimination of ionic impurities is the important goal and the presence of neutral organic substances in trace amounts is not objectionable. Therefore, deionized water obtained from columns containing ion-exchange resins will serve well for almost all purposes. Dissolved air is objectionable for some applications and can be removed by boiling for a short period.

For conductance work, ions other than those resulting from the ionization of water itself must be reduced to the lowest possible concentrations. If distilled water is used, it should be triply distilled (ordinary distilled water distilled a second time from dilute acidified permanganate to oxidize organic impurities and a third time with a block-tin condenser from dilute barium hydroxide to remove volatile acids and CO_2). Conductivity water should be stored in polyethylene bottles in order to prevent leaching of soluble constituents from the surface of a glass bottle. Exposure to air should be minimized to prevent contamination with carbon dioxide.

Organic liquids. Very brief comments are given here on the purification of a few common compounds. The principal contamination of *benzene* is thiophene, which cannot be removed from benzene by distillation but can be removed by treatment with concentrated sulfuric acid. Water can be removed by fractional distillation, a binary azeotrope coming over first. *Ethanol* cannot be freed of water by a simple fractionation, since an azeotrope with 5 percent water forms. Benzene can be added, permitting the water to be removed in a ternary azeotrope by fractionation. Alternatively, the 95 percent ethanol can be digested with calcium oxide to remove water and then distilled. For highest purity, this should be preceded by treatment with silver oxide to remove aldehydes. Absolute alcohol should be protected from exposure to air in order to prevent contamination by water. *Hydrocarbons* can be purified by extensive fractionation combined with treatments with molecular sieves.

Sodium hydroxide. For use as a reagent in acidimetric titration, sodium hydroxide must be freed from contamination by sodium carbonate, which forms rapidly on exposure of NaOH to air. Commercially obtainable reagent sodium hydroxide needs no further purification if available in stick form, since any carbonate that forms on exposure to air can be quickly washed off with distilled water before dissolving the sticks in CO_2-free water (distilled water, reboiled if necessary) to make up the desired solution. This probably cannot be done effectively with the usual pellets. However, sodium carbonate is virtually insoluble in a saturated solution (15 to 18 M) of sodium hydroxide. Such a solution can be made up by stirring the solid hyroxide with cracked ice and allowing the resulting hot solution to cool. This syrupy liquid can be stored in a polyethylene bottle. When the insoluble carbonate has settled, the clear hyroxide solution can be drawn off as needed by pipette or siphon and diluted with CO_2-free water to the desired concentration. Sodium hydroxide solutions should not be stored for long in untreated glass flasks or bottles. Polyethylene bottles are satisfactory.

Handling of sodium hydroxide solutions at the high concentrations involved here entails extreme hazard of the destruction of the cornea if any solution gets into the eye. **Wear a face shield. Carefully avoid skin contact**.

Other procedures. It is seldom necessary in physical chemistry laboratory work to purify chemical compounds beyond the purity attainable commercially. For special purposes such techniques as recrystallization, fractional distillation, chromatographic separation, and zone refining may be used.[3]

GAS-HANDLING PROCEDURES

Many of the experiments in this book involve the use of one or more gases such as oxygen, nitrogen, hydrogen, helium, argon, and carbon dioxide. We shall be concerned here with procedures for handling these gases.

Cylinders, gas regulators, reducing valves. Although some gases can be prepared in chemical generators, it is far more convenient and desirable to obtain gases commercially in steel cylinders. These cylinders are available in several types and sizes. The large size ordinarily used in the laboratory for oxygen, nitrogen, hydrogen, etc. is illustrated in Fig. 3. A typical cylinder is

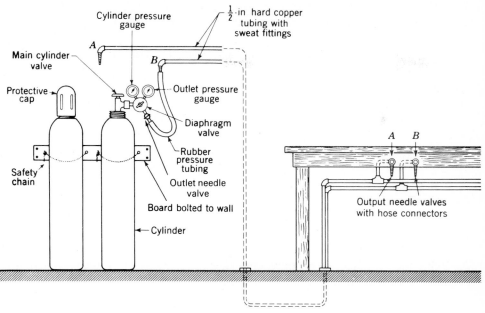

FIGURE 3
Design for laboratory gas lines.

51 inches in height and 9 inches in diameter; at 2200 psi and 70°F it contains approximately 280 moles of gas (~7000 liters STP). In addition, smaller cylinders and very small lecture bottles are useful for supplying special gases such as hydrogen chloride, acetylene, sulfur dioxide. Note that most gas pressure gauges read in psi (pound/square inch), and that 14.7 psi equals 1 atm.

Each cylinder is delivered with a protective cap, which should be removed only when the cylinder has been chained against a laboratory table or a wall. **Cylinders should always be chained** to prevent upset, which has been known to cause violent release of the gas or even bursting of the cylinder, with serious consequences.

At the top of the cylinder are a needle valve and a threaded outlet. To clear the outlet of dust the needle valve should be barely opened for an instant and reclosed. In most laboratory work a *regulator* is attached to the cylinder, as shown in Fig. 3. This usually comprises a Bourdon gauge to indicate the cylinder pressure (up to 4000 psi), another gauge to indicate outlet pressure (ordinarily up to 60 psi), an adjustable diaphragm valve to regulate the outlet pressure, and an outlet needle valve to control the flow rate. The diaphragm valve is a simple form of manostat (see p. 783) in which the needle of a needle valve is attached to a flexible diaphragm. When the outlet pressure exceeds the desired value (controlled by a spring between the diaphragm and an adjusting screw), the motion of the diaphragm is such as to close this needle valve, shutting off the flow of gas from the high-pressure side; when the outlet pressure is too low, the diaphragm opens this needle valve.

The fittings and threads on the cylinder outlet are of several types, depending on the kind of gas, in order to prevent the wrong regulator or other fitting from being used with a given cylinder; e.g., an H_2 regulator on an O_2 cylinder.

The regulator fastens to the cylinder with a metal-to-metal contact, no gasket ordinarily being required except in the case of CO_2 cylinders. The connection is made tight with a large wrench. To verify that the seal is gastight, close the diaphragm valve, then open and close the cylinder valve and watch the cylinder pressure gauge for a few minutes.

Regulator operation. When it is desired to use gas from the cylinder, follow the procedure given below.

1. Close the outlet needle valve and the diaphragm valve (turn screw handle counterclockwise until it rotates freely).
2. Open the cylinder valve fully and check the cylinder pressure gauge to make sure that the cylinder is not empty.
3. *Make sure that the outlet needle valve is closed,* and then slowly turn the handle on the diaphragm valve clockwise until the outlet gauge reads about 2 psi. Keep in mind that the outlet pressure gauge reads overpressure, i.e., psi *above* atmospheric pressure. 1 psi = 0.068 atm = 51.7 Torr.
4. In order to purge the system and the regulator chamber of air, open a vent valve and then *very slowly* open the outlet needle valve enough to achieve a reasonable flow rate. (If the gas is toxic or expensive, one can connect a vacuum pump to the vent valve and pump out the system and regulator.) After a brief purge period, close the vent valve, allow the system to fill, and then close the outlet needle valve. When one wishes to fill the system to an overpressure greater than 2 psi, *slowly* turn the diaphragm valve handle until the outlet pressure rises to the desired value.
5. When gas is no longer needed (or at the end of a laboratory period), close the outlet needle valve and the cylinder valve.

Do not allow a cylinder to be emptied down to 1 atm; there should be some overpressure left when the cylinder is returned. Unneeded cylinders should be returned to the supplier to avoid needless demurrage charges.

A simple *reducing valve,* consisting of a needle valve with fittings to match the cylinder and a hose connector for the low-pressure side, can be used in place of a regulator for some purposes when gas flow through an open system is desired. This device is less expensive than a regulator, but its use requires more care because if the system should accidentally be closed, the pressure will build up to whatever is required to burst it at its weakest point if this is less than the full cylinder pressure. It is advisable to connect the system to the reducing valve with a lightweight rubber tube that will easily blow off the hose connector if the pressure becomes too high, or to provide some other safety valve near the inlet end of the system.

Warning: Never connect a gas cylinder directly to a closed system that is not specially designed to withstand high pressures (such as a "combustion bomb" for heats of combustion). Even when a regulator is used, remember that the diaphragm valve may slowly leak; arrange for some kind of safety pressure release, if only a rubber tubing connection that can be easily blown off.

Most gases undergo a Joule–Thomson cooling when they expand in a regulator or reducing valve. In the case of CO_2 the cooling at high flow rates is often large enough to be troublesome, causing frosting of apparatus or even clogging of the regulator or the reducing valve with solid. In addition to this effect, there may be a compressional heating of the gas if a significant pressure is built up quickly in some part of the system. If a gas must be maintained at a constant temperature, it should be passed through a long coil of copper tubing immersed in a constant-temperature bath. One hundred feet of $\frac{1}{4}$-inch copper tubing is adequate for flow rates up to about $5 \, \text{L min}^{-1}$.

Needle valves. For control of gas flow at ordinary pressures, needle valves give much better control than stopcocks. The hard-steel tapered needle, at the end of a screw-threaded shaft, seats in a cylindrical hole so that the area of open space for gas flow is gradually increased or decreased on rotating the shaft. These usually do *not* provide a reliable shut-off and are used for flow control only. Persons whose acquaintance with valves is limited to water faucets often damage needle valves by needlessly overtightening them when shutting off the flow. This results in a decrease in the sensitivity of control of the gas flow. It is important not to exert any more force than necessary.

Gas-distribution lines. For an individual experiment, the cylinder may be chained at the laboratory table and the gas carried from the regulator to the experimental apparatus by a length of rubber or plastic tubing. If the same gas is needed simultaneously in several experiments, a gas-distribution line (see Fig. 3) is a great convenience. The cylinders are chained against a wall, and the gas is conducted through $\frac{1}{2}$-inch copper tubes to the laboratory tables, where they service as many outlets as are needed. Each outlet consists of a needle valve with a convenient knob and a nipple to which a length of rubber tubing may be attached. If two or more gas lines are available, they should be clearly distinguished by color coding. When a line is changed from one gas to another, it should be flushed out thoroughly before use.

Hoses. Gases for open systems may be carried by ordinary $\frac{1}{4}$- or $\frac{5}{16}$-inch gum-rubber tubing. Closed systems may require heavy-wall rubber or plastic pressure tubing, which can safely be used with pressures up to several atmospheres. **Warning:** Do not subject glass apparatus containing bulbs more than 2 or 3 inches in diameter to internal pressures of more than 1 atm above the outside pressure.

Gas purification. Although in many cases the gas from the cylinder is sufficiently pure for direct use, for certain purposes it should be subjected to one or more purification procedures. Hydrogen is frequently contaminated with small amounts of oxygen, which should be removed if the gas is to be used in a hydogen electrode. The most convenient procedure is to use a catalytic purifier such as Deoxo, which contains palladium; the oxygen combines with hydrogen to form water, which is subsequently removed with a drying tube if it is objectionable.

Another frequent contaminant is water vapor. This can be removed by passing the gas through a U tube filled with a suitable drying agent. A tube of this type is shown in Fig. 4-4. Gas flow should not be too fast, or the drying will be incomplete. Use of two or more drying tubes in series may provide better drying. A list of commonly used adsorbents for filling desiccators or drying tubes is given in Table 2.

Water saturation. Gases to be used in systems containing water or aqueous solutions should be saturated with water before they are admitted to the sample cell. For this purpose a bubbler containing a fritted disk that disperses the gas in the form of very small bubbles is far superior to the ordinary laboratory bubbler. The temperature of the bubbler should be the same as that

TABLE 2
Common drying agents

Agent (trade name)	Hydrated form	Comments
$Mg(ClO_4)_2$ (Anhydrone)	$Mg(ClO_4)_2 \cdot 6H_2O$	**Caution:** This is an oxidizing agent. Cloth and other organic materials impregnated with perchlorate can be flammable.
$MgSO_4$	$MgSO_4 \cdot 7H_2O$	Chemically inert. Commonly used for drying organic solvents; rapid, and high capacity.
$CaSO_4$ (Drierite)	$CaSO_4 \cdot H_2O$	Inert, very rapid and efficient (H_2O vapor pressure 0.004 Torr above hydrate at 25°C); low capacity. Available as "indicating Drierite", containing some $CoCl_2$ which changes from blue to pink when hydrated. Relatively expensive but regenerated by heating in air to 250°C.
$CaCl_2$	$CaCl_2 \cdot 6H_2O$	Inexpensive drying agent with large capacity. Will liquefy on absorbing sufficient water.
P_2O_5	$(H_3PO_4)_n$	Rapid and efficient agent as long as P_2O_5 surface is exposed (not coated with H_3PO_4). **Caution:** P_2O_5 is toxic.
Molecular sieves (Linde type 5A)	Pores filled	Useful for drying gases. Can be reactivated by heating.

of the system, and the connection to the system should be as short as possible to avoid condensation of water from the gas before it enters the system.

Flowmeters. A wide variety of instruments are available for measuring the flow rates of liquids and gases in closed-tube systems.[4,5] Four general types are (1) differential pressure devices, (2) variable area devices, (3) velocity meters, and (4) mass meters. Important examples of each type are listed in Table 3, which contains information on the range of liquid volume flow rates Q covered by a given style of instrument as well as the accuracy and range for any individual meter.

Differential pressure devices consist of two elements—one causes a *change* in the flow rate of a flowing fluid, which creates a pressure difference (Bernoulli effect) between two sections of the tube or pipe, and the second element measures the resultant Δp. Such a device is nonlinear, since Q is proportional to $(\Delta p)^{1/2}$. As a result, the range of Q values that can be covered by a given instrument is limited. A familiar example of a differential pressure device is the Venturi tube, which consists of a pipe tapered down to a straight

TABLE 3
Flowmeters for liquids and gases[a]

These devices usually measure the volume flow rate Q. The mass flow rate $\dot{m}$ is given by ρQ and is directly measured by mass meters.

Type	Range of Q (in L min^{-1})	Max / Min	Accuracy (reproducibility)	Viscosity effects	Gas use
1. Differential pressure					
Venturi tube	100–10k	4	1% FS	High	Y
Pitot tube	400–20k	5	1–2% R (0.1% R)	Low	Y
2. Variable area meter					
Rotameter	0–20	10(high Q) 100(low Q)	2% R (0.5% R)	Medium	Y
3. Velocity meter					
Turbine/vane	0.25–2.5k	15	1% FS (0.25% FS)	High	Y
Vortex	25–20k	10	1% R (0.2% R)	Medium	Y
Electromagnetic	0–5k	40	0.5–1% R	None	N
Ultrasonic	0–6k	20	2–3% FS (0.2% FS)	None	N
4. Mass meter					
Thermal	—	10	1% FS (0.2% FS)	None	Y
Coriolis	—	10	0.4% R	None	N

Notes. The ratio max/min indicates the range of Q measurable with a given individual instrument. FS = full scale, R = reading, Y = yes, N = no.

[a] Adapted from ref. 4.

"throat" (small-diameter section) and then flared out to its original diameter. Another variant is the Pitot tube, which measures the difference between the static fluid pressure at the downstream wall of the probe and the impact pressure where the flowing fluid impinges directly onto the pressure port (mounted in the upstream wall of the probe). In this device, $\Delta p \propto \rho Q^2$ and one needs to know the mass density ρ.

Variable-area devices, such as rotameters, also involve a differential pressure; but they represent a distinct type of device, since the differential pressure is constant during operation. In the rotameter, a small "float" is contained in a vertical glass tube having a tapered inside diameter (the cross-sectional area increasing with distance from the bottom of the tube).† When there is no flow the float rests at the bottom of the tube. When gas or liquid is flowing upward in the tube, the float rises until the net force due to the fluid flow just balances the force of gravity. Thus the steady-state height h of the float depends on the rate of flow. This device can be designed to yield an almost linear response (i.e., $h \propto Q$), and rotameters are available in a wide variety of flow-rate ranges below 20 liter min^{-1} for liquids or 600 liter min^{-1} for gases. Rotameters are delivered with a calibration for air or water; information about the density ρ and viscosity η is needed in order to recalibrate the device if other gases or liquids are used.

Velocity m ters measure the velocity v of fluid flow in a pipe of known cross section, thus yielding a signal linearly proportional to the volume flow rate Q. Mass meters provide signals directly proportional to the mass flow rate $\dot{m} = \rho Q$, where ρ is the mass density. Coriolis meters, which are true mass meters, can be used only for liquids. Thermal-type flowmeters use a heating element and determine the rate of heat transfer, which is proportional to the mass flow rate. This type of device is used mostly for gas measurements, but liquid flow designs are also available. For gases, a mass flow rate $Q_s = \dot{m}/\rho_s = (\rho/\rho_s)Q$ is frequently used. Since ρ_s is the gas density at STP (1 bar, 273.15 K), Q_s is a standard volume flow rate, usually expressed in units of liter(STP) per min (denoted as SLM). There are two styles of thermal mass meters: the heated tube and the immersion probe. In the heated tube design, all the gas passes through a sensor tube element with an external heater and a pair of precision temperature sensors. The accuracy is high (± 1 per cent full scale), but the flow rate range is low ($Q_s < 10$ SLM), the response time is 1 s, and very clean gases are required. Immersion probes are somewhat less accurate ($\pm 2\%$ of reading or $\pm 0.5\%$ of full scale), but higher flow rates up to 50 SLM and more rapid response (~ 100 ms) can be achieved even with "dirty" gas samples.

Figure 4 shows a simple gas flow manometer that can be easily constructed in the laboratory. In this device a pressure difference, caused by viscous flow in a capillary tube (see Exp. 4), is measured by a simple

---

† A related simple device in which a plastic ball moves around a circular course is often used as a semiquantitative water flow indicator to monitor the flow of cooling water through a diffusion pump or bath circulating coil.

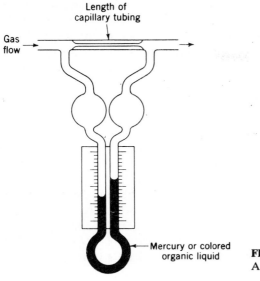

FIGURE 4
A typical flow manometer.

manometer containing mercury or a colored organic liquid (such as dibutyl phthalate with added eosin). The response of this flowmeter is very nearly linear over the range in which the gas flow in the capillary is laminar and may be roughly linear over a useful range even when the flow is turbulent. By variation of the capillary length and diameter and of the liquid density a wide range of flow rates can be measured.

A steady gas flow calibrating a flowmeter of this kind can be obtained from a needle valve attached to a regulator set to 5–10 psi gauge pressure or more (to avoid perturbations due to small variations in outlet pressure). At small flow rates the volume of gas flow over an interval of time measured with a timer can be determined with a gas burette; a three-way stopcock can be used to switch the gas burette in and out of the system. For larger flow rates a water-filled inverted graduated cylinder or volumetric flask, its mouth held under the surface of a water bath, can be used to collect gas from a rubber tube held underneath it for a time interval measured with a timer. For precise work a correction should be made for the partial pressure of water vapor in the gas collected.

Pressure gauges.[5,6] The most commonly used gauge for moderately precise measurement of pressures up to ~1000 bar is the *Bourdon gauge*. This gauge makes use of a curved metal tube of oval or flattened cross section; a higher pressure inside the tube than outside tends to straighten out the tube by a slight amount, producing a motion that is converted to a rotation of the indicating needle by a rack and pinion. In the best designs, the Bourdon tube is made of stainless steel and there is a mirror ring around the outer perimeter of the gauge dial so that the pointer can be read without parallax errors. Such gauges have an accuracy of ±0.25 percent of full scale. The typical Bourdon gauge measures pressure relative to the ambient air pressure and has a scale in PSIG (pounds per square inch gauge). Thus a reading of 14.7 psi means an overpressure of 1 atm or an absolute pressure of 2 atm. Some gauges have scales indicating absolute pressure in psi (PSIA) or kilopascal (100 kPa = 1 bar = 0.987 atm). For low-pressure measurements (0–600 Torr) on dry gases, there are direct-reading gauges using flexible bronze diaphragms, usually corrugated, with appropriate mechanical linkages.

For precise work, other devices may be used. The most familiar is the closed-tube mercury manometer (see Chapter XVII). For very precise manometer readings, the meniscus positions can be determined with the aid of a cathetometer (see Chapter XVIII). Clearly, the mercury manometer is convenient only for pressures that do not exceed ~1000 Torr (1.33 bar). Furthermore, manometer readings are time consuming and do not provide a convenient analog or digital output signal.

In recent years, electrical pressure transducers based on the use of strain gauges have been developed for use over ranges varying from 0–1 PSIG to 0–10 000 PSIG. The sensor is usually a thin diaphragm with a strain gauge mounted on it to detect deformations caused by a pressure differential across the diaphragm. The best stability is achieved using a stainless-steel diaphragm with a foil-type strain gauge bonded to it with epoxy cement (accuracy ±0.25 percent in the electrical output). This sensor is part of a complex compensation network needed to set the null (zero balance), correct for nonlinear characteristics of the strain gauge, and compensate for ambient temperature changes. The output can be displayed directly on an analog recorder or sent via an interface to a microcomputer. With microprocessor instruments, one can achieve conversion accuracy of ±0.05 percent full scale (including nonlinearity, hysteresis, and reproducibility errors) over ranges as small as 0–10 PSIG or as large as 0–10 000 PSIG. Moderate-cost instruments will perform almost as well, although with less flexibility in range selection and calibration checking. Within the last few years, piezoresistive silicon diaphragms have been introduced. In these integrated-circuit devices, piezoresistors are diffused into a homogeneous single-crystal silicon diaphragm. A bend in this elastic diaphragm causes a change in the resistance of the embedded resistor, and the resulting accuracy can be as good as ±0.1 percent. Optical pressure transducers seem likely to become widely used in the future. In such devices, the elastic deformation due to an applied pressure is detected optically, eliminating the friction-induced errors associated with conventional strain-gauge diaphragms.

Another type of electrical pressure transducer is a variable reluctance device such as those manufactured by Validyne Engineering Corp. The deflections of a diaphragm of magnetic stainless steel are sensed by two inductance pick-up coils. With ac-bridge circuitry, this transducer delivers a full-scale output of 40 mV per volt for a variety of ranges from 0–4 Torr (0–0.08 PSIA) to 0–3200 PSIA.

For pressures in the range 500–10 000 bar, wire-wound *manganin resistance gauges* are widely used. The resistance change with pressure is rather low ($d \ln R/dp \cong 2.4 \times 10^{-6}\,\text{bar}^{-1}$) but quite linear and stable. Absolute gas pressure measurements over a very wide range can be made with a *deadweight gauge*. This gauge measures the force exerted by a gas on a highly polished piston with a close sliding fit in a highly polished vertical cylinder, the mean diameter of which is known accurately. The effect of friction is largely eliminated by the presence of a lubricant and by a reciprocating mechanism that rotates the piston or the cylinder back and forth a few degrees around the vertical axis. The force due to gas pressure is balanced by weights placed on a pan supported by the floating piston. Pressures up to several thousand bar can be measured with high absolute accuracy (±0.01 percent). The deadweight gauge is used mostly for the calibration of more convenient secondary pressure gauges. Other pressure devices, such as capacitance manometers (which are treated in Chapter XVII), will not be discussed here.[5]

Manostats. A manostat is a device for maintaining the pressure in a system at a constant value. A gas regulator valve like that described on p. 775, operating in conjunction with an outlet leak, constitutes a simple manostat that prevents the system pressure exceeding a preset value. With the outlet open to the atmosphere, control pressures lie in the range 1–5 atm. More precise manostats, also operating on the feedback principle, can be based on mercury manometers. An electrical contact, made or broken when a mercury meniscus reaches a preset level, can be made to cause the opening or closing of a solenoid-operated valve. Pressure controllers can also be based on the use of bellows (0–300 bar), diaphragms (0.3–100 bar), or pistons (30–200 bar) as the sensing element. The motion of the sensor is then coupled to an electrical control device. In the simplest designs this is a single-pole double-throw (ON–OFF) switch, but proportionating controls can also be used.

Control of pressures less than 1 bar is sometimes desirable, as in a reduced-pressure distillation or in pumping on a refrigerant bath. Figure 5(*a*) shows a mercury manometer in which the control point is established by a needle making electrical contact with the mercury meniscus. The electrical signal is used to open or close either a valve connecting the system to the pump or a leak admitting air to the system. A controlled leak can be constructed from a simple electrical relay in which the armature, padded with rubber or neoprene, covers or uncovers the end of a capillary tube; see Fig. 5(*a*).

Less complicated and more ingenious devices can be made to control the pressure. Some of these depend on the fact that surface tension prevents liquid mercury from penetrating a fritted-glass disk without considerable applied pressure, while gas can penetrate it easily. The gas penetrates such a disk against a given hydrostatic head of mercury on the other side; see Fig. 5(*b*).

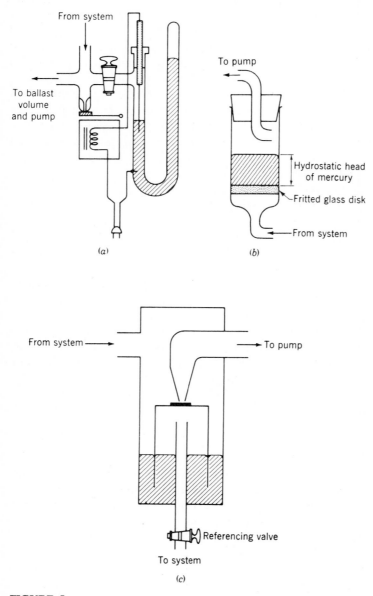

FIGURE 5
Manostats (schematic): (*a*) feedback manostat employing mercury manometer with adjustable wire contact and leak operated by a converted relay; (*b*) device using fritted glass disk and adjustable head of mercury; (*c*) Cartesian diver manostat.

A popular form of manostat makes use of the "Cartesian diver" principle; see Fig. 5(c). The "diver" is a float consisting of an inverted cup in a fluid such as mercury. The position of the diver depends on the difference between the system pressure and the reference pressure inside the cup (which can be adjusted to any desired value). When the system pressure is less than the reference pressure, the diver will rise until its top surface bears against the orifice to the vacuum pumping line. This cuts off the gas flow from the system until the pressure builds up again. When the system pressure becomes greater than the reference value, the diver sinks and the orifice to the pump is open. Thus the system pressure cycles back and forth over a narrow range centered at the preset value. These devices are available commercially from Greiner Scientific Corp.

Precautions. In addition to several general precautions given above for gas handling, certain particular precautions should be followed in handling hydrogen and oxygen.

Hydrogen is very flammable and also forms explosive mixtures with air over a wide range of compositions. If used in any quantity, the effluent hydrogen should be vented out of doors by a tube through a window. This is not essential for the slow rates of flow necessary with a hydrogen electrode, but good ventilation is important; there should be no open flames, and no smoking permitted.

Oxygen is hazardous when in contact with flammable substances, particularly oil. *No oil or other organic substance should be allowed to come into contact with oxygen under pressure.* Oxygen lines, valves, and regulators must be kept scrupulously free of oil.

ELECTRODES FOR ELECTROCHEMICAL CELLS

We describe below the electrodeposition procedures required in the preparation of the platinum and silver–silver chloride electrodes used in the electrochemical experiments described in this book. We shall not attempt to give a general treatment of electroplating or of electrode preparation. Further details can be found in various monographs.[7]

Platinum electrodes. These are used in conductivity cells and in hydrogen electrodes. For these purposes they are usually covered with a deposit of platinum black to increase the surface area. For certain other purposes, such as use in redox electrodes, this is usually not necessary.

Platinum can be used in the form of thin sheet or screen. Platinum wire can be welded to a small square of sheet or screen by placing it in position on a metal surface, heating to red heat with a torch, and striking lightly with a small ball peen hammer. The platinum wire can then be sealed into the end of a piece of soft-glass tubing, and electrical contact can be made through mercury placed in this tube.

Platinum can be plated onto other metals by use of a solution containing 1 g platinum diammino nitrate, 1 g sodium nitrate, 10 g ammonium nitrate, and 5 mL concentrated ammonium hydroxide in enough water to make 100 mL of solution. The cleaned metal is made the cathode, and a strip of platinum metal is the anode; a 4.5-V battery, a rheostat, and a milliammeter are connected between the electrodes, and the plating is carried out at a current density of 50 to 100 mA per cm^2 of cathode surface.

For depositing platinum black, a solution of 3 g platinic chloride and 0.2 g lead acetate in 100 mL of distilled water is prepared. This can be kept in a glass-stoppered bottle and used repeatedly. The platinum electrode should be treated with warm aqua regia (one part concentrated HNO$_3$ to three parts concentrated HCl) to clean the surface and, if necessary, to remove old platinum black. It is then rinsed thoroughly with distilled water. While still wet the electrode is immersed in (or the cell is filled with) the platinizing solution. If there is only one electrode to be treated, platinum wire will serve as the anode. A 3-V battery and a rheostat in series are connected between the electrodes, the electrode to be platinized being the cathode (negative). The rheostat is adjusted so that gas is produced only slowly. If both electrodes are to be platinized, the polarity is reversed every 30 s. The electrolysis should be stopped as soon as the electrodes are sooty black; an excessive deposit should be avoided. The platinic chloride solution is then returned to its stock bottle; the electrodes are rinsed thoroughly in distilled water, and electrolysis is continued with a very dilute solution of sulfuric acid in order to remove traces of chlorine. After a final washing with distilled water the electrodes are ready for use. Pending use they should be stored in contact with water; *they should never be allowed to dry out.*

Silver–silver chloride electrodes. These electrodes may be made from thin sheet silver of high purity, but it is perhaps better to plate silver onto a clean square of platinum sheet or screen. This is made the cathode and a strip of very pure sheet silver (at least 99.95 percent pure) the anode in a plating bath containing 41 g silver cyanide, 40 g potassium cyanide, 11 g potassium hydroxide, and 62 g potassium carbonate per liter. Use a 4.5-V battery connected in series with a rheostat and milliammeter. Set the rheostat to about 1000 Ω and make electrical connections to the electrodes *before* immersion; then adjust the rheostat so as to obtain a current density of about 5 mA per cm^2 of electrode area. Plate for a few hours. Remove the electrodes, and wash very thoroughly with distilled water to remove all traces of cyanides.

The silver-plated electrode is "aged" in an acidified solution of silver nitrate and is then made the anode (and a platinum wire the cathode) in a 1 M HCl solution at a current density of 5 to 10 mA per cm^2 of electrode area. In a few minutes a brownish coat of silver chloride will appear and the electrolysis can be stopped. The electrode should be aged for a few days in distilled water before use, and it *should never be allowed to dry out.* After long periods the potential of the electrode changes, probably owing to crystal growth. The old silver chloride coating can be removed with ammonia or cyanide; after washing, a fresh coating can be prepared.

In use, care should be taken to avoid exposure of a silver–silver chloride electrode to bromides. Best results are obtained when oxygen is removed from the solution by a stream of nitrogen.

MATERIALS FOR CONSTRUCTION

> "They should know enough not to build a bridge out of sodium."
> *Anonymous MIT Civil Engineering professor*

This section describes a wide variety of materials that can be used successfully in the construction and repair of laboratory apparatus. Considerable details about the construction of scientific apparatus are given in monographs[8] and specialized journals.[9]

Glass.[10] Glasses and glassblowing are discussed briefly in Chapter XVII in connection with vacuum systems. The transparency, low thermal conductivity, and chemical inertness (except to HF and strongly basic solutions) of glass make it a valuable material for constructing many pieces of apparatus. Because of its low thermal expansion coefficient, Pyrex glass can be blown into many complex shapes without too great a danger of cracking due to internal strains. A major disadvantage of glass is its breakability, but this problem may be reduced in the future by the application of Lexan polycarbonate resin to glass objects.

Glasses do not have a sharp melting point but soften over a broad temperature range. The ranges for typical glassy materials are:

soft glass (soda lime)	400–700°C
Pyrex (borate)	500–800°C
Vycor	800–1000°C
fused quartz (silica)	1200–1700°C

A glass object can be used for indefinite periods at the low end of its softening range but only for very short intermittent periods at the high end. Fused silica is obviously very useful for high-temperature apparatus; and Vycor, which contains a few percent of other oxides, is almost as good. At elevated temperatures, fused quartz will devitrify, i.e., crystallize and become opaque.

Soft glass has a higher expansion coefficient than Pyrex and is much more subject to thermal stress. It is mainly useful as a glass that will produce a vacuum-tight seal with platinum. It cannot be sealed directly to Pyrex, but "graded seals" are available commercially. Pyrex does not make satisfactory glass-to-metal seals with most metals (tungsten is an exception if it is cleaned with sodium nitrite in a flame); however, vacuum-tight Pyrex–Kovar seals are readily available. Various thermosetting cements, such as quartz cement and "solder glass" for soft glass, are also available commercially. Fused silica has high tensile strength and very low mechanical hysteresis; these properties make quartz fibers useful for certain applications. Glass transmits light from about 320 nm to 2 μm, thus covering the entire visible range. If ultraviolet transmittance is required, fused silica can be used for wavelengths down to 200 nm.

Metals.[11] Hot-rolled *steel* is the least expensive metal for construction purposes and is available in sheet, strip, rod, girder, and also in angle form, all of which are useful in the laboratory for constructing stands and frameworks. To prevent rusting, it should be galvanized (zinc-dipped), cadmium plated, or at least painted. Several commercial primers are effective for the retardation of rusting; they can be covered with additional coats of ordinary paint if desired. Cold-rolled (low-carbon) steel is also used for heavy-duty structural purposes, and owing to its excellent machining properties it has many varied uses in apparatus construction. It can be galvanized or plated with cadmium, copper, nickel, chromium, or any of several other metals. Copper should be plated on steel before nickel and both before chromium. Steel parts can be joined by welding, silver soldering, soft soldering, or copper brazing (if previously copper- or nickel-plated). High-carbon steels, such as tool steels (e.g., drill rod), are less easily machinable but have the advantage that they can be hardened by heat-treatment.

Stainless steels are in a special class, differing from other steels in being nonmagnetic and essentially free from rusting and corrosion. A typical stainless steel (type 316) contains 17 percent chromium, 12 percent nickel, and 3 percent molybdenum. The chemical resistance of stainless steel makes it very attractive for many purposes, but its high cost and the difficulty of machining it limit its use somewhat. It can be silver-soldered or welded; soft soldering is difficult.

Aluminum and its alloys are excellent structural metals, with good machinability, fair corrosion resistance, good electrical and thermal conductivity. It is readily sand-cast or die-cast. Aluminum is seldom plated; it can be anodized (to produce a thick oxide layer) and then dyed or painted. Aluminum is difficult to weld owing to its flammability. Soldering is also difficult but can be aided by tinning with indium metal. The low melting point of aluminum (660°C) somewhat restricts its application.

Aluminum and its alloys are often used in the laboratory in sheet form for making electrical chassis and panels and in rod form for constructing frames for apparatus support. Aluminum foil is useful for heat-reflecting shields in low-temperature work. A common structural alloy is Duralumin (Dural), which contains about 4 percent copper and traces of manganese and magnesium. Duralumin that has been heated to 530°C and quenched in water is ductile for $\frac{1}{2}$ hr or more and can be readily cold-worked; thereafter the alloy hardens and attains considerable strength. The hardening can be delayed for long periods (for rivets, etc.) by storage at "dry ice" temperatures.

Copper is used where its high electrical and thermal conductivity, its malleability and ductility, and its ease of soft soldering confer advantages. OFHC (oxygen-free, high-conductivity) copper should be used where highest conductivity is required or when employed in a vacuum system, particularly when soldering or welding is to be done in a hydrogen atmosphere. Soft copper is exceedingly difficult to machine. It is subject to oxidation and should not be heated above 100 to 200°C for any length of time except in a reducing atmosphere or unless adequately plated. Copper can be joined by brazing, silver soldering, or soft soldering. Copper amalgamates readily with mercury.

Brass is basically a copper–zinc alloy; *bronze* is a copper–tin alloy. In practice both often contain many other metals. Their high machinability, resistance to corrosion, and ease of soft soldering make them very useful in apparatus construction. Owing to the volatility of zinc, brass should not be used in vacuum components that must be baked out or operated hot. Certain bronzes such as phosphor bronze are useful for springs and diaphragms; beryllium copper is also useful in these applications.

Monel and *Inconel* are basically Ni–Cu–Co alloys containing small amounts of iron and manganese. These alloys have fair machinability and excellent corrosion resistance even at elevated temperatures; some are magnetic, others are not. They can be brazed, silver-soldered, and soft-soldered. Monel and cupronickel are very useful in the construction of apparatus for low-temperature work owing to their low thermal conductivities. Inconel can be used for heating elements; Nichrome (Ni–Cr or Ni–Cr–Fe–Mn) and Chromel (Ni–Cr–Fe) are also useful for this purpose.

Invar (64 percent iron, 36 percent nickel) has a very low coefficient of thermal expansion ($1 \times 10^{-6}\,K^{-1}$), which makes it a useful material for laser resonators and other devices in which dimensional stability is important. It is magnetic and only moderately corrosion resistant. *Kovar* (53.7 percent iron, 29 percent nickel, 17 percent cobalt, 0.3 percent manganese) has been mentioned in Chapter XVII as a glass-sealing metal.

Silver is an excellent conductor of heat and electricity and a good reflector of light. It is relatively immune to oxidation but becomes tarnished by exposure to sulfur compounds in exceedingly small concentrations. It is an excellent electroplating metal and can also be deposited in thin films by evaporation. In Dewar flasks and other vacuum glassware it is deposited from an aqueous medium by the Brashear process. Silver is an excellent brazing material and an important constituent of "silver solder." The term silver is often applied to alloys of silver with copper; for example, "Sterling" silver contains 7.5 percent copper. "Fine" silver is 99.9+ percent silver.

Platinum and *palladium* are useful because of their chemical inertness, electrical conductivity, high reflectivity, and high melting point; their very high cost restricts their use to applications in which only small amounts are employed: electrical contacts, suspension wires, heating elements, radiation shields, etc. They absorb hydrogen; palladium is very permeable to it and may be mechanically damaged by exposure to it. These metals easily spot-weld to themselves and to each other. *Gold* is also very useful because of its inertness. It is an excellent plating material and can be deposited by vacuum evaporation or chemical deposition.

Tungsten has the highest melting point of any known metal (3380°C) and is useful for heating elements and various vacuum-cell components. It is difficult to work and is usually handled in the form of wire or ribbon. Tungsten wires tend to have a fibrous structure; lead-through tungsten–Pyrex seals may not be vacuum-tight unless one end of the wire is coated with nickel. Tungsten spot-welds to itself and to nickel and tantalum. Tungsten wires will oxidize if heated in air. Molybdenum and tantalum are much more easily machinable and workable, are chemically resistant, and also have high melting points.

The mechanical properties of a number of commonly used metals are summarized in Table 4.

Polymeric materials.[12] There now exists a vast array of polymeric materials, both natural and synthetic. We shall concern ourselves here only with a few which are of particular usefulness in laboratory apparatus construction.

Rubber is the most commonly encountered of such materials. In its vulcanized form it is used in rubber tubing and rubber stoppers. A sulfur

TABLE 4
Mechanical properties of metals

Metal	Hardness (Brinell)	Tensile strength (units of 10^3 lb/in.^{2})	Young's modulus of elasticity (units of 10^6 lb/in.^{2})	Melting point (°C)	Density (g/cm^3)
Aluminum (pure)	16	8.5	8–11	660	2.7
Duralumin	125	88.2	10	~640	3.0
Brass (67 Cu–33 Zn)	145	66.8	13	940	8.4
Copper		33	18	1082	8.5
Cast iron	77	16	12–14	1200	⎫ 7.7
Mild steel	16	50–60			⎬ to
Carbon steel	400	200	28	1430	⎪ 7.9
Stainless steel	~500	250–300		~1500	⎭
Gold		36	11.4	1063	19.3
Silver		40	11.2	961	10.5
Platinum	64	48	24	1773	21.5
Monel	130–300	75–170	24–26	1350	8.9
Tantalum	105	50	27	2996	16.6
Tungsten	365	175	60	3410	19.4

coating may appear on the surface of such stoppers in the course of time. For tubing (other than pressure tubing), pure gum rubber is preferable, although it must be replaced more frequently. Gum rubber can be used up to ~100°C; vulcanized rubber is useful up to ~150°C.

Neoprene (du Pont) is a rubber-like material that is a polymer of 2-chloro-1,3-butadiene. Somewhat less flexible than natural rubber, it has greater resistance to oils, greases, hydrocarbon solvents, and other chemicals. Neoprene is useful for gaskets, O rings, and tubing.

A convenient substitute for rubber tubing in the laboratory is transparent vinyl-plastic tubing such as *Tygon,* a compounding of polyvinylidene chloride with certain liquid plasticizers. This tubing is tough and flexible and makes a very good seal with glass fittings. It tends to become yellow with age, and after long exposure to water it may become somewhat milky. It is attacked by many organic solvents but has fairly good resistance to most other ordinary chemicals.

Lucite and *Plexiglas* (polymethyl methacrylate as marketed by du Pont and by Rohm and Haas, respectively) and *polystyrene* are transparent thermoplastic materials. Their machinability is fairly good, but somewhat limited by their thermoplasticity. They are strongly attacked by solvents such as acetone. They can be cemented with solvents alone (trichloroethylene) or with such cements as Duco. Over long periods cracks may develop at points of strain, and discoloration may result from prolonged exposure to strong light.

Nylon (du Pont polyhexamethylenediamine adipic polyamide) is available in solid form as well as fiber and sheet. It has high strength and mechanical stability, excellent machinability, and low surface friction; it is excellent for bearings, small gears and cams, etc., in which it can be used with minimal lubrication or none at all. Nylon fibers and threads are useful in apparatus construction.

Teflon (du Pont polytetrafluoroethylene) is a somewhat more flexible solid that is virtually unsurpassed in chemical inertness, electrical insulating properties, and self-lubricating qualities. It is available in the form of rod, tube, tape, and sheet and is readily machinable. It is useful for gaskets and bushings, unlubricated vacuum seals for rotating shafts, etc. It can be used in dynamic high-vacuum systems if these are not baked out much above 130°C. **Caution:** When Teflon is heated to decomposition, it reportedly gives off fumes that are extremely toxic. After machining it, clean up all chips and scraps at once. *Viton* (du Pont polydifluorodichloroethylene) has properties very similar to those of Teflon and is used for O rings and greaseless stopcock bores.

Saran (Dow polyvinylidene dichloride) is a tough, chemically resistant plastic available in a variety of forms that are useful in the laboratory. Saran pipe or tubing can easily be welded to itself or sealed to glass and is useful for handling corrosive solutions. Thin Saran film, available commercially as a packaging material, is useful for windows, support films, etc. *Mylar* (du Pont polyethylene terephthalate) film and other polyester films are also useful for these purposes. Mylar is chemically inert and has excellent electrical properties for electrical insulation and for use as a dielectric medium in capacitors. Much thinner than these are films that can be made in the laboratory by allowing a dilute ethylene dichloride solution of Formvar (polyvinyl acetal) to spread on a water surface and dry.

Delrin (du Pont polyformaldehyde) has excellent chemical, mechanical, and electrical properties. It is easily machineable, tough, and resilient even at low temperatures. Although there are no good adhesives for joining Delrin pieces, it is weldable and easily molded. Chemically stable enough for continuous use at temperatures up to 85°C, it is also inert to most organic solvents.

Polyethylene and *polypropylene* are somewhat similar to Teflon but are inferior in chemical resistance and many other respects. They are useful in the form of bottles, flasks, and beakers for containing such reagents as hydrofluoric acid, strong bases, etc., that attack glass. Polyethylene tubing is much less flexible than rubber or Tygon but more flexible than Saran; it can be used for handling caustics, corrosive gases, etc. Polyethylene film has better chemical resistance than Saran and Mylar but lower strength and poorer optical properties.

Mechanical and chemical properties of a number of commonly used plastics are summarized in Table 5.

Thermal insulating materials. Vermiculite (expanded mica) is useful for insulating ovens and furnaces when the maximum temperatures are not too high. For high-temperature applications, alumina (Al_2O_3) or magnesia (MgO) in the form of compacted powders can be used for insulation up to 1550°C and 1675°C, respectively. For uncompacted powders, the upper limits are approximately 100°C higher. Braided fiberglass is suitable for use up to 480°C, and special high-temperature glass braids can be used up to 720°C. Teflon sleeves or tapes will provide insulation up to 260°C. Fabricated high-temperature parts can be made from boron nitride and other machinable ceramics.

At room temperature and below, foamed plastics are very effective insulators. Styrofoam, a rigid prefoamed material, is useful for making "dry-ice" cooling chambers and is very easily cut into any desired shape. Silicone foam is generally available in a soft and flexible form. Foam-in-place insulation is also handy for awkwardly shaped applications. Typically, this is based on a polyurethane resin containing a "bubbling agent," i.e., a pair of reactants which, when mixed, release N_2 or CO_2 gas into the polymerizing resin and thus produce a foam.

At very low cryogenic temperatures, the best insulation is a vacuum jacket that is silvered to eliminate radiative heat leaks. Various kinds of

TABLE 5
Properties of common plastics

	Plexiglas or Lucite	Saran	Polystyrene	Polyethylene	Teflon	Silicone rubber
Composition	Polymethyl-methacrylate	Polyvinylidene chloride	Polystyrene	Polyethylene	Polytetra-fluoroethylene	Organo-polysiloxane
High-temperature limit, °C	60	66	72	conventional, 60 linear, 120	260	320
Chemically resistant to:	Aqueous acids and alkalis, most oils	Aqueous acids and alkalis, most oils, some organic solvents	Aqueous acids and alkalis, alcohols	Aqueous acids and alkalis, most oils, hydroxylic solvents	Almost everything	Aqueous solutions, alcohols, ketones, some oils
Chemically sensitive to:	Alcohols, ketones, esters, aromatic hydrocarbons, halogenated hydrocarbons	Ketones, aromatic hydrocarbons, oxidizing acids	Most organic solvents	Hydrocarbon solvents, some ketones and esters	Almost nothing	Strong acids and alkalis, hydrocarbon solvents
Machinability	Good	Poor	Fair	Poor	Good	—
Tensile strength (units of 10^3 lb/in.^{2})	4–10	0.85–9	3–10	1.3	1.8	0.3–0.6
Young's modulus (units of 10^6 lb/in.^{2})	0.5	0.2–2.0	0.2–0.6	0.02	0.06	—
Other properties	High optical quality, high electrical resistivity, good mechanical strength	Fireproof (but gives off toxic HCl fumes), poor mechanical strength		Poor mechanical strength, flexible	Fireproof (but gives off toxic fumes), low friction coefficient	Poor mechanical strength (rubbery)

"superinsulation" exist and are used to store and transport volatile cyrogens like liquid helium. In one type, a fine insulating powder is placed in the vacuum jacket; in another type, a metallized variety of Kapton film is used.

Lubricants and coatings. Most of the lubrication problems encountered in the laboratory can be solved with SAE 10 or SAE 20 machine oil, except for ball bearings subject to heavy loads, for which cup grease should be used.

Teflon self-lubricating plastic bearings can be used in situations where a lubricant might contaminate the system. There are also a variety of nonoily lubricants such as silicone lubricant (usually dispensed as a spray) and molybdenum disulfide (suspended in petroleum distillates that evaporate to leave a dry coating). Molybdenum disulfide is an excellent alternative to graphite for lubrication of moving parts in a high-vacuum system.

Most of the materials that we have discussed—glass, stainless steel, aluminum, brass, plastics—require no protective coating. Parts made of mild steel, welded angle-iron, or plywood should be given a coat of paint, which is most conveniently done using du Pont Krylon acrylic aerosol sprays. Clear polymer varnishes and Teflon coatings can also be applied in spray or liquid form.

SOLDERS AND ADHESIVES

The joining together of various parts of an experimental apparatus is a frequently encountered task. This section describes two major approaches to this task: the soldering together of metal parts and the cementing together with adhesives of a wide variety of materials.

Solders. Soldering is the technique of joining two metals with a fused metal of lower melting point that wets both surfaces.[11,13] *Soft solder* is a low-melting alloy of lead, tin, and often other metals (e.g., bismuth or antimony). It is commonly used in making electrical connections, making connections between copper or brass tubes or fixtures by means of "sweat fittings" (such as sleeves, elbows, tees, etc.), assembling small parts of an apparatus, and occasionally assembling large parts (when reliance is not placed on it for much mechanical strength). It can be used for ordinary vacuum work but is not satisfactory for high-vacuum work, since its low melting point ($\sim$200°C) is lower than most bakeout temperatures. Indium solders are especially handy for low-temperature use ($\sim$120–155°C).

Silver solder is a common form of *hard solder*; it is an alloy of silver and copper, usually with zinc and sometimes tin or cadmium added. Silver solders melt in the range 600–700°C and have high mechanical strength.

The three requirements for successful soldering are clean surfaces, the correct flux, and sufficient heat. The first and crucial step is to clean thoroughly the parts that are to be soldered. A solder joint is easier to make and more reliable if the surfaces are initially clean and bright (free from dirt, oil, and oxide coatings). It is also important to use plenty of soldering flux. The function of this flux is to free the surface of any residual contamination and to protect both the surface and the solder from oxidation.

In electrical wiring, soft solder is commonly used with noncorrosive rosin fluxes and soldering pastes. Soldering of electrical connections with rosin-core solder is easiest when both parts to be joined have been "tinned," i.e., wetted with a coating of solder. Hookup wire is often pretinned, as are the leads of resistors and capacitors, the lugs on terminal strips, etc. Copper wire usually requires no tinning but may require scraping to remove Formvar or other lacquer coatings (which are often invisible) or oxide and dirt. A tinned soldering iron or soldering gun carrying a drop of solder is applied to the connection and held there until the solder spreads over the metal by capillary action; more solder can be applied if necessary, but the minimum required amount should be used.

Soft-soldering of copper, iron, steel, and brass objects of large size is usually accomplished by the use of a burner or hand torch, with an acid flux—frequently a concentrated aqueous solution of zinc chloride and ammonium chloride (2:1 ratio)—brushed onto the hot metal concurrently with the addition of solder. Initially, heat should be applied around the area to be soldered but not directly on the region to be soldered. If the region to be soldered is heated too much without the application of flux, an oxide coating will form that can make proper surface adhesion impossible. The surfaces to be joined should be pretinned if possible, and excess solder is shaken off or wiped off with a cloth. The two surfaces to be joined are then placed in contact and heated with the torch until the solder begins to flow; more solder is then added as required, and the pieces are allowed to cool undisturbed. The finished work should be washed thoroughly with water to remove the flux.

Hard-soldering, or silver-soldering, requires the use of a hand torch and a soldering paste of borax and boracic acid. A fluoride flux "Aircosil" can also be used. The pieces to be joined are placed in contact, and the surfaces are brushed with the flux during heating. Solder is applied as soon as the work is hot enough to melt it and encouraged to spread if necessary by further application of flux. Silver solders containing zinc or cadmium should not be used in vacuum systems that must be baked out or operated hot. **Caution:** Zinc and cadmium fumes are toxic; make sure there is adequate ventilation.

The cleanest and most reliable joints in vacuum systems are made by *hydrogen furnace brazing.* The parts to be assembled are clamped together with a thin gasket or sheet of the brazing alloy (silver or gold solder) between them. They are heated in a hydrogen atmosphere until the brazing alloy melts, runs, and wets both metal surfaces. No flux is required.

Welding techniques are many and varied[13] and will not be discussed in detail. The ideal construction material and joining technique for vacuum apparatus and many other systems involves stainless-steel parts connected by heliarc welding (on the *inside* of the joint). However, stainless steel is difficult to machine, and the welding must be done by an experienced welder. A completely different welding technique, called cold-welding, involves the use of pressure rather than heat. This requires a soft welding material that flows easily and "wets" the metal surface well. Indium, a soft metal that melts at 155°C, can be cold-welded by the application of modest loads with a handpress and makes a convenient vacuum-tight seal without heating the apparatus above room temperature.

Adhesives. The use of adhesives to cement together pieces of apparatus has become a common technique for a wide variety of situations. A general class of adhesive material is *epoxy cement,* consisting of a reactive epoxide resin liquid that must be mixed with a "curing agent" activator (generally an alkylamine) before being applied. These cements set irreversibly to form tough, adherent solids that bond well to metals (including aluminum, which is difficult to solder), glass, wood, and some plastics. Epoxy seals are relatively inert to chemicals and solvents and are usually poor conductors of heat and electricity. However, many special varieties of epoxy are available: thermally conducting epoxies, electrically conducting epoxies, optical epoxies that are transparent, vacuum epoxies with very low residual vapor pressures, metal-filled epoxies that can be machined. A wide range of strength and hardness can be achieved by varying the resin/activator ratio and the curing time. It is advisable to roughen the surfaces to be joined slightly with emery paper or sandblasting prior to applying the adhesive. The best seals are obtained when the gap is between 0.002 inch and 0.006 inch; use shims if necessary. It should be noted that epoxy resin cements sold in stationary stores often contain fillers and are not necessarily as good as those available for laboratory use such as Araldite (Ciba–Geigy Co.), F-88 (American Consolidated Dental Co.), and Tygoweld (U.S. Stoneware Co.). **Caution:** Epoxy resins should be considered to be toxic until thoroughly set; skin contact should be avoided.

A special-purpose adhesive that forms an exceedingly strong bond to metals and many other materials, including careless experimenters, is cyanoacrylate *contact cement,* available as Eastman 910 Adhesive and Techni-Tool Permabond. It is expensive, but only small amounts are needed. In this cement, the material polymerizes rapidly without an added activator. It is not void filling, and a 0.001 inch-thick film yields the best bond. Do not buy more than you will need in a few months since it slowly deteriorates; store in the refrigerator.

For affixing devices such as heaters, strain gauges, and thermometers to experimental cells, one wants a thin adhesive layer that is stable over the temperature range of projected use but need not be very strong mechanically. *GE varnish* (General Electric product no. 7031) is an excellent adhesive that is widely used over the range from helium temperatures (4 K = −269°C) to above room temperature (~150°C). A hot-curing epoxy adhesive EP 310, available from Omega Engineering,[6] is useful over the range −270°C to +310°C. Although expensive, it gives a strong and rigid bond that can meet the special needs of strains gauges.

Silicone rubbers, such as General Electric RTV, represent a useful type of flexible adhesive. They can be clear or opaque, are fairly inert chemically, and some are usable up to 300°C or above. These silicones are just poured or extruded from a tube, and no heat cure is required since they cure by air drying. The resulting seals are not mechanically strong but will provide protection against water and oil.

In cases where vacuum leaks cannot be properly repaired, sealants like *Apiezon W* and *Glyptal* can be useful. Apiezon W (black waxes of petroleum origin) softens or "melts" on warming to 50–150°C, flows readily on warm surfaces, and sticks well to clean glass or metal surfaces. It has a fairly low vapor pressure, but its exposure on the surfaces of vacuum systems should be kept to a minimum. Glyptal (General Electric glycol phthalate), a lacquer with or without added pigment and with a solvent such as xylene, is useful for a wide variety of vacuum-sealing applications. It adheres well to clean glass or metal and has a relatively low vapor pressure after baking. It should not be heated above 150°C. Glyptal hardens very slowly because it forms a surface film that retards evaporation of the solvent from beneath the surface. It should be allowed to dry for several days at room temperature or for 12 hours under an infrared lamp. For short-term emergency repair of a gross leak, the plumbing product Duct-Seal can save an experiment that might otherwise have to be terminated.

For high-temperature applications, *sauereisen cement* (Omega CC cement) and *zinc oxychloride* (dental cement) are useful irreversible cements. Sauereisen cement is made by suspending ceramic powders in sodium silicate solution ("water-glass"). This cement sets very hard and withstands temperatures up to 1000°C. Zinc oxychloride is made by mixing calcined zinc oxide powder with concentrated zinc chloride solution. One can also use a ceramic putty (Omega CR 760), which must be cured at 180°C and is then serviceable up to 850°C.

TUBING CONNECTIONS

It is often necessary to seal joints between dissimilar materials against leakage. A number of materials are available for doing this.

Metal tubing is often terminated in a pipe thread. This is an accurately tapered and threaded unit, and an assembly must be well tightened with a pair

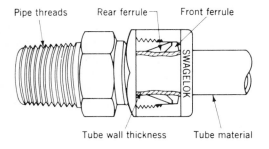

FIGURE 6
Typical Swagelok compression fitting.

TABLE 6
Recommended materials for Swagelok fittings

Tubing	Threaded body	Front ferrule	Rear ferrule	Nut
Glass	Brass or stainless	Teflon	Nylon	Brass or stainless
Teflon	Teflon	Teflon	Stainless	Brass
Polyethylene	Brass	Nylon	Nylon	Brass
Tygon	(Either metal, Nylon, or polyethylene fittings can be used, but a special tube insert must be included.)			

of wrenches in order to seat the joint. Small leaks can be sealed by wrapping the inner member with a thin film of Teflon tape (available as Strip-eeze or Tape Dope) before assembling the joint.

Various types of compression fittings are used as an alternative to pipe threads. The Swagelok fittings (Fig. 6) and Gyrolok fittings of similar design are more expensive than the standard types but can be disassembled more easily. They can also be used to couple wide varieties of tubing by using the combinations of materials shown in Table 6. It should be emphasized that inner and outer elements of the different types of fittings are *not* interchangeable.

Ultra-Torr fittings (Fig. 7) seal by means of an O ring, usually made of Viton, and are especially useful in making glass–metal–plastic seals in simple vacuum systems. Viton is a rubber-like material with a very low vapor pressure, and Viton O rings can be baked to ~200°C. Teflon is also occasionally used as a material for O rings.

Glass pipe and fittings are available for applications where clean inert surfaces are required or where one wants to see what is going on inside the system. Pyrex "Double Tough" pipes, tees, crosses, and reducers are made by Corning Glass. The sections may be connected together or to metal apparatus using special flanges and O rings. Pieces with 2-, 3-, or 4-inch diameters are the most convenient. Teflon O rings may be obtained from Corning Glass but they must be used with great care, since it is very easy to break the glass pipe when the bolts are tightened enough to seal the Teflon joint. Very good results can be obtained by using Viton O rings or gaskets cut from 1/8-inch-thick silicone rubber.

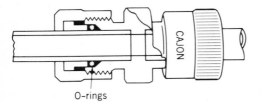

O-rings

FIGURE 7
Cajon Ultra-Torr compression fitting.

SHOPWORK

Experimental work in physical chemistry is not limited to the assembly of standard pieces of commercial apparatus and the making of measurements with them. New or modified apparatus is frequently needed. This must be designed and must often be constructed by the experimenter. One must therefore be acquainted not only with the appropriate materials for constructing apparatus but also with many of the techniques for operating machine tools.

The most important machine tools are the drill press, the band saw, the grinding wheel, the lathe, and the milling machine. The first three of these are very simple basic tools that every experimenter should learn how to use. The lathe and the milling machine are more complex and more expensive tools of great value in constructing special apparatus; if possible, anyone interested in physical chemistry research work should learn how to operate these tools also. At the very least, it is necessary to understand the principles of their operation and to appreciate what they can and cannot do. Indeed, a familarity with machine tools is a vital part of the ability to design complex apparatus properly. No attempt will be made here to describe the operation of any machine tools. Written descriptions are available elsewhere,[8] but it is of great importance to obtain basic instruction in the use of these tools from a qualified machinist in order to avoid risk of personal injury to the operator and damage to the machines.

REFERENCES

1. D. A. Skoog and D. M. West, "Fundamentals of Analytical Chemistry," 4th ed., Holt, New York (1982); I. M. Kolthoff and P. J. Elving (eds.), "Treatise on Analytical Chemistry," Wiley-Interscience, New York (1959 *et seq.*, a multivolume continuing series).
2. E. L. Peffer and G. C. Mulligan, "Testing of Glass Volumetric Apparatus," U.S. Natl. Bur. Stand. Circ. 602, U.S. Government Printing Office, Washington, D.C. (1959).
3. K. Lark-Horovitz and V. A. Johnson (eds.), "Methods of Experimental Physics," Vol. 6A. "Solid State Physics," chap. 2, Academic Press, New York (1959).
4. "Flow and Level Measurement Handbook and Encyclopedia," Omega Engineering, Inc., Stamford, Conn. (1987).
5. R. P. Benedict, "Fundamentals of Temperature, Pressure, and Flow Measurements," 3d ed., Wiley, New York (1984).
6. "Pressure, Strain and Force Measurement Handbook and Encyclopedia," Omega Engineering, Inc., Stamford Conn. (1987).
7. D. J. G. Ives and G. J. Janz (eds.), "Reference Electrodes," Academic Press, New York (1969).
8. J. H. Moore, C. C. Davis, and M. A. Coplan, "Building Scientific Apparatus," Addison-Wesley, Reading, Mass. (1983); R. F. Eisenberg, R. R. Kraybill and R. B. Seymour, "Selection of Materials for Construction of Equipment", in E. S. Perry and A. Weissberger (eds.), "Techniques of Chemistry," vol. XIII, chap. 1, Wiley-Interscience, New York (1979).
9. See, for example, *Review of Scientific Instruments* and *Journal of Physics E: Scientific Instruments*.
10. A. Paul, "Chemistry of Glasses," Chapman and Hall, London (1982); D. Holloway, "The Physical Properties of Glass," Springer, New York (1973).

11. H. E. Boyer and T. L. Gall (eds.), "Metals Handbook: Desk Edition," American Society for Metals, Metals Park, Ohio (1985).
12. C. A. Harper, "Handbook of Plastics and Elastomers," McGraw-Hill. New York (1975).
13. "Metals Handbook: Vol. 6. Welding, Brazing, and Soldering," American Society for Metals, Metals Park, Ohio (1983).

GENERAL READING

R. P. Benedict, *op. cit.*
J. H. Moore, C. C. Davis, and M. A. Coplan, *op. cit.*

CHAPTER
XX

LEAST-SQUARES
FITTING
PROCEDURES

Among the most common computational operations in experimental science is the fitting of experimental data to a theoretical function or model. Traditionally, this has often been done graphically when there is only one independent variable, as in fitting a straight line to a number of experimental points (x, y), with reliance on visual judgement for maximizing the quality of the fit. This is still widely done, particularly when the experimental results are of only modest quality. When the number of variables or the number of fitting parameters is large, or when it is important to make the best possible use of the data, the fitting must be done numerically. The computational method most commonly used is the method of least squares.

INTRODUCTION

The method of least squares has been used for many years, but the computations were once slow and onerous. Today, the wide availability of computers has made least-squares fitting a routine operation. Indeed, *least squares has become so easy to do that it is possible to be seduced into a false sense of security concerning the validity of the results obtained.* Least-squares fitting is not magic and is subject to the maxim of computer science, "garbage in, garbage out." It is important to realize that a set of experimental points can often be fitted quite well with an analytical expression bearing no relation whatsoever to any valid physical theory. There is nothing particularly wrong with fitting a data set with an artificial function for purely empirical purposes, such as to obtain an analytical representation of the data convenient for computational purposes (interpolation, differentiation, integration). For example, one might use a least-squares fit to the guassian error function, Eq. (II-11), in various contexts having little or nothing to do with the gaussian error function in principle; atomic electron densities, for instance. However, it is important to keep in mind which features of the model being tested are physically meaningful and which are not. One should resist cluttering up a theoretical model with features of doubtful physical significance merely to improve the fit. Furthermore, in trying to decide between two theoretical models on the basis of the "best" least-squares fit, one must be very careful in assessing the magnitude and the character of the experimental errors.

We will not undertake to present here a comprehensive and rigorous treatment of least squares. Our aim is to give a succinct and practical description of least squares as used in physical chemistry.

It must be kept in mind that the treatment of uncertainties given in this chapter is concerned with uncertainty in the data caused by random errors and not with that due to possible systematic errors, concerning which those considerations discussed in Chapter II apply.

FOUNDATIONS OF LEAST SQUARES

The maximum likelihood criterion. Let it be assumed that for m different values of an independent variable x (x_i; $i = 1, \ldots, m$) there are corresponding measured values of a dependent variable y (y_i; $i = 1, \ldots, m$). Let us further assume that there is a theoretical model predicting the way in which y is expected to depend on x, and that this model may be represented by an analytical function f:

$$y = f(\alpha_1, \ldots, \alpha_n; x) \tag{1}$$

The quantities α_j ($j = 1, \ldots, n$)† are independently adjustable parameters that are initially unknown or known only approximately. The initial trial values of these parameters will be denoted as $\alpha_1^0, \ldots, \alpha_n^0$. It should be pointed out that we might just as easily have more than one independent variable; let there be p of them (x^k; $k = 1, \ldots, p$). The model is now represented by

$$y = f(\alpha_1, \ldots, \alpha_n; x^1, \ldots, x^p) \tag{2}$$

---

† In appropriate circumstances the index j can be chosen to run instead from 0 through $n - 1$, for example when the model function is a polynomial

$$y = \alpha_0 + \alpha_1 x + \alpha_2 x^2 + \cdots + \alpha_{n-1} x^{n-1}$$

There are virtually no additional complexities or difficulties of any consequence resulting from having more than one independent variable, but for the
sake of simplicity of expression we will assume only one in the treatment
presented below.

Least-squares fitting deals with the problem of determining the "best"
values of the n adjustable parameters $\alpha_1, \ldots, \alpha_n$ so as to maximize the
agreement between the m observed y values and the values of y calculated with
Eq. (1). In effect, we are seeking the "best" (though generally not exact)
solutions to a set of m simultaneous equations (called *observational equations*
or *equations of condition*) in n unknowns:

$$y_i = f(\alpha_1, \ldots, \alpha_n; x_i) \tag{3}$$

The number m of observations exceeds (usually by an order of magnitude or
more) the number n of adjustable parameters. Thus, the mathematical
problem is overdetermined. [For a situation in which $m = n$, the corresponding
set of simultaneous equations could in principle be solved exactly for the
parameters α_j. But this "exact" fit to the data would provide no test at all of
the validity of the model.] In least squares as properly applied, the number of
observations is made large compared to the number of parameters in order (1)
to sample adequately a domain of respectable size for testing the validity of the
model, (2) to increase the accuracy and precision of the parameter determinations, and (3) to obtain statistical information as to the quality of the
parameter determination and the applicability of the model.

The least-squares criterion of best fit depends upon the concept known as
maximum likelihood: the best set of parameters α_j is one that maximizes the
probability function for the full set of measurements y_i. The probability P_i for
making at $x = x_i$ a single measurement of y equal to y_i is [see Eq. (II-11)]

$$P_i = \frac{1}{\sqrt{2\pi}\,\sigma_i} \exp\left\{ -\frac{[y_i - f(\alpha_1, \ldots, \alpha_n; x_i)]^2}{2\sigma_i^2} \right\}$$

assuming that the measurements y_i are normally distributed (i.e., are samples
of a gaussian distribution) with standard deviations σ_i. For making the entire
set of measurements y_i the probability is

$$P = \prod_i P_i = \prod_i \frac{1}{\sqrt{2\pi}\,\sigma_i} \exp\left\{ -\sum_i \frac{[y_i - f(\alpha_1, \ldots, \alpha_n; x_i)]^2}{2\sigma_i^2} \right\}$$

Consider now the variation of P with respect to the α_j values. This affects only
the sum in the exponent; maximizing P is the same thing as minimizing this
sum. Thus, maximum likelihood for P becomes the *least-squares principle*

$$\delta_{\alpha_1} X^2 = \delta_{\alpha_2} X^2 = \cdots = \delta_{\alpha_n} X^2 = 0 \tag{4}$$

where δ signifies variation with respect to infinitesimal and independent variations of the α_j,† and X^2 is defined by

$$X^2 \equiv \sum_i \left[\frac{y_i - f(\alpha_1, \ldots, \alpha_n; x_i)}{\sigma_i} \right]^2 \tag{5}$$

Actually, least squares is often applied in cases where it is not known with any certainty that measurements of y_i conform to a normal distribution, or even when it is in fact known that they do *not* conform to a normal distribution. Does this destroy the applicability of the maximum-likelihood criterion? The answer is: not necessarily. The central-limit theorem is discussed briefly in Chapter II.[2] Simply stated, it says that the sum (or average) of a large number of measurements conforms very nearly to a normal distribution irrespective of the distributions of the individual measurements *provided* that no one measurement contributes more than a small fraction to the sum (or average) and that the variations in the widths of the individual distributions are within reasonable bounds. (As we shall see, the average of a group of numbers is a special case of a least-squares determination.)

The factor $1/\sigma_i^2$ in Eq. (5) has the significance of the weight of the ith observation; i.e., it determines how heavily that observation contributes to the sum. However, it is not always possible to know even approximately the values of σ_i at the time the measurements are made. In this case, it is important and usually possible to assign *relative* weights w_i^R which are estimates related to the *true* weights $w_i^T \equiv \sigma_i^{-2}$ by

$$w_i^R \approx k w_i^T = k \sigma_i^{-2} \tag{6}$$

where k is an unknown scale factor. Thus we will rewrite Eq. (5) as

$$X^2 = \sum_i w_i [y_i - f(\alpha_1, \ldots, \alpha_n; x_i)]^2 \tag{7}$$

and use this form in what follows on the assumption that true weights w_i^T are available. If relative weights w_i^R are used, this will only introduce an unknown constant multiplicative factor into X^2 and will have no effect on minimizing this quantity.

If by reason of difficulty or lack of time, problems of weighting are to be ignored, all factors w_i can be set equal to one in the equations that follow. In that case, however, subsequent sections dealing with evaluation of uncertainties and goodness of fit lose much of their validity unless applied to a case in which all the observations just happen to be of equal weight.

Normal equations. For X^2 to be a minimum, we require that

$$\frac{\partial X^2}{\partial \alpha_1} = \frac{\partial X^2}{\partial \alpha_2} = \cdots = \frac{\partial X^2}{\partial \alpha_n} = 0$$

---

† The validity of Eq. (4) depends on the function f being well behaved, i.e., possessing no discontinuities in the function itself or in the first derivative with respect to any α.

For the derivative with respect to α_1 we have

$$-2 \sum_i w_i[y_i - f(\alpha_1, \ldots, \alpha_n; x_i)] \frac{\partial f}{\partial \alpha_1} = 0 \qquad (8)$$

For the general case that f is not a linear function of the α_j, let us expand it in a Taylor series around the values

$$y_i^0 \equiv f(\alpha_1^0, \ldots, \alpha_n^0; x_i)$$

calculated with the trial values α_i^0 of the parameters. *Keeping only terms to first order,*

$$f(\alpha_1, \ldots, \alpha_n; x_i) = y_i^0 + \frac{\partial f_i}{\partial \alpha_1} \Delta\alpha_1 + \frac{\partial f_i}{\partial \alpha_2} \Delta\alpha_2 + \cdots + \frac{\partial f_i}{\partial \alpha_n} \Delta\alpha_n, \qquad (9)$$

we obtain after rearranging Eq. (8)

$$\sum_i w_i \left(\frac{\partial f_i}{\partial \alpha_1}\right)^2 \Delta\alpha_1 + \sum_i w_i \frac{\partial f_i}{\partial \alpha_1} \frac{\partial f_i}{\partial \alpha_2} \Delta\alpha_2 + \cdots$$

$$+ \sum_i w_i \frac{\partial f_i}{\partial \alpha_1} \frac{\partial f_i}{\partial \alpha_n} \Delta\alpha_n = \sum_i w_i \frac{\partial f_i}{\partial \alpha_1} (y_i - y_i^0) \qquad (10)$$

where by $\Delta\alpha_j$ we mean $\alpha_j - \alpha_j^0$ and by $\partial f_i/\partial\alpha_j$ we mean $\partial f/\partial\alpha_j$ evaluated at $\alpha_1 = \alpha_1^0, \ldots, \alpha_n = \alpha_n^0, x = x_i$. To make the notation more compact, we define

$$A_{jk} \equiv \sum_i w_i \frac{\partial f_i}{\partial \alpha_j} \frac{\partial f_i}{\partial \alpha_k} \qquad (11a)$$

$$h_j \equiv \sum_i w_i \left(\frac{\partial f_i}{\partial \alpha_j}\right)(y - y_i^0) \qquad (11b)$$

Then Eq. (10) takes the form

$$A_{11} \Delta\alpha_1 + A_{12} \Delta\alpha_2 + \cdots + A_{1n} \Delta\alpha_n = h_1 \qquad (12a)$$

Similarly
$$A_{21} \Delta\alpha_1 + A_{22} \Delta\alpha_2 + \cdots + A_{2n} \Delta\alpha_n = h_2$$

$$\cdots$$
$$\cdots \qquad (12b)$$

$$A_{n1} \Delta\alpha_1 + A_{n2} \Delta\alpha_2 + \cdots + A_{nn} \Delta\alpha_n = h_n$$

Eqs. (12), constituting a set of n simultaneous linear equations in n unknowns, are known as the *normal equations*. They can be solved to yield a set of equations in the form

$$\Delta\alpha_1 = B_{11}h_1 + B_{12}h_2 + \cdots + B_{1n}h_n$$
$$\Delta\alpha_2 = B_{21}h_1 + B_{22}h_2 + \cdots + B_{2n}h_n$$
$$\cdots$$
$$\cdots \qquad (13)$$
$$\Delta\alpha_n = B_{n1}h_1 + B_{n2}h_2 + \cdots + B_{nn}h_n$$

The details of the solution process do not concern us here. The quantities A_{ij} are the elements of an $n \times n$ square matrix $\mathbf{A}$, and the B_{ij} are the elements of another $n \times n$ square matrix $\mathbf{B}$ that is the *inverse* of matrix A. The inversion of a matrix is a routine task for a computer.[1]

Now a new and improved set of α's can be obtained:

$$
\begin{aligned}
\alpha_1^1 &= \alpha_1^0 + \Delta\alpha_1 \\
\alpha_2^1 &= \alpha_2^0 + \Delta\alpha_2 \\
&\vdots \\
\alpha_n^1 &= \alpha_n^0 + \Delta\alpha_n
\end{aligned}
\tag{14}
$$

If the function f is linear in the parameters α_j the fitting procedure is now complete and the values α_j^1 are the best values in a least-squares sense. If f is nonlinear in terms of any of the α_j values, it is usually necessary to improve the α_j values by carrying out another "cycle" of minimization where α_j^1 plays the role previously played by α_j^0. The iteration process is repeated as many times as necessary to obtain convergence of the α_j values to some predetermined level of accuracy.

The normal-equations algorithm here described is generally the fitting method of choice when (1) the errors in the observations conform to a normal distribution (see discussion of this distribution in Chapter II), and (2) the observational equations are linear in the adjustable parameters. As a matter of convenience, this algorithm is often used (especially in "canned" least-squares computer programs) when one or both of these conditions is not fulfilled. This is not always bad practice, but one should be aware of the hazards discussed below.

The occasional large deviation of a y_i from its population mean can have a large effect on the refined parameters α_j because the quantity $[y_i - f(\cdots \alpha_j \cdots, x_i)]$ enters X^2 as the square. This is tolerable with a normal distribution because deviations become more and more unlikely as they increase in magnitude. However, if the error distribution has tails containing a higher than normal proportion of the total probability, the normal equations algorithm is significantly less reliable. Modifications of the normal equations algorithm for cases of abnormal distributions have been proposed by Tukey and Andrews.[3]

Major difficulties often encountered with the normal equations approach for *nonlinear* least squares are poor convergence to the "best" values of the parameters, and possible convergence to a set of bad values of the parameters corresponding to a false minimum in X^2. A number of other algorithms are available that have advantages over the normal-equations algorithm for many circumstances.[4] In the "steepest descent" algorithm the parameter shifts follow the negative gradient of the function X^2 in parameter space to maximize the rate of decrease in X^2 in the early stages, but final convergence with this algorithm is often very slow. A mixed algorithm due to Marquardt[4-6] combines the virtues of the normal-equations and steepest-descent algorithms and is widely used. The Marquardt algorithm is equivalent to replacing Eq. (11*a*) by

$$A_{jk} = \sum_i (1 + \delta_{jk}\lambda)\frac{\partial f_i}{\partial \alpha_j}\frac{\partial f_i}{\partial \alpha_k} \tag{15}$$

where δ_{jk} is the Kronecker delta (equal to unity when $j = k$ and equal to zero otherwise) and λ is a parameter that is adjusted after each cycle to obtain optimum performance. When λ is large, as is typical in the early cycles, the procedure resembles that of steepest descents; when it is small, as it typically becomes in the later cycles, it resembles the normal equations procedure. This adjustment of λ is done automatically in computer programs that utilize this algorithm.[5]

Simple examples. A trivial example of a least-squares calculation is the calculation of the arithmetic mean. In this case the independent variable x does not appear and there is only one parameter α, which is the desired average value of y. The observational equations are of the form

$$\Delta\alpha = \alpha = y_i - 0$$

since we may take $\alpha^0 = 0$. The single normal equation is

$$\sum_i w_i(1)^2\alpha = \sum_i (1)w_i y_i$$

whence

$$\alpha = \bar{y} = \frac{\sum w_i y_i}{\sum w_i} \tag{16}$$

When the weights are all taken as unity this becomes

$$\alpha = \bar{y} = \frac{1}{m}\sum_i y_i \tag{17}$$

A more significant example is that of a linear relationship

$$y = f(\alpha_0, \alpha_1; x) = \alpha_0 + \alpha_1 x \tag{18}$$

This is the equation of a straight line, where α_1 is the slope and α_0 is the y intercept. Here again, because of linearity, we may take $\alpha_0^0 = \alpha_1^0 = 0$. The m observational equations are of the form

$$\alpha_0 + \alpha_1 x_i = y_i$$

and the two normal equations are

$$\sum_i w_i(1)^2\alpha_0 + \sum_i w_i(1)x_i\alpha_1 = \sum_i w_i(1)y_i$$

$$\sum_i w_i x_i(1)\alpha_0 + \sum_i w_i x_i^2\alpha_1 = \sum_i w_i x_i y_i \tag{19}$$

Solving, we obtain

$$\alpha_0 = \frac{1}{D}\left(\sum_i w_i x_i^2 \sum_i w_i y_i - \sum_i w_i x_i \sum_i w_i x_i y_i\right) \tag{20a}$$

$$\alpha_1 = \frac{1}{D}\left(-\sum_i w_i x_i \sum_i w_i y_i + \sum_i w_i \sum_i w_i x_i y_i\right) \tag{20b}$$

where

$$D = \sum_i w_i \sum_i w_i x_i^2 - \left(\sum_i w_i x_i\right)^2 \tag{21}$$

is the determinant $|A|$ of the matrix

$$\mathbf{A} = \begin{pmatrix} \sum w_i & \sum w_i x_i \\ \sum w_i x_i & \sum w_i x_i^2 \end{pmatrix} \tag{22}$$

(If unit weights are employed, all w_i are deleted and $\sum w_i$ is replaced by m.) These equations may be useful when a simple straight-line fit ("linear regression") is being done with a pocket calculator. Many such calculators accumulate most or all of the sums required in Eqs. (20)–(22); some complete the calculation and offer both the refined parameters and their estimated standard deviations.

    The student may find it illuminating to derive expressions for the three-parameter linear least-squares case with two independent variables:

$$z = \alpha_0 + \alpha_{1x} x + \alpha_{1y} y$$

Note that if the independent variable y is replaced by x^2 we have a fitting function that is nonlinear in x but that can be treated with linear least squares since it is linear in the adjustable parameters. Similarly, a general polynomial can be fitted by linear least squares:

$$y = \sum_{j=0}^{n-1} \alpha_j x^j$$

WEIGHTS

In general, a weight w_i must be assigned for each measurement y_i. If estimates of the standard deviations σ_i are available, the true weights $w_i^T = \sigma_i^{-2}$ can be used. If σ_i are not known but it is manifest that all measurements should have the same uncertainty, the weights are all equal and may be set equal to unity for convenience in finding the best fit. There may be other circumstances in which the standard deviations are not known, but in which *relative weights* w_i^R can be assigned on an arbitrary scale by judgement based on experience or by common-sense criteria. (For example, a value of y_i, which is the mean of two or three measurements of y at the same x_i, has a weight twice or three times that of a single measurement at that x_i.) Such relative weights are related to the true weights by Eq. (6).

The best basis for assigning weights is to make several measurements y_{ik} ($k = 1, \ldots, N$) at the same x_i, and determine the experimental variance S_i^2 of each measurement with Eq. (II-4). Since the mean of those N measurements will enter the least-squares calculations as y_i, that variance divided by N will constitute the first approximation to σ_i^2. Since N is likely to be small (less than 6), σ_i^2 will be rather uncertain and consequently a certain amount of pooling or smoothing of the σ_i^2 values may be appropriate. This may be done by plotting the σ_i^2 against x_i and drawing a smooth curve to obtain the values to be used in the least-squares calculation. Alternatively, the σ_i^2 may be averaged in blocks of data having in each block uniform conditions of measurement. We may summarize by writing

$$\frac{1}{w_i} = \sigma_i^2(\text{est}) = \left[\frac{1}{N(N-1)} \sum_{k=1}^{N} (y_{ik} - y_i)^2\right]_{\text{smoothed}} \tag{23}$$

where y_i is the mean of the y_{ik}.

Since y is functionally dependent on x, the uncertainty in y must contain a contribution from any uncertainty in x. If in the above-described procedure the variable x is set separately and independently to its assigned value x_i for each of the N measurements y_i, then presumably the contribution of the uncertainty in x_i will automatically be reflected in the uncertainty of y_i and Eq. (23) will apply directly. However, if x is set only once to x_i for the N measurements y_i, so that the error contributed by x_i is the same for all N measurements, then the uncertainty in both must be reflected in the weight:

$$w_i = \frac{1}{\sigma_i^2} = \frac{1}{\sigma_{iy}^2 + \left(\dfrac{\partial f}{\partial x}\right)_{x_i}^2 \sigma_{ix}^2} \tag{24}$$

where σ_{iy} and σ_{ix} are the independent standard deviations in y and x, respectively, and $(\partial f / \partial x)_{x_i}$ is evaluated with $\alpha_j = \alpha_j^0$ (the trial values of the parameters). Equation (24) can also be used in other cases where the uncertainties in y and x are independent and have been estimated separately. In principle, the least-squares method assumes that all the uncertainty is in the y_i value at some exactly known x_i value, but Eq. (24) is a good approximation for adjusting the weights to reflect the experimental fact that both x and y are usually subject to experimental error.

It should be stressed that in any experiment for which there is to be a least-squares refinement of parameters, intended to yield not only the best possible parameter values but also respectable estimates of their uncertainties *and* a test of the validity of the model, it is vital to take the trouble to analyze the methods and the circumstances of the experiment carefully in order to get the best possible values of the *a priori* weights.

REJECTION OF DISCORDANT DATA

There may be one or more y_i values that deviate markedly from the trend of the others. Not only may y_i values be in error because of the kinds of occurrences discussed in Chapter II (reading errors, transposition of digits, power-line transients), there may also be deviations occurring at discrete x_i values arising from physical effects not taken into account in the model. (Examples: in spectroscopy, an unexpected resonance happening at a certain frequency v_i, enhancing or reducing the spectral response relative to that envisaged by the model; in x-ray crystallography, multiple reflection or thermal diffuse scattering producing abnormal intensity for a particular Bragg reflection.)

Sometimes the deviation is so gross that the offending measurement can be discarded at once. More often, it is necessary to carry out one or more cycles of least squares and to obtain *weighted residuals* δ_i *based on properly scaled weights* [defined below by Eqs. (27a) and (27b)]. It must then be decided by inspection of the δ_i values whether the apparently discordant y_i should be rejected and one or more additional cycles of least squares carried out. This question is seldom discussed adequately, and there seem to be no well-defined and generally accepted criteria. We present some suggestions based on the magnitudes of the weighted residuals.

For values of m no larger than 10, one can use a Q test criterion (see Chapter II) for considering the rejection of *one* outlying datum y_i. In the present context, the quantity Q is defined by

$$Q = (\delta_{max} - \delta_{near})/(\delta_{max} - \delta_{far}) \tag{25}$$

The quantities appearing in this equation may be best understood with the aid of an example. Let the residuals for a set of 10 measurements be arranged from the largest negative to the largest positive:

$$-1.0 \quad -0.9 \quad -0.7 \quad -0.7 \quad -0.3 \quad 0.0 \quad 0.3 \quad 0.4 \quad 0.4 \quad 2.0$$

Here δ_{max} (the δ with the largest absolute magnitude) is 2.0, δ_{near} (the δ nearest to δ_{max}) is 0.4, and δ_{far} (the δ farthest from δ_{max}) is -1.0. Thus $Q = (2.0 - 0.4)/[2.0 - (-1.0)] = 0.53$. From Table II-1 the critical value Q_c is 0.41. Therefore the y_i point that yields $\delta_i = 2.0$ should be rejected.

For values of m ranging from 11 to 50 we suggest that a seemingly discordant datum for which $\delta_i > 2.6$ be rejected. (A value of 2.58 corresponds to an *a priori* estimate that a given measurement has about a 1 percent chance of being valid.) For larger values of m it is difficult or impossible to distinguish data having large weighted residuals because of faulty measurement from those having large weighted residuals because they happen to be in the tail of the normal distribution. However, data with "wild" values of δ_i (i.e., exceeding 4) may always be rejected.

GOODNESS OF FIT

The χ^2 test. Having completed the least-squares computations (with enough cycles to obtain convergence if the function is nonlinear in the α_j) and having determined the best values α_j^* of the parameters, it remains to determine how good they are and how good the model is. In this and the following sections a number of equations are presented without proof; detailed developments can be found in the books referred to at the end of this chapter.

At this point the observed y_i are not in perfect agreement with the calculated y_i^* given by the model with the refined parameters

$$y_i^* = f(\alpha_i^*, \ldots, \alpha_n^*; x_i) \tag{26}$$

There remain residuals $y_i - y_i^*$, which we may subject to statistical analysis. To reduce these to the same statistical population we define a *weighted residual* δ_i:

$$\delta_i \equiv \frac{y_i - y_i^*}{\sigma_i} \tag{27a}$$

where the σ_i are the *a priori* standard deviations. In practice, the weighted residuals are approximated by

$$\delta_{i0} = \sqrt{w_i}\,(y_i - y_i^*) \tag{27b}$$

where w_i are the *a priori* estimated weights. If the true weights were used, these two quantities would be identical; however, in the case where the weights are estimated on an arbitrary relative basis [see Eq. (6)] it is necessary to make a distinction between them.

Analogous to Eq. (II-4) there is the following expression for the *variance of an observation of unit weight:*

$$S_{(1)}^2 = \frac{\sum \delta_{i0}^2}{m - n} = \frac{X_{min}^2}{m - n} \tag{28}$$

where m is the number of observations and n is the number of adjustable parameters. The square root of this quantity, $S_{(1)}$, is an *estimate* of the standard deviation of an observation of unit weight. If $S_{(1)}$ were indeed equal to the standard deviation of an observation of unit weight, it would have the value unity. Presumably such a value would be obtained if there were an infinite number of observations, the model were rigorously correct, and the correct values $w_i = \sigma_i^{-2}$ were used in Eq. (27b). In general, $S_{(1)}$ differs from unity. For an error-free model, $S_{(1)}$ may be greater than or less than unity depending on whether the weights are overestimated or underestimated.† If the weights are correct, errors in the model tend to make $S_{(1)}$ greater than unity. Thus we see the reason for the emphasis on assigning *a priori* weights

---

† One can rescale the weights *a posteriori* on the approximate basis that $w_i(\text{new}) = w_i(\text{old})/S_{(1)}^2$ if one is confident the model is correct.

realistically and accurately. If one has confidence in the weights, the value of $S_{(1)}$ can provide a real test of the validity of the model. On the other hand, if weights have been assigned only on a relative basis, $S_{(1)}$ does not afford an effective test of a single model but may be of value in comparing two models (see below).

The test of a fit can be put into the form of the following question: "At a specified level of confidence, is the value of $S_{(1)}^2$ consistent with the assumption that the residuals δ_{i0} are representative of a normal distribution as they would be if the model were correct and the weights properly assigned?" Here we make use of the fact that in principle the minimized quantity X^2, known as χ^2, should conform to the so-called *chi-square distribution*. The probability distribution function for χ^2, with $v = (m - n)$ degrees of freedom, is[6,7]

$$P(\chi^2, v) = \frac{(\chi^2)^{(v-2)/2} e^{-\chi^2/2}}{2^{v/2} \Gamma(v/2)} \tag{29}$$

where $\Gamma(v/2)$ is the gamma function. The probability that χ^2 will exceed a certain limiting value is

$$P_{\text{int}}(\chi^2, v) = \int_{\chi^2}^{\infty} P(x^2, v) \, dx^2 \tag{30}$$

where x^2 is a variable of integration corresponding to χ^2.

At this point it is convenient to make a change in notation. It has become common practice to cite the quantity "reduced chi square" χ_v^2:

$$\chi_v^2 \equiv \frac{\chi^2}{v} \tag{31a}$$

where
$$\equiv \frac{1}{v} \sum_i \frac{(y_i - y_i^*)^2}{\sigma_i^2} \simeq \frac{1}{v} \sum_i w_i (y_i - y_i^*)^2 \tag{31b}$$

$$v = m - n \tag{31c}$$

is the number of degrees of freedom when m data points are fitted with a form involving n adjustable parameters. The quantity χ_v^2 is defined by the first expression in Eq. (31b) and, in practice, calculated with the second expression. Thus, in the literature, χ_v^2 is the symbol often given to what we have been calling $S_{(1)}^2$. One can refer to published tables[8-10] of χ_v^2 for given probability P_{int} and a number of degrees of freedom v. An abridged table for $P_{\text{int}} = 0.95$ and $P_{\text{int}} = 0.05$ is given in Table 1.

If the model is known in advance to be above reproach, the value of $S_{(1)}^2$ can be used to answer the question: "At a specified level of confidence is the value of $S_{(1)}^2$ consistent with the assumption that the weighted residuals δ_{i0} represent a normal distribution with mean zero and standard deviation unity?" This could be crudely translated into the question: "Is the value of $S_{(1)}^2$ consistent with the assumption that the weights have been properly assigned?"

TABLE 1
Limiting values of $\chi_v^2 \equiv \chi^2(v)/v$, $F(1, v)$, and $F(v, v)$ for stated probability of exceeding these values[a]

v	Limiting χ_v^2		Limiting $F(1, v)$	Limiting $F(v, v)$
	$P_{int} = 0.95$	$P_{int} = 0.05$	$P_{int} = 0.05$	$P_{int} = 0.05$
1	0.0039	3.84	161	161
2	0.0515	3.00	18.5	19.0
3	0.117	2.60	10.13	9.28
4	0.178	2.37	7.71	6.39
5	0.229	2.21	6.61	5.05
6	0.273	2.10	5.99	4.28
8	0.342	1.94	5.32	3.44
10	0.394	1.83	4.96	2.98
12	0.436	1.75	4.75	2.59
15	0.484	1.67	4.54	2.40
20	0.543	1.57	4.35	2.12
30	0.616	1.46	4.17	1.84
40	0.662	1.39	4.08	1.69
60	0.720	1.32	4.00	1.53
120	0.798	1.22	3.92	1.35
∞	1.000	1.00	3.84	1.00

[a] Adapted from Refs. 9 and 10.

Here $S_{(1)}^2$ can be either less than or greater than unity; for 90-percent confidence limits, $S_{(1)}^2$ should lie between the *limiting* χ_v^2 values corresponding to $P_{int} = 0.05$ and $P_{int} = 0.95$. For example, when $v = m - n = 30$, we find from Table 1 that $S_{(1)}^2$ should lie between 0.616 and 1.46 if the above question is to be answered in the affirmative. An affirmative answer is *not* a guarantee that the weights have been assigned correctly, but only an assertion that they cannot be criticized on the basis of the statistics available.

If the assignment of weights is known in advance to be above reproach, the value of $S_{(1)}^2$ can be used to answer the more interesting question: "At a specified level of confidence is the value of $S_{(1)}^2$ consistent with the assumption that the data set conforms to the assumed model function?" For the above example of $v = 30$ and 90-percent confidence limits, $S_{(1)}^2$ should lie between 0.616 and 1.46 if the question is to be answered positively. That is, if the model is valid, there is only a 5 percent *a priori* statistical probability that $S_{(1)}^2$ would lie below 0.616 and the same probability that it might lie above 1.46.

Good fits require χ_v^2 values that lie between the limiting values specified (in Table 1, for example) for a given confidence level and v value. To continue with our example of a fit to a data set with $m = 32$ and $n = 2$, let us assume that χ_v^2 calculated with Eq. (31b) has the value 1.7. At the 90-percent confidence level, the validity of the model, the assignment of weights, or both are suspect. However, one must be quite certain about the weights before this χ_v^2 value be used to reject the model (within the specified confidence limits). Let us now assume that the fit to our data set had given $\chi_v^2 = 1.2$. For this χ_v^2 value, the validity of the model and the assignment of weights cannot be challenged on a statistical basis at this confidence level. However, this does not guarantee that the model is sound unless one can show that artifically low weights have not been used.

If the model function is a nonlinear function of the parameters α_j, and the trial values α_j^0 are not quite close to the true values, the least-squares treatment may converge on a false minimum for χ_v^2. Usually when this happens, $\chi_v^2 \simeq S_{(1)}^2 \gg 1$. However, in problems where m and n are both very large (as in x-ray crystallography), there are a great many false minima, some of which are not far from the true minimum in n-dimensional parameter space and yield χ_v^2 values not greatly different from unity. This is one additional argument for making careful *a priori* estimates of σ_i. When the weights for the observations are properly estimated, it is unlikely (although occasionally possible) that a false minimum will satisfy the χ^2 test within the appropriate confidence limits.

Distribution of residuals. When the number of measurements is large (preferably more than 100) one can carry out a χ^2 test of the frequency distribution of the δ_i values.[2] This frequency test can be more instructive than the χ^2 test of $S_{(1)}^2$ since it may allow a diagnosis of defects in the model or defects in the weight distribution apart from a mere scaling error in the weights.

Some workers have introduced a more detailed statistical test of the least-squares fit and the weighting scheme by using *normal probability plots*.[11,12] This test compares the actual distribution of the observed weighted residuals δ^{obs} to the ideal values δ^{ideal} expected for a normal distribution of mean zero and standard deviation unity. A plot of δ^{obs} versus δ^{ideal} should yield points close to a straight line with slope unity that passes through the origin, and most of the scatter should occur at the ends. If the plotted points look linear but with a slope considerably different from unity, the weights may be relatively correct but either overestimated or underestimated. If the plotted points exhibit a pronounced deviation from linearity, either the model is defective or erroneous relative weights have been assigned.

In the spirit of the above, it is helpful to make a qualitative inspection for trends in the residuals even if a formal analysis is not undertaken. In some fitting situations, one can see that dropping a few points will allow a change in the adjustable fitting parameters that will appreciably decrease all the remaining δ_i values and thus significantly lower χ_v^2. Such a procedure may or may not be defensible on purely statistical grounds, but it can at least lead one to carefully inspect the experimental validity of possibly errant points. Even when there are no experimentally suspect points with peculiar residuals, it is important to be aware of the trend in the residuals across the data set. Models that give a best fit with systematic trends in the residuals (say a block of negative residuals at each end of the data set with a central block of positive residuals) are suspect even if the χ_v^2 values appear to be satisfactory.

COMPARISON OF MODELS

It often happens that a decision has to be made between two somewhat different models for describing the data. Both models have physical significance but they are based on different physical hypotheses. If both appear to provide reasonably good fits to the data, can preference be given for one model over the other?

The judgement may be based on the reduced chi-square values χ_v^2 for the fits with the two respective models. However, even if fitting with the two models involves the same number of degrees of freedom v, it should not be said that model 2 is significantly preferable to model 1 just because χ_v^2 for model 2 is smaller than χ_v^2 for model 1. One must answer the question: "Is the difference *significant*?" Here we need to look at the probability distribution for the ratio of reduced chi squares. Such a ratio, χ_{v1}^2/χ_{v2}^2, should conform to a distribution known as the F distribution, for which the probability distribution function is[8]

$$P_F(F, v_1, v_2) = \frac{\Gamma\left(\dfrac{v_1 + v_2}{2}\right)}{\Gamma\left(\dfrac{v_1}{2}\right)\Gamma\left(\dfrac{v_2}{2}\right)} \left(\frac{v_1}{v_2}\right)^{v_1/2} \frac{F^{(v_1-1)/2}}{\left(1 + F\dfrac{v_1}{v_2}\right)^{(v_1+v_2)/2}} \tag{32}$$

The test is made with the integral probability

$$P_{int}(F, v_1, v_2) = \int_F^\infty P_F(f, v_1, v_2)\, df \tag{33}$$

(where f is an integration variable corresponding to F), which expresses the probability that F obtained from a random data set exceeds a certain value. We make use of the fact that

$$F \equiv \frac{\chi_{v1}^2}{\chi_{v2}^2} = \frac{\chi_1^2/(m - n_1)}{\chi_2^2/(m - n_2)} = \frac{S_{(1)1}^2}{S_{(1)2}^2} \tag{34}$$

and look up in Table 1 or published tables[9,10] the value of

$$P_{int}(F, v_1, v_2) = P_{int}\left(\frac{S_{(1)1}^2}{S_{(1)2}^2}, m - n_1, m - n_2\right)$$

or else look up the limiting value of $F(v_1, v_2)$ for a given probability level, say 5 percent.

For example, suppose that there are two models both with the same number of adjustable parameters. Each fit will involve the same number of degrees of freedom, say 20. Reference to statistical tables gives an $F(20, 20)$ value of 2.12 for a probability level of 0.05. That means that if $\chi^2_{v1}/\chi^2_{v2} > 2.12$ it can be said at the 95 percent confidence level that model 1 may be rejected in favor of model 2. Note the important feature that an accurate value for the ratio χ^2_{v1}/χ^2_{v2} does *not* require *a priori* knowledge of the σ_i values and true weights. If good relative weights are used [see Eq. (6)], the reduced chi-square ratio will be correct.

This F test can also be used when the number of degrees of freedom is different for the two models. The most frequently encountered circumstance of this kind results from least-squares fitting with two models, one having n parameters and the other having the same n parameters plus p additional parameters associated with an elaboration of the first model. As a simple example, consider a unimolecular gas-phase kinetics experiment like Exp. 24. Strict first-order kinetics predicts that the partial pressure of the decaying molecular species will vary as

$$\ln p = \ln p_0 - kt \quad \text{(model 1)} \tag{35}$$

However, at low pressures, owing to decreased probability of collisional deactivation of the activated molecule, the behavior may be better represented by

$$\ln p = \ln p_0 - kt + \frac{b}{p_0}(e^{kt} - 1) \quad \text{(model 2)} \tag{36}$$

Equation (35) has two adjustable parameters: p_0 and k. Equation (36) has three: p_0, k, and b. The experimental data consist of measurements of p as a function of t over a range in which p decreases by a factor of about e^2. Least-squares parameter determinations are carried out separately with each of the two models. They give slightly different sets of values of p_0 and k. It is found that $S^2_{(1)}$ for model 2 is less than $S^2_{(1)}$ for model 1. Is model 2 to be preferred? That is: does a nonzero coefficient b really arise from a physically better model, or does it merely provide a cosmetic improvement in the fit by adding an extra and nonphysical adjustable parameter?

The F test in this case can be cast in a somewhat different form:[8]

$$F_\chi(p, m - n - p) \equiv \frac{\chi^2(m - n) - \chi^2(m - n - p)}{p(m - n - p)^{-1}\chi^2(m - n - p)}$$

Substituting in the $S^2_{(1)}$ values, we have

$$\frac{(m - n)S^2_{(1)}(m - n) - (m - n - p)S^2_{(1)}(m - n - p)}{pS^2_{(1)}(m - n - p)} = \frac{v_1 \chi^2_{v_1}}{p \chi^2_{v_2}} - \frac{v_2}{p} = F_\chi(p, v_2) \tag{38}$$

In the present case $n = 2$, $p = 1$; let us suppose that $m = 20$. For model 1,

$v_1 = m - n = 18$; for model 2, $v_2 = m - n - p = 17$. Therefore

$$\frac{18S^2_{(1)1} - 17S^2_{(1)2}}{1 \cdot S^2_{(1)2}} = 18\frac{\chi^2_{v_1}}{\chi^2_{v_2}} - 17 = F_\chi(1, 17) \tag{39}$$

It can be shown that F_χ follows the F distribution. At $P = 0.05$, the limiting value of $F(1, 17)$ is found from Table 1 to be 4.5. The corresponding limiting ratio of reduced chi squares, $\chi^2_{v_1}/\chi^2_{v_2}$, is found from Eq. (39) to be $(4.5 + 17)/18 = 1.19$. Suppose in our hypothetical example we find that the actual ratio is 1.24. Then at the 95-percent confidence level model 2 is to be preferred, and the hypothesis that decay of the activated species is competing significantly with deactivation is confirmed. If, on the other hand, this ratio is found to be only 1.12, the 'better" fit with model 2 is not statistically significant. We might have strong theoretical reasons for preferring model 2, but this set of data on this particular reacting system does not allow one to reject model 1.

More detailed comparison of two models could involve the use of normal probability plots[11,12] or at least inspection of residuals for systematic trends as discussed earlier. If two models produce fits of equivalent quality in the sense of an F test, one might still prefer the model that gave the more random sequence of residuals (or, put differently, be suspicious of a model that gave blocks of residuals with alternating signs). Another way to compare models that yield almost equivalent fits is *range shrinking*, in which the data set is systematically truncated from either end of the range in the variable x and the stability of the parameter value is inspected.

Finally, one should pay considerable attention to the physical reasonableness of the adjustable parameters α_j before choosing one model over another. To embrace model 2 and reject model 1 on a purely statistical analysis of one set of data is dangerous. If possible, one should vary the experimental conditions, analyze several sets of data, and look at the behavior of the α_j values. In our gas kinetics example, one could vary the temperature at which the rate of decomposition is studied and test the temperature dependence of the k values obtained from least-squares fits at each temperature. What if model 2 were statistically better for fitting the data at each T but gave k (and probably b) values that were erratic functions of T, while model 1 gave k values that smoothly varied with T in a way that was theoretically pleasing? One might then prefer model 1 but report the statistical problems associated with the data. In particular, one should then look carefully for *a posteriori* evidence that the relative weights might be reassigned. Unsuspected systematic errors present over part of the range could have influenced the quality of the overall fit.

UNCERTAINTIES IN THE PARAMETERS

We now assume that the least squares refinement has converged satisfactorily, that any necessary rejection of discordant data has taken place before the final cycles were carried out, and that statistical tests on the weighted residuals have given reassuring results. It is now appropriate to estimate the uncertainties in the determined values of the adjustable parameters α_j.†

A propagation-of-error treatment of Eq. (13) yields for the estimated standard deviation in parameter α_j the expression:

$$S(\alpha_j) = B_{jj}^{1/2} S_{(1)} \tag{40}$$

where B_{jj} is the jth diagonal element in the inverse normal equation matrix $\mathbf{B} = \mathbf{A}^{-1}$ and $S_{(1)}$ is the estimated standard deviation of an observation of unit weight given by Eq. (28). In the unusual case where (1) the number m of observations is so small (less that 6 or 8, say) that $S_{(1)}$ cannot be determined with great reliability from Eq. (28) and (2) the reliance that can be placed on the *a priori* σ_j values is uncommonly high, $S_{(1)}$ can be simply set equal to unity. Equation (40) is the basis of the calculation of estimated standard deviations that are automatically given by packaged least-squares computer programs (and all too often are uncritically accepted by the users of such programs without much concern for the validity of the weighting system or of the model).

For the trivial one-parameter case in which the parameter to be. determined is the arithmetic mean,

$$B_{11} = \frac{1}{\displaystyle\sum_i w_i}$$

Making use of Eqs. (28) and (40) we obtain

$$S(\bar{y}) = \left[\frac{\sum \delta_i^2}{(N-1)\sum w_i} \right]^{1/2}$$

If all weights are taken as unity, $\sum_i w_i = m = N$ and

$$S(\bar{y}) = \left[\frac{\sum (y_i - y_i^*)^2}{N(N-1)} \right]^{1/2} \tag{41}$$

which corresponds to the result cited in Chapter II, Eqs. (II-1) and (II-8).

For the two-parameter case of the linear relationship we may apply Eqs. (28) and (40) to the estimation of the standard deviations in the intercept α_0 and the slope α_1 of the corresponding straight line. From Eq. (22) we have

$$B_{00} = \frac{1}{D} \sum_i w_i x_i^2 \qquad B_{11} = \frac{1}{D} \sum_i w_i$$

---

† If the results of statistical tests were *not* reassuring, only qualified confidence in the estimated uncertainties is justified; if they are to be quoted at all, it may be appropriate to multiply them by a factor of at least 2 or 3. In any such case, the circumstances should be fully reported along with the parameter values and the quoted uncertainties, if any.

where D is given by Eq. (21). Thus

$$S(\alpha_0) = \left(\frac{1}{D}\sum_i w_i x_i^2\right)^{1/2}\left(\frac{\sum \delta_i^2}{m-n}\right)^{1/2} \tag{42a}$$

$$S(\alpha_1) = \left(\frac{1}{D}\sum_i w_i\right)^{1/2}\left(\frac{\sum \delta_i^2}{m-n}\right)^{1/2} \tag{42b}$$

As in the case of Eqs. (20)–(22), when unit weights are employed the w_i are deleted and $\sum w_i$ is replaced by m.

Knowledge of estimated standard deviations is not by itself very useful, especially when they are unthinkingly accepted without regard to the validity of the model and of the weighting scheme. Assuming that these aspects are assuredly satisfactory, we are usually more interested in confidence limits for the parameters at a stated probability level P, as in the one-parameter case of the algebraic mean discussed in the section on confidence limits in Chapter II. Equation (II-33) for the one-parameter case applies also to the n-parameter case; for 95-percent confidence

$$\Delta_{0.95,j} = t_{0.95,v}S_j \tag{43}$$

but here we must take as the number of degrees of freedom $v = m - n$.

This method is usually satisfactory for routine work, particularly when the model function is linear in the adjustable parameters. However, there may arise circumstances in nonlinear least squares for which more care is needed. Consider a situation in which one parameter, say α_k, is of special importance. It may be that it was the whole point of the experiment to determine that one parameter, while the other parameters are of minor importance apart from their being necessary parts of the model. Let the "best" value of this parameter, determined by a convergent least-squares procedure, be α_k^*. Let us step away from α_k^* in both directions by a constant interval to obtain several values of α_k for the procedure to be described. [The interval chosen might be, for example, S_k as calculated with Eq. (40), and the number of steps might be three or more on each side.] We then *fix* α_k at each of these trial values in turn, refine the remaining parameters to convergence, and evaluate χ_v^2 with Eq. (31b), using in each case the current fixed value of α_k and the refined values of the other parameters. Let us now plot χ_v^2 against α_k and draw a smooth curve through the points. In a linear least-squares case the plot may be expected to be parabolic around α_k^*, but in a nonlinear case it may deviate considerably from parabolic shape and is likely to be unsymmetrical. We now use the F test to establish a ratio

$$F(v, v) = \frac{\chi_{v\,max}^2}{\chi_{v\,min}^2} \tag{44}$$

that corresponds to a probability level of, say, 5 percent. For example, with 15 degrees of freedom, we find from Table 1 that $F_{0.05}(15, 15) = 2.40$. We multiply the observed $\chi_{v\,min}^2$, evaluated at α_k^*, by the appropriate value of F to obtain a value for $\chi^2_{\nu\,\text{max}}$, and we mark the two opposite points on the curve where $\chi^2_\nu = \chi^2_{\nu\,\text{max}}$. The corresponding values of α_k are the maximum and minimum limiting values, and their differences from α_k^* may be quoted as the confidence limits at the stated probability level, say 95 percent. Note that the two limits may not necessarily be equal; instead of quoting a result in the customary form 3.154 ± 0.021 we may have to use the form

$$3.154^{+0.027}_{-0.015}$$

A plot of χ^2_ν against a parameter that is fixed at selected values as described above is useful also for diagnostic purposes. It may, for example, indicate that the value obtained for α_k^* is at a false minimum of X^2. If this situation is suspected, the plot should cover a much wider range than is necessary to determine confidence limits of error.

It frequently happens that one wishes to obtain an estimated standard deviation for a quantity F that is calculated from two or more of the α_j according to a functional relationship of some sort, say $F(\alpha_1, \ldots, \alpha_n)$. It would *not* be correct to estimate separately the $\sigma(\alpha_j)$ with Eqs. (28) and (40) and apply Eq. (II-51). The reason is that the errors in the α_j are usually not independent; in general they are *correlated*. A simple example will show how errors can be correlated. Imagine three points a, b, c along a straight line. Let the distance ab be α_1, and let the distance bc be α_2. If the estimated standard deviation (e.s.d.) in the position of any of the points is *independently* σ_p, the e.s.d. of α_1 is $\sqrt{2}\,\sigma_p$ and the e.s.d. of α_2 is also $\sqrt{2}\,\sigma_p$. What is the estimated standard deviation of $F = \alpha_1 + \alpha_2 =$ the distance ac? If we use the propagation-of-errors treatment without allowing for correlation, we get the answer that $\sigma(F) = [\sigma^2(\alpha_1) + \sigma^2(\alpha_2)]^{1/2} = 2\sigma_p$, while on going back to first principles we see that the true answer is $\sqrt{2}\,\sigma_p$. The two distances α_1 and α_2 are correlated; when point b moves toward a, it is moving away from c. For a general discussion of this problem and methods for handling it with the correlation matrix see Refs. 1 and 7. The correlation matrix may also provide useful information for the design of an experiment.

SUMMARY OF PROCEDURES

This section will provide a brief "road map" of the steps recommended in carrying out a least-squares fit.

1. Establish the weights w_i to be assigned to each observed value y_i. It is vital to have good relative weights, and preferable to have true weights based on the standard deviations σ_i. Think carefully about your measurements before taking the easy way out and adopting equal weights (see Weights). It may be necessary to smooth the observed variances over the entire data set (see Sample Least-Squares Calculation).
2. Reject any obviously bad points (see Rejection of Discordant Data).
3. Decide on the model function to be used. This choice may be guided by a

theoretical prediction or by a rough empirical assessment of the type of dependence likely.

4. Choose a reasonable set of trial values α_j^0 for the adjustable parameters of the model. In nonlinear fitting, a good choice of α_j^0 is very important to reduce computational time and to avoid false minima (see Goodness of Fit, towards the end of the section on χ^2 tests).

5. Carry out the least-squares minimization of the quantity X^2 in Eq. (7) according to an appropriate algorithm (presumably normal equations if the observational equations are linear in the parameters to be determined; otherwise some other such as Marquardt's[4,5]). Convergence should not be assumed in the nonlinear case until successive cycles produce no significant change in any of the parameters.

6. After convergence has been achieved, check the residuals $y_i - y_i^*$ to see if any data points should be deleted from the data set (see end of the section on Goodness of Fit). If so, rerun the least-squares fit on the new, edited data set.

7. Obtain the reduced chi square value χ_ν^2 given by Eq. (31b) and carry out the appropriate statistical tests for goodness of fit, including inspection of the weighted residuals for systematic trends (see Goodness of Fit).

8. If either theory or clear systematic trends in the residuals suggests an alternative model, carry out a new least-squares fit to the same data set with the new model. Make a statistical analysis to help decide whether one of the models can be rejected in favor of the other. Be careful to consider the physical reality of the adjustable parameters in the model that you finally choose (see Comparison of Models).

9. Evaluate the statistical uncertainties in the adjustable parameters obtained from the best fit (see Uncertainties in the Parameters).

10. *Think about what you are doing. Do not treat least-squares fitting as a magical mathematical game,* and do not accept without searching examination the results of packaged least-squares computer programs.

SAMPLE LEAST-SQUARES CALCULATION

We will illustrate here as many as possible of the principles and techniques discussed in this chapter with a sample "curve fitting" by least squares.

For values of x_i ranging from 2.1 to 5.0 ($i = 1$ through 30) in intervals of 0.1, measurements of y were made, four y_{ik} for each x_i value. The measured values are listed in Table 2 and plotted in Fig. 1. For each x_i, the four y_{ik} were averaged to give y_i, and the variance S_i^2 was calculated. Since the measurements were made under uniform conditions and since the variances, when plotted against x, showed (apart from their natural statistical scatter) a smooth variation suggestive of quadratic behavior, the variances were smoothed by least-squares fitting with a quadratic using unit weights. The variance function

TABLE 2
Sample least-squares calculation: data and least-squares fits with two and three parameters

i	x_i	$y_{ik}, k=1,\ldots,4$				$\overline{y}_i$ (mean)	s_i^2	$s_{sm_i}^2$	$s_{sm_i}^{2(mean)}$	w_i	y_i^{cal}	δ_i^2	δ_i	y_i^{cal}	δ_i^2	δ_i
		THE DATA SET									2 Parameters[a]			3 Parameters[b]		
1	2.1	17.15	16.73	16.80	17.24	16.98	.0638	.0372	.0093	107	17.13	2.41	-1.55	16.85	1.81	1.34
2	2.2	16.62	16.46	16.62	16.60	16.57	.0060	.0342	.0086	116	16.81	6.68	-2.58	16.58	.01	-.11
3	2.3	16.33	15.94	16.35	16.26	16.22	.0363	.0315	.0079	127	16.48	7.32	-2.70	16.30	.81	-.90
4	2.4	16.14	15.70	15.77	15.97	15.89	.0396	.0291	.0073	137	16.16	9.99	-3.16	16.03	2.69	-1.64
5	2.5	15.64	15.87	15.90	15.61	15.75	.0228	.0270	.0068	147	15.83	.94	-.97	15.74	.01	.12
6	2.6	15.66	15.84	15.55	15.62	15.62	.0303	.0251	.0063	159	15.50	2.29	1.51	15.46	4.07	2.02
7	2.7	15.10	15.28	15.20	15.32	15.22	.0094	.0235	.0059	169	15.18	.27	.52	15.16	.61	.78
8	2.8	14.70	14.97	14.67	15.06	14.85	.0355	.0222	.0056	179	14.85	.00	.00	14.87	.07	-.27
9	2.9	14.64	14.40	14.44	14.58	14.52	.0129	.0212	.0053	189	14.53	.02	-.14	14.57	.47	-.69
10	3.0	14.23	14.11	14.12	14.18	14.16	.0031	.0204	.0051	196	14.20	.31	-.56	14.27	2.37	-1.54
11	3.1	14.03	14.06	14.06	14.04	14.05	.0002	.0199	.0050	200	13.88	5.78	2.40	13.96	1.62	1.27
12	3.2	13.66	13.75	13.52	13.57	13.63	.0103	.0197	.0049	204	13.55	1.31	1.14	13.65	.08	-.29
13	3.3	13.33	13.28	13.05	13.28	13.24	.0158	.0197	.0049	204	13.23	.02	.14	13.34	2.04	-1.43
14	3.4	12.81	13.11	12.88	13.16	12.99	.0293	.0200	.0050	200	12.90	1.62	1.27	13.02	.18	-.42
15	3.5	12.79	12.48	12.55	12.74	12.64	.0221	.0206	.0052	192	12.58	.69	.83	12.69	.48	-.69
16	3.6	12.13	12.62	12.67	12.25	12.42	.0718	.0215	.0054	185	12.25	6.35	2.31	12.37	.46	.68
17	3.7	12.27	12.08	11.91	11.97	12.06	.0250	.0226	.0056	179	11.93	3.03	1.74	12.04	.07	.27
18	3.8	11.67	11.66	11.81	11.75	11.72	.0050	.0240	.0060	167	11.60	2.40	1.55	11.70	.07	.26
19	3.9	11.34	11.42	11.66	11.47	11.47	.0185	.0257	.0064	156	11.28	5.63	2.37	11.36	1.89	1.37
20	4.0	11.26	11.24	11.03	11.28	11.20	.0135	.0276	.0069	145	10.95	9.06	3.01	11.02	4.70	2.17
21	4.1	10.67	10.53	10.64	10.98	10.70	.0372	.0298	.0074	135	10.63	.66	.81	10.67	.12	.35
22	4.2	10.39	10.40	10.08	10.33	10.30	.0225	.0323	.0081	123	10.30	.00	.00	10.32	.05	-.22
23	4.3	9.96	10.00	9.65	9.93	9.88	.0254	.0357	.0088	114	9.98	1.14	-1.07	9.96	.73	-.85
24	4.4	9.31	9.68	9.21	9.84	9.51	.0893	.0381	.0095	105	9.65	2.06	-1.43	9.61	1.05	-1.02
25	4.5	9.09	8.95	9.16	9.25	9.09	.0294	.0414	.0104	96	9.33	5.53	-2.35	9.24	2.16	-1.47
26	4.6	9.07	8.86	8.86	8.53	8.83	.0498	.0450	.0112	89	9.00	2.57	-1.60	8.88	.22	-.47
27	4.7	8.71	8.39	8.58	8.34	8.50	.0294	.0489	.0122	82	8.68	2.66	-1.63	8.50	.00	.00
28	4.8	7.94	8.52	8.17	8.04	8.17	.0641	.0529	.0132	76	8.35	2.46	-1.57	8.13	.12	.35
29	4.9	7.74	7.88	7.71	7.84	7.79	.0065	.0573	.0143	70	8.03	4.03	-2.01	7.75	.11	.33
30	5.0	7.01	7.37	7.70	7.62	7.42	.0963	.0620	.0155	65	7.70	5.10	-2.26	7.37	.16	.40

$$^a \quad y_i^{cal} = 23.957 - 3.251\, x_i$$

$$^b \quad y_i^{cal} = 21.618 - 1.850\, x_i - .2000\, x_i^2$$

$$\Sigma \delta_i^2 = 91.33 \qquad\qquad \Sigma \delta_i^2 = 29.23$$
$$\div 28 \qquad\qquad\qquad \div 27$$
$$S_{(0)}^2 = 3.262 \qquad\qquad S_{(0)}^2 = 1.083$$
$$S_{(1)} = 1.806 \qquad\qquad S_{(1)} = 1.040$$

obtained was $S_{sm}^2 = 0.162 - 0.0879x + 0.01357x^2$. The values of $S_{sm_i}^2$ calculated with this equation are also listed in Table 2.

No measurements were rejected as being discordant. For example, let us look at the measurements at $x_i = 2.2$, where the observed y_{ik} values were

16.62   16.46   16.62   16.60

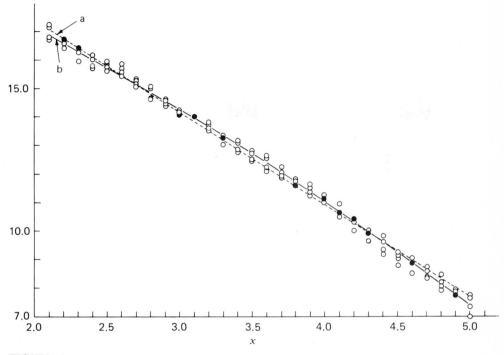

FIGURE 1
Plot of experimental points and least-squares lines: (*a*) straight line fit (2 parameters), (*b*) quadratic fit (3 parameters). A filled-in circle represents two or more nearly coincident points.

The value of Q is $24/26 = 0.923$, which exceeds $Q_c(90\%)$ for $N = 4$. Should the value 16.46 be rejected? If these four measurements were the only ones, one should certainly reject that value. Here, however, we have the benefit of much more information, and a glance at the rest of the data show that the range of y_{ik} values at $x_i = 2.2$ is reasonably representative. In applying rejection criteria or other statistical criteria one must always take into account the context in which the criteria are given, and be careful to make the best possible use of *all* the data at hand.

There being four measurements averaged, the value of the smoothed variance for y_i, the mean of the y_{ik}, was taken as $S_{sm_i}^2/N$ where $N = 4$. This now is our best estimate of σ_i^2. Its reciprocal, rounded to an integer for computational convenience, was taken as the weight w_i. In many other circumstances, there will not be as many as four measurements of each y_i to form a basis for estimating w_i values. Often only one y_i value is available at each x_i. If this is the case, *it is important to carry out multiple measurements at least for some x_i values to provide a feel for "typical" σ_i values.* Also, one should use all one's general knowledge and pay careful attention to the range of y_i, x_i values in estimating the best weights.

Since the plot in Fig. 1 appears to confirm roughly a theoretical expectation of linear behavior, a two-parameter least-squares fit to the function

$$y = \alpha_0 + \alpha_1 x$$

was undertaken. This fit yielded $\alpha_0 = 23.9567$, $\alpha_1 = -3.2507$. Values of y_i^{cal} calculated with these parameters are given in Table 2. The value $\chi_v^2 \simeq S_{(1)}^2 = 3.262$ is quite different from unity; for $P_{int} = 0.05$ and $v = m - n = 30 - 2 = 28$, the limiting χ_v^2 is 1.476. Therefore, at the 95-percent confidence level, our χ_v^2 value is *not* consistent with the assumption that the residuals δ_i are representative of a normal distribution. Either our weighting scheme is wrong or the data are not well represented by a linear fit.

Examination of the fit of the straight line to the data points (as characterized by the residuals) shows that the points tend to be a little below the line near the ends and a little above near the middle, as if the correct fitting function should have a small amount of curvature. Let us now consider a quadratic model, which may be justified by a more refined theory or may be purely empirical. Accordingly, a new fit was made with the function

$$y = \alpha_0 + \alpha_1 x + \alpha_2 x^2$$

This fit yielded $\alpha_0 = 21.6183$, $\alpha_1 = -1.8505$, $\alpha_2 = -0.2000$. The value of $\chi_v^2 = S_{(1)}^2$ was 1.083 for this three-parameter fit, clearly within the bounds of the χ^2 test. A decrease in χ^2 from that for the two-parameter fit was to be expected; χ^2 never increases and practically always decreases when an additional parameter is added to an existing model (provided that the new model reverts to the previous one for some value of the added parameter). While the fact that $S_{(1)}^2$ is so close to unity is gratifying, we may apply an additional test to confirm that the improvement resulting from the addition of the quadratic parameter is statistically significant at the 95-percent confidence level. With the F test in the form given by Eq. (38) we obtain

$$F(1, 27) = \frac{28 S_{(1)2}^2 - 27 S_{(1)3}^2}{1 \cdot S_{(1)3}^2} = \frac{28 \times 3.262 - 27 \times 1.083}{1.083} = 57.34$$

At $P = 0.05$, the limiting value of $F(1, 27)$ is 4.22; thus, the improvement is significant at the 95-percent confidence level; indeed, it is still significant at about the 99.99-percent confidence level.

Having satisfied ourselves of the statistical adequacy of the fit, we are now entitled to estimate the uncertainties in the refined parameters. We apply Eqs. (28), (40), and (43) to obtain estimated standard deviations and 95-percent confidence limits for both the two-parameter and three-parameter determinations:

2-parameter model $(v = 30 - 2 = 28)$

$$\alpha_0 = 23.96(13) \qquad \text{or} \qquad 23.96 \pm 0.22$$
$$\alpha_1 = -3.25(4) \qquad \text{or} \qquad -3.25 \pm 0.06$$

3-parameter model ($v = 30 - 3 = 27$)

$$\alpha_0 = 21.6(3) \qquad \text{or} \qquad 21.6 \pm 0.5$$
$$\alpha_1 = -1.85(18) \qquad \text{or} \qquad -1.85 \pm 0.30$$
$$\alpha_2 = -0.20(3) \qquad \text{or} \qquad -0.20 \pm 0.04$$

The uncertainty values calculated for the two-parameter model are quoted only for illustrative purposes, since we have good reason to believe from our statistical tests that they are not to be relied upon. Note that they are much smaller than the uncertainty values calculated for the three-parameter model. Note also that the differences between the two models in the values of α_0 and α_1 are far outside the limits of error. Had we lazily accepted the two-parameter fit and its estimated error limits, and attached physical significance to either of the two parameters, we would have badly deceived ourselves. On the other hand, in view of the excellent statistical performance of the three-parameter fit, we may accept the parameter values and the estimated uncertainty values with considerable confidence.

Coda. Now it can be told: The "experimental" y_i values in Table 2 were actually generated from *known* "true" values of the parameters and *known* standard deviations! Thus we have here a demonstration of how well the method of least squares works.

We started out with

$$y_i^{\text{true}} = 21.33 - 1.697x_i - 0.2232x_i^2$$

and the y_i values were assigned a standard deviation of

$$\sigma_i^{\text{true}} = 0.8 - 0.4x_i + 0.06x_i^2$$

Numbers from a random-number table were used to generate normal deviates. These were multiplied by σ_i^{true} as given above and the resulting synthetic "errors" were added to the y_i^{true} to produce the "experimental" data.

Let us see how well the least squares worked. The smoothed variance came out to be $S_i^2 = 0.162 - 0.088x_i + 0.136x_i^2$. This cannot correspond exactly with $(\sigma_i^{\text{true}})^2$, which is a quartic. The least-squares smoothing does not look particularly good, especially near the ends of the range. In the following comparison, results of a completely independent visual smoothing (not used) are also included:

x_i	$(\sigma_i^{\text{true}})^2$	S_i^2(l.s.)	S_i^2(visual)
2.1	0.0504	0.0372	0.041
3.0	0.0196	0.0204	0.017
4.0	0.0256	0.0276	0.024
5.0	0.0900	0.0620	0.061

The least-squares smoothing in this case cannot be said to be really superior to visual smoothing; although neither the least squares not the visual variances are particularly satisfying compared to the true variances, they are the best we have. Certainly either of them is better than equal variances, and very much better than the individual variances, for the purpose of assigning weights. (For example, the variance determined for the four measurements at $x_i = 3.1$ is 0.0002, corresponding to a weight of 5000, about 25 times the maximum weight eventually assigned.) However, the least-squares method is not very sensitive to modest errors in the weights. If such errors are quite large, statistical tests should indicate the existence of problems with the assigned weights.

Now let us look at the parameter values with their 95-percent confidence limits, tabulated below.

	2-Parameter	3-Parameter	True
α_0	23.96 ± 0.22	21.6 ± 0.5	21.33
α_1	-3.25 ± 0.06	-1.85 ± 0.30	-1.697
α_2	—	-0.20 ± 0.04	-0.2232

The values from the three-parameter determination agree with the true values well within the 95-percent confidence limits quoted for the former. However, the values from the two-parameter determination deviated from the true values by 12 and 26 times the confidence limits! This provides further illustration of the danger of relying on estimated uncertainties when the model may be defective.

Finally, it should be noted that all of the calculations for this example were done by hand on a \$30 calculator. They were later repeated on a computer for checking, but this example illustrates that one does not always have to be dependent on the availability of a high-speed and expensive computer.

REFERENCES

1. P. R. Bevington, "Data Reduction and Error Analysis for the Physical Sciences," pp. 92–118, 134–186, 204–245, 297–303, McGraw-Hill, New York (1969).
2. A. I. Khinchin, "Mathematical Foundations of Statistical Mechanics," p. 166, Dover, New York (1949).
3. E. Prince, "Mathematical Techniques in Crystallography and Materials Science," pp. 77–82, Springer-Verlag, New York (1982).
4. F. S. Acton, "Numerical Methods that Work," chap. 17, Harper & Row, New York (1970).
5. R. Annino and R. Driver, "Scientific and Engineering Applications with Personal Computers," pp. 336–340, 533–536, Wiley, New York (1986).
6. P. R. Bevington, *op. cit.,* pp. 81–88, 235–236, 313–316.
7. W. C. Hamilton, "Statistics in Physical Science," pp. 49–60, 124–144, Ronald Press, New York, (1964).
8. P. R. Bevington, *op. cit.,* pp. 196–203, 317–323.
9. "CRC Handbook for Probability and Statistics," 2d ed., pp. 296–298, 306, CRC Press, Cleveland, Ohio (1976).

10. "CRC Handbook of Chemistry and Physics," 65th ed., pp. 105–109, CRC Press, Boca Raton, Fla. (1984).
11. S. C. Abrahams and E. T. Keve, *Acta Crystallogr.* **A27,** 157 (1971).
12. W. C. Hamilton and S. C. Abrahams, *Acta Crystallogr.* **A28,** 215 (1972).

GENERAL READING

P. R. Bevington, *op. cit.*
R. de Levie, "When, Why, and How to use Weighted Least Squares," *J. Chem. Educ.* **63,** 10 (1986).
W. C. Hamilton, *op. cit.*
E. Whittaker and G. Robinson, "The Calculus of Observations," 4th ed., Blackie, Glasgow (1944).

CHAPTER
XXI

USE OF COMPUTERS

In most fields of physical chemistry and chemical physics, the use of digital computers is considered indispensable. Many things are done today that would be impossible without modern computers. These include, for example, Hartree–Fock *ab initio* quantum-mechanical calculations, least-squares refinements of x-ray crystal structure determinations with hundreds of adjustable parameters and many thousands of observational equations, and Monte Carlo calculations of statistical mechanics, to mention only a few. Moreover, computers are now commonly used to control commercial instruments such as mass spectrometers, x-ray single-crystal diffractometers, and NMR spectrometers, as well as to control specialized devices that are part of an independently designed experimental apparatus. In this role a computer may give all necessary instructions to the apparatus, and record and process the experimental data produced, with relatively little human intervention.

Accordingly, acquaintance with computers and with the basics of programming is essential to a sound foundation in physical chemistry. The student interested in either experimental or theoretical physical chemistry should obtain some experience with computers early in his or her training. A laboratory course in physical chemistry is an excellent place to start, if a start has not already been made. The purpose of this chapter is to provide a brief introduction to computers, especially microcomputers, which have become essential tools in research and industrial laboratories.†

The field of computers is large and growing rapidly. No one book, much less a single chapter like this one, can provide all the information that a beginner might need in setting up calculations for experiments such as the ones in this book. Much of the necessary information, which includes not only the principles and techniques of programming but also many practical aspects of dealing with a specific machine, is best gained from the operating manuals for that machine, from operating-system manuals and programming manuals, and from experience. If one is dealing with an institutional computer, it is also necessary to gain familiarity with the local operating system: such operating systems are often idiosyncratic. The present chapter is intended only as a prelude and a stimulus.

Before proceeding further, it would be well to sound a warning. Computers may be essential to modern science, but they also can be a tremendous time sink. The feeling of power that one may get from programming a complicated calculation can be intoxicating, but one should not lose sight of the physical problem that is being solved. Serious physical scientists might do well to hang a sign over their desks saying something like

WARNING
EXCESSIVE COMPUTER PROGRAMMING MAY BE
HAZARDOUS TO YOUR SCIENTIFIC HEALTH

However, this warning should not be used as an excuse to do a "quick and dirty job." A careful and disciplined approach to programming is essential. A program should be written with adequate documentation, including liberal comment statements, so that it will be clear to any later user of the program how the computation is being carried out.

For the purpose of this chapter, the word computer will mean a machine that can store a large number of digital data, and can operate on those data with stored instructions to perform the required arithmetical and/or logical operations. The digital computer stores all of its information in the form of binary numbers, for which the only digits are 0 and 1. A single binary digit is known as a *bit*; a sequence of eight bits is known as a *byte*. It is awkward to deal with computers at the level of individual bits; therefore it is common to describe computer memory and computer data in terms of bytes. A sequence

---

† The arithmetic calculations for nearly all of the experiments in this book can be done with nothing more than a good pocket calculator. In some cases, however, the use of a digital computer may be beneficial in avoiding excessive tedium (which is conducive to errors), saving time (if not too much time is spent on programming), gaining reliability, and in some cases enabling the direct recording of experimental data in digital form. Examples are Exps. 6 to 8 (calorimetry; time–temperature curves), Exp. 9 (partial molar volume), Exp. 15 (liquid–solid phase diagram; cooling curves), and Exps. 20 to 25 (chemical kinetics). Other examples may be found by the student.

of bytes (usually 2 bytes in microcomputers, 4 or more in larger computers) upon which the computer can operate as a unit is known as a *word*. The word may be an instruction, or a number, or a component of either. A byte may represent an alphanumeric character; each of its $2^8 = 256$ values may be assigned to a different alphabetical letter, decimal digit, punctuation mark, or control characer. A particular set of assigned character codes, known as ASCII character codes (American Standard Code for Information Interchange) is in almost universal use in computer work. A list of ASCII codes may be found in many programming manuals.

Within the computer's memory there is no distinction between instructions and data; the distinction exists only in the manner in which the memory contents are treated in the operation of the program. The instructions include arithmetic operations (add, subtract, multiply, divide, change sign), certain logical operations, word-modification instructions (shift left or right, truncate, etc.), control instructions (such as conditional jumps), and input–output (I/O) instructions (read, write).

This chapter will deal principally with the class of computers known as *microcomputers*. Before we undertake to discuss microcomputers and their use, it would be well to point out that there is so much variety among the available microcomputers and microcomputer operating systems that it is impossible in this chapter to present discussions applicable to all; yet, for illustrative purposes, it is useful to discuss features of some particular types of microcomputers and systems. The discussions in this chapter apply most accurately to the family of computers represented by the International Business Machines IBM PC and the so-called "IBM compatibles," and to the operating system known as PC-DOS/MS-DOS,[1,2] which was developed by the Microsoft Corporation for this family of computers.

TYPES OF INFORMATION HANDLED BY COMPUTERS

Computer instructions. Generally, a computer instruction consists of two main parts: the operation code (add, multiply, jump, etc.) and the operand specification (register or registers involved, and/or data address). The structure of an instruction in memory is too complicated to merit discussion here; fortunately, it is seldom necessary to deal with an instruction in the form in which it is stored in the computer ("machine language"), since an instruction in both of its parts can be represented symbolically in what is called "assembly language," which will be discussed briefly later. In a microcomputer an instruction occupies 1 to 6 bytes in memory.

Integers. Integers are stored in computers in the form of binary numbers. Typically, in microcomputers, an integer is stored in one 16-bit, 2-byte word.

The decimal arithmetic notation of common usage represents integers in the form

$$N \equiv d_m d_{m-1} \cdots d_2 d_1 d_0 \tag{1}$$

where each of the d_i represents a digit 0 through 9. In this notation, the given array of digits represents the numerical value given by

$$N = \sum_{i=0}^{m} d_i B^i \tag{2}$$

where B is the base of the numbering system; for decimal numbers, $B = 10$. For example, the decimal number 137 has the value $1 \times 10^2 + 3 \times 10^1 + 7 \times 10^0$. There is nothing particularly special about the number 10 for the base of a number system; presumably it arose historically from the fact that the human being has ten fingers. The basic memory devices of a digital computer have in effect only two fingers, represented by the two states of a bistable micro-electronic circuit, and therefore can accommodate only the digits 0 and 1. Equations (1) and (2) apply as well to binary (base 2) notation as to decimal notation.

Long strings of binary numbers are tedious for humans to write and to read. Since 8 and 16 are powers of 2, they represent bases that can be converted easily to and from binary. In the octal system (base 8), each octal digit 0 through 7 represents a binary number of up to three digits; one can simply partition the sequence of digits in a binary number into groups of three to obtain conversion to octal. In the hexadecimal system (base 16), now much more widely used than octal in computer work, each digit 0 through 9 and A through F represents a binary number of up to four digits. The binary expressions for the digits in the octal, decimal, and hexadecimal systems are given in Table 1. Arithmetic operations in these various bases are intrinsically identical except that the addition and multiplication tables are different for the different bases.

A negative integer is usually represented as the "2's complement" of the corresponding positive integer. If an integer is stored as one 16-bit word (two bytes), then the 2's complement is obtained by subtracting the number from 2^{16}. This is equivalent to changing all 1's to 0's and all 0's to 1's, and then adding 1 to the resulting number. Example:

Decimal	Binary	Hexadecimal
425	0000 0001 1010 1001	1A9
−425	1111 1110 0101 0111	FE57

The left-most digit is used only to indicate the sign of the number: $0 = +$, $1 = -$. The magnitude of the number itself is thus limited to the right-most 15 binary digits. Thus a two-byte signed integer is restricted to the range $-32\,768$ to $+32\,767$.

TABLE 1
Number systems used with computers

Binary B = 2	Octal B = 8	Decimal B = 10	Hexadecimal B = 16	Expression in binary form
0	0	0	0	0
1	1	1	1	1
	2	2	2	10
	3	3	3	11
	4	4	4	100
	5	5	5	101
	6	6	6	110
	7	7	7	111
		8	8	1000
		9	9	1001
			A	1010
			B	1011
			C	1100
			D	1101
			E	1110
			F	1111

Reals. These are nonintegers that can cover an extremely wide range of magnitudes, for they are represented in storage in *floating-point* form:

$$\text{mantissa} \times 2^{\text{exponent}}$$

The format of the real may vary. Typically a floating-point real occupies 4 or 6 bytes in storage, depending on the programming language; details of the form in which the exponent and mantissa are accommodated in this space are beyond the scope of this discussion. The usual formats of 4- and 6-byte reals allow respectively about 7 and 11 decimal digits of precision, with a magnitude range of approximately 10^{-38} to 10^{38} in either case. Most programming languages allow also a "double precision" real, typically stored in 8 bytes. This can provide, for example, 16 decimal digits of precision with magnitudes ranging approximately from 10^{-308} to 10^{308}.

For input into the computer, several formats are available for reals, for example:

> 56.239
>
> −0.00030012
>
> 4.50652E15 ($\equiv 4.50652 \times 10^{15}$)
>
> −2.115436E-28 ($\equiv -2.115436 \times 10^{-28}$)
>
> −5.4117037821332D-87 ($\equiv -5.4117037821332 \times 10^{-87}$)

The last three examples are in what is called exponential notation. D has the same meaning as E, except that E is intended to indicate the normal (single precision) real while D is intended to indicate the double-precision real. The same formats can be specified for output of reals from the computer.

Arrays. An array is a sequence of subscripted variables, such as the components of a vector or the components of a matrix of two or more dimensions. Examples:

$$a(j) \; (\equiv a_j), \quad j = 1, \ldots, 10$$
$$b(j, k), \quad j = 1, \ldots, 10, \; k = 1, \ldots, 20$$
$$c(j, k, l), \quad j = 1, \ldots, 10, \; k = 1, \ldots, 20, \; l = 1, \ldots, 50$$

The components of an array are usually reals, while the subscripts are integers.

Strings. A string is a sequence of bytes representing characters, usually ASCII characters. When specified within a program a string is usually identified by enclosing it in quotation marks:

"This is a string."

"Pi has the value 3.14159."

"Press any key to continue this program."

A string occupies in memory a number of bytes equal to the number of bytes in the string plus one or two for the length of the string.

Boolean quantities. Boolean variables, used in logical operations, are limited to two values, namely "TRUE" and "FALSE." These may be represented in the computer by a single bit, with value 1 and 0 respectively, but in practice a byte is usually reserved for it. Boolean variables may be manipulated by the operations of logical arithmetic: "NOT", "AND", "OR", "XOR", etc.

Among Boolean variables of interest to programmers are arithmetical statements which must be either true or false:

$$A = 0$$
$$X < Y$$
$$Z \geqslant 10 \quad \text{(rendered in a program as } Z > = 10\text{)}$$

These statements are found as arguments of "IF" statements used with high-level programming languages for program branching; e.g.,

IF $X < Y$ THEN (statement A) ELSE (statement B)

If the Boolean $X < Y$ is TRUE then statement A is executed; if FALSE, then statement B is executed.

Files. Files are extremely important in computer work, for they represent information stored outside the computer on auxiliary devices (magnetic tapes, disks, etc.). Apart from entering data on the keyboard or printing the results out on the printer, virtually all input–output (I/O) operations involve files. In the case of a microcomputer, files are recorded on "floppy disks" (also known as "diskettes") and/or on a hard disk if the computer has one.

Files are of many different kinds. An important classification is "files of byte", which are usually text files. In many cases the bytes are ASCII characters, and the files may be data files (input data, output data), documents for communication, etc. Text files are not always in ASCII; most word processors produce documents that are not in ASCII, and program files in older versions of the BASIC programming language are usually not ASCII, but non-ASCII files can be converted to ASCII files if necessary. The computer operating system generally has commands that display the contents of an ASCII file on the computer screen or print it out on the printer; in the operating system PC-DOS/MS-DOS[1,2] these commands are respectively "TYPE" and "PRINT" followed by the file specification. However many data files are not "files of byte"; data are usually stored in binary form, as they are in the computer, often in units known as "records", each of which may contain several pieces of data. At the end of each file there may be an "end-of-file" (EOF) character; for ASCII files it is ASCII 26 (or 1A hexadecimal) produced on the keyboard for text files as Ctrl-Z or ˆZ.

Every file has a "filename", with which it is referenced by the computer. In the PC-DOS/MS-DOS operating system it has the form

$$xxxxxxxx.yyy$$

in which xxxxxxxx is a string of no more than eight alphanumeric characters, and .yyy is an (optional) *extension* generally indicating the type of file. Examples are .EXE and .COM, executable program files; .ASM, .FOR, .BAS, and .PAS for program source files in assembly language, FORTRAN, BASIC, and Pascal respectively, .DOC or .TXT for text files, and .DAT for data files. To call up a file on a computer with several disk storage drives (often named A, B, C, . . .) it is necessary to precede the filename with a drive or "path" designation, unless the current or intended residence of the file is the "default drive". In its simplest form the drive or path designation is simply the name of the disk drive followed by a colon. Thus the complete "filespec" for a BASIC program source file on drive B might be, for example:

$$B:MYPROG.BAS$$

HARDWARE

So far, we have treated the computer as if it were a black box, with only passing reference to some of the best-known I/O components. It is time now to describe the various elements of a computer system. These can be divided into two major categories, known as *hardware* and *software*. The hardware consists of the electronic and mechanical equipment. The software consists of the information that tells the computer what to do.

The organization of the hardware of a microcomputer is illustrated in Fig. 1. The essential elements, apart from I/O devices, are VLSI (Very-Large-Scale Integrated) circuits on silicon or gallium-arsenide wafers ("chips") mounted on a printed-circuit board often called the "mother board". Those chips that have processing functions are called *microprocessors*. The principal chips in a microcomputer are the central processing unit (CPU) and possibly a separate arithmetic coprocessor, the random access memory (RAM), a read-only memory (ROM) chip known as BIOS (Basic Input–Output System) as well as perhaps other ROM chips (such as the BASIC interpreter ROM in Fig. 1), and controllers for the I/O devices. The mother board also has "slots" for plugging in "extension boards" or "cards" that may contain additional memory, interfaces for external equipment or communication, a calendar-clock, and/or other devices. Plugged into the mother board directly or via chassis terminals are the necessary I/O devices for getting information into and out of the computer: the video monitor, the keyboard, one or more magnetic storage devices, and the printer. The various components of the mother board communicate with each other via a set of parallel conductors called the *bus*. Many of these components will be discussed briefly.

Central processing unit (CPU). This unit contains a number of registers, the most important of which are an accumulator, which is principally involved in arithmetic and logical operations, and instruction registers that contain the current instruction and the address from which it came in memory. The CPU, which is usually on a single chip, reads an instruction from a location in memory determined by the sequential position of that instruction in the stored program, or, in certain cases, by a previously executed jump instruction. Then, in accord with the instruction read, the CPU performs some arithmetic, logical, or control function, referring to the contents of one or more of its own registers and/or to the contents of a register in memory identified by the address part of the instruction. Built into the CPU is the "hard-wired" circuitry that represents the *instruction set*—the translation of the binary instruction code into the elementary step or steps that must be performed in order to execute the instruction. The separate arithmetic coprocessor, if one is present, assists the CPU by handling, at high speed, operations involving floating-point real numbers; if one is not present such operations are performed (at much lower speed) in the CPU through the use of routines provided by the compiler software or programmed in assembly language.

Random access memory (RAM). The RAM constitutes storage for a large number of bytes of information; the modifier "random access" indicates that any location in memory can be accessed in no more time than is required for any other location, typically one or a few microseconds in a microcomputer. Usually the number of locations (bytes of storage) is expressed in units of K ($1\,K = 2^{10} = 1024$), and very frequently it is $2^n\,K$; 512 K is a common size for a RAM. With 16 bits for the address of an instruction, any of $2^{16} = 65\,536$

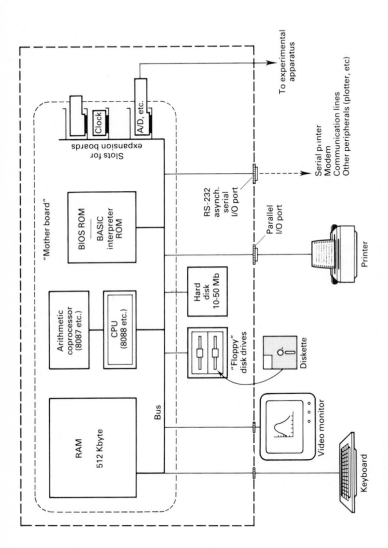

FIGURE 1

Schematic diagram of a hypothetical microcomputer as used in laboratory work. Components inside the heavy dashed line are typically within the computer chassis, as are also the power supply and I/O controller cards (not shown). The light dashed line encloses the "mother board", a printed-circuit board containing slots and sockets into which are plugged the various internal components. Among these may be "expansion boards", such as a calender-clock, a module containing A/D and D/A converters and a timer clock for interacting with experimental apparatus, additional RAM, "firmware" (ROM or EPROM boards; e.g., a BASIC interpreter), and a telephone modem.

locations (64 K) may be addressed. A larger number of bits is required for accessing a RAM larger than 64 K. This problem is solved in the computer architecture either by using more than two bytes for the address, or (as is the case for many microcomputers) by storing the additional needed bits (the "segment") in a separate register in the CPU, which may be changed by special instructions when necessary. At the start of a computation the RAM generally contains only the operating system, which will assist the reading into the RAM of the user's program. The program, with the assistance of the operating system, reads the input data into the RAM as needed, carries out the intended calculations, and reads the results out of the RAM into the output devices. Information stored in the RAM is "volatile", meaning that it will be lost when the computer is shut off or when there is a power failure, unless (as in the case of some portable machines) the contents are preserved with the aid of a rechargeable battery.

Read-only memory (ROM). The ROM is nonvolatile memory, the contents of which are stored permanently in the process of the manufacture of the ROM chip. Information can be read by the CPU from a ROM but cannot be written into the ROM by the CPU. One particular ROM that is an essential part of the modern microcomputer is the BIOS, or Basic Input–Output System. (The word "Basic" in this context has nothing to do with the programming language called "BASIC".) It contains routines for initiating the initial "bootstrap operations" of loading the operating system into the computer, and for operating the I/O devices, particularly the magnetic disk drives. Other essential ROM and RAM chips serve the video monitor and the keyboard. Some microcomputers contain the BASIC interpreter on a ROM chip; other special purpose ROMs may also be present. A variant of the ROM, the programmable read-only memory (PROM) can be written into *once* by a special procedure; what is written into it is permanently stored. More popular at present is the erasable programmable read-only memory (EPROM), the contents of which can be erased by exposure of the chip to ultraviolet light through a special window so that new information can be written into it. The EPROM is particularly valuable for storing routines used to control an instrument (see below), as the contents are nonvolatile and not inadvertently erasable. There now exists also the EEPROM (electronically erasable programmable read-only memory).

Auxiliary memory: the hard disk. Usually the capacity of the RAM in a microcomputer is no more than 512 K or 1024 K; larger capacities are becoming more common as the prices of memory chips fall. However, vastly larger memory capacity can be available in devices that are not random access, time of access being measured at best in milliseconds and at worst in many seconds. The permanently installed rapidly spinning magnetic disk, usually referred to as the "hard disk", is the most important high-capacity device with access times of small fractions of a second (typically about 40 ms). Time of access is limited by the time necessary to position the read-write head over the appropriate track and the time necessary for the sector of the disk containing the desired data to rotate to the head. Hard disks in microcomputers usually have capacities of from 10 to 50 megabytes (MB); the disks in large computers have larger capacities. Magnetic tape offers even higher capacities, but takes much more time to access. This latter medium is often used as a means for *backing up* a hard disk, providing archival copy of critical files. (Floppy disks may also be used to back up a hard disk.)

I/O devices. In a microcomputer the absolutely essential I/O components are (1) a keyboard for manual entry of information into the computer; (2) a video monitor for displaying the status of the computer, operating system, and working program, as well as input data and calculated results; and (3) a device for receiving data from or delivering data to a magnetic storage medium such as a magnetic disk or magnetic tape. (The hard disk has already been discussed as an auxiliary memory device but it is usually treated by the computer system as an I/O device since it is external to the central computer circuitry.)

The monitor is very similar to a TV screen. Most often, information is displayed on the monitor in alphanumeric form; however, the computer can usually be programmed to produce graphic displays (bar graphs and pie charts for business use; data plots and mathematical functions for scientific applications). The monitor display may be monochrome or in color. Color is useful in games and business applications, but for serious scientific computing the monochrome monitor is satisfactory for most purposes. Moreover, it often has higher resolution than the color monitor, and thus clearer alphanumeric characters and sharper graphics.

The magnetic medium in virtually universal use for microcomputers is the flexible or floppy disk. This is usually $5\frac{1}{4}$ inches in diameter, although a $3\frac{1}{2}$-inch disk is used in some machines. The disks used in microcomputers are sometimes called "diskettes" to distinguish them from the 8-inch diameter disks used in larger computers, but we will use the term "disk" in our description, generally referring to the $5\frac{1}{4}$-inch size. The disk is contained in a special protective envelope within which it is caused to spin by the action of a spindle that is inserted in the center hole when the disk drive is operated; a slot provides access for the magnetic read–write head or heads. Most floppy disks are coated with magnetic material on both sides and are read from and written to on both sides. Before the disk can be used, it must be *formatted*. Typically each side of the disk is considered to be divided into 40 concentric circular tracks, and each track is divided into nine sectors each with a capacity of 512 bytes; thus the disk has a total capacity of $360\,K = 368\,640$ bytes. The formatting, which is done on command by the operating system, marks the tracks and the origins of the sectors. The floppy disks must be handled carefully. Since the envelope has openings for the drive spindle and the magnetic heads, *it must be protected from dust and fingerprints*. For this purpose each disk is provided with a specially treated paper sleeve, in which it should always be stored. It must also be protected from high temperatures, bending, contact with heavy *or pointed* objects, and *magnetic fields*. In particular, any writing on a disk label must be done with a soft felt-tip pen and not with a pencil or ball-point pen. Since accidents can happen with disks, including inadvertent erasure or modification in the computer itself, all disks containing information that one cannot afford to lose should be backed up by copying the contents onto another disk which should be stored in a separate location.

The printer is essential for obtaining "hard copy" of the information created or processed by the computer. For scientific work the dot-matrix printer is the most useful because of its speed, flexibility, and reliability. The more sophisticated models can be programmed to print Greek and mathematical characters, as well as characters designed by the user, and to plot graphics output (graphs, curves, etc.) at somewhat limited resolution. Dot-matrix printers are durable, but noisy. Usually they receive their input in parallel (entire bytes being transmitted simultaneously, rather than a bit at a time) from the designated parallel printer port (Centronics or similar standard). Only passing mention can be given here to other types of printers: letter-quality printers using "daisy wheels" (noisy), ink jets (quiet), or lasers and electrostatic transfer (quiet, versatile, and expensive). Mention should be made of X–Y digital plotters, which are essential for high-resolution graphics.

Other I/O devices are the telephone modem, which permits communication between computers or terminals via telephone lines, and devices for communicating with scientific instruments and experimental apparatus. The latter devices will be described more in detail later.

Types of computers. Table 2 illustrates the wide range of computer types and capabilities. *Supercomputers,* such as the Cray models and forthcoming parallel multiprocessor computers of potentially even greater computing power, are very expensive; those that exist are located in widely scattered centers and are accessible to users virtually everywhere in the country via telephone, microwave, and satellite links. Their hourly charges are very high but for huge computations they are more economical than smaller computers; moreover, they are the only computers available that can do certain types of computations, such as real-time computer modeling of complex physical systems. The traditional *mainframe* central institutional computers are still in considerable use, but as they become obsolete they may not be replaced, since most of their applications are becoming more appropriate to supercomputers on the one hand or smaller computers like the Digital Equipment Corporation (DEC) VAX on the other. The emergence of the VAX family of computers has caused a revolution in scientific computing, since it is now possible for individual science departments to have such computers, to be operated directly by their staff members and students. Also of considerable importance is the DEC PDP-11 series of *minicomputers* which have come into considerable use in laboratories.[3]

TABLE 2
Representative computers

Type	Examples	Speed[a]	Typical cost	Typical scientific applications
Ultrafast multiprocessor computers	Cray X-MP	400	$10–20M	Mathematical modeling (e.g., aeronautics); high-energy physics; exceptional statistical and quantum-mechanical calculations; regional computing
Ultrafast computers	Cray 1 & 2	100		
Mainframe computers	IBM 360 and 370 IBM 3081D CDC 7600 CDC 6300–6600	1–20	$1–5M	Institutional computer center; general and time-share computing
Mini-maxi computers	DEC VAX HP 3000	0.1–1	$50–500K	Departmental or research group computer center; laboratory computing
Mini-computers	DEC MICROVAX DEC PDP-11	0.01–1	$20–100K	Laboratory computing; instrument control
Micro-computers ("personal computers")	IBM PC/AT (w 80287) IBM PC (w 8087) IBM PC (w/o 8087) IBM compatibles Apple IIc Apple Macintosh HP 150 HP Vectra	0.01 0.005 0.001	$2–4K $1–2K	Personal computing; laboratory computing; instrument control

[a] Speeds are given in "MFLOPS"—million floating-point operations per second. The actual speed depends very much on the computer configuration and the conditions of operation. These speed and cost estimates are very rough and are intended only to indicate orders of magnitude for comparison purposes.

It is likely that most undergraduate students for the next few years will get the bulk of their computing experience with *microcomputers,* although some will probably have access to VAX- and PDP-11-type computers as students. Discussion of details of the use of these and larger computers is beyond the scope of this chapter. In our discussion of hardware and software in this chapter we emphasize general concepts of microcomputers, applicable to those produced by IBM, DEC, Apple, Hewlett-Packard, etc. Our discussions tend to apply most closely to one particular family of computers: that represented by the IBM PC, PC/XT, and PC/AT computers, and the so-called "IBM compatible" computers exemplified by Compaq, AT & T 6300, Zenith ZF-151-22, and others. These are built around the Intel 8086/8088 microprocessors and their successors. The main reasons for this are (1) that the relatively open architecture of this group of computers permits great flexibility for specialized applications, such as data communications and interfacing with external systems (such as experimental apparatus); and (2) this group of computers, operating under the PC-DOS/MS-DOS operating system, has attracted by far the most attention from software producers and makers of accessories such as experimental interfaces. We believe that this situation may exist for some time to come. However, computers built around the Motorola 68000 microprocessor—e.g., the Apple Macintosh, the Commodore Amiga, and their successors—may provide increasing competition, particularly since later versions of the very popular Macintosh, for example, also have an open architecture.

The Intel 8088 microprocessor,[4] found in many of the IBM family microcomputers, is a 16-bit processor; however, the final "8" signifies that it connects only to an 8-bit bus, and thus it can only receive or send data one byte at a time. Some microcomputers have the 8086 microprocessor, which connects to a 16-bit bus. Some newer machines have the updated 80286 microprocessor, or the 32-bit 80386 that can perform at about 20 times the speed of the 8088. All members of this family have provision for installing an optional arithmetic coprocessor, the Intel 8087 or 80287, which handles floating-point arithmetic, in double precision if desired.

Many of the components of the IBM/compatible family have become standard throughout the industry. However, the compatibility is not absolute. The BIOS ROM chip is generally proprietary and is copyrighted by the individual manufacturers. The video systems show marked differences throughout this family; it is perhaps here that the most trouble is encountered with incompatibility when it is found that graphics software written for one computer will not work on others. This can be overcome to some degree with a special color/graphics extension board plugged into an expansion slot to replace the function of the existing video controller chip. Graphics aside, almost all software written for the IBM PC will run on any of the compatibles. Moreover, most expansion boards manufactured for the IBM PC, including those used for interfacing, will work on the compatibles, which generally follow the IBM system of memory allocation and interrupt assignments.

SOFTWARE

The software consists of information, in the form of programs, that tells the computer what to do. The individual user is able to generate only a part of the software needed to operate the computer—namely, his own application program. However, so much more than that is necessary to exploit the computer fully that a whole industry—comparable in magnitude to the computer hardware manufacturing industry—has been built up to produce computer software, which is generally distributed in the form of floppy disks. Software is of several types, which will be discussed briefly here.

Operating systems. The operating system combines facilities of two main types. One type contains the indispensable mechanics of processing information from the keyboard, outputting information to the printer, reading files from a disk or writing them to a disk, and assigning memory locations to programs and data read into the computer. The most essential of these functions are resident in the computer at all times during its operation; others may be automatically read in from a disk as needed. The other type contains "utilities." Some of these, such as copying or renaming files, or exhibiting their contents on the monitor screen or printer, are in some systems (e.g., PC-DOS/MS-DOS) also resident at all times. Utilities that require large amounts of memory space generally reside on the operating-system floppy disk or in the hard disk, from which they are read into the computer on command. These include disk formatting and disk copying and comparing functions, a text editor for creating and modifying text files, a binary file editor, perhaps an assembler, a linker for linking together binary files (e.g., user program and library subroutine files), and perhaps a compiler, among others.

CP/M is one of the oldest microcomputer operating systems, having been developed for 8-bit microcomputers. It has been updated to the 16-bit IBM and compatibles, but the system in greatest current use for these is PC-DOS/MS-DOS (DOS = Disk Operating System), developed by the Microsoft Corporation.[1,2] PC-DOS is the version written for IBM; MS-DOS is a very similar version for the compatibles. Hereafter we will refer to both as DOS.

UNIX[5] is an operating system developed by AT & T Bell Laboratories in 1969, originally as a multitasking, multiuser system for the Digital Equipment Corporation's PDP series of computers. (Multitasking is the use of a computer to handle several jobs concurrently; multiuser means accessibility by more than one user at the same time.) UNIX has gone through much evolution and now is the primary system recommended for some of the newest microcomputers, notably the newest IBM and AT & T computers, as well as being available for other machines such as the IBM PC/XT and the IBM compatibles. It is complex and not very easy to learn, but is more versatile and powerful than DOS. It has spawned variants such as Venturcom's VENIX for 8088 CPU machines and Microsoft's XENIX for the 68000 and other CPUs (now including the 8088). In the opinion of many, UNIX represents the trend of operating systems of the future.

Programming languages. The only programming language that the central processing unit understands is *machine language,* consisting of binary numerical codes, which are next to impossible for human users to deal with on any appreciable scale. At a level above machine language is *assembly language,*[6,7,8] in which the various instructions are coded with alphabetical mnemonics and the addresses are replaced by human-assigned alphanumeric symbolic codes. A brief discussion of the latter will be given in a later section.

Nearly all programming by computer users is done with one or another of several high-level programming languages.[9,10] These languages bear a limited relationship to an ordinary written language (usually English) but have their own structure and syntax. A high-level language frees the programmer from being concerned with assigning storage locations and with individual arithmetical steps in the evaluation of mathematical formulae; while assembly language requires that every machine operation be specified, a high-level language enables a program to be composed of "statements," each of which may be the equivalent of several machine operations. Further, the high-level programming systems include extensive diagnostics that alert the user to syntactical and run-time errors.

Machine language and assembly language are very machine dependent because they are intimately dependent on the details of the architecture of the CPU. High-level languages have the very great advantage of being essentially machine independent. However, to use a particular language on a particular machine requires that the programming system (interpreter and/or compiler; see below) has been adapted to that machine. Since the major high-level programming language systems have been adapted to practically all widely used computers, programs written in them are generally transportable, i.e., are capable of being run on all such computers with little or no changes.

Historically, FORTRAN (= FORmula TRANslation) is the oldest of these high-level languages, having been developed in the 1950s by IBM. After the *source code* is written in FORTRAN, it is *compiled*; in this operation, analogous to assembly, the program is converted to executable machine language. For many years FORTRAN has been the principal high-level language used in institutional mainframe computers, and has gone through periodic revisions and extensions; it is now available for microcomputers. Although FORTRAN is used less for microcomputer programming than some other languages (and for that reason will not be treated further in this chapter), it remains of importance to microcomputer users because programs written in FORTRAN for larger computers can often be adapted for compilation on microcomputers, and because software written in FORTRAN for microcomputers continues to be available. The current version of FORTRAN is FORTRAN 77.[11,12]

For microcomputers, on the other hand, high-level languages began with BASIC (Beginners' All-purpose Symbolic Instruction Code), developed in 1965 at Dartmouth College by John G. Kemeny and Thomas E. Kurtz to respond to the needs of students and users of smaller computers. BASIC originally had a simpler structure and smaller instruction set than FORTRAN, but has evolved through the years to a level of sophistication greater than that of FORTRAN. In contrast to FORTRAN, which is a compiled language, BASIC was originally designed as an *interpreted* language and is still usually employed in that form. During the interpretative execution of a BASIC program, the BASIC interpreter must be resident in the computer memory. The source text is read character by character by the interpreter, and as each statement is assimilated it is translated into machine language commands that are immediately executed. This process is followed every time the program is run; when the program contains a loop, the interpretation process is repeated throughout every journey through the loop. Thus, an interpreted BASIC program runs very much more slowly than the equivalent program in FORTRAN (by about a factor of 10), since a FORTRAN program needs to be compiled only once. However, if speed is not important, as in relatively short and straightforward programs, interpreted BASIC may be preferred because the program can be run as soon as written without the intervention of the compilation step. The current forms of interpreted BASIC are IBM's BASICA and Microsoft's MBASIC and GWBASIC; these are very similar.[13,14] Interpreted BASIC is so widely used that a BASIC ROM chip is a standard component in some microcomputers; furthermore, some extension boards and peripheral equipment have BASIC ROM chips.

In recent years, compilers have appeared for BASIC that produce compiled programs approximately equivalent in speed to those written in FORTRAN. An example is Microsoft's QuickBASIC,[15] which also provides considerable relaxation from some of the programming constraints of interpreted BASIC. For example, while in interpreted BASIC it is necessary that all of the lines of a program be numbered, with the numbers increasing monotonically from beginning to end, this is not necessary in QuickBASIC; moreover, in QuickBASIC alphanumeric labels are allowed, as in FORTRAN. Fortunately, programs written in BASICA or GWBASIC can be compiled by QuickBASIC with only minor changes or none at all.

Another compilable language, Pascal, developed originally by Niklaus Wirth at the Eidgenossische Technische Hochschule in Zürich, Switzerland, in 1970–1971, was originally intended primarily as a teaching tool; it has undergone some evolution since that time.[16] Unlike nearly all other high-level computer languages, its name is not conventionally rendered in all capital letters, perhaps because it was named after a person, Blaise Pascal (1623–1662), a French mathematician who as a young man conceived and constructed a machine for performing arithmetic calculations. Its most important ancestor is ALGOL (ALGOrithmic Language), developed several years earlier. Pascal is a language developed for what is called *structured programming*. The characteristics that make programming structured are largely beyond the scope of this book, but some of them are the following. The program produced is completely modular (while those produced by languages like traditional BASIC and FORTRAN are only haphazardly so). Much use is made of "procedures" (analogous to "subroutines"), which are called by name. There is a wide variety of data types (integer, real, string, byte, char, set, array, record, pointer, file, user-created types, etc.), some of which are themselves structured. The language syntax and command structure make the program development largely self-documenting. A Pascal source program, by reason of the organization forced upon it by the strict syntax and structure, tends to be more readable than source programs in traditional FORTRAN and BASIC; however, FORTRAN 77 and QuickBASIC are largely capable of supporting structured programming and thus permit improved readability.

Pascal's most popular present form under DOS, Turbo Pascal[17] (developed by Borland International), differs in a few minor respects from more standard forms of Pascal, but at least in the opinion of its many users has significant advantages over them. It compiles very much faster than FORTRAN or other Pascals, and produces more compact machine code. The very high compiling speed enables rapid debugging and makes Turbo Pascal competitive with interpreted BASIC in convenience for writing short programs, while the running times are competitive with those of the other compiled languages and superior to those that do not support the 8087/80287 arithmetic coprocessor, which is supported by Turbo Pascal.

A relatively recent-comer to the field of high-level languages is called C.[18,19] It has roots in ALGOL and Pascal but has features that make it more powerful and at the same time more difficult to learn. The source code is more terse, and is less readable than that of Pascal. The compiler imposes fewer constraints than Pascal, and places much more responsibility on the programmer, but with this responsibility comes more freedom for innovative and highly efficient programming. C is closely associated with the UNIX operating system used in the newer computers; indeed the UNIX operating system is written in C, which gives UNIX marked advantages in "transportability" over other operating systems, which have had to be written in assembly languages for all of the types of computer on which they are intended to be run.

User software. In addition to programs that may be written by the computer user for specific applications, the computer user will frequently need programs or packages of programs for various kinds of standard calculations, such as least-squares fitting, statistical treatments, matrix manipulations, generation of special functions (Bessel functions, etc.), and handling of graphics. These, generally written in high-level languages for compilation by the user, are available from other computer users, commercial vendors, computer "bulletin boards," computer magazines, and books.[20,21,22]

Other software. Word processors and other business software need no treatment here with the exception of *spreadsheets*, such as Lotus 1–2–3™ and Symphony™. Besides their applications to business, these have important scientific applications when long columns of experimental data need to be processed in a uniform way. If the data (such as spectroscopic data) are entered into one or more columns of a spreadsheet, they can be manipulated simply by specifying simple formulae that relate the columns; thus spectra may be added or subtracted or scaled, all without doing any computer programming. The spreadsheets are, however, expensive enough to rule out their purchase for limited or casual purposes.

There are a variety of small packages of utilities on the software market, of which Superkey and Prokey are examples; these can be used to greatly enhance the power of the keyboard by permitting a combination of functions to be exercised through a single keystroke.

COMPUTER PROGRAMMING

Computers are often characterized in the popular press as wonderful electronic brains. Wonderful they may be, but brains they are not in any real sense. In truth, the computer is a literal-minded idiot. It will do anything in its power that you tell it to do, no matter how stupid. It makes no allowance for human frailties; even the diagnostics it gives you for the mechanical errors you may make in programming it have themselves been programmed by some human being. The computer has, however, important saving graces: whatever you tell it to do will be done very quickly and accurately, and it never quits out of boredom or makes errors due to fatigue (although it may suffer occasional nervous breakdowns).

Since "to err is human," no one writing a program of any length should expect it to work properly in all respects the first time. Something as trivial as a missing or misplaced punctuation mark will prevent the successful compilation of a program. Fortunately, something this trivial is easy to correct with the aid of diagnostic messages given by the compiler or operating system. The hardest mistakes to avoid, and the ones the computer will usually be unable to point out to you, are *mistakes in the logical structure of the program.* The logical structuring of the program is the aspect of programming that requires the greatest care and thought; the rest is merely mechanical.

A program is an algorithm, i.e., a rigorously defined computational procedure. As such it can be diagrammed on paper, before the actual coding begins. When written out, the program is basically a linear rendering of the algorithmic flow diagram. In such a rendering, program branching (represented two-dimensionally on the flow diagram much as a road is branched on a map) is accomplished with conditional jump statements. The construction of a flow diagram (or its equivalent) for a new program, although often neglected, is strongly recommended for beginners, and even for experienced programmers if the program is very complicated.

We will give here, as an example of a simple computation to be programmed for a computer, the problem of finding the greatest common divisor of two positive integers. We first describe the algorithm with words, then present a flow diagram, and then show how the computation is accomplished by means of a program written in Turbo Pascal.

We employ Euclid's algorithm. Let the two integers be A and B. Euclid tells us to divide the larger by the smaller; *if* A is the larger we will interchange the two so that A is the smaller, B the larger. We examine the remainder R in B/A. *If* R is zero, the answer is almost trivial since A is already the greatest common divisor. *If* R is not zero, we rename the quantity A as B and rename the quantity R as A, repeat the division and again examine the remainder R. The procedure is repeated until we find a value of R that is zero. Then the last value of A is the greatest common divisor.

A flow diagram for this algorithm is presented in Fig. 2. We add an extra feature permitting this algorithm to be applied successively to any number of

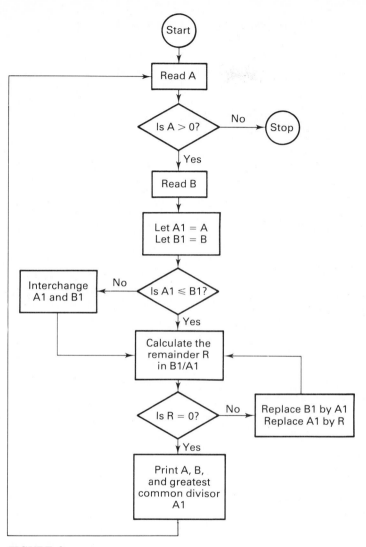

FIGURE 2
Flow diagram for Euclid's algorithm for finding the greatest common divisors of pairs of positive integers *A* and *B*.

A,B integer pairs. *A* is tested on entry to make sure it is a nonzero positive integer; if it is not, this is taken as a signal to end the computation. Figure 3 shows the program in Pascal. (Information contained between curly brackets, {...}, is treated by the compiler as comment and ignored during compilation.) Here input values of *A* and *B* are entered by keyboard, and displayed on the screen as keyed in; the great common divisor is displayed on the screen at the end of the computation, and both input data and output data are output on

```
Program EUCLID;   {to find greatest common divisor of two positive integers}

var
     A, B, a1, b1, t, r, gcd: integer;

begin
     writeln('Program EUCLID');
     writeln('for finding greatest common divisor of two positive integers');
     writeln;
     writeln('NOTE: Integers may not exceed 32767'); writeln;
     writeln(LST,'Program EUCLID Output'); writeln(LST);
     writeln(LST,'    FOR INTEGERS    GREATEST COMMON');
     writeln(LST,'      A        B            DIVISOR IS',^M^J); {Return+linefeed}
     repeat
          writeln('Enter first positive integer, A: (to QUIT, enter zero)');
          readln(A);
          writeln;
          if A>0 then
          begin
               writeln('Now enter second positive integer, B (NOT ZERO): ');
               readln(B);
               writeln; writeln;
               a1:=A;
               b1:=B;
               if a1>b1 then
               begin      {interchange a1 and b1}
                    t:=a1;
                    a1:=b1;
                    b1:=t;
               end;       {a1 is now less than or equal to b1}
               repeat
                    r:=b1 mod a1;   {find remainder r in b1 divided by a1}
                    gcd:=a1; {greatest common divisor if r=0}
                    b1:=a1;
                    a1:=r;     {replace b1 by a1, and then a1 by r}
               until r=0;
               writeln('For integers A = ', A, ' and B = ', B, ',');
               writeln('     the greatest common divisor is ', gcd);
               writeln(LST, A:6, B:7, gcd:14);
               writeln; writeln('********************');
               writeln; writeln;
          end;
     until A=0;   {go back and do another run unless zero was entered for A}
     writeln(LST, ^L);    {advance printer paper to top of next page}
end.
```

```
Program EUCLID Output

   FOR INTEGERS    GREATEST COMMON
      A       B         DIVISOR IS

     25      11              1
     16      64             16
    437     228             19
   1015    2494             29
  13281   18407            233
```

FIGURE 3
Euclid program in Pascal and output of results of sample calculations.

the printer. (The code "LST," in some of the Writeln statements, refers to the printer; in the absence of a device code, output is to the screen.) The structured character of the program is evident; it is built of blocks, each enclosed between the words "begin" and "end," or between the words "repeat" and "until." Note that blocks may be nested within larger blocks. The modular structure is here emphasized, for clarity, by indentation, which is optional and has no bearing on the eventual compilation. (This can also be done in other languages.)

Note the instances of the italicized word *if* in the next to the last paragraph. The first instance corresponds to the first instance of *if* in the program. If the program were written in traditional FORTRAN or BASIC the subsequent instances of *if* would also be matched by an *if* in the program, and the resulting program branching would involve the use of the GOTO statement. In turn the GOTO statement would require that the destination statement carry a line number or label. However, in this Pascal program the *if* and GOTO and label are made unnecessary by the "repeat...until $r = 0$;" construction. (Alternatively, a "while $r > 0$ do begin...end;" construction could have been used.) It is characteristic of the structured programming permitted by Pascal that labels and GOTO statements, while permitted, are used rarely or not at all. The result is a logical, orderly, and highly readable program.

Note that in Pascal all variables must be declared at the beginning of the program, and their types (integer, real, string, etc.) identified. This is not required in FORTRAN or BASIC (except that in those two languages, as in Pascal, the dimensions of arrays must be specified, and some attention must be given in naming the variables to the types they must represent). This predeclaration is often perceived as an inconvenience, but it is a very minor one and contributes to the orderly structure of the program as well as to the speed of compilation.

For the remainder of this section we give instances of some features of BASIC and Pascal that are important in arithmetical computations of the type required by physical chemists, using as examples a couple of the equations in this book. We will use the traditional BASIC and Turbo Pascal.

The Joule–Thomson coefficient of a van der Waals gas is given by Eq. (2–11):

$$\mu = \frac{(2a/RT) - b}{C_p}$$

In the examples below we assume that the variables T, a, b, and C_p have been read in as input data elsewhere in the program and that μ is read out. The value of the gas constant R is specified in the program.

BASIC

.
.
.

    20 R = 8.314 'Gas constant

.
.

    90 REM Joule Thomson Coefficient
    100 MU = (2 * A/(R * T)–B)/CP

.
.

Pascal

```
Program JouleThomson;
var
     mu, a, T, b, Cp: real;
const
     R: Real = 8.314;      {Gas constant}
begin
.
.
     mu : = (2 * a/(R * T)–b)/Cp;      {J.T. coefficient}
.
.
end.
```

Note that the expression of the mathematical formula is almost the same in the two languages. In both, the asterisk ($*$) is used as the multiplication operator, and the slash (/) is used as the division operator, at least for reals. In BASIC a comment may be preceded by the command REM, or by a single apostrophe ('). In Pascal, a comment is enclosed in curly brackets {}. Note again that in Pascal all variables and constants must be predeclared, and their types must be indicated. Notice also the use of ":=" as the assignment operator in Pascal while "=" is used in BASIC, and that all statements are numbered in the conventional BASIC illustrated here.

A powerful feature of all high-level languages is the ability to code repetitive operations. The most common application is in summing. Operations that deal repeatedly with subscripted variables are handled with loops, the format for which varies somewhat among the various languages. As an example of coding repetitive calculations, let us consider Eq. (XX-13), which can be represented in summation form by

$$\Delta \alpha_j = \sum_{k=1}^{n} B_{jk} h_k$$

In matrix terms, this corresponds to the multiplication of an $n \times n$ inverted normal-equations matrix $\mathbf{B}$ by an n-element column vector $\mathbf{h}$ to give the n-element column vector $\Delta \boldsymbol{\alpha}$ representing the shifts in the least-squares parameters.

BASIC

```
10 DIM DEL(20), B(20, 20), H(20)
.
.
50 FOR J = 1 TO N
60 DEL(J) = 0
```

```
 70 FOR K = 1 TO N
 80 DEL(J) = DEL(J) + B(J, K) * H(K)
 90 NEXT K
100 NEXT J
        .
        .
        .
```

Pascal

```
Program XYZ;
var
      del:  array[1..20] of real;
      b:    array[1..20, 1..20] of real;
      h:    array[1..20] of real;
      n, j, k:  integer;
        .
        .
        .
begin
        .
        .
        .
      for j := 1 to n do
      begin
            del[j] := 0.;
            for k := 1 to n do del[j] := del[j] + b[j, k] * h[k];
      end;
        .
        .
        .
end.
```

The DIM statement in the first example and the array declarations in the second are needed to inform the system of the memory storage requirements for the arrays used. These statements must always precede any statements in which the arrays are referenced. In these examples, we have assumed that n or N is no greater than 20, and may be less. To adapt the programs to a problem in which n or N is, say, 30, it is necessary to change the numbers in the dimension statement or array declarations to at least 30.

In the examples above we see that two loops are nested. More than two loops may be nested; the maximum depth of nesting depends on the version being used of the programming language.

Detailed instruction in programming is beyond the scope of this chapter. Programming is best learned by experience with actual computations on a computer, with the aid of appropriate programming manuals.[11-19] There are two levels of concern. First, the syntax and logic must be correct, or the program will not run correctly. Second, even if the program gives a correct result, it may be grossly inefficient and therefore costly in computer time. We will content ourselves here with offering one hint for programming efficiency: if there are operations in a computation that need to be performed only once, keep them out of loops. As a very simple example in BASIC, consider

```
10 F = 0.0
20 FOR I = 1 TO 5000
30 F = F + COS(6.28318 * XJ * X) * SIN(6.28318 * YJ(I) * Y)
40 NEXT I
```

Since the cosine term does not depend on the index I, it needs to be calculated only once. Moreover, Y may be multiplied just once by 6.28318 in advance of the loop. A more efficient version of the above would be

```
10 F = 0.0
15 A = COS(6.28318 * XJ * X)
17 B = 6.28318 * Y
20 FOR I = 1 TO 5000
30 F = F + A * SIN(B * YJ(I))
40 NEXT I.
```

For very short calculations there is no need to be so meticulous, since the additional time invested in programming may not be recovered in execution of the program.

We end this section with a brief discussion of assembly language and its use (under DOS). It is not important for most computer users to be able to use assembly language, but it may be of value for them to have some conception of what assembly language is and what it may be useful for. In an assembly-language program the complete sequence of CPU instructions is specified in minute detail. Instead of the numerical codes for individual CPU instructions, the programmer specifies mnemonics, usually three-letter codes such as *add* (add a byte or a word from one specified CPU register or RAM location to a byte or word in another), *mov* (move a byte or word from one specified CPU register or RAM location to another), *jnz* (cause the CPU to jump out of the normal sequence of instructions to a specified instruction elsewhere in the sequence if the content of a particular register is not zero), etc.; instead of numerical addresses, the programmer assigns convenient alphanumeric codes of his or her devising. The *source program* composed in this manner is stored as an ASCII file with an .ASM extension. This file is processed by a program known as the *assembler*, which converts the source program to the numerical machine language that the CPU understands. The mnemonic instruction codes are replaced by the corresponding numerical codes, and symbolic address codes are replaced by actual addresses relative to an arbitrary origin, which will later be adjusted by DOS to take account of the actual location of the program in RAM. The executable program obtained (often after an additional step of linking with necessary subroutines) is stored as a file with an .EXE or .COM extension. For a more complete understanding the referenced works should be consulted.[6,7,8]

Generally, assembly language is employed to any great extent only by expert programmers. It is much more difficult and time-consuming to use than high-level languages, but in experienced hands it can be much more powerful. Much of the software of commerce is produced in assembly language because the finished machine code will be more compact, and also will operate at higher speed, than the codes produced by high-level languages. (However, the C language produces machine code that is almost as efficient as that produced by optimum assembly language.) A programmer with some familiarity with assembly language can use it to produce small utility programs that will be used again and again, where speed and economy of memory are important considerations, or to produce highly efficient subroutines for programs written in high-level languages. It has already been mentioned that assembly language is to a very considerable degree machine dependent, so that programs written in it have very limited transportability. Fortunately, however, assembly language for IBM microcomputers under DOS is in nearly all respects applicable to all members of the IBM-compatible family.

INTERFACING THE COMPUTER WITH EXPERIMENTS

Just as in modern industry many operations of manufacturing machinery or chemical processing equipment are computer controlled, the control of scientific instruments in the laboratory is now commonly accomplished by computers. There has arisen an industry devoted to the manufacture of equipment for this purpose. A laboratory instrument equipped with an appropriate interface device can be connected directly to an RS-232 serial port on the computer, and can then receive instructions from the computer or on command send data in digital form to the computer. ("Serial" means that bits are sent one at a time rather than in parallel.) If the computer has more than one serial port it can be interfaced in this way to as many separate instruments, each with its own interface device. Alternatively, use may be made of an external "bus" known by the designation IEEE-488,[23] the standards for which have been set up by the Institute for Electrical and Electronics Engineers. This device can interface a computer with up to sixteen separate, properly configured instruments in what amounts to a laboratory network. For example, during a given period of time a computer can control and collect data from an infrared spectrometer, a gas chromatograph, a mass spectrometer, etc., all more or less concurrently. Each instrument will listen only to instructions prefaced by its own address, and will talk only when the IEEE-488 bus is available.

For the remainder of this chapter we will be concerned with a somewhat different approach, namely to equip the computer with an *internal* interface module, in the form of a board that may be plugged into one of the available expansion slots (usually connected by a cable to a box outside the computer that contains additional circuitry and terminals). Manufacturers of such modules include Data Translation, Keithley Instruments (e.g., the Keithley Instruments System 570 Data Acquisition Workstation), Metrabyte Corporation, ICS, Omega Engineering, and Scientific Solutions, Inc., among many others. Features commonly incorporated into interfacing modules include one or more digital-to-analog converters (DAC), one or more analog-to-digital converters (ADC), timers and timing-pulse generators, and various terminals (screw-terminals, BNC connectors, etc.) for making electrical connections to analog measuring devices in instruments. Some modules are designed to be connected directly to specific measuring devices, such as thermocouples and strain gauges; others have much more general applicability. The programming required to communicate with such modules is usually quite different for similar devices made by different manufacturers. It is therefore impossible to give programming examples that are widely applicable; we will content ourselves with discussing, for the purpose of illustration, one device that we have found useful in laboratory instruction.

This device is the LabMaster, manufactured by Scientific Solutions, Inc., of Solon, Ohio. This comes in a number of different models that require only small differences in programming. The one used here converts an external voltage to 12 bits of digital information (11 significant bits plus a sign bit). Other LabMaster models accommodate 14 or 16 bits of digital information. We present in Fig. 4 a program in BASIC for measuring with this device a voltage produced by external equipment. With the LabMaster board discussed here, the allowable voltage input is limited to the range -10.000 to $+9.995$ V. This program deals with a rather simple application, namely analog-to-digital (A/D) conversion only, making use of only one of several analog channels available with the board. No timing functions are carried out by the board; they are all done within the host computer.

In order to understand the program, it is necessary to know a little about the storage registers of the LabMaster board and their functions.[24] The board has 16 one-byte registers addressable by the host computer at locations that we here call ADDRESS to ADDRESS + 15. ADDRESS has for this board a default value of 1808 ($=710$ hexadecimal) but this address may be changed if necessary. These addressable registers often have two functions, one when *written* to the LabMaster by the computer, one when *read* from the LabMaster to the computer. The functions that concern us are described below. It is important to note in advance that the least-significant bit (LSB) in a byte is conventionally numbered bit 0, and the most-significant bit (MSB) is numbered bit 7.

ADDRESS + 4. This is the control byte. When written to by the computer, a 1 in bit 7 is "Computer Load Enable:" autoincrementing by the board or by the peripheral equipment is disabled, and the entire control is in the hands of the computer. The other bits in this control byte deal with various conditions that do not concern us here; we will want them to be zero. Thus, as part of the initialization process, we *write* from computer to the board at ADDRESS + 4 the number 128 ($= 2^7 = 10000000$ binary). When later, during the course of data collection, this byte is *read* from, a 1 in bit 7 informs the host computer that the previous A/D conversion has been finished, so that the computer can now read the two bytes containing the converted digital voltage. After the second of those two bytes has been read, the "done" bit goes back to a zero to await the completion of the subsequent A/D conversion.

ADDRESS + 5. This is *written* to in order to inform the board of the input channel number for the next A/D conversion. When *read* from, this register yields the "Low Data Byte"—the lower 8 bits of the most recent A/D conversion.

ADDRESS + 6. Anything (even a zero) *written* to this register by the computer causes the A/D converter to start a conversion on the specified input channel. When *read* from, this register yields the "High Data Byte"—the upper 8 bits of the result of the most recent A/D conversion. This byte *must* be read *after* the Low Data Byte has been read. (In the model of the board used here, there are only 4 bits of data in the High Data Byte since the effective data word contains only 12 bits.) The two bytes are combined into a *real* result according to the equation

$$A = (\text{High Byte}) \times 256 + (\text{Low Byte})$$

The matter of sign demands some attention here; it exemplifies the kinds of problems that must be faced in programming the application of interface boards like this one. If the voltage is negative, the board stores the converted voltage in 2's complement form, which means that bit 7 of the High Data Byte is a 1 (the structure of that byte being therefore $11111xyz$). However, when each byte is read by the computer, it is stored in the computer as a *two-byte integer* ($00000000\ 11111xyz$); regardless of the content of bit 7, bits 8 through 15 of the High Data Word, *including the sign bit,* are zero, indicating erroneously a *positive* voltage. Thus, if and only if bit 7 in the High Data Byte is 1, the above equation will give for A an erroneous *positive* real result greater than 32 767 ($= 2^{15} - 1$). In this case the real result must be converted into the appropriate negative value by subtracting 65 536! ($= 2^{16}$). The exclamation point in BASIC means that the number 65 536 is real; an integer can have no positive value greater than 32 767.

Once the sign has been taken care of, if necessary, the voltage is given by

$$V = A/204.8$$

for the 12-bit LabMaster considered here. We present here some questions to

```
10 'DVM.BAS DIGITAL VOLTMETER PROGRAM for IBM PC and Compatibles
20 DIM T(2000), V(2000)
30 '   Equipped with 12-Bit Scientific Solutions "LabMaster" ADC/DAC Board
40 GOTO 1080  'Leap over subroutines to start of program
50 '*******************
60 'Subroutine to display voltages on screen as collected, if enough time
70 IF INTVAL<0 THEN 90
80 IF INTVAL<.2 THEN RETURN
90 IF I%=1 THEN PRINT "  Point #     Time (sec)       Volts"
100 PRINT USING "    ####      #####.###      ###.###"; I%, T(I%), V(I%)
110 RETURN
120 '*******************
130 'Data collection subroutine
140 CLS: PRINT "Number of points to collect: if MANUAL mode, enter 1;"
150 PRINT "     if TIMER mode, enter 2-2000": INPUT N%: PRINT
160 IF N%=1 THEN 310
170 IF N%<1 OR N%>2000 THEN BEEP: GOTO 140
180 PRINT "For TIMER control, enter time interval in seconds between successive
points"
190 PRINT "   Data will be displayed on screen as collected if interval >= 0.2 s
econd"
200 PRINT "   (Time indication is unreliable for interval < 0.2 second)": PRINT
210 PRINT "   For high speed set interval = 0. You will be asked for number"
220 PRINT "   of 'CYCLES' (FOR NEXT, to use up time). Number required for given"
230 PRINT "   time interval must be established by trial.  With QUICKBASIC"
240 PRINT "   compiler and a 4.77 MHz computer clock, interval is approximately"
250 PRINT "   (6 + 0.027*CYCLES) milliseconds": PRINT
260 PRINT "   NOW ENTER TIME INTERVAL"
270 INPUT INTVAL: IF INTVAL<0 THEN BEEP: GOTO 270
280 IF INTVAL=0 THEN INPUT "Enter number of CYCLES: ", CYCLES%
290 PRINT "Press spacebar to start    (To interrupt collection, press any key)"
300 GOTO 350
310 INTVAL=-1: N%=2000      'High upper limit for manual mode
320 PRINT "For Manual control, "
330 PRINT "     Press spacebar each time you want a measurement"
340 PRINT "     Press M to quit and return to Menu": PRINT
350 'Write control byte to board: disable auto-incrementing, external start
360 '   conversions, and all interrupts
370 OUT ADDRESS%+4,128     'Bit 7 = 1; all other bits = 0
380 OUT ADDRESS%+5, CHANNEL%   'Write channel # to board
390 X=INP(ADDRESS%+6)      'Reset done flipflop by reading high byte
400 A$=INKEY$: IF A$<>" " THEN 400
410 IF INTVAL>=0 THEN PRINT "Collecting . . ."
420 START!=TIMER
430 FOR I%=1 TO N%         'Data collection loop
440 OUT ADDRESS%+6,0       'Initiate conversion
450 IF INP(ADDRESS%+4)<128 THEN 450   'Await conversion done (bit 7 high)
460 IF INTVAL<>0 THEN TM!=TIMER: IF TM!<START! THEN START!=START!-86400! 'Timer
resets at 86400 sec
470 IF INTVAL<>0 THEN T(I%)=TM!-START!      'Elapsed time
480 LO%=INP(ADDRESS%+5)    'Input low byte
490 HI%=INP(ADDRESS%+6)    'Input high byte
500 LOW!=LO%
510 HIGH!=HI%
520 A=256!*HIGH!+LOW!              'Assemble word
530 IF A>32767! THEN A=A-65536!    'in case voltage is negative
540 V(I%)=A/204.8          'Convert to voltage
550 GOSUB 60   'Display on screen if possible
560 IF INTVAL>0 THEN TM!=TIMER: IF TM!<START! THEN START!=START!-86400!
570 IF INTVAL>0 THEN IF TM!<START!+I%*INTVAL THEN 560    'Wait for proper time
580 IF INTVAL=0 THEN FOR M%=0 TO CYCLES%: NEXT M%
590 IF INTVAL>=0 THEN Y$=INKEY$: IF Y$="" THEN 620 ELSE N%=I%: GOTO 630
600 X$=INKEY$: IF X$="m" OR X$="M" THEN N%=I%: GOTO 630
610 IF X$<>" " THEN 600
620 NEXT I%
630 IF INTVAL<>0 THEN GOTO 670
640 TM!=TIMER: IF TM!<START! THEN START!=START!-86400!
650 INTVAL=(TM!-START!)/N%
660 FOR I%=1 TO N%: T(I%)=(I%-1)*INTVAL: NEXT I%
```

FIGURE 4 (Part 1)

Program DVM in BASIC, for recording voltages from experimental apparatus, written for an IBM PC or compatible computer equipped with a 12-bit Scientific Solutions "LabMaster" expansion board. See text for details.

```
670 PRINT "Press any key to return to Menu"
680 A$=INKEY$: IF A$="" THEN GOTO 680
690 RETURN
700 '********************
710 'Subroutine to display voltages when collection is complete
720 CLS: PRINT "Press spacebar successively to display 20 points at a time"
730 PRINT "Press M to quit and return to Menu after any group": PRINT
740 PRINT "Use Shift/Prt8c to print out any 20-point group if desired": PRINT
750 PRINT "Now enter the number of the point at which you wish to start": PRINT
760 INPUT M%: IF M%<1 OR M%>N% THEN BEEP: GOTO 760
770 PRINT "  Point #      Time (sec)        Volts"
780 FOR I%=M% TO M%+19
790 IF I%>N% THEN 850
800 PRINT USING "    ####        #####.###        ###.###"; I%, T(I%), V(I%)
810 NEXT I%
820 X$=INKEY$: IF X$="m" OR X$="M" THEN 870
830 IF X$<>" " THEN 820
840 M%=I%: IF M%<=N% THEN 770
850 PRINT "Press any key to return to Menu"
860 X$=INKEY$: IF X$="" THEN 860
870 RETURN
880 '********************
890 'Statistics and summary subroutine
900 IF N%<4 THEN PRINT "Not enough points for valid statistics": BEEP: FOR J%=1
TO 2000: NEXT J%: RETURN
910 CLS: TOTAL=0: ST=0
920 FOR I%=1 TO N%: TOTAL=TOTAL+V(I%): ST=ST+V(I%)*V(I%): NEXT I%
930 AVERAGE=TOTAL/N%: DEV=ST-(N%*AVERAGE*AVERAGE)
940 SD=SQR(ABS(DEV/(N%-1)))
950 PRINT "For ";: PRINT USING "####"; N%;: PRINT " points,"
960 PRINT "     Average voltage = ";: PRINT USING "###.###"; AVERAGE
970 PRINT "     Standard deviation = ";: PRINT USING "###.###"; SD: PRINT
980 PRINT "Press any key to return to Menu"
990 X$=INKEY$: IF X$="" THEN 990
1000 RETURN
1010 '********************
1020 'Subroutine to store output on disk
1030 CLS: PRINT "Enter filespec for OUTPUT (recommended format d:filename.DAT; d
= disk drive)"
1040 INPUT F$
1050 OPEN F$ FOR OUTPUT AS #1
1060 FOR I%=1 TO N%: PRINT #1, I%; T(I%); V(I%): NEXT I%
1070 CLOSE #1: RETURN
1080 '*** *** *** START OF MAIN PROGRAM *** *** ***
1090 KEY OFF: CLS
1100 ADDRESS%=1808     '= 710h; starting address of LabMaster board
1110 CHANNEL%=0     'ADC input channel #
1120 CLS: PRINT "RECORDING DIGITAL VOLTMETER using 12-bit LabMaster": PRINT
1130 PRINT "Connect your source of voltage to 'ADC 0',"
1140 PRINT "     being sure to use a ground wire": PRINT
1150 PRINT "The measurable range is -10.000 to +10.000 volts": PRINT
1160 PRINT "You can collect up to 2000 values, in either of two modes:"
1170 PRINT "     MANUAL, or under TIMER control    (Timer limit: 24 hr)": PRINT
1180 PRINT "NOW PRESS ANY KEY TO CONTINUE"
1190 X$=INKEY$: IF X$="" THEN 1190
1200 'Menu
1210 CLS: PRINT "                    M E N U": PRINT: PRINT
1220 PRINT "To collect data, press. . . . . . . . . . . . . . . . . . C": PRINT
1230 PRINT "To display data on screen for viewing, press. . . . . . . V": PRINT
1240 PRINT "To present statistical data (if meaningful), press. . . . S": PRINT
1250 PRINT "To store data on disk, press. . . . . . . . . . . . . . . D": PRINT
1260 PRINT "To QUIT, press. . . . . . . . . . . . . . . . . . . . . . Q": PRINT
1270 PRINT "FOR DATA PRINTOUT, USE Shift/PrtSc": PRINT
1280 X$=INKEY$: IF X$="c" OR X$="C" THEN GOSUB 130: GOTO 1210
1290 IF X$="v" OR X$="V" THEN GOSUB 710: GOTO 1210
1300 IF X$="s" OR X$="S" THEN GOSUB 890: GOTO 1210
1310 IF X$="d" OR X$="D" THEN GOSUB 1020: GOTO 1210
1320 IF X$<>"q" AND X$<>"Q" THEN 1280
1330 END
```

FIGURE 4 (Part 2)

challenge the interested student: Where did the number 204.8 come from? What number would be used for a 16-bit LabMaster? Would the precaution described above for negative voltages have to be followed for the 16-bit device? All of the necessary clues are present in this chapter.

The program presented in Fig. 4 was written in GWBASIC, and it may be run with an interpreter, or it may be compiled with QuickBASIC or another BASIC compiler for higher speed of operation. The essence of the program is contained in the subroutine occupying statements 130 to 690; this subroutine may be modified, simplified, inserted into other BASIC programs, adapted for other interface modules, or translated into other programming languages, as desired by the user. (For Pascal, the INP function and the OUT statement are replaced by statements using the *Port* array.) The program permits time–voltage data to be collected in either of two ways: in a MANUAL mode, in which a voltage measurement is made each time the spacebar is pressed, and a TIMER mode, in which the number of measurements and the time interval between measurements are specified by the user. In either case the times and voltages are stored in arrays, and may be output onto the monitor screen (from which they may be printed with the Shift/PrtSc command), or stored as a file on a disk, from which the data may be read into another program for analysis.

A limitation of the present program is that of timing with the use of the TIMER function in BASIC. Specified time intervals less than 0.2 s do not yield evenly spaced reading times. The clock in the PC runs at 4.77 MHz; for the TIMER function this is scaled down by a factor of 2^{18}, causing the TIMER to "tick" 18.2 times per second; this is one tick every 55 ms. (The LabMaster board contains an AMD 9513 timer chip, but the programming of this chip is exceedingly complicated and beyond the scope of this chapter.) To facilitate reading at very short time intervals, the present program contains an option whereby time measurements are made only at the beginning and at the end of the series of voltage measurements, and the computer is allowed to run in the period between without reference to the TIMER function. The total elapsed time is then divided evenly among the measurements, on the assumption that the computer speed is constant. This is only an approximation, since the operation of the computer is affected slightly by clock and other interrupts and by variation, from one program cycle to the next, in the contents of the instruction queue of the CPU. Tests of this program as compiled with QuickBASIC show that, with a 4.77 MHz computer clock, voltage measurements can be made as fast as every 6 ms, and that the variations from a constant time interval are at most 1 ms.

The program has been written to be self-explanatory in operation; all needed instructions are given on the monitor screen. For further understanding of details of the program, or for dealing with problems in operation, the reader is referred to BASIC programming manuals.[13,15]

A BASIC program for reading the data from a disk file and printing it out is presented in Fig. 5. The disk-reading portion of this program can be

```
10 REM DVMLIST - To read from disk, and print out, previous DVM output
20 CLS
30 DIM V(2000), T(2000)
40 PRINT "DVMLIST: Program to read from disk, and print out, previous DVM output
": PRINT
50 PRINT "Enter filespec ([d:]filename.DAT) for INPUT ": PRINT
60 INPUT F$
70 OPEN F$ FOR INPUT AS #1
80 I%=1
90 IF EOF(1) THEN 120     'Halt read-in on END OF FILE
100 INPUT #1, I%, T(I%), V(I%)
110 I%=I%+1: GOTO 90
120 CLOSE #1
130 N%=I%-1
140 '************
150 'List data on printer
160 LPRINT: LPRINT "  POINT #     Time (sec)        Volts": LPRINT
170 FOR I%=1 TO N%
180 LPRINT USING "    ####        #####.###        ###.###"; I%, T(I%), V(I%)
190 NEXT I%
200 LPRINT CHR$(12)      'Advance form to top of next page
210 END
```

FIGURE 5
Program DVMLIST in BASIC, for reading the disk file created by program DVM and printing out that file.

inserted in any reader-generated BASIC program for processing data as required in any particular experiment.

We leave to the reader the responsibility of going from this limited illustrative example to applications for specific experiments, as well as to programming for other interface boards and more complicated applications involving D/A conversion, external timing, etc.

A device such as the LabMaster can be used with any device that produces output in the form of a voltage: an electrochemical cell, a thermocouple, a thermistor, a strain gauge (in, for example, a Wheatstone bridge circuit), a piezocrystal pressure sensor, etc., but one must reckon with inherent limitations of sensitivity and resolution. A voltage range as wide as -10 to $+10$ volts, combined with a 12-bit A/D converter (giving 11 significant bits) provides a resolution of only 5 mV, which corresponds to the voltage equivalent of the least-significant bit. Thus, the unassisted 12-bit LabMaster model described here is inadequate for use with thermocouples as used in calorimetry, the output of which is measured in microvolts. Other (more expensive) versions of the LabMaster, as well as boards furnished by other manufacturers, incorporate programmable gain by a factor as high as 500; when this is combined with a 16-bit input, the resolution is increased to $0.6\,\mu\text{V}$. When this is used with a copper–constantan thermocouple, the temperature resolution is 0.014 K, perhaps adequate for calorimetry. Thermistors (which must be calibrated) give greater resolution than thermocouples. A larger gain can be accomplished by placing an operational-amplifier circuit between the interface module and the instrument. This has to be done very carefully, with proper attention to electrical grounding, shielding, and electrical isolation requirements, because very low voltage inputs are subject to electrical noise from the host computer. For very low-level input voltages, such as those involved in calorimetry with thermocouples, greater cost-effectiveness and reliability might be obtained by employing more sensitive electronic instrumentation, with an IEEE-488 interface to the computer. An example is the Keithley Model 196 DMM (Digital Multimeter). This has $6\frac{1}{2}$ decimal digits of precision, which at its most sensitive range of 0–300 mV gives a resolution of 0.1 μV, or about 0.002 K with a copper–constantan thermocouple. Such an instrument has many other valuable features, making it of general use in the laboratory.

Application of the kind of interfacing described in this section to some of the experiments in this book will depend on the equipment available in each particular laboratory, and on the ingenuity of the students and teaching staff to deal with problems such as those described in the previous paragraph. With this important reservation, calorimetry experiments are a potential application; they involve reading a temperature at regular intervals. The enterprising student may be challenged to program the analysis of the time–temperature curve, in accordance with the treatment given in Chapter V.

REFERENCES

1. C. DeVoney, "PC-DOS User's Guide and MS–DOS User's Guide," Que Corporation, Indianapolis, Ind. (1984).
2. P. Norton, "MS-DOS and PC-DOS User's Guide," Brady/Prentice-Hall, Old Tappan, N.J. (1985).
3. J. W. Cooper, "The Minicomputer in the Laboratory, with Examples Using the PDP 11," Wiley, New York (1977).
4. P. Norton, "Inside the IBM PC," Brady/Prentice-Hall, Old Tappan, N.J. (1985).
5. K. Christian, "The UNIX Operating System," Wiley, New York (1983).
6. E. J. Desautels, "Assembly Language Programming for PDP-11 and LSI-11 Computers," William C. Brown Company, Dubuque, Iowa (1983).
7. L. J. Scanlon, "IBM PC & XT Assembly Language: A Guide for Programmers," Brady Communications/Prentice-Hall, Bowie, Md. (1983).
8. D. Rollins, "IBM PC 8088 Macro Assembler Programming," Macmillan, New York (1985).
9. J. A. Feldman, "Programming Languages," *Scientific American,* **241,** 94 (December 1979).
10. R. Annino and R. Driver, "Scientific and Engineering Applications with Personal Computers," chap. 4, Wiley, New York (1986).
11. P. Meissner and E. I. Organick, "FORTRAN 77, Featuring Structured Programming," 3d ed., Addison-Wesley, Reading, Mass. (1980).
12. D. D. McCracken, "Computing for Engineers and Scientists with FORTRAN 77," Wiley, New York (1984).
13. V. Kassab, "Technical BASIC," Prentice-Hall, Englewood Cliffs, N.J. (1984).
14. F. R. Ruckdeschel, "BASIC Scientific Subroutines," vol. 1, Byte/McGraw-Hill, Peterborough, N.H. (1981).
15. "Microsoft QuickBASIC Compiler," (with compiler and other software, version 2.0), Microsoft Corporation, Redmond, Wash. (1986).
16. D. Cooper and M. Clancy, "Oh! Pascal!," 2d ed., W. W. Norton, New York (1985).
17. "Turbo Pascal," (with compiler and other software, version 3.0), Borland International, Scotts Valley, Calif. (1985).
18. N. Gehani, "C for Personal Computers," Computer Science Press, Rockville, Md. (1985).

19. B. W. Kernighan and D. M. Ritchie, "The C Programming Language," Prentice-Hall, Englewood Cliffs, N.J. (1978).
20. R. Annino and R. Driver, *op. cit.,* chap. 10, Appendices C-F.
21. W. H. Press, B. T. Flannery, S. A. Teukolsky, and W. T. Vetterling, "Numerical Recipes—The Art of Scientific Computing," Cambridge University Press (1986). [A paperback supplement, accompanied by disks containing the programs given in this book in either FORTRAN or Pascal, is also available.]
22. "ASYST, a Scientific System," (in four volumes, with software), MacMillan Software Company, New York (1985).
23. R. Annino and R. Driver, *op. cit.,* chaps. 12 and 13.
24. "Scientific Solutions LabMaster," (reference manual), Scientific Solutions, Solon, Ohio (1985).

GENERAL READING

D. S. Hussain and K. M. Hussain, "The Computer Challenge," Burgess Communications, Santa Rosa, Calif. (1986).

G. I. Ouchi, "Personal Computers for Scientists: A Byte at a Time," American Chemical Society, Washington, D.C. (1986).

J. A. Feldman, "Programming Languages," *Scientific American,* **241,** 94 (December, 1979).

R. Annino and R. Driver, "Scientific and Engineering Applications with Personal Computers," Wiley, New York (1986).

P. C. Jurs, "Computer Software Applications in Chemistry," Wiley, Somerset, N.J. (1986).

K. L. Ratzlaff, "Introduction to Computer-Assisted Experimentation," Wiley, Somerset, N.J. (1986).

B. W. Rossiter and J. F. Hamilton (eds.), "Physical Methods of Chemistry," vol. I. "Components of Scientific Instruments and Applications of Computers to Chemical Research," 2d ed., chaps. 6–9, Wiley–Interscience, New York (1986).

APPENDIX
A

GLOSSARY
OF SYMBOLS

Listed below are the most common meanings of those symbols that occur frequently in this book; special usages of these symbols and the meanings of any unlisted symbols are defined in the text wherever they occur. Symbols used to represent units for physical quantities are given in Appendix B.

Symbol	Meaning
a	Activity
c	Concentration, molecular speed, speed of light
a, b, c	Crystal unit cell dimensions
d	Diameter (molecular), density, Bragg lattice-plane spacing
e	Electronic charge, base of natural logarithms
f	Force, function, frequency, fugacity, formation (subscript)
g	Acceleration due to gravity, gas, degeneracy
h	Planck's constant, height
i	$\sqrt{-1}$
k	Boltzmann's constant (also k_B), rate constant, force constant
i, j, k	Indices of array elements (in vectors, matrices)
l	Liquid, length
h, k, l	Miller indices for crystal planes
m	Mass, mass of atom or molecule, molality

n	Number of moles, index of refraction
p	Pressure
p_B	Partial pressure of B
q	Heat absorbed by the system, molecular partition function
r	Radius, distance
s	Solid
t	Celsius (centigrade) temperature, time, parameter in Student distribution
u	Root-mean-square speed, molecular potential energy
v	Scalar velocity (speed), vibrational quantum number
w	Work done upon the system
x, y, z	Cartesian coordinates
z	Valence of an ion
A	Helmholtz free energy, area, absorbance
B	Second virial coefficient, scalar magnetic flux density
C	Heat capacity, capacitance, number of components
C_p	Heat capacity at constant pressure
C_v	Heat capacity at constant volume
D	Diffusion constant
D_0, D_e	Dissociation energy referenced to ground state and potential minimum respectively
E	Energy, scalar electric field strength
E_a	Activation energy
G	Gibbs free energy
H	Enthalpy, scalar magnetic field intensity, Hamiltonian
I	Intensity of radiation, moment of inertia, ionic strength, electric current
J	Rotational quantum number, flux
K	Equilibrium constant, thermal conductivity
K_f	Molal freezing constant
L	Conductance
M	Molar mass ("molecular weight"), molarity
N	Number of particles (molecules, atoms, ions), normality
$\bar{N}$	Concentration in molecules per unit volume
N_0	Avogadro's number
P	Number of phases, power, polarization
P_M	Molar polarization
Q	Electric charge, generalized thermodynamic quantity, canonical partition function, vibrational coordinate
R	Gas constant, resistance
R_M	Molar refraction
S	Entropy, estimated standard deviation
T	Absolute temperature, time period, spectroscopic term value, transmittance
U	Potential energy, ionic mobility

V	Volume, voltage (potential difference)
W	Weight
X, Y	Mole fraction
Z	Collision frequency, atomic number, number of formula units per unit cell, compressibility factor, impedance

$\mathbf{a, b, c}$	Crystal lattice vectors	
$\mathbf{p}$	Momentum	
$\mathbf{r}$	Position vector	
$\mathbf{v}$	Velocity	
$\mathbf{B}$	Magnetic flux density	vector quantities
$\mathbf{E}$	Electric field strength	
$\mathbf{H}$	Magnetic field strength	
$\mathbf{M}$	Magnetization	
$\mathbf{P}$	Polarization (dielectric)	

$\mathscr{A}$	Pre-exponential factor in Arrhenius expression
$\mathscr{E}$	Electromotive force (emf)
$\mathscr{F}$	Faraday constant
$\mathscr{R}_\infty$	Rydberg constant

α	Thermal-expansion coefficient, degree of dissociation, polarizability, angle of optical rotation, adjustable parameter (least squares)
α_0	Distortion polarizability
β	$1/kT$
γ	Activity coefficient, surface tension, ratio C_p/C_v
δ	Deviation, chemical shift (NMR)
ε	Molecular energy, permittivity, error, molar absorption coefficient (absorptivity)
ε_0	Permittivity of vacuum
ε_r	Relative permittivity
η	Coefficient of viscosity
θ	Surface coverage, angle (e.g., Bragg angle)
κ	Electrical conductivity, relative permittivity (dielectric constant)
κ_S, κ_T	Adiabatic, isothermal compressibility
λ	Wavelength, ionic equivalent conductance, mean free path
μ	Chemical potential, Joule–Thomson coefficient, scalar dipole moment, reduced mass, permeability
μ_0	Permeability of vacuum
μ_e, μ_N	Bohr magneton, nuclear magneton
ν	Frequency (in Hz)
$\tilde{\nu}$	Wavenumber (in cm^{-1})
ρ	Density, resistivity
σ	Molecular area or dimension, standard error, order parameter, shielding constant (NMR), symmetry number

τ	Relaxation time, time constant
ϕ	Apparent molal volume, angle
χ, χ_e	Magnetic susceptibility, dielectric susceptibility
χ^2	distribution of goodness of fit
ψ	Wavefunction
ω	Circular frequency $(2\pi\nu)$
Δ	Limit of error, difference
Γ	Surface concentration, gamma function (math)
Θ	Characteristic temperature (e.g., Θ_{rot}, Θ_{vib}; Θ_D for Debye temperature)
Λ	Equivalent conductance
Π	Osmotic pressure
Ω	Solid angle
aq	Aqueous solution
ln	Natural logarithm
log	Logaritm to the base 10
pH	$-\log (a_{H^+})$
°C	Degree Celsius (centigrade)
K	Degree kelvin
Q^0	Any thermodynamic property Q of a substance in its standard state
$\bar{Q}$	Molal quantity Q
$\bar{Q}_A$	Partial molal quantity Q for component A

INTERNATIONAL SYSTEM OF UNITS

The International System of Units (SI for Système International) is based on seven dimensionally independent quantities:

Physical quantity	Name	Symbol
Length	meter	m
Mass	kilogram	kg
Time	second	s
Electric current	ampere	A
Thermodynamic temperature	kelvin	K
Amount of substance	mole	mol
Luminous intensity	candela	cd

There are also SI *derived units* with special names and symbols; some of these are listed below.

Physical quantity	Name	Symbol	Definition
Frequency f (*not* $\omega = 2\pi f$)	hertz	Hz	s^{-1}
Force	newton	N	$\mathrm{m\,kg\,s^{-2}}$
Pressure, stress	pascal	Pa	$\mathrm{N\,m^{-2}} = \mathrm{m^{-1}\,kg\,s^{-2}}$
Energy, work, heat	joule	J	$\mathrm{N\,m} = \mathrm{m^2\,kg\,s^{-2}}$
Power, radiant flux	watt	W	$\mathrm{J\,s^{-1}} = \mathrm{m^2\,kg\,s^{-3}}$
Electric charge	coulomb	C	$\mathrm{A\,s}$
Electric potential, electromotive force	volt	V	$\mathrm{J\,C^{-1}} = \mathrm{m^2\,kg\,s^{-3}\,A^{-1}}$
Electric resistance	ohm	Ω	$\mathrm{V\,A^{-1}} = \mathrm{m^2\,kg\,s^{-3}\,A^{-2}}$
Electric capacitance	farad	F	$\mathrm{C\,V^{-1}} = \mathrm{m^{-2}\,kg^{-1}\,s^4\,A^2}$
Magnetic flux density	tesla	T	$\mathrm{V\,s\,m^{-2}} = \mathrm{kg\,s^{-2}\,A^{-1}}$
Magnetic flux	weber	Wb	$\mathrm{V\,s} = \mathrm{m^2\,kg\,s^{-2}\,A^{-1}}$
Inductance	henry	H	$\mathrm{V\,A^{-1}\,s} = \mathrm{m^2\,kg\,s^{-2}\,A^{-2}}$
Luminous flux	lumen	lm	$\mathrm{cd\,sr^{-1}}$
Plane angle	radian	rad	—
Solid angle	steradian	sr	—

SI derived units exist for other physical quantities, but many of these do not have special names of symbols. Most of these are obvious—volume ($\mathrm{m^3}$), mass density ($\mathrm{kg\,m^{-3}}$), velocity ($\mathrm{m\,s^{-1}}$), molar volume ($\mathrm{m^3\,mol^{-1}}$), molar energy ($\mathrm{J\,mol^{-1}}$), etc. A few important but less obvious cases are listed below as a convenient illustration.

Physical quantity	Definition
Heat capacity, entropy	$\mathrm{J\,K^{-1}}$
Surface tension	$\mathrm{N\,m^{-1}}$
Heat flux density, irradiance	$\mathrm{W\,m^{-2}} = \mathrm{kg\,s^{-3}}$
Thermal conductivity	$\mathrm{W\,m^{-1}\,K^{-1}} = \mathrm{m\,kg\,s^{-3}\,K^{-1}}$
Kinematic viscosity, diffusion coefficient	$\mathrm{m^2\,s^{-1}}$
Dynamic viscosity	$\mathrm{N\,s\,m^{-2}} = \mathrm{Pa\,s} = \mathrm{m^{-1}\,kg\,s^{-1}}$
Electric displacement, polarization	$\mathrm{C\,m^{-2}} = \mathrm{m^{-2}\,s\,A}$
Permittivity	$\mathrm{F\,m^{-1}} = \mathrm{m^{-3}\,kg^{-1}\,s^4\,A^2}$
Permeability	$\mathrm{H\,m^{-1}} = \mathrm{m\,kg\,s^{-2}\,A^{-2}}$
Electric field strength	$\mathrm{V\,m^{-1}} = \mathrm{m\,kg\,s^{-3}\,A^{-1}}$
Magnetic field strength	$\mathrm{A\,m^{-1}}$
Luminance	$\mathrm{cd\,m^{-2}}$
Exposure (x- and γ-rays)	$\mathrm{C\,kg^{-1}} = \mathrm{kg^{-1}\,s\,A}$

Note that the symbols for SI units and derived units are all represented by roman letters, with an initial capital letter used when the unit is named after a person. These symbols are never expressed as plurals (with an added s) and are never followed by a period except at the end of a sentence. A sequence of symbols should alternate with spaces as shown above. The format a/b can be used in place of a b^{-1} (e.g., J/K instead of $J K^{-1}$ for entropy), but the exponent format is preferable since it is unambiguous for compound units (e.g., there is some possibility of confusion with W/mK for thermal conductivity but none with $W m^{-1} K^{-1}$).

It is often convenient to handle decimal fractions and multiples of the above SI units by attaching prefixes to the symbols. The commonly used prefixes are listed below.

Fraction	Prefix	Symbol	Multiple	Prefix	Symbol
10^{-1}	deci	d	10	deca	da[a]
10^{-2}	centi	c	10^2	hecto	h[a]
10^{-3}	milli	m	10^3	kilo	k
10^{-6}	micro	μ	10^6	mega	M
10^{-9}	nano	n	10^9	giga	G
10^{-12}	pico	p	10^{12}	tera	T
10^{-15}	femto	f	10^{15}	peta	P[a]
10^{-18}	atto	a	10^{18}	exa	E[a]

[a] These prefixes are rarely used.

CONCENTRATION
UNITS FOR
SOLUTIONS

Name	Symbol	Definition
Weight percent	%	(Grams of solute per grams of solution) × 100
Mole fraction[a]	X_A	Moles of A per total number of moles
Molarity	M	Moles of solute per liter of solution[b]
Normality	N	Equivalents of solute per liter of solution[b]
Formality[c]	F	Formula weights of solute per liter of solution[b]
Molality	m	Moles of solute per kg of solvent
Weight formality[c]	f	Formula weight of solute per kg of solvent

[a] The symbol Y_A is often used for the mole fraction of A in a gas phase that is in equilibrium with a liquid solution.

[b] Note that 1 liter $\equiv 1 \, dm^3$. The SI concentration unit of $mol \, m^{-3}$ is not very attractive for most work in chemistry, and non-SI concentrations based on the liter are widely used.

[c] These units are infrequently used but are of great convenience in expressing the overall composition of a solution when the solute is partially associated or dissociated.

BAROMETER CORRECTIONS

The entries Δ in the table below are calculated from Eq. (XVII-23) on the assumption that the barometer has a *brass* scale graduated to be accurate at 0°C (see Chapter XVIII). These corrections should be **subtracted** from the observed barometer readings; i.e., $p(\text{Torr}) = p_{\text{obs}}(\text{mm}, T) - \Delta$. (If the brass scale is accurate at 20°C, the appropriate corrections are approximately 0.3 mm greater than those given.) Once the barometer reading has been corrected to 0°C, the pressure is referred to as Torr rather than mm Hg.

t, °C	720 mm	740 mm	760 mm	780 mm	800 mm
16	1.88	1.93	1.98	2.03	2.09
17	1.99	2.05	2.10	2.16	2.22
18	2.11	2.17	2.23	2.29	2.35
19	2.23	2.29	2.35	2.41	2.48
20	2.34	2.41	2.47	2.54	2.60
21	2.46	2.53	2.60	2.67	2.73
22	2.58	2.65	2.72	2.79	2.86
23	2.69	2.77	2.84	2.92	2.99
24	2.81	2.89	2.97	3.05	3.12
25	2.93	3.01	3.09	3.17	3.25
26	3.04	3.13	3.21	3.30	3.38
27	3.16	3.25	3.34	3.42	3.51
28	3.28	3.37	3.46	3.55	3.64
29	3.39	3.49	3.58	3.68	3.77
30	3.51	3.61	3.71	3.80	3.90

APPENDIX

E

SAFETY

Given below is a short reminder list of basic safety principles and specific warnings about hazards that can occur in a physical chemistry laboratory. Items 1 and 2 are crucially important. Knowing how to respond effectively in the unlikely event of a serious accident is essential. Safety information should be displayed prominently in the laboratory—find it and read it before beginning any experimental work.

Following the list is a discussion of many safety issues that are pertinent to physical chemistry. A detailed treatment of all aspects of safety is given in *Prudent Practices for Handling Hazardous Chemicals in Laboratories*, prepared by the National Research Council Committee on Hazardous Substances and published by the National Academy Press in 1981.

1. Review experimental procedures and identify possible safety hazards *before* beginning laboratory work.
2. Learn the location and proper use of all safety equipment available in the laboratory as well as the fastest method of obtaining emergency medical assistance.
3. Never work alone in the laboratory.
4. Beware of high-voltage electricity.

5. Never pipette chemicals by mouth; use a pipetting bulb.

6. Wear safety glasses or goggles in the laboratory at all times.

7. Limit the use of open flames and never use them in the presence of flammable materials.

8. Avoid looking directly into any laser beam. A laser beam or even a reflected part of such a beam that enters the eye can cause permanent eye damage.

9. Beware of possible explosions due to gas overpressures, especially in glass systems.

10. Do not eat, drink, or smoke in the laboratory.

Electrical hazards. Several experiments make use of 110-V ac electrical power and employ apparatus in which exposed metal parts are "live." If the laboratory table has a metal surface, cover it with an insulating sheet of plywood or other material before assembling an electrical circuit. Remember that metal fixtures of all kinds and pipes or tubes of any kind that carry water are usually grounded. Turn off all electrical apparatus before altering circuits, if possible; if apparatus must be left on, use properly insulated test prods and leads. Be on the lookout for charged condensers, which may not be discharged owing to a broken circuit or a defective bleeder resistor. Naturally, 220 V represents a greater hazard than 110 V. It should be kept in mind that the laboratory is often served with 220 V in a three-wire system, with 110 V each side of ground. If 110-V outlets are supplied with a ground wire and one side of the 220-V line, as much as a 220-V difference can be obtained in accidental contact between circuits plugged into outlets serviced by opposite sides of the 220-V line. Shock, if it does occur, can be a serious matter; medical help should be summoned at once. Keep the victim quiet and comfortable; administer no stimulants of any kind.

Chemical hazards. These are many and varied. It should be taken for granted that any chemical substance taken by mouth or inhaled is toxic until and unless definite assurance has been given to the contrary. Reactions that evolve toxic fumes or vapors or entail risk of fire should always take place in a fume hood. As a matter of standard safety practice, never pipette any liquid or solution by mouth; use a rubber pipetting bulb. Mercury vapor can attain a hazardous concentration in the laboratory atmosphere. Mercury should be kept in covered vessels at all times. Spills should be carefully cleaned up (a capillary tube attached to a suction flask is convenient for this), and inaccessible droplets in floor cracks and hard-to-reach places should be covered with a light dusting of powdered sulfur. Another insidious hazard is that of vapors from organic solvents. Such solvents should not be used indiscriminately for cleaning purposes, and spills should be avoided. Good ventilation is important.

Environmental exposure to chemical hazards is currently a subject of concern and awareness. Zero exposure or zero risk of exposure is impossible in practice either in the chemical laboratory or elsewhere. Part of the professional role of chemists is to acquire knowledge and to develop judgement as to which precautions are necessary to limit these risks. One should not be blindly afraid of every chemical in the laboratory, nor should one be foolishly fearless. Many chemical hazards can be avoided by simply not eating or drinking in the laboratory, or breathing large volumes of vapors. Most chemical poisons are eliminated from the body, so that the effects of exposure gradually diminish. However, some poisons are not eliminated completely, and they accumulate, usually in particular tissues.

Recently, chronic exposure to low levels of certain chemicals has been shown to increase significantly the incidence of cancer. Such chemicals are referred to as carcinogens. The following list gives those chemicals classified as strong carcinogens by OSHA (Occupational Safety and Health Administration):

1. 2-Acetylaminofluorene
2. Acrylonitrile
3. 4-Aminobiphenyl
4. Benzidine (salts implied)
5. Bis(chloromethyl) ether
6. Chloromethyl methyl ether
7. 3,3'-Dichlorobenzidine
8. 4-Dimethylaminoazobenzene
9. Ethylenimine
10. 4,4'-Methylene-bis(2-chloroaniline)
11. α-Naphthylamine
12. β-Naphthylamine
13. 4-Nitrobiphenyl
14. N-Nitrosodimethylamine
15. β-Propiolactone
16. Vinyl chloride
17. Asbestos
18. Inorganic arsenic

None of these chemicals is used in the experiments described in this book. There are other chemicals thought to be weak carcinogens on the basis of statistical inferences involving data from long-term studies on large numbers of subjects. In these cases, there may be a hazard in industrial settings from chronic exposure to vapors or direct contact with the skin, but the risk from brief use in a research or educational setting is not serious.

Chemical burns. Strong acids (particularly oxidizing acids such as chromic acid cleaning solution) and bases may cause severe burns to the skin. If skin contact is made, wash copiously with water. If the exposure is to a strong acid, washing with a very dilute weak base (ammonia) is helpful; for a strong base use a very dilute weak acid (acetic acid). Particular attention should be directed to eye protection; *safety glasses*, safety goggles, or a face shield must be worn in the laboratory at all times. Prompt and effective action is essential if any chemical agent gets into the eyes; a strong base such as sodium hydroxide can permanently destroy the cornea in a few seconds. *Speed is all-important* in getting the exposed individual to an eyewash fountain or other source of copiously flowing (but low-pressure) water and *thoroughly* bathing the eyeball.

The eyelids should be lifted away from the eyeball to facilitate effective washing. Use nothing but water. *Get medical help promptly.*

Fire and explosion. Any flammable substance provides a potential fire hazard. In experiments that make use of hydrogen gas or other flammable gases, not only open flames but also cigarettes and sparking electrical contacts provide the possibility of explosion. If used in large quantities, such gases must be vented into the open air outside the building. A direct exhaust line from the experiment to a nearby window is best, but an exhaust fan and cross-ventilation will serve. The distillation of flammable liquids must be carried out in the absence of open flames; use a steam bath or electrical heating mantle. If an experiment involves an irreducible risk of fire or explosion, arrange for an adequate barrier. *Safety goggles* are required in all circumstances in which fire or explosion is a possible eventuality. Know the location of the nearest water supply and fire extinguishers (use water only on paper or cloth fires). In the event of serious burns, do not apply ointments or medications; get medical help.

Radiation hazards. Ultraviolet light from a mercury lamp or carbon arc is highly damaging to the eyes. Ordinary glasses give some protection, but the experimental arrangement should be well shielded so as to decrease the possibility of accidental exposure to a minimum. Prolonged exposure of the skin to such radiation can produce a severe "sunburn." An optical laser beam, even from a laser of very low power, that enters directly into the eye or is accidentally reflected into the eye by a surface can cause irreparable damage to the retina as a result of focusing by the lens of the eye. Eyeglasses provide no protection from this hazard, and the appropriate laser safety goggles should be worn when working with such sources (see Exp. 37). Exposure to strong radiofrequency or microwave fields can "cook" tissue and produce deep internal burns. Exposure to x-rays and to the radiation from radioactive materials must be carefully guarded against in experiments dealing with them. Any such experiments should be done under the direct supervision of an experienced research worker who will assume personal responsibility for all required safety measures, and under an appropriate license if radioactive materials are involved.

Mechanical and other hazards. Most mechanical hazards are too clearly apparent to warrant mention here. The danger lies in forgetfulness or casual disregard of risks. Vacuum systems often carry a hazard of collapse or implosion; bulbs more than 1 liter in volume should be surrounded by a metal screen or else wrapped with strong tape to reduce hazard from flying glass particles in the event of implosion. The bursting of a container due to overpressure is a frequent cause of accident or injury. A compressed-air line (usual pressure of the order of 50 psi) should never be connected to a closed system containing rubber tubing or glass bulbs. No closed system, except a properly designed combustion bomb, should be attached to a cylinder of compressed gas (usual maximum pressure: about 3000 psi) unless a suitable reducing valve is attached; even then, a relief valve should be provided to guard against accidental overpressure. Gas cylinders must be chained or strapped to prevent their falling over. The protective cap must be in place whenever a gas cylinder is being moved; cylinders should be moved with an appropriate hand truck, not dragged across the floor. *Mechanical pumps must have belt guards.*

Unattended operations must be planned with automatic safety switches that prevent serious damage (fire, flooding, explosion) in case of accidental equipment failure or interruption of utility services such as electricity, water, or gas supplies. Of special concern are the constant flow of cooling water and the operation of high-temperature baths. In the case of water flow, a device should be installed in the water line to (1) automatically regulate the water pressure (so as to avoid surges that might disconnect or rupture a water hose), and (2) automatically turn off electrical connections and water-supply valves in case of a total loss of water supply. In the case of hot thermostat baths or ovens, a sensor/control device should be installed that automatically turns off the electrical power to all heaters if the temperature exceeds some preset upper limit.

Safety equipment. *Safety glasses* or *goggles* have been mentioned in this appendix and in other places in this book in connection with specific hazards. However, use of safety glasses equipped with side shields, or other approved means of eye protection (plastic goggles alone or over ordinary prescription glasses, plastic face shields) is usually *mandatory at all times in instructional laboratories,* just as it is in industrial research laboratories. The use of safety goggles is strongly recommended as standard laboratory policy. Their use is essential in all circumstances where there exists the possibility of fire, explosion, implosion, spattering of caustic chemicals, or flying fragments from machine-shop operations. Fortunately, very few of these risks exist in the experiments described in this book, and explicit warnings are given in each case where hazards do occur.

The laboratory should be equipped with a conveniently accessible safety shower and an eye-wash fountain; there should be more than one of each in a large laboratory. Increasingly, the fixed type of eye-wash fountain is being superseded by a spray nozzle at the end of an extensible hose; there should be one of these on each laboratory bench. In lieu of such devices—or in addition to them—2- or 3-ft lengths of rubber hose (*not* small-bore pressure tubing) attached with wire or clamps to water faucets are certainly better than nothing.

The laboratory should have convenient access to one or more fume hoods (with a face velocity of at least 100 ft/min) for any operations involving more than insignificant quantities of volatile chemicals in open containers.

An approved fire extinguisher [the "dry chemical" (bicarbonate) type is preferred, but the CO_2 type is satisfactory] should be mounted near at least one exit and refilled after every use, no matter how small. The laboratory should be arranged so as to provide two or more avenues of escape from any experimental setup in case of emergency. A first-aid kit containing Band-Aids, sterile gauze, adhesive tape, petroleum jelly, a mild antiseptic, sterile cotton swabs, tweezers, a set of sewing needles, a packet of razor blades, and a quick-reference first-aid manual will provide adequately for most minor emergencies.

The location of an inhalator, a stretcher, and other rescue equipment, if not in the laboratory itself, should be known. The telephone number of the nearest medical emergency room and the local ambulance service should be posted conspicuously. Instructions for emergency evacuation, including special procedures for evacuating physically handicapped persons, should also be posted. An evacuation drill held near the beginning of each academic term is recommended.

Under no circumstances should a person be allowed to work in the laboratory alone.

Finally, safety depends on habits that must be gained outside the laboratory as well as inside. Thus, on your way to and from the laboratory, look both ways before crossing the street; after finishing the writing of that laboratory report, don't smoke in bed (or anywhere else!).

RESEARCH JOURNALS

Out of the hundreds of different research periodicals currently being published, the 42 listed below describe research of special interest to physical chemists. A very complete tabulation of periodicals, the correct abbreviation of their titles, and their distribution in American and selected foreign libraries can be found in *Chemical Abstracts Service Source Index*.

Abbreviated title	Full title
Accounts Chem. Res.	Accounts of Chemical Research
Acta Crystallogr. Sect. A, B, C	Acta Crystallographica
Bull. Chem. Soc. Jap.	Bulletin of the Chemical Society of Japan (no papers in Japanese!)
Can. J. Chem.	Canadian Journal of Chemistry
Chem. Phys.	Chemical Physics
Chem. Phys. Lett.	Chemical Physics Letters
Discuss. Faraday Soc.	Discussions of the Faraday Society (up to 1971)
Dokl. Phys. Chem.	Doklady Physical Chemistry (English translation of Doklady Akademii Nauk SSSR: Seriya Fiz. Khim.)
Int. J. Quantum Chem.	International Journal of Quantum Chemistry
J. Amer. Chem. Soc.	Journal of the American Chemical Society
J. Appl. Phys.	Journal of Applied Physics
J. Chem. Phys.	Journal of Chemical Physics

(continued)

TABLE (*Continued*)

Abbreviated title	Full title
J. Chem. Soc., Faraday Trans. 1, 2	Journal of the Chemical Society, Faraday Transactions 1 and 2
J. Chem. Thermodyn.	Journal of Chemical Thermodynamics
J. Magn. Reson.	Journal of Magnetic Resonance
J. Mol. Spectrosc.	Journal of Molecular Spectroscopy
J. Phys. A, B, C	Journal of Physics, sections A, B, and C
J. Phys. Chem.	Journal of Physical Chemistry
J. Phys. Chem. Ref. Data	Journal of Physical and Chemical Reference Data
J. Phys. Chem. Solids	Journal of Physics and Chemistry of Solids
J. Phys. Soc. Jap.	Journal of Physical Society of Japan
J. Polym. Sci.	Journal of Polymer Science
J. Quant. Spectrosc. Radiat. Transfer	Journal of Quantitative Spectroscopy and Radiative Transfer
J. Ram. Spectrosc.	Journal of Raman Spectroscopy
J. Solid State Chem.	Journal of Solid State Chemistry
Opt. Spectrosc.	Optics and Spectroscopy (English translation of Optika i Spektroskopiya)
Phase Trans.	Phase Transitions
Phil. Mag.	Philosophical Magazine
Physica A, B + C	Physica, sections A and B + C
Phys. Lett.	Physics Letters
Phys. Rev. A, B	Physical Review, sections A and B
Phys. Rev. Lett.	Physical Review Letters
Proc. Roy. Soc. (*London*), *Ser. A.*	Proceedings of the Royal Society of London, Series A, Mathematical and Physical Sciences
Rev. Sci. Instrum.	Review of Scientific Instruments
Russ. J. Phys. Chem.	Russian Journal of Physical Chemistry (English translation of Zhurnal Fizicheskoi Khimii)
Solid State Commun.	Solid State Communications
Sov. Phys.-JETP	Soviets Physics-JETP (English translation of Zh. Eksp. Teor. Fiz.)
Spectrochim. Acta A	Spectrochimica Acta A: Molecular Spectroscopy
Surface Sci.	Surface Science
Thermochim. Acta	Thermochimica Acta
Trans. Faraday Soc.	Transactions of the Faraday Society (up to 1971)
Z. Phys. Chem.	Zeitschrift für Physikalische Chemie (Frankfurt am Main)

INDEX

Abbe refractometer, 734–736
Absorbance, 313, 332, 443, 592–595, 754
Absorption coefficient, molar, 443–444, 754–755
  of iodine, 596
Absorption spectrum (*see* Spectrum)
Accuracy, 29
Acetal (diethyl), rate of hydrolysis of, 287–297
Acetic acid, ionization constant of, 255–256, 281
Acetylene spectrum, 469–482
Acid-catalyzed reactions, 288, 290–291, 299–300, 305
Activated complex, 317–319
Activation energy, 291, 310, 312, 371
  (*See also* Kinetics)
Activity, 206, 212, 271–272
Activity coefficient:
  from cell measurements, 270–276
  Debye–Hückel equation for, 212, 256, 272–273
  for hydrogen gas, 272
  of ionic species, 212, 256
  of neutral solute species, 212
  for a solvent, 206–207
Adhesives, 796–797
Adiabatic jacket, 159–163
Adiabatic processes:
  in chemical reactions, 153, 157–161
  expansion of gas, 95–103, 107–113
Adsorbed molecules, area of nitrogen, 360
Adsorbents, 355–356
Adsorption:
  BET theory of, 351–353
  of gases, 350–360
    gravimetric method, 353–354

Adsorption (*Cont.*):
  volumetric method, 353–359
  isosteric heat of, 353
  isotherms, 350–353
  at liquid surface, 343–345
  multilayer, 350–353
  phsyical, of gases, 350–360
    (*See also* Isotherm)
Air:
  density of, 720
  index of refraction of, 488, 732, 741n.
  viscosity of, 139
Algorithm, 846–849
Aluminum, 788, 790
Amici prism, 733–735
Amplifiers:
  frequency-selective (lock-in), 737
  instrumentation, 620–621
  operational, 615–621
Analog-to-digital conversion, 621–624
Anharmonic oscillator, energy levels of, 462, 491–492, 498–499
Antibonding orbital, 517
Apparatus lists, 3
Apparent molar volume, 190–191
Area of adsorbed nitrogen, 360
Argon:
  Debye temperature of solid, 575
  lattice energy of, 572–581
  lattice parameter of, 581
  Lennard-Jones potential parameters for, 575–576
  molecular diameter of, 151
  mutual diffusion constant for He–Ar, 151
  thermodynamic properties of, 574

Argon (*Cont.*):
  vapor pressure of solid, 576–581
  virial coefficient for, 94, 573
Arithmetical calculations (*see* Calculations)
Arrhenius activation energy, 291, 310, 312, 371
Ascarite, 146
ASCII code, 830
Aspirator, water, 687
Atomic mass table (*see back endpaper*)
Average deviation, 31
Azeotropes, 234

Balances:
  analytical, 715–720
  corrections and errors, 719–720
  Gouy, 424–426
  Westphal, 764
Ballast bulb, 91, 223
Balmer spectrum, 482–488
Band spectrum (*see* Spectrum)
Barometer, 720–722
  corrections, 870
BASIC, 843–844, 849–852, 856–859
Baths (*see* Temperature control)
Battery control unit, 630–631
Beattie–Bridgeman equation, 98–99
Beckmann thermometer, 647–648
Beer–Lambert law, 308, 312, 443, 754
Benzene:
  compressibility factor for, 221
  heat capacity of gas and liquid, 222
Benzenediazonium ion, rate of decomposition of, 311–316

Benzoic acid, specific energy of combustion, 168
Bernoulli's law, 136
BET (Brunauer, Emmett, and Teller) isotherm, 351–353
Binary notation, 831
Birge–Sponer plot, 499, 505
Bit, 829
Black-body radiation, 661–663, 741–742
Bohr magneton, 420, 508
Boiling point, 219
Boiling-point diagrams, 231–233
Bolometer, 754
Bond energies, 170–173
Bond type and magnetism, 422–423
Bonding:
  orbital, 517
  valence force model, 448, 456, 473–474
Bourdon gauges (see Gauges)
Bragg construction, 536, 554
Bragg equation, 536–537, 553
Bragg reflections, 535–538
  (See also Crystal structure; X-ray diffraction)
Brass, 789–790
  coefficient of thermal expansion, 721
Bravais lattice, 535, 559–560
Brillouin zone, 588–589
Bronze, 789
Bubblers, 147–148, 275, 778
Buerger precession camera, 550, 562–569
Buffers:
  acetate, 281, 293
  temperature dependence of pH in, 291–292
Buoyancy correction, 719–720
Burette, 767, 769
  gas, 353–357
Byte, 829

Cadmium electrode, 268–269
Calculations, 63–72, 74–76
  arithmetical, 63–65
  checking of, 65
  graphical methods, 74–76
  numerical methods, 65–72
    differentiation, 66–67
    integration, 67–68
    polynomial and Fourier series, 68–69
    roots of equation, 70–72
    smoothing of data, 65–66

Calculations (Cont.):
  precision, 63–65
  significant figures, 49–50
  (See also Least-squares method)
Callendar–vanDusen equation, 657–658
Calomel electrode, 725–726
Calorimetry:
  adiabatic jacket, 159–163
  bomb, 161–169
  principles of, 153–161
  solution, 180–186
  (See also Heat)
Capacitance cell:
  for gas, 406–408
  for solution, 397–398
Capacitance measurements:
  bridge method, 398, 405–406, 637–638
  heterodyne-beat method, 398, 636
  LC oscillator, 398–400, 406
  resonance method, 636–637
Capacitors, 602–603
Capillary rise, 345–347
Carbon dioxide:
  C_p of gaseous, 103
  critical point of, 246–252
  molecular diameter of, 151
  van der Waals constants for, 98
  virial coefficients for, 94
Carbon tetrachloride:
  Raman spectrum of, 451–460
  safety warning about, 216–217
Carcinogenic chemicals, 873
Cassia flask, 192–193
Cathetometer, 368, 722
Cell constants, 254
Cellophane, 366
Cells (see Emf cells)
Celsius (centigrade) temperature scale, 87, 640
  (See also Temperature scale)
Cements, 796–797
Center of inversion, 473, 555–557, 569
Central processing unit (CPU), 835
Centrifugal distortion, 474
Chart recorder, 731
Chemical Abstracts, 8–9
Chemical equilibrium, 205–218, 255–256, 467–468, 527–530

Chemical potential, 189, 362
  statistical mechanical calculation, 583–590
Chemical shifts (NMR), 432, 435, 522–525
Chi-square distribution, 811–815
χ^2 test, 811–817, 824
Chlorobenzene, dipole moment of, 401
Chromel, 651
Circuit elements, 599–604
Clapeyron equation (See Clausius–Clapeyron equation)
Classical probability statistics, 42–43
Clausius–Clapeyron equation, 196, 220–222, 572–573, 591
Clausius–Mossotti equation, 393
Cleaning solution, 771
Clock reactions, 277–287
Coexistence curve, 246–252
Cold trap, 682–683, 703–704
Colligative properties, 195–200, 205–208, 361–363
  freezing-point depression, 195–211
  osmotic pressure, 361–369
Collision frequency, 121
Combustion (see Heat)
Comparator, 487
Complexes, transition metal, 422–423
Compressibility factor, 220–221
  of benzene, heptane, water, 221
Computers, 828–861
  files, 833–834
  hardware, 834–841
  interfacing, 853–860
  I/O devices, 838–839
  operating systems, 841–842
  programming, 846–853
  programming languages, 842–845
  software, 841–845
  types of, 839–841
Concentration, 869
  surface, 341, 343–345
Conductance:
  of pumping line, 679–683
  of solutions, 253–266
    equivalent conductance, 254–255
    Λ_0 for HCl, KCl, KAc, 265

Conductance (*Cont.*):
  molar conductance, 254
  strong electrolytes,
    254–255
  weak electrolytes, 255
Conductance bridge, 257–262
Conductivity:
  cell, 262
  electrolytic, 254–255
  of water, 263
  (*See also* Thermal
    conductivity)
Confidence limits, 45–49
Constants, physical (*see front
  endpapers*)
Construction materials,
  787–794
Contact angle, 346–347
Conversion factors (*see front
  endpapers*)
Cooling curves, 201–202,
  241–243
  thermal arrest in, 241
Copper, 789–790
  wavelength of $K\alpha$ radiation,
    545
Corresponding states, law of,
  221
Counters:
  electronic, 763
  radiation, 730–731
Coupling constant (J) for
  spins, 523–527
Coverage, surface, 351–352
Critical exponents, 251–252
Critical opalescence, 248–249
Critical point, 246–252
Critical temperature, 221, 251
Cryogenic:
  baths, 665–666
  thermometers, 663–664
Cryopump, 695
Crystal:
  partition function, 586–590
  vibrations, 586–589
  zero-point energy of, 575,
    586, 590–591
Crystal class, 534, 555–558
Crystal-field complex,
  422–423
Crystal lattice, 534–540, 551,
  561–562
Crystal structure, 532–571
  of CsCl, 541–542
  deduction of, 539–542,
    550–562
  of I_2, 587
  lattice parameter for argon,
    581

Crystal structure (*Cont.*):
  Miller indices, 533,
    537–540, 551–554
  unit cell, 533–535, 587
  (*See also* X-ray diffraction)
Crystal systems, 534, 555–556
Curie constant, 421, 427*n.*,
  437
Curie–Weiss law, 427*n.*
Cyclohexane:
  density of, 204
  freezing point of, 199
  molal freezing-point
    depression constant, 199
  molar heat of fusion, 199
Cyclohexanone, refractive in-
    dex of, 235
Cylopentene, rate of
    decomposition of, 324–326

Data:
  recording of, 5–7, 24–25
  rejection of discordant,
    34–36, 810
  smoothing of, 65–66
  treatment of, 29–30, 51–55,
    80–82
  (*See also* Calculations;
    Errors; Graphs)
Debye–Hückel theory, 212,
  256, 272–273
  limiting law, 272
Debye–Scherrer X-ray
  method, 542–545
  (*See also* X-ray diffraction)
Debye temperature of argon,
  575
Debye unit for dipole mo-
    ment, 396
Degeneracy:
  of d orbitals, 422–423
  electron spin, 508
  nuclear spin, 430, 475
  rotational, 475
  vibrational, 455, 470–472
Degrees of freedom for a
  molecule, 105–107,
    227–228, 446, 587
Density:
  of air, 720
  of cyclohexane, 204
  of dibutyl phthalate, 112,
    697
  flotation method, 570
  of mercury, 112
  of NaCl solutions, 349
  pycnometers for determin-
    ing, 17, 191–193

Density (*Cont.*):
  of water, 193
  Westphal balance for deter-
    mining, 764
Deshielding effects in NMR,
  522–523
Deslandres table, 492–494
Desorption in vacuum sys-
  tems, 685–686
Detectors, 730–731, 747–754
  photodiodes, 750–752
  photographic film, 486–487,
    747
  photomultiplier tubes,
    747–750
  pyroelectric, 753
  radiation counters, 730–731
  thermal devices, 752–754
Deutero-acetylene prepara-
    tion, 478–479
Deuterium chloride prepara-
    tion, 466–467
Dewar flask, 159, 182
Diamagnetism, 419–420,
  432–434
  of water, 428
Diameter (*see* Molecular
    diameter)
Diatomic molecule, energy
  levels, 461–465, 490–494,
    498–501
Dibutyl phthalate, density of,
  112, 697
Dichlorobenzene, dipole mo-
    ment of, 400–401
Dielectric cell for solutions,
  397
Dielectric constant, 390–392
  of gas, 405–411
  of solution, 397–400
  value for reference gases,
    409
Differentiation:
  graphical, 74–75
  numerical, 65–67, 70
Diffraction camera, 555
Diffusion:
  Fick's laws of, 141–143
  of gases, 125–129, 141–152
  Loschmidt apparatus,
    143–146, 149
  self-: in gases, 128–129, 142
Diffusion constant for gases,
  127–129, 141–142, 151
Diffusion pumps, 691–693
Digital to analog conversion,
  623–624
Digital multimeter, 624–626
Dilatometer, 292–295

Diodes, 605–608
Dipole moment:
  of chlorobenzene, 401
  of HCl gas, 402–418
  of solutes, 390–402
  of succinonitrile, 401
Discharge:
  in hydrogen, 482–487
  in nitrogen, 494–495
Dissociation energy, 499–500
Dissociation of weak
    electrolyte, 205–206,
    255–256
Distillation, 233–237
Distribution function:
  for molecular weights
    $P(M)$, 375–376
  normal, of errors, 40–43
Distribution ratio, 213–216
Distribution of residuals, 814
DOS (disk operating system),
    842
Dry Ice (see Temperature
    control)
Drying agents, 778
Drying tubes, 138, 778
Duralumin, 788
Dyes, absorption spectra of,
    440–445

Effusion of gases, 120–122
Electrical heating, 155,
    181–183
  circuit for solution calori-
    meter, 183
Electrical measurements, 615–
    638
Electrodes:
  cadmium, 268–269
  cadmium amalgam, 269
  calomel, 725–726
  glass, 726–727
  hydrogen, 271, 273–275
  platinum, 273–274, 785–786
  silver–silver chloride, 271,
    274–275, 786–787
Electrolytes, 205–211,
    253–266, 270–276
Electromagnetic spectrum,
    739
Electron g value, 508–509
Electron spin resonance
    spectrum (see Spectrum)
Electronic circuit elements,
    599–604
Electronic devices, 604–615
Electronic spectrum (see
    Spectrum)

Emf cells:
  activity coefficients from,
    270–276
  electrodes for, 268–269,
    273–275, 725–727,
    785–787
  standard, 628–629
  temperature dependence of,
    267–270
  thermodynamics of,
    267–270
Emission spectrum (see
    Spectrum)
Energy:
  activation, 291, 310, 312,
    371
  bond, 170–173
  conversion of units (see
    front endpaper)
  of a crystalline solid,
    575–576
  dissociation, 499–500
  strain, 170–180
  zero-point, of crystal, 575,
    586, 590–591
Energy change in chemical
    reactions, 153–161, 168,
    178–179
Energy levels:
  anharmonic oscillator, 462,
    491–492, 498–499
  of atomic hydrogen,
    482–488
  of conjugated dyes,
    440–445
  of diatomic molecules, 461–
    465, 490–494, 498–501
  of nuclei (NMR), 430–431,
    522–525
  of polyatomic molecules,
    470–474
  of unpaired electrons
    (ESR), 508–510
Enthalpy (see Heat)
Enthalpy change in chemical
    reactions, 153–161,
    179–185
  (See also Heat)
Entropy:
  of water, 227–228
  of sublimation, 591
  statistical mechanical, 228,
    591
Enzyme kinetics, 297–311
  Michaelis–Menten mechan-
    ism of, 297–299
Enzyme solution:
  preparation of, 307
  specific activity, 309

Epoxy resins, 796
Equation of state, 88, 97–99
  Beattie–Bridgeman equa-
    tion, 99
  perfect-gas law, 88
  van der Waals', 97, 109, 251
  virial, 88
Equilibria:
  binary liquid–vapor phase,
    229–238
  binary solid–liquid phase,
    238–246
  chemical in gas phase,
    467–468
  chemical in solution, 206,
    211–218, 255–256
  coexisting liquid–vapor,
    246–252
  helix-coil polypeptide,
    380–389
  heterogeneous, 213
  homogeneous, 212–213,
    219–252
  isotopic exchange, 467–468
  keto-enol tautomerism,
    522–531
  osmotic, 361–369
  physical adsorption of gas,
    350–360
  solution-solid solvent, 195–
    211
  vapor pressure of liquid,
    219–229
  vapor pressure of solid,
    572–598
Equilibrium constant, 211–218
  for gas-phase exchange,
    467–468
  for keto-enol tautomerism,
    527–529
  for weak electrolyte ioniza-
    tion, 206, 255–256
Equipartition theorem,
    105–107
Equipment, general labora-
    tory, 3
Equivalent conductance,
    254–255
Error(s), 30–63, 76–80,
    817–820
  in arithmetic mean, 31–32,
    43–44
  confidence limits, 45–48
  error probability function,
    39–40
  estimation of, 48–49
  examples, 36–39, 52–55,
    58–62
  mistakes, 34

Error(s) (*Cont.*):
  normal error function,
    41–43
  precision, 30, 63–65, 77–80
  propagation of, 55–63
    sample treatment, 20–21,
      58–62
  Q test, 34–36
  random, 30–32, 39–43
  range, 31, 47–48
  significance testing, 51–55
  significant figures, 49–50,
    63–65
  in slopes and intercepts,
    76–77
  standard deviation, 32,
    41–44
  Student *t* distribution,
    45–47
  systematic, 32–34, 56
  uncertainty in least-squares
    parameters, 817–820
  variance, 31–32
  (*See also* Least-squares
    method)
Euler's theorem, 188
Eutectic, 239–242
Ewald construction, 553–555,
  565–566
Exchange interactions, 512
Expansion of gas, 107–113
Experimental data (*see* Data)
Extensive variables, 187,
  339–340
Extinction coefficient (*see*
  Absorption coefficient,
  molar)

F distribution, 815
F test, 815–817
Ferric ion hydrolysis constant,
  337
Ferric thiocyanate complex:
  equilibrium constant for
    formation of, 337
  rate of formation of,
    329–338
  Fick's laws of diffusion,
    141–143
Fitting procedures, 68–69,
  801–827
Fixed point (*see* Temperature
  scale)
Flow diagram (computers),
  846–847
Flow method for chemical
  kinetics, 284–287,
  329–338

Flowmeters, 779–781
Fluorescence of iodine, 498
Force constants:
  of acetylene, 470, 473–474
  of CCl_4, 456–457
  of diatomic molecules, 464
  of SO_2, 448
Forepump, 688–690
Formvar, 792
FORTRAN, 843
Fourier series expansion, 69
Franck–Condon factor, 506
Free-electron model of dye
  spectra, 440–443
Free energy:
  changes in electrochemical
    cells, 267–270
  surface, 341
Free radical anions, 510–511
Freedom, degrees of, for a
  molecule, 105–106
Freezing-point depression,
  195–211
  of electrolytes, 205–211
  for molecular weight deter-
    mination, 195–205
  molal, constant, of cyclo-
    hexane and water,
    199
Freezing-point diagrams,
  239–243
Frequency of LC circuit, 398
Friedel's law, 557
FTIR (*see* Spectrometer)
Furnaces, 674

g value:
  of electron, 420
  of nucleus, 430
Galvanometers, 100, 628
Gas burette, 353–354
Gas-handling procedures,
  774–778
Gas law (*see* Equation of
  state)
Gas purification, 778
Gas regulators, 774–776
Gas thermometer, 86–95, 640
Gases:
  adsorption of, 350–360
  degrees of freedom in,
    105–106
  diffusion of, 125–129,
    141–152
  effusion of, 120–122
  expansion of, 107–113
  heat capacity of, 97–99,
    103–107, 222, 449

Gases (*Cont.*):
  Maxwellian velocity
    distribution in, 119,
    120
  partition function, 449,
    467–468, 584–586
  thermal conductivity of,
    124–125
  vibrational heat capacity of,
    106, 449
  viscosity of, 122–124,
    130–140
  water saturation of, 778
Gated integrator, 737
Gauges, 696–703
  Bourdon, 781–782
  ionization, 696, 701–703
  McLeod, 697–700
  manometers, 696–698
  residual-gas analyzer, 703
  Pirani, 700–701
  strain-gauge pressure trans-
    ducers, 782
  thermocouple, 701
  (*See also* Manometer)
Geiger–Müller counter,
  730
Geissler discharge tube,
  485–486, 494
Gibbs–Duhem equation,
  188–189
Gibbs–Helmholtz equation,
  267–268
Gibbs isotherm, 341–343
Glass electrode, 726–727
Glasses, 708, 787–788
  thermal expansion of, 708
Glassware:
  calibration of, 770–771
  cleaning of, 773
  volumetric, 766–770
Glide plane, 558
Globar, 742
Glyptal, 797
Goodness of fit, 811
Gouy balance, 424–426
Graphical differentiation,
  74–75
Graphical integration, 74–75
Graphs:
  errors in slopes and inter-
    cepts, 76–77
  preparation of, 72–74
  presentation of, 14
  (*See also* Calculations)
Gratings, 743–747
Greases, stopcock, 705–707
Guggenheim method, 289–290
  (*see also* Kinetics)

Half-life, 289, 324, 329
(*See also* Kinetics)
Handbooks, 10–12
Harmonic oscillator functions, 470
Hazards, safety, 871–876
HCl mean ion activity coefficient, 246
HCN vibrational spectrum, 476
Heat:
  of adsorption, isosteric, 353
  of combustion, 161–180
    for iron and benzoic acid, 168
  of formation, 171–172
  of fusion, 197, 241
    for cyclohexane and water, 199
  of ionic reaction, 180–186
  of reaction, 268
  of sublimation, 572–597
  of vaporization, 220–227
Heat capacity, 155–156
  of benzene, haptane, water, 222
  of gases, 97–99, 103–107, 222, 449
    vibrational contribution, 449
Heat flow, one-dimensional, 124–125
Heating wire, 675
Helium:
  molecular diameter of, 151
  mutual diffusion constant He-Ar, 151
  van der Waals' constants for, 98
  virial coefficient for, 94
Helix-coil transition, 380–389
Heptane:
  compressibility factor for, 221
  heat capacity of gas and liquid, 222
Hermann–Mauguin symbol, 557, 559–561
Heterodyne-beat method, 398–399, 636
Heterogeneous equilibrium, 213
Hexadecimal notation, 831–832
Homogeneous equilibrium, 212–213, 219–252, 572–573, 583
Hoses, 777–778
Hückel molecular orbital theory, 508, 516–520

Hund's first rule, 420, 423
Hydrochloric acid, Λ_0 value, 265
Hydrogen:
  discharge in, 482–489
  van der Waals' constants for, 98
Hydrogen atom:
  energy levels of, 482–483
  spectrum of, 483, 489
Hydrogen bonding in polypeptides, 380–383
Hydrogen chloride:
  mean ion activity coefficient, 270–276
  spectrum, 461–468
Hydrogen electrode, 271, 273–275
Hydrolysis kinetics, acetal, 287–297
Hyperfine splitting, 510

Ice bath, 92, 666
Ice point, 87, 92, 642
  (*See also* Temperature scale)
Ideal-gas law, 88
Ideal solutions, 196, 230–232
Inconel, 789
Index of refraction, 732–737
  of air, 732
  of cyclohexanone-tetra-chloroethane solutions, 235
Inductors, 600, 604
Inertia, moment of, 461, 474, 585
Infrared spectrum, 446–451, 469–482, 762
  fundamentals, 446–448
  gas cell, 413, 449–450, 594
  HCl, rotational structure of, 461–468
  isotope effect, 460, 464–465, 467, 474–477
  overtone and combination bands, 448, 470, 473
  prism materials, 743
  selection rules for diatomic molecules, 463–464
  vibrational, of SO_2, 446–451
  (*See also* Spectrum)
Infrared spectrometers, 754–763
  detectors, 747–754
  gratings, 744–747
  prisms, 743
  sources, 741–743

Initial rates, method of, 277–287
Instrument control by micro-computers, 853–860
Instrumentation amplifiers, 620
Insulation materials, 792, 794
Integration:
  graphical, 74–76
  numerical, 67–68
Intensive variables, 187–188, 339
  surface quantities, 341
Interfacial tension, 340–341
Interfacing of a computer, 853–860
Interferometer, 411–412, 417
Interpolation, 66–67
Intrinsic viscosity (*see* Viscosity)
Invar, 789
Inversion, center of, 473, 555–558, 569
Inversion of sucrose, 297–311
Inversion temperature, 98
Iodine-clock reaction, 277–287
Iodine crystal structure, 587
Iodine–iodide–triiodide equilibrium, 211–218
Iodine spectrum:
  absorption, 497–498, 501–502, 505, 592–595
  emission (fluorescence), 497–507
Iodine sublimation pressure, 582, 590–591
Ion pump, 693–695
Ionic mobility, 254
Ionic strength, 212–272
Ionization constant of acetic acid, 281
Ionization gauge, 696, 701–703
Iron, specific energy of combustion, 168
Iron arc, 494–495
Isoelectronic ions, 542
Isomorphism, 560, 569
Isoteniscope, 224–226
Isotherm:
  BET adsorption, 351–353
  Gibbs, 341–343
Isotope effect on spectra, 459–460, 464–465, 467, 474–477
Isotope exchange, 467–468

Joule coefficient, 103

Joule heating, 155
Joule–Thomson coefficient, 95–104

Kelvin temperature scale, 87, 640
(*See also* Temperature scale)
Keto–enol tautomerism, 522–531
Kinetic theory of transport phenomena, 119–129
Kinetics, chemical: acid-catalyzed reaction, 288, 290–291, 299–300, 305
activated complex, 317–319
Arrhenius activation energy, 291, 310, 312, 371
chain mechanisms, 320–321
continuous-flow method, 284–287, 331–336
dependence of rate constant on temperature, 291
enzyme, 297–311
fast reactions, 329–338
first-order reaction: half-life of, 324
methods of testing, 288–290, 312–313, 324
gas-phase reaction, 316–329
pressure method, 321–324
wall effects, 321
Guggenheim method, 289–290, 296
of hydrolysis reaction, 287–297
Lineweaver–Burke plot, 299
method of initial rates, 277–284
order of reaction, 278
reaction mechanisms, 278–279, 283
RRK and RRKM theories, 318–319
specific rate constant, 288
spectrophotometric methods, 301, 312–313, 331–335
steady-state mechanisms, 317, 320
transition-state theory, 319n
unimolecular reactions, 317–320
Knudsen flow, 120–121
Kovar seals, 704, 789
Kundt's tube, 115–116

Laminar flow, 130, 135
Laser, 412, 453, 458–459, 503
Lattice constants, 538, 568
Lattice energy, 575–576
Lattice planes, 537–538, 551–552
Lattice types, crystal, 535, 538–539
Laue symmetry, 557
LC circuit, frequency of, 398
Lattice, crystal, 534–540, 551, 559–562
Lattice, reciprocal, 551–555
Lattice vibrations, 586–590
Leak detection, 711–713
Least-squares method, 801–827
chi-square (χ^2) test, 811–815, 824
degrees of freedom, 812
estimated standard deviation of parameters, 817–819
F test, 815–817, 824
goodness of fit, 811–815
foundations of, 802–808
linear fits, 806, 808, 824
nonlinear fits, 806–807, 824–825
normal equations, 804–805, 807
sample calculation, 821–826
summary of procedures, 820–821
uncertainties in the parameters, 817–820
weighting, 608–609, 811–814, 816–817, 820
Lennard-Jones potential, 151, 575–576
parameters for argon, 576
Ligand-field theory, 422–423, 438
Limits of error, 47–49, 519
Liquid–vapor coexistence, 246–252
Liquids:
dielectric constant of, 390–402
surface adsorption in, 343
vapor pressure of, 219–229
Lissajous figures, 724–725
Literature searches, 7–12, 877–878
Lock-in amplifier, 514
Lock-in detector, 737
Lorentz–Lorenz relation 394
Loschmidt diffusion apparatus, 143–146, 149

Lubricants, 774
Lucite, 791–793

Machine tools, 799
Maclaurin series, 198
McLeod gauge, 697–700
Macromolecules:
intrinsic viscosity, 370–380
optical rotation and conformations, 380–389
osmotic pressure, 361–369
Magnetic field, 508
Magnetic induction, 508
Magnetic moment:
electron (orbital), 421
electron (spin), 420–421, 434, 437
nuclear, 430
Magnetic susceptibility, 418–429, 431–432, 437
Manometer:
capacitance, 410, 696, 703
flow, 780–781
mercury, 91–92, 696–698
null, 90–91
oil, 697
open-tube, 110, 112
temperature correction for readings, 226, 697
Manostats, 783–785
Marquardt's algorithm, 806–807, 821
Mass, reduced, 128, 416, 461, 483, 585
Mass action, law of (chemical equilibrium), 205–218, 467–468, 529
Maxwellian distribution of velocities, 119–120
Mean, uncertainty in, 43–49
Mean free path, 120
Mean speed, 119
Mechanical pumps, 688–690
Membranes, 366
Mercury:
density of, 112
purification of, 773
spectrum of, 486
thermal expansion of, 697
thermometers, 645–648
Mercury vapor lamp, 486, 504
Metals for construction, 788–790
Michaelis–Menton mechanism, 298–301
Microprocessor, 835
Miller indices, 533, 537–540, 551–554
Mirror plane, 555–556

Mobility of ions, 254
Molality, 869
Molar refraction, 394, 405
Molarity, 869
Mole fractions, 189
Molecular diameter, 120, 123, 127–129
  values for Ar, CO_2, He, 151
Molecular flow (effusion of gases), 120–122, 681–682
Molecular-orbital calculations, 508, 515–520
Molecular weight:
  distribution function, 376
  number-average, 362–363, 375–376
  viscosity-average, 375–376
  weight-average, 376
Molecular-weight determination of solutes, 195–205
Moment of inertia, 461, 474, 585
Monel, 789–790
Monochromator, 755–756
Morse potential, 500
Mylar, 791–792

Needle valve, 777
Neoprene, 791
Nernst glower, 743
Nichrome, 789
Nickel-chloride solution, susceptibility of, 428
Nicol prism, 728–729
Nitrogen:
  adsorbed, area of, 360
  band spectrum of, 489–497
  gaseous, C_p of, 103
  liquid, vapor pressure of, 580
  pressure variation of normal boiling point, 94
  van der Waals' constants for, 98
  virial coefficient for, 94
Nomenclature, 12
Normal equations, 804–805, 807
Normal error probability function, 41–43
Normal modes of vibration, 447, 455, 469, 474
  of a crystal, 586–590
Normal probability plot, 814, 817
Notebooks, 5–7, 24
Nuclear magnetic resonance (NMR), 431–439, 522–531

Nuclear magnetic resonance (NMR) (Cont.):
  chemical shifts in, 432, 435, 522–525
  deshielding effects, 522–523
  intensities in, 524–526
  spectrometer, 434–436
  spin-spin splitting in, 523–527
Nuclear spin quantum number (I), 430
Number systems, 830–832
Null detectors, 628
Numerical methods, 65–72
Nylon, 791

O-ring gaskets, 705–706, 798
Observational equations, 803
Ohm's law, 253–254
Onsager theory, 255, 266
Operational amplifiers, 600, 615–620
  circuit configurations, 618
Operating systems, computer, 841–842
Optical density (see Absorbance)
Optical pyrometers, 661–663
Optical rotation, 386–389, 728–729
Order of reaction, 278
  (See also Kinetics)
Orthorhombic system, 534, 553, 555–562
Oscillators, LC, 398–399, 406
Oscilloscope, 722–725
Osmometer designs, 363–366
Osmotic coefficient, 207–208
Osmotic pressure, 361–370
  half-sum method, 364
  static method, 364
Ostwald viscosimeter, 372, 378

p–n junction, 604–606
Palladium, 789–790
Paramagnetism, 430–439
  spin magnetic moment, 420–421, 434, 437
Partial molar quantities, 188–189
  free energy, 189
  volume, 187–194
Partition function:
  configurational, 382–386
  for polymer segments, 382–386
  rotational, 585

Partition function (Cont.):
  translational, 584
  vibrational:
    of a crystal, 586–590
    of a gas, 449, 467–468, 584–586
Pascal computer language, 844–851
Pascal's constant, 433–434
Pauli exclusion principle, 420–421, 442
Perfect-gas law, 88
Peritectic transformation, 242
Permittivity of vacuum, 391, 403
pH meter, 725–728
Phase diagram:
  binary liquid–vapor system, 229–238
  binary solid–liquid system, 238–246
  for one-component system, 195
Phase rule, 229
Phonon dispersion curves, 588–589
Phosphorescence, 495
Photodiodes, 750–752
Photography, 747
  spectroscopic plates, 486–487, 506
  X-ray film, 547–548
Photomultipliers, 747–750
Physical absorption of gases, 350–360
Physical constants (see front endpapers)
Pipette, 767–768
Pirani gauge, 696, 700–701
Planck radiation law, 661–662, 741–742
Plastics, 790–793
Plating methods, 786
Platinum, 789–790
  electrodes, 273–274, 785–786
  resistance thermometers, 641, 654–658
  temperature coefficient of resistivity, 655
Plexiglas, 791–793
Plotters, 731
Point groups, 454–455, 555–558
Poiseuille's law, 133, 372
Polar molecules in solution, 390–402
Polarimeter, 386, 728–729
Polarizability, 392
Polarization, 392, 452, 456

Polarization (*Cont.*):
  atomic, 403–405, 415–417
  distortion, 390, 393–394
  of electrodes, 256–257
  molar, 393
    in solution, 395–396
    of a gas, 403–405,
      415–417
  orientation, 390, 393–394
  of solute at infinite dilu-
    tion, 395–396
Polybenzyl glutamate, helix-
  coil transition in, 380–389
Polyethylene, 792–793
Polynomial, fitting with, 68
Polymers, 370
  statistically coiled, 374–375,
    380–389
  (*See also* Macromolecules;
    Plastics)
Polypeptides, 380–383
  hydrogen bonding in,
    381–384
Polyvinyl alcohol:
  chain linkage in, 370–380
  intrinsic viscosity of,
    373–375
Potassium acetate, Λ_0 value
  for, 265
Potaasium chloride:
  Λ_0 value for, 265
  solutions: conductance of,
    265
Potential energy:
  internuclear, 462, 490,
    497–498
  Lennard-Jones, 575–576
  Morse function, 500
Potentiometer:
  basic circuit, 626–628
  L & N Student, 629–631
  procedure for use of, 632
  null detectors, 628
  standardization of, 631
Precession camera (*see*
  Buerger precession
  camera)
Precision, 30, 63–65
  fundamental limitations on,
    77–80
Pressure measurements (*see*
  Gauges; Manometers)
Pressure units, 679
Prisms:
  Amici, 733–735
  infrared, 743
  Nicol, 728–729
  ultraviolet, 484–486, 743
  visible, 484–485, 743
Program languages, 842–845

Programming, 846–853
  flow diagram, 846–849
Projects, special, 27–28
Propagation of errors, 55–63
  sample treatment 20–21,
    58–62
Proportional counter, 730
Pumping speed, 678–684
Pumps:
  cryopump, 695
  diffusion, 691–693
  forepump, 693
  ion, 693–695
  mechanical oil, 688–690
  Roots, 690
  Toepler, 687–688
  turbomolecular, 690
  water aspirator, 687
Purification methods,
  773–774, 778
Pycnometer, 15, 191–193
Pyrex glass, 708, 787–788
Pyrometers, optical, 661–663

Q-test, 35–36
Quartile deviation 43
Quenching, 498

Radiation, blackbody,
  661–662, 741–742
Radiation detectors (*see*
  Detectors)
Radiation sources, 485, 494,
  741–743
Radical anions, 507–521
Raman spectrum, 451–460
  depolarization ratio,
    454–456
  selection rules, 454
  spectrometers, 457–458,
    757–758
  Stokes and anti-Stokes in-
    tensities, 453–454
  (*See also* Spectrum)
Raman spectrometers,
  457–459, 757–758
Random errors, 30–32, 39–43
Range shrinking, 817
Raoult's law, 196, 230–232
  deviations from, 232
Rate laws (*see* Kinetics)
Rayleigh scattering, 452–453
Reaction mechanisms,
  278–279, 283
  (*See also* Kinetics)
Reciprocal lattice, 551–553,
  567–569
Reciprocal space, 551–555
Reciprocity relation, 342

Recorders, 731
Rectifier, 607–608
Rectilinear diameter, law of,
  246
Reduced mass, 128, 461, 483
Reducing valve, 776–777
References:
  abstracts, 8–10
  books, 10–12
  citation of, 23–24
Refraction, molar, 393
Refractive index, 411–415
  of air, 741
  of cyclohexanone-tetra-
    chloroethane solutions,
    235
  of gas, 411–415
  of various optical materials,
    744
Refractometers, 732–737
  principles of operation,
    732–733
  types of: Abbe, 734–736
    immersion, 733
Refrigerant baths, 665–667
  pumping on liquid nitrogen,
    576–577
  (*See also* Temperature
    control)
Regulators, gas, 774–776
  (*See also* Manostats)
Rejection of discordant data,
  34–38, 810
Reports, 12–26
  format of, 13–14
  presentation of figures in,
    14
  sample, 14–26
  style of, 12–13
Resistance thermometers
  654–658
Resistors, 600–601
Resonance energy, 170
Resonance method, 636–637
Reynolds number 135–136
Root-mean-square speed, 119
Rotameter, 780
Rotation, specific, 386–388,
  728
Rotation axis, 555–557
Rotational structure:
  of IR spectra, 446–447,
    461–482
  of vibronic spectra, 492,
    494, 499–501
Rounding off, 49–50
Rubber, 790–791
Rydberg constant, 484, 488

Safety, 5, 28, 871–876

Safety (*Cont.*):
  for vacuum systems,
    713–714
  use of hydrogen and oxygen
    gas, 275, 785
  use of CCl_4, 216–217
  use of lasers, 458
Saran, 791–793
Schoenflies symbol, 557
Science Citation index, 10
Screw axis, 557–559
Seebeck coefficients, 649
Selection rules,
  electronic, 498
  ESR, 508
  IR, 454, 472–473
  NMR, 430, 522
  Raman, 454, 472–473
  rotational, 463, 475, 498
  vibrational, 446–448, 454,
    463, 472–473
Self-diffusion (*see* Diffusion)
Semiconductor devices,
  604–615
Shopwork, 799
Signal averaging devices,
  737–739
Significance testing, 51–55
Significant figures, 49–50,
  63–65
Silver, 788–789
Silver–silver chloride
  electrode, 271–274,
  786–787
Simpson's rules, 68
Smoothing of data, 65–66
Sodium chloride solutions,
  density of, 349
Sodium hydroxide, purifica-
  tion of, 774
Software (computer), 834,
  841–846
Soldering, 794–796
Solid solutions, 239
Solids:
  crystal structure of, 532–572
  lattice energy of, 572–582
  sublimation pressure,
    572–597
Solution:
  chemical equilibrium in,
    211–218
  conductance in, 253–267
  dielectric constant of,
    390–402
  freezing point of, 195–211
  ideal, 196, 230–232
  partial molar volume in,
    187–194

Solution (*Cont.*):
  Raoult's law, 196–197,
    230–232
  solid, 239
  storage of, 772–773
  surface tension of, 339–350
Sound velocity, 113–118
Space group, 550, 558–562,
  567–569
Specific activity of enzyme,
  309
Specific heat (*see* Heat
  capacity)
Specific rotation, 386–388, 728
Spectrographs, 484–485, 743
Spectrometer:
  ESR, 513–514
  Fourier transform IR, 449,
    465, 477, 760–763
  infrared, 758–760
  NMR, 434–435
  Raman, 457–459, 757–758
Spectrophotometer, 313,
  332–334, 443
  Beckman DU, 332–334
  infrared, 758–760
  Raman, 757–758
  visible-ultraviolet, 755–758
Spectroscopic components:
  gratings, 744–747
  prisms, 743
  radiation detectors, 747–754
  radiation sources, 741–743
Spectroscopic plates, 485–487,
  747
Spectrum:
  absorption, of I_2, 497–507,
    592–595
  Balmer, of atomic hydro-
    gen, 482–489
  band, of N_2, 489–497
    rotational structure,
      492–494
    vibrational structure,
      491–492
  band heads, 493, 495, 501
  of conjugated dyes, 440–445
  electron spin resonance
    (ESR), 507–521
  electronic, of I_2, 497–507
  emission:
    of atomic hydrogen,
      482–489
    of I_2, 497–507
    of N_2, 489–497
  infrared (IR) vibrational:
    of acetylene, 469–482,
      762
    of HCN, 476

Spectrum (*Cont.*):
  of SO_2, 446–451
  IR, rotational structure of,
    447, 461–465, 474–477,
    481
  iron-arc reference, 494–
    495
  isotope effect, 460,
    464–465, 467, 474–477
  measurement of spectral
    lines, 487, 495
  mercury reference, 485–486
  Raman vibrational spectrum
    of CCl_4, 451–460
  resonance fluorescence, of
    I_2, 497–507
  rotational–vibrational:
    of acetylene, 469–482
    of HCl, 461–468
    of HCN, 476, 480
  wavelength calibration,
    487–488, 495, 503–505,
    740–741
  (*See also* Infrared spectrum;
    Raman spectrum)
Speed (molecular):
  mean, 119
  root-mean-square, 119
Speed of sound, 113–118
Spin–spin splitting, 523–527
Standard cells, 628–629
Standard deviation, 41–43
  estimated, 32, 43–44,
    818–820
  in least-squares parameters,
    820, 825
Standard electrode emf, 270,
  271
Standard emf (Weston) cell,
  628–629
Standard free-energy change,
  270
Statistical mechanics:
  chemical potential of I_2 gas
    and crystal, 583–590
  classical equipartition
    theorem, 105–106
  model for helix-coil transi-
    tion 382–386
  treatment of errors, 39–43
  vibrational partition func-
    tion, 449, 467–468,
    585–590
Statistical thermodynamics,
  582–591
Steady-state chemical kinetics,
  277–279, 317, 320–321
Steam point, 87, 92–93
Steel, stainless, 788, 790

Stefan–Boltzmann equation, 741
Steric factor, 371
Stopcocks, 706–707
    method of greasing, 706–707
Strain energy, 170–173
Strain gauges, 782
Student t distribution, 45–47
Sucrose inversion, 297–311
Sulfur dioxide:
    bond angle in, 448
    normal modes for, 447–448
    vibrational spectrum of, 446–451
Supercooling, 202–203
Surface adsorption in liquids, 343
Surface concentration, 341, 343–345
Surface coverage, 351
Surface tension, 341
    capillary-rise method, 345–347
    of solutions, 339–350
    of water, 347
Susceptibility, magnetic, 418–429, 431–432, 437
Symmetry:
    in crystals, 555–561
    of normal modes of vibration, 455, 469–471
    of vibrational levels, 469–472
Symmetry species, 454–455, 471–473
Symmorphic space groups, 560
Syringes, 769–770
Systematic errors, 32–34
Systematic extinctions, 561–562, 569

t-butyl peroxide, decomposition of 326–327
t distribution, 45–47
t factor, table, 46
t test, 51–55
Teflon, 791, 793
Temperature, critical, 221, 251
Temperature control, 665–675
    air thermostats, 674
    controllers, 667–671
    fixed-temperature baths, 665–667
    Dry Ice, 666

Temperature control (*Cont.*):
    ice, 644–645, 666–667
    liquid nitrogen, 665–666
    vapor, 667
    liquid baths,
        circulating type, 673
        oil, 672–673
        water, 671–672
    ovens and furnaces, 674–675
    thermostat blocks, 673–674
Temperature scale, 86, 640–641
    Callendar–van Dusen equation, 657–658
    Celsius (centigrade), 89, 640
    characteristic, 575, 585–586
    fixed points, 86, 90, 641
        ice, 87, 92
        steam, 87, 92–93
        triple, of water, 87, 641
    International Practical, 88, 641
    Kelvin, 87, 640
    perfect-gas, 88
Term value, 462, 470, 474, 490–492, 498–499
Tesla coil, 712
Tetrachloroethane, refractive index of, 234–235
Thermal analysis, 242
Thermal arrest in cooling curve, 240–241
Thermal conductivity of gases, 124–125
Thermal expansion:
    of brass, 721
    of glass, 93
    of mercury, 697
Thermal insulation, 792
Thermistors, 658–661
Thermochemistry, 153–186
    (*See also* Calorimetry; Heat)
Thermocouples, 649–654
    amplification, 620–621
    reference junction bath, 650, 661
    single junction, 650
    types of, 650–652
    use of, 99–102, 244, 652–656
Thermocouple gauge, 701
Thermodynamics of cells, 267–276
Thermometer:
    Beckmann, 647–648
    calibration, 645–649
    calorimetric, 647

Thermometer (*Cont.*):
    comparison of different types, 660
    cryogenic, 663–664
    cryoscopic, 200–201, 647
    gas, 86–95, 664
    mercury, 87, 645–647
        stem corrections, 646
    other devices, 664–665
    platinum resistance, 654–658
    quartz, 664
    special liquid, 648
    thermistor, 658–661
    vapor-pressure, 577–580, 664
    (*See also* Pyrometers; Thermocouple)
Thermopile, 752, 758
Thermoregulating circuits, 667–672
Thermostats (*see* Temperature control)
Thyodene, 215
Timing devices, 763–764
Titration, 771–772
    of acids, 210
    of iodine, 214–216
Toepler pump, 687–688
Torr, definition of, 677–678
Transformer, 600, 604
Transitions, vibrational
    combination, difference, 448, 473
    fundamental, 446, 470
    overtones, 448, 470
Transistors, 600–615
Transmittance, 754
Transport properties, kinetic theory of gases, 119–129
    (*See also* Diffusion; Viscosity)
Trapezoidal rule, 67–68
Traps, cold, 682–683
Triple point, 195
    cell, 643–645
    of water, 87, 641
Tubing connections, 797–798
Turbulent flow, 130, 135
Tygon, 791

Uncertainties (*see* Errors)
    in least-squares parameters, 817–821
Uncertainty principle, 79–80
Unimolecular reactions, 316–320
    (*See also* Kinetics)

Units:
  conversion factors (see
    endpapers)
  international system (SI),
    866–868

Vacuum system, general
  purpose, 709–711
Vacuum techniques, 677–714
  cold traps, 682–683,
    703–704
  design of vacuum lines,
    707–711
  joints and seals, 704–706
  leak detection, 711–713
  leaks, 686
  O-rings, 705–706
  (See also Gauges; Pumps)
Valence force model, 448,
    456–457, 473–474
Valves, needle and reducing,
    776–777
van der Waals' equation,
    97–99, 251
van't Hoff equation, 388
Vapor pressure:
  empirical Antoine equation,
    580
  of argon (solid), 572–582
  of iodine (solid), 582–598
  of liquid, 219–229
    boiling-point method,
      223–224
    isotenisope method,
      224–226
  of nitrogen, 580
Vapor-pressure thermometer,
    576–580
Variance, 31–32, 229
Variation principle, 516
Velocities, molecular, Max-
    wellian distribution of,
    119–120
Velocity of sound, 113–118
Vermiculite, 792
Vibration, normal modes of,
    446–448, 455, 469
Vibrational heat capacity of
    gases, 449
Vibrational partition function,
    449, 467–468, 585–590
Vibrational Product Rule, 481
Vibrational spectrum (see
    Spectrum)

Vibrational wavefunctions,
    470
Virial coefficients, 88
  for He, Ar, N_2, and CO_2,
    94
  osmotic second, 362
Virial equation, 88
Viscosimeter, 131, 134
  Ostwald, 372, 378
Viscosity:
  of air, 139
  of gases, 122–124, 130–140
  intrinsic, 373
    of macromolecules,
      370–380
    of polyvinyl alcohol, 375
  kinetic-energy correction,
    136–137
  of several fluids, 130–131
  slip correction, 137
  of a solution of rigid par-
    ticles, 373
  specific, 373
  of water, 379
Viscosity coefficient:
  of gas, 123, 130–131
  of water, 379
Viscous flow, 131–138
Voltage measurement,
    621–632, 722–725
Volume, molar: apparent, 190
  partial, 187–194
  of water, 190
Volumetric procedures,
    766–773

Wagner earthing circuit,
    259–263
Water:
  coefficient of viscosity, 379
  compressibility factor for,
    221
  critical temperature, 221
  density of, 193
  diamagnetism of, 428
  entropy of, 227–228
  heat capacity of gas and
    liquid, 222
  molar freezing-point de-
    pression constant for,
    199
  molar heat of fusion, 199
  molar volume of, 190

Water (Cont.):
  pressure variation of normal
    boiling point, 94
  purification of, 773
  surface tension of, 347
  triple point of, 87, 640
Water baths, 666–667
  (See also Temperature
    control)
Wavefunctions, 470
Wavelengths:
  argon, 740
  copper $K\alpha$ X-ray, 545
  krypton, 740
  mercury, 486, 740
  neon, 740
  selection, 743–747
  sodium D doublet, 732
  in vacuum, 488, 741
  xenon, 740
Wavenumber, 446
Weighing errors, 719–720
Weight of an observation,
    804, 807–809, 811–814,
    818–820, 823
Weighted residual, 811
Weston cell, 628–629
Westphal balance, 764
Wheatstone bridge, 257–262,
    632–636, 655–656
  ac circuit, 257–262, 635–636
  dc circuit, 632–634
Wiens law, 741
Work:
  electrical, 155
  expansion, 97

X-ray diffraction, 532–572
  Bragg reflection, 535–538
  camera, 547, 550, 562–566
  film, 547–548
  intensities of reflections,
    541–542
  powder method, 490–493
  single crystal method,
    562–566
  (See also Crystal structure)

Zeeman energy, 430,
    508–510, 522
Zener diode, 600, 605–608
Zero-point energy of crystal,
    575